# Prealgebra

# Prealgebra

### Third Edition

## K. Elayn Martin-Gay

University of New Orleans

Prentice
Hall

Upper Saddle River, New Jersey 07458

Executive Acquisition Editor: Karin E. Wagner
Editor-in-Chief: Christine Hoag
Project Manager: Ann Marie Jones
Assistant Vice President of Production and Manufacturing: David W. Riccardi
Executive Managing Editor: Kathleen Schiaparelli
Senior Managing Editor: Linda Mihatov Behrens
Project Management: Elm Street Publishing Services, Inc.
Manufacturing Buyer: Alan Fischer
Manufacturing Manager: Trudy Pisciotti
Senior Marketing Manager: Eilish Collins Main
Marketing Assistant: Dan Auld
Director of Marketing: John Tweeddale
Associate Editor, Mathematics/Statistics Media: Audra J. Walsh
Editorial Assistant/Supplements Editor: Kate Marks
Art Director: Maureen Eide
Assistant to the Art Director: John Christiana
Interior Designer: Joseph Sengotta/Elm Street Publishing Service, Inc.
Cover Designer: Joseph Sengotta
Art Editor: Grace Hazeldine
Art Manager: Gus Vibal
Director of Creative Services: Paul Belfanti
Photo Researcher: Beaura Kathy Ringrose
Photo Editor: Beth Boyd
Cover Photo: Chuck Pefley/Stock Boston
Art Studio: Academy Artworks
Compositor: Preparé Inc., Italy

 © 2001 by Prentice-Hall, Inc.
Upper Saddle River, NJ 07458

Printed in the United States of America
10 9 8 7 6 5 4 3 2 1

ISBN 0-13-026038-X(0-13-026037-1 student edition)

Prentice-Hall International (UK) Limited, *London*
Prentice-Hall of Australia Pty. Limited, *Sydney*
Prentice-Hall Canada, Inc., *Toronto*
Prentice-Hall Hispanoamericana, S.A., *Mexico*
Prentice-Hall of India Private Limited, *New Delhi*
Prentice-Hall of Japan, Inc., *Tokyo*
Pearson Education Asia, Pte. Ltd.
Editora Prentice-Hall do Brasil, Ltda., *Rio de Janeiro*

*To Jewett B. Gay, and to the memory*
*of her husband, Jack Gay*

# SERIES LIST

The following list of softcover, hardcover, and computer-based formats are part of a library of developmental mathematics texts and programs by K. Elayn Martin-Gay. An acclaimed series of lecture videos specifically support each book in the Martin-Gay series.

*Prealgebra*, Third Edition
(Softcover, Student Edition 0-13-026037-1; Annotated Instructors's Edition 0-13-026038-X)
*Beginning Algebra*, Third Edition
(Hardcover, Student Edition 0-13-086763-2; Annotated Instructor's Edition 0-13-086764-0)
*Intermediate Algebra*, Third Edition
(Hardcover, Student Edition 0-13-016631-6; Annotated Instructor's Edition 0-13-016632-4)

*Intermediate Algebra: A Graphing Approach*, Second Edition
(Hardcover, Student Edition 0-13-016633-2; Annotated Instructor's Edition 0-13-016634-0)
*Beginning and Intermediate Algebra*, Second Edition
(Hardcover, Student Edition 0-13-016636-7, Annotated Instructor's Edition 0-13-016637-5)

*Basic College Mathematics*
(Softcover, Student Edition 0-13-376823-6, Annotated Instructors's Edition 0-13-080985-3)
*Introductory Algebra*
(Softcover*, Student Edition 0-13-228834-6, Annotated Instructor's Edition 0-13-862467-4)
*Intermediate Algebra*
(Softcover*, Student Edition 0-13-228800-1, Annotated Instructor's Edition 0-13-862376-7)
*Algebra, A Combined Approach*
(Softcover*, Student Edition 0-13-085048-9, Annotated Instructor's Edition 0-13-085049-7)

Prentice Hall Interactive Math, Basic Math
(Computer-based course, Student package, Instructor's package)

Prentice Hall Interactive Math, Introductory Algebra
(Computer-based course, Student package, Instructor's package)

Prentice Hall Interactive Math, Intermediate Algebra
(Computer-based course, Student package, Instructor's package)

* A hardcover version of this text is also available. Ask your local Prentice Hall sales representative for details.

# Contents

# Preface

## ABOUT THE BOOK

This book was written to help students make the transition from arithmetic to algebra. To reach this goal, I introduce algebraic concepts early and repeat them as I treat traditional arithmetic topics, thus laying the groundwork for the next algebra course your students will take. A second goal was to show students the relevancy of the mathematics in everyday life and in the workplace.

In preparing this third edition, I considered the comments and suggestions of colleagues throughout the country and of the many users of previous editions. The numerous features that contributed to the success of the second edition have been retained. This updated revision includes new mathematical content and increased attention to geometric concepts, data interpretation, problem solving, and real-life applications. I have carefully chosen pedagogical features to help students understand and retain concepts. The AMATYC *Crossroads in Mathematics* document and the NCTM Standards (plus Addenda) also influenced the careful reexamination of every section of the text. The key content and pedagogical features are described on this and the following pages.

In addition, the supplements to this text have been enhanced and the range of supplements increased, offering a complete integrated teaching and learning package for maximum support and effectiveness.

## KEY PEDAGOGICAL FEATURES IN THE THIRD EDITION

**Readability and Connections**   Many reviewers of this edition as well as users of the previous editions have commented favorably on the readability and clear, organized presentation. I have tried to make the writing style as clear as possible while still retaining the mathematical integrity of the content. As new topics are presented, efforts have been made to relate the new ideas to those that the students may already know. Constant reinforcement and connections within problem-solving strategies, geometric concepts, pattern recognition, and situations from everyday life can help students gradually master both new and old information.

**Accessible Real-World Applications**   An abundance of new practical applications are found throughout the book in worked-out examples and exercise sets. The applications were carefully chosen to be accessible, to help motivate students, and to strengthen their understanding of mathematics in the real world. They help show connections to a wide range of fields such as agriculture, allied health, astronomy, automotive ownership and maintenance, aviation, biology, business, chemistry, communication, computer technology, construction, consumer affairs, cooking, demographics, earth science, education, entertainment, environmental issues, finance and economics, food service, geography, government, history, hobbies, labor and career issues, life science, medicine, music, nutrition, physics, political science, population, recreation, sports, technology, transportation, travel, weather, and important related mathematical areas such as geometry, statistics, and probability. (See also the Index of Applications on page xx.) Many of the applications are based on recent and interesting real-life data. Sources for data include newspapers, magazines, government publications, publicly held companies, special interest groups, research organizations, and reference books. Opportunities for obtaining your own real data are also included.

**Problem-Solving Process**    This is formally introduced in Chapter 3 with a four-step process that is integrated throughout the text. The four steps are Understand, Translate, Solve, and Interpret. The repeated use of these steps in a variety of examples shows their wide applicability. Reinforcing the steps can increase students' comfort level and confidence when solving problems.

**Unique Exercise Sets**    Each section ends with an Exercise Set, divided in parts. All parts contain graded exercises. Each exercise in the set, except those found in the parts labeled Review and Preview or Combining Concepts, is keyed to at least one of the objectives of the section. As appropriate, exercises are also keyed to one or more specific worked examples in the text. Exercises and examples marked with a video icon ▣ have been worked out step-by-step by the author in the videos that accompany this text.

Throughout the exercise sets, there is an emphasis on *Data and Graphical Interpretation* via tables, charts, and graphs. The ability to interpret data and read and create a variety of types of graphs is developed gradually so students become comfortable with it. In addition, sections in Chapter 8 reinforce and extend these concepts. Similarly, throughout the text there is integration of *Geometric Concepts*, such as perimeter and area. Chapter 9 then provides a concise introduction and further coverage of geometry. A geometry icon △ indicates exercises, examples or full sections involving geometric concepts.

In addition to the approximately 4400 exercises in end-of-section exercise sets, exercises may also be found in the Pretests, Integrated Reviews, Chapter Reviews, Chapter Tests, and Cumulative Reviews. Each exercise set contains one or more of the following features.

*Mental Mathematics*    These problems are found at the beginning of many exercise sets. They are mental warmups that reinforce concepts found in the accompanying section and increase students' confidence before they tackle an exercise set. By relying on their own mental skills, students increase not only their confidence in themselves but also their number sense and estimation ability.

*Review and Preview*    Formerly called Review Exercises, these exercises occur in each exercise set (except for those in Chapter 1). These problems are keyed to earlier sections and review concepts learned earlier in the text that are needed in the next section or in the next chapter. These exercises show the links between earlier topics and later material.

*NEW!*    *Combining Concepts*    These exercises are found at the end of each exercise set after the Review and Preview exercises. New to this edition, Combining Concepts exercises require students to combine several concepts from that section or to take the concepts of the section a step further by combining them with concepts learned in previous sections. For instance, sometimes students are required to combine the concepts of a section with the problem-solving process they learned in Chapter 3 to try their hand at solving an application problem. In addition, there are conceptual exercises that occur outside the Combining Concepts part of the exercise set, and these are keyed with the icon ⌑.

*Writing Exercises*    These exercises, found in almost every exercise set, are keyed with the icon ✎. They require students to show an understanding of a concept learned in the corresponding section. This is accomplished by asking students questions that require them to stop, think, and explain in their own words the concept(s) used in the exercises they have just completed.

Guidelines recommended by the American Mathematical Association of Two-Year Colleges (AMATYC Crossroads in Mathematics guidelines) and other professional groups suggest incorporating writing in mathematics courses to reinforce concepts.

*Internet Excursions*    New to this edition, these exercises occur once per chapter. Internet Excursions require students to use the internet as a source of information or a data-collection tool to complete the exercises, allowing students first-hand experience with manipulating and working with real data.

*NEW!*

**Practice Problems**    Throughout the text, each worked example has a parallel problem called a Practice Problem, found in the margin. Practice Problems invite students to be actively involved in the learning process before beginning the section exercise set. Practice Problems *immediately reinforce* a skill or concept as it is being developed.

**Concept Checks**    These new margin exercises are appropriately placed in many sections of the text. They invite students to *immediately* check their conceptual understanding as material is developed, offering an additional way to actively involve students in the learning process.

*NEW!*

**Integrated Reviews**    These "mid-chapter reviews" are appropriately placed once per chapter. The new Integrated Reviews allow students to review and assimilate the many different skills learned separately over several sections before moving on to related material in the chapter.

*NEW!*

**Helpful Hints**    Helpful Hints, formerly Reminders, contain practical advice on applying mathematical concepts. These are found throughout the text and strategically placed where students are most likely to need immediate reinforcement. Helpful Hints are highlighted for quick reference.

*NEW!*

**Focus On**    Appropriately placed throughout each chapter, these are divided into Focus on Study Skills, Focus on Mathematical Connections, Focus on Business and Career, Focus on the Real World, and Focus on History. They are new to this edition and written to help students develop effective habits for studying mathematics, engage in investigations of other branches of mathematics, and understand the importance of mathematics in various careers and in the world of business. They help show the relevance of mathematics in both the present and past through critical thinking exercises and group activities.

**Calculator Explorations**    These optional explorations offer point-of-use instruction, through examples and exercises, on the proper use of calculators as tools in the mathematical problem-solving process. Placed appropriately throughout the text, Calculator Explorations also reinforce concepts learned in the corresponding section.

Additional exercises building on the skill developed in the Explorations may be found in exercise sets throughout the text. Exercises requiring a calculator are marked with the 🖩 icon. The inside back cover of the text includes a brief description of selected keys on a scientific calculator for reference as desired.

**Chapter Activity**    These features, formerly Group Activity, occur once per chapter at the end of the chapter, often serving as a chapter wrap-up. For individual or group completion, the Chapter Activity, usually hands-on or data-based, complements and extends the concepts of the chapter, allowing students to make decisions and interpretations and to think and write about mathematics.

**Visual Reinforcement of Concepts**    The text contains a wealth of graphics, models, photographs, and illustrations to visually clarify and reinforce concepts. These include bar graphs, line graphs, calculator screens, application illustrations, and geometric figures.

*NEW!*    **Pretests**    New to this edition, each chapter begins with a pretest that is designed to help students identify areas where they need to pay special attention in the upcoming chapter.

**Chapter Highlights**    Found at the end of each chapter, these contain key definitions, concepts, *and examples* to help students understand and retain what they have learned.

**Chapter Review and Test**    The end of each chapter contains a review of topics introduced in the chapter. The Chapter Review offers exercises that are keyed to sections of the chapter. The Chapter Test is a practice test and is not keyed to sections of the chapter.

**Cumulative Review**    These are found at the end of each chapter (except Chapter 1). Each problem contained in the cumulative review is actually an earlier worked example in the text that is referenced in the back of the book along with the answer. Students who need to see a complete worked-out solution, with explanation, can do so by turning to the appropriate example in the text.

**Student Resource Icons**    At the beginning of each section, videotape, software, study guide, and solutions manual icons are displayed. These icons help reinforce that these learning aids are available should students wish to use them to help them review concepts and skills at their own pace. These items have direct correlation to the text and emphasize the text's methods of solution.

*NEW!*    **Functional Use of Color and Design**    Elements of the text are now highlighted with full color or design to make it easier for students to read and study. Color is also used to clarify the problem-solving process in worked examples.

## NEW KEY CONTENT FEATURES IN THE THIRD EDITION

**Greater Emphasis on Geometry**    There is increased emphasis and coverage of geometric concepts. This was accomplished by retaining the early introduction and integration of the concepts of perimeter and area (see Sections 1.2 and 1.5), strengthening the coverage of geometry within sections (see Sections 3.1 and 5.8), and providing a new chapter on Geometry and Measurement together with an Integrated Review dedicated to geometry concepts (see Chapter 9). A geometry icon △ indicates exercises, examples, or full sections involving geometric concepts.

Designed to be as flexible as possible, instructors may elect to use the new chapter as a convenient unit on geometry or use a section or sections in conjunction with earlier chapters. For instance, Section 9.3 on perimeter may be taught anytime after Section 3.5.

There is a geometry appendix containing a review of geometric figures and a new appendix contains coverage of surface area. The inside front cover of this text provides a summary of geometric formulas for easy reference. Formulas include perimeter, area, volume, circumference, and surface area.

The sections on geometry and measurement are written to help increase students' spatial sense, relate general geometric ideas to number and measurement ideas, and make and use estimates of measurement.

**Expanded Coverage of Ratio and Proportion**   The treatment of ratio and proportion now includes separate sections on rates (see Section 6.2) and proportions and problem solving (see Section 6.4).

The chapter on percent now offers full section coverage of solving percent problems with equations and solving percent problems with proportions (see Sections 7.2 and 7.3). The expanded coverage of ratio and proportion allows for an increased number of application problems, and increased flexibility in the amount and use of topics needed for students to succeed in both this course and future courses.

**Increased Attention to Reading Graphs**   The widely praised treatment of reading and interpreting graphs has been expanded to two sections (see Sections 8.1 and 8.2). In addition, there has been a significant increase in the number of examples and exercises *throughout the text* using a graph, table, or chart.

**Probability**   A new section on Counting and Introduction to Probability has been added to this edition (see Section 8.6). It provides for exploring concepts of chance and expanding students' number sense. The topic of probability is included in the AMATYC Crossroads in Mathematics content guidelines.

**Solving Equations**   Special emphasis is given to the addition property of equality and the multiplication property of equality (see Sections 3.2 and 3.3).

**Increased Opportunities to Use Technology**   Optional calculator explorations and exercises are integrated appropriately throughout the text.

**New Examples**   Additional detailed step-by-step examples were added where needed. Many of these reflect real-life situations. Examples are used in two ways—numbered, as formal examples, to check or increase understanding of a topic, and unnumbered, to introduce a topic or informally discuss the topic.

**New Exercises**   A significant amount of time was spent on the exercise sets. New exercises and additional examples help address a wide range of student learning styles and abilities. New kinds of exercises, strategically placed, include Combining Concepts, Concept Checks, Integrated Reviews, Internet Excursions, Pretests, Focus On exercises, and reading more tables, charts, and graphs. In addition, the mental math, computational, and word problems, as well as optional calculator exercises, were refined and enhanced. There is now a total of approximately 7,000 exercises.

## OPTIONS FOR ON-LINE AND DISTANCE LEARNING

For maximum convenience, Prentice Hall offers on-line interactivity and delivery options for a variety of distance learning needs. Instructors may access or adopt these in conjunction with this text, *Prealgebra*, Third Edition.

**Companion Website**   Visit http://www.prenhall.com/martin-gay. The companion Website includes basic distance learning access to provide links to the text's Internet Excursion exercises, dozens of additional links, a syllabus manager, and a selection of on-line self quizzes. E-mail is available.

**WebCT**   WebCT includes distance learning access to content found in the Martin-Gay Companion Website plus more. WebCT provides tools to create, manage, and use on-line course materials. Save time and take advantage of items such as on-line help, communication tools, and access to instructor and student manuals. Your college may already have WebCT's software installed on their server or you may choose to download it. Contact your local Prentice Hall sales representative for details..

## SUPPLEMENTS FOR THE INSTRUCTOR

### Printed Supplements   Annotated Instructor's Edition (0-13-026038-X)

- Answers to all exercises printed on the same text page.
- Teaching Tips throughout the text placed at key points in the margin.

Instructor's Solutions Manual (0-13-026461-X)

- Detailed step-by-step solutions to even-numbered section exercises.
- Solutions to every (odd and even) Mental Math exercise.
- Solutions to every (odd and even) Practice Problem.
- Solutions to every (odd and even) exercise found in the Chapter Pretests, Integrated Reviews (mid-chapter reviews), Chapter Reviews, and Chapter Tests.
- Solution methods reflect those emphasized in the textbook.

Instructor's Resource Manual with Tests (0-13-026450-4)

- Notes to the Instructor that include an introduction to Interactive Learning, Interpreting Graphs and Data, Alternative Assessment, and Helping Students Succeed.
- Two free response Pretests per chapter.
- Eight Chapter Tests per chapter (three multiple choice, five free response).
- Two Cumulative Tests, (one multiple choice, one free response) every two chapters.
- Eight Final Examinations (four multiple choice, four free response).
- Twenty additional exercises per section for additional test exercises or review if needed.
- Answers to all pretests, tests, cumulative tests, additional exercises, and final examinations.

### Media Supplements   Computerized Testing

TestGen-EQ CD Rom (Windows/Macintosh) (0-13-027134-9)

- Algorithmically driven, text-specific testing program.
- Networkable for administering tests and collecting grades on-line.
- Edit and add your own questions to create a nearly unlimited number of tests and drill worksheets.
- Features an Equation Editor and Graphing Tool.
- Tests can be easily exported to html so that they can be posted to the Web for student practice.

Computerized Tutorial Software Course Management Tools
MathPro Explorer 4.0 Network CD Rom (0-13-027136-5)

- Algorithmically driven and fully networkable.
- Enables instructors to create either customized, algorithmically generated practice tests from any section of a chapter of up to 100 items or a test of random items.
- Includes an e-mail function for network users, enabling instructors to send a message to a specific student or to an entire group.
- Network based reports and summaries for a class or student and for cumulative or selected scores are available.

Companion Web site: www.prenhall.com/martin-gay
- Create a customized on-line syllabus with Syllabus Manager.
- Instructors can assign Internet based homework projects integrated with chapter content, such as this text's Internet Excursion exercises.
- Instructors can assign Warm Ups and Quizzes or monitor student self-quizzes by having students e-mail results.
- Provides dozens of additional links to other relevant math sites.

## Supplements for the Student

### Printed Supplements    Student Solutions Manual (0-13-026459-8)

- Detailed step-by-step solutions to odd-numbered section exercises.
- Solutions to every (odd and even) Mental Math exercise.
- Solutions to every (odd and even) Practice Problem.
- Solutions to every (odd and even) exercise found in the Chapter Pretests, Integrated Reviews (mid-chapter reviews), Chapter Reviews, and Chapter Tests.
- Solution methods reflect those emphasized in the textbook.

Student Study Guide (0-13-026462-8)
- Additional step-by-step worked out examples and exercises.
- Chapter Practice Tests.
- Practice Final Examination.
- Includes study skills and note-taking suggestions.
- Each chapter contains a Helpful Hints and Insights section.
- Solutions to all exercises, tests, and final examination.
- Solution methods reflect those emphasized in the textbook.

How to Study Mathematics
- Have your instructor contact the local Prentice Hall sales representative.

Math on the Internet: A Student's Guide
- Have your instructor contact the local Prentice Hall sales representative.

Prentice Hall/New York Times Theme of the Times Newspaper Supplement
- Have your instructor contact the local Prentice Hall sales representative.

### Media Supplements

Computerized Tutorial Software
MathPro Explorer 4.0 Network CD Rom (0-13-027136-5)
MathPro Explorer 4.0 Student CD Rom (0-13-027138-1)

- Keyed to each section of the text for text-specific tutorial exercises and instruction.
- Includes warm-up exercises and graded Practice Problems.
- Includes an easily accessible video "Watch" functionality, providing a problem (similar to the one being attempted) being explained and worked out on the board in a 1-2 minute video clip.
- Provides Explorations, allowing explorations of concepts associated with each objective in more detail.
- Algorithmically generated exercises, and includes bookmark, on-line help, glossary, and summary of scores for the exercises tried.

- Have your instructor contact the local Prentice Hall sales representative—also available for purchase for home use.

Videotape Series (0-13-027142-X)

- Written and presented by textbook author K. Elayn Martin-Gay.
- Keyed to each section of the text.
- Step-by-step solutions to exercises from each section of the text. Exercises that are worked in the videos are marked with a video icon .

Companion Web site: www.prenhall.com/martin-gay

- Links allowing you to collect data to solve the textbook Internet Excursion exercises are provided.
- This companion web site offers Warm-ups, True/False Reading Quizzes, and Chapter Quizzes, offering additional opportunities to test mastery of chapter content.
- Provides dozens of additional links to other relevant math sites.

## ACKNOWLEDGMENTS

First, as usual, I would like to thank my husband, Clayton, for his constant encouragement. I would also like to thank my children, Eric and Bryan, for continuing to eat my cooking and going on adventures with me.

I would also like to thank my extended family for their invaluable help and wonderful sense of humor. Their contributions are too numerous to list. They are Rod and Karen Pasch; Peter, Michael, Christopher, Matthew, and Jessica Callac; Stuart, Earline, Melissa, Mandy, and Bailey Martin; Mark, Sabrina, and Madison Martin; Leo, Barbara, Aaron and Andrea Miller; and Jewett Gay.

A special thank you to all users of the first and second editions of this text who made suggestions for improvements that were incorporated into the third edition. I would also like to thank the following reviewers for their input and suggestions.

Jana Canary, *University of Alaska-Fairbanks*
Gail Carter, *St. Petersburg Junior College-Tarpon Springs Campus*
Terry Cassady, *Austin Community College-Cypress Creek Campus*
Patricia Donovan, *San Joaquin Delta College*
Suellen Gifford, *Ferris State University*
Mark Greenhalgh, *Fullerton College*
Robert Hervey, *Hillsborough Community College*
Patricia Hirschy, *Asnuntuck Community Technical College*
Linda Jones, *Vincennes University*
Maryann Justinger, *Erie Community College-South*
Kathy Kopelousos, *Lewis & Clark Community College*
Walter Labhart, *Northern Michigan University*
Lisa Lindloff, *McLennan Community College*
Sue Little, *North Harris College*
Debra McCandrew, *Florence-Darlington Technical College*
Vincent McGarry, *Austin Community College*
Jean McNeil, *Western Nevada Community College*
Dozier Montgomery, *Florence-Darlington Technical College*
Debra Moses, *University of Alaska-Fairbanks, Tanana Valley Campus*
Dennis Runde, *Manatee Community College*
Robert Secrist, *Kellogg Community College*
John Thoo, *Yuba College*
Joseph Tripp, *Ferris State University*

There were many people who helped me develop this text and I will attempt to thank some of them here. Cheryl Roberts Cantwell was invaluable for contributing to the overall accuracy of this text. Emily Keaton was also invaluable for her many suggestions and contributions during the development and writing of this third edition. Ingrid Mount at Elm Street Publishing Services provided guidance throughout the production process. I thank Terri Bittner, Cindy Trimble, Jeff Rector, and Teri Lovelace at Laurel Technical Services for all their work on the supplements, text, and thorough accuracy check. Lastly, a special thank you to my editors Karin Wagner, Chris Hoag, and project manager Ann Marie Jones for their support and assistance throughout the development and production of this text and to all the staff at Prentice Hall: Linda Behrens, Alan Fischer, Maureen Eide, Grace Hazeldine, Gus Vibal, Audra Walsh, Kate Marks, Eilish Main, Dan Auld, John Tweeddale, Paul Corey, Jerome Grant, and Tim Bozik.

*K. Elayn Martin-Gay*

## ABOUT THE AUTHOR

K. Elayn Martin-Gay has taught mathematics at the University of New Orleans for more than 20 years. Her numerous teaching awards include the local University Alumni Association's Award for Excellence in Teaching, and Outstanding Developmental Educator at University of New Orleans, presented by the Louisiana Association of Developmental Educators.

Elayn is the author of an entire product line of highly successful textbooks. The Martin-Gay library includes the successful worktext series made up of *Basic College Mathematics, Prealgebra* 3e, *Introductory Algebra,* and *Intermediate Algebra.* Martin-Gay's exciting new hardbound series includes *Beginning Algebra* 3e; *Intermediate Algebra* 3e, *Intermediate Algebra, A Graphing Approach* 2e (co-authored with Margaret Greene), and *Beginning and Intermediate Algebra* 2e, a combined algebra text.

Prior to writing textbooks, Elayn developed an acclaimed series of lecture videos to support developmental mathematics students in their quest for success. These highly successful videos originally served as the foundation material for her texts. Today the tapes specifically support each book in the Martin-Gay series.

# Applications Index

# Highlights of Prealgebra, Third Edition

*Prealgebra, Third Edition* has been designed as just one of the tools in a fully integrated learning package to help you develop prealgebra skills. Author K. Elayn Martin-Gay focuses on enhancing the traditional emphasis on mastering the basics with innovative pedagogy and a meaningful learning program. There are three goals that drive her authorship:

▲ **Master and apply skills and concepts**

▲ **Build confidence**

▲ **Increase motivation**

Take a few moments now to examine some of the features that have been incorporated into **Prealgebra**, **Third Edition** to help students excel and to smooth the transition from arithmetic to algebra.

◄ Chapter-opening real-world applications introduce you to everyday situations that are applicable to the mathematics you will learn in the upcoming chapter, showing the relevance of mathematics in daily life.

---

## Solving Equations and Problem Solving

3

Throughout this text, we have been making the transition from arithmetic to algebra. We have said that in algebra letters called variables represent numbers. Using variables is a very powerful method for solving problems that cannot be solved with arithmetic alone. This chapter introduces operations on algebraic expressions and solving variable equations.

**CONTENTS**

3.1  Simplifying Algebraic Expressions

3.2  Solving Equations: The Addition Property

3.3  Solving Equations: The Multiplication Property

Integrated Review— Expressions and Equations

3.4  Solving Linear Equations in One Variable

3.5  Linear Equations in One Variable and Problem Solving

The largest purchase that most people usually make during their lifetimes is a house. Every year, millions of families take the plunge and buy a house. That also means that millions of families sell their houses. Real estate agents can help get sellers and buyers together to make the transaction. In return, the real estate agent usually receives a commission, a payment for his or her services based on the selling price of the house. In Exercises 50–51 on page 208, we will see how an equation can be used to describe the relationship between the selling price of a house, the real estate agent's commission, and the amount the seller receives in a real estate transaction.

163

# Master and Apply Basic Skills and Concepts

K. Elayn Martin-Gay provides patient explanations of key concepts and enlivens the content by integrating successful and innovative pedagogy. *Prealgebra*, *Third Edition* integrates skill building throughout the text and provides problem-solving strategies and hints along the way.

---

✓ **CONCEPT CHECK**

When solving the problem given in Example 3, why should you be skeptical if you obtain an answer of 15,714 votes for the incumbent?

◄ **NEW Concept Checks** are special margin exercises found in most sections. Work these to help gauge your grasp of the concept being developed in the text.

---

**NEW Combining Concepts** exercises are found at the end of each exercise set. Solving these exercises will expose you to the way mathematical ideas build upon each other. ▶

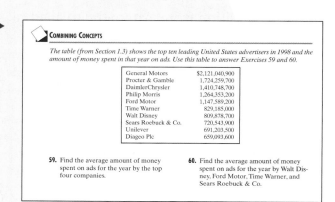

**COMBINING CONCEPTS**

*The table (from Section 1.3) shows the top ten leading United States advertisers in 1998 and the amount of money spent in that year on ads. Use this table to answer Exercises 59 and 60.*

| General Motors | $2,121,040,900 |
|---|---|
| Procter & Gamble | 1,724,259,700 |
| DaimlerChrysler | 1,410,748,700 |
| Philip Morris | 1,264,353,200 |
| Ford Motor | 1,147,589,200 |
| Time Warner | 829,185,000 |
| Walt Disney | 809,878,700 |
| Sears Roebuck & Co. | 720,543,900 |
| Unilever | 691,203,500 |
| Diageo Plc | 659,093,600 |

**59.** Find the average amount of money spent on ads for the year by the top four companies.

**60.** Find the average amount of money spent on ads for the year by Walt Disney, Ford Motor, Time Warner, and Sears Roebuck & Co.

---

## Practice Problem 6

Use the circle graph below to write the ratio of work miles to total miles as a fraction in simplest form.

◄ **Practice Problems** occur in the margins next to every
▼ **Example.** Work these problems after an example to immediately reinforce your understanding.

---

**Example 6**    Writing a Ratio from a Circle Graph

The circle graph in the margin shows the part of a car's total mileage that falls into a particular category. Write the ratio of medical miles to total miles as a fraction in simplest form.

*Solution:*

$$\frac{\text{medical miles}}{\text{total miles}} = \frac{150 \text{ miles}}{15,000 \text{ miles}} = \frac{150}{15,000} = \frac{150}{150 \cdot 100} = \frac{1}{100}$$

# Build Confidence

Good exercise sets and an abundance of worked-out examples are essential for building student confidence. The exercises you will find in this worktext are intended to help you build skills and understand concepts as well as motivate and challenge you. In addition, features like **Chapter Highlights, Chapter Reviews, Chapter Tests** and **Cumulative Reviews** are found at the end of each chapter to help you study and organize your notes.

---

**CHAPTER 3 PRETEST**

*Simplify.*

**1.** $9x - 4 + 6x + 8$　　　　**2.** $3(2x - 1) - (x - 8)$

*Multiply.*

**3.** $8(7b)$　　　　**4.** $-5(2y - 7)$

---

◄ **NEW Pretests** open each chapter. Take a **Pretest** to evaluate where you need the most help before beginning a new chapter.

**NEW Integrated Reviews** serve as mid-chapter reviews and help you to assimilate the new skills you have learned separately over several sections. ▶

**INTEGRATED REVIEW—SUMMARY ON FRACTIONS AND FACTORS**

*Use a fraction to represent the shaded area of each figure or figure group.*

**1.**　　　　**2.**

**3.**

---

**MENTAL MATH**

*State whether the fractions in each list are like or unlike fractions.*

**1.** $\dfrac{7}{8}, \dfrac{7}{10}$　　**2.** $\dfrac{2}{3}, \dfrac{2}{9}$　　**3.** $\dfrac{9}{10}, \dfrac{1}{10}$　　**4.** $\dfrac{8}{11}, \dfrac{2}{11}$

**5.** $\dfrac{2}{31}, \dfrac{30}{31}, \dfrac{19}{31}$　　**6.** $\dfrac{3}{10}, \dfrac{3}{11}, \dfrac{3}{13}$　　**7.** $\dfrac{5}{12}, \dfrac{7}{12}, \dfrac{12}{11}$　　**8.** $\dfrac{1}{5}, \dfrac{2}{5}, \dfrac{4}{5}$

◄ Confidence-building **Mental Math** problems are in many sections.

**Review and Preview** exercises review concepts learned earlier in the text that are needed in the next section or chapter. ▶

**REVIEW AND PREVIEW**

*The trumpeter swan is the largest waterfowl in the United States. Although it was thought to be nearly extinct at the beginning of the twentieth century, recent conservation efforts have been succeeding. Use the bar graph to answer Exercises 63–66. See Section 1.3.*

Source: U.S. Fish and Wildlife Service

**63.** Estimate the number of trumpeter swans in the Great Lakes region in 1998.

**64.** Estimate the number of trumpeter swans in the Great Lakes region in 1996.

# Increase Motivation

Throughout *Prealgebra*, *Third Edition*, K. Elayn Martin-Gay provides interesting real-world applications to strengthen your understanding of the relevance of math in everyday life. When a new topic is presented, an effort has been made to relate the new ideas to those that students may already know. The Third Edition increases emphasis on visualization to clarify and reinforce key concepts.

◀ Real data is integrated throughout the worktext, drawn from current and familiar sources.

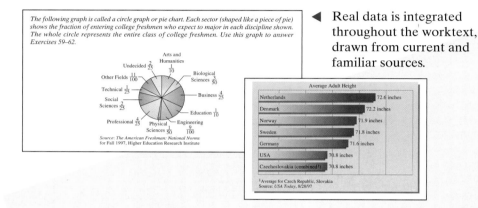

The following graph is called a circle graph or pie chart. Each sector (shaped like a piece of pie) shows the fraction of entering college freshmen who expect to major in each discipline shown. The whole circle represents the entire class of college freshmen. Use this graph to answer Exercises 59–62.

Source: The American Freshman: National Norms for Fall 1997, Higher Education Research Institute

### Average Adult Height

| | |
|---|---|
| Netherlands | 72.6 inches |
| Denmark | 72.2 inches |
| Norway | 71.9 inches |
| Sweden | 71.8 inches |
| Germany | 71.6 inches |
| USA | 70.8 inches |
| Czechoslovakia (combined¹) | 70.8 inches |

¹Average for Czech Republic, Slovakia
Source: USA Today, 8/28/97

▼ Graphics, models, and illustrations provide visual reinforcement.

▼ **Calculator Explorations** and exercises are woven into appropriate sections.

## CALCULATOR EXPLORATIONS
### ENTERING DECIMAL NUMBERS

To enter a decimal number, find the key marked ☐. To enter the number 2.56, for example, press the keys

[ 2 ] [ . ] [ 5 ] [ 6 ]

The display will read [ 2.56 ].

### OPERATIONS ON DECIMAL NUMBERS

Operations on decimal numbers are performed in the same way as operations on whole or signed numbers. For example, to find $8.625 − 4.29$, press the keys

[ 8.625 ] [ − ] [ 4.29 ] [ = ] (or [ ENTER ])

The display will read [ 4.335 ].

(Although entering 8.625, for example, requires pressing more than one key, we group numbers together for easier reading.)

*Use a calculator to perform each indicated operation.*

| | |
|---|---|
| **1.** $315.782 + 12.96$ | **2.** $29.68 + 85.902$ |
| **3.** $6.249 − 1.0076$ | **4.** $5.238 − 0.682$ |
| **5.** 12.555 | **6.** 47.006 |
| 224.987 | 0.17 |
| 5.2 | 313.259 |
| +622.65 | +139.088 |

## CHAPTER 3 ACTIVITY
### USING EQUATIONS

*This activity may be completed by working in groups or individually.*

A hospital nurse working second shift has been keeping track of the fluid intake and output on the chart of one of her patients, Mr. Ramirez. At the end of her shift, she totals the intakes and outputs and returns Mr. Ramirez's chart to the floor nurses' station.

Suppose you are one of the night nurses for this floor and have been assigned to Mr. Ramirez. Shortly after you come on duty, someone at the nurses' station spills coffee over several patients' charts, including the one for Mr. Ramirez. After blotting up the coffee on the chart, you notice that several entries on his chart have been smeared by the coffee spill.

### MEDICAL CHART

Patient: Juan Ramirez     Room: 314

Shift: Second     Nurse's Initials: SRJ     Date: 3/27

| Fluid Intake (cubic centimeters): | | | | Totals |
|---|---|---|---|---|
| Blood | 500 | | | 500 |
| Intravenous | 250 | | | 500 |
| Oral Fluid | 300 | 150 | 100 | 900 |
| Oral Meds | 4 doses of ☐ cc each | | | 200 |
| TOTAL | | | | 2100 |

| Fluid Output (cubic centimeters): | | | Totals |
|---|---|---|---|
| Urine | 240 | 310 | |
| Emesis | 120 | | 120 |
| Irrigation | 50 | 25 | 75 |
| TOTAL | | | 815 |

**1.** On the Fluid Intake chart, is it possible to tell if Mr. Ramirez was given blood more than once? If so, how many times was he given blood and how can you tell?

solve the equation for $x$ to find the obliterated entry.

**4.** For the Oral Meds row of the Fluid Intake chart, notice that the second-shift nurse indicated that

Visit the Internet through the Martin-Gay Companion Website to gather and manipulate data to complete exercises in **NEW Internet Excursions**. ▶

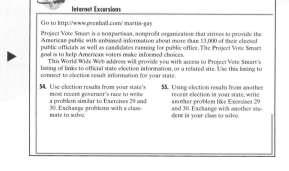

**Internet Excursions**

Go to http://www.prenhall.com/ martin-gay

Project Vote Smart is a nonpartisan, nonprofit organization that strives to provide the American public with unbiased information about more than 13,000 of their elected public officials as well as candidates running for public office. The Project Vote Smart goal is to help American voters make informed choices.

This World Wide Web address will provide you with access to Project Vote Smart's listing of links to official state election information, or a related site. Use this listing to connect to election result information for your state.

**54.** Use election results from your state's most recent governor's race to write a problem similar to Exercises 29 and 30. Exchange problems with a classmate to solve.

**55.** Using election results from another recent election in your state, write another problem like Exercises 29 and 30. Exchange with another student in your class to solve.

# Increase Motivation

**NEW Focus On** boxes found throughout each chapter help you see the relevance of math through critical-thinking exercises and group activities. Try these on your own, or with a classmate.

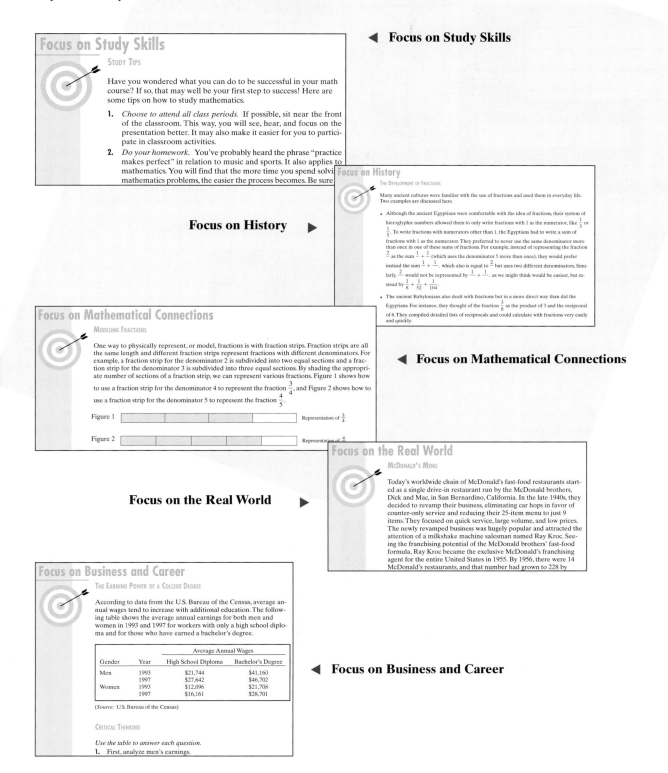

# Enrich Your Learning

Seek out these additional Student Resources to match your personal learning style.

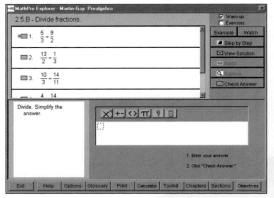

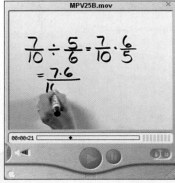

 Multimedia **MathPro Explorer** 4.0 tutorial software is developed around the content and concepts of *Prealgebra*. It provides:

- virtually unlimited practice problems with immediate feedback
- video clips
- step-by-step solutions
- exploratory activities
- on-line help
- summary of progress

Available on CD ROM.

Text-specific videos hosted by the award-winning teacher and author of *Prealgebra* cover each objective in every chapter section as a supplementary review.

## Also Available: ▼

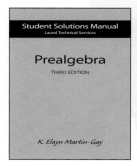

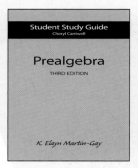

**The New York Times/ Themes of the Times** Newspaper-format supplement

Ask your instructor or bookstore about these additional study aids.

# Whole Numbers and Introduction to Algebra

**M**athematics is an important tool for everyday life. Knowing basic mathematical skills can help simplify many tasks. For example, we use operations on whole numbers to create a monthly budget using the cost of expenses such as rent, utilities, food, car payments, car maintenance, and so on.

This chapter covers basic operations on whole numbers. Knowledge of these operations provides a good foundation on which to build further mathematical skills.

## CONTENTS

**P**rior to the 20th century, criminals and prisoners were identified by physical characteristics such as hair or eye color, weight, height, and certain skeletal measurements. It wasn't until a famous case at Leavenworth Federal Penitentiary in 1903, where two unrelated inmates had the same physical appearance and skeletal measurements, that law enforcement officials realized that these measures were inadequate means of identification. It was at that point that fingerprints, which had been known for some time to be unique to each individual, became the standard for criminal identification. In 1924, the Federal Bureau of Investigation founded its Identification Division using 810,188 fingerprint files recorded at Leavenworth. Today, the FBI maintains a collection of over 250 million sets of fingerprints. In Exercise 20 on page 102, we will see how whole numbers can be used to describe the number of new sets of fingerprints received by the FBI.

**2**

# CHAPTER 1 PRETEST

**1.** Determine the place value of the digit 7 in the whole number 5732.

**2.** Write the whole number 23,490 in words.

**3.** Add: $58 + 29$

**4.** Multiply: $\begin{array}{r} 413 \\ \times \quad 9 \\ \hline \end{array}$

*Subtract. Check by adding.*

**5.** $\begin{array}{r} 857 \\ -231 \\ \hline \end{array}$

**6.** $\begin{array}{r} 51 \\ -19 \\ \hline \end{array}$

*Solve.*

**7.** Karen Lewis is reading a 329-page novel. If she has just finished reading page 193, how many more pages must she read to finish the novel?

**8.** Round 9045 to the nearest ten.

**9.** Round each number to the nearest hundred to find an estimated sum.

$\begin{array}{r} 382 \\ 436 \\ 2084 \\ + \ 176 \\ \hline \end{array}$

**10.** Use the distributive property to rewrite the following expression.

$9(3 + 11)$

△ **11.** Find the perimeter of the figure.

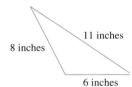

8 inches
11 inches
6 inches

△ **12.** Find the area of the rectangle.

8 yards
23 yards

**13.** The seats in the history lecture hall are arranged in 32 rows with 18 seats in each row. Find how many seats are in this room.

*Divide and then check by multiplying.*

**14.** $2187 \div 9$

**15.** $\dfrac{5361}{12}$

*Solve.*

**16.** Find the average of the following list of numbers. 29, 36, 84, 41, 6, 12, 65

**17.** Write $9 \cdot 9 \cdot 9 \cdot 9 \cdot 9 \cdot 9 \cdot 9$ using exponential notation.

**18.** Evaluate: $7^4$

**19.** Simplify: $36 + 18 \div 6$

**20.** If $a = 4$ and $b = 9$, evaluate $\dfrac{3a + 2b}{5}$.

**21.** Write the phrase "ten decreased by a number" as a variable expression. Use $x$ to represent the number.

# 1.1   PLACE VALUE AND NAMES FOR NUMBERS

The **digits** 0, 1, 2, 3, 4, 5, 6, 7, 8, and 9 can be used to write numbers. For example, the **whole numbers** are

0, 1, 2, 3, 4, 5, 6, 7, 8, 9, 10, 11, …

The three dots (…) after the 11 means that this list continues indefinitely. That is, there is no largest whole number. The smallest whole number is 0.

## A  FINDING THE PLACE VALUE OF A DIGIT IN A WHOLE NUMBER

The position of each digit in a number determines its **place value**. A place-value chart is shown below with the whole number 48,337,000 entered.

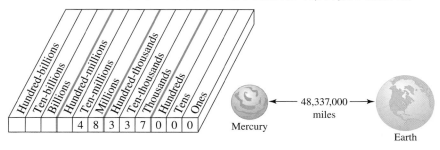

The two 3s in 48,337,000 represent different amounts because of their different placements. The place value of the 3 to the left is hundred-thousands. The place value of the 3 to the right is ten-thousands.

**Examples**   Find the place value of the digit 4 in each whole number.

**1.** 48,761
↑
ten-thousands

**2.** 249
↑
tens

**3.** 524,007,656
↑
millions

## B  WRITING A WHOLE NUMBER IN WORDS AND IN STANDARD FORM

A whole number such as 1,083,664,500 is written in **standard form**. Notice that commas separate the digits into groups of threes, starting from the right. Each group of three digits is called a **period**. The names of the first four periods are shown in red.

> ### WRITING A WHOLE NUMBER IN WORDS
>
> To write a whole number in words, write the number in each period followed by the name of the period. (The ones period is usually not written.) This same procedure can be used to read a whole number.

---

### Objectives

**A** Find the place value of a digit in a whole number.

**B** Write a whole number in words and in standard form.

**C** Write the expanded form of a whole number.

**D** Compare two whole numbers.

**E** Read tables.

| Study Guide | SSM | CD-ROM | Video 1.1 |

#### TEACHING TIP

Before beginning this lesson, ask students how many digits we use to represent numbers. Make sure they understand that we can write all whole numbers with the ten digits 0 through 9. For fun, encourage students to use a dictionary to look up place values that they may be unfamiliar with, such as *zillion* and *googol*.

### Practice Problems 1–3

Find the place value of the digit 7 in each whole number.
1. 72,589,620
2. 67,890
3. 50,722

#### TEACHING TIP

Give students a list of numbers to say aloud. Then ask them to say the whole number that comes immediately before or after each number listed.
Sample List: 63; 25,345; 399,499; 455,699; 7832; 3,000,000; 34,004,000,002; 483

#### TEACHING TIP

Have your students practice writing numbers that you say aloud.
Sample List: 48; 21,924; 7,344,899; 236,341; 345; 876,349; 23,000,426,007; 9342

#### Answers

**1.** ten-millions   **2.** thousands   **3.** hundreds

For example, we write 1,083,664,500 as

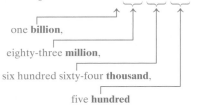

one **billion**,

eighty-three **million**,

six hundred sixty-four **thousand**,

five **hundred**

> ### Helpful Hint
>
> The name of the ones period is not used when reading and writing whole numbers. For example,
>
> 9,265 is read as "nine **thousand**, two **hundred** sixty-five."

## Examples    Write each number in words.

**4.** 85    eighty-five

**5.** 126    one hundred twenty-six

> ### Helpful Hint
>
> The word "and" is *not* used when reading and writing whole numbers. It is used when reading and writing mixed numbers and some decimal values, as shown later in this text.

## Example 6    Write 106,052,447 in words.

*Solution:*          106,052,447 is written as

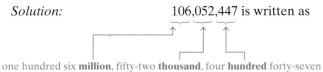

one hundred six **million**, fifty-two **thousand**, four **hundred** forty-seven

### TRY THE CONCEPT CHECK IN THE MARGIN.

### WRITING A WHOLE NUMBER IN STANDARD FORM

To write a whole number in standard form, write the number in each period followed by a comma.

## Examples    Write each number in standard form.

**7.** sixty-one    61

**8.** eight hundred five    805

**9.** two million, five hundred sixty-four thousand, three hundred fifty

     2,564,350

**10.** nine thousand, three hundred eighty-six

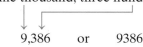

     9,386     or     9386

---

### Practice Problems 4–5

Write each number in words.

4. 67

5. 395

### Practice Problem 6

Write 321,670,200 in words.

### ✓ CONCEPT CHECK

True or false? When writing a check for $30,500, the word name we write for the dollar amount of the check would be "thirty-five hundred." Explain your answer.

### Practice Problems 7–10

Write each number in standard form.

7. twenty-nine

8. seven hundred ten

9. twenty-six thousand, seventy-one

10. six thousand, five hundred seven

#### Answers

**4.** sixty-seven    **5.** three hundred ninety-five
**6.** three hundred twenty-one million, six hundred seventy thousand, two hundred
**7.** 29   **8.** 710   **9.** 26,071   **10.** 6507
**✓ Concept Check:**
False

> **Helpful Hint**
>
> A comma may or may not be inserted in a four-digit number. For example, both
>
> 9,386    and    9386
>
> are acceptable ways of writing nine thousand, three hundred eighty-six.

## C WRITING A WHOLE NUMBER IN EXPANDED FORM

The place value of a digit can be used to write a number in expanded form. The **expanded form** of a number shows each digit of the number with its place value. For example, 5672 is written in expanded form as

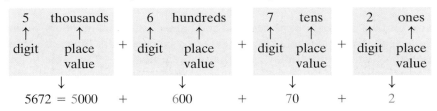

$$5672 = 5000 + 600 + 70 + 2$$

**Example 11** Write 706,449 in expanded form.

*Solution:*    $700,000 + 6000 + 400 + 40 + 9$

## D COMPARING WHOLE NUMBERS

We can picture whole numbers as equally spaced points on a line called the **number line**. The whole numbers are written in "counting" order on the number line beginning with 0. An arrow at the right end of the number line means that the whole numbers continue indefinitely.

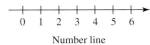

Number line

A whole number is **graphed** by placing a dot on the number line at the point corresponding to the number. For example, 5 is graphed on the number line as shown.

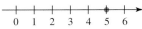

For any two numbers graphed on a number line, the number to the *right* is the **greater number**, and the number to the *left* is the **smaller number**. For example, 3 is *to the left* of 6, so 3 **is less than** 6. Also, 6 is *to the right* of 3, so 6 **is greater than** 3.

We use the symbol < to mean "is less than" and the symbol > to mean "is greater than." For example, the true statement *"3 is less than 6"* can be written in symbols as

   $3 < 6$

Also, *"6 is greater than 3"* is written as

   $6 > 3$

The statements $3 < 6$ and $6 > 3$ are both *true statements*. The statement $5 < 2$ is a *false statement* because 5 is greater than 2, not less than.

**Practice Problem 11**

Write 1,047,608 in expanded form.

**Answer**

**11.** $1,000,000 + 40,000 + 7000 + 600 + 8$

## Practice Problem 12

Insert < or > between each pair of numbers to make a true statement.
a. 0    19        b. 18    32
c. 107    103

TEACHING TIP

To emphasize how prevalent tables are wherever information is given, ask students to bring in a table from a magazine, web site, newspaper, or other printed material. Ask them how difficult it was to find a table. (Once they look for tables, they will notice how many are around.)

**Example 12**  Insert < or > between each pair of numbers to make a true statement.

**a.** 5    50      **b.** 101    0      **c.** 29    27

*Solution:*  **a.** $5 < 50$      **b.** $101 > 0$      **c.** $29 > 27$ ▬▬

> **Helpful Hint**
>
> One way to remember the meaning of the symbols < and > is to think of them as arrowheads "pointing" toward the smaller number. For example, $5 < 11$ and $11 > 5$ are both true statements.

### E  READING TABLES

Now that we know about place value and names for whole numbers, we introduce one way that whole-number data may be presented. **Tables** are often used to organize and display facts that contain numbers. The table below shows the countries that have won the most medals during the Olympic winter games. (Although the medals are won by athletes from the various countries, for simplicity, we will state that countries have won the medals.)

| MOST MEDALS WINTER GAMES (1924–1998) | | | | |
|---|---|---|---|---|
| | Gold | Silver | Bronze | Total |
| USSR[1]/Russia | 107 | 77 | 74 | 258 |
| Norway | 83 | 85 | 68 | 236 |
| U.S. | 59 | 58 | 40 | 157 |
| Germany[2] | 56 | 52 | 44 | 152 |
| Austria | 39 | 53 | 53 | 145 |
| Finland | 37 | 49 | 48 | 134 |
| GDR[3] | 39 | 36 | 35 | 110 |
| Sweden | 39 | 28 | 35 | 102 |
| Switzerland | 29 | 31 | 32 | 92 |
| Canada | 24 | 25 | 28 | 77 |

[1] Includes former USSR to 1992, Russia 1994, 1998.
[2] Includes West Germany 1952, 1968–1988.
[3] GDR (East Germany) 1968–1988.
(*Source:* Winter Games Net)

For example, by reading from left to right along the row marked U.S., we find that the United States has won 59 gold, 58 silver, and 40 bronze medals for the years 1924–1998.

## Practice Problem 13

Use the Winter Games table to answer each question.

a. How many bronze medals has Austria won during the winter games of the Olympics?
b. Which countries shown have won more than 70 gold medals?

**Answers**

**12. a.** <    **b.** <    **c.** >
**13. a.** 53    **b.** USSR/Russia and Norway

**Example 13**  Use the Winter Games table to answer each question.

**a.** How many silver medals has Finland won during the winter games of the Olympics?
**b.** Which country shown has won fewer gold medals than Switzerland?

*Solution:*  **a.** We read from left to right across the line marked Finland until we reach the "Silver" column. We find that Finland has won 49 silver medals.

**b.** Switzerland has won 29 gold medals while Canada has won 24, so Canada has won fewer gold medals than Switzerland.  ▬▬

## EXERCISE SET 1.1

**A** *Determine the place value of the digit 5 in each whole number. See Examples 1 through 3.*

**1.** 352      **2.** 905      **3.** 5890      **4.** 6527

**5.** 62,500,000      **6.** 79,050,000      **7.** 5,070,099      **8.** 51,682,700

**B** *Write each whole number in words. See Examples 4 through 6.*

**9.** 5420      **10.** 3165      **11.** 26,990      **12.** 42,009

**13.** 1,620,000      **14.** 3,204,000      **15.** 53,520,170      **16.** 47,033,107

*Write the number in each sentence in words. See Examples 4 through 6.*

**17.** At this writing, the population of Libya is 4,992,838. (*Source: The World Almanac*)

**18.** Liz Harold has the number 16,820,409

**19.** In a recent year, zinc mines in the United States mined 620,000 metric tons of zinc. (*Source:* U.S. Bureau of Mines)

**20.** The highest point in Montana is at Granite Peak, at an elevation of 12,799 feet. (*Source:* U.S. Geological Survey)

**21.** In a recent year, there were 3893 patients in the United States waiting for a heart transplant. (*Source:* United Network for Organ Sharing)

**22.** Each Home Depot store in the United States and Canada stocks at least 40,000 different kinds of building materials, home improvement supplies, and lawn and garden products. (*Source:* The Home Depot, Inc.)

**Name** _____

*Write each whole number in standard form. See Examples 7 through 10.*

**23.** Six thousand, five hundred eight

**24.** Three thousand, three hundred seventy

**25.** Twenty-nine thousand, nine hundred

**26.** Forty-two thousand, six

**27.** Six million, five hundred four thousand, nineteen

**28.** Ten million, thirty-seven thousand, sixteen

**29.** Three million, fourteen

**30.** Seven million, twelve

*Write the whole number in each sentence in standard form. See Examples 7 through 10.*

**31.** The world's tallest self-supporting structure is the CN Tower in Toronto, Canada. It is one thousand, eight hundred twenty-one feet tall. (*Source: The World Almanac*, 1998)

**32.** The average distance between Earth and the sun is more than ninety-three million miles.

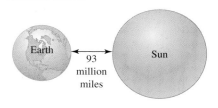

Earth — 93 million miles — Sun

**33.** Grain storage facilities in Toledo, Ohio, have a capacity of sixty-three million, one hundred thousand bushels. (*Source:* Chicago Board of Trade Market Information Department)

**34.** Bobby Hurley of Duke University holds the NCAA Men's Division I record for most career assists. He is credited with one thousand, seventy-six assists from 1990 to 1993.

**35.** One hundred million pounds of non-hazardous waste is recycled at Kodak Park each year. (*Source:* Eastman Kodak Company)

**36.** FedEx employees retrieve packages from over thirty-four thousand FedEx drop boxes around the world each business day. (*Source:* Federal Express Corporation)

Name _____

**C** *Write each whole number in expanded form. See Example 11.*

**37.** 406          **38.** 789          **39.** 5290          **40.** 6040

**41.** 62,407       **42.** 20,215       **43.** 30,680        **44.** 99,032

**45.** 39,680,000   **46.** 47,703,029   **47.** 1006          **48.** 20,304

**D** *Insert < or > to make a true statement. See Example 12.*

**49.** 3     8      **50.** 7     10      **51.** 9     0      **52.** 0     12

**53.** 6     2      **54.** 15    14      **55.** 22    0      **56.** 39    47

**E** *The table shows the five longest rivers in the world. Use this table to answer Exercises 57–61. See Example 13.*

| River | Miles |
|-------|-------|
| Chang jiang-Yangtze (China) | 3964 |
| Amazon (Brazil) | 4000 |
| Tenisei-Angara (Russia) | 3442 |
| Mississippi-Missouri (U.S.) | 3740 |
| Nile (Egypt) | 4145 |

**57.** Write the length of the Amazon River in words.

**58.** Write the length of the Tenisei-Angara River in words.

**59.** Write the length of the Nile River in expanded form.

**60.** Write the length of the Mississippi-Missouri River in expanded form.

**61.** Which river is the longest in the world?

**37.** 400 + 6

**38.** 700 + 80 + 9

**39.** 5000 + 200 + 90

**40.** 6000 + 40

**41.** 60,000 + 2000 + 400 + 7

**42.** 20,000 + 200 + 10 + 5

**43.** 30,000 + 600 + 80

**44.** 90,000 + 9000 + 30 + 2

**45.** 30,000,000 + 9,000,000 + 600,000 + 80,000

**46.** 40,000,000 + 7,000,000 + 700,000 + 3000 + 20 + 9

**47.** 1000 + 6

**48.** 20,000 + 300 + 4

**49.** <

**50.** <

**51.** >

**52.** <

**53.** >

**54.** >

**55.** >

**56.** <

**57.** four thousand

**58.** three thousand, four hundred forty-two

**59.** 4000 + 100 + 40 + 5

**60.** 3000 + 700 + 40

**61.** Nile

**Name** _____

*The table shows the top ten breeds of dogs in 1999 according to the American Kennel Club. Use this table to answer Exercises 62–65. See Example 13.*

| TOP TEN AMERICAN KENNEL CLUB REGISTRATIONS IN 1999 | |
| --- | --- |
| Breed | Number of Registered Dogs |
| Beagle | 53,322 |
| Chihuahua | 43,468 |
| Dachshund | 53,896 |
| German Shepherd | 65,326 |
| Golden Retriever | 65,681 |
| Labrador Retriever | 157,936 |
| Pomeranian | 38,540 |
| Poodle | 51,935 |
| Rottweiler | 55,009 |
| Yorkshire Terrier | 42,900 |

(*Source:* American Kennel Club)

**62.** Which breed has the most American Kennel Club registrations? Write the number of registrations for this breed in words.

**63.** Which breed has the fewest registrations? Write the number of registered dogs for this breed in words.

**64.** Which breed has more dogs registered: Dachshund or Poodle?

**65.** Which breed has fewer dogs registered: Rottweiler or German Shepherd?

◆ **COMBINING CONCEPTS**

**66.** Write the largest four-digit number that can be made from the digits 3, 6, 7, and 2 if each digit must be used once. ___ ___ ___ ___

**67.** Write the largest five-digit number that can be made using the digits 4, 5, and 3 if each digit must be used at least once. ___ ___ , ___ ___ ___

**68.** If a number is given in words, describe the process used to write this number in standard form.

**69.** If a number is written in standard form, describe the process used to write this number in expanded form.

△ **70.** The Pro-Football Hall of Fame was established on September 7, 1963 in this town. Use the information and the diagram to find the name of the town.
▲ Alliance is East of Massillon.
▲ Dover is between Canton and New Philadelphia.
▲ Massillon is not next to Alliance.
▲ Canton is North of Dover.

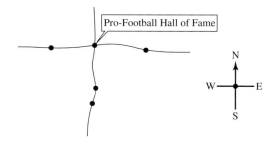

# 1.2   ADDING WHOLE NUMBERS AND PERIMETER

## A   ADDING WHOLE NUMBERS

If one computer in an office has a 2-megabyte memory and a second computer has a 4-megabyte memory, the total memory in the two computers can be found by adding 2 and 4.

$$2 \text{ megabytes} + 4 \text{ megabytes} = 6 \text{ megabytes}$$

The **sum** is 6 megabytes of memory. Each of the numbers 2 and 4 is called an **addend**.

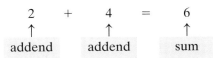

To add whole numbers, we add the digits in the ones place, then the tens place, then the hundreds place, and so on. For example, let's add $2236 + 160$.

```
  2 2 3 6     Line up numbers vertically so that the place values correspond. Then
+   1 6 0     add digits in corresponding place values, starting with the ones place.
---------
  2 3 9 6
```
↑ ↑ ↑ ↑
— sum of ones
— sum of tens
— sum of hundreds
— sum of thousands

### Example 1    Add: $23 + 136$

*Solution:*
```
   23
+ 136
-----
  159
```

When the sum of the digits in corresponding place values is more than 9, "carrying" is necessary. For example, to add $365 + 89$, add the ones-place digits first.

```
   1
  3 6 5     5 ones + 9 ones = 14 ones or 1 ten + 4 ones.
+   8 9
-------
      4     Write the 4 ones in the ones place and carry the 1 ten to the tens place.
```

Next, add the tens-place digits.

```
  1 1
  3 6 5     1 ten + 6 tens + 8 tens = 15 tens or 1 hundred + 5 tens.
+   8 9
-------
    5 4     Write the 5 tens in the tens place and carry the 1 hundred to the
            hundreds place.
```

Next, add the hundreds-place digits.

```
  1 1
  3 6 5     1 hundred + 3 hundreds = 4 hundreds.
+   8 9
-------
  4 5 4     Write the 4 hundreds in the hundreds place.
```

## Practice Problem 1

Add: $7235 + 542$

TEACHING TIP

Remind students that they can and should rewrite problems in a form which makes it easy for them to solve. If an addition problem is given horizontally, they can rewrite it vertically so it is easy to line up numbers with the same place value.

Answer
**1.** 7777

**Practice Problem 2**

Add: 27,364 + 92,977

✓ **CONCEPT CHECK**

What is wrong with the following computation?

$$
\begin{array}{r}
1288 \\
+\ 377 \\
\hline
1555
\end{array}
$$

**Example 2**   Add: 34,285 + 149,761

*Solution:*
$$
\begin{array}{r}
\overset{1\ 1\ \ 1}{34{,}285} \\
+149{,}761 \\
\hline
184{,}046
\end{array}
$$

—————

TRY THE CONCEPT CHECK IN THE MARGIN.

**B**   **USING PROPERTIES OF ADDITION**

Before we continue adding whole numbers, let's review some properties of addition that you may have already discovered. The first property that we will review is the **addition property of 0**. This property reminds us that the sum of 0 and any number is that same number.

> **ADDITION PROPERTY OF 0**
>
> The sum of 0 and any number is that number. For example,
>
> $7 + 0 = 7$
> $0 + 7 = 7$

Next, notice that we can add any two whole numbers in any order and the sum is the same. For example,

$4 + 5 = 9$   and   $5 + 4 = 9$

We call this special property of addition the **commutative property of addition**.

> **COMMUTATIVE PROPERTY OF ADDITION**
>
> Changing the **order** of two addends does not change their sum. For example,
>
> $2 + 3 = 5$   and   $3 + 2 = 5$

Another property that can help us when adding numbers is the **associative property of addition**. This property states that, when adding numbers, the grouping of the numbers can be changed without changing the sum. We use parentheses to group numbers. They indicate what numbers to add first. For example, let's use two different groupings to find the sum of $2 + 1 + 5$.

$2 + \underline{(1 + 5)} = 2 + 6 = 8$

Also,

$\underline{(2 + 1)} + 5 = 3 + 5 = 8$

Both groupings give a sum of 8.

> **ASSOCIATIVE PROPERTY OF ADDITION**
>
> Changing the grouping of addends does not change their sum. For example
>
> $3 + \underline{(5 + 7)} = 3 + 12 = 15$   and   $\underline{(3 + 5)} + 7 = 8 + 7 = 15$

TEACHING TIP

Help students build number flexibility by asking them to notice patterns in the addition table in the appendix. For instance, they might notice that when two odd numbers are added, the result is even. They might also notice that when a two-digit number is the result of adding 9 and another number, the ones digit of the result is one less than the other number.

**Answers**

**2.** 120,341

✓ **Concept Check:**
Forgot to carry 1 ten to the tens place and 1 hundred to the hundreds place.

The commutative and associative properties tell us that we can add whole numbers using any order and grouping that we want.

When adding several numbers, it is often helpful to look for two or three numbers whose sum is 10, 20, and so on.

**Example 3**   Add: $13 + 2 + 7 + 8 + 9$

*Solution:*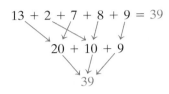

**Example 4**   Add: $1647 + 246 + 32 + 85$

*Solution:*
$$\begin{array}{r} \overset{122}{\phantom{0}} \\ 1647 \\ 246 \\ 32 \\ +\ \ 85 \\ \hline 2010 \end{array}$$

△ **C** FINDING THE PERIMETER OF A POLYGON

A special application of addition is finding the perimeter of a polygon. A **polygon** can be described as a flat figure formed by line segments connected at their ends. (For more review, see Appendix C.) Geometric figures such as triangles, squares. and rectangles are called polygons.

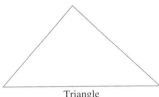

| Triangle | Square | Rectangle |

The **perimeter** of a polygon is the *distance around* the polygon. This means that the perimeter of a polygon is the sum of the lengths of its sides. (For a complete discussion of lengths, see Section 9.2.)

△ **Example 5**   Find the perimeter of the polygon shown.

*Solution:*   To find the perimeter (distance around), we add the lengths of the sides.

2 in. + 3 in. + 1 in. + 3 in. + 4 in. = 13 in.

The perimeter is 13 inches.

---

**Practice Problem 3**

Add: $11 + 7 + 8 + 9 + 13$

**Practice Problem 4**

Add: $19 + 5042 + 638 + 526$

TEACHING TIP

Ask students what they think of when they hear the word "rim." Then point out the word "rim" in perimeter and tell them to use this clue to remember that perimeter is the distance around the "rim" of a figure.

△ **Practice Problem 5**

Find the perimeter of the polygon shown. (A centimeter is a unit of length in the metric system.)

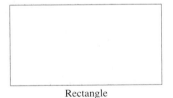

**Answers**

**3.** 48   **4.** 6225   **5.** 27 centimeters

△ **Practice Problem 6**

A new shopping mall has its floor plan in the shape of a triangle. Each of the mall's three sides is 532 feet. Find the perimeter of the building.

532 feet

TEACHING TIP

Ask students if they would know how much fencing material to buy if they were told to enclose a garden with a perimeter of 27. Point out that we need to know the units in which the perimeter was measured. If it was measured in yards, we would buy 27 yards but if it was measured in feet, we would only need 27 feet. Emphasize that units should always be included when giving a perimeter.

△ **Example 6**    Calculating the Perimeter of a Building

The largest commercial building in the world under one roof is the flower auction building of the cooperative VBA in Aalsmeer, Netherlands. The floor plan is a rectangle that measures 776 meters by 639 meters. Find the perimeter of this building. (A meter is a unit of length in the metric system.) (*Source: The Handy Science Answer Book*, Visible Ink Press, 1994)

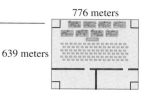

776 meters

639 meters

HELPFUL HINT

The square in each corner denotes that the angle is a right angle (90°).

*Solution:*    Recall that opposite sides of a rectangle have the same length. To find the perimeter of this building, we add the lengths of the sides. The sum of the lengths of its sides is

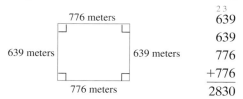

776 meters

639 meters       639 meters

776 meters

$$\begin{array}{r} \overset{2\,3}{639} \\ 639 \\ 776 \\ +776 \\ \hline 2830 \end{array}$$

The perimeter of the building is 2830 meters.

**D**  **SOLVING PROBLEMS BY ADDING**

Often, real-life problems occur that can be solved by writing an addition statement. The first step to solving any word problem is to *understand* the problem by reading it carefully. Descriptions of problems solved through addition *may* include any of these key words or phrases.

| Key Words or Phrases | Example | Symbols |
|---|---|---|
| **added to** | 5 added to 7 | 7 + 5 |
| **plus** | 0 plus 78 | 0 + 78 |
| **increased by** | 12 increased by 6 | 12 + 6 |
| **more than** | 11 more than 25 | 25 + 11 |
| **total** | the total of 8 and 1 | 8 + 1 |
| **sum** | the sum of 4 and 133 | 4 + 133 |

To solve a word problem that involves addition, we first use the facts described to write an addition statement. Then we write the corresponding solution of the real-life problem. It is sometimes helpful to write the statement in words (brief phrases) and then translate to numbers.

**Example 7**    Finding a Salary

The yearly salary for governor of Georgia was recently increased by $12,865. If the old salary was $103,074, find the new salary.

**Practice Problem 7**

The yearly salary for governor of North Dakota was recently increased by $2196. If the old salary was $73,176, find the new salary. (*Source: The World Almanac*, 1998, 1999, 2000)

**Answers**

**6.** 1596 feet   **7.** $75,372

*Solution:*    The key phrase here is "increased by" and suggests that we add. To find the new salary for governor of Georgia, we add the increase, $12,865, to the old salary, $103,074.

| In Words | | Translate to Numbers |
|----------|---|---------------------|
| old salary | $\rightarrow$ | 103,074 |
| + increase | $\rightarrow$ | 12,865 |
| new salary | $\rightarrow$ | 115,939 |

The new salary is $115,939 per year.

## Example 8    Determining the Number of Baseball Cards in a Collection

Alan Mayfield collects baseball cards. He has 109 cards for New York Yankees, 96 for Chicago White Sox, 79 for Kansas City Royals, 42 for Seattle Mariners, 67 for Oakland Athletics, and 52 for California Angels. How many cards does he have in total?

*Solution:*    The key word here is "total." To find the total number of Alan's baseball cards, we find the sum of the quantities from each team.

| In Words | | Translate to Numbers |
|----------|---|---------------------|
| | | 33 |
| New York Yankee cards | $\rightarrow$ | 109 |
| Chicago White Sox cards | $\rightarrow$ | 96 |
| Kansas City Royal cards | $\rightarrow$ | 79 |
| Seattle Mariner cards | $\rightarrow$ | 42 |
| Oakland Athletic cards | $\rightarrow$ | 67 |
| + California Angel cards | $\rightarrow$ | + 52 |
| Total cards | $\rightarrow$ | 445 |

Alan has a total of 445 baseball cards.

**Practice Problem 8**

Elham Abo-Zahrah collects thimbles. She has 42 glass thimbles, 17 steel thimbles, 37 porcelain thimbles, 9 silver thimbles, and 15 plastic thimbles. How many thimbles are in her collection?

Answer

**8.** 120 thimbles

## CALCULATOR EXPLORATIONS
### ADDING NUMBERS

To add numbers on a calculator, find the keys marked $\boxed{+}$ and $\boxed{=}$ (or $\boxed{\text{ENTER}}$).

For example, to add 5 and 7 on a calculator, press the keys $\boxed{5}$ $\boxed{+}$ $\boxed{7}$ $\boxed{=}$ (or $\boxed{\text{ENTER}}$).
The display will read $\boxed{\qquad 12 \quad}$.
Thus, $5 + 7 = 12$.
To add 687 and 981 on a calculator, press the keys
$\boxed{687}$ $\boxed{+}$ $\boxed{981}$ $\boxed{=}$ (or $\boxed{\text{ENTER}}$).
The display will read $\boxed{\qquad 1668 \quad}$.

Thus, $687 + 981 = 1668$. (Although entering 687, for example, requires pressing more than one key, numbers are grouped together for easier reading.)

*Use a calculator to add.*

**1.** $89 + 45$    134

**2.** $76 + 97$    173

**3.** $285 + 55$    340

**4.** $8773 + 652$    9425

**5.**
$$\begin{array}{r} 985 \\ 1210 \\ 562 \\ + \ 77 \\ \hline 2834 \end{array}$$

**6.**
$$\begin{array}{r} 465 \\ 9888 \\ 620 \\ + \ 1550 \\ \hline 12{,}523 \end{array}$$

Name _____ Section _____ Date _____

## MENTAL MATH

*Find each sum.*

**1.** 5 + 7

**2.** 20 + 30

**3.** 5000 + 4000

**4.** 4300 + 26

**5.** 1620 + 0

**6.** 6 + 126 + 4

## EXERCISE SET 1.2

**A** **B** *Add. See Examples 1 through 4.*

**1.**  14
  +22

**2.**  27
  +31

**3.**  62
  +30

**4.**  37
  +42

**5.**  12
  13
  +24

**6.**  23
  45
  +30

**7.**  5267
  + 132

**8.**  236
  +6243

**9.** 53 + 64

**10.** 41 + 74

**11.** 22 + 49

**12.** 35 + 47

**13.** 38 + 79

**14.** 92 + 37

**15.**  8
  9
  2
  5
  +1

**16.**  3
  5
  8
  5
  +7

**17.**  6
  21
  14
  9
  +12

**18.**  12
  4
  8
  26
  +10

**19.**  81
  17
  23
  79
  +12

**20.**  64
  28
  56
  25
  +32

MENTAL MATH ANSWERS

**1.** 12

**2.** 50

**3.** 9000

**4.** 4326

**5.** 1620

**6.** 136

ANSWERS

**1.** 36

**2.** 58

**3.** 92

**4.** 79

**5.** 49

**6.** 98

**7.** 5399

**8.** 6479

**9.** 117

**10.** 115

**11.** 71

**12.** 82

**13.** 117

**14.** 129

**15.** 25

**16.** 28

**17.** 62

**18.** 60

**19.** 212

**20.** 205

**Name** _____

**21.** 62 + 18 + 14          **22.** 23 + 49 + 18          **23.** 40 + 800 + 70

**24.** 30 + 900 + 20          **25.** 7542 + 49 + 682          **26.** 1624 + 1832 + 1976

**27.** 24 + 9006 + 489 + 2407          **28.** 16 + 748 + 1056 + 770

| **29.** | **30.** | **31.** | **32.** |
|---|---|---|---|
| 627 | 427 | 6820 | 6789 |
| 628 | 383 | 4271 | 4321 |
| +629 | +229 | +5626 | +5555 |

| **33.** | **34.** | **35.** | **36.** |
|---|---|---|---|
| 507 | 864 | 4200 | 5000 |
| 593 | 733 | 2107 | 400 |
| + 10 | +356 | +2692 | +3021 |

| **37.** | **38.** | **39.** | **40.** |
|---|---|---|---|
| 49 | 26 | 121,742 | 504,218 |
| 628 | 582 | 57,279 | 321,920 |
| 5762 | 4763 | 6586 | 38,507 |
| +29,462 | +62,511 | +426,782 | +594,687 |

**C** *Find the perimeter of each figure. See Examples 5 and 6.*

△ **41.**

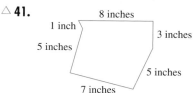

8 inches
1 inch
5 inches
5 inches
7 inches
3 inches

△ **42.**

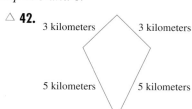

3 kilometers     3 kilometers
5 kilometers     5 kilometers

**18**

**Name** _____

**43.** 25 ft _____

 **43.**

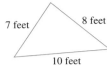

7 feet   8 feet   10 feet

△ **44.**

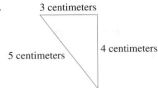

3 centimeters   5 centimeters   4 centimeters

**44.** 12 cm _____

△ **45.**

4 inches

Rectangle   8 inches

△ **46.**

8 miles

Rectangle   4 miles

**45.** 24 in. _____

**46.** 24 mi _____

△ **47.**

2 yards

2 yards   Square

△ **48.**

6 inches
5 inches   5 inches
7 inches
7 inches   3 inches
4 inches

**47.** 8 yd _____

**48.** 37 in. _____

**D** *Solve. See Examples 7 and 8.*

**49.** The distance from Kansas City, Kansas, to Hays, Kansas, is 285 miles. Colby, Kansas, is 98 miles farther from Kansas City than Hays. Find how far it is from Kansas City to Colby.

**50.** The highest point in Kansas is Mt. Sunflower at 4039 feet above sea level. The highest mountain in the world is Mt. Everest in Asia. Its peak is 24,989 feet higher than Mt. Sunflower. Find how high Mt. Everest is. (*Sources:* U.S. Geological Survey and National Geographic Society)

**49.** 383 mi _____

**50.** 29,028 ft _____

**Name** _____

**51.** Leo Callier is installing an invisible fence in his back yard. How many feet of wiring is needed to enclose his yard?

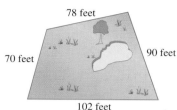

78 feet

70 feet      90 feet

102 feet

△ **52.** A homeowner is considering adding gutters around her home. Find the perimeter of her rectangular home.

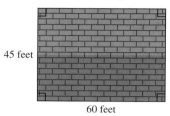

45 feet

60 feet

**53.** The Limited, Inc., had 105,600 employees worldwide in 1995. In 1997, there were 17,500 more Limited employees than in 1995. How many people were employed by the Limited, Inc. in 1997? (*Source:* The Limited, Inc.)

**54.** Papa John's is the fourth-largest pizza chain in the United States. In 1995, Papa John's employed 6762 people. In 1996, employment increased by 2782 employees. How many people did Papa John's employ in 1996? (*Source:* Papa John's International, Inc.)

**55.** As of May 30, 1999, the Broadway show *Cats* had staged 1918 more performances than the Broadway show *Les Miserables*. At that time, *Les Miserables* had staged 5031 performances. How many performances had *Cats* staged by June 1, 1997? (*Source:* The League of American Theatres and Producers)

**56.** Charles Schulz, the creator of the Peanuts comic strip, was born in 1922. Scott Adams, the creator of the Dilbert comic strip, was born 35 years later. In what year was Scott Adams born? (*Source: The World Almanac*, 2000)

**57.** There were 12,166 kidney transplants and 973 kidney-pancreas transplants performed in the United States in 1998. What was the total number of organ transplants performed in 1998 that involved a kidney? (*Source:* United Network for Organ Sharing)

**58.** There were 2345 heart transplants and 47 heart-lung transplants performed in the United States in 1998. What was the total number of organ transplants performed in 1998 that involved a heart? (*Source:* United Network for Organ Sharing)

**Name** _____

**59.** The highest waterfall in the United States is Yosemite Falls in Yosemite National Park in California. Yosemite Falls is made up of three sections, as shown in the table. What is the total height of Yosemite Falls? (*Source:* U.S. Department of the Interior)

| Yosemite Falls | Height |
|---|---|
| Upper Yosemite Falls | 1430 feet |
| Cascades | 675 feet |
| Lower Yosemite Falls | 320 feet |

**60.** Jordan White, a nurse at Mercy Hospital, is recording fluid intake in a patient's medical chart. During his shift, the patient had the following types and amounts of intake measured in cubic centimeters (cc). What amount should Jordan record as the total fluid intake for this patient?

| Oral | Intravenous | Blood |
|---|---|---|
| 240 | 500 | 500 |
| 100 | 200 | |
| 355 | | |

**61.** The state of Alaska has 1795 miles of urban highways and 11,460 miles of rural highways. Find the total highway mileage in Alaska. (*Source:* U.S. Federal Highway Administration)

**62.** The state of Hawaii has 1851 miles of urban highways and 2291 miles of rural highways. Find the total highway mileage in Hawaii. (*Source:* U.S. Federal Highway Administration)

 **COMBINING CONCEPTS**

*The table shows the number of Wal-Mart stores (including Sam's Club) in 14 states. Use this table to answer Exercises 63–67.*

| THE TOP STATES FOR WAL-MART STORES IN 1998 | |
|---|---|
| State | Number of Stores |
| Alabama | 85 |
| Arkansas | 81 |
| California | 124 |
| Florida | 166 |
| Georgia | 103 |
| Illinois | 130 |
| Indiana | 89 |
| Louisiana | 84 |
| Missouri | 121 |
| North Carolina | 100 |
| Ohio | 104 |
| Oklahoma | 84 |
| Tennessee | 99 |
| Texas | 293 |

(*Source:* Wal-Mart Stores, Inc.)

**63.** Which state has the most Wal-Mart stores?

**64.** The mid-South can be defined as Arkansas, Louisiana, Texas, and Oklahoma. What is the total number of Wal-Mart stores in the mid-South?

**65.** What is the total number of Wal-Mart stores located in the three states with the most Wal-Mart stores?

**66.** Which pair of neighboring states has more Wal-Mart stores: Indiana and Illinois or Florida and Georgia?

**67.** 1663 stores

**67.** How many Wal-Mart stores are located in the 14 states given in the table? Use a calculator to check your total.

**68.** answers may vary

**68.** In your own words, explain the commutative property of addition.

**69.** In your own words, explain the associative property of addition.

**69.** answers may vary

**70.** Add: 78,962 + 129,968,350 + 36,462,880

**70.** 166,510,192

**71.** Add: 56,468,980 + 1,236,785 + 986,768,000

**71.** 1,044,473,765

# 1.3  SUBTRACTING WHOLE NUMBERS

## A  SUBTRACTING WHOLE NUMBERS

If you have $5 and someone gives you $3, you have a total of $8 since $5 + 3 = 8$. Similarly, if you have $8, and then someone borrows $3, you have $5 left. **Subtraction** is finding the **difference** of two numbers.

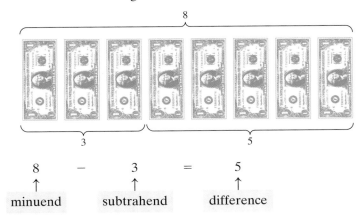

$$8 \quad - \quad 3 \quad = \quad 5$$

minuend    subtrahend    difference

Notice that addition and subtraction are very closely related. In fact, subtraction is defined in terms of addition. For example, $8 - 3$ is the number that when added to 3 gives 8. In other words,

$$8 - 3 = 5 \quad \text{because} \quad 5 + 3 = 8$$

This means that subtraction can be *checked* by addition.

**Example 1**  Subtract. Check each answer by adding.

  **a.** $12 - 9$   **b.** $11 - 6$   **c.** $5 - 5$   **d.** $7 - 0$

*Solution:*   **a.** $12 - 9 = 3$ because $3 + 9 = 12$
  **b.** $11 - 6 = 5$ because $5 + 6 = 11$
  **c.** $5 - 5 = 0$ because $0 + 5 = 5$
  **d.** $7 - 0 = 7$ because $7 + 0 = 7$

Look again at Examples 1(c) and 1(d).

1(c) $5 - 5 = 0$         1(d) $7 - 0 = 7$

same         difference         a number         difference is the
number         is 0         minus 0         same number

These two examples illustrate the **subtraction properties of 0**.

---

**SUBTRACTION PROPERTIES OF 0**

The difference of any number and that same number is 0. For example,

$$11 - 11 = 0$$

The difference of any number and 0 is that same number. For example,

$$45 - 0 = 45$$

---

### Objectives

**A** Subtract whole numbers.
**B** Subtract whole numbers when borrowing is necessary.
**C** Solve problems by subtracting whole numbers.

Study Guide    SSM    CD-ROM    Video 1.3

TEACHING TIP

Ask students to verbalize all the different ways they can think of to state that $8 - 3 = 5$.
For example:
Eight minus 3 equals 5.
Eight take away 3 equals 5.
The difference of 8 and 3 is 5.
Eight decreased by 3 is 5.
Three subtracted from 8 is 5.
Subtracting 3 from 8 results in 5.
Eight less 3 is 5.
Three less than 8 is 5.

## Practice Problem 1

Subtract. Check each answer by adding.
a. $14 - 9$          b. $9 - 9$
c. $4 - 0$

Answers

**1. a.** 5  **b.** 0  **c.** 4

When subtraction involves numbers of two or more digits, it is more convenient to subtract vertically. For example, to subtract $893 - 52$,

$$
\begin{array}{r}
8\;9\;3 \\
-\;\;5\;2 \\
\hline
8\;4\;1
\end{array}
$$

← minuend
← subtrahend
← difference

Line up numbers vertically so that the minuend is on top and the place values correspond. Subtract in corresponding places, starting with the ones place.

$3 - 2$
$9 - 5$
$8 - 0$

To check, add.

$$
\begin{array}{r}
\text{difference} \\
+\ \text{subtrahend} \\
\hline
\text{minuend}
\end{array}
\quad \text{or} \quad
\begin{array}{r}
841 \\
+\ 52 \\
\hline
893
\end{array}
$$

← Since this is the original minuend, the problem checks.

### Example 2

Subtract: $7826 - 505$. Check by adding.

*Solution:*

$$
\begin{array}{r}
7826 \\
-\ 505 \\
\hline
7321
\end{array}
$$

*Check:*

$$
\begin{array}{r}
7321 \\
+\ 505 \\
\hline
7826
\end{array}
$$

### B SUBTRACTING WITH BORROWING

When a digit in the second number (subtrahend) is larger than the corresponding digit in the first number (minuend), **borrowing** is necessary. For example, consider

$$
\begin{array}{r}
81 \\
-\ 63
\end{array}
$$

Since 3 in the ones place of 63 is larger than 1 in the ones place of 81, borrowing is necessary. We borrow 1 ten from the tens place and add it to the ones places.

**Borrowing**

$8 - 1 = 7 \rightarrow$   7   11 ← 1 ten + 1 one = 11 ones

tens   ten   tens

$$
\begin{array}{r}
\cancel{8}\;\cancel{1} \\
-\ 6\;3
\end{array}
$$

Now we subtract the ones-place digits and then the tens-place digits.

$$
\begin{array}{r}
{}^{7}\ {}^{11} \\
\cancel{8}\;\cancel{1} \\
-\ 6\;3 \\
\hline
1\;8
\end{array}
$$

← $11 - 3 = 8$

$7 - 6 = 1$

*Check:*

$$
\begin{array}{r}
18 \\
+\ 63 \\
\hline
81
\end{array}
$$

The original minuend

### Example 3

Subtract $43 - 29$. Check by adding.

*Solution:*

$$
\begin{array}{r}
{}^{3}\ {}^{13} \\
\cancel{4}\;\cancel{3} \\
-\ 2\;9 \\
\hline
1\;4
\end{array}
$$

*Check:*

$$
\begin{array}{r}
14 \\
+\ 29 \\
\hline
43
\end{array}
$$

---

Copyright 2000 Prentice-Hall, Inc.

---

### Practice Problem 2

Subtract. Check by adding.
a. $4689 - 253$     b. $981 - 630$

TEACHING TIP

Some students may find it easy to solve subtraction problems by thinking "what do I add to the bottom number (subtrahend) to get the top number (minuend)?" For example,

$$
\begin{array}{r}
739 \\
-24
\end{array}
$$

ones:     What do I add to 4 to get 9? (5)

tens:     What do I add to 2 to get 3? (1)

hundreds:     What do I add to 0 to get 7? (7)

Answer: 715

### Practice Problem 3

Subtract. Check by adding.

a. $\begin{array}{r} 71 \\ -\ 19 \end{array}$     b. $\begin{array}{r} 1136 \\ -\ 914 \end{array}$

c. $\begin{array}{r} 8627 \\ -\ 4119 \end{array}$

**Answers**

**2. a.** 4436 **b.** 351 **3. a.** 52 **b.** 222 **c.** 4508

Sometimes we may have to borrow from more than one place. For example, to subtract $7631 - 152$, we first borrow from the tens place.

$$
\begin{array}{r}
\overset{2\ \ 11}{7\ 6\ \cancel{3}\ \cancel{1}} \\
-\ 1\ 5\ 2 \\
\hline
9
\end{array}
\quad \leftarrow 11 - 2 = 9
$$

In the tens place, 5 is greater than 2, so we borrow again. This time we borrow from the hundreds place.

$$
\begin{array}{r}
6\ \text{hundreds} - \textbf{1 hundred} = 5\ \text{hundreds} \\
\overset{\ \ 12}{5\ \cancel{2}\ 11} \\
7\ \cancel{6}\ \cancel{3}\ \cancel{1} \\
-\ 1\ 5\ 2 \\
\hline
7\ 4\ 7\ 9
\end{array}
$$

$\left\{ \begin{array}{l} \textbf{1 hundred} + 2\ \text{tens or} \\ 10\ \text{tens} + 2\ \text{tens} = \textbf{12 tens} \end{array} \right.$

*Check:*

$$
\begin{array}{r}
7479 \\
+\ 152 \\
\hline
7631
\end{array}
\quad \text{The original minuend}
$$

## Example 4

Subtract: $900 - 174$. Check by adding.

*Solution:* In the ones place, 4 is larger than 0, so we borrow from the tens place. But the tens place of 900 is 0, so to borrow from the tens place we must first borrow from the hundreds place.

$$
\begin{array}{r}
\overset{8\ \ 10}{\cancel{9}\ \cancel{0}\ 0} \\
-\ 1\ 7\ 4
\end{array}
$$

Now we can borrow from the tens place.

$$
\begin{array}{r}
\overset{\quad 9}{\ \ } \\
\overset{8\ \ \cancel{10}\ \ 10}{\cancel{9}\ \cancel{0}\ \cancel{0}} \\
-\ 1\ 7\ 4 \\
\hline
7\ 2\ 6
\end{array}
$$

*Check:*

$$
\begin{array}{r}
\overset{1\ 1}{726} \\
+\ 174 \\
\hline
900
\end{array}
$$

## C SOLVING PROBLEMS BY SUBTRACTING

Descriptions of real-life problems that suggest solving by subtracting include these key words or phrases.

| Key Words or Phrases | Examples | Symbols |
|---|---|---|
| **subtract** | subtract 5 from 8 | $8 - 5$ |
| **difference** | the difference of 10 and 2 | $10 - 2$ |
| **less** | 17 less 3 | $17 - 3$ |
| **take away** | 14 take away 9 | $14 - 9$ |
| **decreased by** | 7 decreased by 5 | $7 - 5$ |
| **subtracted from** | 9 subtracted from 12 | $12 - 9$ |

**TRY THE CONCEPT CHECK IN THE MARGIN.**

**Practice Problem 4**

Subtract. Check by adding.

a. $\begin{array}{r} 400 \\ -\ 164 \end{array}$     b. $\begin{array}{r} 200 \\ -\ 45 \end{array}$

c. $\begin{array}{r} 1000 \\ -\ 762 \end{array}$

**TEACHING TIP**

Point out to students that the order of the numbers in a subtraction translation is important. Point out that it depends on the meaning of the key word or phrase that is used.

## ✓ CONCEPT CHECK

In each of the following problems, identify which number is the minuend and which number is the subtrahend.

a. What is the result when 9 is subtracted from 20?

b. What is the difference of 15 and 8?

c. Find a number that is 15 fewer than 23.

**Answers**

**4. a.** 236 **b.** 155 **c.** 238

✓ Concept Check:

**a.** minuend: 20, subtrahend: 9

**b.** minuend: 15, subtrahend: 8

**c.** minuend: 23, subtrahend: 15

## Practice Problem 5

The radius of Earth is 6378 kilometers. The radius of Mars is 2981 kilometers less than the radius of Earth. What is the radius of Mars? (*Source:* National Space Science Data Center)

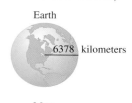

Earth

6378 kilometers

Mars

?

## Practice Problem 6

A new suit originally priced at $92 is now on sale for $47. How much money was taken off the original price?

## Practice Problem 7

Use the graph in Example 7 to answer each question.

a. Find the total number of people who responded to the five favorite sport activities. (Assume that each person could respond to only one activity.)

b. How many more people responded that walking rather than bowling is their favorite activity?

**Answers**

**5.** 3397 kilometers   **6.** $45
**7. a.** 122 people   **b.** 19 people

△ **Example 5**    Finding the Radius of a Planet

The radius of Venus is 6052 kilometers. The radius of Mercury is 3612 kilometers less than the radius of Venus. Find the radius of Mercury. (*Source:* National Space Science Data Center)

> **HELPFUL HINT**
>
> Recall that the radius of a sphere is the distance from its center to any point on the sphere.

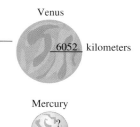

Venus

6052 kilometers

Mercury

?

*Solution:*

| Words | | Translate to Numbers |
|---|---|---|
| radius of Venus | → | $\overset{5\ \ 10}{\cancel{6}\,\cancel{0}\,5\,2}$ |
| less 3612 | → | $-\ 3\ 6\ 1\ 2$ |
| radius of Mercury | → | $2\ 4\ 4\ 0$ |

The radius of Mercury is 2440 kilometers.

**Example 6**    Calculating Miles Per Gallon

A subcompact car gets 42 miles per gallon of gas. A full-size car gets 17 miles per gallon of gas. How many more miles per gallon does the subcompact car get than the full-size car?

*Solution:*

| | Words | | Translate to Numbers |
|---|---|---|---|
| | subcompact miles per gallon | → | $\overset{3\ \ 12}{\cancel{4}\,\cancel{2}}$ |
| − | full-size miles per gallon | → | $-\ 1\ 7$ |
| | more miles per gallon | → | $2\ 5$ |

The subcompact car gets 25 more miles per gallon than the full-size car.

> **Helpful Hint**
>
> Don't forget that a subtraction problem can be checked by adding.

Graphs can be used to visualize data. The graph in Example 7 is called a **bar graph**.

**Example 7**    Reading a Bar Graph

A telephone survey was taken to identify favorite sport activities, and the results from the five most popular activities are shown in the form of a bar graph. In this particular graph, each bar represents a different sport activity, and the height of each bar represents the number of people who responded that the particular sport was their favorite activity. Use this graph to answer each question.

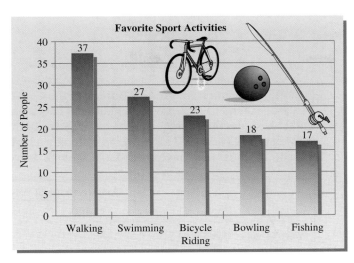

**a.** What activity was preferred by the most people?

**b.** How many more people responded that swimming rather than fishing is their favorite activity?

*Solution:*    **a.** The activity preferred by the most people is the one with the tallest bar, which is walking.

**b.** The number of people who responded with swimming is 27. The number of people who responded with fishing is 17. To find how many more responded with swimming than with fishing, find the difference:

$$27 - 17 = 10$$

Ten more people responded with swimming than with fishing. ■

# CALCULATOR EXPLORATIONS
## SUBTRACTING NUMBERS

To subtract numbers on a calculator, find the keys marked $\boxed{-}$ and $\boxed{=}$ (or $\boxed{\text{ENTER}}$).

For example, to find $83 - 49$ on a calculator, press the keys $\boxed{83}$ $\boxed{-}$ $\boxed{49}$ $\boxed{=}$ (or $\boxed{\text{ENTER}}$).

The display will read $\boxed{\phantom{xxxxxx} 34}$. Thus, $83 - 49 = 34$.

*Use a calculator to subtract.*

**1.** $865 - 95$    770

**2.** $76 - 27$    49

**3.** $147 - 38$    109

**4.** $366 - 87$    279

**5.** $9625 - 647$    8978

**6.** $10,711 - 8925$    1786

# Focus on Business and Career

### THE EARNING POWER OF A COLLEGE DEGREE

According to data from the U.S. Bureau of the Census, average annual wages tend to increase with additional education. The following table shows the average annual earnings for both men and women in 1993 and 1997 for workers with only a high school diploma and for those who have earned a bachelor's degree.

| Gender | Year | Average Annual Wages | |
| | | High School Diploma | Bachelor's Degree |
| --- | --- | --- | --- |
| Men | 1993 | $21,744 | $41,160 |
| | 1997 | $27,642 | $46,702 |
| Women | 1993 | $12,096 | $21,708 |
| | 1997 | $16,161 | $28,701 |

(*Source:* U.S. Bureau of the Census)

### CRITICAL THINKING

*Use the table to answer each question.*

1. First, analyze men's earnings.
   **a.** Did the average annual wages for men with a high school diploma increase or decrease from 1993 to 1997? By how much?
   **b.** Did the average annual wages for men with a bachelor's degree increase or decrease from 1993 to 1997? By how much?
   **c.** How much more could a man with a bachelor's degree earn than a man with a high school diploma in 1993? In 1997?

2. Now analyze women's earnings.
   **a.** Did the average annual wages for women with a high school diploma increase or decrease from 1993 to 1997? By how much?
   **b.** Did the average annual wages for women with a bachelor's degree increase or decrease from 1993 to 1997? By how much?
   **c.** How much more could a woman with a bachelor's degree earn than a woman with a high school diploma in 1993? In 1997?

3. Now compare men's and women's earnings.
   **a.** Find the difference between average annual wages (the "wage gap") for men and women with high school diplomas in 1993.
   **b.** Find the wage gap for men and women with high school diplomas in 1997.
   **c.** Find the wage gap for men and women with bachelor's degrees in 1993.
   **d.** Find the wage gap for men and women with bachelor's degrees in 1997.

4. Write a paragraph summarizing the conclusions that can be drawn from this table of data. Identify any apparent trends.

## MENTAL MATH

*Find each difference.*

**1.** $9 - 2$      **2.** $6 - 6$      **3.** $5 - 0$      **4.** $44 - 22$

**5.** $93 - 93$      **6.** $700 - 400$      **7.** $700 - 300$      **8.** $700 - 700$

**9.** $600 - 100$      **10.** $600 - 0$

# EXERCISE SET 1.3

**A** *Subtract. Check by adding. See Examples 1 and 2.*

**1.**
$$\begin{array}{r} 67 \\ -23 \\ \hline \end{array}$$
**2.**
$$\begin{array}{r} 72 \\ -41 \\ \hline \end{array}$$
**3.**
$$\begin{array}{r} 82 \\ -22 \\ \hline \end{array}$$
**4.**
$$\begin{array}{r} 27 \\ -10 \\ \hline \end{array}$$

**5.**
$$\begin{array}{r} 389 \\ -124 \\ \hline \end{array}$$
**6.**
$$\begin{array}{r} 572 \\ -321 \\ \hline \end{array}$$
**7.**
$$\begin{array}{r} 677 \\ -423 \\ \hline \end{array}$$
**8.**
$$\begin{array}{r} 766 \\ -324 \\ \hline \end{array}$$

**9.**
$$\begin{array}{r} 998 \\ -453 \\ \hline \end{array}$$
**10.**
$$\begin{array}{r} 912 \\ -610 \\ \hline \end{array}$$
**11.**
$$\begin{array}{r} 749 \\ -149 \\ \hline \end{array}$$
**12.**
$$\begin{array}{r} 257 \\ -257 \\ \hline \end{array}$$

**B** *Subtract. Check by adding. See Examples 1 through 4.*

**13.**
$$\begin{array}{r} 62 \\ -37 \\ \hline \end{array}$$
**14.**
$$\begin{array}{r} 55 \\ -29 \\ \hline \end{array}$$
**15.**
$$\begin{array}{r} 70 \\ -25 \\ \hline \end{array}$$
**16.**
$$\begin{array}{r} 80 \\ -37 \\ \hline \end{array}$$

**17.**
$$\begin{array}{r} 938 \\ -792 \\ \hline \end{array}$$
**18.**
$$\begin{array}{r} 436 \\ -275 \\ \hline \end{array}$$
**19.**
$$\begin{array}{r} 922 \\ -634 \\ \hline \end{array}$$
**20.**
$$\begin{array}{r} 674 \\ -299 \\ \hline \end{array}$$

**MENTAL MATH ANSWERS**

**1.** 7
**2.** 0
**3.** 5
**4.** 22
**5.** 0
**6.** 300
**7.** 400
**8.** 0
**9.** 500
**10.** 600

**ANSWERS**

**1.** 44
**2.** 31
**3.** 60
**4.** 17
**5.** 265
**6.** 251
**7.** 254
**8.** 442
**9.** 545
**10.** 302
**11.** 600
**12.** 0
**13.** 25
**14.** 26
**15.** 45
**16.** 43
**17.** 146
**18.** 161
**19.** 288
**20.** 375

**Name** _____

**21.**  600
      −432

**22.**  300
      −149

**23.**  42
      −36

**24.**  73
      −29

**25.**  923
      −476

**26.**  813
      −227

**27.**  6283
      − 560

**28.**  5349
      − 720

**29.**  533
      − 29

**30.**  724
      − 16

**31.**  200
      −111

**32.**  300
      −211

**33.**  1983
      −1904

**34.**  1983
      −1914

**35.**  56,422
      −16,508

**36.**  76,652
      −29,498

**37.** $50,000 - 17,289$     **38.** $40,000 - 23,582$     **39.** $7020 - 1979$

**40.** $6050 - 1878$     **41.** $51,111 - 19,898$     **42.** $62,222 - 39,898$

**43.** Subtract 5 from 9.

**44.** Subtract 9 from 21.

**45.** Find the difference of 41 and 21.

**46.** Find the difference of 16 and 5.

**47.** Subtract 56 from 63.

**48.** Subtract 41 from 59.

**C** *Solve. See Examples 5 through 7.*

**49.** Michelle and Ronnie Brower entered a crawfish-eating contest. Michelle ate 63 crawfish, which was 7 more than Ronnie ate. Find how many crawfish Ronnie ate.

**50.** A stock worth $135 per share on October 10 dropped to $78 per share on October 30 of the same year. Find how much it lost in value from October 10th to the 30th.

**51.** Dyllis King is reading a 503-page book. If she has just finished reading page 239, how many more pages must she read to finish the book?

**52.** When Lou and Judy Zawislak began a trip, the odometer on their car read 55,492. When the trip was over, the odometer read 59,320. How many miles did they drive on their trip?

**53.** The peak of Mt. McKinley in Alaska is 20,320 feet above sea level. The peak of Long's Peak in Colorado is 14,255 feet above sea level. How much higher is the peak of Mt. McKinley than Long's Peak? (*Source:* U.S. Geological Survey)

**54.** On one day in May the temperature in Paddin, Indiana, dropped 27 degrees from 2 p.m. to 4 p.m. If the temperature at 2 p.m. was 73° Fahrenheit, what was the temperature at 4 p.m.?

**55.** Buhler Gomez has a total of $539 in his checking account. If he writes a check for each of the items below, how much money will be left in his account?

| | |
|---|---|
| South Central Bell | $  27 |
| Central LA Electric Co. | $101 |
| Mellon Finance | $236 |

**56.** Pat Salanki's blood cholesterol level is 243. The doctor tells him it should be decreased to 185. How much of a decrease is this?

**57.** The distance from Kansas City to Denver is 645 miles. Hays, Kansas, lies on the road between the two and is 287 miles from Kansas City. What is the distance between Hays and Denver?

**58.** Alan Little is trading his car in on a new car. The new car costs $25,425. His car is worth $7998. How much more money does he need to buy the new car?

**59.** A new VCR with remote control costs $525. Prunella Pasch has $914 in her savings account. How much will she have left in her savings account after she buys the VCR?

**60.** Christopher Columbus landed in America in 1492. The Revolutionary War started in 1776. How many years after Columbus landed did the war begin?

**Name** _____

**61.** A stereo that sells regularly for $547 is discounted by $99 in a sale. What is the sale price?

**62.** A meeting of the board of governors of the Mathematical Association of America was attended by 72 governors. Forty-six of the governors were men. How many were women?

**63.** The numbers of women enrolled in mathematics classes at FHSU during the fall semester were 78 in Basic Mathematics, 185 in College Algebra, and 23 in Calculus. The total number of students enrolled in these mathematics classes was 459. How many men were enrolled in these classes?

**64.** The number of cable TV systems operating in the United States in 1999 was 10,700. In 1990, there were 9575 cable TV systems in operation. How many more cable systems were operating in 1999 than in 1990? (*Source: Television and Cable Factbook*, Warren Publishing, Inc.)

**65.** Jo Keen and Trudy Waterbury were candidates for student government president. Who won the election if the votes were cast as follows? By how many votes did the winner win?

|  | Candidate | |
| Class | Jo | Trudy |
| --- | --- | --- |
| Freshman | 276 | 295 |
| Sophomore | 362 | 122 |
| Junior | 201 | 312 |
| Senior | 179 | 182 |

**66.** Two students submitted advertising budgets for a student government fund-raiser. If $1200 is available for advertising, how much excess would each budget have?

|  | Student A | Student B |
| --- | --- | --- |
| Radio ads | $600 | $300 |
| Newspaper ads | $200 | $400 |
| Posters | $150 | $240 |
| Hand bills | $120 | $170 |

**67.** Until recently, the world's largest permanent maze was located in Ruurlo, Netherlands. This maze of beech hedges covers 94,080 square feet. A new hedge maze using hibiscus bushes at the Dole Plantation in Wahiawa, Hawaii, covers 100,000 square feet. How much larger is the Dole Plantation maze than the Ruurlo maze? (*Source: The 2000 Guinness Book of Records*)

**68.** There were only 27 California condors in the entire world in 1987. By 2000, the number of California condors had increased to 157. How much of an increase was this? (*Source: California Department of Fish and Game*)

**69.** In 1997, the size of North Korea's armed forces was estimated at 1,130,000 soldiers. South Korea's fighting force was estimated at 633,000. How much bigger was the fighting force of North Korea than South Korea? (*Source: The Top 10 of Everything 1997* by Russell Ash)

**70.** The Gap, Inc., had a net income of $1,127,065,000 in 1999. In 1997 The Gap's net income was only $533,901,000. How much did The Gap's net income increase from 1997 to 1999? (*Source:* The Gap, Inc.)

**71.** A patient's white blood cell count can be an indicator of relative health or sickness. Before a surgical procedure, a patient had a white blood count (WBC) of 13,152. After surgery, the same patient had a WBC of 6473. How much did the patient's WBC decline?

**72.** In October 1998, the hole in the Earth's ozone layer over Antarctica was about 26 million square kilometers in size. At the same time the previous year, the hole was only 19 million square kilometers in size. By how much did the ozone hole grow since the previous year? (*Sources:* National Oceanic and Atmospheric Administration and the National Aeronautics and Space Administration)

*The bar graph shows the number of passenger arrivals and departures in 1998 for the top five airports in the United States. Use this graph to answer Exercises 73–76. See Example 7.*

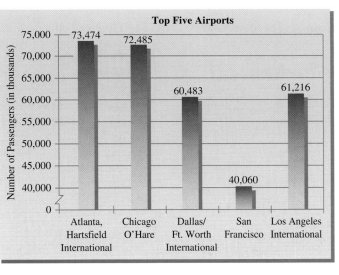

**73.** Which airport is busiest?

**74.** Which airports have more than 70 million passengers per year?

**75.** How many more passengers per year does Los Angeles International Airport have than Dallas/Ft. Worth International Airport?

**76.** How many more passengers per year does Atlanta, Hartsfield International Airport have than Chicago O'Hare Airport?

**69.** 497,000 soldiers

**70.** $593,164,000

**71.** 6679

**72.** 7 million sq. km

**73.** Atlanta, Hartsfield International

**74.** Atlanta, Hartsfield International; Chicago O'Hare

**75.** 733 thousand or 733,000 passengers

**76.** 989 thousand or 989,000 passengers

**COMBINING CONCEPTS**

*The table shows the top ten leading United States advertisers in 1998 and the amount of money spent in that year on ads. Use this table to answer Exercises 77–81. (Source: Competitive Media Reporting and Publishers Information Bureau)*

| | |
|---|---|
| General Motors | $2,121,040,900 |
| Procter & Gamble | 1,724,259,700 |
| DaimlerChrysler | 1,410,748,700 |
| Philip Morris | 1,264,353,200 |
| Ford Motor | 1,147,589,200 |
| Time Warner | 829,185,000 |
| Walt Disney | 809,878,700 |
| Sears Roebuck & Co. | 720,543,900 |
| Unilever | 691,203,500 |
| Diageo Plc | 659,093,600 |

**77.** Which companies spent more than $1500 million on ads?

**78.** Which companies spent less than $700 million on ads?

**79.** How much more money did Philip Morris spend on ads than Walt Disney?

**80.** How much more money did Ford Motor spend on ads than Unilever?

**81.** Find the total amount of money spent by the top five companies on ads.

**82.** The local college library is having a Million Pages of Reading promotion. The freshmen have read a total of 289,462 pages, the sophomores have read a total of 369,477 pages, the juniors have read a total of 218,287 pages, and the seniors have read a total of 121,685 pages. Have they reached a goal of a million pages? If not, how many more pages need to be read?

*Fill in the missing digits in each problem.*

**83.**
```
  526_
 -2_85
 ─────
  28_4
```

**84.**
```
  10,_4_
 -  85_4
 ──────
    _710
```

**85.** When subtracting two numbers, explain how you know when borrowing is necessary.

---

Answers (margin):

**77.** General Motors, Procter & Gamble

**78.** Unilever, Diageo Plc

**79.** $454,474,500

**80.** $456,385,700

**81.** $7,667,991,700

**82.** no, 1089 more pages

**83.**
```
 5269
-2385
─────
 2884
```

**84.**
```
 10,244
-8 534
──────
 1 710
```

**85.** answers may vary

# 1.4 ROUNDING AND ESTIMATING

## A ROUNDING WHOLE NUMBERS

**Rounding** a whole number means approximating it. A rounded whole number is often easier to use, understand, or remember than the precise whole number. For example, instead of trying to remember the Iowa state population as 2,851,792, it is much easier to remember it rounded to the nearest million: 3 million people.

To understand rounding, let's look at the following illustrations. The whole number 36 is closer to 40 than 30, so 36 rounded to the nearest ten is 40.

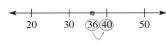

The whole number 52 rounded to the nearest ten is 50, because 52 is closer to 50 than to 60.

In trying to round 25 to the nearest ten, we see that 25 is halfway between 20 and 30. It is not closer to either number. In such a case, we round to the larger ten, that is, to 30.

To round a whole number without using a number line, follow these steps.

---

**ROUNDING WHOLE NUMBERS TO A GIVEN PLACE VALUE**

**Step 1.** Locate the digit to the right of the given place value.

**Step 2.** If this digit is 5 or greater, add 1 to the digit in the given place value and replace each digit to its right by 0.

**Step 3.** If this digit is less than 5, replace it and each digit to its right by 0.

---

**Example 1**   Round 568 to the nearest ten.

*Solution:*   5 6 ⑧   The digit to the right of the tens place is the ones place,
              ↑        which is circled.
           Tens place

           5 6 ⑧   Since the circled digit is 5 or greater, add 1 to the 6 in the
           ↑  ↖   tens place and replace the digit to the right by 0.
        Add 1.  Replace
               with 0.

We find that 568 rounded to the nearest ten is 570.

**Example 2**   Round 278,362 to the nearest thousand.

*Solution:*
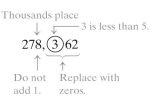

### Objectives

**A** Round whole numbers.

**B** Use rounding to estimate sums and differences.

**C** Solve problems by estimating.

Study   SSM   CD-ROM   Video
Guide                   1.4

The number 278,362 rounded to the nearest thousand is 278,000. ▬▬▬

## Practice Problem 3

Round to the nearest hundred.
a. 2777            b. 38,152
c. 762,955

**Example 3**    Round 248,982 to the nearest hundred.

*Solution:*

Hundreds place
↓ ↓    8 is greater than or equal to 5.
248,9⑧2
↑
Add 1.    9 + 1 = 10, so replace the digit 9 by 0 and carry 1 to the place value to the left.

8+1   0
2   4   8̸   9̸   8   2
       ↗    Replace
Add 1.   with zeros

The number 248,982 rounded to the nearest hundred is 249,000. ▬▬▬

## ✓ CONCEPT CHECK

Round each of the following numbers to the nearest *hundred*. Explain your reasoning.
a. 26            b. 83

**TRY THE CONCEPT CHECK IN THE MARGIN.**

### B ROUNDING TO ESTIMATE SUMS AND DIFFERENCES

By rounding addends, we can estimate sums. An estimated sum is appropriate when an exact sum is not necessary. To estimate the sum shown, round each number to the nearest hundred and then add.

| | | |
|---|---|---|
| 768 | rounds to | 800 |
| 1952 | rounds to | 2000 |
| 225 | rounds to | 200 |
| + 149 | rounds to | + 100 |
| | | 3100 |

The estimated sum is 3100, which is close to the exact sum of 3094.

TEACHING TIP

Check students' understanding of rounding by asking them to give the range of values that would round to a certain number. For example, if a number rounded to the nearest hundred is 2400, the smallest the original number could be is 2350 and the largest it could be is 2449.

## Practice Problem 4

Round each number to the nearest ten to find an estimated sum.

     79
     35
     42
     21
  + 98

**Example 4**    Round each number to the nearest hundred to find an estimated sum.

     294
     625
  1071
  + 349

*Solution:*

| | | |
|---|---|---|
| 294 | rounds to | 300 |
| 625 | rounds to | 600 |
| 1071 | rounds to | 1100 |
| + 349 | rounds to | + 300 |
| | | 2300 |

The estimated sum is 2300. (The exact sum is 2339.) ▬▬▬

**Answers**

**3. a.** 2800   **b.** 38,200   **c.** 763,000   **4.** 280
✓ Concept Check:
**a.** 0   **b.** 100

**Example 5**  Round each number to the nearest hundred to find an estimated difference.

$$4725$$
$$-2879$$

*Solution:*
| | | |
|---|---|---|
| 4725 | rounds to | 4700 |
| −2879 | rounds to | −2900 |
| | | 1800 |

The estimated difference is 1800. (The exact difference is 1846.)

### C SOLVING PROBLEMS BY ESTIMATING

Making estimates is often the quickest way to solve real-life problems when their solutions do not need to be exact.

**Example 6**  **Estimating Distances**

Jose Guillermo is trying to estimate quickly the distance from Temple, Texas, to Brenham, Texas. Round each distance given on the map to the nearest ten to estimate the total distance.

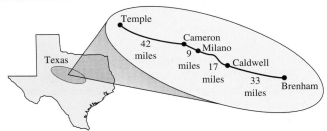

*Solution:*

| Distance | | Estimation |
|---|---|---|
| 42 | rounds to | 40 |
| 9 | rounds to | 10 |
| 17 | rounds to | 20 |
| +33 | rounds to | +30 |
| | | 100 |

It is approximately 100 miles from Temple to Brenham. (The exact distance is 101 miles.)

**Example 7**  **Estimating Data**

In three recent years the numbers of reported cases of mumps in the United States were 906, 1537, and 1692, respectively. Round each number to the nearest hundred to estimate the total number of cases reported over this period. (*Source:* National Centers for Disease Control and Prevention)

*Solution:*

| Number of Cases | | Estimation |
|---|---|---|
| 906 | rounds to | 900 |
| 1537 | rounds to | 1500 |
| +1692 | rounds to | +1700 |
| | | 4100 |

The approximate number of cases reported over this period is 4100.

---

**Practice Problem 5**

Round each number to the nearest thousand to find an estimated difference.

$$4725$$
$$-2879$$

**Practice Problem 6**

Tasha Kilbey is trying to estimate how far it is from Grove, Kansas, to Hays, Kansas. Round each distance given on the map to the nearest ten to estimate the total distance.

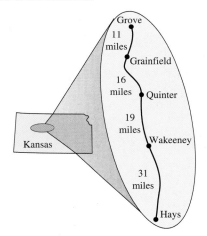

**Practice Problem 7**

In a recent year, there were 120,624 reported cases of chicken pox, 22,866 reported cases of tuberculosis, and 45,970 reported cases of salmonellosis in the United States. Round each number to the nearest ten-thousand to estimate the total number of cases reported for these diseases. (*Source:* National Centers for Disease Control and Prevention)

**Answers**

**5.** 2000  **6.** 80 miles  **7.** 190,000 cases

# Focus on Study Skills

### WHAT IS CRITICAL THINKING?

Although exact definitions often vary, critical thinking usually refers to evaluating, analyzing, and interpreting information to make a decision, draw a conclusion, reach a goal, make a prediction, or form an opinion. It often involves problem solving, communication, and reasoning skills. Critical thinking is more than a technique that helps you pass your courses—critical thinking skills are life skills. Developing these skills can help you solve problems in your workplace and in everyday life. For instance, well-developed critical thinking skills would be useful in the following situation:

> Suppose you work as a medical lab technician. Your lab supervisor has decided that some lab equipment should be replaced. She asks you to collect information on several different models from equipment manufacturers. Your assignment is to study the data and then make a recommendation on which model the lab should buy.

### HOW CAN CRITICAL THINKING BE DEVELOPED?

Just as physical exercise can help to develop and strengthen certain muscles of the body, mental exercise can help to develop critical thinking skills. Mathematics is ideal for helping to develop such skills because it requires using logic and reasoning, recognizing patterns, making conjectures and educated guesses, and drawing conclusions. You will find many opportunities to build your critical thinking skills throughout *Prealgebra*.

▲ In real-life application problems (see Exercise 92 in Section 4.5)
▲ In conceptual exercises found in the Combining Concepts subsection of the exercise sets (see Exercise 77 in Section 1.3)
▲ In the Chapter Activities (see the Chapter 2 Activity)
▲ In the Critical Thinking and Group Activities found in Focus On features like this one throughout the book (see pages 28 and 90)

## EXERCISE SET 1.4

**A** *Round each whole number to the given place value. See Examples 1 through 3.*

**1.** 632 to the nearest ten

**2.** 273 to the nearest ten

**3.** 635 to the nearest ten

**4.** 275 to the nearest ten

**5.** 792 to the nearest ten

**6.** 394 to the nearest ten

**7.** 395 to the nearest ten

**8.** 582 to the nearest ten

**9.** 1096 to the nearest ten

**10.** 2198 to the nearest ten

**11.** 42,682 to the nearest thousand

**12.** 42,682 to the nearest ten-thousand

**13.** 248,695 to the nearest hundred

**14.** 179,406 to the nearest hundred

**15.** 36,499 to the nearest thousand

**16.** 96,501 to the nearest thousand

**17.** 99,995 to the nearest ten

**18.** 39,994 to the nearest ten

**19.** 59,725,642 to the nearest ten-million

**20.** 39,523,698 to the nearest million

**21.** see table

**22.** see table

**23.** see table

**24.** see table

**25.** see table

**26.** see table

**27.** 11,000

**28.** $6600

**29.** 73,000

**30.** 90,500

**31.** 170,000,000

**32.** $14,583,000,000

**33.** $372,000,000

**34.** 490,000

**Name** _____

*Complete the table by estimating the given number to the given place value.*

| | | Ten | Hundred | Thousand |
|---|---|---|---|---|
| **21.** | 5281 | 5280 | 5300 | 5000 |
| **22.** | 7619 | 7620 | 7600 | 8000 |
| **23.** | 9444 | 9440 | 9400 | 9000 |
| **24.** | 7777 | 7780 | 7800 | 8000 |
| **25.** | 14,876 | 14,880 | 14,900 | 15,000 |
| **26.** | 85,049 | 85,050 | 85,000 | 85,000 |

**27.** Round to the nearest thousand the Fall 1999 enrollment of East Tennessee State University: 11,187 (*Source:* East Tennessee State University)

**28.** Round to the nearest hundred the hourly cost of operating a B747-400 aircraft: $6592. (*Source:* Air Transport Association of America)

*Round each number to the indicated place value.*

**29.** In 2000, Superbowl XXXIV was won by the St. Louis Rams over the Tennessee Titans. Attendance at the game was 72,625. Round the attendance figure to the nearest thousand. (*Source:* Sports Illustrated)

**30.** It takes 90,465 days for Pluto to make a complete orbit around the sun. Round this number to the nearest hundred. (*Source:* National Space Science Data Center)

**31.** In 1998, U.S. farms produced 167,122,000 bushels of oats. Round the oat production figure to the nearest ten-million. (*Source:* U.S. Department of Agriculture)

**32.** Revenues collected for Florida public schools during the 1997–1998 school year totaled $14,583,108,000. Round the revenue figure to the nearest million. (*Source:* National Education Association)

**33.** In 1998, JCPenney spent $371,971,500 on advertising. Round the advertising figure to the nearest hundred-thousand. (*Source:* Competitive Media Reporting and Publishers Information Bureau)

**34.** In 1998, there were 491,707 U.S. Army personnel on active duty. Round this personnel figure to the nearest ten-thousand. (*Source:* U.S. Department of Defense)

**B** *Estimate the sum or difference by rounding each number to the nearest ten. See Examples 4 and 5.*

**35.**  29
     35
     42
  +16

**36.**  62
     72
     15
  +19

**37.**  649
  −272

**38.**  555
  −235

*Estimate the sum or difference by rounding each number to the nearest hundred. See Examples 4 and 5.*

**39.**  1812
   1776
 +1945

**40.**  2010
   2001
 +1984

**41.**  1774
 −1492

**42.**  1989
 −1870

**43.**  2995
   1649
 +3940

**44.**  799
  1655
 + 271

---

**Helpful Hint**

Estimation is useful to check for incorrect answers when using a calculator. For example, pressing a key too hard may result in a double digit, while pressing a key too softly may result in the number not appearing in the display.

---

*Two of the given calculator answers below are incorrect. Find them by estimating each sum.*

**45.** 362 + 419     781

**46.** 522 + 785     1307

**47.** 432 + 679 + 198     1139

**48.** 229 + 443 + 606     1278

**49.** 7806 + 5150     12,956

**50.** 5233 + 4988     9011

**51.** 31,439 + 18,781     50,220

**52.** 68,721 + 52,335     121,056

**35.** 130
**36.** 170
**37.** 380
**38.** 320
**39.** 5500
**40.** 6000
**41.** 300
**42.** 100
**43.** 8500
**44.** 2800
**45.** correct
**46.** correct
**47.** incorrect
**48.** correct
**49.** correct
**50.** incorrect
**51.** correct
**52.** correct

**Name** _____

**C** *Solve each problem by estimating. See Examples 6 and 7.*

**53.** Campo Appliance Store advertises three refrigerators on sale at $799, $1299, and $999. Round each cost to the nearest hundred to estimate the total cost.

**54.** Jared Nuss scored 89, 92, 100, 67, 75, and 79 on his calculus tests. Round each score to the nearest ten to estimate his total score.

**55.** Arlene Neville wants to estimate quickly the distance from Stockton to LaCrosse. Round each distance given on the map to the nearest ten miles to estimate the total distance.

**56.** Carmelita Watkins is pricing new stereo systems. One system sells for $1895 and another system sells for $1524. Round each price to the nearest hundred dollars to estimate the difference in price of these systems.

**57.** The peak of Mt. Everest, in Asia, is 29,028 feet above sea level. The top of Mt. Sunflower, in Kansas, is 4039 feet above sea level. Round each height to the nearest thousand to estimate the difference in elevation of these two peaks. (*Sources:* U.S. Geological Survey and National Geographic Society)

**58.** The Gonzales family took a trip and traveled 458, 489, 377, 243, 69, and 702 miles on 6 consecutive days. Round each distance to the nearest hundred to estimate the distance they traveled.

**59.** In 1998, the population of New York City was 7,420,166 and the population of Houston, Texas, was 1,786,691. Round each population to the nearest hundred-thousand to estimate how much larger New York was than Houston. (*Source: The World Almanac, 2000*)

**60.** The distance from Kansas City to Boston is 1429 miles and from Kansas City to Chicago is 530 miles. Round each distance to the nearest hundred to estimate how much farther Boston is from Kansas City than Chicago is.

**61.** In the 1964 presidential election, Lyndon Johnson received 41,126,233 votes and Barry Goldwater received 27,174,898 votes. Round each number of votes to the nearest million to estimate the number of votes by which Johnson won the election.

**62.** Enrollment figures at Normal State University showed an increase from 49,713 credit hours in 1999 to 51,746 credit hours in 2000. Round each number to the nearest thousand to estimate the increase.

**63.** Head Start is a national program that provides developmental and social services for America's low-income preschool children ages three to five. Enrollment figures in Head Start programs showed an increase from 540,930 children in 1990 to 794,000 in 1997. Round each number of children to the nearest thousand to estimate this increase. (*Source:* Head Start Bureau)

**64.** In 1986, Rubbermaid employed 6509 associates on average. By 1996, the average number of Rubbermaid associates had increased to 13,861. Round each number of associates to the nearest hundred to estimate this increase. (*Source:* Rubbermaid, Inc.)

*The table (from the previous section) shows the top ten leading United States advertisers in 1998 and the amount of money spent in that year on ads. Use this table to answer Exercises 65–68. (Source: Competitive Media Reporting and Publishers Information Bureau)*

| | |
|---|---|
| General Motors | $2,121,040,900 |
| Procter & Gamble | 1,724,259,700 |
| DaimlerChrysler | 1,410,748,700 |
| Philip Morris | 1,264,353,200 |
| Ford Motor | 1,147,589,200 |
| Time Warner | 829,185,000 |
| Walt Disney | 809,878,700 |
| Sears Roebuck & Co. | 720,543,900 |
| Unilever | 691,203,500 |
| Diageo Plc | 659,093,600 |

**65.** Approximate the amount of money spent by Philip Morris to the nearest hundred-million.

**66.** Approximate the amount of money spent by General Motors to the nearest hundred-million.

**67.** Approximate the amount of money spent by Daimler Chrysler to the nearest million.

**68.** Approximate the amount of money spent by Ford Motor to the nearest million.

### COMBINING CONCEPTS

*A number rounded to the nearest hundred is 8600.*

**69.** Determine the smallest possible number.

**70.** Determine the largest possible number.

---

**63.** 253,000 children

**64.** 7400 associates

**65.** $1,300,000,000

**66.** $2,100,000,000

**67.** $1,411,000,000

**68.** $1,148,000,000

**69.** 8550

**70.** 8649

**71.** On August 23, 1989, it was estimated that one million five hundred thousand people joined hands in a human chain stretching 370 miles to protest the fiftieth anniversary of the pact that allowed the then Soviet Union to annex the Baltic nations in 1939. If the estimate of the number of people is to the nearest one hundred-thousand, determine the largest possible number of people in the chain.

**72.** In your own words, explain how to round a number to the nearest thousand.

## Internet Excursions

Go to http://www.prenhall.com/martin-gay

This World Wide Web address will provide you with access to summaries of visitor use statistics for the U.S. National Park Service, or a related site. Select the summary report for the most recent year, and use the data given to answer the following questions.

**73.** Scan the visitor use information listed by park type. Which type of park was visited most often? List the five most visited park types along with the number of visitors to each. Estimate the total of these five visitation figures to the nearest million visitors.

**74.** Scan the visitor use information listed by park type. Which type of park was visited least often? List the five least visited park types along with the number of visitors to each. Estimate the total of these five visitation figures to the nearest ten thousand.

# 1.5   MULTIPLYING WHOLE NUMBERS AND AREA

Suppose that we wish to count the number of desks in a classroom. The desks are arranged in 5 rows and each row has 6 desks.

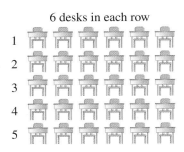

6 desks in each row

Adding 5 sixes gives the total number of desks: $6 + 6 + 6 + 6 + 6 = 30$ desks. When each addend is the same, we refer to this as **repeated addition**. **Multiplication** is repeated addition but with different notation.

$$\underbrace{6 + 6 + 6 + 6 + 6}_{\text{5 sixes}} = 5 \underset{\text{factor}}{\times} 6 = \underset{\text{product}}{30}$$

The $\times$ is called a multiplication sign. The numbers 5 and 6 are called **factors**. The number 30 is called the **product**. The notation $5 \times 6$ is read as "five times six." The symbols $\cdot$ and ( ) can also be used to indicate multiplication.

$$5 \times 6 = 30, \quad 5 \cdot 6 = 30, \quad (5)(6) = 30, \quad \text{and} \quad 5(6) = 30$$

**TRY THE CONCEPT CHECK IN THE MARGIN.**

## A   USING THE PROPERTIES OF MULTIPLICATION

As for addition, we memorize products of one-digit whole numbers and then use certain properties of multiplication to multiply larger numbers. (If necessary, review the multiplication of one-digit numbers in Appendix B.) Notice in the appendix that when any number is multiplied by 0, the result is always 0. This is called the **multiplication property of 0**.

---

**MULTIPLICATION PROPERTY OF 0**

The product of 0 and any number is 0. For example,

$$5 \cdot 0 = 0$$
$$0 \cdot 8 = 0$$

---

Also notice in the appendix that when any number is multiplied by 1, the result is always the original number. We call this result the **multiplication property of 1**.

TEACHING TIP

Show students how to use rectangles to visualize multiplication. For instance, $3 \times 2$ can be represented by the problem of finding how many 1-inch squares are in a rectangle which is 3 inches by 2 inches.

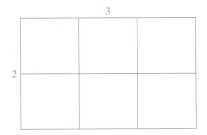

## ✓ CONCEPT CHECK

a. Rewrite $2 + 2 + 2 + 2$ using multiplication.

b. Rewrite $4 \times 10$ as repeated addition. Is there more than one way to do this? If so, show all ways.

TEACHING TIP

Help your students build number flexibility by asking them to notice patterns in the multiplication table in the appendix. For instance, they may notice repeating patterns in the ones digit when multiplying by 2, 4, 5, 6, or 8.

**Answers**

✓ **Concept Check:**
**a.** $4 \times 2 = 8$  **b.** $10 + 10 + 10 + 10 = 40$; yes, $4 + 4 + 4 + 4 + 4 + 4 + 4 + 4 + 4 + 4 = 40$

**MULTIPLICATION PROPERTY OF 1**

The product of 1 and any number is that same number. For example,

$$1 \cdot 9 = 9$$
$$6 \cdot 1 = 6$$

## Practice Problem 1

Multiply.
a. $3 \times 0$          b. $4(1)$
c. $(0)(34)$          d. $1 \cdot 76$

**Example 1**    Multiply.

   **a.** $6 \times 1$     **b.** $0(8)$     **c.** $1 \cdot 45$     **d.** $(75)(0)$

*Solution:*    **a.** $6 \times 1 = 6$          **b.** $0(8) = 0$
              **c.** $1 \cdot 45 = 45$          **d.** $(75)(0) = 0$

Like addition, multiplication is commutative and associative. Notice that, when multiplying two numbers, the order of these numbers can be changed without changing the product. For example,

$$3 \cdot 5 = 15 \quad \text{and} \quad 5 \cdot 3 = 15$$

This property is the **commutative property of multiplication**.

**COMMUTATIVE PROPERTY OF MULTIPLICATION**

Changing the **order** of two factors does not change their product. For example,

$$9 \cdot 2 = 18 \quad \text{and} \quad 2 \cdot 9 = 18$$

Another property that can help us when multiplying is the **associative property of multiplication**. This property states that, when multiplying numbers, the grouping of the numbers can be changed without changing the product. For example,

$$2 \cdot (3 \cdot 4) = 2 \cdot 12 = 24$$

Also,

$$(2 \cdot 3) \cdot 4 = 6 \cdot 4 = 24$$

Both groupings give a product of 24.

**ASSOCIATIVE PROPERTY OF MULTIPLICATION**

Changing the **grouping** of factors does not change their product. For example,

$$5 \cdot (3 \cdot 2) = 5 \cdot 6 = 30 \quad \text{and} \quad (5 \cdot 3) \cdot 2 = 15 \cdot 2 = 30$$

This means $5 \cdot (3 \cdot 2) = (5 \cdot 3) \cdot 2$

With these properties, along with the **distributive property**, we can find the product of any whole numbers. The distributive property says that multiplication **distributes** over addition. For example, notice that $3(2 + 5)$ is the same as $3 \cdot 2 + 3 \cdot 5$.

**Answers**

**1. a.** 0    **b.** 4    **c.** 0    **d.** 76

$$3(2 + 5) = 3(7) = 21$$

$$3 \cdot 2 + 3 \cdot 5 = 6 + 15 = 21$$

Notice in $3(2 + 5) = 3 \cdot 2 + 3 \cdot 5$ that each number inside the parentheses is multiplied by 3.

---

**DISTRIBUTIVE PROPERTY**

Multiplication distributes over addition. For example,

$$2(3 + 4) = 2 \cdot 3 + 2 \cdot 4$$

---

**Example 2**   Rewrite each using the distributive property.

**a.** $3(4 + 5)$      **b.** $10(6 + 8)$      **c.** $2(7 + 3)$

*Solution:*   Using the distributive property, we have

**a.** $3(4 + 5) = 3 \cdot 4 + 3 \cdot 5$

**b.** $10(6 + 8) = 10 \cdot 6 + 10 \cdot 8$

**c.** $2(7 + 3) = 2 \cdot 7 + 2 \cdot 3$

**Practice Problem 2**

Rewrite each using the distributive property.
a. $5(2 + 3)$      b. $9(8 + 7)$
c. $3(6 + 1)$

## B   MULTIPLYING WHOLE NUMBERS

Let's use the distributive property to multiply $7(48)$. To do so, we begin by writing the expanded form of 48 and then applying the distributive property.

$$7(48) = 7(40 + 8)$$
$$= 7 \cdot 40 + 7 \cdot 8 \quad \text{Apply the distributive property.}$$
$$= 280 + 56 \quad \text{Multiply.}$$
$$= 336 \quad \text{Add.}$$

This is how we multiply whole numbers. When multiplying whole numbers, we will use the following notation.

$$\begin{array}{r} 5 \\ 48 \\ \times\ 7 \\ \hline 336 \end{array} \leftarrow 7 \cdot 8 = 56 \qquad \begin{array}{l}\text{Write 6 in the ones} \\ \text{place and carry 5} \\ \text{to the tens place.}\end{array}$$

$7 \cdot 4 = 28$ and $28 + 5 = 33$

**Example 3**   Multiply:   $\begin{array}{r} 25 \\ \times\ 8 \end{array}$

*Solution:*   $\begin{array}{r} 4 \\ 25 \\ \times\ 8 \\ \hline 200 \end{array}$

**Practice Problem 3**

Multiply.
a.   $\begin{array}{r} 36 \\ \times\ 4 \end{array}$      b.   $\begin{array}{r} 92 \\ \times\ 9 \end{array}$

**Answers**

**2. a.** $5(2 + 3) = 5 \cdot 2 + 5 \cdot 3$
**b.** $9(8 + 7) = 9 \cdot 8 + 9 \cdot 7$
**c.** $3(6 + 1) = 3 \cdot 6 + 3 \cdot 1$   **3. a.** 144   **b.** 828

To multiply larger whole numbers, use the following similar notation. Multiply $89 \times 52$.

| Step 1 | Step 2 | Step 3 |
|---|---|---|
| 1 | 4 | |
| 89 | 89 | 89 |
| × 52 | × 52 | × 52 |
| 178 ← Multiply 89 × 2 | 178 | 178 |
| | 4450 ← Multiply 89 × 50 | 4450 |
| | | 4628  Add. |

The numbers 178 and 4450 are called **partial products**. The sum of the partial products, 4628, is the product of 89 and 52.

### Example 4    Multiply: $236 \times 86$

*Solution:*

$$
\begin{array}{r}
236 \\
\times\ 86 \\
\hline
1416 \quad \leftarrow 6(236) \\
18{,}880 \quad \leftarrow 80(236) \\
\hline
20{,}296 \quad \text{Add.}
\end{array}
$$

### Example 5    Multiply: $631 \times 125$

*Solution:*

$$
\begin{array}{r}
631 \\
\times\ 125 \\
\hline
3155 \quad \leftarrow 5(631) \\
12{,}620 \quad \leftarrow 20(631) \\
63{,}100 \quad \leftarrow 100(631) \\
\hline
78{,}875 \quad \text{Add.}
\end{array}
$$

**TRY THE CONCEPT CHECK IN THE MARGIN.**

### C  FINDING THE AREA OF A RECTANGLE

A special application of multiplication is finding the area of a region. Area measures the amount of surface of a region. For example, we measure a plot of land or the living space of a home by area. The figures show two examples of units of area measure. (Recall that a centimeter is a unit of length in the metric system.)

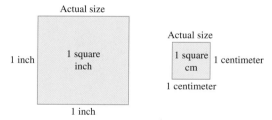

To measure the area of a geometric figure such as the rectangle shown, count the number of square units that cover the region.

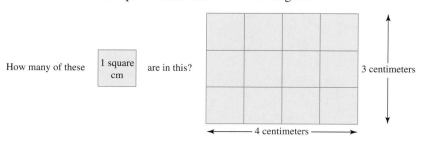

---

## Practice Problem 4

Multiply.

a.  594    b.  306
   × 72       × 81

## Practice Problem 5

Multiply.

a.  726    b.     4
   × 142       × 288

**Answers**

**4. a.** 42,768  **b.** 24,786  **5. a.** 103,092  **b.** 1152

✓ Concept Check:

$$
\begin{array}{r}
314 \\
\times\ 22 \\
\hline
628 \\
6280 \\
\hline
6908
\end{array}
$$

This rectangular region contains 12 square units, each 1 square centimeter. Thus, the area is 12 square centimeters. This total number of squares can be found by counting or by multiplying **4 · 3** (length · width).

$$\begin{aligned}
\text{area of a rectangle} &= \text{length} \cdot \text{width} \\
&= (4 \text{ centimeters})(3 \text{ centimeters}) \\
&= 12 \text{ square centimeters}
\end{aligned}$$

In this section, we find the areas of rectangles only. In later sections, we find the areas of other geometric regions.

---

**Helpful Hint**

Remember that *perimeter* (distance around a plane figure) is measured in units. *Area* (space enclosed by a plane figure) is measured in square units.

5 in.

| Rectangle | 4 in. |

$$\text{Perimeter} = 5 \text{ in.} + 4 \text{ in.} + 5 \text{ in.} + 4 \text{ in.} = 18 \text{ in.}$$
$$\text{Area} = (5 \text{ in.})(4 \text{ in.}) = 20 \text{ square in.}$$

---

△ **Example 6**    Finding the Area of a State

The state of Colorado is in the shape of a rectangle whose length is 380 miles and whose width is 280 miles. Find its area.

*Solution:*    The area of a rectangle is the product of its length and its width.

$$\begin{aligned}
\text{area} &= \text{length} \cdot \text{width} \\
&= (380 \text{ miles})(280 \text{ miles}) \\
&= 106,400 \text{ square miles}
\end{aligned}$$

Colorado

The area of Colorado is 106,400 square miles.

△ **Practice Problem 6**

The state of Wyoming is in the shape of a rectangle whose length is 360 miles and whose width is 280 miles. Find its area.

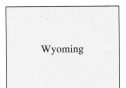

Wyoming

TEACHING TIP

Ask students to name types of index cards, photos, or frames. Then ask them what the type indicates. For example, a 3 × 5 index card, which is stated 3 by 5, indicates the index card is 3 inches wide by 5 inches long.

**Answer**

**6.** 100,800 square miles

---

**D**  **SOLVING PROBLEMS BY MULTIPLYING**

There are several words or phrases that indicate the operation of multiplication. Some of these are as follows.

| Key Words or Phrases | Example | Symbols |
|---|---|---|
| **multiply** | multiply 5 by 7 | $5 \cdot 7$ |
| **product** | the product of 3 and 2 | $3 \cdot 2$ |
| **times** | 10 times 13 | $10 \cdot 13$ |

Many key words or phrases describing real-life problems that suggest addition might be better solved by multiplication instead. For example, to find the *total* cost of 8 shirts, each selling for $27, we can either add $27 + 27 + 27 + 27 + 27 + 27 + 27 + 27$, or we can multiply $8(27)$.

## Practice Problem 7

A computer printer can print 240 characters per second in draft mode. How many total characters can it print in 15 seconds?

### Example 7   Finding Disk Space

A single-density computer disk can hold 370 thousand bytes of information. How many total bytes can 42 such disks hold?

*Solution:*   Forty-two disks will hold $42 \times 370$ thousand bytes.

| Words | | Translate to Numbers | |
|---|---|---|---|
| bytes per disk | $\rightarrow$ | 370 | thousand |
| $\times$     disks | $\rightarrow$ | $\times$ 42 | |
| | | 740 | |
| | | 14,800 | |
| total bytes | | 15,540 | thousand |

Forty-two disks will hold 15,540 thousand bytes. ▬▬▬

## Practice Problem 8

Softball T-shirts come in two styles: plain at $6 each and striped at $7 each. The team orders 4 plain shirts and 5 striped shirts. Find the total cost of the order.

### Example 8   Budgeting Money

Earline Martin agrees to take her children and their cousins to the San Antonio Zoo. The ticket price for each child is $4 and for each adult, $6. If 8 children and 1 adult plan to go, how much money is needed for admission?

*Solution:*   If the price of one child's ticket is $4. the price for 8 children is $8 \cdot 4 = \$32$. The price of one adult ticket is $6, so the total cost is

| Words | | Translate to Numbers |
|---|---|---|
| price of 8 children | $\rightarrow$ | 32 |
| +    price of adult | $\rightarrow$ | + 6 |
| total cost | | 38 |

The total admission cost is $38. ▬▬▬

## Practice Problem 9

If an average page in a book contains 259 words, estimate, rounding to the nearest hundred, the total number of words contained on 195 pages.

### Example 9   Estimating Word Count

The average page of a book contains 259 words. Estimate, rounding to the nearest ten, the total number of words contained on 22 pages.

**Answers**

**7.** 3600 characters   **8.** $59   **9.** 60,000 words

*Solution:* The exact number of words is $259 \times 22$. Estimate this product by rounding each factor to the nearest ten.

$$
\begin{array}{cccr}
259 & \text{rounds to} & & 260 \\
\times\ \ 22 & \text{rounds to} & \times & 20 \\
\hline
& & & 5200
\end{array}
$$

There are approximately 5200 words contained on 22 pages.

## CALCULATOR EXPLORATIONS
### MULTIPLYING NUMBERS

To multiply numbers on a calculator, find the key marked $\boxed{\times}$ and $\boxed{=}$ (or $\boxed{\text{ENTER}}$). For example, to find $31 \cdot 66$ on a calculator, press the keys $\boxed{31}$ $\boxed{\times}$ $\boxed{66}$ $\boxed{=}$ (or $\boxed{\text{ENTER}}$). The display will read $\boxed{\phantom{000}2046}$. Thus, $31 \cdot 66 = 2046$.

*Use a calculator to multiply.*

**1.** $72 \times 48$    3456       **2.** $81 \times 92$    7452

**3.** $163 \cdot 94$    15,322      **4.** $285 \cdot 144$    41,040

**5.** $983(277)$    272,291     **6.** $1562(843)$    1,316,766

# Focus on Study Skills

## HOW TO USE THIS TEXT

One way you can succeed in your Prealgebra course is to under-stand how to best use this text and to take advantage of all its help-ful features. Here are some tips.

1. Each example in the section has a parallel Practice Problem. As you read a section, try each Practice Problem after you've fin-ished the corresponding example. This "learn-by-doing" approach will help you grasp ideas before you move on to other concepts.

2. The main section of exercises in an exercise set are referenced by an objective, such as $\boxed{A}$ or $\boxed{B}$, and also an example(s). Use this referencing if you have trouble completing an assignment from the exercise set.

3. At the back of the book, there is a section of Answers to Select-ed Exercises and Solutions to Selected Exercises. Use these sec-tions as an aid in checking your work.

4. If you need extra help in a particular section, check at the beginning of the section to see what videotapes and software are available.

5. Integrated Reviews in each chapter offer you a chance to prac-tice—in one place—the many concepts that you have learned sep-arately over several sections.

6. There are many opportunities at the end of each chapter to help you understand the concepts of the chapter.
   - ▲ **Highlights** contain chapter summaries with examples.
   - ▲ **Chapter Review** contains review problems organized by section.
   - ▲ **Chapter Test** is a sample test to help you prepare for an exam.
   - ▲ **Cumulative Review** is a review consisting of material from the beginning of the book to the end of the particular chapter.

**Name** _____  **Section** _____ **Date** _____

## MENTAL MATH

*Multiply. See Example 1.*

**1.** $1 \cdot 24$  **2.** $55 \cdot 1$  **3.** $0 \cdot 19$  **4.** $27 \cdot 0$

**5.** $8 \cdot 0 \cdot 9$  **6.** $7 \cdot 6 \cdot 0$  **7.** $87 \cdot 1$  **8.** $1 \cdot 41$

## EXERCISE SET 1.5

**A** *Use the distributive property to rewrite each expression. See Example 2.*

**1.** $4(3 + 9)$  **2.** $5(8 + 2)$  **3.** $2(4 + 6)$

**4.** $6(1 + 4)$  **5.** $10(11 + 7)$  **6.** $12(12 + 3)$

**B** *Multiply. See Example 3.*

| **7.** $\phantom{0}42$ | **8.** $\phantom{0}79$ | **9.** $\phantom{0}624$ | **10.** $\phantom{0}638$ |
|---|---|---|---|
| $\times\ \ 6$ | $\times\ \ 3$ | $\times\ \ \ 3$ | $\times\ \ \ 5$ |

| **11.** $\phantom{0}277$ | **12.** $\phantom{0}882$ | **13.** $\phantom{0}1062$ | **14.** $\phantom{0}9021$ |
|---|---|---|---|
| $\times\ \ 6$ | $\times\ \ 2$ | $\times\ \ \ \ 5$ | $\times\ \ \ \ 3$ |

*Multiply. See Examples 4 and 5.*

| **15.** $\phantom{0}298$ | **16.** $\phantom{0}591$ | **17.** $\phantom{0}231$ | **18.** $\phantom{0}526$ |
|---|---|---|---|
| $\times\ \ 14$ | $\times\ \ 72$ | $\times\ \ 47$ | $\times\ \ 23$ |

| **19.** $\phantom{0}809$ | **20.** $\phantom{0}307$ | **21.** $(620)(40)$ | **22.** $(720)(80)$ |
|---|---|---|---|
| $\times\ \ 14$ | $\times\ \ 16$ | | |

**23.** $(998)(12)(0)$  **24.** $(593)(47)(0)$  **25.** $(590)(1)(10)$  **26.** $(240)(1)(20)$

| **27.** $\phantom{0}1234$ | **28.** $\phantom{0}1357$ | **29.** $\phantom{0}609$ | **30.** $\phantom{0}505$ |
|---|---|---|---|
| $\times\ \ \ 48$ | $\times\ \ \ 79$ | $\times\ \ 234$ | $\times\ \ 127$ |

**Name** _____

| **31.** | 5621 | **32.** | 1234 | **33.** | 1941 | **34.** | 1876 |
|---|---|---|---|---|---|---|---|
| | × 324 | | × 567 | | × 235 | | × 437 |

**35.**    589
   × 110

**36.**    426
   × 110

**D** *Estimate each product by rounding each factor to the nearest hundred. See Example 9.*

**37.** 576 × 354     **38.** 982 × 650     **39.** 604 × 451     **40.** 111 × 999

**C** *Find the area of each rectangle. See Example 6.*

△ **41.**

9 meters

7 meters

△ **42.** 4 inches

12 inches

△ **43.** 13 feet

30 feet

△ **44.** 25 centimeters

20 centimeters

**C** **D** *Solve. See Examples 6 through 9.*

**45.** One tablespoon of olive oil contains 125 calories. How many calories are in 3 tablespoons of olive oil? (*Source: Home and Garden Bulletin No. 72*, U.S. Department of Agriculture)

**46.** One ounce of hulled sunflower seeds contains 14 grams of fat. How many grams of fat are in 6 ounces of hulled sunflower seeds? (*Source: Home and Garden Bulletin No. 72*, U.S. Department of Agriculture)

**54**

**Name** _____

**47.** The textbook for a course in Civil War history costs $37. There are 35 students in the class. Find the total cost of the history books for the class.

**48.** The seats in the mathematics lecture hall are arranged in 12 rows with 6 seats in each row. Find how many seats are in this room.

**49.** A case of canned peas has *two layers* of cans. In each layer are 8 rows with 12 cans in each row. Find how many cans are in a case.

**50.** An apartment building has *three floors*. Each floor has five rows of apartments with four apartments in each row. Find how many apartments are in this building.

△ **51.** A plot of land measures 90 feet by 110 feet. Find its area.

△ **52.** A house measures 45 feet by 60 feet. Find the floor area of the house.

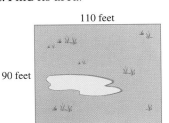

110 feet

90 feet

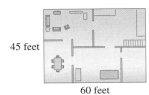

45 feet

60 feet

△ **53.** Recall from an earlier section that the largest commercial building in the world under one roof is the flower auction building of the cooperative VBA in Aalsmeer, Netherlands. The floor plan is a rectangle that measures 776 meters by 639 meters. Find the area of this building. (*Source: The Handy Science Answer Book*, Visible Ink Press, 1994)

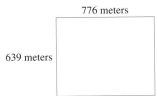

776 meters

639 meters

△ **54.** The largest lobby can be found at the Hyatt Regency in San Francisco, California. It is in the shape of a rectangle that measures 350 feet by 160 feet. Find its area.

**Name** _____

**55.** A pixel is a rectangular dot on a graphing calculator screen. If a graphing calculator screen contains 62 pixels in a row and 94 pixels in a column, find the total number of pixels on a screen.

**56.** A high-density computer disk can hold 1510 thousand bytes of information. Find how many bytes of information 17 disks can hold.

**57.** A line of print on a computer contains 60 characters (letters, spaces, punctuation marks). Find how many characters there are in 25 lines.

**58.** An average cow eats 3 pounds of grain per day. Find how much grain a cow eats in a year. (Assume 365 days in 1 year.)

**59.** One ounce of Planters® Dry Roasted Peanuts has 160 calories. How many calories are in 8 ounces? (*Source:* RJR Nabisco, Inc.)

**60.** One ounce of Planters® Dry Roasted Peanuts has 13 grams of fat. How many grams of fat are in 8 ounces? (*Source:* RJR Nabisco, Inc.)

△ **61.** The diameter of the planet Saturn is 9 times as great as the diameter of Earth. The diameter of Earth is 7927 miles. Find the diameter of Saturn.

Saturn

Earth

**62.** The planet Uranus orbits the sun every 84 Earth years. Find how many Earth days two orbits take. (Assume 365 days in 1 year.)

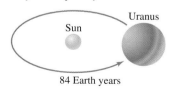

Uranus

Sun

84 Earth years

**56**

63. Suppose a wild elk needs 39 acres of natural grazing area to survive. If there are 576 elk in a certain herd, find how much grazing area the herd needs.

64. A window washer in New York City is bidding for a contract to wash the windows of a 23-story building. To write a bid, the number of windows in the building is needed. If there are 7 windows in each row of windows on 2 sides of the building and 4 windows per row on the other 2 sides of the building, find the total number of windows.

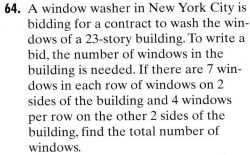

65. Hershey's main chocolate factory in Hershey, Pennsylvania, uses 700,000 quarts of milk each day. How many quarts of milk would be used during the month of March, assuming that chocolate is made at the factory every day of the month? (*Source:* Hershey Foods Corp.)

66. Among older Americans (age 65 years and older), there are 5 times as many widows as widowers. There were 1,700,000 widowers in 1995. How many widows were there in 1995? (*Source:* Administration on Aging)

67. In 1996, the average cost of the Head Start program was $4571 per child. That year 752,077 children participated in the program. Round each number to the nearest thousand and estimate the total cost of the Head Start program in 1996. (*Source:* Head Start Bureau)

68. Approximately 5570 people celebrated their 65th birthday *each day* of 1996 in the United States. Round each number to the nearest hundred and estimate the total number of people who turned 65 *throughout the year* in 1996. (*Source:* Administration on Aging)

◣ **COMBINING CONCEPTS**

*In a survey, college students were asked to name their favorite fruit. The results are shown in the picture graph. Use this graph to answer Exercises 69–72.*

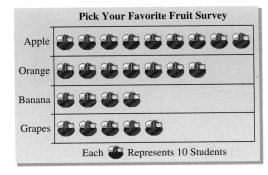

69. How many students chose grapes as their favorite fruit?

70. How many students chose the banana as their favorite fruit?

63. 22,464 acres

64. 506 windows

65. 21,700,000 qt

66. 8,500,000 widows

67. $3,760,000,000

68. 2,240,000 people

69. 50 students

70. 40 students

**71.** Which two fruits were the most popular?

**72.** Which two fruits were chosen by a total of 110 students?

*Fill in the missing digits in each problem.*

**73.**
```
     4_
×   _3
-------
   126
  3780
-------
  3906
```

**74.**
```
     _7
×  6_
-------
   171
  3420
-------
  3591
```

**75.** Explain how to multiply two 2-digit numbers using partial products.

**76.** A slice of enriched white bread has 65 calories. One tablespoon of jam has 55 calories, and one tablespoon of peanut butter has 95 calories. Suppose a peanut butter and jelly sandwich is made with two slices of white bread, one tablespoon of jam, and one tablespoon of peanut butter. How many calories are in two such peanut butter and jelly sandwiches? (*Source: Home and Garden Bulletin No. 72*, U.S. Department of Agriculture)

**77.** In 1999 the WNBA's top scorer was Cynthia Cooper of the Houston Comets. She made 204 free throws, 154 two-point field goals, and 58 three-point field goals. (Free throws are worth one point.) How many points did Cynthia Cooper score during the 1999 season? (*Source:* Women's National Basketball Association)

**78.** In your own words, explain the commutative property of multiplication.

**79.** In your own words, explain the associative property of multiplication.

# 1.6   DIVIDING WHOLE NUMBERS

Suppose three people pooled their money and bought a raffle ticket at a local fund-raiser. Their ticket was the winner and they won a $60 cash prize. They then divided the prize into three equal parts so that each person received $20.

```
        $60
       / | \
    $20 $20 $20
```

## A   DIVIDING WHOLE NUMBERS

The process of separating the quantity into these three equal parts is called **division**. Division can be symbolized by several notations.

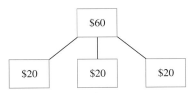

In the notation $\dfrac{60}{3}$, the bar separating 60 and 3 is called a **fraction bar**.

Just as addition and subtraction are closely related, so are multiplication and division. Division is defined in terms of multiplication. For example, $60 \div 3$ is the number that when multiplied by 3 gives 60. In other words,

$$\begin{array}{r} 20 \\ 3\overline{)60} \end{array} \quad \text{because} \quad 20 \cdot 3 = 60$$

This means that division can be *checked* by multiplication.

Since multiplication and division are related in this way, you can use the multiplication table in Appendix B to review quotients of one-digit divisors, if necessary.

**Example 1**    Find each quotient. Check by multiplying.

**a.** $42 \div 7$    **b.** $\dfrac{81}{9}$    **c.** $4\overline{)24}$

*Solution:*    **a.** $42 \div 7 = 6$, because $6 \cdot 7 = 42$.

**b.** $\dfrac{81}{9} = 9$, because $9 \cdot 9 = 81$.

**c.** $4\overline{)24}^{\,6}$, because $6 \cdot 4 = 24$.

**Example 2**    Find each quotient. Check by multiplying.

**a.** $1\overline{)8}$    **b.** $11 \div 1$    **c.** $\dfrac{9}{9}$    **d.** $7 \div 7$    **e.** $\dfrac{10}{1}$    **f.** $6\overline{)6}$

*Solution:*    **a.** $1\overline{)8}^{\,8}$, because $8 \cdot 1 = 8$.

**b.** $11 \div 1 = 11$, because $11 \cdot 1 = 11$.

**Objectives**

**A** Divide whole numbers.
**B** Perform long division.
**C** Solve problems that require dividing by whole numbers.
**D** Find the average of a list of numbers.

Study       SSM    CD-ROM    Video
Guide                        1.6

TEACHING TIP

Ask students to verbalize all the different ways they can think of to state that $60 \div 3 = 20$. For example, 60 divided by 3 equals 20. 3 goes into 60 twenty times. The quotient of 60 and 3 is 20.

**Practice Problem 1**

Find each quotient. Check by multiplying.

a. $8\overline{)48}$    b. $35 \div 5$    c. $\dfrac{49}{7}$

**Practice Problem 2**

Find each quotient. Check by multiplying.

a. $\dfrac{8}{8}$    b. $3 \div 1$    c. $1\overline{)12}$

d. $2 \div 1$    e. $\dfrac{5}{1}$

**Answers**

**1. a.** 6  **b.** 7  **c.** 7  **2. a.** 1  **b.** 3  **c.** 12  **d.** 2  **e.** 5

**c.** $\dfrac{9}{9} = 1$, because $1 \cdot 9 = 9$.

**d.** $7 \div 7 = 1$, because $1 \cdot 7 = 7$.

**e.** $\dfrac{10}{1} = 10$, because $10 \cdot 1 = 10$.

**f.** $6\overline{)6}$, because $1 \cdot 6 = 6$.

Example 2 illustrates important **division properties of 1**.

---

**DIVISION PROPERTIES OF 1**

The quotient of any number and that same number is 1. For example,

$$8 \div 8 = 1, \qquad \dfrac{7}{7} = 1, \qquad \text{and} \qquad 4\overline{)4}$$

The quotient of any number and 1 is that same number. For example,

$$9 \div 1 = 9, \qquad \dfrac{6}{1} = 6, \qquad 1\overline{)3}, \qquad \text{and} \qquad \dfrac{0}{1} = 0$$

---

**Practice Problem 3**

Find each quotient. Check by multiplying.

a. $\dfrac{0}{7}$             b. $5\overline{)0}$

c. $0 \div 6$             d. $9 \div 0$

⬛ **Example 3**     Find each quotient. Check by multiplying.

**a.** $9\overline{)0}$       **b.** $0 \div 12$       **c.** $\dfrac{0}{5}$       **d.** $\dfrac{3}{0}$

*Solution:*     **a.** $9\overline{)0}$, because $0 \cdot 9 = 0$.

**b.** $0 \div 12 = 0$, because $0 \cdot 12 = 0$.

**c.** $\dfrac{0}{5} = 0$, because $0 \cdot 5 = 0$.

**d.** If $\dfrac{3}{0} = a$ *number*, then the *number* $\cdot 0 = 3$. Recall that any number multiplied by 0 is 0 and not 3. We say then that $\dfrac{3}{0}$ is **undefined**.

Example 3 illustrates important **division properties of 0**.

---

**DIVISION PROPERTIES OF 0**

The quotient of 0 and any number (except 0) is 0. For example,

$$0 \div 9 = 0, \qquad \dfrac{0}{5} = 0, \qquad \text{and} \qquad 14\overline{)0}$$

The quotient of any number and 0 is not a number. We say that

$$\dfrac{3}{0}, \qquad 0\overline{)3}, \qquad \text{and} \qquad 3 \div 0$$

are **undefined**.

---

**Answers**

**3. a.** 0   **b.** 0   **c.** 0   **d.** undefined

## B  PERFORMING LONG DIVISION

When dividends are larger, the quotient can be found by a process called **long division**. For example, let's divide 2541 by 3.

$$3\overline{)2541}$$

We can't divide 3 into 2, so we try dividing 3 into the first two digits.

$$\begin{array}{r} 8 \\ 3\overline{)2541} \end{array}$$   $25 \div 3 = 8$ with 1 left, so our best estimate is 8. We place 8 over the 5 in 25.

Next, we multiply 8 and 3 and subtract this product from 25, making sure that this difference is less than the divisor.

$$\begin{array}{r} 8 \\ 3\overline{)2541} \\ -24 \\ \hline 1 \end{array}$$   $8(3) = 24$
$25 - 24 = 1$ and 1 is less than the divisor 3.

Next, we bring down the next digit and start the process again.

$$\begin{array}{r} 84 \\ 3\overline{)2541} \\ -24\downarrow \\ \hline 14 \\ -12 \\ \hline 2 \end{array}$$   $14 \div 3 = 4$ with 2 left

$4(3) = 12$
$14 - 12 = 2$

Again, we bring down the next digit and start the process again.

$$\begin{array}{r} 847 \\ 3\overline{)2541} \\ -24 \\ \hline 14 \\ -12 \\ \hline 21 \\ -21 \\ \hline 0 \end{array}$$   $21 \div 3 = 7$

$7(3) = 21$
$21 - 21 = 0$

The quotient is 847. To check, see that $847 \times 3 = 2541$.

## Example 4    Divide: $3705 \div 5$. Check by multiplying.

*Solution:*

$$\begin{array}{r} 7 \\ 5\overline{)3705} \\ -35\downarrow \\ \hline 20 \end{array}$$   $37 \div 5 = 7$ with 2 left. Place this estimate 7 over the 7 in 37.

$7(5) = 35$
$37 - 35 = 2$ and 2 is less than the divisor 5.
⤴ —— Bring down the 0.

$$\begin{array}{r} 74 \\ 5\overline{)3705} \\ -35 \\ \hline 20 \\ -20\downarrow \\ \hline 05 \end{array}$$   $20 \div 5 = 4$

$4(5) = 20$
$20 - 20 = 0$ and 0 is less than the divisor 5.
⤴ —— Bring down the 5.

### Practice Problem 4

Divide. Check by multiplying.
a. $6\overline{)5382}$          b. $4\overline{)2212}$

**Answers**
**4. a.** 897  **b.** 553

$$
\begin{array}{r}
741 \\
5\overline{)3705} \\
-35 \\
\hline
20 \\
-20 \\
\hline
5 \\
-5 \\
\hline
0
\end{array}
$$

$5 \div 5 = 1$

$1(5) = 5$
$5 - 5 = 0$

*Check:*
$$
\begin{array}{r}
741 \\
\times \quad 5 \\
\hline
3705
\end{array}
$$

---

⌐ **Helpful Hint**

Don't forget that a division problem can be checked by multiplying.

---

**Example 5**    Divide and check: $1872 \div 9$

*Solution:*
$$
\begin{array}{r}
208 \\
9\overline{)1872} \\
-18 \\
\hline
07 \\
-0 \\
\hline
72 \\
-72 \\
\hline
0
\end{array}
$$

$2(9) = 18$
$18 - 18 = 0$, bring down the 7.
$0(9) = 0$
$7 - 0 = 7$, bring down the 2.
$8(9) = 72$
$72 - 72 = 0$

*Check:*    $208 \cdot 9 = 1872$

Naturally, quotients don't always "come out even." Making 4 rows out of 26 chairs, for example, isn't possible if each row is supposed to have exactly the same number of chairs. Each of 4 rows can have 6 chairs, but 2 chairs are still left over.

6 chairs in each row

4 rows

2 chairs left over

We signify "leftovers" or *remainders* in this way:

$$
\begin{array}{r}
6 \text{ R } 2 \\
4\overline{)26}
\end{array}
$$

The **whole-number part of the quotient** is 6; the **remainder part of the quotient** is 2. Checking by multiplying.

---

**Practice Problem 5**

Divide and check.
a. $3\overline{)2397}$    b. $7\overline{)2520}$

| whole-number part | · | divisor | + | remainder part | = | dividend |
|---|---|---|---|---|---|---|
| ↓ | | ↓ | | ↓ | | ↓ |
| 6 | · | 4 | + | 2 | | |
| | | 24 | + | 2 | = | 26 |

**Example 6**   Divide and check: $2557 \div 7$

*Solution:*

$$
\begin{array}{r}
365 \text{ R } 2 \\
7\overline{)2557}
\end{array}
$$

$-21\downarrow$     $3(7) = 21$

$\quad 45$     $25 - 21 = 4$, bring down the 5.

$\;-42\downarrow$     $6(7) = 42$

$\qquad 37$     $45 - 42 = 3$, bring down the 7.

$\quad-35$     $5(7) = 35$

$\qquad\; 2$     $37 - 35 = 2$. The remainder is 2.

*Check:*

| 365 | · | 7 | + | 2 | = | 2557 |
|---|---|---|---|---|---|---|
| ↑ | | ↑ | | ↑ | | ↑ |
| whole-number part | · | divisor | + | remainder part | = | dividend |

**Practice Problem 6**

Divide and check.
a. $5\overline{)949}$      b. $6\overline{)4399}$

**Example 7**   Divide and check: $56{,}717 \div 8$

*Solution:*

$$
\begin{array}{r}
7089 \text{ R } 5 \\
8\overline{)56717}
\end{array}
$$

$-56\downarrow\downarrow$     $7(8) = 56$

$\quad 7$     Subtract and bring down the 7.

$\;-0\downarrow$     $0(8) = 0$

$\quad 71$     Subtract and bring down the 1.

$\;-64\downarrow$     $8(8) = 64$

$\qquad 77$     Subtract and bring down the 7.

$\quad-72$     $9(8) = 72$

$\qquad\; 5$     Subtract. The remainder is 5.

*Check:*

| 7089 | · | 8 | + | 5 | = | 56,717 |
|---|---|---|---|---|---|---|
| ↑ | | ↑ | | ↑ | | ↑ |
| whole-number part | · | divisor | + | remainder part | = | dividend |

**Practice Problem 7**

Divide and check.
a. $5\overline{)40{,}841}$      b. $7\overline{)22{,}430}$

When the divisor has more than one digit, the same pattern applies. For example, let's find $1358 \div 23$.

$$
\begin{array}{r}
5 \\
23\overline{)1358}
\end{array}
$$

$135 \div 23 = 5$ with 20 left over. Our estimate is 5.

$\quad 115\downarrow$     $5(23) = 115$

$\quad 208$     $135 - 115 = 20$. Bring down the 8.

**Answers**

**6. a.** 189 R 4   **b.** 733 R 1
**7. a.** 8168 R 1   **b.** 3204 R 2

Now we continue estimating.

$$
\begin{array}{r}
59\ \text{R}\ 1 \\
23\overline{)1358} \\
-115\phantom{0} \\
\hline
208 \\
-207 \\
\hline
1
\end{array}
$$

$208 \div 23 = 9$ with 1 left over.

$9(23) = 207$

$208 - 207 = 1$. The remainder is 1.

To check, see that $59 \cdot 23 + 1 = 1358$.

**Practice Problem 8**

Divide: $5740 \div 19$

**Example 8**    Divide: $6819 \div 17$

*Solution:*

$$
\begin{array}{r}
401\ \text{R}\ 2 \\
17\overline{)6819} \\
-68\phantom{00} \\
\hline
01\phantom{0} \\
-0\phantom{0} \\
\hline
19 \\
-17 \\
\hline
2
\end{array}
$$

$4(17) = 68$
Subtract and bring down the 1.
$0(17) = 0$
Subtract and bring down the 9.
$1(17) = 17$
Subtract. The remainder is 2.

To check, see that $401 \cdot 17 + 2 = 6819$.

**Example 9**    Divide: $51,600 \div 403$

*Solution:*

$$
\begin{array}{r}
128\ \text{R}\ 16 \\
403\overline{)51600} \\
-403\phantom{00} \\
\hline
1130\phantom{0} \\
-806\phantom{0} \\
\hline
3240 \\
-3224 \\
\hline
16
\end{array}
$$

$1(403) = 403$
Subtract and bring down the 0.
$2(403) = 806$
Subtract and bring down the 0.
$8(403) = 3224$
Subtract. The remainder is 16.

To check, see that $128 \cdot 403 + 16 = 51,600$.

**Practice Problem 9**

Divide: $16,589 \div 247$

## C  SOLVING PROBLEMS BY DIVIDING

Here are some key words and phrases that indicate the operation of division in real-life problems.

| Key Words or Phrases | Examples | Symbols |
| --- | --- | --- |
| **divide** | divide 10 by 5 | $10 \div 5$ or $\dfrac{10}{5}$ |
| **quotient** | the quotient of 64 and 4 | $64 \div 4$ or $\dfrac{64}{4}$ |
| **divided by** | 9 divided by 3 | $9 \div 3$ or $\dfrac{9}{3}$ |
| **divided or shared equally among** | \$100 divided equally among five people | $100 \div 5$ or $\dfrac{100}{5}$ |

✓ **CONCEPT CHECK**

Which of the following statement(s) correctly represent "the quotient of 36 and 6?" Or are they all correct? Explain your answer.

a. $36 \div 6$    b. $6 \div 36$    c. $6$

**Answers**

**8.** 302 R 2    **9.** 67 R 40
✓ **Concept Check:**
**a.** correct    **b.** incorrect    **c.** correct

**TRY THE CONCEPT CHECK IN THE MARGIN.**

## Example 10    Finding Shared Earnings

Zachary, Tyler, and Stephanie McMillan shared a paper route to earn money for college expenses. The total in their fund after expenses was $2895. How much is each person's equal share?

*Solution:*

Each person's equal share is (total) ÷ (number of people) or

$$2895 \quad ÷ \quad 3$$

Then

$$
\begin{array}{r}
965 \\
3)\overline{2895} \\
-27 \phantom{00} \\
\hline
19 \phantom{0} \\
-18 \phantom{0} \\
\hline
15 \\
-15 \\
\hline
0
\end{array}
$$

Each person's share is $965.

### Practice Problem 10

Marina, Manual, and Min Ramerez bought 10 dozen high-density computer diskettes to share equally. How many diskettes did each person get?

## Example 11    Calculating Shipping Needs

How many boxes are needed to ship 56 pairs of Nikes to a shoe store in Texarkana if 9 pairs of shoes will fit in each shipping box?

56 pairs

*Solution:*

| number of boxes | = | total pairs of shoes | ÷ | how many pairs in a box |
|---|---|---|---|---|

$$
\begin{array}{ccccc}
\text{number of boxes} & = & 56 & ÷ & 9
\end{array}
$$

$$
\begin{array}{r}
6\,\text{R}\,2 \\
9)\overline{56} \\
-54 \\
\hline
2
\end{array}
$$

There are 6 full boxes with 2 pairs of shoes left over, so 7 boxes will be needed.

### Practice Problem 11

Peanut butter and cheese cracker sandwiches are packaged 6 sandwiches to a package. How many full packages are formed from 195 sandwiches?

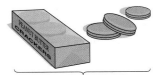

195 cracker sandwiches

## Example 12    Dividing Holiday Favors Among Students

Mary Schultz has 48 kindergarten students. She buys 260 stickers as Thanksgiving Day favors for her students. Can she divide the stickers up equally among her students? If not, how many stickers will be left over?

### Practice Problem 12

Calculators can be packed 24 to a box. If 497 calculators are to be packed, but only full boxes are shipped, how many full boxes will be shipped? How many calculators are left over and not shipped?

**Answers**

**10.** 40 diskettes  **11.** 32 full packages

**12.** 20 full boxes; 17 calculators left over

*Solution:*

| number of stickers | ÷ | number of students |
|:---:|:---:|:---:|
| ↓ | | ↓ |
| 260 | ÷ | 48 |

$$\begin{array}{r} 5\ R\ 20 \\ 48\overline{)260} \\ -240 \\ \hline 20 \end{array}$$

No, the stickers cannot be divided equally among her students since there is a nonzero remainder. There will be 20 stickers left over.

### D   FINDING AVERAGES

A special application of division (and addition) is finding the average of a list of numbers. The **average** of a list of numbers is the sum of the numbers divided by the *number* of numbers.

$$\text{average} = \frac{\text{sum of numbers}}{\textit{number} \text{ of numbers}}$$

### Example 13   Averaging Scores

Liam Reilly's scores in his mathematics class so far are 93, 86, 71, and 82. Find his average score.

*Solution:*   To find his average score, we find the sum of his scores and divide by 4, the number of scores.

$$\begin{array}{r} 93 \\ 86 \\ 71 \\ +\ 82 \\ \hline 332 \end{array} \text{ sum}$$

$$\text{average} = \frac{332}{4} = 83$$

$$\begin{array}{r} 83 \\ 4\overline{)332} \\ -32 \\ \hline 12 \\ -12 \\ \hline 0 \end{array}$$

His average score is 83.

**Practice Problem 13**

To compute a safe time to wait after allergy shots are administered, a lab technician is given a list of elapsed times between administered shots and reactions. Find the average of the times: 5 minutes, 7 minutes, 20 minutes, 6 minutes, 9 minutes, 3 minutes, and 48 minutes.

## CALCULATOR EXPLORATIONS
### DIVIDING NUMBERS

To divide numbers on a calculator, find the keys marked $\div$ and $=$ (or ENTER). For example, to find $435 \div 5$ on a calculator, press the keys $\boxed{435}$ $\boxed{\div}$ $\boxed{5}$ $\boxed{=}$ (or ENTER). The display will read $\boxed{87}$. Thus, $435 \div 5 = 87$.

*Use a calculator to divide.*

**1.** $848 \div 16$   53     **2.** $564 \div 12$   47     **3.** $95\overline{)5890}$   62

**4.** $27\overline{)1053}$   39     **5.** $\dfrac{32{,}886}{126}$   261     **6.** $\dfrac{143{,}088}{264}$   542

**7.** $0 \div 315$   0     **8.** $315 \div 0$   Error

**Answer**

**13.** 14 minutes

## MENTAL MATH

**A** *Find each quotient. See Examples 1 through 3.*

1. $40 \div 8$

2. $72 \div 9$

3. $45 \div 5$

4. $24 \div 3$

5. $0 \div 5$

6. $0 \div 8$

7. $9 \div 1$

8. $12 \div 1$

9. $\dfrac{16}{16}$

10. $\dfrac{49}{49}$

11. $\dfrac{25}{5}$

12. $\dfrac{45}{9}$

13. $6 \div 0$

14. $\dfrac{12}{0}$

15. $7 \div 1$

16. $6 \div 6$

17. $0 \div 4$

18. $7 \div 0$

19. $16 \div 2$

20. $18 \div 3$

# EXERCISE SET 1.6

**B** *Divide and then check by multiplying. See Examples 4 and 5.*

1. $9\overline{)108}$

2. $5\overline{)85}$

3. $6\overline{)222}$

4. $8\overline{)640}$

5. $3\overline{)1014}$

6. $4\overline{)504}$

*Divide and then check by multiplying. See Examples 6 and 7.*

7. $6\overline{)98}$

8. $7\overline{)422}$

9. $2\overline{)1127}$

10. $3\overline{)1240}$

11. $186 \div 5$

12. $167 \div 3$

13. $2121 \div 8$

14. $333 \div 4$

*Divide and then check by multiplying. See Examples 8 and 9.*

15. $23\overline{)1127}$

16. $42\overline{)2016}$

▭ 17. $55\overline{)715}$

18. $32\overline{)1856}$

19. $97\overline{)9449}$

20. $1938 \div 44$

21. $3708 \div 18$

22. $7224 \div 12$

### MENTAL MATH ANSWERS

1. 5
2. 8
3. 9
4. 8
5. 0
6. 0
7. 9
8. 12
9. 1
10. 1
11. 5
12. 5
13. undefined
14. undefined
15. 7
16. 1
17. 0
18. undefined
19. 8
20. 6

### ANSWERS

1. 12
2. 17
3. 37
4. 80
5. 338
6. 126
7. 16 R 2
8. 60 R 2
9. 563 R 1
10. 413 R 1
11. 37 R 1
12. 55 R 2
13. 265 R 1
14. 83 R 1
15. 49
16. 48
17. 13
18. 58
19. 97 R 40
20. 44 R 2
21. 206
22. 602

**Name** _____

**23.** 6578 ÷ 13    **24.** 5670 ÷ 14    **25.** 9299 ÷ 46    **26.** 2539 ÷ 64

**27.** $\dfrac{10,620}{236}$    **28.** $\dfrac{5781}{123}$    **29.** $\dfrac{10,194}{103}$    **30.** $\dfrac{23,048}{240}$

**31.** 20,619 ÷ 102    **32.** 40,803 ÷ 203    **33.** 45,046 ÷ 223    **34.** 164,592 ÷ 543

**C** *Solve. See Examples 10 through 12.*

**35.** Kathy Gomez teaches Spanish lessons for $85 per student for a 5-week session. From one group of students, she collects $4930. Find how many students are in the group.

**36.** Martin Thieme teaches American Sign Language classes for $55 per student for a 7-week session. He collects $1430 from the group of students. Find how many students are in the group.

**37.** Twenty-one people pooled their money and bought lottery tickets. One ticket won a prize of $5,292,000. Find how many dollars each person receives.

**38.** The gravity of Jupiter is 318 times as strong as the gravity of Earth, so objects on Jupiter weigh 318 times as much as they weigh on Earth. If a person would weigh 52,470 pounds on Jupiter, find how much the person weighs on Earth.

**39.** A truck hauls wheat to a storage granary. It carries a total of 5810 bushels of wheat in 14 trips. How much does the truck haul each trip. if each trip it hauls the same amount?

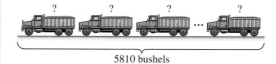

5810 bushels

**40.** The white stripes dividing the lanes on a highway are 25 feet long and the spaces between them are 25 feet long. Find how many whole stripes there are in 1 mile of highway. (A mile is 5280 feet.)

**41.** There is a bridge over highway I-35 every three miles. Find how many bridges there are over 265 miles of I-35.

**42.** An 18-hole golf course is 5580 yards long. If the distance to each hole is the same, find the distance between holes.

**43.** Wendy Holladay has a piece of rope 185 feet long that she cuts into pieces for an experiment in her second-grade class. Each piece of rope is to be 8 feet long. Determine whether she has enough rope for her 22-student class. Determine the amount extra or the amount short.

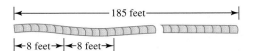

**44.** Jesse White is in the requisitions department of Central Electric Lighting Company. Light poles along a highway are placed 492 feet apart. Find how many poles he should order for a 1-mile strip of highway. (A mile is 5280 feet.)

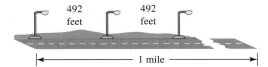

**45.** Edgerrin James of the Indianapolis Colts co-led the NFL in touchdowns during the 1999 football season. James scored a total of 102 points as touchdowns during the season. If a touchdown is worth 6 points, how many touchdowns did James make during 1999? (*Source:* National Football League)

**46.** Hershey's main chocolate factory in Hershey, Pennsylvania, uses 700,000 quarts of milk each day. There are 4 quarts in a gallon. How many gallons of milk are used each day? (*Source:* Hershey Foods Corp.)

**47.** Find how many yards there are in 1 mile. (A mile is 5280 feet; a yard is 3 feet.)

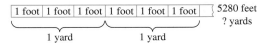

**48.** Find how many whole feet there are in 1 rod. (A mile is 5280 feet; 1 mile is 320 rods.)

**49.** The worldwide wind energy capacity in 1998 was enough to generate approximately 21,000,000,000 kilowatt-hours of electricity, or enough to power 3,500,000 U.S. suburban homes for one year. According to this estimate, how much electricity does a single U.S. suburban home use during one year on average? (*Source:* American Wind Energy Association)

**50.** When geologists study volcanic eruptions, they are often interested in finding the eruption rate. To find the eruption rate, the volume of erupted lava in cubic meters is divided by the length of the episode in hours. In 1983, an eruption episode of the Hawaiian volcano Kilauea produced a volume of 11,000,000 cubic meters of lava in a period of 100 hours. What was the eruption rate for this eruption episode? (*Hint:* The rate will be in cubic meters per hour.) (*Source:* U.S. Geological Survey)

**D** *Find the average of each list of numbers. See Example 13.*

**51.** 14, 22, 45, 18, 30, 27      **52.** 38, 26, 15, 29, 29, 51, 22      **53.** 204, 968, 552, 268

**54.** 121, 200, 185, 176, 163      ▭ **55.** 86, 79, 81, 69, 80      **56.** 92, 96, 90, 85, 92, 79

**43.** Yes, she needs 176 ft; she has 9 ft left over.

**44.** 11 light poles

**45.** 17 touchdowns

**46.** 175,000 gal

**47.** 1760 yd

**48.** 16 ft

**49.** 6000 kilowatt-hours

**50.** 110,000 cu. m per hr

**51.** 26

**52.** 30

**53.** 498

**54.** 169

**55.** 79

**56.** 89

**Name** _____

*The monthly normal temperature in degrees Fahrenheit for Minneapolis, Minnesota, is given in the table. Use this table to answer Exercises 57 and 58. (Source: National Climatic Data Center)*

| | | | |
|---------|-----|--------|-----|
| January | 12° | July | 74° |
| February | 18° | August | 71° |
| March | 31° | Sept. | 61° |
| April | 46° | Oct. | 49° |
| May | 59° | Nov. | 33° |
| June | 68° | Dec. | 18° |

**57.** Find the average temperature for December, January, and February.

**58.** Find the average temperature for the entire year.

### COMBINING CONCEPTS

*The table (from Section 1.3) shows the top ten leading United States advertisers in 1998 and the amount of money spent in that year on ads. Use this table to answer Exercises 59 and 60.*

| | |
|--------------------|----------------|
| General Motors | $2,121,040,900 |
| Procter & Gamble | 1,724,259,700 |
| DaimlerChrysler | 1,410,748,700 |
| Philip Morris | 1,264,353,200 |
| Ford Motor | 1,147,589,200 |
| Time Warner | 829,185,000 |
| Walt Disney | 809,878,700 |
| Sears Roebuck & Co. | 720,543,900 |
| Unilever | 691,203,500 |
| Diageo Plc | 659,093,600 |

**59.** Find the average amount of money spent on ads for the year by the top four companies.

**60.** Find the average amount of money spent on ads for the year by Walt Disney, Ford Motor, Time Warner, and Sears Roebuck & Co.

*In Example 13 in this section, we found that the average of 93, 86, 71, and 82 is 83. Use this information to answer Exercises 61 and 62.*

**61.** If the number 71 is removed from the list of numbers, does the average increase or decrease? Explain why.

**62.** If the number 93 is removed from the list of numbers, does the average increase or decrease? Explain why.

**63.** Without computing it, tell whether the average of 126, 135, 198, and 113 is 86. Explain why it is or why it is not.

△ **64.** If the area of a rectangle is 30 square feet and its width is 3 feet, what is its length?

| 30 square feet | 3 feet |
|:---:|:---|
| ? | |

**70**

# INTEGRATED REVIEW—OPERATIONS ON WHOLE NUMBERS

*Perform each indicated operation.*

**1.**  23
    46
    +79

**2.**  7006
     − 451

**3.**   36
    × 45

**4.** $8\overline{)4496}$

**5.** $1 \cdot 79$

**6.** $\dfrac{36}{0}$

**7.** $9 \div 1$

**8.** $9 \div 9$

**9.** $0 \cdot 13$

**10.** $7 \cdot 0 \cdot 8$

**11.** $0 \div 2$

**12.** $12 \div 4$

**13.**  4219
    −1786

**14.**  1861
    +7965

**15.** $5\overline{)1068}$

**16.**  1259
    ×   63

**17.** $3 \cdot 9$

**18.** $45 \div 5$

**19.**  207
    − 69

**20.**  207
    + 69

**21.** $32\overline{)21{,}222}$

**22.** $65\overline{)70{,}000}$

*Determine the place value of the digit 3 in each whole number.*

**23.** 732

**24.** 23,000

**25.** 8003

**26.** 32,222

**ANSWERS**

**1.** 148

**2.** 6555

**3.** 1620

**4.** 562

**5.** 79

**6.** undefined

**7.** 9

**8.** 1

**9.** 0

**10.** 0

**11.** 0

**12.** 3

**13.** 2433

**14.** 9826

**15.** 213 R 3

**16.** 79,317

**17.** 27

**18.** 9

**19.** 138

**20.** 276

**21.** 663 R 6

**22.** 1076 R 60

**23.** tens

**24.** thousands

**25.** ones

**26.** ten-thousands

**27.** eight hundred fifty

**28.** eight hundred five

**29.** twenty-one thousand, sixty

**30.** four thousand, forty-four

**31.** 5612

**32.** 73,001

**33.** 100,306

**34.** 4,000,020

**35.** see table

**36.** see table

**37.** see table

**38.** see table

**Name** _____

*Write each whole number in words.*

**27.** 850          **28.** 805          **29.** 21,060          **30.** 4044

*Write each whole number in standard form.*

**31.** five thousand, six hundred twelve          **32.** seventy-three thousand, one

**33.** one hundred-thousand, three hund-red six          **34.** four million, twenty

*Fill in the table by rounding each number to the place value indicated.*

|  |  | Tens | Hundreds |
|---|---|---|---|
| **35.** | 3265 | 3270 | 3300 |
| **36.** | 46,817 | 46,820 | 46,800 |
| **37.** | 60,505 | 60,510 | 60,500 |
| **38.** | 671 | 670 | 700 |

# 1.7 EXPONENTS AND ORDER OF OPERATIONS

## A USING EXPONENTIAL NOTATION

An **exponent** is a shorthand notation for repeated multiplication. When the same number is a factor a number of times, an exponent may be used. In the product

$$\underbrace{2 \cdot 2 \cdot 2 \cdot 2 \cdot 2}_{\text{2 is a factor 5 times.}}$$

Using an exponent, this product can be written as

$2^5$ — exponent, base. Read as "two to the fifth power."

Thus,

$$2 \cdot 2 \cdot 2 \cdot 2 \cdot 2 = 2^5$$

This is called **exponential notation**. The **exponent**, 5, indicates how many times the **base**, 2, is a factor.

Certain expressions are used when reading exponential notation.

$5 = 5^1$ is read as "five to the first power."
$5 \cdot 5 = 5^2$ is read as "five to the second power" or "five squared."
$5 \cdot 5 \cdot 5 = 5^3$ is read as "five to the third power" or "five cubed."
$5 \cdot 5 \cdot 5 \cdot 5 = 5^4$ is read as "five to the fourth power."

Usually, an exponent of 1 is not written, so when no exponent appears, we assume that the exponent is 1. For example, $2 = 2^1$ and $7 = 7^1$.

**Examples** Write using exponential notation.

**1.** $4 \cdot 4 \cdot 4 = 4^3$
**2.** $7 \cdot 7 = 7^2$
**3.** $5 \cdot 5 \cdot 5 \cdot 5 = 5^4$
**4.** $6 \cdot 6 \cdot 6 \cdot 8 \cdot 8 \cdot 8 \cdot 8 \cdot 8 = 6^3 \cdot 8^5$

## B EVALUATING EXPONENTIAL EXPRESSIONS

To **evaluate** an exponential expression, we write the expression as a product and then find the value of the product.

**Examples** Evaluate.

**5.** $8^2 = 8 \cdot 8 = 64$
**6.** $7^1 = 7$
**7.** $2^5 = 2 \cdot 2 \cdot 2 \cdot 2 \cdot 2 = 32$
**8.** $5 \cdot 6^2 = 5 \cdot 6 \cdot 6 = 180$

Example 8 illustrates an important property: An exponent applies only to its base. The exponent 2, in $5 \cdot 6^2$ applies only to its base, 6.

### Helpful Hint

An exponent applies only to its base. For example, $4 \cdot 2^3$ means $4 \cdot 2 \cdot 2 \cdot 2$.

### Objectives

**A** Write repeated factors using exponential notation.
**B** Evaluate expressions containing exponents.
**C** Use the order of operations.
**D** Find the area of a square.

Study Guide    SSM    CD-ROM    Video 1.7

TEACHING TIP

Using the multiplication table in the Appendix, consider having students high-light the squared numbers, which fall on the diagonal from the upper left corner to the lower right corner. Encourage them to become familiar with these squared numbers. Ask if they notice any pattern in the numbers. Be sure that they see that the difference between consecutive squared numbers is the sequence of odd numbers beginning with 3.

### Practice Problems 1–4

Write using exponential notation.
1. $2 \cdot 2 \cdot 2$
2. $3 \cdot 3$
3. $10 \cdot 10 \cdot 10 \cdot 10 \cdot 10 \cdot 10$
4. $5 \cdot 5 \cdot 4 \cdot 4 \cdot 4$

### Practice Problems 5–8

Evaluate.
5. $2^3$          6. $5^2$
7. $10^1$         8. $4 \cdot 5^2$

**Answers**
**1.** $2^3$  **2.** $3^2$  **3.** $10^6$  **4.** $5^2 \cdot 4^3$  **5.** 8  **6.** 25  **7.** 10  **8.** 100

---

**Helpful Hint**

Don't forget that $2^4$, for example, is *not* $2 \cdot 4$. Instead, $2^4$ means repeated multiplication of the same factor.

$$2^4 = 2 \cdot 2 \cdot 2 \cdot 2 = 16 \quad \text{whereas} \quad 2 \cdot 4 = 8$$

---

✓ **CONCEPT CHECK**

Which of the following statements is correct?

a. "Four cubed" is the same as $4^2$.

b. $5^3$ is the same as $3 \cdot 3 \cdot 3 \cdot 3 \cdot 3$.

c. $7^2$ is the same as $7 \cdot 7$.

d. "Nine to the third power" is the same as $3^9$.

**TRY THE CONCEPT CHECK IN THE MARGIN.**

**C  USING THE ORDER OF OPERATIONS**

Suppose that you are in charge of taking inventory at a local bookstore. An employee has given you the number of a certain book in stock as the expression

$$3 + 2 \cdot 10$$

To calculate the value of this expression, do you add first or multiply first? If you add first, the answer is 50. If you multiply first, the answer is 23.

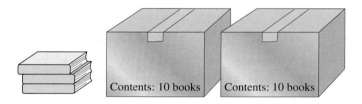

Contents: 10 books        Contents: 10 books

Mathematical symbols wouldn't be very useful if two values were possible for one expression like this. Thus, mathematicians have agreed that, given a choice, we multiply first.

$$3 + 2 \cdot 10 = 3 + 20 \quad \text{Multiply.}$$
$$= 23 \quad \text{Add.}$$

This agreement is one of several **order of operations** agreements.

---

**ORDER OF OPERATIONS**

1. Do all operations within grouping symbols such as parentheses or brackets.

2. Evaluate any expressions with exponents.

3. Multiply or divide in order from left to right.

4. Add or subtract in order from left to right.

---

For example, using the order of operations, let's evaluate $2^3 \cdot 4 - (10 \div 5)$.

$$2^3 \cdot 4 - (10 \div 5) = 2^3 \cdot 4 - 2 \quad \text{Simplify inside parentheses.}$$
$$= 8 \cdot 4 - 2 \quad \text{Write } 2^3 \text{ as 8.}$$
$$= 32 - 2 \quad \text{Multiply } 8 \cdot 4.$$
$$= 30 \quad \text{Subtract.}$$

**Example 9**   Simplify: $2 \cdot 4 - 3 \div 3$

*Solution:*   There are no parentheses and no exponents, so we start by multiplying and dividing, from left to right.

$$2 \cdot 4 - 3 \div 3 = 8 - 3 \div 3 \qquad \text{Multiply.}$$
$$= 8 - 1 \qquad \text{Divide.}$$
$$= 7 \qquad \text{Subtract.}$$

**Example 10**   Simplify: $(8 - 6)^2 + 2^3 \cdot 3$

*Solution:*   $(8 - 6)^2 + 2^3 \cdot 3 = 2^2 + 2^3 \cdot 3 \qquad \text{Simplify inside parentheses.}$

$$= 4 + 8 \cdot 3 \qquad \text{Write } 2^2 \text{ as 4 and } 2^3 \text{ as 8.}$$
$$= 4 + 24 \qquad \text{Multiply.}$$
$$= 28 \qquad \text{Add.}$$

**Example 11**   Simplify: $4^3 + \left[3^2 - (10 \div 2)\right] - 7 \cdot 3$

*Solution:*   Here we begin within the innermost set of parentheses.

$4^3 + \left[3^2 - (10 \div 2)\right] - 7 \cdot 3 = 4^3 + \left[3^2 - 5\right] - 7 \cdot 3 \qquad$ Simplify inside parentheses.

$$= 4^3 + \left[9 - 5\right] - 7 \cdot 3 \qquad \text{Write } 3^2 \text{ as 9.}$$
$$= 4^3 + 4 - 7 \cdot 3 \qquad \text{Simplify inside brackets.}$$
$$= 64 + 4 - 7 \cdot 3 \qquad \text{Write } 4^3 \text{ as 64.}$$
$$= 64 + 4 - 21 \qquad \text{Multiply.}$$
$$= 47 \qquad \text{Add and subtract from left to right.}$$

**Example 12**   Simplify: $\dfrac{7 - 2 \cdot 3 + 3^2}{2^2 + 1}$

*Solution:*   Here, the fraction bar is like a grouping symbol. We simplify above and below the fraction bar separately.

$$\frac{7 - 2 \cdot 3 + 3^2}{2^2 + 1} = \frac{7 - 2 \cdot 3 + 9}{4 + 1} \qquad \text{Evaluate } 3^2 \text{ and } 2^2.$$

$$= \frac{7 - 6 + 9}{5} \qquad \text{Multiply } 2 \cdot 3 \text{ in the numerator and add 4 and 1 in the denominator.}$$

$$= \frac{10}{5} \qquad \text{Add and subtract from left to right.}$$

$$= 2 \qquad \text{Divide.}$$

## △ D FINDING THE AREA OF A SQUARE

Since a square is a special rectangle, we can find its area by finding the product of its length and its width.

area of a rectangle = length · width

Simplify: $16 \div 4 - 2$

**Practice Problem 10**

Simplify: $(9 - 8)^3 + 3 \cdot 2^4$

**Practice Problem 11**

Simplify:
$24 \div \left[20 - (3 \cdot 4)\right] + 2^3 - 5$

TEACHING TIP

If your students are working in groups, ask each group to make a list in 10 minutes of expressions that equal the numbers from 1 to 100 using the basic operations (add, subtract, multiply, divide, and raising to exponents) and a digit of their choice. Their digit must be used four times in each expression. For example, if their digit is 3:

$$1 = \frac{3 + 3}{3 + 3}$$

$$2 = \frac{3}{3} + \frac{3}{3}$$

$$\vdots$$

$$28 = 3^3 + \frac{3}{3}$$

**Practice Problem 12**

Simplify: $\dfrac{60 - 5^2 + 1}{3(1 + 1)}$

Answers
**9.** 2   **10.** 49   **11.** 6   **12.** 6

By recalling that each side of a square has the same measurement, we can use the following to find its area.

$$\text{area of a square} = \text{length} \cdot \text{width}$$
$$= \text{side} \cdot \text{side}$$
$$= (\text{side})^2$$

Square

Side

Side

△ **Practice Problem 13**

Find the area of the square whose side measures 11 centimeters.

11 centimeters

△ **Example 13**    Find the area of the square whose side measures 5 inches.

*Solution:*    $\text{area of a square} = (\text{side})^2$
$$= (5 \text{ inches})^2$$
$$= 25 \text{ square inches}$$

5 inches

The area of the square is 25 square inches.

**Answer**

**13.** 121 square centimeters

# CALCULATOR EXPLORATIONS
## EXPONENTS

To evaluate an exponent such as $4^7$ on a calculator, find the keys marked $\boxed{y^x}$ and $\boxed{=}$ (or $\boxed{\text{ENTER}}$). To evaluate $4^7$, press the keys $\boxed{4}\,\boxed{y^x}\,\boxed{7}\,\boxed{=}$ (or $\boxed{\text{ENTER}}$). The display will read $\boxed{\qquad 16384}$. Thus, $4^7 = 16{,}384$.

*Use a calculator to evaluate.*

| | | |
|---|---|---|
| **1.** $3^6$  729 | **2.** $5^6$  15,625 | **3.** $4^5$  1024 |
| **4.** $7^6$  117,649 | **5.** $2^{11}$  2048 | **6.** $6^8$  1,679,616 |

## ORDER OF OPERATIONS

To see whether your calculator has order of operations built in, evaluate $5 + 2 \cdot 3$ by pressing the keys $\boxed{5}\,\boxed{+}\,\boxed{2}\,\boxed{\times}\,\boxed{3}\,\boxed{=}$ (or $\boxed{\text{ENTER}}$). If the display reads 11, your calculator does have order of operations built in. This means that most of the time you can key in a problem exactly as it is written and the calculator will perform operations in the proper order. When evaluating an expression containing parentheses, key in the parentheses. (If an expression contains brackets, key in parentheses.) For example, to evaluate $2[25 - (8 + 4)] - 11$, press the keys $\boxed{2}\,\boxed{\times}\,\boxed{(}\,\boxed{25}\,\boxed{-}\,\boxed{(}\,\boxed{8}\,\boxed{+}\,\boxed{4}\,\boxed{)}\,\boxed{)}\,\boxed{-}\,\boxed{11}\,\boxed{=}$ (or $\boxed{\text{ENTER}}$). The display will read $\boxed{\qquad 15}$.

## SIMPLIFYING AN EXPRESSION CONTAINING A FRACTION BAR

Even though a calculator follows the order of operations, parentheses must sometimes be inserted. For example, to simplify $\dfrac{8+6}{2}$ on a calculator, enter parentheses about the expression above the fraction bar so that it is simplified separately.

To simplify $\dfrac{8+6}{2}$, press the keys

$\boxed{(}\,\boxed{8}\,\boxed{+}\,\boxed{6}\,\boxed{)}\,\boxed{\div}\,\boxed{2}\,\boxed{=}$ (or $\boxed{\text{ENTER}}$).

The display will read $\boxed{\qquad 7}$.

Thus $\dfrac{8+6}{2} = 7$.

*Use a calculator to evaluate.*   **12.** 300

| | |
|---|---|
| **7.** $7^4 + 5^3$  2526 | **8.** $12^4 - 8^4$  16,640 |
| **9.** $63 \cdot 75 - 43 \cdot 10$  4295 | **10.** $8 \cdot 22 + 7 \cdot 16$  288 |
| **11.** $4(15 \div 3 + 2) - 10 \cdot 2$  8 | **12.** $155 - 2(17 + 3) + 185$ |
| **13.** $\dfrac{720}{4+5}$  80 | **14.** $\dfrac{245}{5+2}$  35 |
| **15.** $\dfrac{78 + 1056}{37 + 5}$  27 | **16.** $\dfrac{869 + 56}{29 + 8}$  25 |

# Focus on Study Skills

## STUDY TIPS

Have you wondered what you can do to be successful in your math course? If so, that may well be your first step to success! Here are some tips on how to study mathematics.

1. *Choose to attend all class periods.* If possible, sit near the front of the classroom. This way, you will see, hear, and focus on the presentation better. It may also make it easier for you to participate in classroom activities.

2. *Do your homework.* You've probably heard the phrase "practice makes perfect" in relation to music and sports. It also applies to mathematics. You will find that the more time you spend solving mathematics problems, the easier the process becomes. Be sure to block out enough time in your schedule to complete your assignments.

3. *Check your work.* Review the steps you made while working a problem. Learn to check your answers in the original problems. You can also compare your answers to the Answers to Selected Exercises listed in the back of the book. If you have made a mistake, figure out what went wrong. Then correct your mistake.

4. *Learn from your mistakes.* Everyone, even your instructor, makes mistakes. You can use your mistakes to become a better math student. The key is finding and understanding your mistakes. Was your mistake a careless one? If so, you can try to work more slowly and make a conscious effort to carefully check your work. Did you make a mistake because you don't understand a concept? If so, take the time to review the concept or ask questions to better understand the concept.

5. *Know how to get help if you need it.* It's OK to ask for help. In fact, it's a good idea to ask for help whenever there is something that you don't understand. Make sure you know when your instructor has office hours and how to find his or her office. Find out if math tutoring services are available on your campus and check out the hours, location, and requirements of the tutoring service. You might also want to find another student in your class that you can call to discuss your assignment.

# EXERCISE SET 1.7

**A** *Write using exponential notation. See Examples 1 through 4.*

**1.** $3 \cdot 3 \cdot 3 \cdot 3$

**2.** $5 \cdot 5 \cdot 5$

**3.** $7 \cdot 7 \cdot 7 \cdot 7 \cdot 7 \cdot 7 \cdot 7 \cdot 7$

**4.** $6 \cdot 6 \cdot 6 \cdot 6 \cdot 6$

**5.** $12 \cdot 12 \cdot 12$

**6.** $10 \cdot 10$

**7.** $6 \cdot 6 \cdot 5 \cdot 5 \cdot 5$

**8.** $4 \cdot 4 \cdot 4 \cdot 3 \cdot 3$

**9.** $9 \cdot 9 \cdot 9 \cdot 8$

**10.** $7 \cdot 7 \cdot 7 \cdot 4$

**11.** $3 \cdot 2 \cdot 2 \cdot 2 \cdot 2 \cdot 2$

**12.** $4 \cdot 6 \cdot 6 \cdot 6 \cdot 6$

**13.** $3 \cdot 2 \cdot 2 \cdot 5 \cdot 5 \cdot 5$

**14.** $6 \cdot 6 \cdot 2 \cdot 9 \cdot 9 \cdot 9 \cdot 9$

**B** *Evaluate. See Examples 5 through 8.*

**15.** $5^2$

**16.** $6^2$

**17.** $5^3$

**18.** $6^3$

**19.** $2^6$

**20.** $2^7$

**21.** $2^{10}$

**22.** $1^{12}$

**23.** $7^1$

**24.** $8^1$

**25.** $3^5$

**26.** $5^4$

ANSWERS

**1.** $3^4$

**2.** $5^3$

**3.** $7^8$

**4.** $6^5$

**5.** $12^3$

**6.** $10^2$

**7.** $6^2 \cdot 5^3$

**8.** $4^3 \cdot 3^2$

**9.** $9^3 \cdot 8$

**10.** $7^3 \cdot 4$

**11.** $3 \cdot 2^5$

**12.** $4 \cdot 6^4$

**13.** $3 \cdot 2^2 \cdot 5^3$

**14.** $6^2 \cdot 2 \cdot 9^4$

**15.** 25

**16.** 36

**17.** 125

**18.** 216

**19.** 64

**20.** 128

**21.** 1024

**22.** 1

**23.** 7

**24.** 8

**25.** 243

**26.** 625

**27.** 256
**28.** 27
**29.** 64
**30.** 256
**31.** 81
**32.** 64
**33.** 729
**34.** 512
**35.** 100
**36.** 1000
**37.** 10,000
**38.** 100,000
**39.** 10
**40.** 14
**41.** 1920
**42.** 6849
**43.** 729
**44.** 1024
**45.** 21
**46.** 42
**47.** 8
**48.** 5
**49.** 29
**50.** 22
**51.** 4
**52.** 0
**53.** 17
**54.** 36
**55.** 46
**56.** 21
**57.** 28
**58.** 42
**59.** 10
**60.** 9
**61.** 7
**62.** 5
**63.** 4
**64.** 6
**65.** 14
**66.** 7
**67.** 72
**68.** 5

**Name** _____

**27.** $2^8$      **28.** $3^3$      **29.** $4^3$      **30.** $4^4$

**31.** $9^2$      **32.** $8^2$      **33.** $9^3$      **34.** $8^3$

**35.** $10^2$      **36.** $10^3$      **37.** $10^4$      **38.** $10^5$

**39.** $10^1$      **40.** $14^1$      **41.** $1920^1$      **42.** $6849^1$

**43.** $3^6$      **44.** $4^5$

**C** *Simplify. See Examples 9 through 12.*

**45.** $15 + 3 \cdot 2$      **46.** $24 + 6 \cdot 3$      **47.** $20 - 4 \cdot 3$      **48.** $17 - 2 \cdot 6$

**49.** $5 \cdot 9 - 16$      **50.** $8 \cdot 4 - 10$      **51.** $28 \div 4 - 3$      **52.** $42 \div 7 - 6$

**53.** $14 + \dfrac{24}{8}$      **54.** $32 + \dfrac{8}{2}$      **55.** $6 \cdot 5 + 8 \cdot 2$      **56.** $3 \cdot 4 + 9 \cdot 1$

**57.** $0 \div 6 + 4 \cdot 7$      **58.** $0 \div 8 + 7 \cdot 6$      **59.** $6 + 8 \div 2$      **60.** $6 + 9 \div 3$

**61.** $(6 + 8) \div 2$      **62.** $(6 + 9) \div 3$      **63.** $\left(6^2 - 4\right) \div 8$      **64.** $\left(7^2 - 7\right) \div 7$

**65.** $\left(3 + 5^2\right) \div 2$      **66.** $\left(13 + 6^2\right) \div 7$      **67.** $6^2 \cdot (10 - 8)$      **68.** $5^3 \div (10 + 15)$

**Name** _____

**69.** $\dfrac{18 + 6}{2^4 - 4}$

**70.** $\dfrac{15 + 17}{5^2 - 3^2}$

**71.** $(2 + 5) \cdot (8 - 3)$

**72.** $(9 - 7) \cdot (12 + 18)$

 **73.** $\dfrac{7(9 - 6) + 3}{3^2 - 3}$

**74.** $\dfrac{5(12 - 7) - 4}{5^2 - 2^3 - 10}$

**75.** $5 \div 0 + 24$

**76.** $18 - 7 \div 0$

**77.** $3^4 - [35 - (12 - 6)]$

**78.** $[40 - (8 - 2)] - 2^5$

 **79.** $(7 \cdot 5) + [9 \div (3 \div 3)]$

**80.** $(18 \div 6) + [(3 + 5) \cdot 2]$

**81.** $8 \cdot [4 + (6 - 1) \cdot 2] - 50 \cdot 2$

**82.** $35 \div [3^2 + (9 - 7) - 2^2] + 10 \cdot 3$

**83.** $7^2 - \{18 - [40 \div (4 \cdot 2) + 2] + 5^2\}$

**84.** $29 - \{5 + 3[8 \cdot (10 - 8)] - 50\}$

**D** *Find the area of each square. See Example 13.*

△ **85.**
20 miles

△ **86.**
4 meters

△ **87.** 8 centimeters

△ **88.**
31 feet

**69.** 2

**70.** 2

**71.** 35

**72.** 60

**73.** 4

**74.** 3

**75.** undefined

**76.** undefined

**77.** 52

**78.** 2

**79.** 44

**80.** 19

**81.** 12

**82.** 35

**83.** 13

**84.** 26

**85.** 400 sq. mi

**86.** 16 sq. m

**87.** 64 sq. cm

**88.** 961 sq. ft

**Name** _____

△ **89.** The Eiffel Tower stands on a square base measuring 100 meters on each side. Find the area of the base.

△ **90.** A square lawn that measures 72 feet on each side is to be fertilized. If 5 bags of fertilizer are available and each bag can fertilize 1000 square feet, is there enough fertilizer to cover the lawn?

### COMBINING CONCEPTS

*Insert grouping symbols (parentheses) so that each given expression evaluates to the given number.*

**91.** $2 + 3 \cdot 6 - 2$; evaluate to 28

**92.** $2 + 3 \cdot 6 - 2$; evaluate to 20

**93.** $24 \div 3 \cdot 2 + 2 \cdot 5$; evaluate to 14

**94.** $24 \div 3 \cdot 2 + 2 \cdot 5$; evaluate to 15

△ **95.** A building contractor is bidding on a contract to install gutters on seven homes in a retirement community all in the shape shown. To estimate his cost of materials, he needs to know the total perimeter of all seven homes. Find the total perimeter.

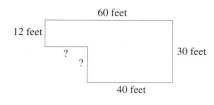

*Simplify.*

▣ **96.** $25^3 \cdot (45 - 7 \cdot 5) \cdot 5$

▣ **97.** $\left(7 + 2^4\right)^5 - \left(3^5 - 2^4\right)^2$

✎ **98.** Explain why $2 \cdot 3^2$ is not the same as $(2 \cdot 3)^2$.

# 1.8 INTRODUCTION TO VARIABLES AND ALGEBRAIC EXPRESSIONS

**Objectives**

**A** Evaluate algebraic expressions given replacement values.

**B** Translate phrases into variable expressions.

Study Guide   SSM   CD-ROM   Video 1.8

## A EVALUATING ALGEBRAIC EXPRESSIONS

Perhaps the most important quality of mathematics is that it is a science of patterns. Communicating about patterns is often made easier by using a letter to represent all the numbers fitting a pattern. We call such a letter a **variable**. For example, in Section 1.2 we presented the addition property of 0, which states that the sum of 0 and any number is that number. We might write

$$0 + 1 = 1$$
$$0 + 2 = 2$$
$$0 + 3 = 3$$
$$0 + 4 = 4$$
$$0 + 5 = 5$$
$$0 + 6 = 6$$
$$\vdots$$

continuing indefinitely. This is a pattern, and all whole numbers fit the pattern. We can communicate this pattern for all whole numbers by letting a letter, such as $a$, represent all whole numbers. We can then write

$$0 + a = a$$

Learning to use variable notation is a primary goal of learning **algebra**. We now take some important beginning steps in learning to use variable notation.

A combination of operations on letters (variables) and numbers is called an **algebraic expression** or simply an **expression**.

### Algebraic Expressions

$$3 + x \qquad 5 \cdot y \qquad 2 \cdot z - 1 + x$$

If two variables or a number and a variable are next to each other, with no operation sign between them, the operation is multiplication. For example,

$$2x \quad \text{means} \quad 2 \cdot x$$

and

$$xy \text{ or } x(y) \quad \text{means} \quad x \cdot y$$

Also, the meaning of an exponent remains the same when the base is a variable. For example,

$$x^2 = \underbrace{x \cdot x}_{2 \text{ factors of } x} \quad \text{and} \quad y^5 = \underbrace{y \cdot y \cdot y \cdot y \cdot y}_{5 \text{ factors of } y}$$

Algebraic expressions such as $3x$ have different values depending on replacement values for $x$. For example, if $x$ is 2, then $3x$ becomes

$$3x = 3 \cdot 2$$
$$= 6$$

If $x$ is 7, then $3x$ becomes

$$3x = 3 \cdot 7$$
$$= 21$$

Replacing a variable in an expression by a number and then finding the value of the expression is called **evaluating the expression** for the variable. When finding the value of an expression, remember to follow the order of operations given in Section 1.7.

**Practice Problem 1**

Evaluate $x - 2$ if $x$ is 5.

**Example 1**     Evaluate $x + 7$ if $x$ is 8.

*Solution:*     Replace $x$ with 8 in the expression $x + 7$.

$$x + 7 = 8 + 7 \qquad \text{Replace } x \text{ with 8.}$$
$$= 15 \qquad \text{Add.}$$

When we write a statement such as "$x$ is 5," we can use an equals symbol to represent "is" so that

$x$ is 5    can be written as    $x = 5$.

**Practice Problem 2**

Evaluate $y(x - 3)$ for $x = 3$ and $y = 7$.

**Example 2**     Evaluate $2(x - y)$ for $x = 8$ and $y = 4$.

*Solution:*     $2(x - y) = 2(8 - 4) \qquad \text{Replace } x \text{ with 8 and } y \text{ with 4.}$
$$= 2(4) \qquad \text{Subtract.}$$
$$= 8 \qquad \text{Multiply.}$$

**Practice Problem 3**

Evaluate $\dfrac{y + 6}{x}$ for $x = 2$ and $y = 8$.

**Example 3**     Evaluate $\dfrac{x - 5y}{y}$ for $x = 21$ and $y = 3$.

*Solution:*     $\dfrac{x - 5y}{y} = \dfrac{21 - 5(3)}{3} \qquad \text{Replace } x \text{ with 21 and } y \text{ with 3.}$

$$= \dfrac{21 - 15}{3} \qquad \text{Multiply.}$$

$$= \dfrac{6}{3} \qquad \text{Subtract.}$$

$$= 2 \qquad \text{Divide.}$$

**Practice Problem 4**

Evaluate $25 - z^3 + x$ for $z = 2$ and $x = 1$.

**Example 4**     Evaluate $x^2 + z - 3$ for $x = 5$ and $z = 4$.

*Solution:*     $x^2 + z - 3 = 5^2 + 4 - 3 \qquad \text{Replace } x \text{ with 5 and } z \text{ with 4.}$
$$= 25 + 4 - 3 \qquad \text{Evaluate } 5^2.$$
$$= 26 \qquad \text{Add and subtract from left to right.}$$

**Answers**

**1.** 3   **2.** 0   **3.** 7   **4.** 18

**Example 5**  The expression $\dfrac{5(F - 32)}{9}$ can be used to write degrees Fahrenheit $F$ as degrees Celsius $C$. Find the value of this expression for $F = 86$.

*Solution:*
$$\frac{5(F - 32)}{9} = \frac{5(86 - 32)}{9}$$
$$= \frac{5(54)}{9}$$
$$= \frac{270}{9}$$
$$= 30$$

Thus 86°F is the same temperature as 30°C.    ▬▬▬

## B  TRANSLATING PHRASES INTO VARIABLE EXPRESSIONS

To aid us in solving problems later, we practice translating verbal phrases into algebraic expressions. Certain key words and phrases suggesting addition, subtraction, multiplication, or division are reviewed next.

| Addition | Subtraction | Multiplication | Division |
|---|---|---|---|
| sum | difference | product | quotient |
| plus | minus | times | divided by |
| added to | subtracted from | multiply | into |
| more than | less than | twice | per |
| increased by | decreased by | of | |
| total | less | double | |

**Example 6**  Write as an algebraic expression. Use $x$ to represent "a number."

   **a.** 7 more than a number

   **b.** 15 decreased by a number

   **c.** The product of 2 and a number

   **d.** The quotient of a number and 5

   **e.** 2 subtracted from a number

*Solution:*

**a.** In words: 7 | more than | a number

Translate: $7 \quad + \quad x$

**b.** In words: 15 | decreased by | a number

Translate: $15 \quad - \quad x$

**c.** In words: The product of

2 | and | a number

Translate: $2 \quad \cdot \quad x \quad$ or $2x$

**d.** In words: The quotient of

a number | and | 5

Translate: $x \quad \div \quad 5 \quad$ or $\dfrac{x}{5}$

**e.** In words: 2 | subtracted from | a number

Translate: $x \quad - \quad 2$

---

## Helpful Hint

Remember that order is important when subtracting. Study the order of numbers and variables below.

| Phrase | Translation |
|---|---|
| a number *decreased by* 5 | $x - 5$ |
| a number *subtracted from* 5 | $5 - x$ |

## Exercise Set 1.8

**A** *Evaluate each following expression for $x = 2$, $y = 5$, and $z = 3$. See Examples 1 through 5.*

**1.** $3 + 2z$

**2.** $7 + 3z$

**3.** $6xz - 5x$

**4.** $4yz + 2x$

**5.** $z - x + y$

**6.** $x + 5y - z$

**7.** $3x - z$

**8.** $2y + 5z$

**9.** $y^3 - 4x$

**10.** $y^3 - z$

**11.** $2xy^2 - 6$

**12.** $3yz^2 + 1$

**13.** $8 - (y - x)$

**14.** $5 + (2x - 1)$

**15.** $y^4 + (z - x)$

**16.** $x^4 - (y - z)$

**17.** $\dfrac{6xy}{z}$

**18.** $\dfrac{8yz}{15}$

**19.** $\dfrac{2y - 2}{x}$

**20.** $\dfrac{6 + 3x}{z}$

**21.** $\dfrac{x + 2y}{z}$

**22.** $\dfrac{2z + 6}{3}$

**23.** $\dfrac{5x}{y} - \dfrac{10}{y}$

**24.** $\dfrac{70}{2y} - \dfrac{15}{z}$

**25.** $2y^2 - 4y + 3$

**26.** $3z^2 - z + 10$

**27.** $(3y - 2x)^2$

**28.** $(4y + 3z)^2$

**29.** $(xy + 1)^2$

**30.** $(xz - 5)^4$

**31.** $2y(4z - x)$

**32.** $3x(y + z)$

**33.** $xy(5 + z - x)$

**34.** $xz(2y + x - z)$

**35.** $\dfrac{7x + 2y}{3x}$

**36.** $\dfrac{6z + 2y}{4}$

ANSWERS

**1.** 9
**2.** 16
**3.** 26
**4.** 64
**5.** 6
**6.** 24
**7.** 3
**8.** 25
**9.** 117
**10.** 122
**11.** 94
**12.** 136
**13.** 5
**14.** 8
**15.** 626
**16.** 14
**17.** 20
**18.** 8
**19.** 4
**20.** 4
**21.** 4
**22.** 4
**23.** 0
**24.** 2
**25.** 33
**26.** 34
**27.** 121
**28.** 841
**29.** 121
**30.** 1
**31.** 100
**32.** 48
**33.** 60
**34.** 54
**35.** 4
**36.** 7

**Name** _____

**37.** The expression $16t^2$ gives the distance in feet that an object falls after $t$ seconds. Complete the table by evaluating $16t^2$ for each given value of $t$.

| $t$ | 1 | 2 | 3 | 4 |
|-----|----|----|-----|-----|
| $16t^2$ | 16 | 64 | 144 | 256 |

**38.** The expression $\dfrac{5(F - 32)}{9}$ gives the equivalent degrees Celsius for $F$ degrees Fahrenheit. Complete the table by evaluating this expression for each given value of $F$.

| $F$ | 50 | 59 | 68 | 77 |
|-----|----|----|----|----|
| $\dfrac{5(F - 32)}{9}$ | 10 | 15 | 20 | 25 |

**B** *Write each phrase as a variable expression. Use x to represent "a number." See Example 6.*

**39.** The sum of a number and five

**40.** Ten plus a number

**41.** The total of a number and eight

**42.** The difference of a number and five hundred

**43.** Twenty decreased by a number

**44.** A number less thirty

**45.** The product of 512 and a number

**46.** A number times twenty

**47.** A number divided by 2

**48.** The quotient of six and a number

**49.** The sum of seventeen and a number added to the product of five and the number

**50.** The difference of twice a number, and four

**51.** The product of five and a number

**52.** The quotient of twenty and a number, decreased by three

**Name** _____

**53.** A number subtracted from 11

**54.** Twelve subtracted from a number

**55.** Fifty decreased by eight times a number

**56.** Twenty decreased by twice a number

 **COMBINING CONCEPTS**

*Use a calculator to evaluate each expression for $x = 23$ and $y = 72$.*

**57.** $x^4 - y^2$

**58.** $2(x + y)^2$

**59.** $x^2 + 5y - 112$

**60.** $16y - 20x + x^3$

**61.** If $x$ is a whole number, which expression is the largest: $2x$, $5x$, or $\dfrac{x}{3}$? Explain your answer.

**62.** If $x$ is a whole number, which expression is the smallest: $2x$, $5x$, or $\dfrac{x}{3}$? Explain your answer.

**63.** In Exercise 37, what do you notice about the value of $16t^2$ as $t$ gets larger?

**64.** In Exercise 38, what do you notice about the value of $\dfrac{5(F - 32)}{9}$ as $F$ gets larger?

**53.** $11 - x$

**54.** $x - 12$

**55.** $50 - 8x$

**56.** $20 - 2x$

**57.** 274,657

**58.** 18,050

**59.** 777

**60.** 12,859

**61.** $5x$

**62.** $\dfrac{x}{3}$

**63.** As $t$ gets larger, $16t^2$ gets larger.

**64.** As $F$ gets larger, $\dfrac{5(F - 32)}{9}$ gets larger.

# Focus On the Real World

## JUDGING DISTANCES

Do you know how to estimate a distance without using a tape measure? One easy way to do this is to use the length of your own stride. First, measure the length of your stride in inches. You can do this with the following steps:

▲ Lay a yardstick on the floor.
▲ Stand next to the yardstick with feet together so that both heels line up with the 0-mark on the yardstick.
▲ Take a normal-sized step forward.
▲ Find the whole-inch mark nearest the toe of the foot farthest from the 0-mark. This is roughly the length of your stride.

To judge a distance, simply pace it off using normal-sized strides. Multiply the number of strides by the length of your stride to get a rough estimate of the distance.

     Suppose you need to measure a distance that can't easily be paced, such as a pond or a busy street. You can easily "transfer" the distance to a more easily paced area by using a baseball cap.

▲ Stand at the edge of the pond or street while wearing a baseball cap.
▲ Bend your head until your chin rests on your chest.
▲ Pull the bill of the cap up or down until it appears to touch the other side of the pond or street.
▲ Without moving your head or the cap, pivot your body to the right until you have a clear path straight ahead for walking.
▲ Notice where the bill seems to be touching the ground now. The distance to this point is the same as the distance across the pond or street you are measuring.
▲ Pace off the distance to this point and find an estimate of the distance as before.

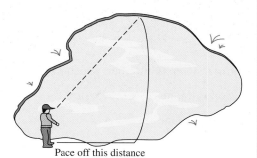

Pace off this distance

Suppose a distance estimate using this method is 1302 inches. To write the distance in terms of feet, divide the estimate by 12. The quotient is the number of whole feet and the remainder is the number of inches. A distance of 1302 inches is the same as 108 feet 6 inches.

$$
\begin{array}{r}
108\,\text{R}\,6 \\
12\,\overline{)1302} \\
\underline{12\phantom{02}} \\
10 \\
\underline{0} \\
102 \\
\underline{96} \\
6
\end{array}
$$

## GROUP ACTIVITY

     Materials: yardstick, baseball cap
     Use the baseball cap procedure to estimate the distance across a river, stream, pond, or busy road on or near your campus. Have each person in your group estimate the same distance using the length of his or her own stride. Compare your results. Write a brief report summarizing your findings. Be sure to include what distance your group estimated, each member's stride length, the number of strides each member paced off, and each member's distance estimate. Conclude by discussing reasons for any differences in estimates.

# Chapter 1 Activity
## Investigating Whole Numbers and Order of Operations

*This activity may be completed by working in groups or individually.*

Try the following activity. You will need 30 index cards.

1. Label each index card with a number from 1 to 30, using each number once.

2. Shuffle the cards. Deal 5 cards face up to each player or team. Then deal a single card face down to each player or team. This single card is the goal card.

3. Play begins when the goal cards are turned over simultaneously. The object is for each player or team to use all 5 number cards (each only once) along with any of the operations +, −, ×, or ÷ to obtain the number on the goal card. The winner is the first team to correctly write out the sequence of numbers and operations equaling the number on the goal card, according to the order of operations. Parentheses or other grouping symbols may be used as needed.

4. If no player/team is able to obtain the number on the goal card, the player/team that is able to come closest to their goal wins.

Example:

Number Cards                    Goal Card

Sequence of numbers and operations to obtain goal:

$$24 \div \left[(15 - 14) + (10 - 9)\right] = 12$$

# CHAPTER 1 HIGHLIGHTS

| DEFINITIONS AND CONCEPTS | EXAMPLES |
|---|---|

### SECTION 1.1   PLACE VALUE AND NAMES FOR NUMBERS

The **whole numbers** are 0, 1, 2, 3, 4, 5,.... The position of each **digit** in a number determines its **place value**. A place-value chart is shown below with the names of the periods shown.

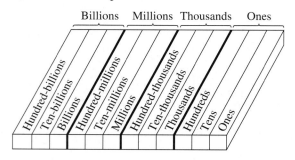

**To write a whole number in words**, write the number in the **period** followed by the name of the period. The name of the ones period is not included.

0,   14,   968,   5,268,619

9,078,651,002 is written as nine billion, seventy-eight million, six hundred fifty-one thousand, two.

### SECTION 1.2   ADDING WHOLE NUMBERS AND PERIMETER

**To add whole numbers**, add the digits in the ones place, then the tens place, then the hundreds place, and so on, carrying when necessary.

The **perimeter** of a polygon is its distance around or the sum of the lengths of its sides.

Find the sum:

```
 211
2689   ← addend
1735   ← addend
+ 662  ← addend
5086   ← sum
```

Find the perimeter of the polygon shown.

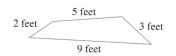

The perimeter is
5 feet + 3 feet + 9 feet + 2 feet = 19 feet.

### SECTION 1.3   SUBTRACTING WHOLE NUMBERS

To **subtract whole numbers**, subtract the digits in the ones place, then the tens place, then the hundreds place, and so on, borrowing when necessary.

Subtract:

```
  815
7954    ← minuend
−5673   ← subtrahend
 2281   ← difference
```

---

### SECTION 1.4　ROUNDING AND ESTIMATING

ROUNDING WHOLE NUMBERS
TO A GIVEN PLACE VALUE

**Step 1.** Locate the digit to the right of the given place value.

**Step 2.** If this digit is 5 or greater, add 1 to the digit in the given place value and replace each digit to its right with 0.

**Step 3.** If this digit is less than 5, replace it and each digit to its right with 0.

Round 15,721 to the nearest thousand.

$$15, \textcircled{7} 21$$

Add 1. ↗ Replace with zeros.

Since the circled digit is 5 or greater, add 1 to the given place value and replace digits to its right with zeros.

15,721 rounded to the nearest thousand is 16,000.

---

### SECTION 1.5　MULTIPLYING WHOLE NUMBERS AND AREA

**To multiply**, for example, 73 and 58, multiply 73 and 8, then 73 and 50. The sum of these partial products is the product of 73 and 58. Use the notation to the right.

$$
\begin{array}{rl}
73 & \leftarrow \text{factor} \\
\times\ 58 & \leftarrow \text{factor} \\
\hline
584 & \leftarrow 73 \times 8 \\
3650 & \leftarrow 73 \times 50 \\
\hline
4234 & \leftarrow \text{product}
\end{array}
$$

To find the **area** of a rectangle, multiply length and width.

Find the area of the rectangle.

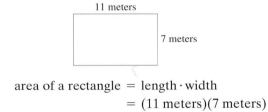

11 meters

7 meters

$$
\begin{aligned}
\text{area of a rectangle} &= \text{length} \cdot \text{width} \\
&= (11 \text{ meters})(7 \text{ meters}) \\
&= 77 \text{ square meters}
\end{aligned}
$$

---

### SECTION 1.6　DIVIDING WHOLE NUMBERS

**To divide larger whole numbers**, use the process called **long division** as shown to the right.

$$
\begin{array}{r}
507 \text{ R } 2 \quad \leftarrow \text{quotient} \\
\text{divisor} \rightarrow\ 14\overline{)7100} \quad \leftarrow \text{dividend} \\
-70\downarrow\quad\quad\ \ 5(14) = 70 \\
\hline
10\quad\quad\ \text{Subtract and bring down the 0.} \\
-\ 0\downarrow\quad\ 0(14) = 0 \\
\hline
100\quad \text{Subtract and bring down the 0.} \\
98\quad\ \ 7(14) = 98 \\
\hline
2\quad\ \ \text{Subtract. The remainder is 2.}
\end{array}
$$

To check, see that $507 \cdot 14 + 2 = 7100$

The **average** of a list of numbers is

$$\text{average} = \frac{\text{sum of numbers}}{\text{number of numbers}}$$

Find the average of 23, 35, and 38.

$$\text{average} = \frac{23 + 35 + 38}{3} = \frac{96}{3} = 32$$

---

### SECTION 1.7   EXPONENTS AND ORDER OF OPERATIONS

An **exponent** is a shorthand notation for repeated multiplication of the same factor.

ORDER OF OPERATIONS

1. Do all operations within grouping symbols such as parentheses or brackets.
2. Evaluate any expressions with exponents.
3. Multiply or divide in order from left to right.
4. Add or subtract in order from left to right.

$$\overset{\text{exponent}}{3^4} = \underbrace{3 \cdot 3 \cdot 3 \cdot 3}_{\substack{4 \text{ factors} \\ \text{of } 3}} = 81$$

base

Simplify: $\dfrac{5 + 3^2}{2(7 - 6)}$

Simplify above and below the fraction bar separately.

$$\frac{5 + 3^2}{2(7 - 6)} = \frac{5 + 9}{2(1)} \quad \text{Evaluate } 3^2.$$
$$\qquad\qquad\qquad\qquad \text{Subtract: } 7 - 6.$$
$$= \frac{14}{2} \quad \text{Add.}$$
$$\qquad\qquad \text{Multiply.}$$
$$= 7 \quad \text{Divide.}$$

The **area of a square** is $(\text{side})^2$.

Find the area of the square shown.

9 inches

$$\begin{aligned} \text{area of the square} &= (\text{side})^2 \\ &= (9 \text{ inches})^2 \\ &= 81 \text{ square inches} \end{aligned}$$

---

### SECTION 1.8   INTRODUCTION TO VARIABLES AND ALGEBRAIC EXPRESSIONS

A letter used to represent a number is called a **variable**.

A combination of operations on variables and numbers is called an **algebraic expression**.

Replacing a variable in an expression by a number, and then finding the value of the expression is called **evaluating the expression** for the variable.

Variables:

$$x, \quad y, \quad z, \quad a, \quad b$$

Algebraic Expressions:

$$3 + x, \quad 7y, \quad x^3 + y - 10$$

Evaluate $2x + y$ for $x = 22$ and $y = 4$.

$$2x + y = 2 \cdot 22 + 4 \quad \text{Replace } x \text{ with 22 and } y \text{ with 4.}$$
$$= 44 + 4 \quad \text{Multiply.}$$
$$= 48 \quad \text{Add.}$$

# CHAPTER 1 REVIEW

**(1.1)** *Determine the place value of the digit 4 in each whole number.*

**1.** 5480   hundreds

**2.** 46,200,120   ten-millions

*Write each whole number in words.*

**3.** 5480   five thousand, four hundred eighty

**4.** 46,200,120   forty-six million, two hundred thousand, one hundred twenty

*Write each whole number in expanded form.*

**5.** 6279   6000 + 200 + 70 + 9

**6.** 403,225,000   400,000,000 + 3,000,000 + 200,000 + 20,000 + 5000

*Write each whole number in standard form.*

**7.** Fifty-nine thousand, eight hundred   59,800

**8.** Six billion, three hundred four million   6,304,000,000

*The following table shows populations of the ten largest cities in the United States. Use this table to answer Exercises 9–12.*

| Rank | City | 1990 | 1980 | 1970 |
|------|------|------|------|------|
| 1 | New York, NY | 7,322,564 | 7,071,639 | 7,895,563 |
| 2 | Los Angeles, CA | 3,485,557 | 2,968,528 | 2,811,801 |
| 3 | Chicago, IL | 2,783,726 | 3,005,072 | 3,369,357 |
| 4 | Houston, TX | 1,629,902 | 1,595,138 | 1,233,535 |
| 5 | Philadelphia, PA | 1,585,577 | 1,688,210 | 1,949,996 |
| 6 | San Diego, CA | 1,110,554 | 875,538 | 697,471 |
| 7 | Detroit, MI | 1,027,974 | 1,203,368 | 1,514,063 |
| 8 | Dallas, TX | 1,007,618 | 904,599 | 844,401 |
| 9 | Phoenix, AZ | 983,403 | 789,704 | 584,303 |
| 10 | San Antonio, TX | 935,393 | 785,940 | 654,153 |

(*Source: The World Almanac*, 2000)

**9.** Find the population of Houston, Texas, in 1980. 1,595,138

**10.** Find the population of Los Angeles, California, in 1970.   2,811,801

**(1.2)**

**11.** Find the increase in population for Phoenix, Arizona, from 1980 to 1990.   193,699

**12.** Find the increase in population for New York, NY from 1980 to 1990.   250,925

**Name** _____

*Add.*

**13.** 7 + 6    13

**14.** 8 + 9    17

**15.** 3 + 0    3

**16.** 0 + 10    10

**17.** 25 + 8 + 5    38

**18.** 27 + 41    68

**19.** 32 + 24    56

**20.** 19 + 21    40

**21.** 47 + 63    110

**22.** 77 + 43    120

**23.** 567 + 383    950

**24.** 463 + 787    1250

**25.** 591 + 623 + 497    1711

**26.** 5982 + 1647 + 2238    9867

**27.** Sean Cruise earned salaries of $62,589, $65,340, and $69,770 during the years of 1997, 1998, and 1999, respectively. Find his total earnings during those three years.    $197,699

**28.** The distance from Chicago to New York City is 714 miles. The distance from New York City to New Delhi, India, is 7318 miles. Find the total distance from Chicago to New Delhi if traveling through New York City.    8032 mi

*Find the perimeter of each figure.*

△ **29.**
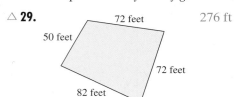
72 feet
50 feet
72 feet
82 feet
276 ft

△ **30.**
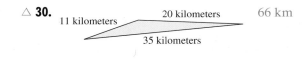
11 kilometers    20 kilometers
35 kilometers
66 km

**(1.3)** *Subtract and then check.*

**31.** 42 − 9    33

**32.** 67 − 24    43

**33.** 93 − 79    14

**34.** 60 − 27    33

**35.** 599 − 237    362

**36.** 462 − 397    65

**37.** 583 − 279    304

**38.** 600 − 124    476

**39.** 4000 − 1886    2114

**40.** 4268 − 3947    321

**41.** Shelly Winters bought a new car listed at $18,425. She received a discount of $1599 and a factory rebate of $1200. Find how much she paid for the car.    $15,626

**42.** Bob Roma is proofreading the Yellow Pages for his county. If he has finished 315 pages of the total 712 pages, how many pages does he have left to proofread?    397 pages

*The bar graph shows the monthly savings account balance for a freshman attending a local community college. Use this graph to answer Exercises 43–46.*

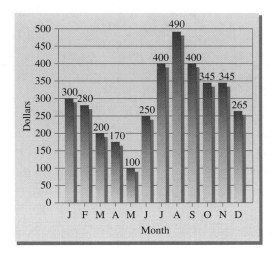

**43.** During what month was the balance the least?
May

**44.** During what month was the balance the greatest?
August

**45.** For what months was the balance greater than $350?   July, August, September

**46.** For what months was the balance less than $200?
April, May

**(1.4)** *Round to the given place value.*

**47.** 93 to the nearest ten   90

**48.** 45 to the nearest ten   50

**49.** 467 to the nearest ten   470

**50.** 493 to the nearest hundred   500

**51.** 4832 to the nearest hundred   4800

**52.** 57,534 to the nearest thousand   58,000

**53.** 49,683,712 to the nearest million   50,000,000

**54.** 768,542 to the nearest hundred-thousand   800,000

*Estimate the sum or difference by rounding each number to the nearest hundred.*

**55.** 4892 + 647 + 1876   7400

**56.** 5925 − 1787   4100

**57.** In 1999, there were 10,506 U.S. commercial radio stations. Round this number to the nearest thousand. (*Source:* M Street Corporation, New York, NY)
11,000

**58.** *The Wall Street Journal* has a circulation of 1,752,693. Round this number to the nearest ten-thousand. (*Source:* Dow Jones & Company, Inc.)   1,750,000

Name _____

**(1.5)** *Multiply.*

**59.** 6 · 7   42

**60.** 8 · 3   24

**61.** 5(0)   0

**62.** 0(9)   0

**63.**   47   1410
       × 30

**64.**   69   2898
       × 42

**65.** 20(8)(5)   800

**66.** 25(9 × 4)   900

**67.**   48   3696
       × 77

**68.**   77   1694
       × 22

**69.** 49 · 49 · 0   0

**70.** 62 · 88 · 0   0

**71.**   586   16,994
        × 29

**72.**   242   8954
        × 37

**73.**   642   113,634
        × 177

**74.**   347   44,763
        × 129

**75.**   1026   411,426
        × 401

**76.**   2107   636,314
        × 302

*Estimate each product by rounding each factor to the given place value.*

**77.** 49 · 32; tens   1500

**78.** 586 · 357; hundreds   240,000

**79.** 5231 · 243; hundreds   1,040,000

**80.** 7836 · 912; hundreds   7,020,000

**81.** There were 5283 students enrolled at Weskan State University in the fall semester. Each paid $927 tuition. Find the total tuition collected.   $4,897,341

**82.** One serving of a particular cake contains 490 calories. If the whole cake contains 8 servings, how many calories are in the whole cake?   3920 calories

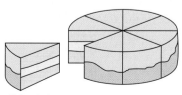

490 calories

**Name** _____

*Find the area of each rectangle.*

△ **83.**

12 miles

5 miles    60 sq. mi

△ **84.** 20 centimeters

25 centimeters    500 sq. cm

**(1.6)** *Divide and then check.*

**85.** $18 \div 6$   3

**86.** $36 \div 9$   4

**87.** $42 \div 7$   6

**88.** $25 \div 5$   5

**89.** $27 \div 5$   5 R 2

**90.** $18 \div 4$   4 R 2

**91.** $16 \div 0$   undefined

**92.** $0 \div 8$   0

**93.** $9 \div 9$   1

**94.** $10 \div 1$   10

**95.** $918 \div 0$   undefined

**96.** $0 \div 668$   0

**97.** $5\overline{)75}$   15

**98.** $8\overline{)159}$   19 R 7

**99.** $26\overline{)626}$   24 R 2

**100.** $6\overline{)336}$   56

**101.** $32\overline{)49}$   1 R 17

**102.** $19\overline{)680}$   35 R 15

**103.** $20\overline{)10,000}$   500

**104.** $43\overline{)909}$   21 R 6

**105.** $47\overline{)23,782}$   506

**106.** $30\overline{)480}$   16

**107.** $16\overline{)3192}$   199 R 8

**108.** $25\overline{)5000}$   200

**109.** One foot is 12 inches. Find how many feet there are in 5496 inches.   458 ft

**110.** Find the average of the numbers 76, 49, 32, and 47.   51

**(1.7)** *Write using exponential notation.*

**111.** $7 \cdot 7 \cdot 7 \cdot 7$   $7^4$

**112.** $3 \cdot 3 \cdot 3$   $3^3$

**113.** $4 \cdot 2 \cdot 2 \cdot 2$   $4 \cdot 2^3$

**114.** $5 \cdot 5 \cdot 7 \cdot 7 \cdot 7$   $5^2 \cdot 7^3$

*Simplify.*

**115.** $7^2$   49

**116.** $2^6$   64

**117.** $5^3 \cdot 3^2$   1125

**118.** $4^1 \cdot 10^2 \cdot 7^2$   19,600

**119.** $18 \div 3 + 7$    13

**120.** $12 - 8 \div 4$    10

**121.** $\dfrac{6^2 - 3}{3^2 + 2}$    3

**122.** $\dfrac{16 - 8}{2^3}$    1

**123.** $2 + 3[1 + (20 - 17) \cdot 3]$    32

**124.** $21 - [2^4 - (7 - 5) - 10] + 8 \cdot 2$    33

*Find the area of each square.*

△ **125.**

49 sq. m

7 meters

△ **126.**

9 sq. in.

3 inches

**(1.8)** *Evaluate each expression for $x = 5$, $y = 0$, and $z = 2$.*

**127.** $\dfrac{2x}{z}$    5

**128.** $4x - 3$    17

**129.** $\dfrac{x + 7}{y}$    undefined

**130.** $\dfrac{y}{5x}$    0

**131.** $x^3 - 2z$    121

**132.** $\dfrac{7 + x}{3z}$    2

**133.** $(y + z)^2$    4

**134.** $\dfrac{100}{x} + \dfrac{y}{3}$    20

*Translate each phrase into a variable expression.*

**135.** Five subtracted from a number
$x - 5$

**136.** Seven more than a number
$x + 7$

**137.** Ten divided by the sum of a number
and 1    $10 \div (x + 1)$

**138.** The product of 5 and the sum of a
number and 3    $5(x + 3)$

**139.** Complete the table by evaluating $8x^2$ for each value of $x$.

| $x$ | 0 | 1 | 2 | 3 |
|------|---|---|----|----|
| $8x^2$ | 0 | 8 | 32 | 72 |

# CHAPTER 1 TEST

*Evaluate.*

**1.** $59 + 82$

**2.** $600 - 487$

**3.** $\begin{array}{r} 496 \\ \times\ 30 \\ \hline \end{array}$

**4.** $52{,}896 \div 69$

**5.** $2^3 \cdot 5^2$

**6.** $6^1 \cdot 2^3$

**7.** $98 \div 1$

**8.** $0 \div 49$

**9.** $62 \div 0$

**10.** $(2^4 - 5) \cdot 3$

**11.** $16 + 9 \div 3 \cdot 4 - 7$

**12.** $2[(6 - 4)^2 + (22 - 19)^2] + 10$

**13.** Round 52,369 to the nearest thousand.

*Estimate the sum or difference by rounding each number to the nearest hundred.*

**14.** $6289 + 5403 + 1957$

**15.** $4267 - 2738$

**16.** Twenty-nine cans of Sherwin Williams paint cost $493. How much was each can?

**17.** Admission to a movie costs $7 per ticket. The Math Club has 17 members who are going to a movie together. What is the total cost of their tickets?

**18.** Jo McElory is looking at two new refrigerators for her apartment. One costs $599 and the other costs $725. How much more expensive is the higher-priced one?

**19.** Aspirin was 100 years old in 1997 and was the first U.S. drug made in tablet form. Today, people take 11 billion tablets a year for heart disease prevention and 4 billion tablets a year for headaches. How many more tablets are taken a year for heart disease prevention? (*Source:* Bayer Market Research)

**1.** 141

**2.** 113

**3.** 14,880

**4.** 766 R 42

**5.** 200

**6.** 48

**7.** 98

**8.** 0

**9.** undefined

**10.** 33

**11.** 21

**12.** 36

**13.** 52,000

**14.** 13,700

**15.** 1600

**16.** $17

**17.** $119

**18.** $126

**19.** 7 billion tablets

**20.** 170,000 cards

**21.** 30

**22.** 1

**23. a.** 17$x$

**b.** 20 − 2$x$

**24.** 20 cm, 25 sq. cm

**25.** 60 yd, 200 sq. yd

**26.** 511

**27.** 1

**20.** The U.S. Federal Bureau of Investigation (FBI) maintains a collection of over 250,000,000 sets of fingerprint records. Each work day, the FBI Identification Division receives over 34,000 standard fingerprint cards to add to its collection. During a standard five-day work week, how many fingerprint cards does the FBI receive? (*Source:* Federal Bureau of Investigation)

**21.** Evaluate $5(x^3 - 2)$ for $x = 2$.

**22.** Evaluate $\dfrac{3x - 5}{2y}$ for $x = 7$ and $y = 8$.

**23.** Translate the following phrases into mathematical expressions. Use $x$ to represent "a number."

**a.** The product of a number and 17

**b.** Twice a number subtracted from 20

*Find the perimeter and the area of each figure.*

△ **24.**

Square | 5 centimeters

△ **25.**

20 yards
Rectangle | 10 yards

*The table shows SAT average scores by state for some selected states. Use this table to answer Exercises 26 and 27.*

| | 1997 | | 1998 | | 1999 | |
|---|---|---|---|---|---|---|
| **SAT Average Scores by State** | Verbal | Math | Verbal | Math | Verbal | Math |
| Louisiana | 560 | 553 | 562 | 558 | 561 | 558 |
| Maine | 507 | 504 | 504 | 501 | 507 | 503 |
| Maryland | 507 | 507 | 506 | 508 | 507 | 507 |
| Massachusetts | 508 | 508 | 508 | 508 | 511 | 511 |
| Michigan | 557 | 566 | 558 | 569 | 557 | 565 |
| Minnesota | 582 | 592 | 585 | 598 | 586 | 598 |
| Mississippi | 567 | 551 | 562 | 549 | 563 | 548 |
| Missouri | 567 | 568 | 570 | 573 | 572 | 572 |
| Montana | 545 | 548 | 543 | 546 | 545 | 546 |
| Nebraska | 562 | 564 | 565 | 571 | 568 | 571 |

(*Source:* The College Board)

**26.** Find the average 1999 SAT verbal score for students from the state of Massachusetts.

**27.** Find the *decrease* in average SAT math scores for students from the state of Michigan for the years 1997 and 1999.

# Integers

**2**

Whole numbers are not sufficient for representing many situations in real life. For example, to express 5° below 0° or $100 in debt, numbers less than 0 are needed. This chapter is devoted to integers, which include numbers less than 0, and to operations on integers.

**CONTENTS**

A lake is a large body of water completely surrounded by land. By this definition, the Caspian Sea, located in southeastern Europe and southwestern Asia, is the world's largest lake with an area of over 143,000 square miles. In comparison, the second largest lake in the world is Lake Superior in North America with an area of only 31,700 square miles. Although the elevation of the Caspian Sea is 92 feet below sea level, it is not the lowest-lying lake. The world's lowest lake surface is the Dead Sea, between Israel and Jordan at 1312 feet below sea level. Nor is the Caspian Sea the deepest lake in the world. That honor goes instead to Lake Baykal in Siberian Russia with a maximum depth of 5315 feet. In Exercises 69–72 on page 128, we will see how integers can be used to describe the elevation of important lakes around the world.

**Name** _____ **Section** _____ **Date** _____

# CHAPTER PRETEST

**1.** The Washington Redskins lost 22 yards on a play. Represent this quantity using an integer.

**2.** Insert < or > between the given pair of numbers to make a true statement.
$-31 \quad -36$

**3.** Simplify: $|-8|$

**4.** Find the opposite of $-12$.

**5.** Add: $-19 + 8$

**6.** Evaluate $x + y$ if $x = -6$ and $y = -11$.

**7.** On February 2 the temperature in Manassas, Virginia, at 7 a.m. was $-9°$ Fahrenheit. By 8 a.m. the temperature had risen by 5°, and then between 8 a.m. and 9 a.m. it rose another 7°. What was the temperature at 9 a.m.?

**8.** Subtract: $-9 - (-14)$

**9.** Simplify: $4 - 6 - (-2) + 8$

**10.** Evaluate $m - n$ if $m = -5$ and $n = 10$.

**11.** Andrea Roberts has $88 in her checking account. She makes a deposit of $35 and writes a check for $72, and then writes another check for $55. Find the amount left in her account. (Write the amount as an integer.)

**12.** Multiply: $(-8)(13)$

**13.** Divide: $\dfrac{-36}{-4}$

**14.** Evaluate $\dfrac{x}{y}$ if $x = -96$ and $y = -2$.

**15.** Evaluate $xy$ if $x = 4$ and $y = -5$.

**16.** A card player had a score of $-20$ for each of three games. Find her total score.

*Simplify.*

**17.** $-5^2$

**18.** $\dfrac{9 - 17}{-6 + 2}$

**19.** $(-2) \cdot |-7| - (-6)$

**20.** Evaluate $x - y^2$ if $x = 2$ and $y = -6$.

**104**

# 2.1   INTRODUCTION TO INTEGERS

## **A** REPRESENTING REAL-LIFE SITUATIONS

Thus far in this text, all numbers have been 0 or greater than 0. Numbers greater than 0 are called **positive numbers**. However, sometimes situations exist that cannot be represented by a number greater than 0. For example,

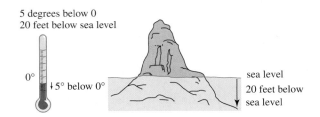

5 degrees below 0
20 feet below sea level

0°   ↓5° below 0°

sea level
20 feet below
sea level

To represent these situations, we need numbers less than 0.

Extending the number line to the left of 0 allows us to picture **negative numbers**, numbers that are less than 0.

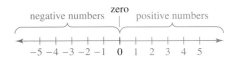

negative numbers   zero   positive numbers

−5 −4 −3 −2 −1  0  1  2  3  4  5

When a single + sign or no sign is in front of a number, the number is a positive number. When a single − sign is in front of a number, the number is a negative number. Together, we call positive numbers, negative numbers, and 0 the **signed numbers**.

   −5 indicates "negative five."
   5 and +5 both indicate "positive five."
   The number 0 is neither positive nor negative.

Some signed numbers are integers. The **integers** consist of the positive numbers, 0, and the negative numbers labeled on the number line above. The integers are

   ..., −3, −2, −1, 0, 1, 2, 3, ...

Now we have numbers to represent the situations previously mentioned.

   5 degrees below 0°      −5°
   20 feet below sea level   −20 feet

## Example 1   Representing Depth with an Integer

Jack Mayfield, a miner for the Molly Kathleen Gold Mine, is presently 150 feet below the surface of the earth. Represent this position using an integer.

*Solution:*   If 0 represents the surface of the earth, then 150 feet below the surface can be represented by −150. ■

TEACHING TIP      Classroom Activity

Ask students to help create a list of real-life situations in which negative numbers are used.

**Practice Problem 1**

a. A deep-sea diver is 800 feet below the surface of the ocean. Represent this position using an integer.

b. A company reports a $2 million loss for the year. Represent this amount using an integer.

**Answers**

**1. a.** −800  **b.** −2 million

## Practice Problem 2

Graph −5, −4, 3, and −3 on the number line.

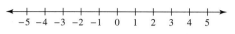

### B GRAPHING INTEGERS

**Example 2**    Graph 0, −3, 2, and −2 on the number line.

*Solution:*

### C COMPARING INTEGERS

We compare integers just as we compare whole numbers. For any two numbers graphed on a number line, the number to the **right** is the **greater number** and the number to the **left** is the **smaller number**. Recall that the symbol > means "is greater than" and the symbol < means "is less than."

Both −5 and −7 are graphed on the number line shown.

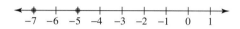

The graph of −7 is **to the left of** −5, so −7 **is less than** −5, written as

$$-7 < -5$$

We can also write

$$-5 > -7$$

since −5 is **to the right** of −7, so −5 **is greater than** −7.

**TRY THE CONCEPT CHECK IN THE MARGIN.**

## ✓ CONCEPT CHECK

Is there a largest positive number? Is there a smallest negative number? Explain.

## Practice Problem 3

Insert < or > between each pair of numbers to make a true statement.
a. 0    −3      b. −5    5
c. −8    −12

**Example 3**    Insert < or > between each pair of numbers to make a true statement.

   **a.** −7      7
   **b.** 0      −4
   **c.** −9      −11

*Solution:*  **a.** −7 is to the left of 7 on a number line, so −7 < 7.

           **b.** 0 is to the right of −4 on a number line, so 0 > −4.

           **c.** −9 is to the right of −11 on a number line, so −9 > −11.

---

**Helpful Hint**

If you think of < and > as arrowheads, notice that in a true statement the arrow always points to the smaller number.

$$5 > -4 \qquad -3 < -1$$

        ↑           ↑
     smaller     smaller
     number     number

---

**Answers**

**2.**

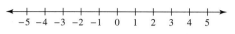

**3. a.** >   **b.** <   **c.** >
✓ **Concept Check:** No.

## D  Finding the Absolute Value of a Number

The **absolute value** of a number is the number's distance from 0 on the number line. The symbol for absolute value is | |. For example, |3| is read as "the absolute value of 3."

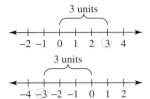

|3| = 3 because 3 is 3 units from 0.

3 units
−2 −1  0  1  2  ③  4

|−3| = 3 because −3 is 3 units from 0.

3 units
−4 ⊖3 −2 −1  0  1  2

## Example 4

Simplify.

a. |−2|        b. |5|        c. |0|

*Solution:*    a. |−2| = 2 because −2 is 2 units from 0.
    b. |5| = 5 because 5 is 5 units from 0.
    c. |0| = 0 because 0 is 0 units from 0.

---

### Helpful Hint

Since the absolute value of a number is that number's *distance* from 0, the absolute value of a number is always 0 or positive. It is never negative.

|0| = 0        |−6| = 6
↑              ↑
zero        a positive number

---

## E  Finding Opposites

Two numbers that are the same distance from 0 on the number line but are on opposite sides of 0 are called **opposites**.

4 and −4 are opposites.

4 units    4 units
−5 ⊖4 −3 −2 −1  0  1  2  3  ④  5

When two numbers are opposites, we say that each is the opposite of the other. Thus **4 is the opposite of − 4** and **− 4 is the opposite of 4.**

The phrase "the opposite of" is written in symbols as " − ." For example,

The opposite of    5    is    −5
      ↓            ↓    ↓     ↓
      −           (5)   =    −5

The opposite of    −3    is    3
      ↓            ↓     ↓     ↓
      −          (−3)   =     3

Notice we just stated that

−(−3) = 3

If *a* is a number, then −(−*a*) = *a*.

---

TEACHING TIP

To help students understand the difference between absolute value and opposite, have them fill out a chart such as the one below.

| Number | Absolute Value | Opposite |
|--------|----------------|----------|
| 5      |                |          |
| −3     |                |          |
| 0      |                |          |
| 27     |                |          |
| −11    |                |          |

Next, ask students:
Is the absolute value of a number ever negative? Is the opposite of a number ever negative?

**Answers**
**4. a.** 4  **b.** 2  **c.** 8

**Practice Problem 5**

Find the opposite of each number.
a. 7                    b. −17

✓ **CONCEPT CHECK**

True or false? The number 0 is the only number that is its own opposite.

**Practice Problem 6**

Simplify.
a. −|−2|                b. −|5|
c. −(−11)

**Example 5**    Find the opposite of each number.

a. 11        **b.** −2        **c.** 0

*Solution:*    **a.** The opposite of 11 is −11.
**b.** The opposite of −2 is −(−2) or 2.
**c.** The opposite of 0 is 0.

> **HELPFUL HINT**
> Remember that 0 is neither positive nor negative.

**TRY THE CONCEPT CHECK IN THE MARGIN.**

**Example 6**    Simplify.

**a.** −(−4)        **b.** −|−5|        **c.** −|6|

*Solution:*    **a.** −(−4) = 4        The opposite of negative 4 is 4.

**b.** −|−5| = −5        The opposite of the absolute value of −5 is the opposite of 5, or −5.

**c.** −|6| = −6        The opposite of the absolute value of 6 is the opposite of 6, or −6.

---

# Focus on the Real World

### USING NEGATIVE NUMBERS

Did you know that in many countries around the world (not including the United States or Russia, for instance) the underground floors in a tall building are numbered using negative numbers? As we have seen in this section and will see throughout this chapter, negative numbers are very useful in many different situations where numbers less than 0 are needed. Another use of negative numbers is in pregnancy and childbirth. Just prior to labor, a baby will drop, or engage, in the mother's pelvis. Doctors and midwives use "stations," ranging from −5 to +5, to describe how much a baby has engaged compared to the 0 station at the entrance to the pelvic canal. A negative station means that a baby is still above the pelvic bones used for reference. A positive station describes how far the baby has moved into the pelvic canal.

### CRITICAL THINKING

Make a list of as many real-world situations as you can think of that use negative numbers in some way. Consider flipping through a newspaper or news magazine for inspiration.

**Answers**
**5. a.** −7  **b.** 17
**6. a.** −2  **b.** −5  **c.** 11
✓ **Concept Check:** True.

## EXERCISE SET 2.1

**A** *Represent each quantity by an integer. See Example 1.*

1. A worker in a silver mine in Nevada works 1445 feet underground.

2. A scuba diver is swimming 35 feet below the surface of the water in the Gulf of Mexico.

3. The peak of Mount Whitney in California is 14,494 feet above sea level. (*Source:* U.S. Geological Survey)

4. The average depth of the Atlantic Ocean is 11,730 feet below the surface of the ocean. (*Source: 2000 World Almanac*)

5. The Virginia Cavaliers football team lost 15 yards on a play.

6. The record high temperature in Arizona is 128 degrees Fahrenheit above zero. (*Source:* National Climatic Data Center)

7. The Dow Jones stock market average fell 317 points in one day.

8. The lowest elevation in the United States is found at Death Valley, California, at an elevation of 282 feet below sea level. (*Source:* U.S. Geological Survey)

9. Converse, Inc., manufactures a wide variety of footwear. In fiscal year 1998, Converse posted a net loss of $5049. (*Source:* Converse, Inc.)

10. For fiscal year 1997, Apple Computer, Inc., reported a net loss of $1045 million. (*Source:* Apple Computer, Inc.)

11. The temperature on one January day in Chicago was 10° below 0° Celsius. Represent this quantity by an integer and tell whether this temperature is cooler or warmer than 5° below 0° Celsius.

12. Two divers are exploring the bottom of a trench in the Pacific Ocean. Joe is at 135 feet below the surface of the ocean and Sara is at 157 feet below the surface. Represent each quantity by an integer and determine who is deeper in the water.

1. $-1445$

2. $-35$

3. $+14{,}494$

4. $-11{,}730$

5. $-15$

6. $+128$

7. $-317$

8. $-282$

9. $-5049$

10. $-1045$ million

11. $-10$; cooler

12. $-135$; $-157$; Sara

**13.** −16

**14.** −8

**15.** see number line

**16.** see number line

**17.** see number line

**18.** see number line

**19.** see number line

**20.** see number line

**21.** see number line

**22.** see number line

**23.** >

**24.** >

**25.** <

**26.** <

**27.** >

**28.** >

**29.** <

**30.** >

**31.** 5

**32.** 7

**33.** 8

**34.** 19

**35.** 0

**36.** 100

**37.** 5

**38.** 10

**Name** _____

**13.** In 1998, the number of music CD singles shipped to retailers reflected a 16 percent loss from the previous year. Write an integer to represent the percent loss in CD singles shipped. (*Source:* Recording Industry Association of America)

**14.** In 1998, the number of music cassettes shipped to retailers reflected an 8 percent loss from the previous year. Write an integer to represent the percent loss of cassettes shipped. (*Source:* Recording Industry Association of America)

**B** *Graph each integer in the list on the same number line. See Example 2.*

**15.** 1, 2, 4, 6

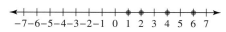

**16.** 3, 5, 2, 0

**17.** 1, −1, −2, −4, −7

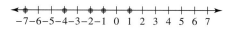

**18.** 2, −2, −4, 6

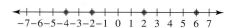

**19.** 0, 2, 5, 7

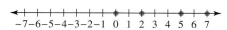

**20.** 0, 3, 6, 10

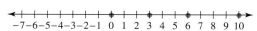

**21.** 0, −2, −7, −5

**22.** 0, −7, 3, −6

**C** *Insert < or > between each pair of integers to make a true statement. See Example 3.*

**23.** 4     0     **24.** 8     0     **25.** −7     −5     **26.** −12     −10

**27.** 0     −3     **28.** 0     −7     **29.** −26     26     **30.** 13     −13

**D** **E** *Simplify. See Example 4.*

**31.** $|5|$     **32.** $|7|$     **33.** $|-8|$     **34.** $|-19|$

**35.** $|0|$     **36.** $|100|$     **37.** $|-5|$     **38.** $|-10|$

**Name** _____

*Find the opposite of each integer. See Example 5.*

**39.** 5                **40.** 8                ▭ **41.** −4              **42.** −6

**43.** 23               **44.** 123              **45.** −10              **46.** −23

*Simplify. See Example 6.*

**47.** |−7|             **48.** |−11|            ▭ **49.** −|20|            **50.** −|43|

▭ **51.** −|−3|          **52.** −|−18|           **53.** −(−8)            **54.** −(−7)

**55.** |−14|            **56.** −(−14)           **57.** −(−29)           **58.** −|−29|

*Insert <, >, or = between each pair of numbers to make a true statement.*

**59.** −3        −5                **60.** −17        −6              **61.** |−9|        |−14|

**62.** |−8|        |−4|            **63.** |−33|        −(−33)        **64.** −|17|        −(−17)

**65.** −|−10|        −(−10)        **66.** |−24|        −(−24)        **67.** 0        −9

**68.** −45        0                **69.** |0|        |−9|          **70.** |−45|        |0|

**71.** −|−2|        −|−10|         **72.** −|−8|        −|−4|

**73.** −(−12)        −(−18)        **74.** −22        −(−38)

| | |
|---|---|
| **39.** −5 | |
| **40.** −8 | |
| **41.** 4 | |
| **42.** 6 | |
| **43.** −23 | |
| **44.** −123 | |
| **45.** 10 | |
| **46.** 23 | |
| **47.** 7 | |
| **48.** 11 | |
| **49.** −20 | |
| **50.** −43 | |
| **51.** −3 | |
| **52.** −18 | |
| **53.** 8 | |
| **54.** 7 | |
| **55.** 14 | |
| **56.** 14 | |
| **57.** 29 | |
| **58.** −29 | |
| **59.** > | |
| **60.** < | |
| **61.** < | |
| **62.** > | |
| **63.** = | |
| **64.** < | |
| **65.** < | |
| **66.** = | |
| **67.** > | |
| **68.** < | |
| **69.** < | |
| **70.** > | |
| **71.** > | |
| **72.** < | |
| **73.** < | |
| **74.** < | |

**Name** _____

## REVIEW AND PREVIEW

*Add. See Section 1.2.*

**75.** $0 + 13$

**76.** $9 + 0$

**77.** $15 + 20$

**78.** $20 + 15$

**79.** $47 + 236 + 77$

**80.** $362 + 37 + 90$

## COMBINING CONCEPTS

*Choose all numbers from each given list that make each statement true.*

**81.** $\boxed{\phantom{xx}} < -8$

    **A.** 0  **B.** $-5$  **C.** 8  **D.** $-12$

**82.** $-3 > \boxed{\phantom{xx}}$

    **A.** 0  **B.** 4  **C.** $-1$  **D.** $-100$

*Answer true or false for each statement.*

**83.** If $a > b$, then $a$ must be a positive number.

**84.** The absolute value of a number is *always* a positive number.

**85.** A positive number is always greater than a negative number.

**86.** Zero is always less than a positive number.

**87.** The number $-a$ is always a negative number. (*Hint:* Read "−" as "the opposite of.")

**88.** Given the number line is it true that $b < a$?

$$a \quad b\,{-}1 \quad 0 \quad 1$$

**89.** Write in your own words how to find the absolute value of an integer.

**90.** Explain how to determine which of two integers is larger.

# 2.2  ADDING INTEGERS

## A  ADDING INTEGERS

Adding integers can be visualized using a number line. A positive number can be represented on the number line by an arrow of appropriate length pointing to the right, and a negative number by an arrow of appropriate length pointing to the left.

Both arrows represent 2 or +2. They both point to the right and they are both 2 units long.

Both arrows represent −3. They both point to the left and they are both 3 units long.

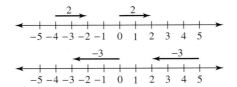

To add integers on a number line, such as $5 + (-2)$, we start at 0 on the number line and draw an arrow representing 5. From the tip of this arrow, we draw another arrow representing −2. The tip of the second arrow ends at their sum, 3.

$$5 + (-2) = 3$$

To add $-1 + (-4)$ on the number line, we start at 0 and draw an arrow representing −1. From the tip of this arrow, we draw another arrow representing −4. The tip of the second arrow ends at their sum, −5.

$$-1 + (-4) = -5$$

**Example 1**  Add using a number line: $-3 + (-4)$

*Solution:*

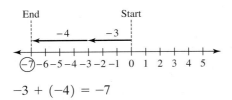

$$-3 + (-4) = -7$$

**Example 2**  Add using a number line: $-7 + 3$

*Solution:*

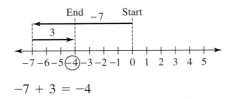

$$-7 + 3 = -4$$

Using a number line each time we add two numbers can be time consuming. Instead, we can notice patterns in the previous examples and write rules for adding signed numbers.

**Objectives**

**A** Add integers.

**B** Evaluate an algebraic expression by adding.

**C** Solve problems by adding integers.

Study Guide   SSM   CD-ROM   Video 2.2

**Practice Problem 1**

Add using a number line:  $5 + (-1)$

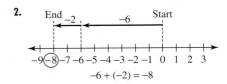

**Practice Problem 2**

Add using a number line:  $-6 + (-2)$

**Answers**

1.

$$5 + (-1) = 4$$

2.

$$-6 + (-2) = -8$$

Rules for adding signed numbers depend on whether we are adding numbers with the same sign or different signs. When adding two numbers with the same sign, notice that the sign of the sum is the same as the sign of the addends.

---

**ADDING TWO NUMBERS WITH THE SAME SIGN**

**Step 1.**   Add their absolute values.

**Step 2.**   Use their common sign as the sign of the sum.

---

**Practice Problem 3**

Add: $(-3) + (-9)$

**Example 3**   Add: $-2 + (-21)$

*Solution:*   **Step 1.** $|-2| = 2, |-21| = 21$, and $2 + 21 = 23$.

**Step 2.** Their common sign is negative, so the sum is negative:

$$-2 + (-21) = -23$$

**Practice Problems 4–5**

Add.
4. $-12 + (-3)$
5. $9 + 5$

**Examples**   Add.

**4.** $-5 + (-1) = -6$

**5.** $2 + 6 = 8$

The rule for adding two numbers with different signs follows.

**TEACHING TIP**

Continue to remind students to keep their work organized and neat. Often students make errors because they simply can't read their own handwriting.

---

**ADDING TWO NUMBERS WITH DIFFERENT SIGNS**

**Step 1.**   Find the larger absolute value minus the smaller absolute value.

**Step 2.**   Use the sign of the number with the larger absolute value as the sign of the sum.

---

**Practice Problem 6**

Add: $-3 + 9$

**Example 6**   Add: $-2 + 5$

*Solution:*   **Step 1.** $|-2| = 2, |5| = 5$, and $5 - 2 = 3$.

**Step 2.** 5 has the larger absolute value and its sign is an understood $+$ :

$$-2 + 5 = +3 \text{ or } 3$$

**Practice Problem 7**

Add: $2 + (-8)$

**Example 7**   Add: $3 + (-7)$

*Solution:*   **Step 1.** $|3| = 3, |-7| = 7$, and $7 - 3 = 4$.

**Step 2.** $-7$ has the larger absolute value and its sign is $-$ :

$$3 + (-7) = -4$$

**Practice Problems 8–10**

Add.
8. $-46 + 20$        9. $8 + (-6)$
10. $-2 + 0$

**Examples**   Add.

**8.** $-18 + 10 = -8$

**9.** $12 + (-8) = 4$

**10.** $0 + (-5) = -5$   The sum of 0 and any number is the number.

**Answers**

**3.** $-12$   **4.** $-15$   **5.** 14   **6.** 6   **7.** $-6$   **8.** $-26$
**9.** 2   **10.** $-2$

TRY THE CONCEPT CHECK IN THE MARGIN.

In the following examples, we add three or more integers. Remember that by the associative and commutative properties for addition, we may add numbers in any order that we wish. In Examples 11 and 12, let's first add the numbers from left to right.

**Example 11**   Add: $(-3) + 4 + (-11)$

*Solution:*   $(-3) + 4 + (-11) = 1 + (-11)$
$$= -10$$

**Example 12**   Add: $1 + (-10) + (-8) + 9$

*Solution:*   $1 + (-10) + (-8) + 9 = -9 + (-8) + 9$
$$= -17 + 9$$
$$= -8$$

The sum will be the same if we add the numbers in any order. To see this, let's add the positive numbers together and then the negative numbers together first.

$1 + 9 = 10$   Add the positive numbers.
$(-10) + (-8) = -18$   Add the negative numbers.
$10 + (-18) = -8$   Add these results.

The sum is $-8$.

> **Helpful Hint**
>
> Don't forget that addition is associative and commutative. In other words, numbers may be added in any order.

## B   EVALUATING ALGEBRAIC EXPRESSIONS

We can continue our work with algebraic expressions by evaluating expressions given integer replacement values.

**Example 13**   Evaluate $x + y$ for $x = 3$ and $y = -5$.

*Solution:*   Replace $x$ with 3 and $y$ with $-5$ in $x + y$.

$x + y = 3 + (-5)$
$$= -2$$

**Example 14**   Evaluate $x + y$ for $x = -2$ and $y = -10$.

*Solution:*   $x + y = (-2) + (-10)$   Replace $x$ with $-2$ and $y$ with $-10$.
$$= -12$$

## Practice Problem 15

If the temperature was −8° Fahrenheit at 6 a.m., and it rose 4 degrees by 7 a.m. and then rose another 7 degrees in the hour from 7 a.m. to 8 a.m., what was the temperature at 8 a.m.?

**C** SOLVING PROBLEMS BY ADDING INTEGERS

Next, we practice solving problems that require adding integers.

### Example 15   Calculating Temperature

On January 6, the temperature in Caribou, Maine, at 8 a.m. was −12° Fahrenheit. By 9 a.m., the temperature had risen 4 degrees, and by 10 a.m. it had risen 6 degrees from the 9 a.m. temperature. What was the temperature at 10 a.m.?

*Solution:*

In words:

| temperature at 10 a.m. | = | 8 a.m. temperature | + | rise of 4° | + | rise of 6° |
|---|---|---|---|---|---|---|

Translate:

$$\text{temperature at 10 a.m.} = -12 + (+4) + (+6)$$
$$= -8 + (+6)$$
$$= -2$$

The temperature was −2°F at 10 a.m.

---

### CALCULATOR EXPLORATIONS
### ENTERING NEGATIVE NUMBERS

To enter a negative number on a calculator, find the key marked ⌈+/−⌋. (Some calculators have a key marked ⌈CHS⌋ and some calculators have a special key ⌈(−)⌋ for entering a negative sign.) To enter the number −2, for example, press the keys ⌈2⌋ ⌈+/−⌋. The display will read ⌈     −2⌋.

To find −32 + (−131), press the keys

⌈32⌋ ⌈+/−⌋ ⌈+⌋ ⌈131⌋ ⌈+/−⌋ ⌈=⌋ or

⌈(−)⌋ ⌈32⌋ ⌈+⌋ ⌈(−)⌋ ⌈131⌋ ⌈ENTER⌋

The display will read ⌈     −163⌋. Thus −32 + (−131) = −163.

*Use a calculator to perform each indicated operation.*

**1.** −256 + 97   −159

**2.** 811 + (−1058)   −247

**3.** 6(15) + (−46)   44

**4.** −129 + 10(48)   351

**5.** −108,650 + (−786,205)
−894,855

**6.** −196,662 + (−129,856)
−326,518

Answer

**15.** 3°F

## Mental Math

*Add.*

**1.** $5 + 0$      **2.** $(-2) + 0$      **3.** $0 + (-35)$      **4.** $0 + 3$

## Exercise Set 2.2

**A**   *Add using a number line. See Examples 1 and 2.*

**1.** $8 + 2$

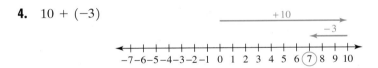

**2.** $9 + (-4)$

**3.** $-4 + 7$

**4.** $10 + (-3)$

**5.** $-13 + 7$

**6.** $(-6) + (-5)$

*Add. See Examples 3 through 10.*

**7.** $23 + 12$      **8.** $15 + 42$      **9.** $-6 + (-2)$      **10.** $-5 + (-4)$

**11.** $-43 + 43$      **12.** $-62 + 62$      ▣ **13.** $6 + (-2)$      **14.** $8 + (-3)$

**15.** $-6 + 8$      **16.** $-8 + 12$      **17.** $3 + (-5)$      **18.** $5 + (-9)$

**Mental Math Answers**

**1.**   5

**2.**   $-2$

**3.**   $-35$

**4.**   3

**Answers**

**1.**   see number line

**2.**   see number line

**3.**   see number line

**4.**   see number line

**5.**   see number line

**6.**   see number line

**7.**   35

**8.**   57

**9.**   $-8$

**10.**   $-9$

**11.**   0

**12.**   0

**13.**   4

**14.**   5

**15.**   2

**16.**   4

**17.**   $-2$

**18.**   $-4$

**19.** $-9$

**20.** $-7$

**21.** $-24$

**22.** $-46$

**23.** $-57$

**24.** $-135$

**25.** $-223$

**26.** $-730$

**27.** $0$

**28.** $0$

**29.** $7$

**30.** $14$

**31.** $-3$

**32.** $-6$

**33.** $-9$

**34.** $-10$

**35.** $30$

**36.** $52$

**37.** $20$

**38.** $40$

**39.** $51$

**40.** $100$

**41.** $-33$

**42.** $-41$

**43.** $-20$

**44.** $-50$

**45.** $-125$

**46.** $-89$

**47.** $-7$

**48.** $-4$

**49.** $-246$

**50.** $-162$

**51.** $16$

**52.** $12$

**53.** $13$

**54.** $224$

**55.** $-33$

**56.** $-17$

**57.** $21$

**58.** $-14$

**59.** $9$

**60.** $-3$

**61.** $0$

**62.** $-44$

**19.** $-2 + (-7)$　　　**20.** $-6 + (-1)$　　　**21.** $-12 + (-12)$　　　**22.** $-23 + (-23)$

**23.** $-25 + (-32)$　　　**24.** $-45 + (-90)$　　　**25.** $-123 + (-100)$

**26.** $-500 + (-230)$　　　**27.** $-7 + 7$　　　**28.** $-10 + 10$

**29.** $12 + (-5)$　　　**30.** $24 + (-10)$　　　**31.** $-6 + 3$

**32.** $-8 + 2$　　　**33.** $-12 + 3$　　　**34.** $-15 + 5$

**35.** $56 + (-26)$　　　**36.** $89 + (-37)$　　　**37.** $-37 + 57$

**38.** $-25 + 65$　　　**39.** $-42 + 93$　　　**40.** $-64 + 164$

**41.** $34 + (-67)$　　　**42.** $42 + (-83)$　　　**43.** $124 + (-144)$

**44.** $325 + (-375)$　　　**45.** $-82 + (-43)$　　　**46.** $-56 + (-33)$

*Add. See Examples 11 and 12.*

**47.** $-4 + 2 + (-5)$　　　　　　**48.** $-1 + 5 + (-8)$

**49.** $-52 + (-77) + (-117)$　　　　　　**50.** $-103 + (-32) + (-27)$

**51.** $12 + (-4) + (-4) + 12$　　　　　　**52.** $18 + (-9) + 5 + (-2)$

**53.** $(-10) + 14 + 25 + (-16)$　　　　　　**54.** $34 + (-12) + (-11) + 213$

**55.** $-8 + (-14) + (-11)$　　　　　　**56.** $-10 + (-6) + (-1)$

**57.** $5 + (-1) + 17$　　　　　　**58.** $3 + (-23) + 6$

**59.** $13 + 14 + (-18)$　　　　　　**60.** $(-45) + 22 + 20$

**61.** $-3 + (-8) + 12 + (-1)$　　　　　　**62.** $-16 + 6 + (-14) + (-20)$

**118**

**Name** _____

**B** *Evaluate x + y for the given replacement values. See Examples 13 and 14.*

**63.** $x = -2$ and $y = 3$     **64.** $x = -7$ and $y = 11$     📷 **65.** $x = -20$ and $y = -50$

**66.** $x = -1$ and $y = -29$     **67.** $x = 3$ and $y = -30$     **68.** $x = 13$ and $y = -17$

**C** *Solve. See Example 15.*

**69.** The temperature at 4 p.m. on February 2 was $-10°$ Celsius. By 11 p.m. the temperature had risen 12 degrees. Find the temperature at 11 p.m.

**70.** Scores in golf can be positive or negative integers. For example, a score of 3 *over* par can be represented by +3 and a score of 5 *under* par can be represented by −5. If Fred Couples had scores of 3 over par, 6 under par, and 7 under par for three games of golf, what was his total score?

**71.** Suppose a deep-sea diver dives from the surface to 165 feet below the surface. He then dives down 16 more feet. Use positive and negative numbers to represent this situation. Then find the diver's present depth.

**72.** Suppose a deep-sea diver dives from the surface to 248 meters below the surface and then swims up 6 meters, down 17 meters, down another 24 meters, and then up 23 meters. Use positive and negative numbers to represent this situation. Then find the diver's depth after these movements.

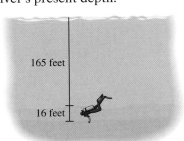

165 feet

16 feet

*In some card games, it is possible to have positive and negative scores. The table shows the scores for two teams playing a series of four card games. Use this table to answer Exercises 73 and 74.*

|        | Game 1 | Game 2 | Game 3 | Game 4 |
|--------|--------|--------|--------|--------|
| Team 1 | −2     | −13    | 20     | 2      |
| Team 2 | 5      | 11     | −7     | −3     |

**73.** Find each team's total score after four games. If the winner is the team with the greater score, find the winning team.

**74.** Find each team's total score after three games. If the winner is the team with the greater score, which team was winning after three games?

**75.** −48°F

**76.** 12°F

**77.** −$581 billion

**78.** −5617 billion cu. ft

**79.** 44

**80.** 91

**81.** 0

**82.** 0

**83.** 28

**84.** 14

**85.** true

**86.** true

**87.** false

**88.** true

**89.** answers may vary

**90.** answers may vary

**75.** The all-time record low temperature for Utah is −69°F, which was recorded on February 1, 1985. Maine's all-time record low temperature is 21°F higher than Utah's record low. What is Maine's record low temperature? (*Source:* National Climatic Data Center)

**76.** The all-time record low temperature for Louisiana is −16°F, which occurred on February 13, 1899. In Hawaii, the lowest temperature ever recorded is 28°F more than Louisiana's all-time low temperature. What is the all-time record low temperature for Hawaii? (*Source:* National Climatic Data Center)

**77.** The difference between a country's exports and imports is called the country's *trade balance.* The United States had a trade balance of −$159 billion in 1995, −$170 billion in 1996, −$181 billion in 1997, and −$230 billion in 1998. What was the total U.S. trade balance for these years? (*Source:* U.S. Department of Commerce)

**78.** The U.S. trade balance for natural gas was −2833 billion cubic feet in 1997 and −2784 billion cubic feet in 1996. What was the total U.S. trade balance for natural gas for these years? (*Source:* U.S. Energy Information Administration)

## REVIEW AND PREVIEW

*Subtract. See Section 1.3.*

**79.** $44 - 0$

**80.** $91 - 0$

**81.** $52 - 52$

**82.** $103 - 103$

**83.** $87 - 59$

**84.** $32 - 18$

## COMBINING CONCEPTS

*For Exercises 85–88, determine whether each statement is true or false.*

**85.** The sum of two negative numbers is always a negative number.

**86.** The sum of two positive numbers is always a positive number.

**87.** The sum of a positive number and a negative number is always a negative number.

**88.** The sum of zero and a negative number is always a negative number.

**89.** In your own words, explain how to add two negative numbers.

**90.** In your own words, explain how to add a positive number and a negative number.

# 2.3 SUBTRACTING INTEGERS

In Section 2.1, we discussed the opposite of an integer.

The opposite of 3 is −3.
The opposite of −6 is 6.

In this section, we use opposites to subtract integers.

## A SUBTRACTING INTEGERS

To subtract integers, we will write the subtraction problem as an addition problem. To see how to do this, study the examples below.

$$10 - 4 = 6$$
$$10 + (-4) = 6$$

Since both expressions simplify to 6, this means that

$$10 - 4 = 10 + (-4) = 6$$

Also,

$$3 - 2 = 3 + (-2) = 1$$
$$15 - 1 = 15 + (-1) = 14$$

Thus, to subtract two numbers, we add the first number to the opposite (also called the *additive inverse*) of the second number.

---

**SUBTRACTING TWO NUMBERS**

If $a$ and $b$ are numbers, then $a - b = a + (-b)$.

---

**Examples**   Subtract.

| subtraction | = | first number | + | opposite of the second number | | |
|---|---|---|---|---|---|---|
| **1.** 8 − 5 | = | 8 | + | (−5) | = | 3 |
| **2.** −4 − 10 | = | −4 | + | (−10) | = | −14 |
| **3.** 6 − (−5) | = | 6 | + | 5 | = | 11 |
| **4.** −11 − (−7) | = | −11 | + | 7 | = | −4 |

**Examples**   Subtract.

**5.** $-10 - 5 = -10 + (-5) = -15$

**6.** $8 - 15 = 8 + (-15) = -7$

**7.** $-4 - (-5) = -4 + 5 = 1$

**Objectives**

**A** Subtract integers.
**B** Add and subtract integers.
**C** Evaluate an algebraic expression by subtracting.
**D** Solve problems by subtracting integers.

Study Guide   SSM   CD-ROM   Video 2.3

**Practice Problems 1–4**

Subtract.
1. 12 − 7
2. −6 − 4
3. 11 − (−14)
4. −9 − (−1)

**Practice Problems 5–7**

Subtract.
5. 5 − 9
6. −12 − 4
7. −2 − (−7)

**Answers**
**1.** 5   **2.** −10   **3.** 25   **4.** −8   **5.** −4   **6.** −16
**7.** 5

✓ **CONCEPT CHECK**

What is wrong with the following calculation?

$$-9 - (-6) = -15$$

**Practice Problem 8**

Subtract 5 from −10.

**Practice Problem 9**

Simplify: $-4 - 3 - 7 - (-5)$

TEACHING TIP

Suggest that students rewrite the expression followed by an equal sign and the first simplification. Then write each additional simplification below the previous one, making sure to align equal signs. This format will help eliminate careless errors.

**Practice Problem 10**

Simplify: $3 + (-5) - 6 - (-4)$

TEACHING TIP

Spend extra time evaluating $x - y$ when $y$ is a negative number. A common mistake is for students to forget to write the subtraction symbol.

**Practice Problem 11**

Evaluate $x - y$ for $x = -2$ and $y = 14$.

**TRY THE CONCEPT CHECK IN THE MARGIN.**

**Example 8**    Subtract 7 from −3.

*Solution:*    To subtract 7 *from* −3, we find

$$-3 - 7 = -3 + (-7) = -10$$

     ▬▬▬

**B** **ADDING AND SUBTRACTING INTEGERS**

If a problem involves adding or subtracting more than two integers, we rewrite differences as sums and add from left to right.

**Example 9**    Simplify: $7 - 8 - (-5) - 1$

*Solution:*

$$\begin{aligned} 7 - 8 - (-5) - 1 &= \underbrace{7 + (-8)} + 5 + (-1) \\ &= \underbrace{-1 + 5} + (-1) \\ &= \underbrace{4 + (-1)} \\ &= 3 \end{aligned}$$

     ▬▬▬

**Example 10**    Simplify: $7 + (-12) - 3 - (-8)$

*Solution:*

$$\begin{aligned} 7 + (-12) - 3 - (-8) &= \underbrace{7 + (-12)} + (-3) + 8 \\ &= -5 + (-3) + 8 \\ &= -8 + 8 \\ &= 0 \end{aligned}$$

     ▬▬▬

**C** **EVALUATING EXPRESSIONS**

Now let's practice evaluating expressions when the replacement values are integers.

**Example 11**    Evaluate $x - y$ for $x = -3$ and $y = 9$.

*Solution:*    Replace $x$ with −3 and $y$ with 9 in $x - y$.

$$\begin{aligned} & x \quad - \quad y \\ & \downarrow \quad \downarrow \quad \downarrow \\ =\;& (-3) - \quad 9 \\ =\;& (-3) + (-9) \\ =\;& -12 \end{aligned}$$

     ▬▬▬

**Answers**

**8.** −15   **9.** −9   **10.** −4   **11.** −16
✓ Concept Check: $-9 - (-6) = -3$

**Example 12**   Evaluate $a - b$ for $a = 8$ and $b = -6$.

*Solution:*   Watch your signs carefully!

$$a \quad - \quad b$$
$$\downarrow \quad \downarrow \quad \downarrow$$
$$= \quad 8 \quad - \quad (-6) \qquad \text{Replace } a \text{ with 8 and } b \text{ with } -6.$$
$$= \quad 8 \quad + \quad 6$$
$$= \quad 14$$

## D  SOLVING PROBLEMS BY SUBTRACTING INTEGERS

Solving problems often requires subtraction of integers.

**Example 13**   **Finding a Change in Elevation**

The highest point in the United States is the top of Mount McKinley, in Denali County, Alaska, at a height of 20,320 feet above sea level. The lowest point is Death Valley, California, which is 282 feet below sea level. How much higher is Mount McKinley than Death Valley? (*Source:* U.S. Geological Survey)

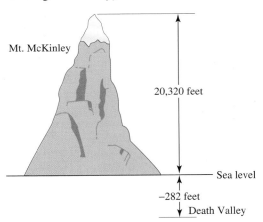

*Solution:*

In words:

| how much higher Mt. McKinley is | = | height of Mt. McKinley | minus | height of Death Valley |
|---|---|---|---|---|
| ↓ | | ↓ | ↓ | ↓ |

Translate:

how much higher = $\quad 20{,}320 \quad - \quad (-282)$
Mt. McKinley is

$$= 20{,}320 + 282 = 20{,}602$$

Mt. McKinley is 20,602 feet higher than Death Valley.

**Practice Problem 12**

Evaluate $y - z$ for $y = -3$ and $z = -4$.

---

**HELPFUL HINT**

Watch carefully when replacing variables in the expression $x - y$. Make sure that all symbols are inserted and accounted for.

---

**Practice Problem 13**

The highest point in Asia is the top of Mount Everest, at a height of 29,028 feet above sea level. The lowest point is the Dead Sea, which is 1312 feet below sea level. How much higher is Mount Everest than the Dead Sea? (*Source:* National Geographic Society)

**Answers**

**12.** 1   **13.** 30,340 feet

# Focus On Study Skills

### STUDYING FOR A MATH EXAM

Remember that one of the best ways to start preparing for an exam is to keep current with your assignments as they are made. Make an effort to clear up any confusion on topics as you cover them.

Begin reviewing for your exam a few days in advance. If you find a topic during your review that you still don't understand, you'll have plenty of time to ask your instructor, another student in your class, or a math tutor for help. Don't wait until the last minute to "cram" for the test.

▲ Reread your notes and carefully review the Chapter Highlights at the end of each chapter.

▲ Try solving a few exercises from each section.

▲ Pay special attention to any new terminology or definitions in the chapter. Be sure you can state the meanings of definitions in your own words.

▲ Find a quiet place to take the Chapter Test found at the end of the chapter to be covered. This gives you a chance to practice taking the real exam, so try the Chapter Test without referring to your notes or looking up anything in your book. Give yourself the same amount of time to take the Chapter Test as you will have to take the exam for which you are preparing. If your exam covers more than one chapter, you should try taking the Chapter Tests for each chapter covered. You may also find working through the Cumulative Reviews helpful when preparing for a multichapter test.

▲ When you have finished taking the Chapter Test, check your answers in the back of the book. Redo any of the problems you missed. Then spend extra time solving similar problems.

▲ If you tend to get anxious while taking an exam, try to visualize yourself taking the exam in advance. Picture yourself being calm, clearheaded, and successful. Picture yourself remembering concepts and definitions with no trouble. When you are well prepared for an exam, a lot of nervousness can be avoided through positive thinking.

▲ Get lots of rest the night before the exam. It's hard to show how well you know the material if your brain is foggy from lack of sleep.

## EXERCISE SET 2.3

**A** *Perform each indicated subtraction. See Examples 1 through 8.*

**1.** $5 - 5$    **2.** $-6 - (-6)$    **3.** $8 - 3$    **4.** $5 - 2$

**5.** $3 - 8$    **6.** $2 - 5$    **7.** $7 - (-7)$    **8.** $12 - (-12)$

**9.** $-5 - (-8)$    **10.** $-25 - (-25)$    **11.** $-14 - 4$    **12.** $-2 - 42$

**13.** $2 - 16$    **14.** $8 - 9$    **15.** $-10 - (-10)$    **16.** $-5 - (-5)$

**17.** $-15 - (-15)$    **18.** $-24 - (-24)$    **19.** $3 - 7$    **20.** $4 - 12$

**21.** $30 - 45$    **22.** $29 - 56$    **23.** $-4 - 10$    **24.** $-5 - 8$

**25.** $-230 - 870$    **26.** $-15 - 26$    **27.** $4 - (-6)$    **28.** $6 - (-9)$

**29.** $-7 - (-3)$    **30.** $-12 - (-5)$    **31.** $-16 - (-23)$    **32.** $-45 - (-16)$

**33.** Subtract 18 from $-20$.        **34.** Subtract 10 from $-22$.

**35.** Find the difference of $-20$ and $-3$.        **36.** Find the difference of $-8$ and $-13$.

**ANSWERS**

**1.** 0
**2.** 0
**3.** 5
**4.** 3
**5.** $-5$
**6.** $-3$
**7.** 14
**8.** 24
**9.** 3
**10.** 0
**11.** $-18$
**12.** $-44$
**13.** $-14$
**14.** $-1$
**15.** 0
**16.** 0
**17.** 0
**18.** 0
**19.** $-4$
**20.** $-8$
**21.** $-15$
**22.** $-27$
**23.** $-14$
**24.** $-13$
**25.** $-1100$
**26.** $-41$
**27.** 10
**28.** 15
**29.** $-4$
**30.** $-7$
**31.** 7
**32.** $-29$
**33.** $-38$
**34.** $-32$
**35.** $-17$
**36.** 5

**125**

**37.** Subtract −11 from 2.          **38.** Subtract −50 from −50.

**B** *Simplify. See Examples 9 and 10.*

**39.** $7 - 3 - 2$          **40.** $8 - 4 - 1$          **41.** $12 - 5 - 7$

**42.** $30 - 7 - 12$          **43.** $-5 - 8 - (-12)$          **44.** $-10 - 6 - (-9)$

**45.** $-10 + (-5) - 12$          **46.** $-15 + (-8) - 4$          **47.** $12 - (-34) + (-6)$

**48.** $23 - (-17) + (-9)$          **49.** $-(-6) - 12 + (-16)$          **50.** $-(-9) - 7 + (-23)$

**51.** $-9 - (-12) + (-7) - 4$          **52.** $-6 - (-8) + (-12) - 7$

**53.** $-3 + 4 - (-23) - 10$          **54.** $5 + (-18) - (-21) - 2$

**C** *Evaluate $x - y$ for the given replacement values. See Examples 11 and 12.*

**55.** $x = -3$ and $y = 5$          **56.** $x = -7$ and $y = 1$          **57.** $x = 6$ and $y = -30$

**58.** $x = 9$ and $y = -2$          **59.** $x = -4$ and $y = -4$          **60.** $x = -8$ and $y = -10$

**61.** $x = 1$ and $y = -18$          **62.** $x = 14$ and $y = -12$

**D** *Solve. See Example 13.*

**63.** The coldest temperature ever record-ed on Earth was −129°F in Antarctica. The warmest temperature ever record-ed was 136°F in the Sahara Desert. How many degrees warmer is 136°F than −126°F? (*Source: Questions Kids Ask*, Grolier Limited, 1991, and *The World Almanac, 2000*)

**64.** The coldest temperature ever record-ed in the United States was −80°F in Alaska. The warmest temperature ever recorded was 134°F in California. How many degrees warmer is 134°F than −80°F? (*Source: The World Al-manac and Book of Facts, 2000*)

**65.** Aaron Aiken has $125 in his checking account. He writes a check for $117, makes a deposit of $45, and then writes another check for $69. Find the amount left in his account. (Write the amount as an integer.)

**66.** In the card game canasta, it is possible to have a negative score. If Juan San-tanilla's score is 15, what is his new score if he loses 20 points?

**67.** The temperature on a February morn-ing is −6° Celsius at 6 a.m. If the tem-perature drops 3 degrees by 7 a.m., rises 4 degrees between 7 a.m. and 8 a.m., and then drops 7 degrees be-tween 8 a.m. and 9 a.m., find the tem-perature at 9 a.m.

**68.** Mt. Elbert in Colorado has an eleva-tion of 14,433 feet above sea level. The Mid-America Trench in the Pacific Ocean has an elevation of 21,857 feet below sea level. Find the difference in elevation between those two points. (*Source:* National Geographic Society and Defense Mapping Agency)

**63.** 265°F

**64.** 214°F

**65.** −$16

**66.** −5 points

**67.** −12°C

**68.** 36,290 ft

**Name** _____

*The bar graph shows heights of selected lakes. For Exercises 69–72, find the difference in elevation for the lakes listed. (Source: U.S. Geological Survey)*

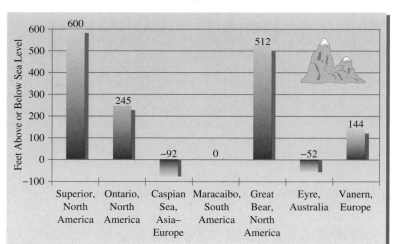

**69.** Lake Superior and Lake Eyre

**70.** Great Bear Lake and Caspian Sea

**71.** Lake Maracaibo and Lake Vanern

**72.** Lake Eyre and Caspian Sea

**73.** The average temperature on the surface of Mercury is 167°C. The average temperature on the surface of Neptune is −215°C. How many degrees warmer is the surface of Mercury than the surface of Neptune? (*Source:* National Space Science Data Center)

**74.** The average temperature on the surface of Earth is 15°C. The average temperature on the surface of Mars is −63°C. How many degrees warmer is the surface of Earth than the surface of Mars? (*Source:* National Space Science Data Center)

**75.** The average temperature on the surface of Saturn is −176°C. The average temperature on the surface of Uranus is −215°C. Which planet has the warmer average temperature? How much warmer is it? (*Source:* National Space Science Data Center)

**76.** The average temperature on the surface of Pluto is −223°C. The average temperature on the surface of Jupiter is −144°C. What is the difference between the average temperature on Jupiter and the average temperature on Pluto? (*Source:* National Space Science Data Center)

**77.** Recall that the difference between a country's exports and imports is called the country's *trade balance*. In 1998, the United States had $663 billion in exports and $912 billion in imports. What was the U.S. trade balance in 1998? (*Source:* U.S. Department of Commerce)

**78.** In 1997, the United States exported 327 million barrels of petroleum products and imported 707 million barrels of petroleum products. What was the U.S. trade balance for petroleum products in 1997? (*Source:* U.S. Energy Information Administration)

## REVIEW AND PREVIEW

*Multiply. See Section 1.5.*

**79.** $8 \cdot 0$

**80.** $0 \cdot 8$

**81.** $1 \cdot 8$

**82.** $8 \cdot 1$

**83.**
$$\begin{array}{r} 23 \\ \times\ 46 \\ \hline \end{array}$$

**84.**
$$\begin{array}{r} 51 \\ \times\ 89 \\ \hline \end{array}$$

## COMBINING CONCEPTS

*Evaluate each expression for the given replacement values.*

**85.** $x - y - z$ for $x = -4$, $y = 3$, and $z = 15$

**86.** $x - y - z$ for $x = -14$, $y = 8$, and $z = -6$

**87.** $a + b - c$ for $a = -16$, $b = 14$, and $c = -22$

**88.** $a + b - c$ for $a = -1$, $b = -1$, and $c = 100$

**89.** $|-3| - |-7|$    **90.** $|-12| - |-5|$    **91.** $|-6| - |6|$    **92.** $|-23| - |-42|$
(Evaluate absolute values first.)

---

**77.** −$249 billion

**78.** −380 million barrels

**79.** 0

**80.** 0

**81.** 8

**82.** 8

**83.** 1058

**84.** 4539

**85.** −22

**86.** −16

**87.** 20

**88.** −102

**89.** −4

**90.** 7

**91.** 0

**92.** −19

**93.** false

**94.** false

**95.** answers may vary

**96.** answers may vary

**97.** answers may vary

*For Exercises 93 and 94, determine whether each statement is true or false.*

**93.** $|-8 - 3| = 8 - 3$

**94.** $|-2 - (-6)| = |-2| - |-6|$

**95.** In your own words, explain how to subtract one integer from another.

**Internet Excursions**

Go to http://www.prenhall.com /martin-gay

This World Wide Web address will provide you with access to a listing of worldwide temperature extremes, or a related site. Use the data given to answer the following questions.

**96.** What are the highest and lowest temperatures (in degrees Fahrenheit) ever recorded for the European continent? Where and when were these temperatures recorded? Find the difference in temperature between these two extremes.

**97.** Locate the extreme temperature information for the state in which your college is located. What are the highest and lowest temperatures (in degrees Fahrenheit) ever recorded for this state? Where and when were these temperatures recorded? How many degrees warmer is the highest temperature than the lowest temperature?

**Name** _____ **Section** _____ **Date** _____

# CHAPTER 2 INTEGRATED REVIEW—INTEGERS

*Represent each quantity by an integer.*

1.  The peak of Mount Everest in Asia is 29,028 feet above sea level. (*Source: U.S. Geological Survey*)

2.  The Mariana Trench in the Pacific Ocean is 35,840 feet below sea level. (*Source: The World Almanac, 2000*)

3.  Graph the signed numbers on the given number line. −4, 0, −1, 3

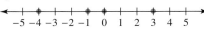

*Insert < or > between each pair of numbers to make a true statement.*

**4.** 0     −3     **5.** −15     −5     **6.** −1     1     **7.** −2     −7

*Simplify.*

**8.** |−1|       **9.** −|−4|       **10.** |−8|       **11.** −(−5)

*Find the opposite of each number.*

**12.** 6       **13.** −3       **14.** 89       **15.** 0

*Add or subtract as indicated.*

**16.** −7 + 12       **17.** −9 + (−11)       **18.** 25 + (−35)

ANSWERS

1.  +29,028

2.  −35,840

3.  see number line

4.  >

5.  <

6.  <

7.  >

8.  1

9.  −4

10. 8

11. 5

12. −6

13. 3

14. −89

15. 0

16. 5

17. −20

18. −10

**131**

**Name** _____

**19.** $1 - 3$　　　　　**20.** $26 - (-26)$　　　　　**21.** $-2 - 1$

**22.** $-18 - (-102)$　　　　　**23.** $-8 + (-6) + 20$

**24.** $-11 - 7 - (-19)$　　　　　**25.** $-4 + (-8) - 16 - (-9)$

**26.** Subtract 14 from 26.　　　　　**27.** Subtract −8 from −12.

*Choose all numbers from each given list that make each statement true.*

**28.** ☐ $< -18$

　　　**A.** 0　**B.** 18　**C.** −3　**D.** −21

**29.** $-5 <$ ☐

　　　**A.** 0　**B.** 3　**C.** −1　**D.** −1000

# 2.4 MULTIPLYING AND DIVIDING INTEGERS

Multiplying and dividing integers is similar to multiplying and dividing whole numbers. One difference is that we need to determine whether the result is a positive number or a negative number.

## A MULTIPLYING INTEGERS

Consider the following pattern of products.

First factor
decreases
by 1
each time.

$3 \cdot 2 = 6$
$2 \cdot 2 = 4$     Product decreases by 2 each time.
$1 \cdot 2 = 2$
$0 \cdot 2 = 0$

This pattern can be continued, as follows.

$-1 \cdot 2 = -2$
$-2 \cdot 2 = -4$
$-3 \cdot 2 = -6$

This suggests that the product of a negative number and a positive number is a negative number.

What is the sign of the product of two negative numbers? To find out, we form another pattern of products. Again, we decrease the first factor by 1 each time, but this time the second factor is negative.

$2 \cdot (-3) = -6$
$1 \cdot (-3) = -3$     Product increases by 3 each time.
$0 \cdot (-3) = 0$

This pattern continues as:

$-1 \cdot (-3) = 3$
$-2 \cdot (-3) = 6$
$-3 \cdot (-3) = 9$

This suggests that the product of two negative numbers is a positive number. Thus we can determine the sign of a product when we know the signs of the factors.

---

**MULTIPLYING NUMBERS**

The product of two numbers having the same sign is a positive number.
The product of two numbers having different signs is a negative number.

---

**Examples**    Multiply.

**1.** $-7 \cdot 3 = -21$
**2.** $-2(-5) = 10$
**3.** $0 \cdot (-4) = 0$
**4.** $10(-8) = -80$

Recall that by the associative and commutative properties for multiplication, we may multiply numbers in any order that we wish. In Example 5, we multiply from left to right.

---

**Objectives**

**A** Multiply integers.
**B** Divide integers.
**C** Evaluate an algebraic expression by multiplying or dividing.
**D** Solve problems by multiplying or dividing integers.

Study   SSM   CD-ROM   Video
Guide                 2.4

---

**Practice Problems 1–4**

Multiply.

1. $-2 \cdot 6$
2. $-4(-3)$
3. $0 \cdot (-10)$
4. $5(-15)$

**Answers**

**1.** −12   **2.** 12   **3.** 0   **4.** −75

## Practice Problem 5

Multiply.
a. $(-2)(-1)(-2)(-1)$
b. $7(2)(-4)$

## ✓ CONCEPT CHECK

What is the sign of the product of five negative numbers? Explain.

## Practice Problem 6

Evaluate $(-3)^4$.

TEACHING TIP

Help students notice a pattern between the number of negative numbers in a product and the sign of the product. Try these examples: $7(-1)$; $2(-3)(-5)$; $(-1)(-4)(-2)$; $5(-2)(-1)(-2)$; and so on.

## Practice Problems 7–9

Divide.
7. $\dfrac{28}{-7}$    8. $-18 \div (-2)$    9. $\dfrac{-60}{10}$

**Answers**

**5. a.** 4  **b.** −56  **6.** 81  **7.** −4  **8.** 9  **9.** −6
✓ **Concept Check:** Negative

## Example 5    Multiply.

a. $7(-6)(-2)$          b. $(-2)(-3)(-4)$

*Solution:*    a. $\overline{7(-6)}(-2) = -42(-2) = 84$

b. $\overline{(-2)(-3)}(-4) = 6(-4) = -24$

**TRY THE CONCEPT CHECK IN THE MARGIN.**

Recall from our study of exponents that $2^3 = 2 \cdot 2 \cdot 2 = 8$. We can now work with bases that are negative numbers. For example,

$$(-2)^3 = (-2)(-2)(-2) = -8$$

## Example 6    Evaluate: $(-5)^2$

*Solution:*    Remember that $(-5)^2$ means 2 factors of −5.

$$(-5)^2 = (-5)(-5) = 25$$

## B  DIVIDING INTEGERS

Division of integers is related to multiplication of integers. The sign rules for division can be discovered by writing a related multiplication problem. For example,

$\dfrac{6}{2} = 3$  because  $3 \cdot 2 = 6$

$\dfrac{-6}{2} = -3$  because  $-3 \cdot 2 = -6$

$\dfrac{6}{-2} = -3$  because  $-3 \cdot (-2) = 6$

$\dfrac{-6}{-2} = 3$  because  $3 \cdot (-2) = -6$

**HELPFUL HINT**
Just as for whole numbers, division can be checked by multiplication.

**DIVIDING NUMBERS**

The quotient of two numbers having the same sign is a positive number. The quotient of two numbers having different signs is a negative number.

## Examples    Divide.

7. $\dfrac{-12}{6} = -2$

8. $-20 \div (-4) = 5$

9. $\dfrac{48}{-3} = -16$

TRY THE CONCEPT CHECK IN THE MARGIN.

**Examples**    Divide, if possible.

**10.** $\dfrac{0}{-5} = 0$   because   $0 \cdot -5 = 0$

**11.** $\dfrac{-7}{0}$ is undefined because there is no number that gives a product of $-7$ when multiplied by $0$.    ▬▬

## C EVALUATING EXPRESSIONS

Next, we practice evaluating expressions given integer replacement values.

**Example 12**   Evaluate $xy$ for $x = -2$ and $y = 7$.

*Solution:*   Recall that $xy$ means $x \cdot y$.
Replace $x$ with $-2$ and $y$ with $7$ in $xy$.

$$xy = -2 \cdot 7$$
$$= -14$$    ▬▬

**Example 13**   Evaluate $\dfrac{x}{y}$ for $x = -24$ and $y = 6$.

*Solution:*   $\dfrac{x}{y} = \dfrac{-24}{6}$   Replace $x$ with $-24$ and $y$ with $6$.

$$= -4$$    ▬▬

## D SOLVING PROBLEMS BY MULTIPLYING AND DIVIDING INTEGERS

Many real-life problems involve multiplication and division of integers.

**Example 14**   Calculating Total Golf Score

A professional golfer finished seven strokes under par $(-7)$ for each of three days of a tournament. What was his total score for the tournament?

*Solution:*   In words:

| golfer's total score | = | number of days | · | score each day |
|:---:|:---:|:---:|:---:|:---:|
| ↓ | | ↓ | | ↓ |

Translate:    golfer's total score   $=$   $3$   $\cdot$   $(-7)$

$$= -21$$

The golfer's total score is $-21$, or $21$ strokes under par.    ▬▬

✓ **CONCEPT CHECK**

What is wrong with the following calculation?

$$\frac{-27}{-9} = -3$$

**Practice Problems 10–11**

Divide, if possible.

10. $\dfrac{-1}{0}$    11. $\dfrac{0}{-2}$

**Practice Problem 12**

Evaluate $xy$ for $x = 5$ and $y = -9$.

**Practice Problem 13**

Evaluate $\dfrac{x}{y}$ for $x = -9$ and $y = -3$.

**Practice Problem 14**

A card player had a score of $-12$ for each of four games. Find her total score.

**Answers**

**10.** undefined   **11.** 0   **12.** $-45$   **13.** 3   **14.** $-48$

✓ Concept Check: $\dfrac{-27}{-9} = 3$

# Focus On Study Skills

### TAKING A MATH EXAM

When it is time to take your math exam, remember these hints.

▲ Make sure you have all the tools you will need to take the exam, including an extra pencil and eraser, paper (if needed), and calculator (if allowed).

▲ Try to relax. Taking a few deep breaths, inhaling and then exhaling slowly before you begin, might help.

▲ Are there any special definitions or solution steps that you'll need to remember during the exam? As soon as you get your exam, write these down at the top, bottom, or on the back of your paper.

▲ Scan the entire test to get an idea of what questions are being asked.

▲ Start with the questions that are easiest for you. This will help build your confidence. Then return to the harder ones.

▲ Read all directions carefully. Make sure that your final result answers the question being asked.

▲ Show all of your work. Try to work neatly.

▲ Don't spend too much time on a single problem. If you get stuck, try moving on to other problems so you can increase your chances of finishing the test. If you have time, you can return to the problem giving you trouble.

▲ Before turning in your exam, check your work carefully if time allows. Be on the lookout for careless mistakes.

# Exercise Set 2.4

**A** *Multiply. See Examples 1 through 4.*

**1.** $-2(-3)$     **2.** $5(-3)$     **3.** $-4(9)$     **4.** $-7(-2)$

**5.** $8(-8)$     **6.** $-9(9)$     **7.** $0(-14)$     **8.** $-6(0)$

*Multiply. See Example 5.*

**9.** $6(-4)(2)$     **10.** $-2(3)(-7)$     **11.** $-1(-2)(-4)$     **12.** $8(-3)(3)$

**13.** $-4(4)(-5)$     **14.** $-2(-5)(-4)$     **15.** $10(-5)(0)$     **16.** $2(-1)(3)(-2)$

**17.** $-5(3)(-1)(-1)$     **18.** $3(0)(-4)(-8)$

*Evaluate. See Example 6.*

**19.** $(-2)^2$     **20.** $(-2)^4$     **21.** $(-3)^3$     **22.** $(-1)^4$

**23.** $(-5)^2$     **24.** $(-4)^3$     **25.** $(-2)^3$     **26.** $(-3)^2$

**B** *Find each quotient. See Examples 7 through 11.*

**27.** $-24 \div 6$     **28.** $90 \div (-9)$     **29.** $\dfrac{-30}{6}$     **30.** $\dfrac{56}{-8}$

**31.** $\dfrac{-88}{-11}$     **32.** $\dfrac{-32}{4}$     **33.** $\dfrac{0}{14}$     **34.** $\dfrac{-13}{0}$

**ANSWERS**

**1.** 6
**2.** $-15$
**3.** $-36$
**4.** 14
**5.** $-64$
**6.** $-81$
**7.** 0
**8.** 0
**9.** $-48$
**10.** 42
**11.** $-8$
**12.** $-72$
**13.** 80
**14.** $-40$
**15.** 0
**16.** 12
**17.** $-15$
**18.** 0
**19.** 4
**20.** 16
**21.** $-27$
**22.** 1
**23.** 25
**24.** $-64$
**25.** $-8$
**26.** 9
**27.** $-4$
**28.** $-10$
**29.** $-5$
**30.** $-7$
**31.** 8
**32.** $-8$
**33.** 0
**34.** undefined

**Name** _____

**35.** $\dfrac{2}{0}$      **36.** $\dfrac{0}{-5}$      **37.** $\dfrac{39}{-3}$      **38.** $\dfrac{-24}{-12}$

**A** **B** *Multiply or divide as indicated.*

**39.** $-12(0)$      **40.** $0(-100)$      **41.** $-4(3)$      **42.** $-6 \cdot 2$

**43.** $-9 \cdot 6$      **44.** $-12(13)$      **45.** $-7(-6)$      **46.** $-9(-5)$

**47.** $-3(-4)(-2)$      **48.** $-7(-5)(-3)$      **49.** $(-4)^2$      **50.** $(-5)^2$

**51.** $-\dfrac{10}{5}$      **52.** $-\dfrac{25}{5}$      **53.** $-\dfrac{56}{8}$      **54.** $-\dfrac{49}{7}$

**55.** $-12 \div 3$      **56.** $-15 \div 3$      **57.** $4(-4)(-3)$      **58.** $6(-5)(-2)$

**59.** $-30(6)(-2)(-3)$      **60.** $-20 \cdot 5 \cdot (-5) \cdot (-3)$      **61.** $3 \cdot (-2) \cdot 0$

**62.** $-5(4)(0)$      **63.** $\dfrac{100}{-20}$      **64.** $\dfrac{45}{-9}$      **65.** $240 \div (-40)$

**66.** $480 \div (-8)$      **67.** $\dfrac{-12}{-4}$      **68.** $\dfrac{-36}{-3}$      **69.** $(-1)^4$

**70.** $(-2)^3$      **71.** $(-3)^5$      **72.** $(-9)^2$      **73.** $-2(3)(5)(-6)$

**74.** $-1(2)(7)(-3)$

**75.** $(-1)^{32}$

**76.** $(-1)^{33}$

**77.** $-2(-2)(-5)$

**78.** $-2(-2)(-3)(-2)$

**79.** $-42 \cdot 23$

**80.** $-56 \cdot 43$

**81.** $25 \cdot (-82)$

**82.** $70 \cdot (-23)$

**C** *Evaluate ab for the given replacement values. See Example 12.*

**83.** $a = -4$ and $b = 7$

**84.** $a = 5$ and $b = -1$

**85.** $a = 3$ and $b = -2$

**86.** $a = -9$ and $b = -6$

**87.** $a = -5$ and $b = -5$

**88.** $a = -8$ and $b = 8$

*Evaluate $\dfrac{x}{y}$ for the given replacement values. See Example 13.*

**89.** $x = 5$ and $y = -5$

**90.** $x = 9$ and $y = -3$

**91.** $x = -12$ and $y = 0$

**92.** $x = -10$ and $y = -10$

**93.** $x = -36$ and $y = -6$

**94.** $x = 0$ and $y = -5$

*Evaluate xy and also $\dfrac{x}{y}$ for the given replacement values.*

**95.** $x = -4$ and $y = -2$

**96.** $x = 20$ and $y = -5$

**97.** $x = 0$ and $y = -6$

**98.** $x = -3$ and $y = 0$

**74.** 42

**75.** 1

**76.** $-1$

**77.** $-20$

**78.** 24

**79.** $-966$

**80.** $-2408$

**81.** $-2050$

**82.** $-1610$

**83.** $-28$

**84.** $-5$

**85.** $-6$

**86.** 54

**87.** 25

**88.** $-64$

**89.** $-1$

**90.** $-3$

**91.** undefined

**92.** 1

**93.** 6

**94.** 0

**95.** 8; 2

**96.** $-100; -4$

**97.** 0; 0

**98.** 0; undefined

**Name** _____

**D** *Solve. See Example 14.*

**99.** A football team lost 4 yards on each of three consecutive plays. Represent the total loss as a product of integers, and find the total loss.

**100.** Joe Norstrom lost $400 on each of seven consecutive days in the stock market. Represent his total loss as a product of integers, and find his total loss.

**101.** A deep-sea diver must move up or down in the water in short steps to keep from getting a physical condition called the bends. Suppose the diver moves down from the surface in five steps of 20 feet each. Represent his movement as a product of integers, and find his final depth.

**102.** A weather forecaster predicts that the temperature will drop 5 degrees each hour for the next six hours. Represent this drop as a product of integers, and find the total drop in temperature.

**103.** During the 2000 Dimension Data Pro Am golf tournament in Sun City, South Africa, David Frost had scores of 1, 0, $-4$, and $-5$ in four rounds of golf. Find his average score per tournament round. (*Source:* Professional Golf Association)

**104.** During the 2000 Phoenix Open golf tournament, Phil Mickelson had scores of $-8$, $+2$, and $-6$ for three rounds of golf. Find his average score per round of golf. (*Source:* Professional Golf Association)

**105.** Pets.com, Inc. is an online retailer of pet products. In 1999, Pets.com posted a net loss of $62 million. If this continued, what would Pets.com's income be after three years? (*Source:* Pets.com. Inc.)

**106.** In 1997, the earnings for a share of Apple Computer stock were approximately $-$8 per share. If an investor owned 15 shares of Apple Computer stock in 1997, what were his total earnings for these shares? (*Source:* Apple Computer, Inc.)

**Name** _____

**107.** In 1979, there were 35 California Condors in the entire world. By 1987, there were only 27 California Condors remaining. (*Source:* California Department of Fish and Game)

   **a.** Find the change in the number of California Condors from 1979 to 1987.

   **b.** Find the average change per year in the California Condor population over this period.

**108.** In 1993, music cassettes with a total value of $2915 million were produced by American music manufacturers. In 1998, this value had dropped to $1420 million. (*Source:* Recording Industry Association of America)

   **a.** Find the change in the value of music cassettes produced from 1993 to 1998.

   **b.** Find the average change per year in the value of music cassettes produced over this period.

## REVIEW AND PREVIEW

*Perform each indicated operation. See Section 1.7.*

**109.** $(3 \cdot 5)^2$

**110.** $(12 - 3)^2(18 - 10)$

**111.** $90 + 12^2 - 5^3$

**112.** $3 \cdot (7 - 4) + 2 \cdot 5^2$

**113.** $12 \div 4 - 2 + 7$

**114.** $12 \div (4 - 2) + 7$

## COMBINING CONCEPTS

*In Exercises 115 through 117, determine whether each statement is true or false.*

**115.** The product of two negative numbers is always a negative number.

**116.** The product of a positive number and a negative number is always a negative number.

**117.** The quotient of two negative numbers is always a positive number.

**107. a.** −8 condors

   **b.** −1 condor per yr

**108. a.** −$1495 million

   **b.** −$299 per yr

**109.** 225

**110.** 648

**111.** 109

**112.** 59

**113.** 8

**114.** 13

**115.** false

**116.** true

**117.** true

**118. a.** −72 radio stations

**b.** −288 radio stations

**c.** 2033 radio stations

**119. a.** −19,024 Saturns

**b.** −38,048 Saturns

**c.** 193,738 Saturns

**120.** true

**121.** false

**122.** true

**123.** false

**124.** answers may vary

**125.** answers may vary

**118.** In 1998, there were 2393 commercial country music radio stations in the United States. By 1999, that number had declined to 2321. (*Source:* M Street Corporation)
  **a.** Find the change in the number of country music radio stations from 1998 to 1999.
  **b.** If this change continues, what will be the total change in the number of country music stations four years after 1999?
  **c.** Based on your answer to part (b), how many country music radio stations will there be in 2003?

**119.** In 1997, Saturn sold a total of 250,810 cars. By 1998, the number of Saturns sold had decreased to 231,786. (*Source:* American Automobile Manufacturers Association)
  **a.** Find the change in the number of Saturns sold from 1997 to 1998.
  **b.** If this change continues, what will be the total change in the number of Saturns sold two years after 1998?
  **c.** Based on your answer to part (b), how many Saturns would have been sold in 2000?

*Let a and b be positive numbers. Answer true or false for each statement.*

**120.** $a(-b)$ is a negative number.

**121.** $(-a)(-b)$ is a negative number.

**122.** $(-a)(-a)$ is a positive number.

**123.** $(-a)(-a)(-a)$ is a positive number.

**124.** In your own words, explain how to multiply two integers.

**125.** In your own words, explain how to divide two integers.

# 2.5  ORDER OF OPERATIONS

## A  SIMPLIFYING EXPRESSIONS

We first discussed the order of operations in Chapter 1. In this section, you are given an opportunity to practice using the order of operations when expressions contain integers. The rules for order of operations from Section 1.7 are repeated here.

If there are no other grouping symbols such as fraction bars or absolute values, perform operations in the following order.

---

**ORDER OF OPERATIONS**

**1.** Do all operations within grouping symbols such as parentheses or brackets.
**2.** Evaluate any expressions with exponents.
**3.** Multiply or divide in order from left to right.
**4.** Add or subtract in order from left to right.

---

**Examples**    Find the value of each expression.

**1.** $(-3)^2 = (-3)(-3) = 9$

**2.** $-3^2 = -(3)(3) = -9$

---

**Helpful Hint**

When simplifying expressions with exponents, notice that parentheses make an important difference.

$(-3)^2$ and $-3^2$ **do not** mean the same thing.
$(-3)^2$ means $(-3)(-3) = 9$.
$-3^2$ means the opposite of $3 \cdot 3$, or $-9$.

Only with parentheses is the $-3$ squared.

---

Spend time studying Examples 1 and 2 and the Helpful Hint above. It is important to be able to determine the base of an exponent. After reading these examples, if you have trouble completing the Practice Problems, review this material again.

**Examples**    Find the value of each expression.

**3.** $2 \cdot 5^2 = 2 \cdot (5 \cdot 5) = 2 \cdot 25 = 50$    The base of the exponent is 5.
**4.** $(-4)^3 = (-4)(-4)(-4) = 16(-4) = -64$    The base of the exponent is $-4$.
**5.** $-2^4 = -(2 \cdot 2 \cdot 2 \cdot 2) = -16$    The base of the exponent is 2.

**Example 6**    Simplify: $\dfrac{-6(2)}{-3}$

*Solution:*    The fraction bar serves as a grouping symbol. First we multiply $-6$ and 2. Then we divide.

$$\frac{-6(2)}{-3} = \frac{-12}{-3}$$
$$= 4$$

---

**Objectives**

**A** Simplify expressions by using the order of operations.

**B** Evaluate an algebraic expression.

Study Guide    SSM    CD-ROM    Video 2.5

**Practice Problems 1–2**

Find the value of each expression.
1. $(-2)^4$
2. $-2^4$

TEACHING TIP

Spend time helping students understand the difference between $(-5)^2$ and $-5^2$, for example. It may help if each time they ask themselves, "The exponent 2 has what base?"

**Practice Problems 3–5**

Find the value of each expression.
3. $3 \cdot 6^2$    4. $(-5)^3$    5. $-3^4$

**Practice Problem 6**

Simplify: $\dfrac{25}{5(-1)}$

**Answers**

**1.** 16  **2.** $-16$  **3.** 108  **4.** $-125$  **5.** $-81$
**6.** $-5$

**Practice Problem 7**

Simplify: $\dfrac{-18 + 6}{-3 - 1}$

**Practice Problem 8**

Simplify: $20 + 50 + (-4)^3$

**Practice Problem 9**

Simplify: $-2^3 + (-4)^2 + 1^5$

**Practice Problem 10**

Simplify: $2(2 - 8) + (-12) - 3$

**Practice Problem 11**

Simplify: $(-5) \cdot |-4| + (-3) + 2^3$

TEACHING TIP   Classroom Activity
See next page.

**Practice Problem 12**

Simplify: $4(-6) \div \left[3(5 - 7)^2\right]$

**✓ CONCEPT CHECK**

True or false? Explain your answer.
The result of
$$-4 \cdot (3 - 7) - 8 \cdot (9 - 6)$$
is positive because there are four negative signs.

**Practice Problem 13**

Evaluate $x^2$ and $-x^2$ for $x = -12$.

**Answers**

**7.** 3  **8.** 6  **9.** 9  **10.** −27  **11.** −15  **12.** −2
**13.** 144; −144
✓ Concept Check:
False; $-4 \cdot (3 - 7) - 8 \cdot (9 - 6) = -8$

**Example 7**   Simplify: $\dfrac{12 - 16}{-1 + 3}$

*Solution:*   We simplify above and below the fraction bar separately. Then we divide.

$$\frac{12 - 16}{-1 + 3} = \frac{-4}{2}$$
$$= -2$$

**Example 8**   Simplify: $60 + 30 + (-2)^3$

*Solution:*
$$60 + 30 + (-2)^3 = 60 + 30 + (-8) \quad \text{Write } (-2)^3 \text{ as } -8.$$
$$= 90 + (-8) \quad \text{Add from left to right.}$$
$$= 82$$

**Example 9**   Simplify: $-4^2 + (-3)^2 - 1^3$

*Solution:*   $-4^2 + (-3)^2 - 1^3 = -16 + 9 - 1$   Simplify expressions with exponents.
$$= -7 - 1 \quad \text{Add or subtract from}$$
$$= -8 \quad \text{left to right.}$$

**Example 10**   Simplify: $3(4 - 7) + (-2) - 5$

*Solution:*
$$3(4 - 7) + (-2) - 5 = 3(-3) + (-2) - 5 \quad \text{Simplify inside parentheses.}$$
$$= -9 + (-2) - 5 \quad \text{Multiply.}$$
$$= -11 - 5 \quad \text{Add or subtract from left to}$$
$$= -16 \quad \text{right.}$$

**Example 11**   Simplify: $(-3) \cdot |-5| - (-2) + 4^2$

*Solution:*
$$(-3) \cdot |-5| - (-2) + 4^2 = (-3) \cdot 5 - (-2) + 4^2 \quad \text{Write } |-5| \text{ as } 5.$$
$$= (-3) \cdot 5 - (-2) + 16 \quad \text{Write } 4^2 \text{ as } 16.$$
$$= -15 - (-2) + 16 \quad \text{Multiply.}$$
$$= -13 + 16 \quad \text{Add or subtract from left to right.}$$
$$= 3$$

**Example 12**   Simplify: $-2\left[-3 + 2(-1 + 6)\right] - 5$

*Solution:*   Here we begin with the innermost set of parentheses.
$$-2\left[-3 + 2(-1 + 6)\right] - 5 = -2\left[-3 + 2(5)\right] - 5 \quad \text{Write } -1 + 6 \text{ as } 5.$$
$$= -2\left[-3 + 10\right] - 5 \quad \text{Multiply.}$$
$$= -2(7) - 5 \quad \text{Add.}$$
$$= -14 - 5 \quad \text{Multiply.}$$
$$= -19 \quad \text{Subtract.}$$

**TRY THE CONCEPT CHECK IN THE MARGIN.**

**B EVALUATING EXPRESSIONS**

Now we practice evaluating expressions.

**Example 13**   Evaluate $x^2$ and $-x^2$ for $x = -11$.

*Solution:*   $x^2 = (-11)^2 = (-11)(-11) = 121$
$-x^2 = -(-11)^2 = -(-11)(-11) = -121$

**Example 14**   Evaluate $6z^2$ for $z = 2$ and $z = -2$.

*Solution:*   $6z^2 = 6(2)^2 = 6(4) = 24$
$6z^2 = 6(-2)^2 = 6(4) = 24$

**Example 15**   Evaluate $x^2 - y$ for $x = -3$ and $y = 10$.

*Solution:*   Replace $x$ with $-3$ and $y$ with 10 and simplify.

$x^2 - y = (-3)^2 - 10$    Let $x = -3$ and $y = 10$.

$\downarrow$

$= 9 - 10$    Replace $(-3)^2$ with 9.
$= -1$    Subtract.

**Example 16**   Evaluate $7 - x^2$ for $x = -4$.

*Solution:*   Replace $x$ with $-4$ and simplify carefully!

$7 - x^2 = 7 - (-4)^2$

$\downarrow \quad \downarrow$

$= 7 - 16$    $(-4)^2 = (-4)(-4) = 16$
$= -9$    Subtract.

---

## CALCULATOR EXPLORATIONS
### SIMPLIFYING AN EXPRESSION CONTAINING A FRACTION BAR

Recall that even though most calculators follow the order of operations, parentheses must sometimes be inserted. For example, to simplify $\dfrac{-8 + 6}{-2}$ on a calculator, enter parentheses about the expression above the fraction bar so that it is simplified separately.

To simplify $\dfrac{-8 + 6}{-2}$, press the keys

| ( | 8 | +/− | + | 6 | ) | ÷ | 2 | +/− | = | or

| ( | (−) | 8 | + | 6 | ) | ÷ | (−) | 2 | ENTER |

The display will read [    1    ].

Thus $\dfrac{-8 + 6}{-2} = 1$.

*Use a calculator to simplify.*

1. $\dfrac{-120 - 360}{-10}$   48

2. $\dfrac{4750}{-2 + (-17)}$   −250

3. $\dfrac{-316 + (-458)}{28 + (-25)}$   −258

4. $\dfrac{-234 + 86}{-18 + 16}$   74

---

**Practice Problem 14**

Evaluate $5y^2$ for $y = 3$ and $y = -3$.

**Practice Problem 15**

Evaluate $x^2 + y$ for $x = -5$ and $y = -2$.

**Practice Problem 16**

Evaluate $4 - x^2$ for $x = -9$.

**TEACHING TIP**   Classroom Activity

For examples 8–12, have students work in groups to form four expressions that are identical except for the placement of parentheses. Have them simplify each expression. Then have them give the expression without parentheses and the four answers to another group to write the expression with parentheses that matches each answer.

**Answers**

**14.** 45; 45   **15.** 23   **16.** −77

# Focus On Study Skills

## TIME MANAGEMENT

As a college student, you know the demands that classes, homework, work, campus life, and family place on your time. Some days you probably wonder how you'll ever get everything done! One key to managing your time is developing a schedule. Here are some hints for making a schedule.

1. Make a list of all the weekly commitments you have for the term. Be sure to include classes, work, regular meetings, extracurricular activities, and so on. You may also find it helpful to list such things as regular workouts, laundry, and grocery shopping.

2. Next, estimate the time needed for each item on the list. Also make a note of how often you will need to do each item. Don't forget to include time estimates for the reading, studying, and homework you do outside your classes. Some instructors suggest spending a minimum of 2 hours studying outside class for every hour you spend in class. You might want to check with your instructor to see if this estimate is realistic. You may also need to consider commuting time.

3. Block out a typical week on a schedule grid like the one below. Start with items with fixed time slots, like classes and work. You might want to shade or draw an X through the boxes on the grid to indicate that these time slots are filled.

4. Add in the items on your list with flexible time slots. You'll need to think carefully about how best to schedule some items such as study time.

5. Don't fill up every time slot on the schedule. Remember: you need to allow time for eating, sleeping, and relaxing! You should also leave a little cushion in case things take longer than planned.

6. You may find it helpful to make a copy of your typical weekly schedule for each week of the term. Then you can add to it or adjust it for the specific events going on each week.

7. If you find that your basic weekly schedule is too full for you to handle, you may need to make some changes in your work load, class load, or in other areas of your life. You may want to talk to your advisor, manager or supervisor at work, or someone in your college's academic counseling center for help with such decisions.

|  | Monday | Tuesday | Wednesday | Thursday | Friday | Saturday | Sunday |
|---|---|---|---|---|---|---|---|
| 7:00 a.m. | | | | | | | |
| 8:00 a.m. | | | | | | | |
| 9:00 a.m. | | | | | | | |
| 10:00 a.m. | | | | | | | |
| 11:00 a.m. | | | | | | | |
| 12:00 p.m. | | | | | | | |
| 1:00 p.m. | | | | | | | |
| 2:00 p.m. | | | | | | | |
| 3:00 p.m. | | | | | | | |
| 4:00 p.m. | | | | | | | |
| 5:00 p.m. | | | | | | | |
| 6:00 p.m. | | | | | | | |
| 7:00 p.m. | | | | | | | |
| 8:00 p.m. | | | | | | | |
| 9:00 p.m. | | | | | | | |

**Name** _____  **Section** _____  **Date** _____

## MENTAL MATH

*Identify the base and exponent of each expression. Do not simplify.*

**1.** $-3^2$     **2.** $(-3)^2$     **3.** $4 \cdot 2^3$     **4.** $9 \cdot 5^6$

**5.** $(-7)^5$     **6.** $-9^4$     **7.** $5^7 \cdot 10$     **8.** $2^8 \cdot 11$

## EXERCISE SET 2.5

**A** *Simplify. See Examples 1 through 12.*

**1.** $-1(-2) + 1$    **2.** $3 + (-8) \div 2$    **3.** $3 - 6 + 2$    **4.** $5 - 9 + 2$

**5.** $9 - 12 - 4$    **6.** $10 - 23 - 12$    **7.** $4 + 3(-6)$    **8.** $8 + 4(3)$

**9.** $5(-9) + 2$    **10.** $7(-6) + 3$    **11.** $(-10) + 4 \div 2$    **12.** $(-12) + 6 \div 3$

**13.** $25 \div (-5) + 12$    **14.** $28 \div (-7) + 10$    **15.** $\dfrac{16 - 13}{-3}$    **16.** $\dfrac{20 - 15}{-1}$

**17.** $\dfrac{24}{10 + (-4)}$    **18.** $\dfrac{88}{-8 - 3}$    **19.** $5(-3) - (-12)$    **20.** $7(-4) - (-6)$

**21.** $(-19) - 12(3)$    **22.** $(-24) - 14(2)$    **23.** $8 + 4^2$    **24.** $12 + 3^3$

**25.** $[8 + (-4)]^2$    **26.** $[9 + (-2)]^3$    **27.** $3^3 - 12$    **28.** $5^2 - 100$

**Name** _____

**29.** $16 - (-3)^4$     **30.** $20 - (-5)^2$     **31.** $|5 + 3| \cdot 2^3$     **32.** $|-3 + 7| \cdot 7^2$

**33.** $7 \cdot 8^2 + 4$     **34.** $10 \cdot 5^4 + 7$     **35.** $5^3 - (4 - 2^3)$     **36.** $8^2 - (5 - 2)^4$

**37.** $(3 - 12) \div 3$     **38.** $(12 - 19) \div 7$     **39.** $5 + 2^3 - 4^2$     **40.** $12 + 5^2 - 2^4$

**41.** $(5 - 9)^2 \div (4 - 2)^2$     **42.** $(2 - 7)^2 \div (4 - 3)^4$     **43.** $|8 - 24| \cdot (-2) \div (-2)$

**44.** $|3 - 15| \cdot (-4) \div (-16)$     **45.** $(-12 - 20) \div 16 - 25$     **46.** $(-20 - 5) \div 5 - 15$

**47.** $5(5 - 2) + (-5)^2 - 6$     **48.** $3 \cdot (8 - 3) + (-4) - 10$     **49.** $(2 - 7) \cdot (6 - 19)$

**50.** $(4 - 12) \cdot (8 - 17)$     **51.** $2 - 7 \cdot 6 - 19$     **52.** $4 - 12 \cdot 8 - 17$

**53.** $(-36 \div 6) - (4 \div 4)$     **54.** $(-4 \div 4) - (8 \div 8)$     **55.** $-5^2 - 6^2$

**56.** $-4^4 - 5^4$     **57.** $(-5)^2 - 6^2$     **58.** $(-4)^4 - (5)^4$

**59.** $(10 - 4^2)^2$     **60.** $(11 - 3^2)^3$     **61.** $2(8 - 10)^2 - 5(1 - 6)^2$

**62.** $-3(4 - 8)^2 + 5(14 - 16)^3$     **63.** $3(-10) \div [5(-3) - 7(-2)]$

**Name** _____

**64.** $12 - [7 - (3 - 6)] + (2 - 3)^3$

**65.** $\dfrac{(-7)(-3) - (4)(3)}{3[7 \div (3 - 10)]}$

**66.** $\dfrac{10(-1) - (-2)(-3)}{2[-8 \div (-2 - 2)]}$

**B** *Evaluate each expression for $x = -2$, $y = 4$, and $z = -1$. See Examples 13 through 16.*

**67.** $x + y + z$   **68.** $x - y - z$   **69.** $2x - y^2$   **70.** $5x - y^2$

**71.** $x^2 - y$   **72.** $x^2 + z$   **73.** $\dfrac{5y}{z}$   **74.** $\dfrac{4x}{y}$

*Evaluate each expression for $x = -3$ and $z = -4$. See Examples 13 through 16.*

**75.** $x^2$   **76.** $z^2$   **77.** $-z^2$   **78.** $-x^2$

**79.** $10 - x^2$   **80.** $3 - z^2$   **81.** $2x^3$   **82.** $3z^2$

## Review and Preview

*Perform each indicated operation. See Sections 1.2, 1.3, 1.5, and 1.6.*

**83.** $45 \cdot 90$   **84.** $90 \div 45$   **85.** $90 - 45$   **86.** $45 + 90$

*Find the perimeter of each figure. See Section 1.2.*

△ **87.**  Square

8 in.

△ **88.**  Parallelogram
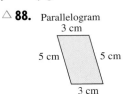
3 cm
5 cm   5 cm
3 cm

△ **89.**  Rectangle
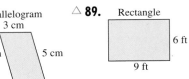
6 ft
9 ft

△ **90.**
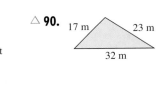
17 m   23 m
32 m

| 64. | 1 |
| 65. | −3 |
| 66. | −4 |
| 67. | 1 |
| 68. | −5 |
| 69. | −20 |
| 70. | −26 |
| 71. | 0 |
| 72. | 3 |
| 73. | −20 |
| 74. | −2 |
| 75. | 9 |
| 76. | 16 |
| 77. | −16 |
| 78. | −9 |
| 79. | 1 |
| 80. | −13 |
| 81. | −54 |
| 82. | 48 |
| 83. | 4050 |
| 84. | 2 |
| 85. | 45 |
| 86. | 135 |
| 87. | 32 in. |
| 88. | 16 cm |
| 89. | 30 ft |
| 90. | 72 m |

**149**

**Name** _____

## COMBINING CONCEPTS

*Insert parentheses where needed so that each expression evaluates to the given number.*

**91.** $2 \cdot 7 - 5 \cdot 3$; evaluates to 12

**92.** $7 \cdot 3 - 4 \cdot 2$; evaluates to 34

**93.** $-6 \cdot 10 - 4$; evaluates to $-36$

**94.** $2 \cdot 8 \div 4 - 20$; evaluates to $-36$

*Evaluate.*

**95.** $(-12)^4$

**96.** $(-17)^6$

**97.** $x^3 - y^2$ for $x = 21$ and $y = -19$

**98.** $3x^2 + 2x - y$ for $x = -18$ and $y = 2868$

**99.** $(xy + z)^x$ for $x = 2$, $y = -5$, and $z = 7$

**100.** $5(ab + 3)^b$ for $a = -2$, $b = 3$

**101.** Discuss the effect parentheses have in an exponential expression. For example, what is the difference between $(-2)^3$ and $-2^3$.

# CHAPTER 2 ACTIVITY
## INVESTIGATING INTEGERS

*Work with a partner or a small group to try the following activity.*

You will need 12 red discs (or small slips of red paper), 12 black discs (or small slips of black paper), a felt-tip marker and/or white correction fluid, and a small paper bag.

| Red: | 1 | 1 | 2 | 2 | 2 | 3 | 3 | 4 | 4 | 4 | 5 | 5 |
|------|---|---|---|---|---|---|---|---|---|---|---|---|
| Black: | 1 | 1 | 1 | 2 | 2 | 3 | 3 | 3 | 4 | 4 | 5 | 5 |

1. Using the marker (or white correction fluid, if necessary), label each red disc with one of the numbers listed in the "red" list above, using each number once. Do the same with the black discs, using the numbers in the "black" list above. (Notice that the red and black lists are slightly different.) Alternatively, write the numbers in each list on small slips of colored paper. Once the discs (or paper slips) are dry, place them in a small paper bag and mix thoroughly.

2. The first player draws two discs from the bag at random. In this game, red discs represent negative numbers and black discs represent positive numbers. Keeping this designation in mind, add the two numbers. For example, a red 3 and a black 2 would represent the sum $-3 + 2 = -1$. Record the sum as the player's score in the Score Sheet shown below and update the Cumulative Score. Replace the discs in the bag and mix thoroughly for the next player.

3. Once each player has taken his or her turn, players take their next turns in order of least to greatest cumulative score.

4. Players continue taking turns, as described in Steps 2 and 3, until the score sheet is filled.

5. The winner is the player with the greatest cumulative score at the end of five turns.

| SCORE SHEET | | | | | | | | | | |
|---|---|---|---|---|---|---|---|---|---|---|
| | Player 1 | | Player 2 | | Player 3 | | Player 4 | | Player 5 | |
| | Score | Cumulative Score | Score | Cumulative Score | Score | Cumulative Score | Score | Cumulative Score | Score | Cumulative Score |
| Turn 1 | | | | | | | | | | |
| Turn 2 | | | | | | | | | | |
| Turn 3 | | | | | | | | | | |
| Turn 4 | | | | | | | | | | |
| Turn 5 | | | | | | | | | | |

# CHAPTER 2 HIGHLIGHTS

| **DEFINITIONS AND CONCEPTS** | **EXAMPLES** |
|---|---|

## SECTION 2.1   INTRODUCTION TO INTEGERS

The **integers** are $\ldots, -3, -2, -1, 0, 1, 2, 3, \ldots$.

The **absolute value** of a number is that number's distance from 0 on the number line. The symbol for absolute value is $|\ \ |$.

Two numbers that are the same distance from 0 on the number line but are on opposite sides of 0 are called **opposites**.

If $a$ is a number, then $-(-a) = a$.

Integers:

$$-432, \quad -10, \quad 0, \quad 15$$

$|-2| = 2$

$|2| = 2$

5 and $-5$ are opposites.

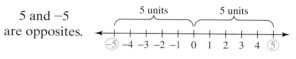

$$-(-11) = 11, \qquad -|-3| = -3$$

## SECTION 2.2   ADDING INTEGERS

ADDING TWO NUMBERS WITH THE SAME SIGN

**Step 1.** Add their absolute values.
**Step 2.** Use their common sign as the sign of the sum.

ADDING TWO NUMBERS WITH DIFFERENT SIGNS

**Step 1.** Find the larger absolute value minus the smaller absolute value.
**Step 2.** Use the sign of the number with the larger absolute value as the sign of the sum.

Add:

$$-3 + (-2) = -5$$

$$-7 + (-15) = -22$$

$$-6 + 4 = -2$$

$$17 + (-12) = 5$$

$$-32 + (-2) + 14 = -34 + 14$$

$$= -20$$

## SECTION 2.3   SUBTRACTING INTEGERS

SUBTRACTING TWO NUMBERS

If $a$ and $b$ are numbers, then $a - b = a + (-b)$.

Subtract:

$$-35 - 4 = -35 + (-4) = -39$$

$$3 - 8 = 3 + (-8) = -5$$

$$-10 - (-12) = -10 + 12 = 2$$

$$7 - 20 - 18 - (-3) = 7 + (-20) + (-18) + (+3)$$

$$= -13 + (-18) + 3$$

$$= -31 + 3$$

$$= -28$$

---

### SECTION 2.4   MULTIPLYING AND DIVIDING INTEGERS

MULTIPLYING NUMBERS

The product of two numbers having the same sign is a positive number.

The product of two numbers having unlike signs is a negative number.

DIVIDING NUMBERS

The quotient of two numbers having the same sign is a positive number.

The quotient of two numbers having unlike signs is a negative number.

Multiply:

$$(-7)(-6) = 42$$
$$9(-4) = -36$$

Evaluate:

$$(-3)^2 = (-3)(-3) = 9$$

Divide:

$$-100 \div (-10) = 10$$

$$\frac{14}{-2} = -7, \quad \frac{0}{-3} = 0, \quad \frac{22}{0} \text{ is undefined.}$$

---

### SECTION 2.5   ORDER OF OPERATIONS

ORDER OF OPERATIONS

1. Do all operations within grouping symbols such as parentheses or brackets.
2. Evaluate any expressions with exponents.
3. Multiply or divide in order from left to right.
4. Add or subtract in order from left to right.

Simplify:

$$3 + 2 \cdot (-5) = 3 + (-10)$$
$$= -7$$

$$\frac{-2(5 - 7)}{-7 + |-3|} = \frac{-2(-2)}{-7 + 3}$$

$$= \frac{4}{-4}$$

$$= -1$$

**Name** _____ **Section** _____ **Date** _____

# CHAPTER 2 REVIEW

**(2.1)** *Represent each quantity by an integer.*

**1.** A gold miner is working 1435 feet down in a mine.   −1435

**2.** A mountain peak is 7562 meters above sea level.   +7562

*Graph each number on the given number line.*

**3.** −2

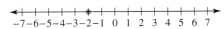

**4.** −7

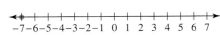

*Insert < or > between each pair of numbers to make a true statement.*

**5.** −18      −20   >

**6.** −5      5   <

**7.** 123      −198   >

*Simplify.*

**8.** |−12|   12

**9.** |0|   0

**10.** −|6|   −6

*Find the opposite of each number.*

**11.** −12   12

**12.** −(−3)   −3

*Answer true or false for each statement.*

**13.** If $a < b$, then $a$ must be a negative number.   false

**14.** The absolute value of an integer is always 0 or a positive number.   true

**15.** A negative number is always less than a positive number.   true

**16.** If $a$ is a negative number, then $-a$ is a positive number.   true

**(2.2)** *Add.*

**17.** 5 + (−3)   2

**18.** 18 + (−4)   14

**19.** −12 + 16   4

**20.** −23 + 40   17

**21.** −8 + (−15)  −23     **22.** −5 + (−17)  −22     **23.** −24 + 3  −21     **24.** −89 + 19  −70

**25.** 15 + (−15)  0     **26.** −24 + 24  0     **27.** −43 + (−108)  −151  **28.** −100 + (−506)  −606

**29.** During the 2000 LPGA Office Depot golf classic, the winner, Karrie Webb had scores of −7, 3, −2, and −1 in four rounds of golf. Find her total score. (*Source:* Ladies Professional Golf Association)  −7

**30.** During the 2000 LPGA Subaru Memorial golf tournament, Meg Mallon had scores of −8, and −1 in her first round. Find her total score in this round. (*Source:* Ladies Professional Golf Association)  −9

**31.** The temperature at 5 a.m. on a day in January was −15°C. By 6 a.m. the temperature had fallen 5 degrees. Find the temperature at 6 a.m.  −20°C

**32.** A diver starts out at 127 feet below the surface and then swims downward another 23 feet. Find the diver's current depth.  150 ft below the surface

**(2.3)** *Subtract.*

**33.** 12 − 4  8     **34.** −12 − 4  −16     **35.** 8 − 19  −11

**36.** −8 − 19  −27     **37.** 7 − (−13)  20     **38.** −6 − (−14)  8

**39.** 16 − 16  0     **40.** −16 − 16  −32     **41.** −12 − (−12)  0

**42.** $|-5| - |-12|$   $-7$

**43.** $-(-5) - 12 + (-3)$   $-10$

**44.** $-8 + |-12| - 10 - |-3|$   $-9$

**45.** Josh Weidner has $142 in his checking account. He writes a check for $125, makes a deposit for $43, and then writes another check for $85. Represent the balance in his account by an integer.   $-25$

**46.** If the elevation of Lake Superior is 600 feet above sea level and the elevation of the Caspian Sea is 92 feet below sea level, find the difference of their elevations.   692 ft

*Answer true or false for each statement.*

**47.** $|-5| - |-6| = 5 - 6$   true

**48.** $|-5 - (-6)| = 5 + 6$   false

**49.** If $b > a$, then $b - a$ is a positive number.   true

**50.** If $b < a$, then $b - a$ is a negative number.   true

**(2.4)** *Multiply.*

**51.** $(-3) \cdot (-7)$   $21$

**52.** $(-6) \cdot (3)$   $-18$

**53.** $-4 \cdot 16$   $-64$

**54.** $(-5) \cdot (-12)$   $60$

*Divide.*

**55.** $-15 \div 3$   $-5$

**56.** $\dfrac{-24}{-8}$   $3$

**57.** $0 \div (-3)$   $0$

**58.** $-2 \div 0$   undefined

**59.** $\dfrac{-38}{-1}$   $38$

**60.** $45 \div (-9)$   $-5$

**61.** A football team lost 5 yards on each of two consecutive plays. Represent the total loss by a product of integers, and find the product.   $(-5)(2) = -10$

**62.** A race horse bettor lost $50 on each of four consecutive races. Represent the total loss by a product of integers, and find the product.   $(-50)(4) = -200$

**(2.5)** *Simplify.*

**63.** $(-7)^2$   49

**64.** $-7^2$   $-49$

**65.** $-2^5$   $-32$

**66.** $(-2)^5$   $-32$

**67.** $5 - 8 + 3$   0

**68.** $-3 + 12 + (-7) - 10$   $-8$

**69.** $-10 + 3 \cdot (-2)$   $-16$

**70.** $5 - 10 \cdot (-3)$   35

**71.** $16 \cdot (-2) + 4$   $-28$

**72.** $3 \cdot (-12) - 8$   $-44$

**73.** $5 + 6 \div (-3)$   3

**74.** $-6 + (-10) \div (-2)$   $-1$

**75.** $16 + (-3) \cdot 12 \div 4$   7

**76.** $(-12) + 25 \cdot 1 \div (-5)$   $-17$

**77.** $4^3 - (8 - 3)^2$   39

**78.** $4^3 - 90$   $-26$

**79.** $-(-4) \cdot |-3| - 5$   7

**80.** $|5 - 1|^2 \cdot (-5)$   $-80$

**81.** $\dfrac{(-4)(-3) - (-2)(-1)}{-10 + 5}$   $-2$

**82.** $\dfrac{4(12 - 18)}{-10 \div (-2 - 3)}$   $-12$

*Evaluate each expression for $x = -2$ and $y = 1$.*

**83.** $2x - y$   $-5$

**84.** $y^2 + x^2$   5

**85.** $\dfrac{3x}{6}$   $-1$

**86.** $\dfrac{5y - x}{-y}$   $-7$

**87.** $x^2$   4

**88.** $-x^2$   $-4$

**89.** $7 - x^2$   3

**90.** $100 - x^3$   108

**Name** _____ **Section** _____ **Date** _____

# CHAPTER 2 TEST

*Simplify each expression.*

**1.** $-5 + 8$      **2.** $18 - 24$      **3.** $5 \cdot (-20)$

**4.** $(-16) \div (-4)$      **5.** $(-18) + (-12)$      **6.** $-7 - (-19)$

**7.** $(-5) \cdot (-13)$      **8.** $\dfrac{-25}{-5}$      **9.** $|-25| + (-13)$

**10.** $14 - |-20|$      **11.** $|5| \cdot |-10|$      **12.** $\dfrac{|-10|}{-|-5|}$

**13.** $(-8) + 9 \div (-3)$      **14.** $-7 + (-32) - 12 + 5$      **15.** $(-5)^3 - 24 \div (-3)$

**16.** $(5 - 9)^2 \cdot (8 - 2)^3$      **17.** $-(-7)^2 \div 7 \cdot (-4)$      **18.** $3 - (8 - 2)^3$

**1.** 3

**2.** $-6$

**3.** $-100$

**4.** 4

**5.** $-30$

**6.** 12

**7.** 65

**8.** 5

**9.** 12

**10.** $-6$

**11.** 50

**12.** $-2$

**13.** $-11$

**14.** $-46$

**15.** $-117$

**16.** 3456

**17.** 28

**18.** $-213$

**19.** $-1$

**20.** $-2$

**21.** $2$

**22.** $-5$

**23.** $-3$

**24.** $5$

**25.** $-1$

**26.** $8$

**27.** $14,893 + (-147) = 14,746; 14,746$ ft

**28.** $41

**29.** $31,642$ ft

**30.** $3820$ ft below sea level

**19.** $-6 + (-15) \div (-3)$

**20.** $\dfrac{4}{2} - \dfrac{8^2}{16}$

**21.** $\dfrac{-3(-2) + 12}{-1(-4 - 5)}$

**22.** $\dfrac{|25 - 30|^2}{2(-6) + 7}$

*Evaluate each expression for $x = 0$, $y = -3$, and $z = 2$.*

**23.** $3x + y$

**24.** $|y| + |x| + |z|$

**25.** $\dfrac{3z}{2y}$

**26.** $5 - y$

**27.** A mountain climber is at an elevation of 14,893 feet and moves down the mountain a distance of 147 feet. Represent his final elevation as a sum, and find the sum.

**28.** Jane Hathaway has $237 in her checking account. She writes a check for $157, she writes another check for $77, and she deposits $38. Represent the balance in her account by an integer.

**29.** Mt. Washington in New Hampshire has an elevation of 6288 feet above sea level. The Romanche Gap in the Atlantic Ocean has an elevation of 25,354 feet below sea level. Represent the difference in elevation between these two points by an integer. (*Source:* National Geographic Society and Defense Mapping Agency)

**30.** Lake Baykal in Siberian Russia is the deepest lake in the world with a maximum depth of 5315 feet. The elevation of the lake's surface is 1495 feet above sea level. What is the elevation (with respect to sea level) of the deepest point in the lake? (*Source:* U.S. Geological Survey)

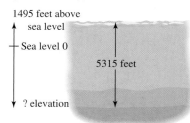

1495 feet above sea level

Sea level 0

5315 feet

? elevation

# Cumulative Review

*Find the place value of the digit 4 in each whole number.*

**1.** 48,761

**2.** 249

**3.** 524,007,656

**4.** Insert < or > to make a true statement.

    **a.** 5     50    **b.** 101    0    **c.** 29    27

**5.** Add: $13 + 2 + 7 + 8 + 9$

**6.** Subtract: $7826 - 505$
Check by adding.

△ **7.** The radius of Venus is 6052 kilometers. The radius of Mercury is 3612 kilometers less than the radius of Venus. Find the radius of Mercury. (*Source:* National Space Science Data Center)

**8.** Round 568 to the nearest ten.

**9.** Round each number to the nearest hundred to find an estimated difference.

    4725
    $-2879$

**10.** Rewrite each using the distributive property.

    **a.** $3(4 + 5)$    **b.** $10(6 + 8)$
    **c.** $2(7 + 3)$

**11.** Multiply: $631 \times 125$

**12.** Find each quotient. Check by multiplying.

    **a.** $42 \div 7$    **b.** $\dfrac{81}{9}$    **c.** $4\overline{)24}$

**Answers**

**1.** ten-thousands (Sec. 1.1, Ex. 1)

**2.** tens (Sec. 1.1, Ex. 2)

**3.** millions (Sec. 1.1, Ex. 3)

**4. a.** <

    **b.** >

    **c.** > (Sec. 1.1, Ex. 12)

**5.** 39 (Sec. 1.2, Ex. 3)

**6.** 7321 (Sec. 1.3, Ex. 2)

**7.** 2440 km (Sec. 1.3, Ex. 5)

**8.** 570 (Sec. 1.4, Ex. 1)

**9.** 1800 (Sec. 1.4, Ex. 5)

**10. a.** $3 \cdot 4 + 3 \cdot 5$

    **b.** $10 \cdot 6 + 10 \cdot 8$

    (Sec. 1.5, Ex. 2)
    **c.** $2 \cdot 7 + 2 \cdot 3$

**11.** 78,875 (Sec. 1.5, Ex. 5)

**12. a.** 6

    **b.** 9

    **c.** 6 (Sec. 1.6, Ex. 1)

**13.** 741 (Sec. 1.6, Ex. 4)

**14.** 7 boxes (Sec. 1.6, Ex. 11)

**15.** 64    (Sec. 1.7, Ex. 5)

**16.** 7    (Sec. 1.7, Ex. 6)

**17.** 180    (Sec. 1.7, Ex. 8)

**18.** 2 (Sec. 1.7, Ex. 12)

**19.** 15 (Sec. 1.8, Ex. 1)

**20. a.** 2

**b.** 5

**c.** 0 (Sec. 2.1, Ex. 4)

**21.** 3 (Sec. 2.2, Ex. 6)

**22.** 14 (Sec. 2.3, Ex. 12)

**23.** −21 (Sec. 2.4, Ex. 1)

**24.** 0 (Sec. 2.4, Ex. 3)

**25.** −16 (Sec. 2.5, Ex. 10)

**Name** _____

**13.** Divide:  $3705 \div 5$
Check by multiplying.

**14.** How many boxes are needed to ship 56 pairs of Nikes to a shoe store in Texarkana if 9 pairs of shoes will fit in each shipping box?

*Evaluate.*

**15.** $8^2$

**16.** $7^1$

**17.** $5 \cdot 6^2$

**18.** Simplify:  $\dfrac{7 - 2 \cdot 3 + 3^2}{2^2 + 1}$

**19.** Evaluate $x + 7$ if $x$ is 8.

**20.** Simplify.

  **a.** $|-2|$    **b.** $|5|$    **c.** $|0|$

**21.** Add:  $-2 + 5$

**22.** Evaluate $a - b$ for $a = 8$ and $b = -6$.

*Multiply.*

**23.** $-7 \cdot 3$

**24.** $0 \cdot (-4)$

**25.** Simplify:  $3(4 - 7) + (-2) - 5$

**162**

# Solving Equations and Problem Solving

Throughout this text, we have been making the transition from arithmetic to algebra. We have said that in algebra letters called variables represent numbers. Using variables is a very powerful method for solving problems that cannot be solved with arithmetic alone. This chapter introduces operations on algebraic expressions and solving variable equations.

3

The largest purchase that most people usually make during their lifetimes is a house. Every year, millions of families take the plunge and buy a house. That also means that millions of families sell their houses. Real estate agents can help get sellers and buyers together to make the transaction. In return, the real estate agent usually receives a commission, a payment for his or her services based on the selling price of the house. In Exercises 50–51 on page 208, we will see how an equation can be used to describe the relationship between the selling price of a house, the real estate agent's commission, and the amount the seller receives in a real estate transaction.

Name _____ Section _____ Date _____

# CHAPTER 3 PRETEST

*Simplify.*

**1.** $9x - 4 + 6x + 8$

**2.** $3(2x - 1) - (x - 8)$

*Multiply.*

**3.** $8(7b)$

**4.** $-5(2y - 7)$

△ **5.** Find the perimeter of the triangle.

△ **6.** Find the area of the rectangle.

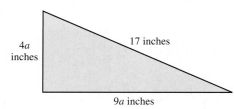

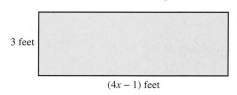

**7.** Determine whether $-4$ is a solution of the equation $2y - 3 = -5$.

*Solve.*

**8.** $-7 = x + 5$

**9.** $9n + 3 - 8n = 8 - 14$

**10.** $-6x = 42$

**11.** $2m - 9m = -77$

*Translate each phrase to an algebraic expression. Let x be the unknown number.*

**12.** Twice the difference of a number and 12

**13.** The product of a number and 3

*Solve.*

**14.** $18 - 3x = -9$

**15.** $8a - 5 = 2a + 7$

**16.** $4(x + 1) = 9x - 1$

**17.** $-2(3x + 4) - 10 = 0$

*Translate each sentence into an equation. Use x to represent "a number."*

**18.** The quotient of 54 and $-6$ is $-9$.

**19.** Five less than a number is 12.

*Solve.*

**20.** Sixty-eight is 5 more than 9 times a number. Find the number.

**164**

# 3.1   SIMPLIFYING ALGEBRAIC EXPRESSIONS

Just as we can add, subtract, multiply, and divide numbers, we can add, subtract, multiply, and divide algebraic expressions. In previous sections we evaluated algebraic expressions like $x + 3$, $4x$, and $x + 2y$ for particular values of the variables. In this section, we explore working with variable expressions without evaluating them. We begin with a definition of a term.

## A   COMBINING LIKE TERMS

The addends of an algebraic expression are called the **terms** of the expression.

$$x + 3$$
↑   ↑ —— 2 terms

$$3y^2 + (-6y) + 4$$
↑     ↑     ↑ —— 3 terms

A term that is only a number has a special name. It is called a **constant term**, or simply a **constant**. A term that contains a variable is called a **variable term**.

|  |  |  |  |
|---|---|---|---|
| $x$ | $+$ | $3$ | |
| ↑ | | ↑ | |
| variable term | | constant term | |

|  |  |  |
|---|---|---|
| $3y^2 + (-6y) +$ | | $4$ |
| ↑ | ↑ | ↑ |
| variable terms | | constant term |

The number factor of a variable term is called the **numerical coefficient**. A numerical coefficient of 1 is usually not written.

| $5x$ | $x$ or $1x$ | $3y^2$ | $-6y$ |
|---|---|---|---|
| ↑ | ↑ | ↑ | ↑ |
| Numerical coefficient is 5. | Understood numerical coefficient is 1. | Numerical coefficient is 3. | Numerical coefficient is $-6$. |

Terms that are exactly the same, except that they may have different numerical coefficients, are called **like terms**.

| Like Terms | Unlike Terms |
|---|---|
| $3x, \dfrac{1}{2}x$ | $5x, x^2$ |
| $-6y, 2y, y$ | $7x, 7y$ |

**TRY THE CONCEPT CHECK IN THE MARGIN.**

A sum or difference of like terms can be simplified using the **distributive property**. Recall from Chapter 1 that the distributive property says that multiplication distributes over addition (and subtraction). Using variables, we can write the distributive property as follows.

---

**DISTRIBUTIVE PROPERTY**

If $a$, $b$, and $c$ are numbers, then

$$ac + bc = (a + b)c$$

Also,

$$ac - bc = (a - b)c$$

---

**Objectives**

A   Use properties of numbers to combine like terms.

B   Use properties of numbers to multiply expressions.

C   Simplify expressions by multiplying and then combining like terms.

D   Find the perimeter and area of figures.

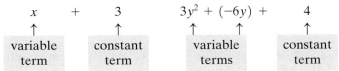

Study Guide     SSM     CD-ROM     Video 3.1

TEACHING TIP

Continue to remind students that $x$ means $1x$ and $-x$ means $-1x$.

TEACHING TIP

Remind students that order of variables does not matter when deciding whether terms are like terms or not. In other words, $4xy$ and $2yx$ are like terms.

## ✓ CONCEPT CHECK

True or false? The terms $-7xz^2$ and $3z^2x$ are like terms. Explain.

**Answer**

✓ Concept Check: True.

By the distributive property then,

$$7x + 2x = (7 + 2)x$$
$$= 9x$$

We have combined like terms and the expression $7x + 2x$ simplifies to $9x$.

**Practice Problem 1**

Combine like terms.
a. $8m - 11m$  b. $5a + a$
c. $-y^2 + 3y^2 + 7$

**Example 1**  Simplify by combining like terms.

a. $3x + 2x$
b. $y - 7y$
c. $3x^2 + 5x^2 - 2$

*Solution:*  Add or subtract like terms.

a. $3x + 2x = (3 + 2)x$
$$= 5x$$
b. $y - 7y = 1y - 7y$
$$= (1 - 7)y$$
$$= -6y$$
c. $3x^2 + 5x^2 - 2 = (3 + 5)x^2 - 2$
$$= 8x^2 - 2$$

The commutative and associative properties of addition and multiplication can also help us simplify expressions. We presented these properties in Sections 1.2 and 1.5 and state them again using variables.

> **PROPERTIES OF ADDITION AND MULTIPLICATION**
>
> If $a$, $b$, and $c$ are numbers, then
>
> $a + b = b + a$    Commutative property of addition
> $a \cdot b = b \cdot a$    Commutative property of multiplication
>
> That is, the **order** of adding or multiplying two numbers can be changed without changing their sum or product.
>
> $(a + b) + c = a + (b + c)$    Associative property of addition
> $(a \cdot b) \cdot c = a \cdot (b \cdot c)$    Associative property of multiplication
>
> That is, the **grouping** of numbers in addition or multiplication can be changed without changing their sum or product.

**Practice Problem 2**

Simplify: $8m + 5 + m - 4$

**Example 2**  Simplify: $2y - 6 + 4y + 8$

*Solution:*  We begin by writing subtraction as the opposite of addition.

$2y - 6 + 4y + 8 = 2y + (-6) + 4y + 8$    Apply the commutative
$\qquad\qquad\qquad = 2y + 4y + (-6) + 8$    property of addition.
$\qquad\qquad\qquad = (2 + 4)y + (-6) + 8$    Apply the distributive property.
$\qquad\qquad\qquad = 6y + 2$    Simplify.

**Answers**

1. a. $-3m$  b. $6a$  c. $2y^2 + 7$
2. $9m + 1$

**Examples**  Simplify each expression by combining like terms.

3.    $6x + 2x - 5 = 8x - 5$

**4.** $4x + 2 - 5x + 3 = 4x + 2 + (-5x) + 3$
$$= 4x + (-5x) + 2 + 3$$
$$= -1x + 5 \quad \text{or} \quad -x + 5$$

**5.** $2x - 5 + 3y + 4x - 10y + 11$
$$= 2x + (-5) + 3y + 4x + (-10y) + 11$$
$$= 2x + 4x + 3y + (-10y) + (-5) + 11$$
$$= 6x - 7y + 6$$

**Practice Problems 3–5**

Simplify each expression by combining like terms.
3. $7y + 11y - 8$
4. $2y - 6 + y + 7$
5. $-9y + 2 - 4y - 8x + 12 - x$

As we practice combining like terms, keep in mind that some of the steps may be performed mentally.

## B    MULTIPLYING EXPRESSIONS

We can also use properties of numbers to multiply expressions such as $3(2x)$. By the associative property of multiplication, we can write the product $3(2x)$ as $(3 \cdot 2)x$, which simplifies to $6x$.

**Examples**     Multiply.

**6.** $5(3y) = (5 \cdot 3)y$     Apply the associative property of multiplication.
$$= 15y$$     Multiply.

**7.** $-2(4x) = (-2 \cdot 4)x$     Apply the associative property of multiplication.
$$= -8x$$     Multiply.

**Practice Problems 6–7**

Multiply.
6. $7(8a)$
7. $-5(9x)$

We can use the distributive property to combine like terms, which we have done, and also to multiply expressions such as $2(3 + x)$. By the distributive property, we have that

$$2\overparen{(3 + x)} = 2 \cdot 3 + 2 \cdot x$$     Apply the distributive property.
$$= 6 + 2x$$     Multiply.

**Example 8**     Use the distributive property to multiply: $6(x + 4)$

*Solution:*     By the distributive property,

$$6\overparen{(x + 4)} = 6 \cdot x + 6 \cdot 4$$     Apply the distributive property.
$$= 6x + 24$$     Multiply.

**TRY THE CONCEPT CHECK IN THE MARGIN.**

**Practice Problem 8**

Use the distributive property to multiply: $7(y + 2)$

**✓ CONCEPT CHECK**

What's wrong with the following?
$8(a - b) = 8a - b$

**Example 9**     Multiply: $-3(5a + 2)$

*Solution:*     By the distributive property,

$$-3\overparen{(5a + 2)} = -3(5a) + (-3)(2)$$     Apply the distributive property.
$$= (-3 \cdot 5)a + (-6)$$     Use the associative property and multiply.
$$= -15a - 6$$     Multiply.

To simplify expressions containing parentheses, we first use the distributive property and multiply.

**Practice Problem 9**

Multiply: $4(7a - 5)$

**Answers**

**3.** $18y - 8$   **4.** $3y + 1$   **5.** $-13y - 9x + 14$
**6.** $56a$   **7.** $-45x$   **8.** $7y + 14$   **9.** $28a - 20$
**✓ Concept Check:** Did not distribute the 8;
$8(a - b) = 8a - 8b$

## Practice Problem 10

Multiply: $6(5 - y)$

TEACHING TIP

Before Example 11, you may want to try the following: Ask students to identify the terms that the 5 distributes to in each expression.

$$y + 5(x - 2)$$
$$5(3x + 6z) - 4y$$
$$4(y + 3) + 5(x + 1)$$

## Practice Problem 11

Simplify: $5(y - 3) - 8 + y$

## Practice Problem 12

Simplify: $5(2x - 3) + 7(x - 1)$

## Practice Problem 13

Simplify: $-(y + 1) + 3y - 12$

---

### Example 10    Multiply: $8(x - 4)$

*Solution:* 
$$8(x - 4) = 8 \cdot x - 8 \cdot 4$$
$$= 8x - 32$$

### ◼C SIMPLIFYING EXPRESSIONS

Next we will **simplify** expressions by first multiplying and then **combining** any like terms.

### Example 11    Simplify: $2(3 + x) - 15$

*Solution:*   First we use the distributive property to remove parentheses.

$$2(3 + x) - 15 = 2 \cdot 3 + 2 \cdot x + (-15) \qquad \text{Apply the distributive property.}$$
$$= 6 + 2x + (-15) \qquad \text{Multiply.}$$
$$= 2x + (-9) \quad \text{or} \quad 2x - 9 \qquad \text{Combine like terms.}$$

> **HELPFUL HINT**
>
> 2 is *not* distributed to the $-15$ since $-15$ is not within the parentheses.

### Example 12    Simplify: $-2(x - 5) + 4(2x + 2)$

*Solution:*   First we use the distributive property to remove parentheses.

$$-2(x - 5) + 4(2x + 2) = -2 \cdot x + (-2)(-5) + 4 \cdot 2x + 4 \cdot 2$$
$$= -2x + 10 + 8x + 8$$
$$= 6x + 18$$

### Example 13    Simplify: $-(x + 4) + 5x + 16$

*Solution:*   The expression $-(x + 4)$ means $-1(x + 4)$.

$$-(x + 4) + 5x + 16 = -1(x + 4) + 5x + 16$$
$$= -1 \cdot x + (-1)(4) + 5x + 16 \qquad \text{Apply the distributive property.}$$
$$= -x + (-4) + 5x + 16 \qquad \text{Multiply.}$$
$$= 4x + 12 \qquad \text{Combine like terms.}$$

---

**Answers**

**10.** $30 - 6y$    **11.** $6y - 23$    **12.** $17x - 22$
**13.** $2y - 13$

△ **Example 14**   Find the perimeter of the triangle.

*Solution:*   Recall that the perimeter of a figure is the distance around the figure. To find the perimeter, then, we find the sum of the lengths of the sides.

$$\text{perimeter} = 2z + 3z + 5z$$
$$= 10z$$

> **HELPFUL HINT**
> Don't forget to insert proper units.

The perimeter is $10z$ feet.

△ **Example 15**   Find the area of the rectangular deck.

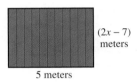

*Solution:*   Recall how to find the area of a rectangle.

$$A = \text{length} \cdot \text{width}$$
$$= 5(2x - 7) \qquad \text{Let length} = 5 \text{ and width} = (2x - 7).$$
$$= 10x - 35 \qquad \text{Multiply.}$$

The area is $(10x - 35)$ *square* meters.

---

△ **Practice Problem 14**

Find the perimeter of the square.

△ **Practice Problem 15**

Find the area of the garden in the shape of a rectangle.

**Answers**
**14.** $8x$ centimeters   **15.** $(36y + 27)$ square yards

# Focus on History

## THE EQUAL SIGN

The most important symbol in an equation is the equal sign, $=$. It indicates that a statement is, in fact, an equation and not just an algebraic expression. However, the equal sign has not always been around. In fact, the $=$ symbol has been used only since 1557 when it first appeared in a book called *The Whetstone of Witte* by English physician and mathematician Robert Recorde. Recorde's original symbol used much longer lines, $\Equal$, and can be seen in this page from *The Whetstone of Witte*.

In his book, Recorde gave an interesting reason for his choice of symbol. He explained, "I will sette as I doe often in woorke use, a paire of parralles, or Gemowe lines of one lengthe, thus: $=$, bicause noe 2, thynges, can be moare equalle."

Prior to the appearance of the equal sign, authors of mathematics texts often used words, such as "aequales" or "gleich" to indicate equality. These words continued to be used regularly until the use of the $=$ symbol became widely accepted in the 1700s.

**Name** _____ **Section** _____ **Date** _____

MENTAL MATH ANSWERS

1. 5

2. −2

3. 1

4. 3

5. 11

6. −1

## MENTAL MATH

*Find the numerical coefficient of each variable term.*

**1.** $5y$        **2.** $-2z$        **3.** $z$

**4.** $3xy^2$        **5.** $11a$        **6.** $-x$

## EXERCISE SET 3.1

**A** *Simplify each expression by combining like terms. See Examples 1 through 5.*

**1.** $3x + 5x$      **2.** $8y + 3y$      **3.** $5n - 9n$

**4.** $7z - 10z$      **5.** $4c + c - 7c$      **6.** $5b - 8b - b$

**7.** $5x - 7x + x - 3x$      **8.** $8y + y - 2y - y$      **9.** $4a + 3a + 6a - 8$

**10.** $5b - 4b + b - 15$

**B** *Multiply. See Examples 6 and 7.*

**11.** $6(5x)$      **12.** $4(4x)$      **13.** $-2(11y)$

**14.** $-3(21z)$      **15.** $12(6a)$      **16.** $9(7b)$

*Multiply. See Examples 8 through 10.*

**17.** $2(y + 2)$      **18.** $3(x + 1)$      **19.** $5(a - 8)$

**20.** $4(y - 6)$      **21.** $-4(3x + 7)$      **22.** $-8(8y + 10)$

ANSWERS

1. $8x$

2. $11y$

3. $-4n$

4. $-3z$

5. $-2c$

6. $-4b$

7. $-4x$

8. $6y$

9. $13a - 8$

10. $2b - 15$

11. $30x$

12. $16x$

13. $-22y$

14. $-63z$

15. $72a$

16. $63b$

17. $2y + 4$

18. $3x + 3$

19. $5a - 40$

20. $4y - 24$

21. $-12x - 28$

22. $-64y - 80$

**23.** $2x + 15$

**24.** $28 - 5y$

**25.** $-21n + 20$

**26.** $2b - 15$

**27.** $15c + 3$

**28.** $24d + 2$

**29.** $7w + 15$

**30.** $13z + 50$

**31.** $11x - 8$

**32.** $21x - 4$

**33.** $-5x - 9$

**34.** $-2y + 16$

**35.** $-2y$

**36.** $13x$

**37.** $-7z$

**38.** $4x$

**39.** $8d - 3c$

**40.** $8r - 6s$

**41.** $6y - 14$

**42.** $-6a - 1$

**43.** $-q$

**44.** $2m$

**45.** $2x + 22$

**46.** $5x + 13$

**47.** $-3x - 35$

**48.** $-8x + 6$

**49.** $-3z - 15$

**50.** $-8 - 2v$

**51.** $-6x + 6$

**52.** $13y + 11$

**53.** $3x - 30$

**54.** $6x - 16$

**55.** $-r + 8$

**Name** _____

**C** *Simplify each expression. Use the distributive property to remove parentheses first. See Examples 11 through 13.*

**23.** $2(x + 4) + 7$

**24.** $5(6 - y) - 2$

**25.** $-4(6n - 5) + 3n$

**26.** $-3(5 - 2b) - 4b$

**27.** $5(3c - 1) + 8$

**28.** $4(6d - 2) + 10$

**29.** $3 + 6(w + 2) + w$

**30.** $8z + 5(6 + z) + 20$

**31.** $2(3x + 1) + 5(x - 2)$

**32.** $3(5x - 2) + 2(3x + 1)$

**33.** $-(5x - 1) - 10$

**34.** $-(2y - 6) + 10$

**A** **B** **C** *Simplify each expression. See Examples 1 through 13.*

**35.** $18y - 20y$

**36.** $x + 12x$

**37.** $z - 8z$

**38.** $12x - 8x$

**39.** $9d - 3c - d$

**40.** $8r + s - 7s$

**41.** $2y - 6 + 4y - 8$

**42.** $a + 4 - 7a - 5$

**43.** $5q + p - 6q - p$

**44.** $m - 8n + m + 8n$

**45.** $2(x + 1) + 20$

**46.** $5(x - 1) + 18$

**47.** $5(x - 7) - 8x$

**48.** $3(x + 2) - 11x$

**49.** $-5(z + 3) + 2z$

**50.** $-8(1 + v) + 6v$

**51.** $8 - x + 4x - 2 - 9x$

**52.** $5y - 4 + 9y - y + 15$

**53.** $-7(x + 5) + 5(2x + 1)$

**54.** $-2(x + 4) + 8(x - 1)$

**55.** $3r - 5r + 8 + r$

**Name** _____

**56.** $6x - 4 + 2x - x + 3$      **57.** $-3(n - 1) - 4n$      **58.** $5(c + 2) + 7c$

**59.** $4(z - 3) + 5z - 2$      **60.** $8(m + 3) - 20 + m$      **61.** $6(2x - 1) - 12x$

**62.** $5(2a + 3) - 10a$      **63.** $-(4x - 5) + 5$      **64.** $-(7y - 2) + 6$

**65.** $-(4xy - 10) + 2(3xy + 5)$      **66.** $-(12ab - 10) + 5(3ab - 2)$

**67.** $3a + 4(a + 3)$      **68.** $b + 2(b - 5)$

**69.** $5y - 2(y - 1) + 3$      **70.** $3x - 4(x + 2) + 1$

**D**   *Find the perimeter of each figure. See Example 14.*

△ **71.**
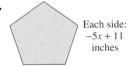
Each side:
$-5x + 11$
inches

△ **72.**

Each side:
$9y + 1$
kilometers

△ **73.**
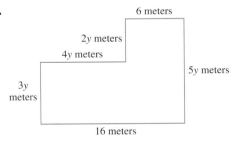
6 meters
$2y$ meters
$4y$ meters
$5y$ meters
$3y$ meters
16 meters

△ **74.**
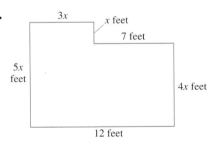
$3x$
$x$ feet
7 feet
$5x$ feet
$4x$ feet
12 feet

△ **75.**
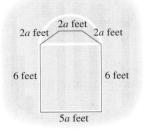
$2a$ feet
$2a$ feet
$2a$ feet
6 feet
6 feet
$5a$ feet

△ **76.**
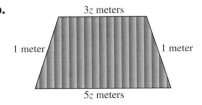
$3z$ meters
1 meter
1 meter
$5z$ meters

**Name** _____

*Find the area of each figure. See Example 15.*

△ **77.**

Square | $(4z)$ centimeters

**78.**
△

12 feet

$(x + 3)$ feet | Rectangle

## REVIEW AND PREVIEW

*Perform each indicated operation. See Sections 2.2 and 2.3.*

**79.** $-13 + 10$

**80.** $-7 - (-4)$

**81.** $-4 - (-12)$

**82.** $-15 + 23$

**83.** $-4 + 4$

**84.** $8 + (-8)$

## COMBINING CONCEPTS

*Simplify.*

**85.** $9684q - 686 - 4860q + 12,960$

**86.** $76(268x + 592) - 2960$

**87.** If $x$ is a whole number, which expression is the greatest: $-2x$ or $-5x$? Explain your answer.

**88.** If $x$ is a whole number, which expression is the least: $-2x$ or $-5x$? Explain your answer.

*Find the area of each figure.*

△ **89.**

$(2x + 1)$ miles

7 miles | Rectangle | $(2x + 3)$ miles

Rectangle | 3 miles

△ **90.**

12 kilometers

$(3x - 5)$ kilometers | Rectangle

$(5x - 1)$ kilometers | Rectangle

4 kilometers

**91.** Explain what makes two terms "like terms."

**92.** Explain how to combine like terms.

# 3.2 SOLVING EQUATIONS: THE ADDITION PROPERTY

Frequently in this book, we have written statements like $7 + 4 = 11$ or area = length·width. Each of these statements is called an **equation**. An equation is of the form

**expression = expression**

An equation can be labeled as

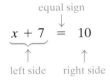

left side    right side

## A DETERMINING WHETHER A NUMBER IS A SOLUTION

When an equation contains a variable, deciding which values of the variable make an equation a true statement is called **solving** an equation for the variable. A **solution** of an equation is a value for the variable that makes an equation a true statement. For example, 2 is a solution of the equation $x + 5 = 7$, since replacing $x$ with 2 results in the *true* statement $2 + 5 = 7$. Similarly, 3 is not a solution of $x + 5 = 7$, since replacing $x$ with 3 results in the *false* statement $3 + 5 = 7$.

**Example 1**   Determine whether 6 is a solution of the equation $4(x - 3) = 12$.

*Solution:*   We replace $x$ with 6 in the equation.

$$4(x - 3) = 12$$
$$4(6 - 3) \stackrel{?}{=} 12 \quad \text{Replace } x \text{ with 6.}$$
$$4(3) \stackrel{?}{=} 12$$
$$12 = 12 \quad \text{True.}$$

Since $12 = 12$ is a true statement, 6 *is* a solution of the equation.

**Example 2**   Determine whether $-1$ is a solution of the equation $3y + 1 = 3$.

*Solution:*

$$3y + 1 = 3$$
$$3(-1) + 1 \stackrel{?}{=} 3$$
$$-3 + 1 \stackrel{?}{=} 3$$
$$-2 = 3 \quad \text{False.}$$

Since $-2 = 3$ is false, $-1$ is *not* a solution of the equation.

## B USING THE ADDITION PROPERTY TO SOLVE EQUATIONS

To solve an equation, we will use properties of equality to write simpler equations, all equivalent to the original equation, until the final equation has the form

$$x = \textbf{number} \quad \text{or} \quad \textbf{number} = x$$

---

**Objectives**

**A** Determine whether a given number is a solution of an equation.

**B** Use the addition property of equality to solve equations.

Study Guide    SSM    CD-ROM    Video 3.2

TEACHING TIP

Remind students that throughout algebra, it's important for them to know whether they are working with an equation or an expression. Continually remind them that an equation has an equal sign and an expression does not.

**Practice Problem 1**

Determine whether 4 is a solution of the equation $3(y - 6) = 6$.

**Practice Problem 2**

Determine whether $-2$ is a solution of the equation $-4x - 3 = 5$.

Answers

**1.** no  **2.** yes

Equivalent equations have the same solution, so the word "number" in the equations on the previous page represents the solution of the original equation. The first property of equality to help us write simpler equations is the **addition property of equality**.

---

**ADDITION PROPERTY OF EQUALITY**

Let $a$, $b$, and $c$ represent numbers.
If $a = b$, then

$$a + c = b + c \quad \text{and} \quad a - c = b - c$$

---

In other words, the same number may be added to or subtracted from both sides of an equation without changing the solution of the equation.

A good way to visualize a true equation is to picture a balanced scale. Since the scale is balanced, each side weighs the same amount. Similarly, in a true equation the expressions on each side have the same value. Picturing our balanced scale, if we add the same weight to each side, the scale remains balanced.

**Practice Problem 3**

Solve for $y$: $y - 5 = -3$

TEACHING TIP

When discussing Example 3, ask students why 2 was chosen to be added to both sides.

**Example 3**    Solve for $x$: $x - 2 = 1$

*Solution:*    To solve the equation for $x$, we need to rewrite the equation in the form $x = $ number. In other words, our goal is to get $x$ alone on one side of the equation. To do so, we add 2 to both sides of the equation.

$$x - 2 = 1$$
$$x - 2 + 2 = 1 + 2 \qquad \text{Add 2 to both sides of the equation.}$$
$$x = 3 \qquad \text{Simplify.}$$

To check, we replace $x$ with 3 in the *original* equation.

$$x - 2 = 1 \qquad \text{Original equation}$$
$$3 - 2 \stackrel{?}{=} 1 \qquad \text{Replace } x \text{ with 3.}$$
$$1 = 1 \qquad \text{True.}$$

Since $1 = 1$ is a true statement, 3 is the solution of the equation.

---

**Helpful Hint**

Remember to check the solution in the *original* equation to see that it makes the equation a true statement.

---

Let's visualize how we used the addition property of equality to solve the equation in Example 3. Picture the original equation $x - 2 = 1$ as a balanced scale. The left side of the equation has the same value as the right side.

$x - 2$          $1$

If the same weight is added to each side of a scale, the scale remains balanced. Likewise, if the same number is added to each side of an equation, the left side continues to have the same value as the right side.

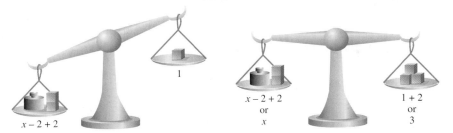

$$x - 2 + 2$$

$$x - 2 + 2$$
or
$$x$$

$$1 + 2$$
or
$$3$$

**Example 4**   Solve: $-8 = x + 1$

*Solution:*   To get $x$ alone on one side of the equation, we subtract 1 from both sides of the equation.

$$-8 = x + 1$$
$$-8 - 1 = x + 1 - 1 \qquad \text{Subtract 1 from both sides.}$$
$$-8 + (-1) = x + 1 + (-1)$$
$$-9 = x \qquad \text{Simplify.}$$

*Check:*   $-8 = x + 1$
$$-8 \stackrel{?}{=} -9 + 1 \qquad \text{Replace } x \text{ with } -9.$$
$$-8 = -8 \qquad \text{True.}$$

The solution is $-9$.

---

**Practice Problem 4**

Solve: $z + 9 = 1$

---

┌─────────────────────────────────────┐

**Helpful Hint**

Remember that we can get the variable alone on either side of the equation. For example, the equations $-9 = x$ and $x = -9$ both have the solution of $-9$.

└─────────────────────────────────────┘

**Example 5**   Solve: $7x = 6x + 4$

*Solution:*   Subtract $6x$ from both sides so that variable terms will be on the same side of the equation.

$$7x = 6x + 4$$
$$7x - 6x = 6x + 4 - 6x \qquad \text{Subtract } 6x \text{ from both sides.}$$
$$1x = 4 \quad \text{or} \quad x = 4 \qquad \text{Simplify.}$$

Check to see that the solution is 4.

If one or both sides of an equation can be simplified, do that first.

**Practice Problem 5**

Solve: $10x = -2 + 9x$

---

**Example 6**   Solve: $y - 5 = -2 - 6$

*Solution:*   First we simplify the right side of the equation.

$$y - 5 = -2 - 6$$
$$y - 5 = -8 \qquad \text{Combine like terms.}$$

Next we get $y$ alone by adding 5 to both sides of the equation.

**Practice Problem 6**

Solve: $x + 6 = 1 - 3$

**Answers**

**4.** $-8$  **5.** $-2$  **6.** $-8$

Copyright 2000 Prentice-Hall, Inc.

## ✓ CONCEPT CHECK

What number should be added to or subtracted from both sides of the equation in order to solve the equation $-3 = y + 2$?

### Practice Problem 7

Solve: $-6y - 1 + 7y = 17$

### Practice Problem 8

Solve: $13x = 4(3x - 1)$

$$y - 5 + 5 = -8 + 5 \quad \text{Add 5 to both sides.}$$
$$y = -3 \quad \text{Simplify.}$$

Check to see that $-3$ is the solution of the equation.

## TRY THE CONCEPT CHECK IN THE MARGIN.

### Example 7

Solve: $5x + 2 - 4x = 7 - 9$

*Solution:*   First we simplify each side of the equation separately.

$$5x + 2 - 4x = 7 - 9$$
$$\underbrace{5x - 4x} + 2 = \underbrace{7 - 9}$$
$$1x + 2 = -2 \quad \text{Combine like terms.}$$

To get $x$ alone on the left side, we subtract 2 from both sides.

$$1x + 2 - 2 = -2 - 2 \quad \text{Subtract 2 from both sides.}$$
$$1x = -4 \quad \text{or} \quad x = -4 \quad \text{Simplify.}$$

Check to verify that $-4$ is the solution.

### Example 8

Solve: $3(3x - 5) = 10x$

*Solution:*   First we multiply on the left side.

$$3(3x - 5) = 10x$$
$$3 \cdot 3x - 3 \cdot 5 = 10x \quad \text{Use the distributive property.}$$
$$9x - 15 = 10x$$

Now we subtract $9x$ from both sides.

$$9x - 15 - 9x = 10x - 9x \quad \text{Subtract } 9x \text{ from both sides.}$$
$$-15 = 1x \quad \text{or} \quad x = -15 \quad \text{Simplify.}$$

**Answers**

**7.** 18    **8.** −4

✓ **Concept Check:** Subtract 2 from both sides.

## Exercise Set 3.2

**A** *Decide whether the given number is a solution of the given equation. See Examples 1 and 2.*

**1.** Is 10 a solution of $x - 8 = 2$?

**2.** Is 9 a solution of $y - 2 = 7$?

**3.** Is 6 a solution of $z + 8 = 14$?

**4.** Is 5 a solution of $n + 11 = 16$?

**5.** Is $-5$ a solution of $x + 12 = 7$?

**6.** Is $-7$ a solution of $a + 23 = 16$?

**7.** Is 8 a solution of $7f = 64 - f$?

**8.** Is 3 a solution of $12 - k = 9$?

**9.** Is 0 a solution of $h - 8 = -8$?

**10.** Is $-2$ a solution of $3 + d = 0$?

**11.** Is 3 a solution of $4c + 2 - 3c = -1 + 6$?

**12.** Is 1 a solution of $2(b - 3) = 10$?

**B** *Solve and check the solution. See Examples 3 through 5.*

**13.** $a + 5 = 23$

**14.** $s - 7 = 15$

**15.** $d - 9 = 17$

**16.** $f + 4 = -6$

**17.** $7 = y - 2$

**18.** $-10 = z - 15$

**19.** $-12 = x + 4$

**20.** $1 = y + 7$

**21.** $3x = 2x + 11$

**22.** $5y = 4y + 12$

**23.** $-4 + y = 2y$

**24.** $8x - 4 = 9x$

*Solve and check the solution. See Examples 6 through 8.*

**25.** $x - 3 = -1 + 4$

**26.** $x + 7 = 2 + 3$

**27.** $y + 1 = -3 + 4$

| ANSWERS | |
|---|---|
| **1.** | yes |
| **2.** | yes |
| **3.** | yes |
| **4.** | yes |
| **5.** | yes |
| **6.** | yes |
| **7.** | yes |
| **8.** | yes |
| **9.** | yes |
| **10.** | no |
| **11.** | yes |
| **12.** | no |
| **13.** | 18 |
| **14.** | 22 |
| **15.** | 26 |
| **16.** | $-10$ |
| **17.** | 9 |
| **18.** | 5 |
| **19.** | $-16$ |
| **20.** | $-6$ |
| **21.** | 11 |
| **22.** | 12 |
| **23.** | $-4$ |
| **24.** | $-4$ |
| **25.** | 6 |
| **26.** | $-2$ |
| **27.** | 0 |

**28.** $y - 8 = -5 - 1$     📼 **29.** $-7 + 10 = m - 5$     **30.** $1 - 8 = n + 2$

**31.** $-2 - 3 = -4 + x$     **32.** $7 - (-10) = x - 5$     **33.** $2(5x - 3) = 11x$

**34.** $6(3x + 1) = 19x$     📼 **35.** $3y = 2(y + 12)$     **36.** $17x = 4(4x - 6)$

**37.** $-8x + 4 + 9x = -1 + 7$     **38.** $3x - 2x + 5 = 5$

**39.** $2 - 2 = 5x - 4x$     **40.** $11 - 15 = 6x - 4 - 5x$

**41.** $7x + 14 - 6x = -4 - 10$     **42.** $-10x + 11x + 5 = 9 - 5$

*Solve each equation. See Examples 3 through 8.*

**43.** $57 = y - 16$     **44.** $56 = -45 + x$     **45.** $67 = z + 67$

**46.** $66 = 8 + z$     **47.** $x + 5 = 4 - 3$     **48.** $x - 7 = -14 + 10$

**49.** $z - 23 = -88$     **50.** $x + 3 = -8$     **51.** $7a + 7 - 6a = 20$

**52.** $-8a - 11 + 9a = -5$     **53.** $-12 + x = -15$     **54.** $10 + x = 12$

📼 **55.** $-8 - 9 = 3x + 5 - 2x$     **56.** $-7 + 10 = 4x - 6 - 3x$

---

**28.** 2

**29.** 8

**30.** −9

**31.** −1

**32.** 22

**33.** −6

**34.** 6

**35.** 24

**36.** −24

**37.** 2

**38.** 0

**39.** 0

**40.** 0

**41.** −28

**42.** −1

**43.** 73

**44.** 101

**45.** 0

**46.** 58

**47.** −4

**48.** 3

**49.** −65

**50.** −11

**51.** 13

**52.** 6

**53.** −3

**54.** 2

**55.** −22

**56.** 9

Name _____

**57.** $8(3x - 2) = 25x$

**58.** $9(4x + 5) = 37x$

**59.** $7x + 7 - 6x = 10$

**60.** $-3 + 5x - 4x = 13$

**61.** $50y = 7(7y + 4)$

**62.** $65y = 8(8y - 9)$

## REVIEW AND PREVIEW

*The trumpeter swan is the largest waterfowl in the United States. Although it was thought to be nearly extinct at the beginning of the twentieth century, recent conservation efforts have been succeeding. Use the bar graph to answer Exercises 63–66. See Section 1.3.*

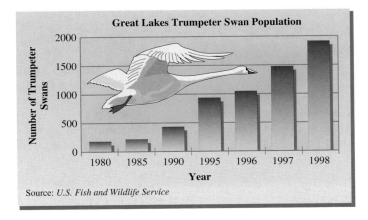

**Great Lakes Trumpeter Swan Population**

Number of Trumpeter Swans

Year

Source: *U.S. Fish and Wildlife Service*

**63.** Estimate the number of trumpeter swans in the Great Lakes region in 1998.

**64.** Estimate the number of trumpeter swans in the Great Lakes region in 1996.

**65.** How many more trumpeter swans in the Great Lakes region were there in 1995 than in 1985?

**66.** Describe any trends shown in this graph.

*Simplify. See Section 1.6.*

**67.** $\dfrac{8}{8}$

**68.** $\dfrac{11}{11}$

---

**57.** −16

**58.** 45

**59.** 3

**60.** 16

**61.** 28

**62.** −72

**63.** 1900

**64.** 1050

**65.** about 700

**66.** answers may vary

**67.** 1

**68.** 1

**69.** 1

**69.** $\dfrac{-3}{-3}$

**70.** $\dfrac{-7}{-7}$

**70.** 1

**71.** In your own words, explain the addition property of equality.

**72.** Write an equation that can be solved using the addition property of equality.

**71.** answers may vary

 **COMBINING CONCEPTS**

*Solve.*

**72.** answers may vary

**73.** $x - 76{,}862 = 86{,}102$

**74.** $-968 + 432 = 86y - 508 - 85y$

*A football team's total offense T is found by adding the total passing yardage P to the total rushing yardage R: T = P + R.*

**73.** 162,964

**75.** During the 1999 football season, the Green Bay Packers' total offense was 5419 yards. The Packers' rushing yardage for the season was 1519 yards. How many yards did the Packers gain by passing during the season? (*Source:* National Football League)

**76.** During the 1999 football season, the Denver Broncos' total offense was 5283 yards. The Broncos' passing yardage for the season was 3419 yards. How many yards did the Broncos gain by rushing during the season? (*Source:* National Football League)

**74.** −28

*In accounting, a company's annual net income I can be computed using the relation I = R − E, where R is the company's total revenues for a year and E is the company's total expenses for the year.*

**75.** 3900 yd

**77.** At the end of fiscal year 1997, Wal-Mart had a net income of $3056 million. During the year, Wal-Mart incurred a total of $103,090 million in expenses. What was Wal-Mart's total revenues for the year? (*Source:* Wal-Mart Stores, Inc.)

**78.** At the end of fiscal year 1997, Home Depot had a net income of $938 million. During the year, Home Depot had a total of $18,615 million in expenses. What was Home Depot's total revenues for the year? (*Source:* Home Depot, Inc.)

**76.** 1864 yd

**77.** $106,146 million

**78.** $19,553 million

**182**

# 3.3 SOLVING EQUATIONS: THE MULTIPLICATION PROPERTY

**Objectives**

**A** Use the multiplication property to solve equations.

**B** Translate word phrases into algebraic expressions.

Study    SSM    CD-ROM    Video
Guide                            3.3

## A USING THE MULTIPLICATION PROPERTY TO SOLVE EQUATIONS

Although the addition property of equality is a powerful tool for helping us solve equations, it cannot help us solve all types of equations. For example, it cannot help us solve an equation such as $2x = 6$. To solve this equation, we use a second property of equality called the **multiplication property of equality**.

---

**MULTIPLICATION PROPERTY OF EQUALITY**

Let $a$, $b$, and $c$ represent numbers and let $c \neq 0$.
If $a = b$, then

$$a \cdot c = b \cdot c \quad \text{and} \quad \frac{a}{c} = \frac{b}{c}$$

---

In other words, both sides of an equation may be multiplied or divided by the same nonzero number without changing the solution of the equation.

Picturing again our balanced scale, if we multiply or divide the weight on each side by the same nonzero number, the scale (or equation) remains balanced.

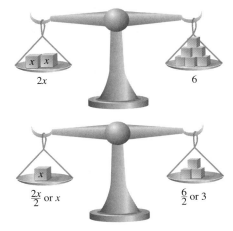

In this section, we will be using the division part of this property only. In other words, we will be dividing both sides of an equation by the same nonzero number only. In the next chapter, we will use the multiplication part of the property.

To solve $2x = 6$ for $x$, we use the multiplication property of equality to divide both sides of the equation by 2, and simplify as follows:

$$2x = 6$$
$$\frac{2 \cdot x}{2} = \frac{6}{2} \qquad \text{Divide both sides by 2.}$$

Then it can be shown that an expression such as $\dfrac{2 \cdot x}{2}$ is equivalent to $\dfrac{2}{2} \cdot x$, so

$$\frac{2 \cdot x}{2} = \frac{6}{2} \quad \text{can be written as} \quad \frac{2}{2} \cdot x = \frac{6}{2}$$
$$1 \cdot x = 3 \quad \text{or} \quad x = 3$$

## Practice Problem 1

Solve for $y$: $3y = -18$

**TEACHING TIP**

Help students remember both the addition and multiplication properties by asking them which property is used to solve each equation below.

$3x = 15$
$3 + x = 15$
$-5x = 20$
$x - 5 = 20$

## Practice Problem 2

Solve for $x$: $-16 = 8x$

## Practice Problem 3

Solve for $y$: $-3y = -27$

## ✓ CONCEPT CHECK

Which operation is appropriate for solving each of the following equations, addition or division?

a. $12 = x - 3$
b. $12 = 3x$

**Answers**

**1.** $-6$  **2.** $-2$  **3.** $9$
✓ **Concept Check: a.** Addition.  **b.** Division.

---

**Example 1**     Solve for $x$: $-5x = 15$

*Solution:*     To get $x$ alone, divide both sides by $-5$.

$$-5x = 15 \qquad \text{Original equation}$$

$$\frac{-5x}{-5} = \frac{15}{-5} \qquad \text{Divide both sides by } -5.$$

$$\frac{-5}{-5} \cdot x = \frac{15}{-5}$$

$$1x = -3 \quad \text{or} \quad x = -3 \qquad \text{Simplify.}$$

To check, replace $x$ with $-3$ in the original equation.

$$-5x = 15 \qquad \text{Original equation}$$

$$-5(-3) \stackrel{?}{=} 15 \qquad \text{Let } x = -3.$$

$$15 = 15 \qquad \text{True.}$$

The solution is $-3$.     ▬▬▬

**Example 2**     Solve for $y$: $-8 = 2y$

*Solution:*     To get $y$ alone, divide both sides of the equation by 2.

$$-8 = 2y$$

$$\frac{-8}{2} = \frac{2y}{2} \qquad \text{Divide both sides by 2.}$$

$$\frac{-8}{2} = \frac{2}{2} \cdot y$$

$$-4 = 1y \quad \text{or} \quad y = -4$$

Check to see that $-4$ is the solution.     ▬▬▬

**Example 3**     Solve for $x$: $-12x = -36$

*Solution:*     Divide both sides of the equation by the coefficient of $x$, which is $-12$.

$$-12x = -36$$

$$\frac{-12x}{-12} = \frac{-36}{-12}$$

$$\frac{-12}{-12} \cdot x = \frac{-36}{-12}$$

$$x = 3$$

To check, replace $x$ with 3 in the original equation.

$$-12x = -36$$

$$-12(3) \stackrel{?}{=} -36 \qquad \text{Replace } x \text{ with 3.}$$

$$-36 = -36 \qquad \text{True.}$$

Since $-36 = -36$ is a true statement, the solution is 3.     ▬▬▬

**TRY THE CONCEPT CHECK IN THE MARGIN.**

We often need to simplify one or both sides of an equation before applying the properties of equality to get the variable alone.

**Example 4**    Solve: $3y - 7y = 12$

*Solution:*    First combine like terms.

$3y - 7y = 12$

$-4y = 12$    Combine like terms.

$\dfrac{-4y}{-4} = \dfrac{12}{-4}$    Divide both sides by $-4$.

$y = -3$    Simplify.

*Check:*    Replace $y$ with $-3$.

$3y - 7y = 12$

$3(-3) - 7(-3) \overset{?}{=} 12$

$-9 + 21 \overset{?}{=} 12$

$12 = 12$    True.

The solution is $-3$.

■

**Example 5**    Solve: $-z = 11 - 5$

*Solution:*    Simplify the right side of the equation first.

$-z = 11 - 5$

$-z = 6$

Next, recall that $-z$ means $-1z$. Divide both sides by $-1$.

$\dfrac{-1z}{-1} = \dfrac{6}{-1}$    Divide both sides by $-1$.

$z = -6$

Check to see that $-6$ is the solution.

■

**Example 6**    Solve: $8x - 3x = 12 - 17$

*Solution:*    First combine like terms on each side of the equation.

$8x - 3x = 12 - 17$

$5x = -5$

$\dfrac{5x}{5} = \dfrac{-5}{5}$    Divide both sides by 5.

$x = -1$    Simplify.

Check to see that the solution is $-1$.

■

**Practice Problem 4**

Solve for $m$: $10 = 2m - 4m$

**Practice Problem 5**

Solve for $a$: $-8 + 6 = -a$

**Practice Problem 6**

Solve for $y$: $-4 - 10 = 4y + 3y$

## B  TRANSLATING WORD PHRASES INTO EXPRESSIONS

A section later in this chapter contains a formal introduction to problem solving. To prepare for this section, let's review writing phrases as algebraic expressions.

**Answers**
**4.** $-5$  **5.** $2$  **6.** $-2$

## Practice Problem 7

Translate each phrase into an algebraic expression. Let $x$ be the unknown number.

a. The product of 5 and a number, decreased by 25

b. Twice the sum of a number and 3

c. The quotient of 39 and twice a number

TEACHING TIP   Classroom Activity

Have students work in groups to translate each expression into a phrase.

$x - 3$

$5(x + 2)$

$4x - 7$

$\dfrac{x}{9}$

## Example 7

Translate each phrase into an algebraic expression. Let $x$ be the unknown number.

a. Twice a number, increased by $-9$

b. Three times the difference of a number and 11

c. The quotient of 5 times a number and 17

*Solution:*

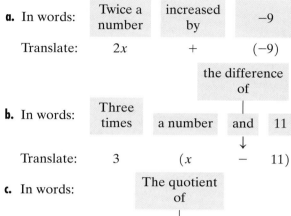

a. In words:  | Twice a number | increased by | $-9$

Translate:   $2x$    $+$    $(-9)$

b. In words:

Three times | a number | the difference of | and | 11

Translate:   $3$   $(x$   $-$   $11)$

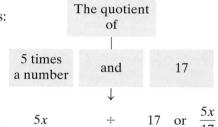

c. In words:  The quotient of

5 times a number | and | 17

$5x$    $\div$    $17$   or   $\dfrac{5x}{17}$

**Answers**

**7. a.** $5x - 25$  **b.** $2(x + 3)$

**c.** $39 \div 2x$ or $\dfrac{39}{2x}$

# Exercise Set 3.3

**A** *Solve each equation. See Examples 1 through 3.*

**1.** $5x = 20$

**2.** $6y = 48$

**3.** $-3z = 12$

**4.** $-2x = 26$

**5.** $4y = 0$

**6.** $8x = -8$

**7.** $2z = -34$

**8.** $7y = 21$

**9.** $-3x = -15$

**10.** $-4z = -12$

*Solve each equation. First combine any like terms on each side of the equation. See Examples 4 through 6.*

**11.** $2w - 12w = 40$

**12.** $8y + y = 45$

**13.** $16 = 10t - 8t$

**14.** $100 = 15y + 5y$

**15.** $2z = 12 - 14$

**16.** $-3x = 11 - 2$

**17.** $4 - 10 = -3z$

**18.** $20 - 2 = -4x + 10$

**19.** $-3x - 3x = 50 - 2$

**20.** $5y - 9y = -14 + (-14)$

*Solve each equation. See Examples 1 through 6.*

**21.** $-10x = 10$

**22.** $2z = -14$

**23.** $5x = -35$

**24.** $-3y = -27$

**25.** $0 = 3x$

**26.** $-3 = -3x$

**27.** $24 = t + 3t$

**28.** $35 = 3r + 2r$

ANSWERS

1. 4

2. 8

3. −4

4. −13

5. 0

6. −1

7. −17

8. 3

9. 5

10. 3

11. −4

12. 5

13. 8

14. 5

15. −1

16. −3

17. 2

18. −2

19. −8

20. 7

21. −1

22. −7

23. −7

24. 9

25. 0

26. 1

27. 6

28. 7

**Name** _____

**29.** $10z - 3z = -63$       **30.** $x + 3x = -48$       **31.** $3z - 10z = -63$

**32.** $6y - y = 20$       **33.** $12 = 13y - 10y$       **34.** $20 = y - 6y$

**35.** $-4x = 20 - (-4)$       **36.** $6x = 5 - 35$       **37.** $18 - 11 = 7x$

**38.** $9 - 14 = 5x$       **39.** $11 - 18 = 7x$       **40.** $14 - 9 = 5x$

**41.** $10p - 11p = 25$       **42.** $5q - 6q = 21$       **43.** $6x - x = 4 - 14$

**44.** $3x + x = 12 - 4$       **45.** $10 = 7t - 12t$       **46.** $-30 = t + 9t$

**47.** $5 - 5 = 3x + 2x$       **48.** $-42 + 20 = -2x + 13x$       **49.** $4r - 9r = -20$

**50.** $11v + 3v = 28$       **51.** $-3x + 4x = -1 + 8$       **52.** $7x + 8x = 12 + (-12)$

**53.** $3w - 12w = -27$       **54.** $5t + 3t = -72$       **55.** $-36 = 9u - 10u$

**56.** $-50 = 4y - 5y$       **57.** $23x - 25x = 7 - 9$       **58.** $8x - 6x = 12 - 22$

**B** *Translate each phrase into an algebraic expression. See Example 7.*

**59.** Seven added to the product of 4 and a number

**60.** The product of 7 and a number, added to 100

**61.** Twice a number, decreased by 17

**62.** A number decreased by 14 and increased by 5 times the number

**63.** The product of −6 and the sum of a number and 15

**64.** Twice the sum of a number and −5

**65.** The quotient of 45 and the product of a number and −5

**66.** The quotient of ten times a number, and −4

## REVIEW AND PREVIEW

*Evaluate each expression for x = −5. See Section 1.8.*

**67.** $3x + 10$

**68.** $40x$

**69.** $\dfrac{x - 5}{2}$

**70.** $7x - 20$

**71.** $\dfrac{3x + 4}{x + 4}$

**72.** $\dfrac{2x - 4}{x - 2}$

## COMBINING CONCEPTS

**73.** Why does the multiplication property of equality not allow us to divide both sides of an equation by zero?

**74.** Is the equation $-x = 6$ solved for the variable? Explain why or why not.

**75.** Solve: $-25x = 900$

**76.** Solve: $3y = -1259 - 3277$

**59.** $4x + 7$

**60.** $100 + 7x$

**61.** $2x - 17$

**62.** $x - 14 + 5x$

**63.** $-6(x + 15)$

**64.** $2(x + (-5))$

**65.** $\dfrac{45}{-5x}$

**66.** $\dfrac{10x}{-4}$

**67.** $-5$

**68.** $-200$

**69.** $-5$

**70.** $-55$

**71.** $11$

**72.** $2$

**73.** answers may vary

**74.** answers may vary

**75.** $-36$

**76.** $-1512$

*The equation d = r · t describes the relationship between distance d in miles, rate r in miles per hour, and time t in hours.*

**77.** The distance between New Orleans, Louisiana, and Kansas City, Missouri, by road, is approximately 780 miles. How long will it take to drive from New Orleans to Kansas City if the driver maintains a speed of 65 miles per hour? (*Source: 2000 World Almanac*)

**78.** The distance between Boston, Massachusetts, and Memphis, Tennessee, by road, is approximately 1320 miles. How long will it take to drive from Boston to Memphis if the driver maintains a speed of 60 miles per hour? (*Source: 2000 World Almanac*)

**79.** The distance between Toledo, Ohio, and Chicago, Illinois, by road, is 232 miles. At what speed should a driver drive if he or she would like to make the trip in 4 hours? (*Source: 2000 World Almanac*)

**80.** The distance between St. Louis, Missouri, and Minneapolis, Minnesota, by road, is approximately 549 miles. If it took 9 hours to drive from St. Louis to Minneapolis, what was the driver's average speed? (*Source: 2000 World Almanac*)

**77.** 12 hr

**78.** 22 hr

**79.** 58 mph

**80.** 61 mph

**Name** _____  **Section** _____ **Date** _____

## INTEGRATED REVIEW—EXPRESSIONS AND EQUATIONS

*Evaluate each expression for $x = -1$ and $y = 3$.*

**1.** $y - x$

**2.** $\dfrac{y}{x}$

**3.** $5x + 2y$

**4.** $\dfrac{y^2 + x}{2x}$

*Simplify each expression by combining like terms.*

**5.** $7x + x$

**6.** $6y - 10y$

**7.** $2a + 5a - 9a - 2$

*Multiply and simplify if possible.*

**8.** $-2(4x + 7)$

**9.** $-(y - 4) + 5(y + 2)$

△ **10.** Find the area of the rectangle.

Rectangle | 3 meters

$(4x - 2)$ meters

*Solve and check.*

**11.** $x + 7 = 20$

**12.** $-11 = x - 2$

**1.** 4

**2.** $-3$

**3.** 1

**4.** $-4$

**5.** $8x$

**6.** $-4y$

**7.** $-2a - 2$

**8.** $-8x - 14$

**9.** $4y + 14$

**10.** $(12x - 6)$ sq. m

**11.** 13

**12.** $-9$

**13.** 5

**14.** 0

**15.** −4

**16.** 5

**17.** −10

**18.** −3

**19.** 6

**20.** 8

**21.** −1

**22.** −24

**Name** _____

**13.** $11x = 55$

**14.** $-7y = 0$

**15.** $12 = 11x - 14x$

**16.** $-3x = -15$

**17.** $x - 12 = -45 + 23$

**18.** $8y + 7y = -45$

**19.** $6 - (-5) = x + 5$

**20.** $-2m = -16$

**21.** $13x = -1 + 12x$

**22.** $6(3x - 4) = 19x$

# 3.4  SOLVING LINEAR EQUATIONS IN ONE VARIABLE

In this chapter, the equations we are solving are called **linear equations in one variable**. For example, an equation such as $5x - 2 = 6x$ is a linear equation in one variable. It is called linear because the exponent on each $x$ is 1 and there is no variable below a fraction bar. It is an equation in one variable because it contains one variable, $x$.

## A  SOLVING EQUATIONS USING THE ADDITION AND MULTIPLICATION PROPERTIES

We will now solve linear equations in one variable using more than one property of equality. To solve an equation such as $2x - 6 = 18$, we first get the variable term $2x$ alone on one side of the equation.

**Example 1**    Solve: $2x - 6 = 18$

*Solution:*    We start by adding 6 to both sides to get the variable term $2x$ alone.

$$2x - 6 = 18$$
$$2x - 6 + 6 = 18 + 6 \quad \text{Add 6 to both sides.}$$
$$2x = 24 \quad \text{Simplify.}$$

To finish solving, we divide both sides by 2.

$$\frac{2x}{2} = \frac{24}{2} \quad \text{Divide both sides by 2.}$$
$$x = 12 \quad \text{Simplify.}$$

> **HELPFUL HINT**
>
> Don't forget to check the proposed solution in the *original* equation.

*Check:*
$$2x - 6 = 18$$
$$2(12) - 6 \stackrel{?}{=} 18 \quad \text{Replace } x \text{ with 12 and simplify.}$$
$$24 - 6 \stackrel{?}{=} 18$$
$$18 = 18 \quad \text{True.}$$

The solution is 12.

**Example 2**    Solve: $20 - x = 21$

*Solution:*    First we get the variable term alone on one side of the equation.

$$20 - x = 21$$
$$20 - x - 20 = 21 - 20 \quad \text{Subtract 20 from both sides.}$$
$$-1x = 1 \quad \text{Simplify. Recall that } -x \text{ means } -1x.$$
$$\frac{-1x}{-1} = \frac{1}{-1} \quad \text{Divide both sides by } -1.$$
$$x = -1 \quad \text{Simplify.}$$

*Check:*
$$20 - x = 21$$
$$20 - (-1) \stackrel{?}{=} 21$$
$$21 = 21 \quad \text{True.}$$

The solution is $-1$.

---

## Objectives

**A** Solve linear equations using the addition and multiplication properties.

**B** Solve linear equations containing parentheses.

**C** Write sentences as equations.

Study Guide | SSM | CD-ROM | Video 3.4

**Practice Problem 1**

Solve: $5y + 2 = 17$

**Practice Problem 2**

Solve: $45 = -10 - y$

TEACHING TIP

You may want to point out to students that their work will be easier to follow and they will make fewer errors if they model their work on the solutions in the text.

**Answers**

**1.** 3   **2.** $-55$

If an equation contains variable terms on both sides, we use the addition property of equality to get all the variable terms on one side and all the constants or numbers on the other side.

### Practice Problem 3

Solve: $7x + 12 = 3x - 4$

**Example 3**   Solve: $3a - 6 = a + 4$

*Solution:*

$$3a - 6 = a + 4$$
$$3a - 6 + 6 = a + 4 + 6 \qquad \text{Add 6 to both sides.}$$
$$3a = a + 10 \qquad \text{Simplify.}$$
$$3a - a = a + 10 - a \qquad \text{Subtract } a \text{ from both sides.}$$
$$2a = 10 \qquad \text{Simplify.}$$
$$\frac{2a}{2} = \frac{10}{2} \qquad \text{Divide both sides by 2.}$$
$$a = 5 \qquad \text{Simplify.}$$

Check to see that the solution is 5.

**TRY THE CONCEPT CHECK IN THE MARGIN.**

### ✓ Concept Check

In Example 3, the solution is 5. If you wanted to check this solution, what equation should you use?

**B**   **SOLVING EQUATIONS CONTAINING PARENTHESES**

If an equation contains parentheses, we must first use the distributive property to remove them.

**Example 4**   Solve: $7(x - 2) = 9x - 6$

*Solution:*   First we apply the distributive property.

$$7(\overset{\frown}{x - 2}) = 9x - 6$$
$$7x - 14 = 9x - 6 \qquad \text{Apply the distributive property.}$$

Next we move variable terms to one side of the equation and constants to the other side.

$$7x - 14 - 9x = 9x - 6 - 9x \qquad \text{Subtract } 9x \text{ from both sides.}$$
$$-2x - 14 = -6 \qquad \text{Simplify.}$$
$$-2x - 14 + 14 = -6 + 14 \qquad \text{Add 14 to both sides.}$$
$$-2x = 8 \qquad \text{Simplify.}$$
$$\frac{-2x}{-2} = \frac{8}{-2} \qquad \text{Divide both sides by } -2.$$
$$x = -4 \qquad \text{Simplify.}$$

Check to see that $-4$ is the solution.

### Practice Problem 4

Solve: $6(a - 5) = 4(a + 1)$

You may want to use the following steps to solve equations.

---

**STEPS FOR SOLVING AN EQUATION**

**Step 1.**   If parentheses are present, use the distributive property.

**Step 2.**   Combine any like terms on each side of the equation.

**Step 3.**   Use the addition property of equality to rewrite the equation so that variable terms are on one side of the equation and constant terms are on the other side.

**Step 4.**   Use the multiplication property of equality to divide both sides by the numerical coefficient of the variable to solve.

**Step 5.**   Check the solution in the *original equation*.

---

Answers

**3.** $-4$   **4.** 17

✓ Concept Check: $3a - 6 = a + 4$

**Example 5**   Solve: $3(2x - 6) + 6 = 0$

*Solution:*   $3(2x - 6) + 6 = 0$

**Step 1.**   $6x - 18 + 6 = 0$   Apply the distributive property.

**Step 2.**   $6x - 12 = 0$   Combine like terms on the left side of the equation.

**Step 3.**   $6x - 12 + 12 = 0 + 12$   Add 12 to both sides.

$6x = 12$   Simplify.

**Step 4.**   $\dfrac{6x}{6} = \dfrac{12}{6}$   Divide both sides by 6.

$x = 2$   Simplify.

*Check:*

**Step 5.**   $3(2x - 6) + 6 = 0$

$3(2 \cdot 2 - 6) + 6 \stackrel{?}{=} 0$

$3(4 - 6) + 6 \stackrel{?}{=} 0$

$3(-2) + 6 \stackrel{?}{=} 0$

$-6 + 6 \stackrel{?}{=} 0$

$0 = 0$   True.

The solution is 2.

## C  WRITING SENTENCES AS EQUATIONS

Next we practice translating sentences into equations. Below are key words and phrases that translate to an equal sign.

| Key Words or Phrases | Examples | Symbols |
|---|---|---|
| **equals** | 3 equals 2 plus 1 | $3 = 2 + 1$ |
| **gives** | the quotient of 10 and $-5$ gives $-2$ | $\dfrac{10}{-5} = -2$ |
| **is/was** | $x$ is 5 | $x = 5$ |
| **yields** | $y$ plus 2 yields 13 | $y + 2 = 13$ |
| **amounts to** | twice $x$ amounts to $-30$ | $2x = -30$ |
| **is equal to** | $-24$ is equal to 2 times $-12$ | $-24 = 2(-12)$ |

**Example 6**   Translate each sentence into an equation.

**a.** The product of 7 and 6 is 42.
**b.** Twice the sum of 3 and 5 is equal to 16.
**c.** The quotient of $-45$ and 5 yields $-9$.

*Solution:*   **a.** In words:

| The product of 7 and 6 | is | 42 |
|---|---|---|
| ↓ | ↓ | ↓ |

Translate:   $7 \cdot 6$   $=$   $42$

---

**Practice Problem 5**

Solve: $4(x + 3) = 12$

TEACHING TIP

Before going over the textbook steps for solving an equation, help students develop these steps on their own. Alternatively, you may want to have students work in groups to make their own steps. Then have them compare their steps to the ones in the book.

**Practice Problem 6**

Translate each sentence into an equation.

a. The difference of 110 and 80 is 30.
b. The product of 3 and the sum of $-9$ and 11 amounts to 6.
c. The quotient of twice 12 and $-6$ yields $-4$.

**Answers**

**5.** 0   **6. a.** $110 - 80 = 30$   **b.** $3(-9 + 11) = 6$
**c.** $\dfrac{2(12)}{-6} = -4$

TEACHING TIP    Classroom Activity

Have students work in groups to translate each equation into a sentence.

$14 \cdot 6 - 3 = 81$

$7(12 - 4) = 84 - 28$

$\dfrac{45}{360} = 0.125$

**b.** In words:    | Twice | the sum of 3 and 5 | is equal to | 16 |

Translate:    $2$        $(3 + 5)$        $=$        $16$

**c.** In words:    | The quotient of −45 and 5 | yields | −9 |

Translate:    $\dfrac{-45}{5}$        $=$        $-9$

---

# CALCULATOR EXPLORATIONS
## CHECKING EQUATIONS

A calculator can be used to check possible solutions of equations. To do this, replace the variable by the possible solution and evaluate each side of the equation separately. For example, to see whether 7 is a solution of the equation $52x = 15x + 259$, replace $x$ with 7 and use your calculator to evaluate each side separately.

Equation:    $52x = 15x + 259$

$52 \cdot 7 \overset{?}{=} 15 \cdot 7 + 259$    Replace $x$ with 7.

Evaluate left side:    | 52 | × | 7 | = |(or | ENTER |).    Display:    | 364 |.

Evaluate right side:    | 15 | × | 7 | + | 259 | = |(or | ENTER |).    Display:    | 364 |.

Since the left side equals the right side, 7 is a solution of the equation $52x = 15x + 259$.

*Use a calculator to determine whether the numbers given are solutions of each equation.*

**1.** $76(x - 25) = -988$;   12   yes

**2.** $-47x + 862 = -783$;   35   yes

**3.** $x + 562 = 3x + 900$;   −170   no

**4.** $55(x + 10) = 75x + 910$;   −18   yes

**5.** $29x - 1034 = 61x - 362$;   −21   yes

**6.** $-38x + 205 = 25x + 120$;   25   no

# EXERCISE SET 3.4

**A** *Solve each equation. See Examples 1 through 3.*

**1.** $2x - 6 = 0$  **2.** $3y - 12 = 0$  **3.** $5n + 10 = 0$  **4.** $4z + 8 = 0$

**5.** $6 - n = 10$  **6.** $7 - y = 9$  **7.** $10x + 15 = 6x + 3$

**8.** $5x - 3 = 2x - 18$  **9.** $3x - 7 = 4x + 5$  **10.** $3x + 1 = 8x - 4$

**B** *Solve each equation. See Examples 4 and 5.*

**11.** $3(x - 1) = 12$  **12.** $2(x + 5) = -8$  **13.** $-2(y + 4) = 2$

**14.** $-1(y + 3) = 10$  **15.** $35 = 17 + 3(x - 2)$  **16.** $22 - 42 = 4(x - 1)$

**A** **B** *Solve each equation. See Examples 1 through 5.*

**17.** $8 - t = 3$  **18.** $6 - x = 4$  **19.** $0 = 4x - 4$

**20.** $0 = 5y + 5$  **21.** $2n + 8 = 0$  **22.** $8w - 40 = 0$

**23.** $7 = 4c - 1$  **24.** $9 = 2b - 5$  **25.** $3r + 4 = 19$

**ANSWERS**

1. 3
2. 4
3. −2
4. −2
5. −4
6. −2
7. −3
8. −5
9. −12
10. 1
11. 5
12. −9
13. −5
14. −13
15. 8
16. −4
17. 5
18. 2
19. 1
20. −1
21. −4
22. 5
23. 2
24. 7
25. 5

**26.** 9

**27.** −3

**28.** −3

**29.** 2

**30.** 4

**31.** −2

**32.** −3

**33.** 3

**34.** 6

**35.** 6

**36.** 5

**37.** −10

**38.** −4

**39.** −4

**40.** −2

**41.** −1

**42.** −1

**43.** 4

**44.** 5

**45.** −4

**46.** 3

**47.** 3

**48.** −2

**49.** −1

**50.** 1

**51.** 0

**52.** 0

**Name** _____

**26.** $5m + 1 = 46$

**27.** $2x - 1 = -7$

**28.** $3t - 2 = -11$

**29.** $2 = 3z - 4$

**30.** $4 = 4p - 12$

**31.** $5x - 2 = -12$

**32.** $7y - 3 = -24$

**33.** $-7c + 1 = -20$

**34.** $-2b + 5 = -7$

**35.** $9 = 4x - 15$

**36.** $16 = 6y - 14$

**37.** $8m + 79 = -1$

**38.** $9a + 29 = -7$

**39.** $10 + 4v = -6$

**40.** $13 + 6b = 1$

**41.** $-5 = -13 - 8k$

**42.** $-7 = -17 - 10d$

**43.** $4x + 3 = 2x + 11$

**44.** $6y - 8 = 3y + 7$

**45.** $-2y - 10 = 5y + 18$

**46.** $7n + 5 = 12n - 10$

**47.** $-8n + 1 = -6n - 5$

**48.** $10w + 8 = w - 10$

**49.** $9 - 3x = 14 + 2x$

**50.** $4 - 7m = -3m$

**51.** $2(y - 3) = y - 6$

**52.** $3(z + 2) = 5z + 6$

**53.** $2t - 1 = 3(t + 7)$

**54.** $4 + 3c = 2(c + 2)$

**55.** $3(5c - 1) - 2 = 13c + 3$

**56.** $4(3t + 4) - 20 = 3 + 5t$

**57.** $10 + 5(z - 2) = 4z + 1$

**58.** $14 + 4(w - 5) = 6 - 2w$

**59.** $7(6 + w) = 6(2 + w)$

**60.** $6(5 + c) = 5(c - 4)$

**C**  *Write each sentence as an equation. See Example 6.*

**61.** The sum of $-42$ and 16 is $-26$.

**62.** The difference of $-30$ and 10 equals $-40$.

**63.** The product of $-5$ and $-29$ gives 145.

**64.** The quotient of $-16$ and 2 yields $-8$.

**65.** Three times the difference of $-14$ and 2 amounts to $-48$.

**66.** Negative 2 times the sum of 3 and 12 is $-30$.

**67.** The quotient of 100 and twice 50 is equal to 1.

**68.** Seventeen subtracted from $-12$ equals $-29$.

---

**53.** $-22$

**54.** $0$

**55.** $4$

**56.** $1$

**57.** $1$

**58.** $2$

**59.** $-30$

**60.** $-50$

**61.** $-42 + 16 = -26$

**62.** $-30 - 10 = -40$

**63.** $-5(-29) = 145$

**64.** $\dfrac{-16}{2} = -8$

**65.** $3(-14 - 2) = -48$

**66.** $-2(3 + 12) = -30$

**67.** $\dfrac{100}{2(50)} = 1$

**68.** $-12 - 17 = -29$

**Name** _____

## REVIEW AND PREVIEW

*Evaluate each expression for $x = 3$, $y = -1$, and $z = 0$. See Section 2.5.*

**69.** $x^3 - 2xy$    **70.** $y^3 + 3xyz$    **71.** $y^5 - 4x^2$    **72.** $(-y)^3 + 3xyz$

**73.** $(2x - y)^2$    **74.** $(6z - 5y)^3$    **75.** $4xy + 6y^2$    **76.** $2x^2 - 7z$

## COMBINING CONCEPTS

*Solve.*

**77.** $(-8)^2 + 3x = 5x + 4^3$    **78.** $3^2 \cdot x = (-9)^3$

**79.** $2^3(x + 4) = 3^2(x + 4)$    **80.** $x + 45^2 = 54^2$

**81.** A classmate tries to solve $3x = 39$ by subtracting 3 from both sides of the equation. Is this correct? Why or why not?

## 3.5   LINEAR EQUATIONS IN ONE VARIABLE AND PROBLEM SOLVING

### A   WRITING SENTENCES AS EQUATIONS

Now that we have practiced solving equations for a variable, we can extend these problem-solving skills considerably. We begin by reviewing how to translate sentences into algebraic equations using the following key words and phrases as a guide.

| Addition | Subtraction | Multiplication | Division | Equal Sign |
|---|---|---|---|---|
| sum | difference | product | quotient | equals |
| plus | minus | times | divided by | gives |
| added to | subtracted from | multiply | into | is/was |
| more than | less than | twice | per | yields |
| increased by | decreased by | of | | amounts to |
| total | less | double | | is equal to |

**Example 1**   Write each sentence as an equation. Use $x$ to represent "a number."

   **a.** Nine more than a number is 5.
   **b.** Twice a number equals −10.
   **c.** A number minus 6 amounts to 168.
   **d.** Three times the sum of a number and 5 is −30.
   **e.** The quotient of 8 and twice a number is equal to 2.

*Solution:*   **a.** In words:   Nine   more than   a number   is   5

            Translate:   9   +   $x$   =   5

   **b.** In words:   Twice a number   equals   −10

            Translate:   $2x$   =   −10

   **c.** In words:   A number   minus   6   amounts to   168

            Translate:   $x$   −   6   =   168

   **d.** In words:   Three times   the sum of a number and 5   is   −30

            Translate:   3   $(x + 5)$   =   −30

   **e.** In words:   The quotient of

            Translate:   8   and   twice a number   is equal to   2

                  8   ÷   $2x$   =   2

$$\text{or } \frac{8}{2x} = 2$$

**Objectives**

 **A** Translate sentences into mathematical equations.

 **B** Use problem-solving steps to solve problems.

Study Guide   SSM   CD-ROM   Video 3.5

**Practice Problem 1**

Write each sentence as an equation. Use $x$ to represent "a number."

a. Five times a number is 20.

b. The sum of a number and −5 yields 14.

c. Ten subtracted from a number amounts to −23.

d. Five times a number added to 7 is equal to −8.

e. The quotient of 6 and the sum of a number and 4 gives 1.

**Answers**

**1. a.** $5x = 20$   **b.** $x + (-5) = 14$
**c.** $x - 10 = -23$   **d.** $7 + 5x = -8$
**e.** $\dfrac{6}{x + 4} = 1$

## B USE PROBLEM-SOLVING STEPS TO SOLVE PROBLEMS

Our main purpose for studying arithmetic and algebra is to solve problems. In previous sections, we have prepared for problem solving by writing phrases as algebraic expressions and sentences as equations. We now draw upon this experience as we solve problems. The following problem-solving steps will be used throughout this text.

---

**PROBLEM-SOLVING STEPS**

1.  UNDERSTAND the problem. During this step, become comfortable with the problem. Some ways of doing this are:
    *   Read and reread the problem.
    *   Choose a variable to represent the unknown.
    *   Construct a drawing.
    *   Propose a solution and check it. Pay careful attention to how you check your proposed solution. This will help when writing an equation to model the problem.
2.  TRANSLATE the problem into an equation.
3.  SOLVE the equation.
4.  INTERPRET the results. *Check* the proposed solution in the stated problem and *state* your conclusion.

---

**Practice Problem 2**

Translate "the sum of a number and 2 equals 6 added to three times the number" into an equation and solve.

### Example 2    Finding an Unknown Number

Twice a number, added to 3, is the same as the number minus 6. Find the number.

*Solution:*

1.  UNDERSTAND the problem. To do so, we read and reread the problem. To help us understand the problem better, let's propose a solution and check it. Suppose that the unknown number is 10. From the stated problem, "twice 10, added to 3" is 23, but "10 minus 6" is 4. Since 23 is not equal to 4, the solution is not 10. The purpose of proposing a solution is not to guess correctly but to better understand the problem.

    Now, let's assign a variable to the unknown. We will let

    $x$ = the unknown number.

2.  TRANSLATE the problem into an equation.

| In words: | Twice a number | added to 3 | is the same as | the number minus 6 |
|---|---|---|---|---|
| | ↓ | ↓ | ↓ | ↓ |
| Translate: | $2x$ | $+ 3$ | $=$ | $x - 6$ |

3.  SOLVE the equation. To solve the equation, we first subtract $x$ from both sides.

$$2x + 3 = x - 6$$
$$2x + 3 - x = x - 6 - x \qquad \text{Subtract } x \text{ from both sides.}$$
$$x + 3 = -6 \qquad \text{Simplify.}$$
$$x + 3 - 3 = -6 - 3 \qquad \text{Subtract 3 from both sides.}$$
$$x = -9 \qquad \text{Simplify.}$$

**Answer**

**2.** −2

**4.** INTERPRET the results. First, *check* the proposed solution in the stated problem. Twice "−9" is −18 and −18 + 3 is −15. This is equal to the number minus 6 or "−9" − 6 or −15. Then *state* your conclusion: The unknown number is −9.

**TRY THE CONCEPT CHECK IN THE MARGIN.**

**Example 3**    Determining Voter Counts

In an election, the incumbent city council member, Laura Hartley, received 275 *more* votes than her challenger. If a total of 7857 votes were cast in the election, find how many votes the incumbent, Laura Hartley, received.

*Solution:*    **1.** UNDERSTAND the problem. We read and reread the problem. Then we assign a variable to an unknown. We use this variable to represent any other unknown quantities. We let

$x$ = the number of challenger votes

Then

$x + 275$ = the number of incumbent votes

since she received 275 more votes.

**2.** TRANSLATE the problem into an equation.

In words:
| challenger votes | + | incumbent votes | = | total votes |
|---|---|---|---|---|
| ↓ | | ↓ | | ↓ |
| $x$ | + | $x + 275$ | = | 7857 |

Translate:

**3.** SOLVE the equation.

$$x + x + 275 = 7857$$
$$2x + 275 = 7857 \qquad \text{Combine like terms.}$$
$$2x + 275 - 275 = 7857 - 275 \qquad \text{Subtract 275 from both sides.}$$
$$2x = 7582 \qquad \text{Simplify.}$$
$$\frac{2x}{2} = \frac{7582}{2} \qquad \text{Divide both sides by 2.}$$
$$x = 3791 \qquad \text{Simplify.}$$

**4.** INTERPRET the results. First *check* the proposed solution in the stated problem. Since $x$ represents the number of votes the challenger received, the challenger received 3791 votes. The incumbent received $x + 275 = 3791 + 275 = 4066$ votes. To check, notice that the total number of challenger votes and incumbent votes is $3791 + 4066 = 7857$ votes, the given total of votes cast. Also, 4066 is 275 more votes than 3791, so the solution checks. Then, *state* your conclusion: The incumbent, Laura Hartley, received 4066 votes.

✓ **CONCEPT CHECK**

Suppose you have solved an equation involving perimeter to find the length of a rectangular table. Explain why you would want to recheck your math if you obtain the result of −5.

**Practice Problem 3**

At a recent United States/Japan summit meeting, 121 delegates attended. If the United States sent 19 more delegates than Japan, find how many the United States sent.

**Answers**

**3.** 70 delegates

✓ **Concept Check:** Length cannot be negative.

## ✓ CONCEPT CHECK

When solving the problem given in Example 3, why should you be skeptical if you obtain an answer of 15,714 votes for the incumbent?

### Practice Problem 4

A woman's $21,000 estate is to be divided so that her husband receives twice as much as her son. How much will each receive?

TEACHING TIP    Classroom Activity

Have students work in groups to write word problems similar to Example 2, Example 3, and Example 4. Then have them exchange their word problems with another group to solve.

TRY THE CONCEPT CHECK IN THE MARGIN.

**Example 4**    Calculating Separate Costs

Leo Leal sold a used computer system and software for $2100, receiving four times as much money for the computer system as for the software. Find the price of each.

*Solution:*

1. UNDERSTAND the problem. We read and reread the problem. Then we assign a variable to an unknown. We use this variable to represent any other unknown quantities. We let

   $x$ = the software price

   $4x$ = the computer system price

2. TRANSLATE the problem into an equation.

| In words: | Software price | and | computer price | is | 2100 |
|-----------|---------------|-----|----------------|-----|------|
|           | ↓             | ↓   | ↓              | ↓   | ↓    |
| Translate: | $x$          | $+$ | $4x$           | $=$ | 2100 |

3. SOLVE the equation.

   $x + 4x = 2100$

   $5x = 2100$    Combine like terms.

   $\dfrac{5x}{5} = \dfrac{2100}{5}$    Divide both sides by 5.

   $x = 420$    Simplify.

4. INTERPRET the results. *Check* the proposed solution in the stated problem. The software sold for $420. The computer system sold for $4x = 4(\$420) = \$1680$. Since $\$420 + \$1680 = \$2100$, the total price, and $1680 is four times $420, the solution checks. *State* your conclusion: The software sold for $420 and the computer system sold for $1680.

### Answers

**4.** husband: $14,000; son $7000

✓ **Concept Check:** Answers may vary.

# EXERCISE SET 3.5

**A** *Write each sentence as an equation. Use x to represent "a number." See Example 1.*

1. A number added to −5 is −7.

2. Five subtracted from a number equals 10.

3. Three times a number yields 27.

4. The quotient of 8 and a number is −2.

5. A number subtracted from −20 amounts to 104.

6. Two added to twice a number gives −14.

7. Twice the sum of a number and −1 is equal to 50.

8. Three times the difference of 7 and a number amounts to −40.

**B** *Solve. See Example 2.*

9. Three times a number, added to 9, is 33. Find the number.

10. Twice a number, subtracted from 60, is 20. Find the number.

11. The product of 9 and a number gives 54. Find the number.

12. The product of 7 and a number is 14. Find the number.

13. The sum of 3, 4, and a number amounts to 16. Find the number.

14. A number less 5 is 11. Find the number.

15. Seventy-two is 8 times a number added to 24. Find the number.

16. The product of 11 and a number is 121. Find the number.

17. The difference of a number and 3 is the same as 45 less the number. Find the number.

18. Sixty-four less half of 64 is equal to the sum of some number and 20. Find the number.

**19.** Three times the difference of some number and 5 amounts to 9. Find the number.

**20.** Five times the sum of some number plus 2 is equal to 8 times the number, minus 11. Find the number.

**21.** Eight decreased by some number equals the quotient of 15 and 5. Find the number.

**22.** Ten less some number is twice the sum of that number and 5. Find the number.

**23.** Thirteen added to the product of 3 and some number amounts to 5 times the number added to 3.

**24.** The product of 4 and a number is the same as 30 less twice that same number.

**25.** Five times a number, less 40, is 8 more than the number.

**26.** Thirty less a number is equal to 3 times the sum of the number and 6.

**27.** The difference of a number and 3 is equal to the quotient of 10 and 5.

**28.** The product of a number and 3 is twice the sum of that number and 5.

*Solve. See Examples 3 and 4*

**29.** In the 1996 presidential election, Bill Clinton received 220 more electoral votes than Bob Dole. If a total of 538 electoral votes were cast for the two candidates, find how many votes each candidate received. (*Source:* Voter News Service)

**30.** California has 22 more electoral votes for president than Texas. If the total number of electoral votes for these two states is 86, find the number for each state. (*Source: The World Almanac 2000*)

**31.** Mark and Stuart Martin collect foreign coins. Mark has twice the number of coins Stuart has. Together they have 120 foreign coins. Find how many coins Mark has.

**32.** Heather and Mary Gamber collect baseball cards. Heather's collection is three times as large as Mary's. Together they have 400 baseball cards. Find the number of baseball cards in Heather's collection.

**Name** _____

**33.** A Toyota Camry is traveling twice as fast as a Dodge truck. If their combined speed is 105 miles per hour, find the speed of the car and find the speed of the truck.

**34.** A crow will eat five more ounces of food a day than a finch. If together they eat 13 ounces of food, find how many ounces of food the crow consumes and how many ounces of food the finch consumes.

**35.** Anthony Tedesco sold his used mountain bike and accessories for $270. If he received five times as much money for the bike as he did for the accessories, find how much money he received for the bike.

**36.** A tractor and a plow attachment are worth $1200. The tractor is worth seven times as much money as the plow. Find the value of the tractor and the value of the plow.

**37.** The two NCAA stadiums with the largest capacities are Michigan Stadium (Univ. of Michigan) and Neyland Stadium (Univ. of Tennessee). Michigan Stadium has a capacity of 4647 more than Neyland. If the combined capacity for the two stadiums is 210,355, find the capacity for each stadium. (*Source:* National Collegiate Athletic Association)

**38.** A NHRA top fuel dragster has a top speed of 95 mph faster than an Indy Racing League car. If the combined top speeds for these two cars is 565 mph, find the top speed of each car. (*Source:* USA Today, 9/2/99)

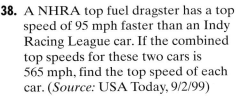

**39.** California contains the largest state population of native Americans. This population is three times the native American population of Washington state. If the total of these two populations is 412 thousand find the native American population in each of these two states. (*Source:* U.S. Census Bureau)

**40.** In 2020, China is projected to be the country with the greatest number of visiting tourists. This number is twice the number of tourists projected for Spain. If the total number of tourists for these two countries is projected to be 210 million, find the number projected for each. (*Source: The State of the World Atlas* by Dan Smith)

**33.** truck, 35 mph; car, 70 mph

**34.** crow, 9 oz; finch, 4 oz

**35.** $225

**36.** plow, $150; tractor, $1050

**37.** Michigan Stadium, 107,501; Neyland Stadium, 102,854

**38.** Top fuel dragster, 330 mph; Indy Racing League, car 235 mph

**39.** California, 309 thousand; Washington, 103 thousand

**40.** China, 140 million; Spain, 70 million

**Name** _____

**41.** During the 2000 Women's NCAA Division I basketball championship game, the Connecticut Huskies scored 19 more points than the Tennessee Lady Volunteers. Together, both teams scored a total of 123 points. How many points did the 2000 Champion Connecticut Huskies score during this game? (*Source:* National Collegiate Athletic Association)

**42.** During the 2000 Men's NCAA Division I basketball championship game, the Florida Gators scored 13 fewer points than the Michigan State Spartans. The Gators scored 76 points. How many points did the 2000 Champion Michigan State Spartans score during this game? (*Source:* National Collegiate Athletic Association)

## REVIEW AND PREVIEW

*Round each whole number to the given place. See Section 1.4.*

**43.** 586 to the nearest ten

**44.** 82 to the nearest ten

**45.** 1026 to the nearest hundred

**46.** 52,333 to the nearest thousand

**47.** 2986 to the nearest thousand

**48.** 101,552 to the nearest hundred

## COMBINING CONCEPTS

**49.** Solve Example 3 again, but this time let $x$ be the number of incumbent votes. Did you get the same results? Explain why or why not.

*In real estate, a house's selling price P is found by adding the real estate agent's commission C to the amount A that the seller of the house receives: $P = A + C$.*

**50.** Brianna Morley's house sold for $230,000. Her real estate agent received a commission of $13,800. How much did Brianna receive? (*Hint:* Substitute the known values into the equation, then solve the equation for the remaining unknown.)

**51.** Duncan Bostic plans to use a real estate agent to sell his house. He hopes to sell the house for $165,000 and keep $156,750 of that. If everything goes as he has planned, how much will his real estate agent receive as a commission?

_In retailing, the retail price P of an item can be computed using the equation $P = C + M$, where C is the wholesale cost of the item and M is the amount of markup._

**52.** The retail price of a computer system is $999 after a markup of $450. What is the wholesale cost of the computer system? (_Hint:_ Substitute the known values into the equation, then solve the equation for the remaining unknown.)

**53.** A retailer paid $12 for a blouse she plans to sell for a retail price of $29. What is the markup on the blouse?

## Internet Excursions

Go to http://www.prenhall.com/ martin-gay

Project Vote Smart is a nonpartisan, nonprofit organization that strives to provide the American public with unbiased information about more than 13,000 of their elected public officials as well as candidates running for public office. The Project Vote Smart goal is to help American voters make informed choices.

This World Wide Web address will provide you with access to Project Vote Smart's listing of links to official state election information, or a related site. Use this listing to connect to election result information for your state.

**54.** Use election results from your state's most recent governor's race to write a problem similar to Exercises 29 and 30. Exchange problems with a classmate to solve.

**55.** Using election results from another recent election in your state, write another problem like Exercises 29 and 30. Exchange with another student in your class to solve.

# Focus on the Real World

### SCALES

In this chapter, we have been using a scale to help illustrate true equations and the properties of equality. Scales of different types are used everyday to measure the mass or weight of objects. The scale shown is called an equal-arm balance. It was developed by the ancient Egyptians around 2500 B.C. to assist in everyday business. To use it, an object with an unknown mass is placed in one pan and objects with known masses are placed in the other pan one at a time, until the arm to which the pans are attached is perfectly level. When the pans are balanced in this way, the user knows that the object with the unknown mass has the same mass as the sum of the known masses in the other pan.

The ancient Romans invented a different kind of scale, called a steelyard, about 2000 years ago. The Romans' steelyard used a pan to hold the unknown mass on one end of the scale's arm and a known mass that slid along the other end of the arm. The position of the sliding mass over markings along the arm indicated the mass of the object in the pan. A variation of the steelyard can still be seen today in many doctors' offices.

Today, most of the scales we use on a daily basis measure weights and masses using springs, pendulums, or changes in electrical resistance. However, the most accurate scales in the world are based on the same balancing principles used in the Egyptians' equal-arm balance and the Romans' steelyard. These highly precise scales can measure accurately to a millionth of a gram.

### GROUP ACTIVITY

Borrow an equal-arm balance or construct a simple one of your own. Use the scale to investigate and demonstrate the properties of equality.

# CHAPTER 3 ACTIVITY
## USING EQUATIONS

*This activity may be completed by working in groups or individually.*

A hospital nurse working second shift has been keeping track of the fluid intake and output on the chart of one of her patients, Mr. Ramirez. At the end of her shift, she totals the intakes and outputs and returns Mr. Ramirez's chart to the floor nurses' station.

Suppose you are one of the night nurses for this floor and have been assigned to Mr. Ramirez. Shortly after you come on duty, someone at the nurses' station spills coffee over several patients' charts, including the one for Mr. Ramirez. After blotting up the coffee on the chart, you notice that several entries on his chart have been smeared by the coffee spill.

---

### MEDICAL CHART

Patient: Juan Ramirez                                    Room: 314

Shift: Second                    Nurse's Initials: SRJ                    Date: 3/27

*Fluid Intake* (cubic centimeters):

| | | | | | Totals |
|---|---|---|---|---|---|
| Blood | 500 | | | | 500 |
| Intravenous | 250 | | | | 500 |
| Oral Fluid | 300 | 150 | | 100 | 900 |
| Oral Meds | 4 doses of __ cc each | | | | 200 |
| **TOTAL** | | | | | 2100 |

*Fluid Output* (cubic centimeters):

| | | | | Totals |
|---|---|---|---|---|
| Urine | 240 | 310 | | |
| Emesis | 120 | | | 120 |
| Irrigation | 50 | 25 | | 75 |
| **TOTAL** | | | | 815 |

---

1. On the Fluid Intake chart, is it possible to tell if Mr. Ramirez was given blood more than once? If so, how many times was he given blood and how can you tell?

2. On the Fluid Intake chart, is it possible to tell if Mr. Ramirez was given intravenous fluid more than once? Is it possible to tell exactly how many times Mr. Ramirez was given intravenous fluid? Explain.

3. For the Oral Fluid row of the Fluid Intake chart, notice that the third entry in that row is obliterated by the coffee stain. Let the variable *x* represent this third entry. Write an equation that describes the total amount of oral fluids Mr. Ramirez received during the second shift. Then solve the equation for *x* to find the obliterated entry.

4. For the Oral Meds row of the Fluid Intake chart, notice that the second-shift nurse indicated that Mr. Ramirez was given four equal doses of oral medication. We cannot see the size of each dose, but the chart shows that he received a total of 200 cc. Write and solve an equation to find the size of each dose.

5. Using the Fluid Output chart, write and solve an equation to find the total urine output.

6. Using the Fluid Output chart and the result from Question 5, write and solve an equation to find the obliterated entry in the Urine output row.

# CHAPTER 3 HIGHLIGHTS

| DEFINITIONS AND CONCEPTS | EXAMPLES |
|---|---|

## SECTION 3.1 SIMPLIFYING ALGEBRAIC EXPRESSIONS

The addends of an algebraic expression are called the **terms** of the expression.

$$5x^2 + (-4x) + (-2)$$

3 terms

The number factor of a variable term is called the **numerical coefficient**.

| Term | Numerical Coefficient |
|---|---|
| $7x$ | 7 |
| $-6y$ | $-6$ |
| $x$ or $1x$ | 1 |

Terms that are exactly the same, except that they may have different numerical coefficients, are called **like terms**.

$$5x + 11x = (5 + 11)x = 16x$$

like terms

An algebraic expression is **simplified** when all like terms have been **combined**.

$$y - 6y = (1 - 6)y = -5y$$

Use the **distributive property** to multiply an algebraic expression by a term.

Simplify: $-4(x + 2) + 3(5x - 7)$
$$= -4(x) + (-4)(2) + 3(5x) + 3(-7)$$
$$= -4x + (-8) + 15x + (-21)$$
$$= 11x + (-29) \quad \text{or} \quad 11x - 29$$

## SECTION 3.2 SOLVING EQUATIONS: THE ADDITION PROPERTY

ADDITION PROPERTY OF EQUALITY

Let $a$, $b$, and $c$ represent numbers.

If $a = b$, then

$a + c = b + c$ and $a - c = b - c$

In other words, the same number may be added to or subtracted from both sides of an equation without changing the solution of the equation.

Solve for $x$: $\quad x + 8 = 2 + (-1)$
| | |
|---|---|
| $x + 8 = 1$ | Combine like terms. |
| $x + 8 - 8 = 1 - 8$ | Subtract 8 from both sides. |
| $x = -7$ | Simplify. |

The solution is $-7$.

## SECTION 3.3 SOLVING EQUATIONS: THE MULTIPLICATION PROPERTY

MULTIPLICATION PROPERTY OF EQUALITY

Let $a$, $b$, and $c$ represent numbers and let $c \neq 0$.

If $a = b$, then

$a \cdot c = b \cdot c$ and $\dfrac{a}{c} = \dfrac{b}{c}$

In other words, both sides of an equation may be multiplied or divided by the same nonzero number without changing the solution of the equation.

Solve for $y$: $\quad y - 7y = 30$
| | |
|---|---|
| $-6y = 30$ | Combine like terms. |
| $\dfrac{-6y}{-6} = \dfrac{30}{-6}$ | Divide by $-6$. |
| $y = -5$ | Simplify. |

The solution is $-5$.

| SECTION 3.4    SOLVING LINEAR EQUATIONS IN ONE VARIABLE | |
|---|---|
| STEPS FOR SOLVING AN EQUATION IN $x$ | Solve for $x$:    $5(3x - 1) + 15 = -5$ |

**STEPS FOR SOLVING AN EQUATION IN $x$**

**Step 1.** If parentheses are present, use the distributive property to remove them.

**Step 2.** Combine any like terms on each side of the equation.

**Step 3.** Use the addition property of equality to rewrite the equation so that variable terms are on one side of the equation and constant terms are on the other side.

**Step 4.** Use the multiplication property of equality to divide both sides by the numerical coefficient of $x$ to solve.

**Step 5.** Check the solution in the *original equation*.

Solve for $x$:    $5(3x - 1) + 15 = -5$

**Step 1.**    $15x - 5 + 15 = -5$    Apply the distributive property.

**Step 2.**    $15x + 10 = -5$    Combine like terms.

**Step 3.**    $15x + 10 - 10 = -5 - 10$    Subtract 10.

$15x = -15$

**Step 4.**    $\dfrac{15x}{15} = \dfrac{-15}{15}$    Divide by 15.

$x = -1$

**Step 5.**    Check to verify that $-1$ is the solution.

---

**SECTION 3.5    LINEAR EQUATIONS IN ONE VARIABLE AND PROBLEM SOLVING**

PROBLEM-SOLVING STEPS

The incubation period for a golden eagle is three times the incubation period for a hummingbird. If the total of their incubation periods is 60 days, find the incubation period for each bird. (*Source: Wildlife Fact File*, International Masters Publishers)

1. UNDERSTAND the problem. Some ways of doing this are:
   - Read and reread the problem.
   - Construct a drawing.
   - Assign a variable to an unknown in the problem.
   - Propose a solution and check.

2. TRANSLATE the problem into an equation.

1. UNDERSTAND the problem. Then assign a variable. Let

   $x =$ incubation period of a hummingbird
   $3x =$ incubation period of a golden eagle

2. TRANSLATE.

| Incubation of hummingbird | + | incubation of golden eagle | is | 60 |
|---|---|---|---|---|
| ↓ | | ↓ | ↓ | ↓ |
| $x$ | $+$ | $3x$ | $=$ | $60$ |

3. SOLVE the equation.

3. SOLVE.

$x + 3x = 60$
$4x = 60$
$\dfrac{4x}{4} = \dfrac{60}{4}$
$x = 15$

4. INTERPRET the results. *Check* the proposed solution in the stated problem and *state* your conclusion.

4. INTERPRET the solution in the stated problem. The incubation period for the hummingbird is 15 days. The incubation period for the golden eagle is $3x = 3 \cdot 15 = 45$ days. Since 15 days + 45 days = 60 days and 45 is 3(15), the solution checks. *State* your conclusion: The incubation period for the hummingbird is 15 days. The incubation period for the golden eagle is 45 days.

---

### SECTION 3.5 LINEAR EQUATIONS IN ONE VARIABLE AND PROBLEM SOLVING (CONTINUED)

| | |
|---|---|
| As a convenient summary and for quick reference, the problem-solving steps are below. | PROBLEM-SOLVING STEPS<br><br>1. UNDERSTAND the problem. During this step, become comfortable with the problem. Some ways of doing this are:<br>  • Read and reread the problem.<br>  • Choose a variable to represent the unknown.<br>  • Construct a drawing.<br>  • Propose a solution and check it. Pay careful attention to how you check your proposed solution. This will help when writing an equation to model the problem.<br>2. TRANSLATE the problem into an equation.<br>3. SOLVE the equation.<br>4. INTERPRET the results. *Check* the proposed solution in the stated problem and *state* your conclusion. |

**Name** _____ **Section** _____ **Date** _____

# Chapter 3 Review

**(3.1)** *Simplify each expression by combining like terms.*

**1.** $3y + 7y - 15$ $\quad 10y - 15$

**2.** $2y - 10 - 8y$ $\quad -6y - 10$

**3.** $8a + a - 7 - 15a$ $\quad -6a - 7$

**4.** $y + 3 - 9y - 1$ $\quad -8y + 2$

*Multiply.*

**5.** $2(x + 5)$ $\quad 2x + 10$

**6.** $-3(y + 8)$ $\quad -3y - 24$

*Simplify.*

**7.** $7x + 3(x - 4) + x$ $\quad 11x - 12$

**8.** $-(3m + 2) - m - 10$ $\quad -4m - 12$

**9.** $3(5a - 2) - 20a + 10$ $\quad -5a + 4$

**10.** $6y + 3 + 2(3y - 6)$ $\quad 12y - 9$

*Find the perimeter of each figure.*

△ **11.**

$2x$ yards

3 yards

Rectangle

$(4x + 6)$ yd

△ **12.**

Square $\quad$ $5y$ meters

$20y$ m

*Find the area of each figure.*

△ **13.**

$(2x - 1)$ yards

3 yards

Rectangle

$(6x - 3)$ sq. yd

△ **14.**

$(x - 2)$ centimeters

$(5x + 4)$ centimeters

10 centimeters

Rectangle

Rectangle

7 centimeters

$(45x + 8)$ sq. cm

*Simplify.*

📱 **15.** $85(7068x - 108) + 42x$   $600{,}822x - 9180$

📱 **16.** $-4268y + 120(63y - 32)$   $3292y - 3840$

**(3.2)**

**17.** Is 4 a solution of $5(2 - x) = -10$?   yes

**18.** Is 0 a solution of $6y + 2 = 23 + 4y$?   no

*Solve each equation.*

**19.** $z - 5 = -7$   $-2$

**20.** $3x + 10 = 4x$   $10$

**21.** $n + 18 = 10 - (-2)$   $-6$

**22.** $c - 5 = -13 + 7$   $-1$

**23.** $7x + 5 - 6x = -20$   $-25$

**24.** $17x = 2(8x - 4)$   $-8$

**(3.3)** *Solve each equation.*

**25.** $-3y = -21$   $7$

**26.** $-8x = 72$   $-9$

**27.** $-5n = -5$   $1$

**28.** $-3a = 15$   $-5$

**29.** $0 = 12x$   $0$

**30.** $-14 = 7x$   $-2$

**31.** $-5t + 32 + 4t = 32$   $0$

**32.** $3z + 72 - 2z = -56$   $-128$

*Translate each phrase into an algebraic expression.*

**33.** Eleven added to twice a number   $2x + 11$

**34.** The product of $-5$ and a number, decreased by 50
$-5x - 50$

**35.** The quotient of 70 and the sum of a number and 6
$$\frac{70}{x + 6}$$

**36.** Twice the difference of a number and 13

$2(x - 13)$

**(3.4)** *Solve each equation.*

**37.** $3x - 4 = 11$   5

**38.** $6y + 1 = 73$   12

**39.** $14 - y = -3$   17

**40.** $7 - z = 0$   7

**41.** $4z - z = -6$   $-2$

**42.** $t - 9t = -64$   8

**43.** $5(n - 3) = 7 + 3n$   11

**44.** $7(2 + x) = 4x - 1$   $-5$

**45.** $2x + 7 = 6x - 1$   2

**46.** $5x - 18 = -4x$   2

*Write each sentence as an equation.*

**47.** The difference of 20 and $-8$ is 28.   $20 - (-8) = 28$

**48.** Five times the sum of 2 and $-6$ yields $-20$.
$5(2 + (-6)) = -20$

**49.** The quotient of −75 and the sum of 5 and 20 is equal to −3. $\dfrac{-75}{5+20} = -3$

**50.** Nineteen subtracted from −2 amounts to −21. $-2 - 19 = -21$

**(3.5)** *Translate each sentence into an equation using x for the unknown number.*

**51.** Twice a number minus 8 is 40. $2x - 8 = 40$

**52.** Twelve subtracted from the quotient of a number and 2 is 10. $\dfrac{x}{2} - 12 = 10$

**53.** The difference of a number and 3 is the quotient of the number and 4. $x - 3 = x \div 4$

**54.** The product of some number and 6 is equal to the sum of the number and 2. $6x = x + 2$

*Solve.*

**55.** Five times a number subtracted from 40 is the same as three times the number. Find the number. 5

**56.** The product of a number and 3 is twice the difference of that number and 8. Find the number. −16

**57.** In an election, the incumbent received 14,000 votes of the 18,500 votes cast. Of the remaining votes, the Democratic candidate received 272 more than the Independent candidate. Find how many votes the Democratic candidate received. 2386 votes

**58.** Rajiv Puri has twice as many compact discs as he has cassette tapes. Find the number of CDs if he has a total of 126 music recordings. 84 CDs

**218**

**Name** _____ **Section** _____ **Date** _____

# Chapter 3 Test

1. Simplify $7x - 5 - 12x + 10$ by combining like terms.

2. Multiply: $-2(3y + 7)$

3. Simplify: $-(3z + 2) - 5z - 18$

△ 4. Find the perimeter of the equilateral triangle. (A triangle with three sides of equal length.)

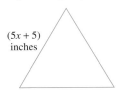

$(5x + 5)$ inches

△ 5. Find the area of the rectangle.

3 meters

Rectangle | $(3x - 1)$ meters

*Solve each equation.*

6. $9x + x = -60$

7. $12 = y - 3y$

8. $2n - 3 = 17$

9. $5 + 4z = 37$

10. $5x + 12 - 4x - 14 = 22$

11. $2 - c + 4c = 5$

12. $3x - 5 = -11$

13. $-4x + 7 = 15$

14. $2(x - 6) = 0$

15. $3(4 + 2y) = 12$

16. $5x - 2 = x - 10$

17. $10y - 1 = 7y + 20$

18. $6 + 2(3n - 1) = 28$

19. $4(5x + 3) = 2(7x + 6)$

1. $-5x + 5$

2. $-6y - 14$

3. $-8z - 20$

4. $(15x + 15)$ in.

5. $(9x - 3)$ sq. m

6. $-6$

7. $-6$

8. $10$

9. $8$

10. $24$

11. $1$

12. $-2$

13. $-2$

14. $6$

15. $0$

16. $-2$

17. $7$

18. $4$

19. $0$

**Name** _____

△ **20.** A lawn is in the shape of a trapezoid with a height of 60 feet and bases of 70 feet and 130 feet. Find the area of the lawn. Use

$$A = \frac{1}{2} \cdot h \cdot (B + b).$$

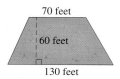

70 feet

60 feet

130 feet

**21.** Translate the following phrases into mathematical expressions. Use $x$ to represent "a number."
   **a.** The product of a number and 17
   **b.** Twice a number subtracted from 20

*Solve.*

**22.** The difference of three times a number and five times the same number is 4. Find the number.

**23.** In a championship basketball game, Paula Zimmerman made twice as many free throws as Maria Kaminsky. If the total number of free throws made by both women was 12, find how many free throws Paula made.

**24.** In a 10-kilometer race, there are 112 more men entered than women. Find the number of female runners if the total number of runners in the race is 600.

**Name** _____ **Section** _____ **Date** _____

# CUMULATIVE REVIEW

**1.** Write 106, 052, 447 in words.

△ **2.** Find the perimeter of the polygon shown.

2 inches  3 inches  1 inch  4 inches  3 inches

**3.** Subtract: $900 - 174$. Check by adding.

**4.** Round 248, 982 to the nearest hundred.

**5.** Multiply: $25 \times 8$

**6.** Divide and check: $1872 \div 9$

**7.** Simplify: $2 \cdot 4 - 3 \div 3$

**8.** Evaluate $x^2 + z - 3$ for $x = 5$ and $z = 4$.

**9.** Insert $<$ or $>$ between each pair of numbers to make a true statement.
  **a.** $-7$    $7$

  **b.** $0$    $-4$

  **c.** $-9$    $-11$

**10.** Add using a number line: $-3 + (-4)$

_Add._

**11.** $-5 + (-1)$

**12.** $2 + 6$

**Name** _____

*Subtract.*

**13.** $-4 - 10$

**14.** $6 - (-5)$

**15.** $-11 - (-7)$

*Divide.*

**16.** $\dfrac{-12}{6}$

**17.** $-20 \div (-4)$

**18.** $\dfrac{48}{-3}$

*Find the value of each expression.*

**19.** $(-3)^2$

**20.** $-3^2$

**21.** Simplify:
$2y - 6 + 4y + 8$

**22.** Determine whether $-1$ is a solution of the equation $3y + 1 = 3$.

**23.** Solve for $x$: $-12x = -36$

**24.** Solve: $2x - 6 = 18$

**25.** Leo Leal sold a used computer system and software for $2100, receiving four times as much money for the computer system as for the software. Find the price of each.

# Fractions

4

Fractions are numbers and, like whole numbers, they can be added, subtracted, multiplied, and divided. Fractions are very useful and appear frequently in everyday language, in common phrases such as "half an hour," "quarter of a pound," and "third of a cup." This chapter reviews the concept of fractions and demonstrates how to add, subtract, multiply, and divide fractions. In this chapter we also solve linear equations containing fractions.

The European honeybee (EHB) is used throughout the United States and Europe to make honey and to pollinate crops such as melons. EHBs have worked harmoniously with humans for centuries and are known for their gentleness. In the 1950s, beekeepers in Brazil were disappointed with EHB performance in the tropical Brazilian climate. Hoping to obtain a strain of bees better suited to the hotter climate, scientists crossbred European honeybees with a strain of African bees. The result was the Africanized honeybee (AHB). Unfortunately, AHBs were accidentally released before the crossbreeding program could eliminate their extremely aggressive traits, for which they are better known today as "killer bees." As it turned out, AHBs were extremely well suited to the tropical climate and spread rapidly through South and Central America. They first entered the United States in 1990. AHBs have since colonized parts of Texas, New Mexico, Arizona, and California, where the entire county of Los Angeles was officially declared to be colonized by killer bees in April 1999. In Exercise 89 on page 285, we will see how fractions can be used to compare the distances that killer bees and European honeybees will chase a perceived threat.

1. $\dfrac{3}{8}$; (4.1B)

2. $\dfrac{15}{18}$; (4.1D)

3. $0$; (4.1E)

4. $2 \cdot 2 \cdot 5 \cdot 7$ or $2^2 \cdot 5 \cdot 7$; (4.2A)

5. $\dfrac{5}{9}$; (4.2B)

6. $\dfrac{9}{20}$; of a gram; (4.2C)

7. $\dfrac{6}{5}$; (4.3A)

8. $\dfrac{1}{14x}$; (4.3C)

9. $\dfrac{5}{11}$; (4.4A)

10. $-\dfrac{1}{10}$; (4.4A)

11. $-\dfrac{8}{27}$; (4.3B)

12. $-\dfrac{1}{3}$; (4.3D)

13. $\dfrac{23}{36}$; (4.5A)

14. $\dfrac{1}{14}$; (4.5B)

15. $\dfrac{3x}{7}$; (4.6A)

16. $-\dfrac{46}{25}$; (4.6B)

17. $-\dfrac{23}{2}$; (4.7A)

18. $\dfrac{13}{5}$; (4.8B)

19. $\dfrac{44}{5}$ or $8\dfrac{4}{5}$; (4.8D)

20. $9\dfrac{5}{6}$; (4.8E)

# CHAPTER PRETEST

1. Write a fraction to represent the shaded area of the figure.

2. Write $\dfrac{5}{6}$ as an equivalent fraction whose denominator is 18.

3. Simplify: $\dfrac{0}{-4}$

4. Write the prime factorization of 140.

5. Simplify: $\dfrac{30}{54}$

6. There are 1000 milligrams in 1 gram. What fraction of a gram is 450 milligrams?

*Perform each indicated operation and simplify.*

7. $\dfrac{3}{4} \cdot \dfrac{24}{15}$

8. $\dfrac{5x}{7} \div 10x^2$

9. $\dfrac{8}{11} - \dfrac{3}{11}$

10. $-\dfrac{3}{10} + \dfrac{2}{10}$

11. Evaluate: $\left(-\dfrac{2}{3}\right)^3$

12. Evaluate $x \div y$ if $x = \dfrac{2}{9}$ and $y = -\dfrac{2}{3}$.

13. Add: $\dfrac{5}{9} + \dfrac{1}{12}$

14. Evaluate $x - y$ if $x = -\dfrac{3}{14}$ and $y = -\dfrac{2}{7}$.

*Simplify.*

15. $\dfrac{\dfrac{x}{3}}{\dfrac{7}{9}}$

16. $\left(\dfrac{2}{5}\right)^2 - 2$

17. Solve: $\dfrac{x}{4} + 3 = \dfrac{1}{8}$

18. Write $2\dfrac{3}{5}$ as an improper fraction.

*Perform each indicated operation and simplify.*

19. $3\dfrac{1}{5} \cdot 2\dfrac{3}{4}$

20. $5\dfrac{2}{3} + 4\dfrac{1}{6}$

# 4.1 Introduction to Fractions and Equivalent Fractions

## A Identifying Numerators and Denominators

Whole numbers are used to count whole things or units, such as cars, ball games, horses, dollars, and people. To refer to a part of a whole, fractions are used. For example, a whole baseball game is divided into nine parts called innings. If a player pitched five complete innings, the fraction $\frac{5}{9}$ can be used to show the part of a whole game he or she pitched. The 9 in the fraction $\frac{5}{9}$ is called the **denominator**, and it refers to the total number of equal parts (innings) in the whole game. The 5 in the fraction $\frac{5}{9}$ is called the **numerator**, and it tells how many of those equal parts (innings) the pitcher pitched.

$$\frac{5}{9} \quad \begin{array}{l} \leftarrow \text{how many of the parts being considered} \\ \leftarrow \text{number of equal parts in the whole} \end{array}$$

**Examples**    Identify the numerator and the denominator of each fraction.

**1.** $\dfrac{3}{7} \quad \begin{array}{l} \leftarrow \text{numerator} \\ \leftarrow \text{denominator} \end{array}$

**2.** $\dfrac{13x}{5} \quad \begin{array}{l} \leftarrow \text{numerator} \\ \leftarrow \text{denominator} \end{array}$

> **Helpful Hint**
>
> $\dfrac{3}{7} \leftarrow$ Remember that the bar in a fraction means division. Since division by 0 is undefined, a fraction with a denominator of 0 is undefined.

## B Writing Fractions to Represent Shaded Areas of Figures

One way to become familiar with the concept of fractions is to visualize fractions with shaded figures. We can then write a fraction to represent the shaded area of the figure.

**Examples**    Write a fraction to represent the shaded area of each figure.

**3.**

The figure is divided into 5 equal parts and 2 of them are shaded.

Thus, $\frac{2}{5}$ of the figure is shaded.

$\dfrac{2}{5} \quad \begin{array}{l} \leftarrow \text{parts shaded} \\ \leftarrow \text{equal parts} \end{array}$

**4.**

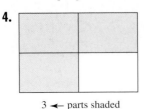

The figure is divided into 4 equal parts and 3 of them are shaded.

Thus, $\frac{3}{4}$ of the figure is shaded.

$\dfrac{3}{4} \quad \begin{array}{l} \leftarrow \text{parts shaded} \\ \leftarrow \text{equal parts} \end{array}$

---

### Objectives

**A** Identify the numerator and the denominator of a fraction.

**B** Write a fraction to represent the shaded area of a figure.

**C** Graph fractions on a number line.

**D** Write equivalent fractions.

**E** Simplify fractions of the form $\frac{a}{a}$, $\frac{a}{1}$, and $\frac{0}{a}$.

Study Guide    SSM    CD-ROM    Video 4.1

### Practice Problems 1–2

Identify the numerator and the denominator of each fraction.

**1.** $\dfrac{9}{2}$      **2.** $\dfrac{10y}{17}$

**TEACHING TIP**

Ask students to also represent the unshaded area of each figure. Help them notice any relationships between the fraction for the unshaded areas and the fraction for the shaded areas. (For example, the sum of the numerators is the denominator.)

### Practice Problems 3–4

Write a fraction to represent the shaded area of each figure.

**3.**

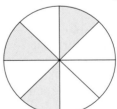

**4.**

### Answers

**1.** $\dfrac{9}{2} \begin{array}{l} \leftarrow \text{numerator} \\ \leftarrow \text{denominator} \end{array}$   **2.** $\dfrac{10y}{17} \begin{array}{l} \leftarrow \text{numerator} \\ \leftarrow \text{denominator} \end{array}$

**3.** $\dfrac{3}{8}$    **4.** $\dfrac{1}{6}$

$$\frac{2}{3}$$

A **proper fraction** is a fraction whose numerator is less than its denominator. Proper fractions are less than 1. The shaded portion of the triangle's area is represented by $\frac{2}{3}$.

An **improper fraction** is a fraction whose numerator is greater than or equal to its denominator. Improper fractions are greater than or equal to 1. The area of the shaded part of the group of circles is $\frac{9}{4}$. The shaded part of the rectangle's area is $\frac{6}{6}$.

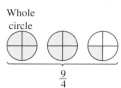

Whole circle

$$\frac{9}{4}$$

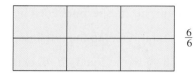

$$\frac{6}{6}$$

## ✓ CONCEPT CHECK

Identify each as a proper fraction or improper fraction.

a. $\frac{6}{7}$  b. $\frac{13}{12}$  c. $\frac{2}{2}$  d. $\frac{99}{101}$

**TRY THE CONCEPT CHECK IN THE MARGIN.**

### Practice Problems 5–6

Represent the shaded part of each figure group with an improper fraction.

5.

6.

**Examples**   Represent the shaded part of each figure group with an improper fraction.

5.

Whole object

improper fraction: $\frac{4}{3}$

mixed number: $1\frac{1}{3}$

6.
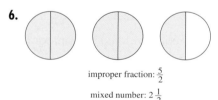

improper fraction: $\frac{5}{2}$

mixed number: $2\frac{1}{2}$

### Practice Problem 7

Of the nine planets in our solar system, seven are further from the sun than Venus is. What fraction of the planets are farther from the sun than Venus is?

## Example 7   Writing Fractions from Real-Life Data

Of the nine planets in our solar system, two are closer to the sun than Earth is. What fraction of the planets are closer to the sun than Earth is?

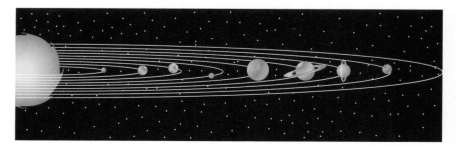

*Solution:*   The fraction closer to the sun is:

$$\frac{2}{9}$$   ← number of planets that are closer

← number of planets in our solar system

**Answers**

**5.** $\frac{8}{3}$  **6.** $\frac{5}{4}$  **7.** $\frac{7}{9}$

✓ **Concept Check: a.** proper  **b.** improper
**c.** improper  **d.** proper

Thus, $\frac{2}{9}$ of our solar system planets are closer to the sun than Earth is.

## C  GRAPHING FRACTIONS ON A NUMBER LINE

Another way to visualize fractions is to graph them on a number line. To do this, think of 1 unit on the number line as a whole. To graph $\frac{2}{5}$, for example, divide the distance from 0 to 1 into 5 equal parts. Then start at 0 and count over 2 parts.

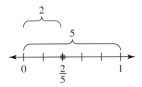

**Example 8**    Graph each proper fraction on a number line.

  **a.** $\frac{3}{4}$      **b.** $\frac{1}{2}$      **c.** $\frac{3}{6}$

*Solution:*    **a.** Divide the distance from 0 to 1 into 4 parts. Then start at 0 and count over 3 parts.

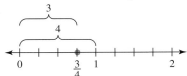

  **b.**

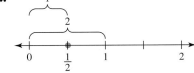

  **c.**

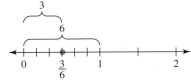

The fractions in Example 8 are all proper fractions. Notice that the graph of each is less than 1. This is always true since the numerator of a proper fraction is less than the denominator.

**Example 9**    Graph each improper fraction on a number line.

  **a.** $\frac{7}{6}$      **b.** $\frac{9}{5}$      **c.** $\frac{6}{6}$      **d.** $\frac{3}{1}$

*Solution:*    **a.**

**Practice Problem 8**

Graph each proper fraction on a number line.

  **a.** $\frac{5}{7}$      **b.** $\frac{2}{3}$      **c.** $\frac{4}{6}$

**Practice Problem 9**

Graph each improper fraction on a number line.

  **a.** $\frac{8}{3}$      **b.** $\frac{5}{4}$      **c.** $\frac{7}{7}$

**Answers**

**8. a.**   (number line) 0 ... $\frac{5}{7}$ 1

**b.**   (number line) 0 ... $\frac{2}{3}$ 1

**c.**   (number line) 0 ... $\frac{4}{6}$ 1

**9. a.**   (number line) 0 1 2 $\frac{8}{3}$ 3

**b.**   (number line) 0 1 $\frac{5}{4}$ 2

**c.**   (number line) 0 1 or $\frac{7}{7}$

**b.**

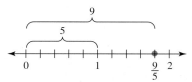

**c.**

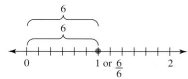

**d.** Each 1-unit distance has 1 equal part. Count over 3 parts.

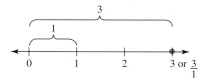

The fractions in Example 9 are all improper fractions. Notice that the graph of each is greater than or equal to 1. This is always true since the numerator of an improper fraction is greater than or equal to the denominator.

## D WRITING EQUIVALENT FRACTIONS

When we graph $\frac{1}{3}$ and $\frac{2}{6}$ on a number line,

notice that both $\frac{1}{3}$ and $\frac{2}{6}$ correspond to the same point. These fractions are called **equivalent fractions**, and we write $\frac{1}{3} = \frac{2}{6}$.

Another way to visualize equivalent fractions is to use shaded figures.

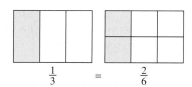

EQUIVALENT FRACTIONS

Fractions that represent the same portion of a whole or the same point on the number line are called **equivalent fractions**.

To write equivalent fractions, we use the **fundamental property of fractions**. This property guarantees that, if we multiply or divide both the numerator and the denominator by the same nonzero number, the result is an equivalent fraction. For example, if we multiply the numerator and denominator of $\frac{1}{3}$ by the same number, 2, the result is the equivalent fraction $\frac{2}{6}$.

$$\frac{1 \cdot 2}{3 \cdot 2} = \frac{2}{6}$$

---

**FUNDAMENTAL PROPERTY OF FRACTIONS**

If $a$, $b$, and $c$ are numbers, then

$$\frac{a}{b} = \frac{a \cdot c}{b \cdot c} \quad \text{and also} \quad \frac{a}{b} = \frac{a \div c}{b \div c}$$

as long as $b$ and $c$ are not 0. In other words, if the numerator and denominator of a fraction are multiplied or divided by the **same** nonzero number, the result is an equivalent fraction.

---

**Example 10**  Write $\frac{2}{5}$ as an equivalent fraction whose denominator is 15.

*Solution:*  Since $5 \cdot 3 = 15$, we use the fundamental property of fractions and multiply the numerator and denominator of $\frac{2}{5}$ by 3.

$$\frac{2}{5} = \frac{2 \cdot 3}{5 \cdot 3} = \frac{6}{15}$$

Then $\frac{2}{5}$ is equivalent to $\frac{6}{15}$. They both represent the same part of a whole.

**Example 11**  Write $\frac{9x}{11}$ as an equivalent fraction whose denominator is 44.

*Solution:*  Since $11 \cdot 4 = 44$, we multiply the numerator and denominator by 4.

$$\frac{9x}{11} = \frac{9x \cdot 4}{11 \cdot 4} = \frac{36x}{44}$$

Then $\frac{9x}{11}$ is equivalent to $\frac{36x}{44}$.

**Example 12**  Write 3 as an equivalent fraction whose denominator is 7.

*Solution:*  Recall that $3 = \frac{3}{1}$. Since $1 \cdot 7 = 7$, multiply the numerator and denominator by 7.

$$\frac{3}{1} = \frac{3 \cdot 7}{1 \cdot 7} = \frac{21}{7}$$

In this chapter, we will perform operations on fractions containing variables. When the denominator of a fraction contains a variable, such as $\frac{8}{3x}$, we will assume that the variable does not represent 0. Recall that the denominator of a fraction cannot be 0.

**Practice Problem 10**

Write $\frac{1}{4}$ as an equivalent fraction whose denominator is 20.

TEACHING TIP
Show students that they can check their work by using cross products.

**Practice Problem 11**

Write $\frac{3x}{7}$ as an equivalent fraction with a denominator of 42.

**Practice Problem 12**

Write 4 as an equivalent fraction whose denominator is 6.

Answers

**10.** $\frac{5}{20}$  **11.** $\frac{18x}{42}$  **12.** $\frac{24}{6}$

## Practice Problem 13

Write $\dfrac{9}{4x}$ as an equivalent fraction with a denominator of $20x$.

## ✓ CONCEPT CHECK

What is the first step in writing $\dfrac{3}{10}$ as an equivalent fraction whose denominator is 100?

## Practice Problems 14–19

Simplify by dividing the numerator by the denominator.

14. $\dfrac{9}{9}$     15. $\dfrac{-6}{-6}$     16. $\dfrac{-1}{-1}$

17. $\dfrac{4}{1}$     18. $\dfrac{-30}{1}$     19. $\dfrac{13}{1}$

## Practice Problems 20–22

20. $\dfrac{0}{14}$     21. $\dfrac{0}{-9}$     22. $\dfrac{6}{0}$

**Answers**

13. $\dfrac{45}{20x}$   14. 1   15. 1   16. 1   17. 4   18. −30

19. 13   20. 0   21. 0   22. undefined

✓ **Concept Check:** Answers may vary.

---

**Example 13**   Write $\dfrac{8}{3x}$ as an equivalent fraction whose denominator is $12x$.

*Solution:*    $3x \cdot 4 = 12x$. Multiply the numerator and denominator by 4.

$$\frac{8}{3x} = \frac{8 \cdot 4}{3x \cdot 4} = \frac{32}{12x}$$
      ▬▬

**TRY THE CONCEPT CHECK IN THE MARGIN.**

**E**   **SIMPLIFYING** $\dfrac{a}{a}$, $\dfrac{a}{1}$, **AND** $\dfrac{0}{a}$.

The graphs of $\dfrac{6}{6}$ and $\dfrac{3}{1}$ are shown next.

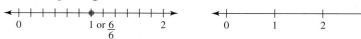

Notice that the graph of $\dfrac{6}{6}$ is the same as the graph of 1, and the graph of $\dfrac{3}{1}$ is the same as 3. This makes sense if we recall that the notation for a fraction, the fraction bar, means the same as the notation for a quotient. They both mean division.

$$\frac{6}{6} = 6 \div 6 = 1 \qquad \text{and} \qquad \frac{3}{1} = 3 \div 1 = 3$$

**Examples**   Simplify by dividing the numerator by the denominator.

14. $\dfrac{5}{5} = 1$     15. $\dfrac{-2}{-2} = 1$     16. $\dfrac{100}{100} = 1$

17. $\dfrac{7}{1} = 7$     18. $\dfrac{-5}{1} = -5$     19. $\dfrac{41}{1} = 41$   ▬▬

In general, if the numerator and the denominator are the same, the fraction is equivalent to 1. Also, if the denominator of a fraction is 1, the fraction is equivalent to the numerator.

---
If $a$ is any number other than 0, then $\dfrac{a}{a} = 1$.

Also, if $a$ is any number, $\dfrac{a}{1} = a$.

---

Recall from Chapter 1 that if the numerator of a fraction is 0 and the denominator is not, the fraction simplifies to 0. Also, recall that we may not divide by 0. We say that $\dfrac{a}{0}$ is undefined. (Here $a$ is not 0.)

---
If $a$ is not 0, then

$$\frac{0}{a} = 0 \qquad \text{and} \qquad \frac{a}{0} \text{ is undefined.}$$

---

**Examples**   Simplify.

20. $\dfrac{0}{5} = 0$     21. $\dfrac{0}{-3} = 0$     22. $\dfrac{19}{0}$ is undefined

        ▬▬

**Name** _____ **Section** _____ **Date** _____

## MENTAL MATH

**A** *Identify the numerator and the denominator of each fraction. See Examples 1 and 2.*

**1.** $\dfrac{1}{2}$

**2.** $\dfrac{1}{4}$

**3.** $\dfrac{10}{3}$

**4.** $\dfrac{53}{21}$

**5.** $\dfrac{3z}{7}$

**6.** $\dfrac{11x}{15}$

## EXERCISE SET 4.1

**B** *Write a fraction to represent the shaded area of each figure. See Examples 3 and 4.*

**1.**

**2.**

**3.**

**4.**

**5.**

**6.**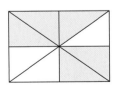

**7.**

**8.**

**231**

**9.** $\dfrac{4}{9}$

**9.**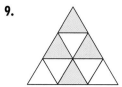

**10.** $\dfrac{7}{8}$

**10.**

**11.** $\dfrac{5}{8}$

**12.** $\dfrac{3}{8}$

*Use the pizza illustration for Exercises 11 and 12.*

**13.** $\dfrac{11}{4}$

**11.** What fraction of the pizza is gone?  **12.** What fraction of the pizza remains?

**14.** $\dfrac{3}{2}$

*Represent the shaded area in each figure group with an improper fraction. See Examples 5 and 6.*

**13.**   **14.**

**15.** $\dfrac{11}{3}$

**16.** $\dfrac{10}{4}$

**15.**   **16.**

**17.** $\dfrac{3}{2}$

**18.** $\dfrac{5}{3}$

**17.**   **18.**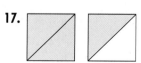

**19.** $\dfrac{4}{3}$

**19.** 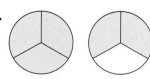  **20.**

**20.** $\dfrac{6}{5}$

**Name** _____

**21.**

**22.**

**23.**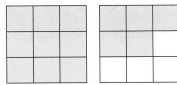

**24.**

*Write each fraction. See Example 7.*

**25.** Of the 131 students at a small private school, 42 are freshmen. What fraction of the students are freshmen?

**26.** Of the 78 executives at a private accounting firm, 61 are men. What fraction of the executives are men?

**27.** From Exercise 25, how many students are *not* freshmen? What fraction of the students are *not* freshmen?

**28.** From Exercise 26, how many of the executives are women? What fraction of the executives are women?

**29.** According to a recent study, 4 out of 10 visits to U.S. hospital emergency rooms were for an injury. What fraction of emergency room visits are injury-related? (*Source:* National Center for Health Statistics)

**30.** The average American driver spends about 1 hour each day in a car or truck. What fraction of a day does an average American driver spend in a car or truck? (*Hint:* How many hours are in a day?) (*Source:* U.S. Department of Transportation—Federal Highway Administration)

**21.** $\dfrac{17}{6}$ _____

**22.** $\dfrac{7}{5}$ _____

**23.** $\dfrac{14}{9}$ _____

**24.** $\dfrac{25}{9}$ _____

**25.** $\dfrac{42}{131}$ of the students _____

**26.** $\dfrac{61}{78}$ of the executives _____

**27.** $89; \dfrac{89}{131}$ of the students _____

**28.** $17; \dfrac{17}{78}$ of the executives _____

**29.** $\dfrac{4}{10}$ of the visits _____

**30.** $\dfrac{1}{24}$ of a day _____

**31.** $\frac{8}{42}$ of the presidents

**32.** $\frac{4}{9}$ of the planets

**33.** $\frac{5}{12}$ of a foot

**34.** $\frac{1}{2}$ of the memory

**35.** $\frac{11}{24}$ of a day

**36.** $\frac{37}{60}$ of an hour

**37.** $\frac{7}{11}$ of the children

**38.** $\frac{10}{31}$ of the class

**39. a.** $\frac{40}{50}$ of the states

**b.** 10 states

**c.** $\frac{10}{50}$ of the states

**40. a.** $\frac{33}{50}$ of the states

**b.** 17 states

**c.** $\frac{17}{50}$ of the states

**Name** _____

**31.** As of 1999, the United States has had 42 different presidents. A total of eight U.S. presidents were born in the state of Virginia, more than any other state. What fraction of U.S. presidents were born in Virginia? (*Source: 2000 World Almanac and Book of Facts*)

Eight U.S. Presidents

**32.** Of the nine planets in our solar system, four have days that are longer than the 24-hour Earth day. What fraction of the planets have longer days than Earth has? (*Source:* National Space Science Data Center)

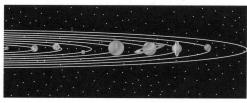

**33.** There are 12 inches in a foot. What fractional part of a foot does 5 inches represent?

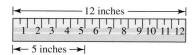

**34.** A 3M high-density disk holds 2 megabytes of memory. What fractional part of the memory on this disk does 1 megabyte represent?

**35.** There are 24 hours in a day. What fraction of a day does 11 hours represent?

**36.** There are 60 minutes in an hour. What fraction of an hour does 37 minutes represent?

**37.** In a family with 11 children, there are 4 boys and 7 girls. What fraction of the children are girls?

**38.** In a prealgebra class containing 31 students, there are 18 freshmen, 10 sophomores, and 3 juniors. What fraction of the class is sophomores?

**39.** Consumer fireworks are legal in 40 states in the United States.
  **a.** In what fraction of the states are consumer fireworks legal?
  **b.** In how many states are consumer fireworks illegal?
  **c.** In what fraction of the states are consumer fireworks illegal? (*Source:* United States Fireworks Safety Council)

**40.** Thirty-three states in the United States contain federal Indian reservations.
  **a.** What fraction of the states contain Indian reservations?
  **b.** How many states do not contain Indian reservations?
  **c.** What fraction of the states do not contain Indian reservations? (*Source:* Tiller Research, Inc., Albuquerque, NM)

**51.** $\dfrac{20}{35}$ _____

**52.** $\dfrac{12}{20}$ _____

**53.** $\dfrac{14}{21}$ _____

**54.** $\dfrac{4}{24}$ _____

**55.** $\dfrac{10y}{25}$ _____

**56.** $\dfrac{63a}{70}$ _____

**57.** $\dfrac{15}{30}$ _____

**58.** $\dfrac{10}{30}$ _____

**59.** $\dfrac{30}{21x}$ _____

**60.** $\dfrac{35}{21b}$ _____

**61.** $\dfrac{10}{5}$ _____

**62.** $\dfrac{40}{8}$ _____

**C** *Graph each fraction on a number line. See Examples 8 and 9.*

**41.** $\dfrac{1}{4}$

**42.** $\dfrac{1}{3}$

**43.** $\dfrac{4}{7}$

**44.** $\dfrac{5}{6}$

**45.** $\dfrac{8}{5}$

**46.** $\dfrac{9}{8}$

**47.** $\dfrac{7}{3}$

**48.** $\dfrac{15}{7}$

**49.** $\dfrac{3}{8}$

**50.** $\dfrac{8}{3}$

**D** *Write each fraction as an equivalent fraction with the given denominator. See Examples 10 through 13.*

**51.** $\dfrac{4}{7}$; denominator of 35

**52.** $\dfrac{3}{5}$; denominator of 20

**53.** $\dfrac{2}{3}$; denominator of 21

**54.** $\dfrac{1}{6}$; denominator of 24

**55.** $\dfrac{2y}{5}$; denominator of 25

**56.** $\dfrac{9a}{10}$; denominator of 70

**57.** $\dfrac{1}{2}$; denominator of 30

**58.** $\dfrac{1}{3}$; denominator of 30

**59.** $\dfrac{10}{7x}$; denominator of 21x

**60.** $\dfrac{5}{3b}$; denominator of 21b

**61.** 2; denominator of 5

**62.** 5; denominator of 8

**235**

**Name** _____

*Write each fraction as an equivalent fraction whose denominator is 12. See Examples 10 and 11.*

63. $\frac{3}{4}$      64. $\frac{4}{6}$      ▣ 65. $\frac{2y}{3}$

66. $\frac{2}{3}$      67. $\frac{1}{2}$      68. $\frac{3x}{2}$

*Write each fraction as an equivalent fraction whose denominator is 36x. See Example 13.*

69. $\frac{4}{3}$      70. $\frac{3}{4}$      ▣ 71. $\frac{5}{9}$

72. $\frac{7}{6}$      73. 1      74. 2

*The table shows the fraction of the population in each country that used cell phones at the beginning of 1999. Use this table to answer Exercises 75–78.*

75. Complete the table by writing each fraction as an equivalent fraction with a denominator of 100.

76. Which of these countries has the largest fraction of cell phone users?

77. Which of these countries has the smallest fraction of cell phone users?

78. In which of these countries does over half of the population use cell phones?

| Country | Fraction of Population Using Cell Phones | Equivalent fraction with a Denominator of 100 |
|---|---|---|
| Denmark | $\frac{7}{20}$ | $\frac{35}{100}$ |
| Finland | $\frac{57}{100}$ | $\frac{57}{100}$ |
| Hong Kong | $\frac{43}{100}$ | $\frac{43}{100}$ |
| Israel | $\frac{37}{100}$ | $\frac{37}{100}$ |
| Italy | $\frac{9}{25}$ | $\frac{36}{100}$ |
| Japan | $\frac{31}{100}$ | $\frac{31}{100}$ |
| Norway | $\frac{12}{25}$ | $\frac{48}{100}$ |
| Singapore | $\frac{8}{25}$ | $\frac{32}{100}$ |
| South Korea | $\frac{3}{10}$ | $\frac{30}{100}$ |
| Sweden | $\frac{23}{50}$ | $\frac{46}{100}$ |
| United States | $\frac{13}{50}$ | $\frac{26}{100}$ |

(*Source:* Dataquest)

**Name** _____

79. 1

80. 1

81. −5

82. −10

83. 0

84. 0

85. 1

86. 1

87. undefined

88. undefined

89. 3

90. 1

91. 9

92. 64

93. 125

94. 81

95. 49

96. 625

97. 24

98. 80

99. $\dfrac{464}{2088}$

100. $\dfrac{1767}{6479}$

101. answers may vary

102. answers may vary

**E** *Simplify by dividing. See Examples 14 through 22.*

**79.** $\dfrac{12}{12}$     **80.** $\dfrac{-3}{-3}$     **81.** $\dfrac{-5}{1}$     **82.** $\dfrac{-10}{1}$

**83.** $\dfrac{0}{-2}$     **84.** $\dfrac{0}{-8}$     **85.** $\dfrac{-8}{-8}$     **86.** $\dfrac{-14}{-14}$

**87.** $\dfrac{-9}{0}$     **88.** $\dfrac{-7}{0}$     **89.** $\dfrac{3}{1}$     **90.** $\dfrac{5}{5}$

## REVIEW AND PREVIEW

*Simplify. See Section 1.7.*

**91.** $3^2$     **92.** $4^3$     **93.** $5^3$     **94.** $3^4$

**95.** $7^2$     **96.** $5^4$     **97.** $2^3 \cdot 3$     **98.** $4^2 \cdot 5$

## COMBINING CONCEPTS

**99.** Write $\dfrac{2}{9}$ as an equivalent fraction whose denominator is 2088.

**100.** Write $\dfrac{3}{11}$ as an equivalent fraction whose denominator is 6479.

**101.** In your own words, explain how to write equivalent fractions.

**102.** In your own words, explain why $\dfrac{0}{10} = 0$ and $\dfrac{10}{0}$ is undefined.

**Name** _____

*Write each fraction.*

**103.** Habitat for Humanity is a nonprofit organization that helps provide affordable housing to families in need. Habitat for Humanity does its work of building and renovating houses through 1300 local affiliates in the United States and 285 international affiliates. What fraction of the total Habitat for Humanity affiliates are located in the United States? (*Source:* Habitat for Humanity International)

**104.** The United States Marine Corps (USMC) has five principal training centers in California, three in North Carolina, two in South Carolina, one in Arizona, one in Hawaii, and one in Virginia. What fraction of the total USMC principal training centers are located in California? (*Source:* U.S. Department of Defense)

**105.** The Public Broadcasting Service (PBS) provides programming to the noncommercial public TV stations of the United States. The table shows a breakdown of the public television licensees by type. Each licensee operates one or more PBS member TV stations. What fraction of the public television licensees are universities or colleges?

| PUBLIC TELEVISION LICENSEES | |
|---|---|
| Type | Number |
| Community organizations | 87 |
| Universities/colleges | 56 |
| State authorities | 21 |
| Local education/ municipal authorities | 8 |

(*Source:* The Public Broadcast Service)

**106.** The table shows the number of Wendy's restaurants operated in various regions. What fraction of Wendy's restaurants are located in the United States?

| WENDY'S RESTAURANTS | |
|---|---|
| Region | Number |
| United States | 4377 |
| Canada | 229 |
| Asia/Pacific | 201 |
| Latin America | 68 |
| Europe & other | 58 |

(*Source:* Wendy's International, Inc.)

238

# 4.2 FACTORS AND SIMPLEST FORM

## A WRITING A NUMBER AS A PRODUCT OF PRIME NUMBERS

Of all the equivalent ways to write a particular fraction, one special way is called **simplest form** or **lowest terms**. To help us write a fraction in simplest form, we first practice writing a number as a product of prime numbers.

> A **prime number** is a whole number greater than 1 whose only divisors are 1 and itself. The first few prime numbers are 2, 3, 5, 7, 11, 13, 17, 19, 23, 29, . . . .
> A **composite number** is a whole number greater than 1 that is not prime.

> **┌ Helpful Hint**
>
> The natural number 1 is neither prime nor composite.

When a number is written as a product of prime numbers, this product is called the **prime factorization** of the number. For example, the prime factorization of 12 is $2 \cdot 2 \cdot 3$ because

$$12 = \underbrace{2 \cdot 2 \cdot 3}$$

This product is 12 and each number is a prime number.

Because multiplication is commutative, the order of the factors is not important. We can write the factorization $2 \cdot 2 \cdot 3$ as $2 \cdot 3 \cdot 2$ or $3 \cdot 2 \cdot 2$. Any of these is called the prime factorization of 12.

> Every whole number greater than 1 has exactly one prime factorization.

Recall from Section 1.5 that since $12 = 2 \cdot 2 \cdot 3$, the numbers 2 and 3 are called *factors* of 12. A **factor** is any number that divides a number evenly (with a remainder of 0).

One method for finding the prime factorization of a number is by using a factor tree, as shown in the next example.

**Example 1**    Write the prime factorization of 45.

*Solution:*    We can begin by writing 45 as the product of two numbers, say 5 and 9.

$$\begin{array}{c} 45 \\ \diagup \diagdown \\ 5 \cdot 9 \end{array}$$

The number 5 is prime but 9 is not, so we write 9 as $3 \cdot 3$.

$$\left.\begin{array}{c} 45 \\ \diagup \diagdown \\ 5 \cdot 9 \\ \diagup \diagdown \\ 5 \cdot 3 \cdot 3 \end{array}\right\} \text{A factor tree}$$

Each factor is now a prime number, so the prime factorization of 45 is $3 \cdot 3 \cdot 5$ or $3^2 \cdot 5$.

---

### Objectives

**A** Write a number as a product of prime numbers.

**B** Write a fraction in simplest form.

**C** Solve problems by writing fractions in simplest form.

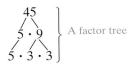

Study Guide    SSM    CD-ROM    Video 4.2

TEACHING TIP

Students may be interested to know that mathematicians are always looking for larger and larger prime numbers. Computers are used to find them and when a new large prime number is found, it is usually reported in the newspaper.

---

**Practice Problem 1**

Write the prime factorization of 28.

**Answer**

**1.** $28 = 2 \cdot 2 \cdot 7$ or $2^2 \cdot 7$

## Practice Problem 2

Write the prime factorization of 60.

TEACHING TIP

To check the prime factorization of a number, students need to ask themselves two questions:

1. Is the prime factorization product equal to the original number?
2. Are all numbers in the product prime?

## Example 2

Write the prime factorization of 80.

*Solution:*    Write 80 as a product of two numbers. Continue this process until all factors are prime.

All factors are now prime, so the prime factorization of 80 is

$$2 \cdot 2 \cdot 2 \cdot 2 \cdot 5 \quad \text{or} \quad 2^4 \cdot 5.$$

There are a few quick **divisibility tests** to determine whether a number is divisible by the primes 2, 3, or 5.

> **DIVISIBILITY TESTS**
>
> A whole number is divisible by
> - **2** if the ones digit is 0, 2, 4, 6, or 8.
>   ↓
>      132 is divisible by 2.
> - **3** if the sum of the digits is divisible by 3.
>      144 is divisible by 3 since $1 + 4 + 4 = 9$ is divisible by 3.
> - **5** if the ones digit is 0 or 5.
>   ↓
>      1115 is divisible by 5.

When finding the prime factorization of larger numbers, you may want to use the procedure shown in Example 3.

## Practice Problem 3

Write the prime factorization of 297.

## Example 3

Write the prime factorization of 252.

*Solution:*    Since the ones digit of 252 is 2, we know that 252 is divisible by 2.

$$\begin{array}{r} 126 \\ 2\overline{)252} \end{array}$$

126 is divisible by 2 also.

$$\begin{array}{r} 63 \\ 2\overline{)126} \\ 2\overline{)252} \end{array}$$

63 is not divisible by 2 but is divisible by 3. Divide 63 by 3 and continue in this same manner until the quotient is a prime number.

$$\begin{array}{r} 7 \\ 3\overline{)21} \\ 3\overline{)63} \\ 2\overline{)126} \\ 2\overline{)252} \end{array}$$

The prime factorization of 252 is $2 \cdot 2 \cdot 3 \cdot 3 \cdot 7$ or $2^2 \cdot 3^2 \cdot 7$.

**Answers**

**2.** $60 = 2 \cdot 2 \cdot 3 \cdot 5$ or $2^2 \cdot 3 \cdot 5$

**3.** $297 = 3 \cdot 3 \cdot 3 \cdot 11$ or $3^3 \cdot 11$

TRY THE CONCEPT CHECK IN THE MARGIN.

### B    WRITING A FRACTION IN SIMPLEST FORM

We can use the prime factorization of a number to help us write a fraction in **simplest form** or **lowest terms**.

> #### SIMPLEST FORM
>
> A fraction is in **simplest form**, or **lowest terms**, when the numerator and the denominator have no common factors other than 1.

For example, the fraction $\dfrac{6}{10}$ is *not* in simplest form because 6 and 10 both have a factor of 2. That is, 2 is a common factor of 6 and 10.

To write a fraction in simplest form, write the prime factorization of the numerator and the denominator and then use the fundamental principle of fractions to divide both by all common factors.

For example,

$$\frac{6}{10} = \frac{2 \cdot 3}{2 \cdot 5} = \frac{2 \cdot 3 \div 2}{2 \cdot 5 \div 2} = \frac{3}{5}$$

The fraction $\dfrac{3}{5}$ is in lowest terms, since the numerator and the denominator have no common factors (other than 1).

In the future, we will use the following notation to show dividing the numerator and the denominator by common factors.

$$\frac{6}{10} = \frac{2 \cdot 3}{2 \cdot 5} = \frac{\cancel{2} \cdot 3}{\cancel{2} \cdot 5} = \frac{3}{5}$$

> #### WRITING A FRACTION IN SIMPLEST FORM
>
> To write a fraction in simplest form, write the prime factorization of the numerator and the denominator and then divide both by all common factors.

The process of writing a fraction in simplest form is called **simplifying** the fraction.

**Example 4**    Simplify: $\dfrac{12}{20}$

*Solution:*    First write the prime factorization of the numerator and the denominator.

$$\frac{12}{20} = \frac{2 \cdot 2 \cdot 3}{2 \cdot 2 \cdot 5}$$

Next divide the numerator and the denominator by all common factors.

$$\frac{12}{20} = \frac{\cancel{2} \cdot \cancel{2} \cdot 3}{\cancel{2} \cdot \cancel{2} \cdot 5} = \frac{3}{5}$$

✓ **CONCEPT CHECK**

True or false? The prime factorization of 72 is $2 \cdot 4 \cdot 9$. Explain your reasoning.

**Practice Problem 4**

Simplify: $\dfrac{30}{45}$

**Answers**

**4.** $\dfrac{2}{3}$

✓ **Concept Check:** No; $4 = 2 \cdot 2$ and $9 = 3 \cdot 3$.

**Practice Problem 5**

Simplify: $\dfrac{39}{51}$

**Practice Problem 6**

Simplify: $\dfrac{45}{105y}$

**Practice Problem 7**

Simplify: $\dfrac{9a}{50a}$

**Practice Problem 8**

Simplify: $\dfrac{38}{4}$

**Practice Problem 9**

Simplify: $\dfrac{7a^3}{56a^2}$

**Answers**

**5.** $\dfrac{13}{17}$  **6.** $\dfrac{3}{7y}$  **7.** $\dfrac{9}{50}$  **8.** $\dfrac{19}{2}$  **9.** $\dfrac{a}{8}$

**Example 5**   Simplify: $\dfrac{42}{66}$

*Solution:*   $\dfrac{42}{66} = \dfrac{2 \cdot 3 \cdot 7}{2 \cdot 3 \cdot 11} = \dfrac{7}{11}$

**Example 6**   Simplify: $\dfrac{84x}{90}$

*Solution:*   $\dfrac{84x}{90} = \dfrac{2 \cdot 2 \cdot 3 \cdot 7 \cdot x}{2 \cdot 3 \cdot 3 \cdot 5} = \dfrac{14x}{15}$

**Example 7**   Simplify: $\dfrac{10y}{27y}$

*Solution:*   $\dfrac{10y}{27y} = \dfrac{2 \cdot 5 \cdot y}{3 \cdot 3 \cdot 3 \cdot y} = \dfrac{10}{27}$

**Example 8**   Simplify: $\dfrac{36}{28}$

*Solution:*   $\dfrac{36}{28} = \dfrac{2 \cdot 2 \cdot 3 \cdot 3}{2 \cdot 2 \cdot 7} = \dfrac{9}{7}$

For the fraction in Example 8, $\dfrac{36}{28}$, you may immediately notice that a common fator of 36 and 28 is 4. If so, you may simply divide out that common factor.

$$\frac{36}{28} = \frac{4 \cdot 9}{4 \cdot 7} = \frac{9}{7}$$

The result is in simplest form. If it were not, we would repeat the same procedure until it was.

**Example 9**   Simplify: $\dfrac{6x^2}{60x^3}$

*Solution:*   Notice that 6 and 60 have a common factor of 6.

$$\frac{6x^2}{60x^3} = \frac{6 \cdot x \cdot x}{6 \cdot 10 \cdot x \cdot x \cdot x} = \frac{1}{10x}$$

> **Helpful Hint**
>
> When all the factors of the numerator or denominator are divided out, don't forget that the result is 1 in that numerator or denominator. If it helps, use the following notation.
>
> $$\frac{6x^2}{60x^3} = \frac{\overset{1}{\cancel{6}} \cdot \overset{1}{\cancel{x}} \cdot \overset{1}{\cancel{x}}}{\underset{1}{\cancel{6}} \cdot 10 \cdot \underset{1}{\cancel{x}} \cdot \underset{1}{\cancel{x}} \cdot x} = \frac{1}{10x} \quad \begin{matrix} \leftarrow 1 \cdot 1 \cdot 1 \text{ is } 1. \\ \leftarrow 1 \cdot 10 \cdot 1 \cdot 1 \cdot x \text{ is } 10x. \end{matrix}$$

TRY THE CONCEPT CHECK IN THE MARGIN.

## C SOLVING PROBLEMS BY WRITING FRACTIONS IN SIMPLEST FORM

Many real-life problems can be solved by writing fractions. To make the answers more clear, these fractions should be written in simplest form.

**Example 10**   Calculating Fraction of Parks in Washington State

As of 1999, there were 54 national parks in the United States. Three of these parks are located in the state of Washington. What fraction of the United States' national parks can be found in Washington state? Write the fraction in simplest form. (*Source:* National Park Service)

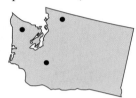

*Solution:*   First we determine the fraction of parks found in Washington state.

$$\frac{3}{54} \quad \leftarrow \text{national parks in Washington} \\ \leftarrow \text{total national parks}$$

Next we simplify the fraction.

$$\frac{3}{54} = \frac{3}{3 \cdot 18} = \frac{1}{18}$$

Thus, $\frac{1}{18}$ of the United States' national parks are in Washington state.

■■■■

✓ **CONCEPT CHECK**

Why is the following way to simplify the fraction $\frac{17}{75}$ incorrect?

$$\frac{1\,7}{7\,5} = \frac{1}{5}$$

**Practice Problem 10**

Eighty pigs were used in a recent study of olestra, a calorie-free fat substitute. A group of 12 of these pigs were fed a diet high in fat. What fraction of the pigs were fed the high-fat diet in this study? Write your answer in simplest form. (*Source:* From a study conducted by the Procter & Gamble Company)

Answers

**10.** $\frac{3}{20}$

✓ Concept Check:   Answers may vary.

# CALCULATOR EXPLORATIONS
## SIMPLIFYING FRACTIONS

### SCIENTIFIC CALCULATOR

Many calculators have a fraction key, such as , that allows you to simplify a fraction on the calculator. For example, to simplify $\frac{324}{612}$, enter

| 3 | 2 | 4 | $a^{b/c}$ | 6 | 1 | 2 | = |

The display will read

| 9 ⌐17 |

which represents $\frac{9}{17}$, the original fraction simplified.

### GRAPHING CALCULATOR

Graphing calculators also allow you to simplify fractions. The fraction option on a graphing calculator may be found under the MATH menu.

To simplify $\frac{324}{612}$, enter

| 3 | 2 | 4 | ÷ | 6 | 1 | 2 | MATH | ENTER | ENTER |

The display will read

| 324/612 ▶ FRAC 9/17 |

---

**Helpful Hint**

The Calculator Explorations boxes in this chapter provide only an introduction to fraction keys on calculators. Any time you use a calculator, there are both advantages and limitations to its use. Never rely solely on your calculator. It is very important that you understand how to perform all operations on fractions by hand in order to progress through later topics. For further information, talk to your instructor.

---

*Use your calculator to simplify each fraction.*

1. $\frac{128}{224}$  $\frac{4}{7}$     2. $\frac{231}{396}$  $\frac{7}{12}$     3. $\frac{340}{459}$  $\frac{20}{27}$     4. $\frac{999}{1350}$  $\frac{37}{50}$

5. $\frac{810}{432}$  $\frac{15}{8}$     6. $\frac{315}{225}$  $\frac{7}{5}$     7. $\frac{243}{54}$  $\frac{9}{2}$     8. $\frac{689}{455}$  $\frac{53}{35}$

**Name** _____ **Section** _____ **Date** _____

## MENTAL MATH

**1.** Is 2430 divisible by 2? By 3? By 5?

*Write the prime factorization of each number.*

**2.** 15      **3.** 10      **4.** 6      **5.** 21

**6.** 4      **7.** 9      **8.** 14

## EXERCISE SET 4.2

**A** *Write the prime factorization of each number. See Examples 1 through 3.*

**1.** 20      **2.** 12      📼 **3.** 48      **4.** 80

**5.** 64      **6.** 45      📼 **7.** 240      **8.** 128

**B** *Simplify each fraction. See Examples 4 through 9.*

**9.** $\dfrac{3}{12}$      **10.** $\dfrac{5}{20}$      **11.** $\dfrac{7x}{35}$      **12.** $\dfrac{9}{48z}$

📼 **13.** $\dfrac{14}{16}$      **14.** $\dfrac{18}{4}$      **15.** $\dfrac{24a}{30a}$      **16.** $\dfrac{70y}{80xy}$

**17.** $\dfrac{35}{42}$      **18.** $\dfrac{25}{55}$      📼 **19.** $\dfrac{30x^2}{36x}$      **20.** $\dfrac{45b}{80b^2}$

**21.** $\dfrac{16}{24}$      **22.** $\dfrac{18}{45}$      **23.** $\dfrac{45xz}{60z}$      **24.** $\dfrac{22a}{99ab}$

**25.** $\dfrac{3b}{2a}$

**26.** $\dfrac{7x}{4y}$

**27.** $\dfrac{7}{8}$

**28.** $\dfrac{7}{8}$

**29.** $\dfrac{3}{7}$

**30.** $\dfrac{2}{5}$

**31.** $\dfrac{3y}{5}$

**32.** $\dfrac{2}{3x}$

**33.** $\dfrac{4}{7}$

**34.** $\dfrac{3}{4}$

**35.** $\dfrac{4x^2y}{5}$

**36.** $\dfrac{4a}{9b^2}$

**37.** $\dfrac{4}{5}$

**38.** $\dfrac{3}{4}$

**39.** $\dfrac{5x}{8}$

**40.** $\dfrac{6y}{7}$

**41.** $\dfrac{3x}{10}$

**42.** $\dfrac{2}{5z}$

**43.** $\dfrac{3a^2}{2b^3}$

**44.** $\dfrac{5}{3xy^2z^2}$

**45.** $\dfrac{5}{8z}$

**46.** $\dfrac{7b}{15}$

**47.** $\dfrac{3}{4}$ of a shift

**48.** $\dfrac{1}{10}$ of the caps

**49.** $\dfrac{1}{2}$ mi

**50.** $\dfrac{1}{5}$ m

**246**

**Name** _____

**25.** $\dfrac{39ab}{26a^2}$

**26.** $\dfrac{42x^2}{24xy}$

**27.** $\dfrac{63}{72}$

**28.** $\dfrac{56}{64}$

**29.** $\dfrac{21}{49}$

**30.** $\dfrac{14}{35}$

**31.** $\dfrac{24y}{40}$

**32.** $\dfrac{36}{54x}$

**33.** $\dfrac{36z}{63z}$

**34.** $\dfrac{39b}{52b}$

**35.** $\dfrac{72x^3y^2}{90xy}$

**36.** $\dfrac{24a^2b}{54ab^3}$

**37.** $\dfrac{12}{15}$

**38.** $\dfrac{18}{24}$

**39.** $\dfrac{25x^2}{40x}$

**40.** $\dfrac{36y^2}{42y}$

**41.** $\dfrac{27xy}{90y}$

**42.** $\dfrac{60y}{150yz}$

**43.** $\dfrac{36a^3bc^2}{24ab^4c^2}$

**44.** $\dfrac{60x^2yz}{36x^3y^3z^3}$

**45.** $\dfrac{40xy}{64xyz}$

**46.** $\dfrac{28abc}{60ac}$

**C** *Solve. Write each fraction in simplest form. See Example 10.*

**47.** A work shift for an employee at Mc-Donald's consists of 8 hours. What fraction of the employee's work shift is represented by 6 hours?

**48.** Two thousand baseball caps were sold one year at the U.S. Open golf tournament. What fractional part of this total does 200 caps represent?

**49.** There are 5280 feet in a mile. What fraction of a mile is represented by 2640 feet?

**50.** There are 100 centimeters in 1 meter. What fraction of a meter is 20 centimeters?

**51.** As of the end of 1999, a total of 408 individuals from around the world had flown in space. Of these, 257 were citizens of the United States. What fraction of individuals who have flown in space were Americans? (*Source:* Congressional Research Service)

**52.** Hallmark Cards employs 20,100 full-time employees worldwide. About 5900 employees work at the Hallmark headquarters in Kansas City, Missouri. What fraction of Hallmark employees work in Kansas City? (*Source:* Hallmark Cards, Inc.)

**53.** There are 16,000 students at a local university. If 8800 are females, what fraction of the students are *male*?

**54.** Four out of 10 marbles are red. What fraction of marbles are *not red*?

**55.** Fifteen states in the United States have Ritz-Carlton hotels. (*Source:* Marriott International)
  **a.** What fraction of states can claim at least one Ritz-Carlton hotel?
  **b.** How many states do not have a Ritz-Carlton hotel?
  **c.** Write the fraction of states without a Ritz-Carlton hotel.

**56.** As of 1999, there were 54 national parks in the United States. Eight of these parks are located in Alaska. (*Source:* National Park Service)
  **a.** What fraction of the United States' national parks can be found in Alaska?
  **b.** How many of the United States' national parks are found outside Alaska?
  **c.** Write the fraction of national parks found in states other than Alaska.

**57.** The outer wall of the Pentagon is 24 inches wide. Ten inches is concrete, 8 inches is brick and 6 inches is limestone. What fraction of the width is concrete? (*Source:* USA Today 1/28/2000)

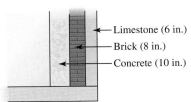

Limestone (6 in.)
Brick (8 in.)
Concrete (10 in.)

**58.** There are 35 students in a biology class. If 10 students made an A on the first test, what fraction of the students made an A?

---

**51.** $\dfrac{257}{408}$ of individuals

**52.** $\dfrac{59}{201}$ of Hallmark employees

**53.** $\dfrac{9}{20}$ of the students

**54.** $\dfrac{3}{5}$ of the marbles

**55. a.** $\dfrac{3}{10}$ of the states

  **b.** 35 states

  **c.** $\dfrac{7}{10}$ of the states

**56. a.** $\dfrac{4}{27}$ of the parks

  **b.** 46 parks

  **c.** $\dfrac{23}{27}$ of the parks

**57.** $\dfrac{5}{12}$ of the width

**58.** $\dfrac{2}{7}$ of the students

**Name** _____

*The following graph is called a circle graph or pie chart. Each sector (shaped like a piece of pie) shows the fraction of entering college freshmen who expect to major in each discipline shown. The whole circle represents the entire class of college freshmen. Use this graph to answer Exercises 59–62.*

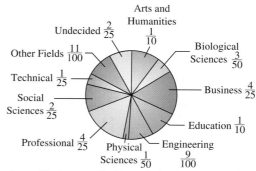

Source: *The American Freshman: National Norms for Fall 1997*, Higher Education Research Institute

**59.** What fraction of entering college freshmen plan to major in education?

**60.** What fraction of entering college freshmen plan to major in social sciences?

**61.** Why is the Professional sector the same size as the Business sector?

**62.** Why is the Physical Sciences sector smaller than the Biological Sciences sector?

## REVIEW AND PREVIEW

*Evaluate each expression using the given replacement numbers. See Section 2.5.*

**63.** $\dfrac{x^3}{9}$ when $x = -3$

**64.** $\dfrac{y^3}{5}$ when $y = -5$

**65.** $2y$ when $y = -7$

**66.** $-5a$ when $a = -4$

**67.** $3z - y$ when $z = 2$ and $y = 6$

**68.** $7a - b$ when $a = 1$ and $b = -5$

**69.** $a^2 + 2b + 3$ when $a = 4$ and $b = 5$

**70.** $yx - z^2$ when $y = 6$, $x = 6$, and $z = 6$

**COMBINING CONCEPTS**

**71.** Which fraction is closest to 0? Explain your answer.

**a.** $\dfrac{1}{2}$  **b.** $\dfrac{1}{3}$

**c.** $\dfrac{1}{4}$  **d.** $\dfrac{1}{5}$

**72.** Which fraction is closest to 1? Explain your answer.

**a.** $\dfrac{1}{2}$  **b.** $\dfrac{2}{3}$

**c.** $\dfrac{3}{4}$  **d.** $\dfrac{5}{6}$

*Answer true or false for each statement.*

**73.** $\dfrac{14}{42} = \dfrac{2 \cdot 7}{2 \cdot 3 \cdot 7} = \dfrac{0}{3}$

**74.** The fractions $\dfrac{8}{20}$, $\dfrac{4}{5}$, and $\dfrac{16}{60}$ are all equivalent.

**75.** A proper fraction cannot be equivalent to an improper fraction.

**76.** A fraction whose numerator and denominator are two different prime numbers cannot be simplified.

*Simplify each fraction.*

**77.** $\dfrac{372}{620}$

**78.** $\dfrac{9506}{12,222}$

*There are generally considered to be eight basic blood types. The table shows the number of people with the various blood types in a typical group of 100 blood donors. Use this table to answer Exercises 79–83. Write each answer in simplest form.*

**79.** What fraction of blood donors have A Rh-positive blood type?

**80.** What fraction of blood donors have O blood type?

**81.** What fraction of blood donors have AB blood type?

**82.** What fraction of blood donors have B blood type?

**83.** What fraction of blood donors have the negative Rh factor?

| DISTRIBUTION OF BLOOD TYPES IN BLOOD DONORS | |
|---|---|
| Blood Type | Number of People |
| O Rh-positive | 37 |
| O Rh-negative | 7 |
| A Rh-positive | 36 |
| A Rh-negative | 6 |
| B Rh-positive | 9 |
| B Rh-negative | 1 |
| AB Rh-positive | 3 |
| AB Rh-negative | 1 |

(*Source:* American Red Cross Biomedical Services)

**71.** d _____

**72.** d _____

**73.** false _____

**74.** false _____

**75.** true _____

**76.** true _____

**77.** $\dfrac{3}{5}$ _____

**78.** $\dfrac{7}{9}$ _____

**79.** $\dfrac{9}{25}$ of the donors

**80.** $\dfrac{11}{25}$ of the donors

**81.** $\dfrac{1}{25}$ of the donors

**82.** $\dfrac{1}{10}$ of the donors

**83.** $\dfrac{3}{20}$ of the donors

# Focus on Mathematical Connections

## MODELING FRACTIONS

One way to physically represent, or model, fractions is with fraction strips. Fraction strips are all the same length and different fraction strips represent fractions with different denominators. For example, a fraction strip for the denominator 2 is subdivided into two equal sections and a fraction strip for the denominator 3 is subdivided into three equal sections. By shading the appropriate number of sections of a fraction strip, we can represent various fractions. Figure 1 shows how to use a fraction strip for the denominator 4 to represent the fraction $\frac{3}{4}$, and Figure 2 shows how to use a fraction strip for the denominator 5 to represent the fraction $\frac{4}{5}$.

Figure 1    [fraction strip diagram]    Representation of $\frac{3}{4}$

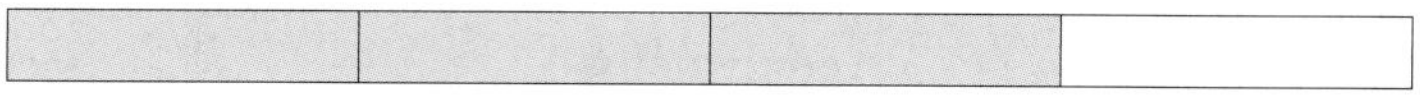

Figure 2    [fraction strip diagram]    Representation of $\frac{4}{5}$

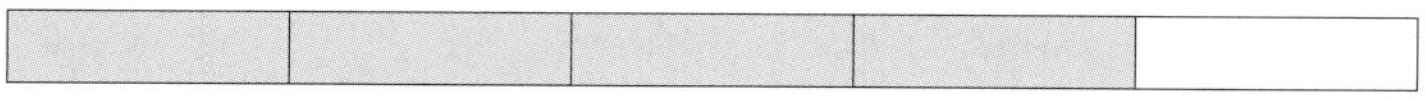

Notice that because the fraction strips are the same length, we can use the representations to compare fractions. We see that shaded sections in Figure 2, when taken together, are longer than the shaded sections in Figure 1. We can conclude from this that $\frac{3}{4} < \frac{4}{5}$.

## CRITICAL THINKING

**1.** Make a set of fraction strips of equal length for the denominators 2, 3, 4, 5, 6, 8, and 12.

**2.** Use the fraction strips to write $<$, $=$, or $>$ between each pair of fractions.

**a.** $\frac{5}{12}$    $\frac{3}{5}$    $<$

**b.** $\frac{1}{2}$    $\frac{4}{8}$    $=$

**c.** $\frac{5}{6}$    $\frac{3}{4}$    $>$

**d.** $\frac{3}{5}$    $\frac{2}{3}$    $<$

# 4.3   MULTIPLYING AND DIVIDING FRACTIONS

## A   MULTIPLYING FRACTIONS

Let's use a diagram to discover how fractions are multiplied. For example, to multiply $\frac{1}{2}$ and $\frac{3}{4}$, we find $\frac{1}{2}$ of $\frac{3}{4}$. To do this, we begin with a diagram showing $\frac{3}{4}$ of rectangle's area shaded.

    $\frac{3}{4}$ of the rectangle's area is shaded.

To find $\frac{1}{2}$ of $\frac{3}{4}$, we heavily shade $\frac{1}{2}$ of the part that is already shaded.

By counting smaller rectangles, we see that $\frac{3}{8}$ of the larger rectangle is now heavily shaded, so that $\frac{1}{2}$ of $\frac{3}{4}$ is $\frac{3}{8}$. This means that

$$\frac{1}{2} \cdot \frac{3}{4} = \frac{3}{8}$$   Notice that $\frac{1}{2} \cdot \frac{3}{4} = \frac{1 \cdot 3}{2 \cdot 4} = \frac{3}{8}$.

Notice that the numerator of the product is equal to the product of the numerators and that the denominator of the product is equal to the product of the denominators. This is how we multiply fractions.

> **MULTIPLYING TWO FRACTIONS**
>
> If $a$, $b$, $c$, and $d$ are numbers and $b$ and $d$ are not 0, then
>
> $$\frac{a}{b} \cdot \frac{c}{d} = \frac{a \cdot c}{b \cdot d}$$
>
> In other words, to multiply two fractions, multiply the numerators and multiply the denominators.

**Examples**   Multiply.

**1.** $\frac{2}{3} \cdot \frac{5}{11} = \frac{2 \cdot 5}{3 \cdot 11} = \frac{10}{33}$   ← Product of numerators
← Product of denominators

**2.** $\frac{1}{4} \cdot \frac{1}{2} = \frac{1 \cdot 1}{4 \cdot 2} = \frac{1}{8}$

**Example 3**   Multiply and simplify: $\frac{6}{7} \cdot \frac{14}{27}$

*Solution:*   $\frac{6}{7} \cdot \frac{14}{27} = \frac{6 \cdot 14}{7 \cdot 27}$

Next, simplify by factoring into primes and dividing out common factors.

$$\frac{6 \cdot 14}{7 \cdot 27} = \frac{2 \cdot 3 \cdot 2 \cdot 7}{7 \cdot 3 \cdot 3 \cdot 3}$$

$$= \frac{4}{9}$$

---

TEACHING TIP

To help students understand multiplication of fractions, you may want to ask: "If you were to eat $\frac{1}{2}$ of $\frac{1}{4}$ of a pie, how much of the whole pie would you eat?" Draw a diagram and show them that this means

$$\frac{1}{2} \cdot \frac{1}{4} = \frac{1}{8}$$

**Practice Problems 1–2**

Multiply.

1. $\frac{3}{8} \cdot \frac{5}{7}$              2. $\frac{1}{3} \cdot \frac{1}{6}$

**Practice Problem 3**

Multiply and simplify: $\frac{6}{11} \cdot \frac{5}{8}$

**Answers**

**1.** $\frac{15}{56}$   **2.** $\frac{1}{18}$   **3.** $\frac{15}{44}$

### Helpful Hint

In simplifying a product, it may be possible to identify common factors without actually writing the prime factorizations. For example,

$$\frac{10}{11} \cdot \frac{1}{20} = \frac{10 \cdot 1}{11 \cdot 20} = \frac{10 \cdot 1}{11 \cdot 10 \cdot 2} = \frac{1}{22}$$

---

## Practice Problem 4

Multiply and simplify: $\dfrac{4}{15} \cdot \dfrac{3}{8}$

---

**Example 4**   Multiply and simplify: $\dfrac{23}{32} \cdot \dfrac{4}{7}$

*Solution:*   Notice that 4 and 32 have a common factor of 4.

$$\frac{23}{32} \cdot \frac{4}{7} = \frac{23 \cdot 4}{32 \cdot 7} = \frac{23 \cdot 4}{4 \cdot 8 \cdot 7} = \frac{23}{56}$$

After multiplying two fractions, *always* check to see whether the product can be simplified.

---

## Practice Problem 5

Multiply: $\dfrac{1}{22} \cdot \left(-\dfrac{11}{28}\right)$

---

**Example 5**   Multiply: $-\dfrac{1}{4} \cdot \dfrac{1}{2}$

*Solution:*   Recall that the product of a negative number and a positive number is a negative number.

$$-\frac{1}{4} \cdot \frac{1}{2} = -\frac{1 \cdot 1}{4 \cdot 2} = -\frac{1}{8}$$

---

## Practice Problem 6

Multiply: $\dfrac{9}{5} \cdot \dfrac{20}{12}$

---

**Example 6**   Multiply: $\dfrac{13}{6} \cdot \dfrac{30}{26}$

*Solution:*   $\dfrac{13}{6} \cdot \dfrac{30}{26} = \dfrac{13 \cdot 6 \cdot 5}{6 \cdot 2 \cdot 13} = \dfrac{5}{2}$

We multiply fractions in the same way if variables are involved.

---

## Practice Problem 7

Multiply: $\dfrac{2}{3} \cdot \dfrac{3y}{2}$

---

**Example 7**   Multiply: $\dfrac{3x}{4} \cdot \dfrac{8}{5x}$

*Solution:*   $\dfrac{3x}{4} \cdot \dfrac{8}{5x} = \dfrac{3 \cdot x \cdot 8}{4 \cdot 5 \cdot x} = \dfrac{3 \cdot 4 \cdot 2}{4 \cdot 5} = \dfrac{6}{5}$

### Helpful Hint

Recall that when the denominator of a fraction contains a variable, such as $\dfrac{8}{5x}$, we assume that the variable does not represent 0.

---

## Practice Problem 8

Multiply: $\dfrac{a^3}{b^2} \cdot \dfrac{b}{a^2}$

**Answers**

**4.** $\dfrac{1}{10}$   **5.** $-\dfrac{1}{56}$   **6.** 3   **7.** $y$   **8.** $\dfrac{a}{b}$

---

**Example 8**   Multiply: $\dfrac{x^2}{y} \cdot \dfrac{y^3}{x}$

*Solution:*   $\dfrac{x^2}{y} \cdot \dfrac{y^3}{x} = \dfrac{x^2 \cdot y^3}{y \cdot x} = \dfrac{x \cdot x \cdot y \cdot y \cdot y}{y \cdot x} = \dfrac{x \cdot y \cdot y}{1} = xy^2$

## B EVALUATING EXPRESSIONS WITH FRACTIONAL BASES

The base of an exponential expression can also be a fraction.

$$\left(\frac{1}{3}\right)^4 = \underbrace{\frac{1}{3} \cdot \frac{1}{3} \cdot \frac{1}{3} \cdot \frac{1}{3}} = \frac{1 \cdot 1 \cdot 1 \cdot 1}{3 \cdot 3 \cdot 3 \cdot 3} = \frac{1}{81}$$

$\frac{1}{3}$ is a factor 4 times.

**Example 9**   Evaluate.

a. $\left(\frac{2}{5}\right)^4$   b. $\left(-\frac{1}{4}\right)^2$

*Solution:*   a. $\left(\frac{2}{5}\right)^4 = \frac{2}{5} \cdot \frac{2}{5} \cdot \frac{2}{5} \cdot \frac{2}{5} = \frac{2 \cdot 2 \cdot 2 \cdot 2}{5 \cdot 5 \cdot 5 \cdot 5} = \frac{16}{625}$

b. $\left(-\frac{1}{4}\right)^2 = \left(-\frac{1}{4}\right) \cdot \left(-\frac{1}{4}\right) = \frac{1 \cdot 1}{4 \cdot 4} = \frac{1}{16}$   ▬▬▬

**Practice Problem 9**

Evaluate.

a. $\left(\frac{3}{4}\right)^3$   b. $\left(-\frac{4}{5}\right)^2$

## C DIVIDING FRACTIONS

Before we can divide fractions, we need to know how to find the **reciprocal** of a fraction.

---

**RECIPROCAL OF A FRACTION**

Two numbers are **reciprocals** of each other if their product is 1. The reciprocal of the fraction $\frac{a}{b}$ is $\frac{b}{a}$ because $\frac{a}{b} \cdot \frac{b}{a} = \frac{a \cdot b}{b \cdot a} = 1$.

---

For example,

The reciprocal of $\frac{2}{5}$ is $\frac{5}{2}$ because $\frac{2}{5} \cdot \frac{5}{2} = \frac{10}{10} = 1$.

The reciprocal of 5 is $\frac{1}{5}$ because $5 \cdot \frac{1}{5} = \frac{5}{1} \cdot \frac{1}{5} = \frac{5}{5} = 1$.

The reciprocal of $-\frac{7}{11}$ is $-\frac{11}{7}$ because $-\frac{7}{11} \cdot -\frac{11}{7} = \frac{77}{77} = 1$.

---

**Helpful Hint**

Every number has a reciprocal except 0. The number 0 has no reciprocal because there is no number as such that $0 \cdot a = 1$.

---

Division of fractions has the same meaning as division of whole numbers. For example,

$10 \div 5$ means: How many 5s are there in 10?

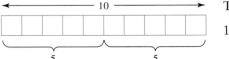

5     5

There are two 5s in 10, so

$10 \div 5 = 2$

Answers

**9. a.** $\frac{27}{64}$   **b.** $\frac{16}{25}$

$\dfrac{3}{4} \div \dfrac{1}{8}$ means: How many $\dfrac{1}{8}$s are there in $\dfrac{3}{4}$?

There are six $\dfrac{1}{8}$s in $\dfrac{3}{4}$, so

$$\dfrac{3}{4} \div \dfrac{1}{8} = 6$$

We can use reciprocals to divide fractions.

---

**DIVIDING FRACTIONS**

If $b$, $c$, and $d$ are not 0, then $\dfrac{a}{b} \div \dfrac{c}{d} = \dfrac{a}{b} \cdot \dfrac{d}{c} = \dfrac{a \cdot d}{b \cdot c}$.

In other words, to divide fractions, multiply the first fraction by the reciprocal of the second fraction.

---

For example,

$$\overset{\text{multiply by reciprocal}}{\dfrac{3}{4} \div \dfrac{1}{8}} = \dfrac{3}{4} \cdot \dfrac{8}{1} = \dfrac{3 \cdot 8}{4 \cdot 1} = \dfrac{3 \cdot 2 \cdot 4}{4 \cdot 1} = \dfrac{6}{1} \quad \text{or} \quad 6$$

**Examples**    Divide and simplify.

**10.** $\dfrac{7}{8} \div \dfrac{2}{9} = \dfrac{7}{8} \cdot \dfrac{9}{2} = \dfrac{7 \cdot 9}{8 \cdot 2} = \dfrac{63}{16}$

**11.** $\dfrac{2}{5} \div \dfrac{1}{2} = \dfrac{2}{5} \cdot \dfrac{2}{1} = \dfrac{2 \cdot 2}{5 \cdot 1} = \dfrac{4}{5}$

After dividing two fractions, *always* check to see whether the result can be simplified.

---

**Helpful Hint**

When dividing by a fraction, do not look for common factors to divide out until you rewrite the division as multiplication.

$$\underset{\downarrow}{\overset{\text{Do not try to divide out these two 2s.}}{\dfrac{1}{2} \div \dfrac{2}{3}}} = \dfrac{1}{2} \cdot \dfrac{3}{2} = \dfrac{3}{4}$$

---

**Example 12**    Divide: $-\dfrac{5}{16} \div -\dfrac{3}{4}$

*Solution:*    Recall that the quotient (or product) of two negative numbers is a positive number.

$$-\dfrac{5}{16} \div -\dfrac{3}{4} = -\dfrac{5}{16} \cdot -\dfrac{4}{3} = \dfrac{5 \cdot 4}{4 \cdot 4 \cdot 3} = \dfrac{5}{12}$$

---

**Practice Problems 10–11**

Divide and simplify.

10. $\dfrac{3}{2} \div \dfrac{14}{5}$

11. $\dfrac{4}{9} \div \dfrac{1}{2}$

TEACHING TIP

Make sure that students read and understand this helpful hint.

**Practice Problem 12**

Divide: $\dfrac{10}{4} \div \dfrac{2}{9}$

**Answers**

**10.** $\dfrac{15}{28}$   **11.** $\dfrac{8}{9}$   **12.** $\dfrac{45}{4}$

**Example 13**   Divide: $\dfrac{2x}{3} \div 3x^2$

*Solution:*   $\dfrac{2x}{3} \div 3x^2 = \dfrac{2x}{3} \div \dfrac{3x^2}{1} = \dfrac{2x}{3} \cdot \dfrac{1}{3x^2} = \dfrac{2 \cdot x \cdot 1}{3 \cdot 3 \cdot x \cdot x} = \dfrac{2}{9x}$

## TRY THE CONCEPT CHECK IN THE MARGIN.

**Example 14**   Simplify: $\left( \dfrac{4}{7} \cdot \dfrac{3}{8} \right) \div -\dfrac{3}{4}$

*Solution:*   Remember to perform the operations inside the ( ) first.

$$\left( \dfrac{4}{7} \cdot \dfrac{3}{8} \right) \div -\dfrac{3}{4} = \left( \dfrac{4 \cdot 3}{7 \cdot 2 \cdot 4} \right) \div -\dfrac{3}{4} = \dfrac{3}{14} \div -\dfrac{3}{4}$$

Now divide.

$$\dfrac{3}{14} \div -\dfrac{3}{4} = \dfrac{3}{14} \cdot -\dfrac{4}{3} = -\dfrac{3 \cdot 2 \cdot 2}{2 \cdot 7 \cdot 3} = -\dfrac{2}{7}$$

## D   MULTIPLYING AND DIVIDING WITH FRACTIONAL REPLACEMENT VALUES

**Example 15**   If $x = \dfrac{7}{8}$ and $y = -\dfrac{1}{3}$, evaluate (**a**) $xy$ and (**b**) $x \div y$.

*Solution:*   Replace $x$ with $\dfrac{7}{8}$ and $y$ with $-\dfrac{1}{3}$.

**a.**   $xy = \dfrac{7}{8} \cdot -\dfrac{1}{3}$        **b.**   $x \div y = \dfrac{7}{8} \div -\dfrac{1}{3}$

$\phantom{xy} = -\dfrac{7 \cdot 1}{8 \cdot 3}$              $\phantom{x \div y} = \dfrac{7}{8} \cdot -\dfrac{3}{1}$

$\phantom{xy} = -\dfrac{7}{24}$                 $\phantom{x \div y} = -\dfrac{7 \cdot 3}{8 \cdot 1}$

$\phantom{x \div y xxxxxxxxxxxxxxxxxxx} = -\dfrac{21}{8}$

**Example 16**   Is $-\dfrac{2}{3}$ a solution of the equation $-\dfrac{1}{2}x = \dfrac{1}{3}$?

*Solution:*   To check whether a number is a solution of an equation, recall that we replace the variable with the given number and see if a true statement results.

$-\dfrac{1}{2} \cdot x = \dfrac{1}{3}$   Recall that $-\dfrac{1}{2}x$ means $-\dfrac{1}{2} \cdot x$.

$-\dfrac{1}{2} \cdot -\dfrac{2}{3} = \dfrac{1}{3}$   Replace $x$ with $-\dfrac{2}{3}$.

$\dfrac{1 \cdot 2}{2 \cdot 3} = \dfrac{1}{3}$   The product of two negative numbers is a positive number.

$\dfrac{1}{3} = \dfrac{1}{3}$   True.

Since we have a true statement, $-\dfrac{2}{3}$ is a solution.

---

**Practice Problem 13**

Divide: $\dfrac{3y}{4} \div 5y^3$

## ✓ CONCEPT CHECK

Which is the correct way to divide $\dfrac{3}{5}$ by $\dfrac{5}{12}$? Explain.

a.   $\dfrac{3}{5} \div \dfrac{5}{12} = \dfrac{3}{5} \cdot \dfrac{12}{5}$

b.   $\dfrac{3}{5} \div \dfrac{5}{12} = \dfrac{5}{3} \cdot \dfrac{5}{12}$

**Practice Problem 14**

Simplify: $\left( -\dfrac{2}{3} \cdot \dfrac{9}{14} \right) \div \dfrac{7}{15}$

**Practice Problem 15**

If $x = -\dfrac{3}{4}$ and $y = \dfrac{9}{2}$, evaluate (a) $xy$, and (b) $x \div y$.

**Practice Problem 16**

Is $-\dfrac{9}{8}$ a solution of the equation
$$2x = -\dfrac{9}{4}?$$

**Answers**

**13.** $\dfrac{3}{20y^2}$   **14.** $-\dfrac{45}{49}$   **15. a.** $-\dfrac{27}{8}$   **b.** $-\dfrac{1}{6}$

**16.** yes

✓ **Concept Check:** a

## E  SOLVING APPLICATIONS BY MULTIPLYING AND DIVIDING FRACTIONS

To solve real-life problems that involve multiplying and dividing fractions, we will use our four problem-solving steps from Chapter 3. In Example 17, a new key word that implies multiplication is used. That key word is "of."

### Practice Problem 17

About $\frac{1}{3}$ of all plant and animal species in the United States are at risk of becoming extinct. There are 20,439 known species of plants and animals in the United States. How many species are at risk of extinction? (*Source:* The Nature Conservancy)

**Example 17**   Finding Number of Roller Coasters in an Amusement Park

Cedar Point is an amusement park located in Sandusky, Ohio. Its collection of 60 rides is the largest in the world. One-fifth of Cedar Point's rides are roller coasters. How many roller coasters are in Cedar Point's collection of rides? (*Source:* Cedar Fair, L.P.)

*Solution:*

1. UNDERSTAND the problem. To do so, read and reread the problem. We are told that $\frac{1}{5}$ of Cedar Point's rides are roller coasters. The word "of" here means multiplication.

2. TRANSLATE.

In words:

| Number of roller coasters | is | $\frac{1}{5}$ | of | total rides at Cedar Point |
|---|---|---|---|---|
| ↓ | ↓ | ↓ | ↓ | ↓ |

Translate:

$$\text{Number of roller coasters} = \frac{1}{5} \cdot 60$$

3. SOLVE.

$$\frac{1}{5} \cdot 60 = \frac{1}{5} \cdot \frac{60}{1} = \frac{1 \cdot 60}{5 \cdot 1} = \frac{1 \cdot 5 \cdot 12}{5 \cdot 1} = \frac{12}{1} \quad \text{or} \quad 12$$

4. INTERPRET. *Check* your work. *State* your conclusion: The number of roller coasters at Cedar Point is 12.

**Answer**

**17.** 6813 species

**Name** _____ **Section** _____ **Date** _____

## MENTAL MATH

*Find each product.*

**1.** $\dfrac{1}{3} \cdot \dfrac{2}{5}$

**2.** $\dfrac{2}{3} \cdot \dfrac{4}{7}$

**3.** $\dfrac{6}{5} \cdot \dfrac{1}{7}$

**4.** $\dfrac{7}{3} \cdot \dfrac{2}{3}$

**5.** $\dfrac{3}{1} \cdot \dfrac{3}{8}$

**6.** $\dfrac{2}{1} \cdot \dfrac{7}{11}$

## EXERCISE SET 4.3

**A** *Multiply. Write the product in simplest form. See Examples 1 through 8.*

**1.** $\dfrac{7}{8} \cdot \dfrac{2}{3}$

**2.** $\dfrac{5}{9} \cdot \dfrac{7}{4}$

**3.** $-\dfrac{2}{7} \cdot \dfrac{5}{8}$

**4.** $\dfrac{5}{8} \cdot -\dfrac{1}{3}$

**5.** $-\dfrac{1}{2} \cdot -\dfrac{2}{15}$

**6.** $-\dfrac{3}{8} \cdot -\dfrac{5}{12}$

**7.** $\dfrac{18x}{20} \cdot \dfrac{36}{99}$

**8.** $\dfrac{5}{32} \cdot \dfrac{64y}{100}$

**9.** $3a^2 \cdot \dfrac{1}{4}$

**10.** $-\dfrac{2}{3} \cdot 6y^3$

**11.** $\dfrac{x^3}{y^3} \cdot \dfrac{y^2}{x}$

**12.** $\dfrac{a}{b^3} \cdot \dfrac{b}{a^3}$

**B** *Evaluate. See Example 9.*

**13.** $\left(\dfrac{1}{5}\right)^3$

**14.** $\left(-\dfrac{1}{2}\right)^4$

**15.** $\left(-\dfrac{2}{3}\right)^2$

**16.** $\left(\dfrac{8}{9}\right)^2$

**17.** $\left(-\dfrac{2}{3}\right)^3 \cdot \dfrac{1}{2}$

**18.** $\left(-\dfrac{3}{4}\right)^3 \cdot \dfrac{1}{3}$

**Name** _____

**C** *Divide. Write all quotients in simplest form. See Examples 10 through 13.*

**19.** $\dfrac{2}{3} \div \dfrac{5}{6}$

**20.** $\dfrac{5}{8} \div \dfrac{2}{3}$

**21.** $-\dfrac{6}{15} \div \dfrac{12}{5}$

**22.** $-\dfrac{4}{15} \div -\dfrac{8}{3}$

**23.** $\dfrac{8}{9} \div \dfrac{x}{2}$

**24.** $\dfrac{10}{11} \div -\dfrac{4}{5}$

**25.** $\dfrac{11y}{20} \div \dfrac{3}{11}$

**26.** $\dfrac{9z}{20} \div \dfrac{2}{9}$

**27.** $-\dfrac{2}{3} \div 4$

**28.** $-\dfrac{5}{6} \div 10$

**29.** $\dfrac{1}{5x} \div \dfrac{5}{x^2}$

**30.** $\dfrac{3}{y^2} \div \dfrac{9}{y}$

**A** **B** **C** *Perform each indicated operation. See Examples 1 through 14.*

**31.** $\dfrac{2}{3} \cdot \dfrac{5}{9}$

**32.** $\dfrac{8}{15} \cdot \dfrac{5}{32}$

**33.** $\dfrac{3x}{7} \div \dfrac{5}{6x}$

**34.** $\dfrac{16}{27y} \div \dfrac{8}{15y}$

**35.** $-\dfrac{5}{28} \cdot \dfrac{35}{25}$

**36.** $\dfrac{24}{45} \cdot -\dfrac{5}{8}$

**37.** $-\dfrac{3}{5} \div -\dfrac{4}{5}$

**38.** $-\dfrac{11}{16} \div -\dfrac{13}{16}$

**39.** $\left(-\dfrac{3}{4}\right)^2$

**40.** $\left(-\dfrac{1}{2}\right)^5$

**41.** $\dfrac{x^2}{y} \cdot \dfrac{y^3}{x}$

**42.** $\dfrac{b}{a^2} \cdot \dfrac{a^3}{b^3}$

**43.** $7 \div \dfrac{2}{11}$

**44.** $-100 \div \dfrac{1}{2}$

**45.** $-3x \div \dfrac{x^2}{12}$

**46.** $7x \div \dfrac{14x}{3}$

**47.** $\left(\dfrac{2}{7} \div \dfrac{7}{2}\right) \cdot \dfrac{3}{4}$  **48.** $\dfrac{1}{2} \cdot \left(\dfrac{5}{6} \div \dfrac{1}{12}\right)$  **49.** $-\dfrac{19}{63y} \cdot 9y^2$  **50.** $16a^2 \cdot -\dfrac{31}{24a}$

**51.** $-\dfrac{2}{3} \cdot -\dfrac{6}{11}$  **52.** $-\dfrac{1}{5} \cdot -\dfrac{6}{7}$  ▭ **53.** $\dfrac{4}{8} \div \dfrac{3}{16}$  **54.** $\dfrac{9}{2} \div \dfrac{16}{15}$

**55.** $\dfrac{21x^2}{10y} \div \dfrac{14x}{25y}$  **56.** $\dfrac{17y^2}{24x} \div \dfrac{13y}{18x}$  **57.** $\left(1 \div \dfrac{3}{4}\right) \cdot \dfrac{2}{3}$  **58.** $\left(33 \div \dfrac{2}{11}\right) \cdot \dfrac{5}{9}$

**59.** $\dfrac{a^3}{2} \div 30a^3$  **60.** $15c^3 \div \dfrac{3c^2}{5}$  **61.** $\dfrac{ab^2}{c} \cdot \dfrac{c}{ab}$  **62.** $\dfrac{ac}{b} \cdot \dfrac{b^3}{a^2c}$

▭ **63.** $\left(\dfrac{1}{2} \cdot \dfrac{2}{3}\right) \div \dfrac{5}{6}$  **64.** $\left(\dfrac{3}{4} \cdot \dfrac{8}{9}\right) \div \dfrac{2}{5}$  **65.** $-\dfrac{4}{7} \div \left(\dfrac{4}{5} \cdot \dfrac{3}{7}\right)$  **66.** $\dfrac{5}{8} \div \left(\dfrac{4}{7} \cdot -\dfrac{5}{16}\right)$

**D** *Given the following replacement values, evaluate (**a**) xy and (**b**) x ÷ y. See Example 15.*

**67.** $x = \dfrac{2}{5}$ and $y = \dfrac{5}{6}$  **68.** $x = \dfrac{8}{9}$ and $y = \dfrac{1}{4}$

---

**47.** $\dfrac{3}{49}$

**48.** $5$

**49.** $-\dfrac{19y}{7}$

**50.** $-\dfrac{62a}{3}$

**51.** $\dfrac{4}{11}$

**52.** $\dfrac{6}{35}$

**53.** $\dfrac{8}{3}$

**54.** $\dfrac{135}{32}$

**55.** $\dfrac{15x}{4}$

**56.** $\dfrac{51y}{52}$

**57.** $\dfrac{8}{9}$

**58.** $\dfrac{605}{6}$

**59.** $\dfrac{1}{60}$

**60.** $25c$

**61.** $b$

**62.** $\dfrac{b^2}{a}$

**63.** $\dfrac{2}{5}$

**64.** $\dfrac{5}{3}$

**65.** $-\dfrac{5}{3}$

**66.** $-\dfrac{7}{2}$

**67. a.** $\dfrac{1}{3}$

**b.** $\dfrac{12}{25}$

**68. a.** $\dfrac{2}{9}$

**b.** $\dfrac{32}{9}$

**Name** _____

**69.** $x = -\dfrac{4}{5}$ and $y = \dfrac{9}{11}$

**70.** $x = \dfrac{7}{6}$ and $y = -\dfrac{1}{2}$

*Determine whether the given replacement values are solutions of the given equations. See Example 16.*

**71.** Is $-\dfrac{5}{18}$ a solution to $3x = -\dfrac{5}{6}$?

**72.** Is $\dfrac{9}{11}$ a solution to $\dfrac{2}{3}y = \dfrac{6}{11}$?

**73.** Is $\dfrac{2}{5}$ a solution to $-\dfrac{1}{2}z = \dfrac{1}{10}$?

**74.** Is $\dfrac{3}{5}$ a solution to $5x = \dfrac{1}{3}$?

**E** *Solve. See Example 17.*

**75.** A veterinarian's dipping vat holds 36 gallons of liquid. She normally fills it $\dfrac{5}{6}$ full of a medicated flea dip solution. Find how many gallons of solution are normally in the vat.

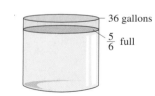

36 gallons

$\dfrac{5}{6}$ full

**76.** The winnings of a race horse are distributed among the owners of the horse according to the amount of ownership each has in the horse. Mr. Gamble has a $\dfrac{3}{25}$ interest in the horse Rainbow, who won $285,000 in the Rye Grass Stakes. Find Mr. Gamble's winnings.

**77.** Each turn of a screw sinks it $\dfrac{3}{16}$ of an inch deeper into a piece of wood. Find how deep the screw is after 8 turns.

**78.** The O'Neill family has a weekly income of $450, $\dfrac{2}{15}$ of which must be budgeted for utilities. Find how many dollars are budgeted each week to pay the utilities.

**79.** As part of his research, famous tornado expert Dr. T. Fujita studied approximately 31,050 tornadoes that occurred in the United States between 1916 and 1985. He found that roughly $\frac{7}{10}$ of these tornadoes occurred during April, May, June, and July. How many of these tornadoes occurred during these four months? (*Source: U.S. Tornadoes Part 1*, T. Fujita, University of Chicago)

**80.** Campbell Soup Company ships its soup in boxes containing 24 cans. If each can weighs $\frac{5}{4}$ pounds, find how much the contents of a box weighs.

**81.** An estimate for the measure of an adult's wrist is $\frac{1}{4}$ of the waist size. If Jorge has a 34-inch waist, estimate the size of his wrist.

**82.** An estimate for an adult's waist measurement is found by dividing the neck size (in inches) by $\frac{1}{2}$. Jock's neck measures 18 inches. Estimate his waist measurement.

*When setting a post for building a deck or fence, it is recommended that $\frac{1}{3}$ of the total length of the post be buried in the ground. Find the amount of post to be buried in the ground for the given post lengths.*

**83.** a 9-foot post

**84.** a 12-foot post

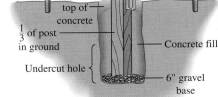

Setting post
Pressure-treated post
Temporary braces
Level
Slope the top of concrete
$\frac{1}{3}$ of post in ground
Concrete fill
Undercut hole
6" gravel base

*Source: Southern Living, May, 1995.*

**79.** 21,735 tornadoes

**80.** 30 lb

**81.** $\frac{17}{2}$ in.

**82.** 36 in.

**83.** 3 ft

**84.** 4 ft

**Name** _____

*Find the area of each rectangle. Recall that area = length · width, or lw.*

△ **85.**

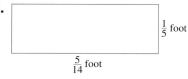

$\frac{1}{5}$ foot

$\frac{5}{14}$ foot

△ **86.**

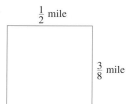

$\frac{1}{2}$ mile

$\frac{3}{8}$ mile

## REVIEW AND PREVIEW

*Write the prime factorization of each number. See Section 4.2.*

**87.** 90

**88.** 42

**89.** 65

**90.** 72

**91.** 126

**92.** 112

## COMBINING CONCEPTS

**93.** In a recent year, approximately $\frac{7}{10}$ of U.S. households that have on-line service made on-line purchases. If 10,300,000 U.S. households have on-line services, how many of these made on-line purchases? (*Source:* Inteco Corp.)

**94.** Approximately $\frac{3}{25}$ of the U.S. population is in the state of California. If the U.S. population is 270,299,000, find the population of California. (*Source:* U.S. Bureau of the Census, 1998)

**95.** In your own words, describe how to multiply fractions.

**96.** In your own words, describe how to divide fractions.

**97.** One-third of all native flowering plant species in the United States are at risk of becoming extinct. That translates into 5144 at-risk flowering plant species. Based on this data, how many flowering plant species are native to the United States overall? (*Source:* The Nature Conservancy) (*Hint:* How many $\frac{1}{3}$s are in 5144?)

**98.** The FedEx fleet of aircraft includes 264 Cessnas. These Cessnas make up $\frac{66}{149}$ of the FedEx fleet. What is the size of the entire FedEx fleet of aircraft? (*Source:* Federal Express Corp.)

# 4.4   ADDING AND SUBTRACTING LIKE FRACTIONS AND LEAST COMMON DENOMINATOR

Fractions that have the same or a common denominator are called **like fractions**. Fractions that have different denominators are called **unlike fractions**.

**Like Fractions**

$\dfrac{2}{5}$ and $\dfrac{3}{5}$

$\dfrac{5}{21}, \dfrac{16}{21}$, and $\dfrac{7}{21}$

$-\dfrac{9}{15}$ and $\dfrac{13}{15}$

**Unlike Fractions**

$\dfrac{2}{5}$ and $\dfrac{3}{4}$

$-\dfrac{5}{7}$ and $\dfrac{5}{9}$

$\dfrac{3}{4}, \dfrac{9}{12}$, and $\dfrac{18}{24}$

**Objectives**

**A** Add or subtract like fractions.

**B** Add and subtract given fractional replacement values.

**C** Solve problems by adding or subtracting like fractions.

**D** Find the least common denominator of a list of fractions.

Study Guide    SSM    CD-ROM    Video 4.4

## A   ADDING OR SUBTRACTING LIKE FRACTIONS

We can add like fractions on a number line just as we added whole numbers and integers on a number line.

To add $\dfrac{1}{5} + \dfrac{3}{5}$, start at 0 and draw an arrow $\dfrac{1}{5}$ of a unit long pointing to the right. From the tip of this arrow, draw an arrow $\dfrac{3}{5}$ of a unit long also pointing to the right. The tip of the second arrow ends at their sum, $\dfrac{4}{5}$.

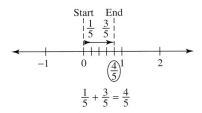

$$\frac{1}{5} + \frac{3}{5} = \frac{4}{5}$$

Notice that the numerator of the sum is the sum of the numerators. Also, the denominator of the sum is their common denominator. This is how we add fractions. A similar method is used to subtract fractions.

---

**ADDING OR SUBTRACTING LIKE FRACTIONS (FRACTIONS WITH THE SAME DENOMINATOR)**

If $a$, $b$, and $c$, are numbers and $b$ is not 0, then

$$\frac{a}{b} + \frac{c}{b} = \frac{a+c}{b} \qquad \text{and also} \qquad \frac{a}{b} - \frac{c}{b} = \frac{a-c}{b}$$

In other words, to add or subtract fractions with the same denominator, add or subtract their numerators and write the sum or difference over the **common** denominator.

---

For example,

$$\frac{1}{4} + \frac{2}{4} = \frac{1+2}{4} = \frac{3}{4}$$   Add the numerators.
Keep the denominator.

$$\frac{4}{5} - \frac{2}{5} = \frac{4-2}{5} = \frac{2}{5}$$   Subtract the numerators.
Keep the denominator.

### Practice Problems 1–3

Add and simplify.

1. $\dfrac{5}{9} + \dfrac{2}{9}$    2. $\dfrac{5}{8} + \dfrac{1}{8}$

3. $\dfrac{10}{11} + \dfrac{1}{11} + \dfrac{7}{11}$

### ✓ CONCEPT CHECK

Find and correct the error in the following.

$$\frac{3}{7} + \frac{3}{7} = \frac{6}{14}$$

### Practice Problems 4–5

Subtract and simplify.

4. $\dfrac{7}{12} - \dfrac{2}{12}$    5. $\dfrac{9}{10} - \dfrac{1}{10}$

**TEACHING TIP**

Remind students that when adding fractions with the same denominator, they add the numerators and *keep* the common denominator. A common mistake is for students to also add denominators.

### Practice Problem 6

Add: $-\dfrac{8}{5} + \dfrac{4}{5}$

### Practice Problem 7

Subtract: $\dfrac{2}{5} - \dfrac{3y}{5}$

**Answers**

1. $\dfrac{7}{9}$   2. $\dfrac{3}{4}$   3. $\dfrac{18}{11}$   4. $\dfrac{5}{12}$   5. $\dfrac{4}{5}$   6. $-\dfrac{4}{5}$

7. $\dfrac{2 - 3y}{5}$

✓ **Concept Check:** $\dfrac{3}{7} + \dfrac{3}{7} = \dfrac{6}{7}$; keep the common denominator.

---

**Helpful Hint**

As usual, don't forget to write all answers in simplest form.

**Examples**   Add and simplify.

1. $\dfrac{2}{7} + \dfrac{3}{7} = \dfrac{2 + 3}{7} = \dfrac{5}{7}$   ← Add the numerators.
   ← Keep the common denominator.

2. $\dfrac{3}{16} + \dfrac{7}{16} = \dfrac{3 + 7}{16} = \dfrac{10}{16} = \dfrac{2 \cdot 5}{2 \cdot 8} = \dfrac{5}{8}$

3. $\dfrac{7}{8} + \dfrac{6}{8} + \dfrac{3}{8} = \dfrac{7 + 6 + 3}{8} = \dfrac{16}{8} = 2$

**TRY THE CONCEPT CHECK IN THE MARGIN.**

**Examples**   Subtract and simplify.

4. $\dfrac{8}{9} - \dfrac{1}{9} = \dfrac{8 - 1}{9} = \dfrac{7}{9}$   ← Subtract the numerators.
   ← Keep the common denominator.

5. $\dfrac{7}{8} - \dfrac{5}{8} = \dfrac{7 - 5}{8} = \dfrac{2}{8} = \dfrac{2}{2 \cdot 4} = \dfrac{1}{4}$

From our earlier work, we know that $\dfrac{-12}{6} = -2$ and that $\dfrac{12}{-6} = -2$. Also, $-\dfrac{12}{6} = -2$. Since all these fractions simplify to $-2$, we have that

$$\frac{-12}{6} = \frac{12}{-6} = -\frac{12}{6}$$

In general, the following is true:

$$\boxed{\dfrac{-a}{b} = \dfrac{a}{-b} = -\dfrac{a}{b} \text{ as long as } b \text{ is not } 0.}$$

For example, $\dfrac{-3}{4} = \dfrac{3}{-4} = -\dfrac{3}{4}$.

**Example 6**   Add: $-\dfrac{11}{8} + \dfrac{6}{8}$

*Solution:*   $-\dfrac{11}{8} + \dfrac{6}{8} = \dfrac{-11 + 6}{8}$

$$= \frac{-5}{8} \text{ or } -\frac{5}{8}$$

**Example 7**   Subtract: $\dfrac{3x}{4} - \dfrac{7}{4}$

*Solution:*   $\dfrac{3x}{4} - \dfrac{7}{4} = \dfrac{3x - 7}{4}$

Recall from Section 3.1 that the terms in the numerator are unlike terms and cannot be combined.

**Example 8**   Subtract: $\dfrac{3}{7} - \dfrac{6}{7} - \dfrac{3}{7}$

*Solution:*   $\dfrac{3}{7} - \dfrac{6}{7} - \dfrac{3}{7} = \dfrac{3 - 6 - 3}{7} = \dfrac{-6}{7}$ or $-\dfrac{6}{7}$

> **HELPFUL HINT**   Recall that
> $\dfrac{-6}{7} = -\dfrac{6}{7}$ and also $\dfrac{6}{-7}$, if needed.

**Practice Problem 8**

Subtract: $\dfrac{4}{11} - \dfrac{6}{11} - \dfrac{3}{11}$

### B   ADDING AND SUBTRACTING GIVEN FRACTIONAL REPLACEMENT VALUES

**Example 9**   Evaluate $x - y$ if $x = -\dfrac{3}{10}$ and $y = \dfrac{2}{10}$.

*Solution:*   $x - y = -\dfrac{3}{10} - \dfrac{2}{10}$   Replace $x$ with $-\dfrac{3}{10}$ and $y$ with $\dfrac{2}{10}$.

$= \dfrac{-3 - 2}{10}$

$= \dfrac{-5}{10} = \dfrac{-1 \cdot 5}{2 \cdot 5} = \dfrac{-1}{2}$ or $-\dfrac{1}{2}$

**Practice Problem 9**

Evaluate $x + y$ if $x = -\dfrac{10}{12}$ and $y = \dfrac{5}{12}$.

**Example 10**   Solve: $x - \dfrac{1}{5} = \dfrac{3}{5}$

*Solution:*   To solve, add $\dfrac{1}{5}$ to both sides of the equation.

$x - \dfrac{1}{5} = \dfrac{3}{5}$

$x - \dfrac{1}{5} + \dfrac{1}{5} = \dfrac{3}{5} + \dfrac{1}{5}$

$x = \dfrac{4}{5}$

To check, replace $x$ with $\dfrac{4}{5}$ in the original equation.

$x - \dfrac{1}{5} = \dfrac{3}{5}$

$\dfrac{4}{5} - \dfrac{1}{5} = \dfrac{3}{5}$   Replace $x$ with $\dfrac{4}{5}$.

$\dfrac{4 - 1}{5} = \dfrac{3}{5}$

$\dfrac{3}{5} = \dfrac{3}{5}$   True.

The solution of $x - \dfrac{1}{5} = \dfrac{3}{5}$ is $\dfrac{4}{5}$.

**Practice Problem 10**

Solve: $\dfrac{7}{10} = x - \dfrac{1}{10}$

### C   SOLVING PROBLEMS BY ADDING AND SUBTRACTING LIKE FRACTIONS

We can combine our skills in adding and subtracting like fractions with our four problem-solving steps from Chapter 3 to solve many kinds of real-life problems.

**Answers**

**8.** $-\dfrac{5}{11}$   **9.** $-\dfrac{5}{12}$   **10.** $\dfrac{4}{5}$

## Practice Problem 11

If a piano student practices the piano $\frac{3}{4}$ of an hour in the morning and $\frac{1}{4}$ of an hour in the evening, how long did she practice that day?

## Example 11     Total Amount of an Ingredient in a Recipe

A recipe calls for $\frac{1}{3}$ of a cup of flour at the beginning and $\frac{2}{3}$ of a cup of flour later. How much total flour is needed to make that recipe?

$\frac{1}{3}$ cup     $\frac{2}{3}$ cup

*Solution:*     **1.** UNDERSTAND the problem. To do so, read and reread the problem. Since we are finding total flour, we will add.

**2.** TRANSLATE.

In words:

| Total flour | is | flour at the beginning | added to | flour later |
|---|---|---|---|---|
| ↓ | ↓ | ↓ | ↓ | ↓ |

Translate:     Total flour $=$     $\frac{1}{3}$     $+$     $\frac{2}{3}$

**3.** SOLVE.     $\frac{1}{3} + \frac{2}{3} = \frac{1+2}{3} = \frac{3}{3} = 1$

**4.** INTERPRET. *Check* your work. *State* your conclusion: The total flour needed for the recipe is 1 cup. ■

## Practice Problem 12

A walker attends a gym that has a $\frac{1}{8}$-mile track. If he walks 9 laps on the track on Monday and 3 laps on Wednesday, how much farther did he walk on Monday than on Wednesday?

## Example 12     Calculating Distance

The distance from home to the World Gym is $\frac{7}{8}$ of a mile and from home to the Post Office is $\frac{5}{8}$ of a mile. How much farther is it from home to the World Gym than from home to the Post Office?

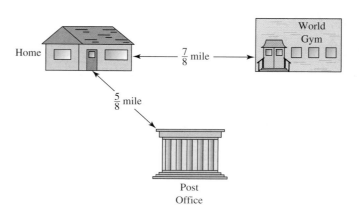

Home     $\frac{7}{8}$ mile     World Gym

$\frac{5}{8}$ mile

Post Office

*Solution:*     **1.** UNDERSTAND. Read and reread the problem. The phrase "How much farther" tells us to subtract distances.

**Answers**

**11.** 1 hour   **12.** 1 mile   $\frac{3}{4}$ mile

**2.** TRANSLATE.

In words:

| Distance farther | is | home to World Gym distance | minus | home to Post Office distance |
|:---:|:---:|:---:|:---:|:---:|
| ↓ | ↓ | ↓ | ↓ | ↓ |

Translate:     $\text{Distance farther} = \dfrac{7}{8} - \dfrac{5}{8}$

**3.** SOLVE:  $\dfrac{7}{8} - \dfrac{5}{8} = \dfrac{7-5}{8} = \dfrac{2}{8} = \dfrac{2}{2 \cdot 4} = \dfrac{1}{4}$

**4.** INTERPRET. *Check* your work. *State* your conclusion:

The distance from home to the World Gym is $\dfrac{1}{4}$ mile farther than from home to the Post Office.  ▬▬▬

## D  FINDING THE LEAST COMMON DENOMINATOR

In the next section, we will add and subtract fractions that have different denominators. To add or subtract fractions that have unlike or different denominators, we first write them as equivalent fractions with a common denominator.

Although any common denominator can be used to add or subtract unlike fractions, we will use the **least common denominator (LCD)** or the **least common multiple (LCM)** of the denominators. Why? Since the LCD is the *smallest* of all common denominators, operations are usually less tedious with this number.

> The **least common denominator (LCD)** of a list of fractions is the smallest positive number divisible by all the denominators in the list. (The least common denominator is also called the **least common multiple (LCM).**)

For example, the LCD of $\dfrac{1}{4}$ and $\dfrac{3}{10}$ is 20 because 20 is the smallest positive number divisible by both 4 and 10.

### Finding the LCD: Method 1

One way to find the LCD is to see whether the larger denominator is divisible by the smaller denominator. If so, the larger number is the LCD. If not, then check consecutive multiples of the larger denominator until the LCD is found.

To find the LCD for $\dfrac{1}{4}$ and $\dfrac{3}{10}$, we check to see whether 10 is a multiple of 4. No, it is not, so we check consecutive multiples of 10.

$2 \cdot 10 = 20$    20 is divisible by 4,    so LCD $= 20$.

**Example 13**  Find the LCD of $\dfrac{3}{8}$ and $\dfrac{1}{6}$.

*Solution:*    Is 8 divisible by 6? No, so we check multiples of 8.

$2 \cdot 8 = 16$    16 is not divisible by 6.
$3 \cdot 8 = 24$    24 is divisible by 6, so LCD $= 24$.  ▬▬▬

**Practice Problem 13**

Find the LCD of $\dfrac{7}{12}$ and $\dfrac{2}{15}$.

**Answer**

**13.** 60

## Finding the LCD: Method 2

Another way to find the LCD is to first write each denominator as a product of primes. To find the LCD of $\frac{1}{4}$ and $\frac{3}{10}$, we write

$$4 = 2 \cdot 2$$
$$10 = 2 \cdot 5$$

If the LCD is divisible by 4, it must contain the factors $2 \cdot 2$. If the LCD is divisible by 10, it must contain the factors $2 \cdot 5$. Since 4 and 10 will divide into the LCD separately, the LCD needs to contain a factor the greatest number of times that the factor appears in any **one** prime factorization.

$$\text{LCD} = 2 \cdot 2 \cdot 5 = 20$$

factors of 4

factors of 10

---

**HELPFUL HINT**

The number 2 is a factor twice since that is the greatest number of times that 2 is a factor in the prime factorization of any one denominator.

---

**Practice Problem 14**

Find the LCD of $\frac{9}{14}$ and $\frac{11}{35}$.

---

**Example 14**    Find the LCD of $\frac{5}{24}$ and $\frac{1}{18}$.

*Solution:*    First write each denominator as a product of primes.

$$24 = 2 \cdot 2 \cdot 2 \cdot 3$$
$$18 = 2 \cdot 3 \cdot 3$$

Write each factor the greatest number of times that it appears in any **one** prime factorization.

The greatest number of times that 2 appears is **3** times:
$$24 = 2 \cdot 2 \cdot 2 \cdot 3$$

The greatest number of times that 3 appears is **2** times:
$$8 = 2 \cdot 3 \cdot 3$$

$$\text{LCD} = 2 \cdot 2 \cdot 2 \cdot 3 \cdot 3 = 72$$

Notice that 72 is the smallest positive number that is divisible by both 18 and 24.

---

**Practice Problem 15**

Find the LCD of $\frac{7}{4}, \frac{7}{15}$, and $\frac{3}{10}$.

**Practice Problem 16**

Find the LCD of $\frac{7}{y}$ and $\frac{6}{11}$.

---

**Example 15**    Find the LCD of $-\frac{2}{5}, \frac{1}{6}$, and $\frac{5}{12}$.

*Solution:*    To help find the LCD, we write the prime factorization of each denominator and circle each different factor the greatest number of times that it appears in any one factor.

$$5 = \boxed{5}$$
$$6 = 2 \cdot \boxed{3}$$
$$12 = \boxed{2 \cdot 2} \cdot 3$$
$$\text{LCD} = 2 \cdot 2 \cdot 3 \cdot 5 = 60$$

---

**✓ CONCEPT CHECK**

True or false? The LCD of the fractions $\frac{1}{6}$ and $\frac{1}{8}$ is 48.

---

**Example 16**    Find the LCD of $\frac{3}{5}$ and $\frac{2}{x}$.

*Solution:*   
$$5 = \boxed{5}$$
$$x = \boxed{x}$$
$$\text{LCD} = 5 \cdot x = 5x$$

---

**Answers**

**14.** 70   **15.** 60   **16.** $11y$
✓ **Concept Check:** False; it is 24.

**TRY THE CONCEPT CHECK IN THE MARGIN.**

**Name** _____ **Section** _____ **Date** _____

## MENTAL MATH

*State whether the fractions in each list are like or unlike fractions.*

**1.** $\dfrac{7}{8}, \dfrac{7}{10}$

**2.** $\dfrac{2}{3}, \dfrac{2}{9}$

**3.** $\dfrac{9}{10}, \dfrac{1}{10}$

**4.** $\dfrac{8}{11}, \dfrac{2}{11}$

**5.** $\dfrac{2}{31}, \dfrac{30}{31}, \dfrac{19}{31}$

**6.** $\dfrac{3}{10}, \dfrac{3}{11}, \dfrac{3}{13}$

**7.** $\dfrac{5}{12}, \dfrac{7}{12}, \dfrac{12}{11}$

**8.** $\dfrac{1}{5}, \dfrac{2}{5}, \dfrac{4}{5}$

**A** *Add or subtract as indicated. See Examples 1 through 8.*

**9.** $\dfrac{3}{7} + \dfrac{2}{7}$

**10.** $\dfrac{5}{9} + \dfrac{2}{9}$

**11.** $\dfrac{10}{11} - \dfrac{4}{11}$

**12.** $\dfrac{9}{13} - \dfrac{5}{13}$

**13.** $\dfrac{5}{11} + \dfrac{2}{11}$

**14.** $\dfrac{4}{7} + \dfrac{2}{7}$

**15.** $\dfrac{9}{15} - \dfrac{1}{15}$

**16.** $\dfrac{3}{15} - \dfrac{1}{15}$

## EXERCISE SET 4.4

*Add or subtract as indicated. See Examples 1 through 8.*

**1.** $-\dfrac{1}{2} + \dfrac{1}{2}$

**2.** $-\dfrac{3}{x} + \dfrac{1}{x}$

**3.** $\dfrac{2}{9x} + \dfrac{4}{9x}$

**4.** $\dfrac{3}{10y} + \dfrac{2}{10y}$

**5.** $-\dfrac{4}{13} + \dfrac{2}{13} + \dfrac{1}{13}$
**6.** $-\dfrac{5}{11} + \dfrac{1}{11} + \dfrac{2}{11}$
**7.** $\dfrac{7}{18} + \dfrac{3}{18} + \dfrac{2}{18}$
**8.** $\dfrac{2}{15} + \dfrac{4}{15} + \dfrac{9}{15}$

**9.** $\dfrac{-\dfrac{3}{y}}{}$

**10.** $\dfrac{-\dfrac{3}{z}}{}$

**11.** $\dfrac{7a-3}{4}$

**12.** $\dfrac{18b-3}{5}$

**13.** $-\dfrac{3}{4}$

**14.** $-\dfrac{2}{3}$

**15.** $-\dfrac{1}{3}$

**16.** $-\dfrac{3}{14}$

**17.** $\dfrac{2x}{3}$

**18.** $\dfrac{2x-7}{15}$

**19.** $-\dfrac{x}{2}$

**20.** $\dfrac{11b}{8}$

**21.** $\dfrac{3}{4z}$

**22.** $\dfrac{11}{8a}$

**23.** $-\dfrac{3}{10}$

**24.** $-\dfrac{3}{10}$

**25.** 2

**26.** $-\dfrac{3}{2}$

**27.** $-\dfrac{2}{3}$

**28.** 2

**29.** $\dfrac{3x}{4}$

**30.** $\dfrac{3y}{2}$

**31.** $\dfrac{5}{4}$

**32.** $-\dfrac{1}{4}$

**Name** _____

**9.** $\dfrac{1}{y} - \dfrac{4}{y}$ 　　　**10.** $\dfrac{4}{z} - \dfrac{7}{z}$ 　　　**11.** $\dfrac{7a}{4} - \dfrac{3}{4}$ 　　　**12.** $\dfrac{18b}{5} - \dfrac{3}{5}$

**13.** $\dfrac{1}{8} - \dfrac{7}{8}$ 　　　**14.** $\dfrac{1}{6} - \dfrac{5}{6}$ 　　　**15.** $\dfrac{20}{21} - \dfrac{10}{21} - \dfrac{17}{21}$ 　　　**16.** $\dfrac{27}{28} - \dfrac{5}{28} - \dfrac{28}{28}$

**17.** $\dfrac{9x}{15} + \dfrac{1x}{15}$ 　　　**18.** $\dfrac{2x}{15} - \dfrac{7}{15}$ 　　　▭ **19.** $\dfrac{7x}{16} - \dfrac{15x}{16}$ 　　　**20.** $\dfrac{15b}{16} + \dfrac{7b}{16}$

**21.** $\dfrac{15}{16z} - \dfrac{3}{16z}$ 　　　**22.** $\dfrac{7}{16a} + \dfrac{15}{16a}$ 　　　**23.** $\dfrac{3}{10} - \dfrac{6}{10}$ 　　　**24.** $-\dfrac{6}{10} + \dfrac{3}{10}$

**25.** $\dfrac{15}{17} + \dfrac{5}{17} + \dfrac{14}{17}$ 　　　**26.** $\dfrac{1}{8} - \dfrac{15}{8} + \dfrac{2}{8}$ 　　　▭ **27.** $\dfrac{9}{12} - \dfrac{7}{12} - \dfrac{10}{12}$

**28.** $\dfrac{9}{13} + \dfrac{10}{13} + \dfrac{7}{13}$ 　　　**29.** $\dfrac{x}{4} + \dfrac{3x}{4} - \dfrac{2x}{4} + \dfrac{x}{4}$ 　　　**30.** $\dfrac{9y}{8} + \dfrac{2y}{8} + \dfrac{5y}{8} - \dfrac{4y}{8}$

**B** *Evaluate each expression for the given replacement values. See Example 9.*

**31.** $x + y$; $x = \dfrac{3}{4}$, $y = \dfrac{2}{4}$ 　　　　　　**32.** $x - y$; $x = \dfrac{7}{8}$, $y = \dfrac{9}{8}$

**33.** $x - y$; $x = -\dfrac{1}{5}$, $y = \dfrac{3}{5}$

**34.** $x + y$; $x = -\dfrac{1}{6}$, $y = \dfrac{5}{6}$

**35.** $x - y + z$; $x = \dfrac{3}{12}$, $y = \dfrac{5}{12}$, $z = -\dfrac{7}{12}$

**36.** $x + y - z$; $x = \dfrac{2}{14}$, $y = \dfrac{3}{14}$, $z = \dfrac{8}{14}$

*Solve and check. See Example 10.*

**37.** $x + \dfrac{1}{3} = -\dfrac{1}{3}$

**38.** $x + \dfrac{1}{9} = -\dfrac{7}{9}$

**39.** $y - \dfrac{3}{13} = -\dfrac{2}{13}$

**40.** $z - \dfrac{5}{14} = \dfrac{4}{14}$

**41.** $3x - \dfrac{1}{5} - 2x = \dfrac{1}{5} + \dfrac{2}{5}$

**42.** $5x + \dfrac{1}{11} - 4x = \dfrac{2}{11} - \dfrac{5}{11}$

**C** *Find the perimeter of each figure. Recall that the perimeter of a figure is the distance around a figure.*

△ **43.** $\dfrac{4}{20}$ inch   $\dfrac{7}{20}$ inch   $\dfrac{9}{20}$ inch

△ **44.** Square   $\dfrac{1}{6}$ centimeter

△ **45.** $\dfrac{5}{12}$ meter   Rectangle   $\dfrac{7}{12}$ meter

△ **46.** $\dfrac{3}{13}$ foot   $\dfrac{2}{13}$ foot   $\dfrac{6}{13}$ foot   $\dfrac{3}{13}$ foot   $\dfrac{4}{13}$ foot

**33.** $-\dfrac{4}{5}$

**34.** $\dfrac{2}{3}$

**35.** $-\dfrac{3}{4}$

**36.** $-\dfrac{3}{14}$

**37.** $-\dfrac{2}{3}$

**38.** $-\dfrac{8}{9}$

**39.** $\dfrac{1}{13}$

**40.** $\dfrac{9}{14}$

**41.** $\dfrac{4}{5}$

**42.** $-\dfrac{4}{11}$

**43.** $1$ in.

**44.** $\dfrac{2}{3}$ cm

**45.** $2$ m

**46.** $\dfrac{18}{13}$ ft

**Name** _____

*Solve. Write each answer in simplest form. See Examples 11 and 12.*

**47.** Emil Vasquez, a body builder, worked out $\dfrac{7}{8}$ of an hour one morning before school and $\dfrac{5}{8}$ of an hour that evening. How long did he work out that day?

**48.** A recipe for Heavenly Hash cake calls for $\dfrac{3}{4}$ cup of flour and later $\dfrac{1}{4}$ cup of flour. How much flour is needed to make the recipe?

*The chart shows the breakdown of all U.S. employees covered by health benefits in 1997 by type of health plan. Use this chart to answer Exercises 49–52.*

**49.** Put the health plans in order from the smallest fraction of employees covered to the largest fraction of employees covered.

**50.** Find the fraction of employees that are *not* covered by a Health Maintenance Organization.

| Type of Health Plan | Fraction of Employees with Health Benefits |
|---|---|
| Health Maintenance Organization | $\dfrac{6}{20}$ |
| Point-of-Service | $\dfrac{4}{20}$ |
| Preferred Provider Organization | $\dfrac{7}{20}$ |
| Traditional fee-for-service | $\dfrac{3}{20}$ |

(*Source:* William M. Mercer, Inc.)

**51.** Find the fraction of the employees that are *not* covered by a Preferred Provider Organization.

**52.** Which type of health plan is the most popular?

**53.** As of September 1999, the fraction of states in the United States with maximum interstate highway speed limits up to and including 70 mph was $\dfrac{39}{50}$. The fraction of states with 70 mph speed limits was $\dfrac{16}{50}$. What fraction of states had speed limits that were less than 70 mph? (*Source:* National Motorists Association)

**54.** When people take aspirin, $\dfrac{31}{50}$ of the time it is used to treat some type of pain. Approximately $\dfrac{7}{50}$ of all aspirin use is for treating headaches. What fraction of aspirin use is for treating pain other than headaches? (*Source:* Bayer Market Research)

**D** *Find the LCD of each list of fractions. See Examples 13 through 16.*

**55.** $\dfrac{1}{3}, \dfrac{3}{4}$  **56.** $\dfrac{1}{4}, \dfrac{5}{6}$  **57.** $-\dfrac{2}{9}, \dfrac{6}{15}$  **58.** $-\dfrac{7}{12}, \dfrac{3}{20}$

**59.** $\dfrac{5}{12}, \dfrac{5}{18}$  **60.** $\dfrac{7}{12}, \dfrac{7}{15}$  **61.** $-\dfrac{7}{24}, -\dfrac{5}{x}$  **62.** $-\dfrac{11}{y}, -\dfrac{13}{70}$

**63.** $\dfrac{2}{25}, \dfrac{3}{15}, \dfrac{5}{6}$  **64.** $\dfrac{3}{4}, \dfrac{1}{6}, \dfrac{13}{18}$  **65.** $\dfrac{23}{18}, \dfrac{1}{21}$  **66.** $\dfrac{45}{24}, \dfrac{2}{45}$

**67.** $-\dfrac{16}{15}, -\dfrac{11}{25}$  **68.** $-\dfrac{22}{21}, \dfrac{3}{14}$  **69.** $\dfrac{1}{8}, \dfrac{1}{24}$  **70.** $\dfrac{1}{15}, \dfrac{1}{90}$

**71.** $\dfrac{8}{25}, \dfrac{7}{10}$  **72.** $\dfrac{7}{8}, \dfrac{13}{12}$  **73.** $-\dfrac{1}{a}, -\dfrac{7}{12}$  **74.** $-\dfrac{1}{9}, -\dfrac{80}{b}$

**75.** $\dfrac{4}{3}, \dfrac{8}{21}, \dfrac{3}{56}$  **76.** $\dfrac{6}{70}, \dfrac{11}{80}, \dfrac{15}{90}$  **77.** $\dfrac{12}{11}, \dfrac{20}{33}, \dfrac{12}{121}$  **78.** $\dfrac{7}{10}, \dfrac{8}{15}, \dfrac{9}{100}$

**55.** 12

**56.** 12

**57.** 45

**58.** 60

**59.** 36

**60.** 60

**61.** $24x$

**62.** $70y$

**63.** 150

**64.** 36

**65.** 126

**66.** 360

**67.** 75

**68.** 42

**69.** 24

**70.** 90

**71.** 50

**72.** 24

**73.** $12a$

**74.** $9b$

**75.** 168

**76.** 5040

**77.** 363

**78.** 300

**Name** _____

## REVIEW AND PREVIEW

*Perform each indicated operation. See Sections 4.3 and 2.5.*

**79.** $\dfrac{4}{5} \cdot \dfrac{3}{7}$

**80.** $\dfrac{5}{3} \cdot \dfrac{4}{9}$

**81.** $-2 + 10$

**82.** $-2(10)$

**83.** $\dfrac{2}{5} \div \dfrac{1}{2}$

**84.** $\dfrac{4}{7} \div \dfrac{1}{3}$

**85.** $-12 - 16$

**86.** $-18 - (-2)$

## COMBINING CONCEPTS

*Perform each indicated operation.*

**87.** $\dfrac{4}{11} + \dfrac{5}{11} - \dfrac{3}{11} + \dfrac{2}{11}$

**88.** $\dfrac{9}{12} + \dfrac{1}{12} - \dfrac{3}{12} - \dfrac{5}{12}$

*Solve. Write each answer in simplest form.*

**89.** A person's marital status is either single (never married), married, widowed, or divorced. Of American men over the age of 65, $\dfrac{38}{50}$ are married and $\dfrac{7}{50}$ are widowed. What fraction of American men over age 65 are either single or divorced? (*Source:* Based on data from the U.S. Bureau of the Census)

**90.** Mike Cannon jogged $\dfrac{3}{8}$ of a mile from home and then rested. Then he continued jogging for another $\dfrac{3}{8}$ of a mile until he discovered his watch had fallen off. He walked back along the same path for $\dfrac{4}{8}$ of a mile until he found his watch. Find how far he was from his *starting point*.

**91.** In your own words, explain how to add like fractions.

**92.** In your own words, explain how to subtract like fractions.

# 4.5   ADDING AND SUBTRACTING UNLIKE FRACTIONS

## A  ADDING AND SUBTRACTING UNLIKE FRACTIONS

In this section we add and subtract fractions with different denominators. To add or subtract these unlike fractions, first write the fractions as equivalent fractions with a common denominator and then add or subtract the like fractions. The common denominator we will use is the least common denominator (LCD).

For example, add the following unlike fractions: $\frac{3}{4} + \frac{1}{6}$. The LCD of the denominators 4 and 6 is 12. Write each fraction as an equivalent fraction with a denominator of 12.

$$\frac{3}{4} = \frac{3 \cdot 3}{4 \cdot 3} = \frac{9}{12} \qquad \text{and} \qquad \frac{1}{6} = \frac{1 \cdot 2}{6 \cdot 2} = \frac{2}{12}$$

Then

$$\frac{3}{4} + \frac{1}{6} = \frac{9}{12} + \frac{2}{12} = \frac{11}{12}$$

---

### ADDING OR SUBTRACTING UNLIKE FRACTIONS

**Step 1.**  Find the LCD of the denominators of the fractions.
**Step 2.**  Write each fraction as an equivalent fraction whose denominator is the LCD.
**Step 3.**  Add or subtract the like fractions.
**Step 4.**  Write the sum or difference in simplest form.

---

**Example 1**   Add: $\frac{2}{5} + \frac{4}{15}$

*Solution:*   **Step 1.**  The LCD of the denominators 5 and 15 is 15.

**Step 2.**  $\frac{2}{5} = \frac{2 \cdot 3}{5 \cdot 3} = \frac{6}{15},$    $\frac{4}{15} = \frac{4}{15}$   ← This fraction already has a denominator of 15.

**Step 3.**  $\frac{2}{5} + \frac{4}{15} = \frac{6}{15} + \frac{4}{15} = \frac{10}{15}$

**Step 4.**  Write in simplest form.

$$\frac{10}{15} = \frac{2 \cdot 5}{3 \cdot 5} = \frac{2}{3}$$

---

**Helpful Hint**

Remember that $-\frac{a}{b} = \frac{a}{-b} = \frac{-a}{b}$. For example, $-\frac{2}{3} = \frac{2}{-3} = \frac{-2}{3}$.

---

### Objectives

**A**  Add or subtract unlike fractions.
**B**  Evaluate expressions given fractional replacement values.
**C**  Solve equations containing fractions.
**D**  Solve problems by adding or subtracting unlike fractions.

Study Guide   SSM   CD-ROM   Video 4.5

---

**Practice Problem 1**

Add: $\frac{4}{7} + \frac{3}{14}$

---

**Answer**

**1.** $\frac{11}{14}$

## Practice Problem 2

Subtract: $\dfrac{3}{7} - \dfrac{9}{10}$

## Practice Problem 3

Add: $-\dfrac{1}{5} + \dfrac{3}{20}$

**TEACHING TIP**

As a review, ask students how they multiply, divide, add, and subtract fractions. Ask them, "For which operation(s) must the fractions have the same denominator before performing the operation?"

## Practice Problem 4

Add: $5 + \dfrac{3y}{4}$

**Answers**

**2.** $-\dfrac{33}{70}$   **3.** $-\dfrac{1}{20}$   **4.** $\dfrac{20 + 3y}{4}$

**Example 2**   Subtract: $\dfrac{2}{3} - \dfrac{10}{11}$

*Solution:*   **Step 1.** The LCD of the denominators 3 and 11 is 33.

**Step 2.** $\dfrac{2}{3} = \dfrac{2 \cdot 11}{3 \cdot 11} = \dfrac{22}{33}$   and   $\dfrac{10}{11} = \dfrac{10 \cdot 3}{11 \cdot 3} = \dfrac{30}{33}$

**Step 3.** Subtract.

$$\dfrac{2}{3} - \dfrac{10}{11} = \dfrac{22}{33} - \dfrac{30}{33}$$

$$= \dfrac{22 - 30}{33}$$

$$= \dfrac{-8}{33} \quad \text{or} \quad -\dfrac{8}{33}$$

**Step 4.** $-\dfrac{8}{33}$ is in simplest form.

**Example 3**   Add: $-\dfrac{1}{6} + \dfrac{1}{2}$

*Solution:*   The LCD of the denominators 6 and 2 is 6.

$$-\dfrac{1}{6} + \dfrac{1}{2} = \dfrac{-1}{6} + \dfrac{1 \cdot 3}{2 \cdot 3}$$

$$= \dfrac{-1}{6} + \dfrac{3}{6}$$

$$= \dfrac{2}{6}$$

Next, simplify $\dfrac{2}{6}$.

$$\dfrac{2}{6} = \dfrac{2}{2 \cdot 3} = \dfrac{1}{3}$$

When the fractions contain variables, we add and subtract the same way.

**Example 4**   Subtract: $2 - \dfrac{x}{3}$

*Solution:*   Recall that $2 = \dfrac{2}{1}$. The LCD of the denominators 1 and 3 is 3.

$$\dfrac{2}{1} - \dfrac{x}{3} = \dfrac{2 \cdot 3}{1 \cdot 3} - \dfrac{x}{3}$$

$$= \dfrac{6}{3} - \dfrac{x}{3}$$

$$= \dfrac{6 - x}{3}$$

The numerator $6 - x$ cannot be simplified further since 6 and $-x$ are unlike terms.

**Example 5**   Find: $-\dfrac{3}{4} - \dfrac{1}{14} + \dfrac{6}{7}$

*Solution:*   The LCD of 4, 14, and 7 is 28.

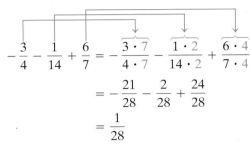

$$-\frac{3}{4} - \frac{1}{14} + \frac{6}{7} = -\frac{3 \cdot 7}{4 \cdot 7} - \frac{1 \cdot 2}{14 \cdot 2} + \frac{6 \cdot 4}{7 \cdot 4}$$

$$= -\frac{21}{28} - \frac{2}{28} + \frac{24}{28}$$

$$= \frac{1}{28}$$

**TRY THE CONCEPT CHECK IN THE MARGIN.**

### B   EVALUATING EXPRESSIONS GIVEN FRACTIONAL REPLACEMENT VALUES

**Example 6**   Evaluate $x - y$ if $x = \dfrac{7}{18}$ and $y = \dfrac{2}{9}$.

*Solution:*   Replace $x$ with $\dfrac{7}{18}$ and $y$ with $\dfrac{2}{9}$ in the expression $x - y$.

$$x - y = \frac{7}{18} - \frac{2}{9}$$

The LCD of the denominators 18 and 9 is 18. Then

$$\frac{7}{18} - \frac{2}{9} = \frac{7}{18} - \frac{2 \cdot 2}{9 \cdot 2}$$

$$= \frac{7}{18} - \frac{4}{18}$$

$$= \frac{3}{18} = \frac{1}{6} \quad \text{Simplified.}$$

### C   SOLVING EQUATIONS CONTAINING FRACTIONS

**Example 7**   Solve: $x - \dfrac{3}{4} = \dfrac{1}{20}$

*Solution:*   To get $x$ by itself, add $\dfrac{3}{4}$ to both sides.

$$x - \frac{3}{4} = \frac{1}{20}$$

$$x - \frac{3}{4} + \frac{3}{4} = \frac{1}{20} + \frac{3}{4} \quad \text{Add } \frac{3}{4} \text{ to both sides.}$$

$$x = \frac{1}{20} + \frac{3 \cdot 5}{4 \cdot 5} \quad \text{The LCD of 20 and 4 is 20.}$$

$$x = \frac{1}{20} + \frac{15}{20}$$

$$x = \frac{16}{20}$$

$$x = \frac{4 \cdot 4}{4 \cdot 5} = \frac{4}{5} \quad \text{Write } \frac{16}{20} \text{ in simplest form.}$$

**Practice Problem 5**

Find: $\dfrac{5}{8} - \dfrac{1}{3} - \dfrac{1}{12}$

✓ **CONCEPT CHECK**

Find and correct the error in the following: $\dfrac{7}{12} - \dfrac{3}{4} = \dfrac{4}{8} = \dfrac{1}{2}$.

**Practice Problem 6**

Evaluate $x + y$ if $x = \dfrac{5}{11}$ and $y = \dfrac{4}{9}$.

**Practice Problem 7**

Solve: $y - \dfrac{2}{3} = \dfrac{5}{12}$

**Answers**

**5.** $\dfrac{5}{24}$   **6.** $\dfrac{89}{99}$   **7.** $\dfrac{13}{12}$

✓ **Concept Check:**

$\dfrac{7}{12} - \dfrac{3}{4} = \dfrac{7}{12} - \dfrac{9}{12} = -\dfrac{2}{12} = -\dfrac{1}{6}$

*Check:*    To check, replace $x$ with $\dfrac{4}{5}$ in the original equation.

$$x - \frac{3}{4} = \frac{1}{20}$$

$$\frac{4}{5} - \frac{3}{4} = \frac{1}{20} \qquad \text{Replace } x \text{ with } \tfrac{4}{5}.$$

$$\frac{4 \cdot 4}{5 \cdot 4} - \frac{3 \cdot 5}{4 \cdot 5} = \frac{1}{20} \qquad \text{The LCD of 5 and 4 is 20.}$$

$$\frac{16}{20} - \frac{15}{20} = \frac{1}{20}$$

$$\frac{1}{20} = \frac{1}{20} \qquad \text{True.}$$

Thus $\dfrac{4}{5}$ is the solution of $x - \dfrac{3}{4} = \dfrac{1}{20}$.

## D SOLVING PROBLEMS BY ADDING OR SUBTRACTING UNLIKE FRACTIONS

Very often, real-world problems involve adding or subtracting unlike fractions.

### Practice Problem 8

To repair her sidewalk, a homeowner must pour small amounts of cement in three different locations. She needs $\dfrac{3}{5}$ of a cubic yard, $\dfrac{2}{10}$ of a cubic yard, and $\dfrac{2}{15}$ of a cubic yard for these locations. Find the total amount of concrete the homeowner needs. If she bought enough cement to mix 1 cubic yard, did she buy enough?

### Example 8    Finding Total Weight

A freight truck has $\dfrac{1}{4}$ ton of computers, $\dfrac{1}{3}$ ton of televisions, and $\dfrac{3}{8}$ ton of small appliances. Find the total weight of its load.

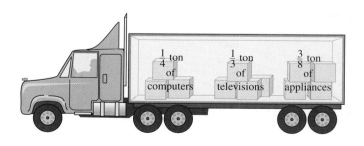

*Solution:*    **1.** UNDERSTAND. Read and reread the problem. The phrase "total weight" tells us to add.

**2.** TRANSLATE.

| In words: | Total weight | is | weight of computers | plus | weight of televisions | plus | weight of appliances |
|---|---|---|---|---|---|---|---|
| | ↓ | ↓ | ↓ | ↓ | ↓ | ↓ | ↓ |
| Translate: | Total weight | = | $\dfrac{1}{4}$ | + | $\dfrac{1}{3}$ | + | $\dfrac{3}{8}$ |

**Answer**

**8.** $\dfrac{14}{15}$ cubic yard; yes, she bought enough.

**3.** SOLVE. The LCD is 24.

$$\frac{1}{4} + \frac{1}{3} + \frac{3}{8} = \frac{1 \cdot 6}{4 \cdot 6} + \frac{1 \cdot 8}{3 \cdot 8} + \frac{3 \cdot 3}{8 \cdot 3}$$

$$= \frac{6}{24} + \frac{8}{24} + \frac{9}{24}$$

$$= \frac{23}{24}$$

**4.** INTERPRET. *Check* the solution. *State* your conclusion: The total weight of the truck's load is $\frac{23}{24}$ ton.

## Example 9    Calculating Flight Time

A flight from Tucson, Arizona, to Phoenix, Arizona, requires $\frac{5}{12}$ of an hour. If the plane has been flying $\frac{1}{4}$ of an hour, find how much time remains before landing.

*Solution:*   **1.** UNDERSTAND. Read and reread the problem. The phrase "how much time remains" tells us to subtract.

**2.** TRANSLATE.

| In words: | Time remaining | is | flight time from Tucson to Phoenix | minus | flight time already passed |
|---|---|---|---|---|---|
| | ↓ | ↓ | ↓ | ↓ | ↓ |
| Translate: | Time remaining | = | $\frac{5}{12}$ | − | $\frac{1}{4}$ |

**3.** SOLVE. The LCD is 12.

$$\frac{5}{12} - \frac{1}{4} = \frac{5}{12} - \frac{1 \cdot 3}{4 \cdot 3}$$

$$= \frac{5}{12} - \frac{3}{12}$$

$$= \frac{2}{12} = \frac{2}{2 \cdot 6} = \frac{1}{6}$$

**4.** INTERPRET. *Check* the solution. *State* your conclusion: The flight time remaining is $\frac{1}{6}$ of an hour.

---

**Practice Problem 9**

Find the difference in length of two boards if one board is $\frac{4}{5}$ of a foot long and the other is $\frac{2}{3}$ of a foot long.

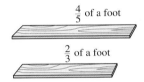

TEACHING TIP    Classroom Activity

Ask students to work in groups to make up their own word problems involving the addition and subtraction of fractions with unlike denominators. Then have them give their problems to another group to solve.

**Answer**

**9.** $\frac{2}{15}$ of a foot

# CALCULATOR EXPLORATIONS
## PERFORMING OPERATIONS ON FRACTIONS

### SCIENTIFIC CALCULATOR

Many calculators have a fraction key, such as $\boxed{a^{b/c}}$, that allows you to enter fractions, perform operations on fractions, and will give the result as a fraction. If your calculator has a fraction key, use it to calculate

$$\frac{3}{5} + \frac{4}{7}$$

Enter the keystrokes

$\boxed{3}$ $\boxed{a^{b/c}}$ $\boxed{5}$ $\boxed{+}$ $\boxed{4}$ $\boxed{a^{b/c}}$ $\boxed{7}$ $\boxed{=}$

The display should read

$$\boxed{1\_6\lrcorner35}$$

which represents the mixed number $1\frac{6}{35}$. Until we discuss mixed numbers in Section 4.8, let's write the result as a fraction. To convert from mixed number notation to fractional notation, press

$\boxed{2^{nd}}$ $\boxed{d/c}$.

The display now reads

$$\boxed{41\lrcorner35}$$

which represents $\frac{41}{35}$, the sum in fractional notation.

### GRAPHING CALCULATOR

Graphing calculators also allow you to perform operations on fractions and will give exact fractional results. The fraction option on a graphing calculator may be found under the $\boxed{\text{MATH}}$ menu. To perform the addition above, try the keystrokes.

$\boxed{3}$ $\boxed{\div}$ $\boxed{5}$ $\boxed{+}$ $\boxed{4}$ $\boxed{\div}$ $\boxed{7}$ $\boxed{\text{MATH}}$ $\boxed{\text{ENTER}}$ $\boxed{\text{ENTER}}$

The display should read

$$\boxed{3/5+4/7 \blacktriangleright \text{FRAC } 41/35}$$

*Use a calculator to add the following fractions. Give each sum as a fraction.*

**1.** $\frac{1}{16} + \frac{2}{5}$ $\quad\frac{37}{80}$

**2.** $\frac{3}{20} + \frac{2}{25}$ $\quad\frac{23}{100}$

**3.** $\frac{4}{9} + \frac{7}{8}$ $\quad\frac{95}{72}$

**4.** $\frac{9}{11} + \frac{5}{12}$ $\quad\frac{163}{132}$

**5.** $\frac{10}{17} + \frac{12}{19}$ $\quad\frac{394}{323}$

**6.** $\frac{14}{31} + \frac{15}{21}$ $\quad\frac{253}{217}$

## MENTAL MATH

*Find the LCD of each pair of fractions.*

**1.** $\dfrac{1}{2}, \dfrac{2}{3}$    **2.** $\dfrac{1}{2}, \dfrac{3}{4}$    **3.** $\dfrac{1}{6}, \dfrac{5}{12}$    **4.** $\dfrac{2}{5}, \dfrac{7}{10}$

**5.** $\dfrac{4}{7}, \dfrac{1}{8}$    **6.** $\dfrac{23}{24}, \dfrac{1}{3}$    **7.** $\dfrac{11}{12}, \dfrac{3}{4}$    **8.** $\dfrac{2}{3}, \dfrac{3}{11}$

# EXERCISE SET 4.5

**A**  *Add or subtract as indicated. See Examples 1 through 4.*

**1.** $\dfrac{2}{3} + \dfrac{1}{6}$    **2.** $\dfrac{5}{6} + \dfrac{1}{12}$    **3.** $\dfrac{1}{2} - \dfrac{1}{3}$    **4.** $\dfrac{2}{3} - \dfrac{1}{4}$

**5.** $-\dfrac{2}{11} + \dfrac{2}{33}$    **6.** $-\dfrac{5}{9} + \dfrac{1}{3}$    **7.** $\dfrac{3x}{14} - \dfrac{3}{7}$    **8.** $\dfrac{2y}{5} - \dfrac{2}{15}$

**9.** $\dfrac{11}{35} + \dfrac{2}{7}$    **10.** $\dfrac{2}{5} + \dfrac{3}{25}$    **11.** $2y - \dfrac{5}{12}$    **12.** $5y - \dfrac{3}{20}$

**13.** $\dfrac{5}{12} - \dfrac{1}{9}$    **14.** $\dfrac{7}{12} - \dfrac{5}{18}$    **15.** $\dfrac{5}{7} + 1$    **16.** $-10 + \dfrac{7}{10}$

**17.** $\dfrac{5a}{11} + \dfrac{4a}{9}$    **18.** $\dfrac{7x}{18} + \dfrac{2x}{9}$    ▣ **19.** $\dfrac{2y}{3} - \dfrac{1}{6}$    **20.** $\dfrac{5}{6} - \dfrac{1}{12}$

**Name** _____

**21.** $\dfrac{1}{2} + \dfrac{3}{x}$

**22.** $\dfrac{2}{5} + \dfrac{3}{x}$

**23.** $-\dfrac{2}{11} - \dfrac{2}{33}$

**24.** $-\dfrac{5}{9} - \dfrac{1}{3}$

**25.** $\dfrac{9}{14} - \dfrac{3}{7}$

**26.** $\dfrac{4}{5} - \dfrac{2}{15}$

**27.** $\dfrac{11y}{35} - \dfrac{2}{7}$

**28.** $\dfrac{2b}{5} - \dfrac{3}{25}$

**29.** $\dfrac{1}{9} - \dfrac{5}{12}$

**30.** $\dfrac{5}{18} - \dfrac{7}{12}$

**31.** $\dfrac{7}{15} - \dfrac{5}{12}$

**32.** $\dfrac{5}{8} - \dfrac{3}{20}$

**33.** $\dfrac{5}{7} - \dfrac{1}{8}$

**34.** $\dfrac{10}{13} - \dfrac{7}{10}$

**35.** $\dfrac{7}{8} + \dfrac{3}{16}$

**36.** $-\dfrac{7}{18} - \dfrac{2}{9}$

**37.** $\dfrac{5}{9} + \dfrac{3}{9}$

**38.** $\dfrac{4}{13} - \dfrac{1}{13}$

**39.** $\dfrac{5}{11} + \dfrac{y}{3}$

**40.** $\dfrac{5z}{13} + \dfrac{3}{26}$

**41.** $-\dfrac{5}{6} - \dfrac{3}{7}$

**42.** $\dfrac{1}{2} - \dfrac{3}{29}$

**43.** $\dfrac{7}{9} - \dfrac{1}{6}$

**44.** $\dfrac{9}{16} - \dfrac{3}{8}$

**45.** $\dfrac{2a}{3} + \dfrac{6a}{13}$     **46.** $\dfrac{3y}{4} + \dfrac{y}{7}$     **47.** $\dfrac{7}{30} - \dfrac{5}{12}$     **48.** $\dfrac{7}{30} - \dfrac{3}{20}$

**49.** $\dfrac{5}{9} + \dfrac{1}{y}$     **50.** $\dfrac{1}{12} - \dfrac{5}{x}$     **51.** $\dfrac{4}{5} + \dfrac{4}{9}$     **52.** $\dfrac{11}{12} - \dfrac{7}{24}$

**53.** $\dfrac{5}{9x} + \dfrac{1}{8}$     **54.** $\dfrac{3}{8} + \dfrac{5}{12x}$

*Perform each indicated operation. See Example 5.*

**55.** $\dfrac{9}{16} + \dfrac{5}{16} + \dfrac{2}{16}$     **56.** $\dfrac{4}{7} + \dfrac{1}{7} + \dfrac{2}{7}$     **57.** $-\dfrac{2}{5} + \dfrac{1}{3} - \dfrac{3}{10}$     **58.** $-\dfrac{1}{3} - \dfrac{1}{4} + \dfrac{2}{5}$

**59.** $\dfrac{x}{2} + \dfrac{x}{4} + \dfrac{2x}{16}$     **60.** $\dfrac{z}{4} + \dfrac{z}{8} + \dfrac{2z}{16}$     **61.** $-\dfrac{1}{2} - \dfrac{1}{4} - \dfrac{1}{16}$     **62.** $\dfrac{1}{4} - \dfrac{1}{8} - \dfrac{1}{16}$

**63.** $\dfrac{6}{5} - \dfrac{3}{4} + \dfrac{1}{2}$     **64.** $\dfrac{6}{5} + \dfrac{3}{4} - \dfrac{1}{2}$     **65.** $-\dfrac{9}{12} + \dfrac{17}{24} - \dfrac{1}{6}$     **66.** $-\dfrac{5}{14} + \dfrac{3}{7} - \dfrac{1}{2}$

**67.** $\dfrac{3}{16} + \dfrac{7x}{18} + \dfrac{5}{24}$     **68.** $\dfrac{1}{10} + \dfrac{4y}{15} + \dfrac{9}{25}$     **69.** $\dfrac{3x}{8} + \dfrac{2x}{7} - \dfrac{5}{14}$     **70.** $\dfrac{9x}{10} - \dfrac{1}{2} + \dfrac{x}{5}$

---

**45.** $\dfrac{44a}{39}$

**46.** $\dfrac{25y}{28}$

**47.** $-\dfrac{11}{60}$

**48.** $\dfrac{1}{12}$

**49.** $\dfrac{5y + 9}{9y}$

**50.** $\dfrac{x - 60}{12x}$

**51.** $\dfrac{56}{45}$

**52.** $\dfrac{5}{8}$

**53.** $\dfrac{40 + 9x}{72x}$

**54.** $\dfrac{9x + 10}{24x}$

**55.** $1$

**56.** $1$

**57.** $-\dfrac{11}{30}$

**58.** $-\dfrac{11}{60}$

**59.** $\dfrac{7x}{8}$

**60.** $\dfrac{z}{2}$

**61.** $-\dfrac{13}{16}$

**62.** $\dfrac{1}{16}$

**63.** $\dfrac{19}{20}$

**64.** $\dfrac{29}{20}$

**65.** $-\dfrac{5}{24}$

**66.** $-\dfrac{3}{7}$

**67.** $\dfrac{57 + 56x}{144}$

**68.** $\dfrac{69 + 40y}{150}$

**69.** $\dfrac{37x - 20}{56}$

**70.** $\dfrac{11x - 5}{10}$

**284**

---

**Name** _____

**B** *Evaluate each expression if* $x = \dfrac{1}{3}$ *and* $y = \dfrac{3}{4}$. *See Example 6.*

**71.** $x + y$      **72.** $x - y$      **73.** $xy$

**74.** $x \div y$      **75.** $2y + x$      **76.** $2x + y$

**C** *Solve and check. See Example 7.*

**77.** $x - \dfrac{1}{12} = \dfrac{5}{6}$      **78.** $y - \dfrac{8}{9} = \dfrac{1}{3}$      **79.** $\dfrac{2}{5} + y = -\dfrac{3}{10}$

**80.** $\dfrac{1}{2} + a = -\dfrac{3}{8}$      **81.** $7z + \dfrac{1}{16} - 6z = \dfrac{3}{4}$      **82.** $9x - \dfrac{2}{7} - 8x = \dfrac{11}{14}$

**83.** $-\dfrac{2}{9} = x - \dfrac{5}{6}$      **84.** $-\dfrac{1}{4} = y - \dfrac{7}{10}$

**D** *Find the perimeter of each geometric figure. (Hint: Recall that perimeter means distance around.)*

△ **85.**

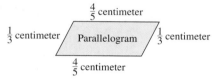

$\frac{4}{5}$ centimeter

$\frac{1}{3}$ centimeter   Parallelogram   $\frac{1}{3}$ centimeter

$\frac{4}{5}$ centimeter

△ **86.**

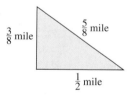

$\frac{3}{8}$ mile   $\frac{5}{8}$ mile

$\frac{1}{2}$ mile

△ **87.**

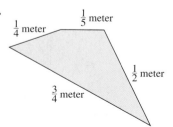

$\frac{1}{4}$ meter   $\frac{1}{5}$ meter

$\frac{1}{2}$ meter

$\frac{3}{4}$ meter

△ **88.**

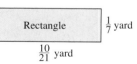

Rectangle   $\frac{1}{7}$ yard

$\frac{10}{21}$ yard

**Name** _____

*Solve. See Examples 8 and 9.*

**89.** Killer bees have been known to chase people for up to $\dfrac{1}{4}$ of a mile, while domestic European honeybees will normally chase a person for no more than 100 feet, or $\dfrac{5}{264}$ of a mile. How much farther will a killer bee chase a person than a domestic honeybee? (*Source:* Coachella Valley Mosquito & Vector Control District)

**90.** The slowest mammal is the three-toed sloth from South America. The sloth has an average ground speed of $\dfrac{1}{10}$ mph. In the trees, it can accelerate to $\dfrac{17}{100}$ mph. How much faster can a sloth travel in the trees? (*Source: The Guiness Book of World Records*, 1999)

**91.** About $\dfrac{13}{20}$ of American students ages 10 to 17 name math, science, or art as their favorite subject in school. Art is the favorite subject for about $\dfrac{4}{25}$ of these students. For what fraction of students this age is math or science their favorite subject? (*Source:* Peter D. Hart Research Associates for the National Science Foundation)

**92.** Together, the United States' and Japan's postal services handle $\dfrac{49}{100}$ of the world's mail volume. Japan's postal service alone handles $\dfrac{3}{50}$ of the world's mail. What fraction of the world's mail is handled by the postal service of the United States? (*Source:* United States Postal Service)

*The table gives the fraction of Americans who eat pasta at various intervals. Use this table to answer Exercises 93 and 94.*

| HOW OFTEN AMERICANS EAT PASTA ||
| Frequency | Fraction |
| --- | --- |
| 3 times per week | $\dfrac{31}{100}$ |
| 1 or 2 times per week | $\dfrac{23}{50}$ |
| 1 or 2 times per month | $\dfrac{17}{100}$ |
| Less often | $\dfrac{3}{50}$ |

(*Source:* Princeton Survey Research)

**93.** What fraction of Americans eat pasta 1, 2, or 3 times a week?

**94.** What fraction of Americans eat pasta 1 or 2 times a month or less often?

**95.** $\frac{29}{100}$ of adults _____

**96.** $\frac{8}{25}$ of adults _____

**97.** $\frac{25}{36}$ _____

**98.** $\frac{1}{4}$ _____

**99.** $\frac{25}{36}$ _____

**100.** $\frac{1}{4}$ _____

**101.** 57,200 _____

**102.** 600 _____

**103.** 330 _____

**104.** 2330 _____

**105.** $\frac{49}{44}$ _____

**106.** $\frac{5}{338}$ _____

*The circle graph shows the fraction of American adults who drive various distances in miles in an average week. Use this graph to answer Exercises 95 and 96.*

**Miles Driven by American Adults in an Average Week**

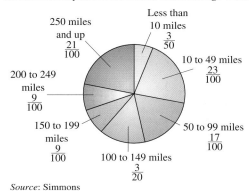

*Source*: Simmons

**95.** What fraction of American adults drive less than 50 miles in an average week?

**96.** What fraction of American adults drive between 50 and 149 miles in an average week?

## REVIEW AND PREVIEW

*Evaluate. See Section 4.3.*

**97.** $\left(\frac{5}{6}\right)^2$     **98.** $\left(\frac{1}{2}\right)^2$     **99.** $\left(-\frac{5}{6}\right)^2$     **100.** $\left(-\frac{1}{2}\right)^2$

*Round each number to the given place value. See Section 1.4.*

**101.** 57,236 to the nearest hundred

**102.** 576 to the nearest hundred

**103.** 327 to the nearest ten

**104.** 2333 to the nearest ten

## COMBINING CONCEPTS

*Perform each indicated operation.*

**105.** $\frac{30}{55} + \frac{1000}{1760}$; the LCD of 55 and 1760 is 1760.

**106.** $\frac{19}{26} - \frac{968}{1352}$; the LCD of 26 and 1352 is 1352.

**107.** In your own words, describe how to add two fractions with different denominators.

*The table shows the fraction of the world's land area occupied by each continent. Use this table to answer Exercises 108–112.*

| Continent | Fraction of World's Land Area |
|-----------|--------------|
| Africa | $\dfrac{39}{193}$ |
| Antarctica | $\dfrac{18}{193}$ |
| Asia | $\dfrac{58}{193}$ |
| Australia | $\dfrac{11}{193}$ |
| Europe | $\dfrac{38}{579}$ |
| North America | $\dfrac{94}{579}$ |
| South America | $\dfrac{23}{193}$ |

(*Source: 2000 World Almanac*)

**108.** What fraction of the world's land area is accounted for by North and South America?

**109.** What fraction of the world's land area is accounted for by Asia and Europe?

**110.** If the total land area of Earth's surface is 57,900,000 square miles, what is the combined land area of the North and South American continents?

**111.** If the total land area of Earth's surface is 57,900,000 square miles, what is the combined land area of the European and Asian continents?

**112.** Antarctica is generally considered to be uninhabited. What fraction of the world's land area is accounted for by inhabited continents?

# Focus on History

### THE DEVELOPMENT OF FRACTIONS

Many ancient cultures were familiar with the use of fractions and used them in everyday life. Two examples are discussed here.

▲ Although the ancient Egyptians were comfortable with the idea of fractions, their system of hieroglyphic numbers allowed them to only write fractions with 1 as the numerator, like $\frac{1}{3}$ or $\frac{1}{5}$. To write fractions with numerators other than 1, the Egyptians had to write a sum of fractions with 1 as the numerator. They preferred to never use the same denominator more than once in one of these sums of fractions. For example, instead of representing the fraction $\frac{2}{5}$ as the sum $\frac{1}{5} + \frac{1}{5}$ (which uses the denominator 5 more than once), they would prefer instead the sum $\frac{1}{3} + \frac{1}{15}$, which also is equal to $\frac{2}{5}$ but uses two different denominators. Similarly, $\frac{2}{13}$ would not be represented by $\frac{1}{13} + \frac{1}{13}$, as we might think would be easiest, but instead by $\frac{1}{8} + \frac{1}{52} + \frac{1}{104}$.

▲ The ancient Babylonians also dealt with fractions but in a more direct way than did the Egyptians. For instance, they thought of the fraction $\frac{3}{8}$ as the product of 3 and the reciprocal of 8. They compiled detailed lists of reciprocals and could calculate with fractions very easily and quickly.

Although many ancient cultures used fractions, they didn't represent them the way we do today. Fractions did not evolve into anything similar to our present-day format until the seventh century A.D. Around that time, the Hindus became the first to write fractions with the numerator above the denominator. However, Hindu mathematicians, such as Brahmagupta in A.D. 628, wrote these fractions without the fraction bar we use today. Thus, the fraction $\frac{5}{8}$ would have been written by the Hindus as $\frac{5}{8}$.

It wasn't until between A.D. 1100 and 1200 that the next giant step in fraction notation was made. Up to this point, Arab mathematicians had copied Hindu notation for fractions. However, they then improved this notation by inserting a horizontal bar, called a *vinculum*, between the numerator and denominator to separate the two parts of the fraction.

Today we see fractions in the form ⁴/₉, or 4/9 just as often as in the $\frac{4}{9}$ form, developed by the Arabs. The use of the diagonal fraction bar, called a *solidus*, was introduced with the appearance of the printing press. Typesetting fractions with a horizontal fraction bar was difficult and space-consuming because three lines of type were required for the fraction. Typesetting fractions with a diagonal fraction bar was far easier for printers because the entire fraction fit on a single line of type.

# INTEGRATED REVIEW—SUMMARY ON FRACTIONS AND FACTORS

*Use a fraction to represent the shaded area of each figure or figure group.*

**1.**

**2.**

**3.**

**4.** In a survey, 73 people out of 85 get less than 8 hours of sleep each night. What fraction of people in the survey get less than 8 hours of sleep?

*Write the prime factorization of each number. Write any repeated factors using exponents.*

**5.** 6            **6.** 70            **7.** 252

*Write each fraction in simplest form.*

**8.** $\dfrac{2}{14}$         **9.** $\dfrac{20}{24}$         **10.** $\dfrac{18}{38}$

**11.** $\dfrac{42}{110}$         **12.** $\dfrac{32}{64}$         **13.** $\dfrac{72}{80}$

**14.** Of the 50 United States, 2 states are not adjacent to any other states. What fraction of the states are not adjacent to other states? Write the fraction in simplest form.

**15.** $\dfrac{4}{5}$

**16.** $-\dfrac{2}{5}$

**17.** $\dfrac{3}{25}$

**18.** $\dfrac{1}{3}$

**19.** $\dfrac{4}{5}$

**20.** $\dfrac{5}{9}$

**21.** $-\dfrac{1}{6}$

**22.** $\dfrac{3}{2}$

**23.** $\dfrac{1}{18}$

**24.** $\dfrac{4}{21}$

**25.** $-\dfrac{7}{48}$

**26.** $-\dfrac{9}{50}$

**27.** $\dfrac{37}{40}$

**28.** $\dfrac{11}{36}$

**29.** $\dfrac{5}{33}$

**30.** $\dfrac{5}{18}$

**31.** $\dfrac{11}{18}$

**32.** $\dfrac{37}{50}$

**Name** _____

*The following summary will help you with this review of operations on fractions.*

---

**OPERATIONS ON FRACTIONS**

Let $a$, $b$, $c$, and $d$ be integers.

Addition: $\dfrac{a}{b} + \dfrac{c}{b} = \dfrac{a + c}{b}$

$(b \neq 0)$ $\uparrow$ $\uparrow$
**common denominators**

Subtraction: $\dfrac{a}{b} - \dfrac{c}{b} = \dfrac{a - c}{b}$

$(b \neq 0)$ $\uparrow$ $\uparrow$
**common denominators**

Multiplication: $\dfrac{a}{b} \cdot \dfrac{c}{d} = \dfrac{a \cdot c}{b \cdot d}$
$(b \neq 0, d \neq 0)$

Division: $\dfrac{a}{b} \div \dfrac{c}{d} = \dfrac{a}{b} \cdot \dfrac{d}{c} = \dfrac{a \cdot d}{b \cdot c}$
$(b \neq 0, d \neq 0, c \neq 0)$

---

*Perform each indicated operation.*

**15.** $\dfrac{1}{5} + \dfrac{3}{5}$     **16.** $\dfrac{1}{5} - \dfrac{3}{5}$     **17.** $\dfrac{1}{5} \cdot \dfrac{3}{5}$     **18.** $\dfrac{1}{5} \div \dfrac{3}{5}$

**19.** $\dfrac{2}{3} \div \dfrac{5}{6}$     **20.** $\dfrac{2}{3} \cdot \dfrac{5}{6}$     **21.** $\dfrac{2}{3} - \dfrac{5}{6}$     **22.** $\dfrac{2}{3} + \dfrac{5}{6}$

**23.** $-\dfrac{1}{7} \cdot -\dfrac{7}{18}$     **24.** $-\dfrac{4}{9} \cdot -\dfrac{3}{7}$     **25.** $-\dfrac{7}{8} \div 6$     **26.** $-\dfrac{9}{10} \div 5$

**27.** $\dfrac{7}{8} + \dfrac{1}{20}$     **28.** $\dfrac{5}{12} - \dfrac{1}{9}$     **29.** $\dfrac{9}{11} - \dfrac{2}{3}$     **30.** $\dfrac{2}{9} + \dfrac{1}{18}$

**31.** $\dfrac{2}{9} + \dfrac{1}{18} + \dfrac{1}{3}$     **32.** $\dfrac{3}{10} + \dfrac{1}{5} + \dfrac{6}{25}$

## 4.6   COMPLEX FRACTIONS AND REVIEW OF ORDER OF OPERATIONS

### A   SIMPLIFYING COMPLEX FRACTIONS

Thus far, we have studied operations on fractions. We now practice simplifying fractions whose numerators or denominators themselves contain fractions. These fractions are called **complex fractions**.

> **COMPLEX FRACTION**
>
> A fraction whose numerator or denominator or both numerator and denominator contain fractions is called a **complex fraction**.

Examples of complex fractions are

$$\frac{\frac{x}{4}}{\frac{3}{2}} \qquad \frac{\frac{1}{2} + \frac{3}{8}}{\frac{3}{4} - \frac{1}{6}} \qquad \frac{\frac{y}{5} - 2}{\frac{3}{10}}$$

### Method 1 for Simplifying Complex Fractions

Two methods are presented to simplify complex fractions. The first method makes use of the fact that a fraction bar means division.

**Example 1**   Simplify: $\dfrac{\frac{x}{4}}{\frac{3}{2}}$

*Solution:*   Since a fraction bar means division, the complex fraction

$\dfrac{\frac{x}{4}}{\frac{3}{2}}$ can be written as $\dfrac{x}{4} \div \dfrac{3}{2}$. Then divide as usual to simplify.

$$\frac{x}{4} \div \frac{3}{2} = \frac{x}{4} \cdot \frac{2}{3} \qquad \text{Multiply by the reciprocal.}$$

$$= \frac{x \cdot 2}{2 \cdot 2 \cdot 3}$$

$$= \frac{x}{6}$$

**Example 2**   Simplify: $\dfrac{\frac{1}{2} + \frac{3}{8}}{\frac{3}{4} - \frac{1}{6}}$

*Solution:*   Recall the order of operations. Before performing the division, we simplify the numerator and the denominator of the complex fraction separately.

$$\frac{\frac{1}{2} + \frac{3}{8}}{\frac{3}{4} - \frac{1}{6}} = \frac{\frac{1 \cdot 4}{2 \cdot 4} + \frac{3}{8}}{\frac{3 \cdot 3}{4 \cdot 3} - \frac{1 \cdot 2}{6 \cdot 2}} = \frac{\frac{4}{8} + \frac{3}{8}}{\frac{9}{12} - \frac{2}{12}} = \frac{\frac{7}{8}}{\frac{7}{12}}$$

Thus,

$$\frac{\dfrac{1}{2} + \dfrac{3}{8}}{\dfrac{3}{4} - \dfrac{1}{6}} = \frac{\dfrac{7}{8}}{\dfrac{7}{12}}$$

$$= \frac{7}{8} \div \frac{7}{12} \qquad \text{Rewrite the quotient using the} \div \text{sign.}$$

$$= \frac{7}{8} \cdot \frac{12}{7} \qquad \text{Multiply by the reciprocal.}$$

$$= \frac{7 \cdot 3 \cdot 4}{2 \cdot 4 \cdot 7} \qquad \text{Multiply.}$$

$$= \frac{3}{2} \qquad \text{Simplify.}$$

## Method 2 for Simplifying Complex Fractions

The second method for simplifying complex fractions is to multiply the numerator and the denominator of the complex fraction by the LCD of all the fractions in its numerator and its denominator. Since this LCD is divisible by all denominators, this has the effect of leaving sums and differences of integers in the numerator and the denominator. Let's use this second method to simplify the complex fraction in Example 2 again.

### Practice Problem 3

Use Method 2 to simplify: $\dfrac{\dfrac{1}{2} + \dfrac{1}{6}}{\dfrac{3}{4} - \dfrac{2}{3}}$

**Example 3**   Simplify: $\dfrac{\dfrac{1}{2} + \dfrac{3}{8}}{\dfrac{3}{4} - \dfrac{1}{6}}$

*Solution:*   The complex fraction contains fractions with denominators 2, 8, 4, and 6. The LCD is 24. By the fundamental property of fractions, we can multiply the numerator and the denominator of the complex fraction by 24.

$$\frac{\dfrac{1}{2} + \dfrac{3}{8}}{\dfrac{3}{4} - \dfrac{1}{6}} = \frac{24\left(\dfrac{1}{2} + \dfrac{3}{8}\right)}{24\left(\dfrac{3}{4} - \dfrac{1}{6}\right)}$$

$$= \frac{\left(24 \cdot \dfrac{1}{2}\right) + \left(24 \cdot \dfrac{3}{8}\right)}{\left(24 \cdot \dfrac{3}{4}\right) - \left(24 \cdot \dfrac{1}{6}\right)} \quad \text{Apply the distributive property.}$$

$$= \frac{12 + 9}{18 - 4} \qquad \text{Multiply.}$$

$$= \frac{21}{14}$$

$$= \frac{7 \cdot 3}{7 \cdot 2} = \frac{3}{2} \qquad \text{Simplify.}$$

The simplified result is the same, of course, no matter which method is used.

**Example 4**    Simplify: $\dfrac{\dfrac{y}{5} - 2}{\dfrac{3}{10}}$

*Solution:*    Use the second method and multiply the numerator and the denominator of the complex fraction by the LCD of all fractions. Recall that $2 = \dfrac{2}{1}$. The LCD of the denominators 5, 1, and 10 is 10.

$$\frac{\dfrac{y}{5} - \dfrac{2}{1}}{\dfrac{3}{10}} = \frac{10\left(\dfrac{y}{5} - \dfrac{2}{1}\right)}{10\left(\dfrac{3}{10}\right)}$$    Multiply the numerator and denominator by 10.

$$= \frac{\left(10 \cdot \dfrac{y}{5}\right) - (10 \cdot 2)}{10 \cdot \dfrac{3}{10}}$$    Apply the distributive property.

$$= \frac{2y - 20}{3}$$    Multiply.

**Practice Problem 4**

Simplify: $\dfrac{\dfrac{3}{4}}{\dfrac{x}{5} - 1}$

> **HELPFUL HINT**
>
> Don't forget to multiply the numerator and the denominator of the complex fraction by the same number—the LCD.

**TEACHING TIP**

Students often want to now use this method when adding and subtracting fractions. Remind students that they may not multiply through and "cancel" when adding and subtracting fractions.

**B**  **REVIEWING THE ORDER OF OPERATIONS**

At this time, it is probably a good idea to review the order of operations on expressions containing fractions.

> **ORDER OF OPERATIONS**
>
> 1.  Do all operations within grouping symbols such as parentheses or brackets.
> 2.  Evaluate any expressions with exponents and find any square roots.
> 3.  Multiply or divide in order from left to right.
> 4.  Add or subtract in order from left to right.

**Example 5**    Simplify: $\left(\dfrac{4}{5}\right)^2 - 1$

*Solution:*    According to the order of operations, first evaluate $\left(\dfrac{4}{5}\right)^2$.

$$\left(\frac{4}{5}\right)^2 - 1 = \frac{16}{25} - 1$$    Write $\left(\dfrac{4}{5}\right)^2$ as $\dfrac{16}{25}$.

Next, combine the fractions. The LCD of 25 and 1 is 25.

$$\frac{16}{25} - 1 = \frac{16}{25} - \frac{25}{25}$$    Write 1 as $\dfrac{25}{25}$.

$$= \frac{-9}{25} \text{ or } -\frac{9}{25}$$    Subtract.

**Practice Problem 5**

Simplify: $\left(2 - \dfrac{2}{3}\right)^3$

**Answers**

**4.** $\dfrac{15}{4x - 20}$    **5.** $\dfrac{64}{27}$

## Practice Problem 6

Simplify: $\left(-\dfrac{1}{2} + \dfrac{1}{5}\right)\left(\dfrac{7}{8} + \dfrac{1}{8}\right)$

## ✓ CONCEPT CHECK

What should be done first to simplify the expression $\dfrac{1}{5} \cdot \dfrac{5}{2} - \left(\dfrac{2}{3} + \dfrac{4}{5}\right)^2$?

## Practice Problem 7

Evaluate $-\dfrac{3}{5} - xy$ if $x = \dfrac{3}{10}$ and $y = \dfrac{2}{3}$.

---

**Example 6**    Simplify: $\left(\dfrac{1}{4} + \dfrac{2}{3}\right)\left(\dfrac{11}{12} + \dfrac{1}{4}\right)$

*Solution:*    First perform operations inside parentheses. Then multiply.

$$\left(\dfrac{1}{4} + \dfrac{2}{3}\right)\left(\dfrac{11}{12} + \dfrac{1}{4}\right) = \left(\dfrac{1 \cdot 3}{4 \cdot 3} + \dfrac{2 \cdot 4}{3 \cdot 4}\right)\left(\dfrac{11}{12} + \dfrac{1 \cdot 3}{4 \cdot 3}\right) \quad \text{The LCD is 12.}$$

$$= \left(\dfrac{3}{12} + \dfrac{8}{12}\right)\left(\dfrac{11}{12} + \dfrac{3}{12}\right)$$

$$= \left(\dfrac{11}{12}\right)\left(\dfrac{14}{12}\right) \quad \text{Add.}$$

$$= \dfrac{11 \cdot 2 \cdot 7}{2 \cdot 6 \cdot 12} \quad \text{Multiply.}$$

$$= \dfrac{77}{72} \quad \text{Simplify.}$$

**TRY THE CONCEPT CHECK IN THE MARGIN.**

**Example 7**    Evaluate $2x + y^2$ if $x = -\dfrac{1}{2}$ and $y = \dfrac{1}{3}$.

*Solution:*    Replace $x$ and $y$ with the given values and simplify.

$$2x + y^2 = 2\left(-\dfrac{1}{2}\right) + \left(\dfrac{1}{3}\right)^2 \quad \text{Replace } x \text{ with } -\dfrac{1}{2} \text{ and } y \text{ with } \dfrac{1}{3}.$$

$$= 2\left(-\dfrac{1}{2}\right) + \dfrac{1}{9} \quad \text{Write } \left(\dfrac{1}{3}\right)^2 \text{ as } \dfrac{1}{9}.$$

$$= -1 + \dfrac{1}{9} \quad \text{Multiply.}$$

$$= -\dfrac{9}{9} + \dfrac{1}{9} \quad \text{The LCD is 9.}$$

$$= -\dfrac{8}{9} \quad \text{Add.}$$

---

**Answers**

**6.** $-\dfrac{3}{10}$    **7.** $-\dfrac{4}{5}$

✓ **Concept Check:** Add inside parentheses.

## EXERCISE SET 4.6

**A** *Simplify each complex fraction. See Examples 1 through 4.*

**1.** $\dfrac{\frac{1}{8}}{\frac{3}{4}}$

**2.** $\dfrac{\frac{2}{3}}{\frac{2}{7}}$

**3.** $\dfrac{\frac{9}{10}}{\frac{21}{10}}$

**4.** $\dfrac{\frac{14}{5}}{\frac{7}{5}}$

**5.** $\dfrac{\frac{2x}{27}}{\frac{4}{9}}$

**6.** $\dfrac{\frac{3y}{11}}{\frac{1}{2}}$

**7.** $\dfrac{\frac{3}{4}+\frac{2}{5}}{\frac{1}{2}+\frac{3}{5}}$

**8.** $\dfrac{\frac{7}{6}+\frac{2}{3}}{\frac{3}{2}-\frac{8}{9}}$

**9.** $\dfrac{\frac{3x}{4}}{5-\frac{1}{8}}$

**10.** $\dfrac{\frac{3}{10}+2}{\frac{2}{5y}}$

**B** *Simplify. See Examples 5 and 6.*

**11.** $2^2 - \left(\dfrac{1}{3}\right)^2$

**12.** $3^2 - \left(\dfrac{1}{2}\right)^2$

**13.** $\left(\dfrac{2}{9}+\dfrac{4}{9}\right)\left(\dfrac{1}{3}-\dfrac{9}{10}\right)$

**14.** $\left(\dfrac{1}{5}-\dfrac{1}{10}\right)\left(\dfrac{1}{5}+\dfrac{1}{10}\right)$

**15.** $\left(\dfrac{7}{8}-\dfrac{1}{2}\right) \div \dfrac{3}{11}$

**16.** $\left(-\dfrac{2}{3}-\dfrac{7}{3}\right) \div \dfrac{4}{9}$

*Evaluate each expression if $x = -\dfrac{1}{3}$, $y = \dfrac{2}{5}$, and $z = \dfrac{5}{6}$. See Example 7.*

**17.** $5y - z$

**18.** $2z - x$

**19.** $\dfrac{x}{z}$

**20.** $\dfrac{y+x}{z}$

**21.** $x^2 - yz$

**22.** $(1 + x)(1 + z)$

ANSWERS

**1.** $\dfrac{1}{6}$

**2.** $\dfrac{7}{3}$

**3.** $\dfrac{3}{7}$

**4.** $2$

**5.** $\dfrac{x}{6}$

**6.** $\dfrac{6y}{11}$

**7.** $\dfrac{23}{22}$

**8.** $3$

**9.** $\dfrac{2x}{13}$

**10.** $\dfrac{23y}{4}$

**11.** $\dfrac{35}{9}$

**12.** $\dfrac{35}{4}$

**13.** $-\dfrac{17}{45}$

**14.** $\dfrac{3}{100}$

**15.** $\dfrac{11}{8}$

**16.** $-\dfrac{27}{4}$

**17.** $\dfrac{7}{6}$

**18.** $2$

**19.** $-\dfrac{2}{5}$

**20.** $\dfrac{2}{25}$

**21.** $-\dfrac{2}{9}$

**23.** $\dfrac{5}{2}$ _____

**24.** $\dfrac{5}{4}$ _____

**25.** $\dfrac{7}{2}$ _____

**26.** $\dfrac{11}{21}$ _____

**27.** $\dfrac{4}{9}$ _____

**28.** $\dfrac{15}{16}$ _____

**29.** $-\dfrac{13}{2}$ _____

**30.** $\dfrac{25}{42}$ _____

**31.** $\dfrac{9}{25}$ _____

**32.** $-\dfrac{5}{16}$ _____

**33.** $-\dfrac{5}{32}$ _____

**34.** $-\dfrac{1}{5}$ _____

**35.** 1 _____

**36.** $\dfrac{1}{81}$ _____

**37.** $\dfrac{1}{10}$ _____

**38.** $\dfrac{24}{31}$ _____

**39.** $-\dfrac{11}{40}$ _____

**40.** $-\dfrac{45}{28}$ _____

**41.** $\dfrac{x+6}{16}$ _____

**42.** $\dfrac{8-2x}{19}$ _____

**A** **B**  *Simplify the following. See Examples 1 through 6.*

**23.** $\dfrac{\frac{5}{24}}{\frac{1}{12}}$

**24.** $\dfrac{\frac{7}{10}}{\frac{14}{25}}$

**25.** $\left(\dfrac{3}{2}\right)^3 + \left(\dfrac{1}{2}\right)^3$

**26.** $\left(\dfrac{5}{21} \div \dfrac{1}{2}\right) + \left(\dfrac{1}{7} \cdot \dfrac{1}{3}\right)$

**27.** $\left(-\dfrac{1}{3}\right)^2 + \dfrac{1}{3}$

**28.** $\left(-\dfrac{3}{4}\right)^2 + \dfrac{3}{8}$

**29.** $\dfrac{2 + \frac{1}{6}}{1 - \frac{4}{3}}$

**30.** $\dfrac{3 - \frac{1}{2}}{4 + \frac{1}{5}}$

**31.** $\left(1 - \dfrac{2}{5}\right)^2$

**32.** $\left(-\dfrac{1}{2}\right)^2 - \left(\dfrac{3}{4}\right)^2$

**33.** $\left(\dfrac{3}{4} - 1\right)\left(\dfrac{1}{8} + \dfrac{1}{2}\right)$

**34.** $\left(\dfrac{1}{10} + \dfrac{3}{20}\right)\left(\dfrac{1}{5} - 1\right)$

**35.** $\left(-\dfrac{2}{9} - \dfrac{7}{9}\right)^4$

**36.** $\left(\dfrac{5}{9} - \dfrac{2}{3}\right)^2$

**37.** $\dfrac{\frac{1}{2} - \frac{3}{8}}{\frac{3}{4} + \frac{1}{2}}$

**38.** $\dfrac{\frac{7}{10} + \frac{1}{2}}{\frac{4}{5} + \frac{3}{4}}$

**39.** $\left(\dfrac{3}{4} \div \dfrac{6}{5}\right) - \left(\dfrac{3}{4} \cdot \dfrac{6}{5}\right)$

**40.** $\left(\dfrac{1}{2} \cdot \dfrac{2}{7}\right) - \left(\dfrac{1}{2} \div \dfrac{2}{7}\right)$

**41.** $\dfrac{\frac{x}{3} + 2}{5 + \frac{1}{3}}$

**42.** $\dfrac{1 - \frac{x}{4}}{2 + \frac{3}{8}}$

**296**

**Name** _____

## REVIEW AND PREVIEW

*Evaluate. See Section 1.7.*

**43.** $2^3$      **44.** $3^2$      **45.** $5^2$      **46.** $2^5$

*Multiply. See Section 2.4.*

**47.** $\dfrac{1}{3}(3x)$      **48.** $\dfrac{1}{5}(5y)$      **49.** $\dfrac{2}{3}\left(\dfrac{3}{2}a\right)$      **50.** $-\dfrac{9}{10}\left(-\dfrac{10}{9}m\right)$

## COMBINING CONCEPTS

*Evaluate each expression if $x = \dfrac{3}{4}$ and $y = -\dfrac{4}{7}$.*

**51.** $\dfrac{2+x}{y}$      **52.** $4x + y$      **53.** $x^2 + 7y$      **54.** $\dfrac{\frac{9}{14}}{x + y}$

*Recall that to find the average of two numbers, find their sum and divide by 2. For example, the average of $\dfrac{1}{2}$ and $\dfrac{3}{4}$ is $\dfrac{\frac{1}{2} + \frac{3}{4}}{2}$. Find the average of each pair of numbers.*

**55.** $\dfrac{1}{2}, \dfrac{3}{4}$      **56.** $\dfrac{3}{5}, \dfrac{9}{10}$      **57.** $\dfrac{1}{4}, \dfrac{2}{14}$      **58.** $\dfrac{5}{6}, \dfrac{7}{9}$

**59.** Two positive numbers, $a$ and $b$, are graphed below. Where should the graph of their average lie?

| | |
|---|---|
| **43.** | 8 |
| **44.** | 9 |
| **45.** | 25 |
| **46.** | 32 |
| **47.** | $1x$ or $x$ |
| **48.** | $1y$ or $y$ |
| **49.** | $1a$ or $a$ |
| **50.** | $1m$ or $m$ |
| **51.** | $-\dfrac{77}{16}$ |
| **52.** | $\dfrac{17}{7}$ |
| **53.** | $-\dfrac{55}{16}$ |
| **54.** | $\dfrac{18}{5}$ |
| **55.** | $\dfrac{5}{8}$ |
| **56.** | $\dfrac{3}{4}$ |
| **57.** | $\dfrac{11}{56}$ |
| **58.** | $\dfrac{29}{36}$ |
| **59.** | halfway between $a$ and $b$ |

*Answer true or false for each statement.*

**60.** It is possible for the average of two numbers to be greater than both numbers.

**61.** It is possible for the average of two numbers to be less than both numbers.

**62.** The sum of two negative fractions is always a negative number.

**63.** The sum of a negative fraction and a positive fraction is always a positive number.

**64.** It is possible for the sum of two fractions to be a whole number.

**65.** It is possible for the difference of two fractions to be a whole number.

# 4.7 SOLVING EQUATIONS CONTAINING FRACTIONS

## A SOLVING EQUATIONS CONTAINING FRACTIONS

In Chapter 3, we solved linear equations in one variable. In this section, we practice this skill by solving linear equations containing fractions. To help us solve these equations, let's review the multiplication property of equality and the division property of equality.

---

**MULTIPLICATION AND DIVISION PROPERTIES OF EQUALITY**

Let $a$, $b$, and $c$ be numbers, and $c \neq 0$.

$$\text{If } a = b, \text{ then } a \cdot c = b \cdot c \text{ and also } \frac{a}{c} = \frac{b}{c}.$$

---

In other words, both sides of an equation may be multiplied or divided by the **same** nonzero number without changing the solution of the equation.

**Example 1**    Solve for $x$: $\frac{1}{3}x = 7$

*Solution:*    Recall that isolating $x$ means that we want the coefficient of $x$ to be 1. To do so, we use the multiplication property of equality and multiply both sides of the equation by the reciprocal of $\frac{1}{3}$, or 3. Since $\frac{1}{3} \cdot 3 = 1$, we have isolated $x$.

$$\frac{1}{3}x = 7$$

$$3 \cdot \frac{1}{3}x = 3 \cdot 7 \qquad \text{Multiply both sides by 3.}$$

$$1 \cdot x = 21 \text{ or } x = 21 \qquad \text{Simplify.}$$

To check, replace $x$ with 21 in the original equation.

$$\frac{1}{3}x = 7 \qquad \text{Original equation}$$

$$\frac{1}{3} \cdot 21 = 7 \qquad \text{Replace } x \text{ with 21.}$$

$$7 = 7 \qquad \text{True.}$$

Since $7 = 7$ is a true statement, 21 is the solution of $\frac{1}{3}x = 7$.

---

### Objectives

**A** Solve equations containing fractions.

**B** Review adding and subtracting fractions.

Study    SSM    CD-ROM    Video
Guide                              4.7

**Practice Problem 1**

Solve for $y$: $\frac{1}{5}y = 2$

**Answer**
**1.** 10

## Practice Problem 2

Solve for $b$: $\dfrac{5}{7} b = 25$

## Practice Problem 3

Solve for $x$: $-\dfrac{7}{10} x = \dfrac{2}{5}$

**Answers**

**2.** $35$   **3.** $-\dfrac{4}{7}$

---

**Example 2**   Solve for $a$: $\dfrac{3}{5} a = 9$

*Solution:*   Multiply both sides by $\dfrac{5}{3}$, the reciprocal of $\dfrac{3}{5}$, so that the coefficient of $a$ is 1.

$$\frac{3}{5} a = 9$$

$$\frac{5}{3} \cdot \frac{3}{5} a = \frac{5}{3} \cdot 9 \qquad \text{Multiply both sides by } \tfrac{5}{3}.$$

$$1a = \frac{5 \cdot 9}{3} \qquad \text{Multiply.}$$

$$a = 15 \qquad \text{Simplify.}$$

To check, replace $a$ with 15 in the original equation.

$$\frac{3}{5} a = 9$$

$$\frac{3}{5} \cdot 15 = 9 \qquad \text{Replace } a \text{ with 15.}$$

$$\frac{3 \cdot 15}{5} = 9 \qquad \text{Multiply.}$$

$$9 = 9 \qquad \text{True.}$$

Since $9 = 9$ is true, 15 is the solution of $\dfrac{3}{5} a = 9$. ▬▬▬

**Example 3**   Solve for $x$: $\dfrac{3}{4} x = -\dfrac{1}{8}$

*Solution:*   Multiply both sides of the equation by $\dfrac{4}{3}$, the reciprocal of $\dfrac{3}{4}$.

$$\frac{3}{4} x = -\frac{1}{8}$$

$$\frac{4}{3} \cdot \frac{3}{4} x = \frac{4}{3} \cdot -\frac{1}{8} \qquad \text{Multiply both sides by } \tfrac{4}{3}.$$

$$1x = -\frac{4 \cdot 1}{3 \cdot 8} \qquad \text{Multiply.}$$

$$x = -\frac{1}{6} \qquad \text{Simplify.}$$

To check, replace $x$ with $-\dfrac{1}{6}$ in the original equation.

$$\frac{3}{4} x = -\frac{1}{8} \qquad \text{Original equation}$$

$$\frac{3}{4} \cdot -\frac{1}{6} = -\frac{1}{8} \qquad \text{Replace } x \text{ with } -\tfrac{1}{6}.$$

$$-\frac{1}{8} = -\frac{1}{8} \qquad \text{True.}$$

Since we arrived at a true statement, $-\dfrac{1}{6}$ is the solution of $\dfrac{3}{4} x = -\dfrac{1}{8}$.   ▬▬▬

**Example 4**   Solve for $y$: $3y = -\dfrac{2}{11}$

*Solution:*   We can either divide both sides by 3 or multiply both sides by the reciprocal of 3.

$$3y = -\frac{2}{11}$$

$$\frac{1}{3} \cdot 3y = \frac{1}{3} \cdot -\frac{2}{11} \qquad \text{Multiply both sides by } \frac{1}{3}.$$

$$1y = -\frac{1 \cdot 2}{3 \cdot 11} \qquad \text{Multiply.}$$

$$y = -\frac{2}{33} \qquad \text{Simplify.}$$

Check to see that the solution is $-\dfrac{2}{33}$.

Solving equations with fractions can be tedious. If an equation contains fractions, it is often helpful to first multiply both sides of the equation by the LCD of the fractions. This has the effect of eliminating the fractions in the equation, as shown in the next example.

**Example 5**   Solve for $x$: $\dfrac{x}{6} + 1 = \dfrac{4}{3}$

*Solution:*   First multiply both sides of the equation by the LCD of the fractions. The LCD of the denominators 6 and 3 is 6.

$$\frac{x}{6} + 1 = \frac{4}{3}$$

$$6\left(\frac{x}{6} + 1\right) = 6\left(\frac{4}{3}\right) \qquad \text{Multiply both sides by 6.}$$

$$6\left(\frac{x}{6}\right) + 6(1) = 6\left(\frac{4}{3}\right) \qquad \text{Apply the distributive property.}$$

$$x + 6 = 8 \qquad \text{Simplify.}$$

$$x + 6 + (-6) = 8 + (-6) \qquad \text{Add } -6 \text{ to both sides.}$$

$$x = 2 \qquad \text{Simplify.}$$

To check, replace $x$ with 2 in the original equation.

$$\frac{x}{6} + 1 = \frac{4}{3} \qquad \text{Original equation}$$

$$\frac{2}{6} + 1 = \frac{4}{3} \qquad \text{Replace } x \text{ with 2.}$$

$$\frac{1}{3} + \frac{3}{3} = \frac{4}{3} \qquad \text{Simplify } \frac{2}{6}. \text{ The LCD of 3 and 1 is 3.}$$

$$\frac{4}{3} = \frac{4}{3} \qquad \text{True.}$$

Since we arrived at a true statement, 2 is the solution of $\dfrac{x}{6} + 1 = \dfrac{4}{3}$.

**Practice Problem 4**

Solve for $x$: $5x = -\dfrac{3}{4}$

**Practice Problem 5**

Solve for $y$: $\dfrac{y}{8} + \dfrac{3}{4} = 2$

TEACHING TIP

Remind students that we may multiply through by the LCD and divide out denominators because we have an equation. Here we are using the multiplication property of equality. They may only use this property if there is an equation.

**Answers**

**4.** $-\dfrac{3}{20}$   **5.** 10

Let's review the steps for solving equations in $x$. An extra step is now included to handle equations containing fractions.

---

**SOLVING AN EQUATION IN $x$**

**Step 1.** If fractions are present, multiply both sides of the equation by the LCD of the fractions.

**Step 2.** If parentheses are present, use the distributive property.

**Step 3.** Combine any like terms on each side of the equation.

**Step 4.** Use the addition property of equality to rewrite the equation so that variable terms are on one side of the equation and constant terms are on the other side.

**Step 5.** Divide both sides of the equation by the numerical coefficient of $x$ to solve.

**Step 6.** Check the answer in the **original equation**.

---

**Practice Problem 6**

Solve for $x$: $\dfrac{x}{5} - x = \dfrac{1}{5}$

**HELPFUL HINT**

Don't forget to multiply *both* sides of the equation by the LCD.

**Example 6**    Solve for $z$: $\dfrac{z}{5} - \dfrac{z}{3} = 6$

*Solution:*

$$\frac{z}{5} - \frac{z}{3} = 6$$

$$15\left(\frac{z}{5} - \frac{z}{3}\right) = 15(6) \qquad \text{Multiply both sides by the LCD, 15.}$$

$$15\left(\frac{z}{5}\right) - 15\left(\frac{z}{3}\right) = 15(6) \qquad \text{Apply the distributive property.}$$

$$3z - 5z = 90 \qquad \text{Simplify.}$$

$$-2z = 90 \qquad \text{Combine like terms.}$$

$$\frac{-2z}{-2} = \frac{90}{-2} \qquad \text{Divide both sides by } -2, \text{ the coefficient of } z.$$

$$z = -45 \qquad \text{Simplify.}$$

To check, replace $z$ with $-45$ in the **original equation** to see that a true statement results.

**Practice Problem 7**

Solve: $\dfrac{y}{2} = \dfrac{y}{5} + \dfrac{3}{2}$

**Example 7**    Solve: $\dfrac{x}{2} = \dfrac{x}{3} + \dfrac{1}{2}$

*Solution:*    First multiply both sides by the LCD, 6.

$$\frac{x}{2} = \frac{x}{3} + \frac{1}{2}$$

$$6\left(\frac{x}{2}\right) = 6\left(\frac{x}{3} + \frac{1}{2}\right) \qquad \text{Multiply both sides by the LCD, 6.}$$

$$6\left(\frac{x}{2}\right) = 6\left(\frac{x}{3}\right) + 6\left(\frac{1}{2}\right) \qquad \text{Apply the distributive property.}$$

$$3x = 2x + 3 \qquad \text{Simplify.}$$

$$3x + (-2x) = 2x + 3 + (-2x) \qquad \text{Add } (-2x) \text{ to both sides.}$$

$$x = 3 \qquad \text{Simplify.}$$

To check, replace $x$ with 3 in the original equation to see that a true statement results.

**Answers**

**6.** $-\dfrac{1}{4}$    **7.** 5

## B REVIEW OF ADDING AND SUBTRACTING FRACTIONS

Make sure you understand the difference between **solving an equation** containing fractions and **adding or subtracting two fractions**. To solve an equation containing fractions, we use the multiplication property of equality and multiply both sides by the LCD of the fractions, thus eliminating the fractions. This method does not apply to adding or subtracting fractions. The multiplication property of equality applies only to equations. To add or subtract unlike fractions, we write each fraction as an equivalent fraction using the LCD of the fractions as the denominator. See the next example for a review.

**Example 8**   Add: $\dfrac{x}{3} + \dfrac{2}{5}$

*Solution:*   This expression is not an equation. Here, we are adding two unlike fractions. To add unlike fractions, we need to find the LCD. The LCD of the denominators 3 and 5 is 15. Write each fraction as an equivalent fraction with a denominator of 15.

$$\frac{x}{3} + \frac{2}{5} = \frac{x \cdot 5}{3 \cdot 5} + \frac{2 \cdot 3}{5 \cdot 3}$$
$$= \frac{5x}{15} + \frac{6}{15}$$
$$= \frac{5x + 6}{15}$$

**TRY THE CONCEPT CHECK IN THE MARGIN.**

### Practice Problem 8

Subtract: $\dfrac{9}{10} - \dfrac{y}{3}$

TEACHING TIP

Ask students to explain the difference between Example 7 and Example 8. Why may we multiply through by the LCD in Example 7 and not in Example 8?

### ✓ CONCEPT CHECK

Which of the following are equations and which are expressions?

a. $\dfrac{1}{2} + 3x = 5$

b. $\dfrac{2}{3}x - \dfrac{x}{5}$

c. $\dfrac{x}{12} + \dfrac{5x}{24}$

d. $\dfrac{x}{5} = \dfrac{1}{10}$

**Answers**

8. $\dfrac{27 - 10y}{30}$

✓ Concept Check:  equations: a, d
expressions: b, c

# Focus on Business and Career

## INTENDED MAJORS

Table 1 shows the breakdown of entering college freshmen who plan to major in a professional field by specific professional major. Table 2 shows a similar breakdown for technical majors.

| TABLE 1 | |
|---|---|
| Professional Majors | Fraction |
| Architecture or urban planning | $\frac{11}{155}$ |
| Home economics | $\frac{1}{155}$ |
| Health technology | $\frac{13}{155}$ |
| Nursing | $\frac{7}{31}$ |
| Pharmacy | $\frac{9}{155}$ |
| Pre-dental, pre-medical, pre-veterinary | $\frac{39}{155}$ |
| Therapy (occupational, physical, speech) | $\frac{36}{155}$ |
| Other | $\frac{11}{155}$ |

(*Source:* Higher Education Research Institute)

| TABLE 2 | |
|---|---|
| Technical Majors | Fraction |
| Building trades | $\frac{1}{8}$ |
| Data processing, computer programming | $\frac{3}{8}$ |
| Drafting or design | $\frac{1}{8}$ |
| Electronics | $\frac{3}{40}$ |
| Mechanics | $\frac{7}{40}$ |
| Other | $\frac{1}{8}$ |

(*Source:* Higher Education Research Institute)

## CRITICAL THINKING

1. Interpret the meaning of the first line of Table 1.
2. Interpret the meaning of the third line of Table 2.
3. Which of the professional majors is most popular? Explain your reasoning.
4. Which of the technical majors is most popular? Explain your reasoning.

## GROUP ACTIVITY

Research the number of majors offered through your discipline area (such as Arts and Humanities, Biological Sciences, Business, Education, Engineering, etc.) at your school. How many entering students plan to take each major? Make a table similar to Tables 1 and 2 to display your results. Show the fraction of entering students taking each major. Do additional research to find the number of students that graduate with each major. Make another table to display your results. How does the breakdown of majors (those intended by entering students and those actually completed by graduating) change?

# EXERCISE SET 4.7

**A** *Solve each equation. See Examples 1 through 4.*

**1.** $7x = 2$    **2.** $-5x = 4$    📼 **3.** $\dfrac{1}{4}x = 3$    **4.** $\dfrac{1}{3}x = 6$

**5.** $\dfrac{2}{9}y = -6$    **6.** $\dfrac{4}{7}x = -8$    📼 **7.** $-\dfrac{4}{9}z = -\dfrac{3}{2}$    **8.** $-\dfrac{11}{10}x = -\dfrac{2}{7}$

**9.** $7a = \dfrac{1}{3}$    **10.** $2z = -\dfrac{5}{12}$    **11.** $-3x = -\dfrac{6}{11}$    **12.** $-4z = -\dfrac{12}{25}$

*Solve each equation. See Examples 5 through 7.*

**13.** $\dfrac{x}{3} + 2 = \dfrac{7}{3}$    **14.** $\dfrac{x}{5} - 1 = \dfrac{7}{5}$    **15.** $\dfrac{x}{5} - \dfrac{5}{10} = 1$

**16.** $\dfrac{x}{3} - x = -6$    **17.** $\dfrac{1}{2} - \dfrac{3}{5} = \dfrac{x}{10}$    **18.** $\dfrac{2}{3} - \dfrac{1}{4} = \dfrac{x}{12}$

**19.** $\dfrac{x}{3} = \dfrac{x}{5} - 2$    **20.** $\dfrac{a}{2} = \dfrac{a}{7} + \dfrac{5}{2}$

**ANSWERS**

1. $\dfrac{2}{7}$

2. $-\dfrac{4}{5}$

3. $12$

4. $18$

5. $-27$

6. $-14$

7. $\dfrac{27}{8}$

8. $\dfrac{20}{77}$

9. $\dfrac{1}{21}$

10. $-\dfrac{5}{24}$

11. $\dfrac{2}{11}$

12. $\dfrac{3}{25}$

13. $1$

14. $12$

15. $\dfrac{15}{2}$

16. $9$

17. $-1$

18. $5$

19. $-15$

20. $7$

**306**

---

**Name** _____

**B** *Add or subtract as indicated. See Example 8.*

**21.** $\dfrac{x}{7} - \dfrac{4}{3}$

**22.** $-\dfrac{5}{9} + \dfrac{y}{8}$

**23.** $\dfrac{y}{2} + 5$

**24.** $2 + \dfrac{7x}{3}$

**25.** $\dfrac{3x}{10} + \dfrac{x}{6}$

**26.** $\dfrac{9x}{8} - \dfrac{5x}{6}$

**A B** *Solve. If no equation is given, perform the indicated operation.*

**27.** $\dfrac{3}{8}x = \dfrac{1}{2}$

**28.** $\dfrac{2}{5}y = \dfrac{3}{10}$

**29.** $\dfrac{2}{3} - \dfrac{x}{5} = \dfrac{4}{15}$

**30.** $\dfrac{4}{5} + \dfrac{x}{4} = \dfrac{21}{20}$

**31.** $\dfrac{9}{14}z = \dfrac{27}{20}$

**32.** $\dfrac{5}{16}a = \dfrac{5}{6}$

**33.** $-3m - 5m = \dfrac{4}{7}$

**34.** $30n - 34n = \dfrac{3}{20}$

**35.** $\dfrac{x}{4} + 1 = \dfrac{1}{4}$

**36.** $\dfrac{y}{7} - 2 = \dfrac{1}{7}$

**37.** $\dfrac{1}{5}y = 10$

**38.** $\dfrac{1}{4}x = -2$

**39.** $\dfrac{5}{9} - \dfrac{2}{3}$

**40.** $\dfrac{8}{11} - \dfrac{1}{2}$

**41.** $-\dfrac{3}{4}x = \dfrac{9}{2}$

**42.** $\dfrac{5}{7}y = -\dfrac{15}{49}$

**43.** $\dfrac{x}{2} - x = -2$

**44.** $\dfrac{y}{3} = -4 + y$

**45.** $-\dfrac{5}{8}y = \dfrac{3}{16} - \dfrac{9}{16}$

**46.** $-\dfrac{7}{9}x = -\dfrac{5}{18} - \dfrac{4}{18}$

**47.** $17x - 25x = \dfrac{1}{3}$

**48.** $27x - 30x = \dfrac{4}{9}$

**49.** $\dfrac{7}{6}x = \dfrac{1}{4} - \dfrac{2}{3}$

**50.** $\dfrac{5}{4}y = \dfrac{1}{2} - \dfrac{7}{10}$

**▣ 51.** $\dfrac{b}{4} = \dfrac{b}{12} + \dfrac{2}{3}$

**52.** $\dfrac{a}{6} = \dfrac{a}{3} + \dfrac{1}{2}$

**53.** $\dfrac{x}{3} + 2 = \dfrac{x}{2} + 8$

**54.** $\dfrac{y}{5} - 2 = \dfrac{y}{3} - 4$

**39.** $-\dfrac{1}{9}$

**40.** $\dfrac{5}{22}$

**41.** $-6$

**42.** $-\dfrac{3}{7}$

**43.** $4$

**44.** $6$

**45.** $\dfrac{3}{5}$

**46.** $\dfrac{9}{14}$

**47.** $-\dfrac{1}{24}$

**48.** $-\dfrac{4}{27}$

**49.** $-\dfrac{5}{14}$

**50.** $-\dfrac{4}{25}$

**51.** $4$

**52.** $-3$

**53.** $-36$

**54.** $15$

55. $\dfrac{7}{2}$

56. $\dfrac{8}{3}$

57. $\dfrac{59}{10}$

58. $\dfrac{15}{8}$

59. $\dfrac{49}{6}$

60. $\dfrac{19}{5}$

61. answers may vary

62. $\dfrac{112}{11}$

63. $-2$

## REVIEW AND PREVIEW

*Perform each indicated operation. See Section 4.5.*

**55.** $3 + \dfrac{1}{2}$      **56.** $2 + \dfrac{2}{3}$      **57.** $5 + \dfrac{9}{10}$

**58.** $1 + \dfrac{7}{8}$      **59.** $9 - \dfrac{5}{6}$      **60.** $4 - \dfrac{1}{5}$

## COMBINING CONCEPTS

**61.** Explain why the method for eliminating fractions in an **equation** does not apply to simplifying **expressions** containing fractions.

*Solve.*

**62.** $\dfrac{14}{11} + \dfrac{3x}{8} = \dfrac{x}{2}$      **63.** $\dfrac{19}{53} = \dfrac{353x}{1431} + \dfrac{23}{27}$

# 4.8   OPERATIONS ON MIXED NUMBERS

## A   ILLUSTRATING MIXED NUMBERS

Recall the graph of the improper fraction $\frac{9}{5}$.

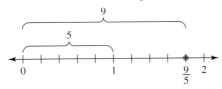

Notice that the graph of $\frac{9}{5}$ is $\frac{4}{5}$ past the number 1. Another way to name the improper fraction $\frac{9}{5}$ is by the mixed number $1\frac{4}{5}$ (read as "one and four fifths"). This **mixed number** contains a whole number and a proper fraction. The mixed number $1\frac{4}{5}$ means $1 + \frac{4}{5}$. Examples of mixed numbers are

$$2\frac{1}{4} = 2 + \frac{1}{4} \quad \text{and} \quad 5\frac{3}{8} = 5 + \frac{3}{8}$$

We can write a mixed number as an improper fraction by adding.

$$1\frac{4}{5} = 1 + \frac{4}{5} = \frac{1 \cdot 5}{1 \cdot 5} + \frac{4}{5} = \frac{5}{5} + \frac{4}{5} = \frac{9}{5}$$

We can also see that $1\frac{4}{5} = \frac{9}{5}$ by studying the following illustration.

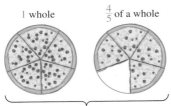

1 whole        $\frac{4}{5}$ of a whole

$\frac{9}{5}$  Number of pieces of pizza
    Number of equal parts in a whole

**Examples**   Represent the shaded area of each figure group with both an improper fraction and a mixed number.

**1.**   Whole object

improper fraction: $\frac{4}{3}$

mixed number: $1\frac{1}{3}$

**2.**

improper fraction: $\frac{5}{2}$

mixed number: $2\frac{1}{2}$

## Objectives

**A**  Illustrate mixed numbers.
**B**  Write a mixed number as an improper fraction.
**C**  Write an improper fraction as a mixed or whole number.
**D**  Multiply or divide mixed or whole numbers.
**E**  Add or subtract mixed numbers.
**F**  Solve problems containing mixed numbers.

Study     SSM     CD-ROM     Video
Guide                          4.8

**Practice Problems 1–2**

Represent the shaded area of each figure group with both an improper fraction and a mixed number.

1.

2.

**Answers**

**1.** $\frac{8}{3}$ or $2\frac{2}{3}$   **2.** $\frac{5}{4}$ or $1\frac{1}{4}$

## ✓ CONCEPT CHECK

Is $2\frac{1}{8}$ closer to the number 2 or the number 3? If you were to estimate $2\frac{1}{8}$ by a whole number, which number would you choose? Why?

TEACHING TIP

Tell students that they will need to know how to write a mixed number as an improper fraction in order to multiply and divide mixed numbers.

## Practice Problem 3

Write each mixed number as an improper fraction.

a. $2\frac{5}{7}$             b. $5\frac{1}{3}$

c. $9\frac{3}{10}$             d. $1\frac{1}{5}$

## Answers

3. a. $\frac{19}{7}$   b. $\frac{16}{3}$   c. $\frac{93}{10}$   d. $\frac{6}{5}$

✓ Concept Check: 2; Answers may vary.

---

TRY THE CONCEPT CHECK IN THE MARGIN.

### B  WRITING MIXED NUMBERS AS IMPROPER FRACTIONS

Notice from Examples 1 and 2 that mixed numbers and improper fractions were both used to represent the shaded area of the figure groups. For example,

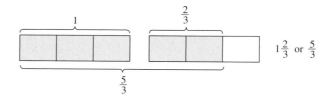

By studying these examples, we can see that a pattern occurs when writing mixed numbers as improper fractions. This pattern leads to the following method for writing a mixed number as an improper fraction.

> ### WRITING A MIXED NUMBER AS AN IMPROPER FRACTION
>
> To write a mixed number as an improper fraction,
> **Step 1.**  Multiply the whole number by the denominator of the fraction.
> **Step 2.**  Add the numerator of the fraction to the product from Step 1.
> **Step 3.**  Write the sum from Step 2 as the numerator of the improper fraction over the original denominator.

For example,

$$1\frac{2}{3} = \frac{3 \cdot 1 + 2}{3} = \frac{3 + 2}{3} = \frac{5}{3}$$

### Example 3     Write each mixed number as an improper fraction.

a. $4\frac{2}{9}$             b. $1\frac{8}{11}$

*Solution:*   a. $4\frac{2}{9} = \frac{9 \cdot 4 + 2}{9} = \frac{36 + 2}{9} = \frac{38}{9}$

b. $1\frac{8}{11} = \frac{11 \cdot 1 + 8}{11} = \frac{11 + 8}{11} = \frac{19}{11}$

### C  WRITING IMPROPER FRACTIONS AS MIXED NUMBERS OR WHOLE NUMBERS

Just as there are times when an improper fraction is preferred, sometimes a mixed or a whole number better suits a situation. To write improper fractions as mixed or whole numbers, just remember that the fraction bar means division.

> **WRITING AN IMPROPER FRACTION AS A MIXED NUMBER OR A WHOLE NUMBER**
>
> To write an improper fraction as a mixed number or a whole number,
> **Step 1.** Divide the denominator into the numerator.
> **Step 2.** The whole-number part of the mixed number is the quotient. The fraction part of the mixed number is the remainder over the original denominator.
>
> $$\text{quotient } \frac{\text{remainder}}{\text{original denominator}}$$

For example,

$$\frac{5}{3} = 3\overline{)5} = 1\frac{2}{3} \quad \begin{array}{l}\leftarrow \text{remainder}\\ \leftarrow \text{original denominator}\end{array}$$
$$\frac{3}{2} \quad \uparrow \text{ quotient}$$

**Example 4**   Write each improper fraction as a mixed number or a whole number.

a. $\dfrac{30}{7}$   b. $\dfrac{16}{15}$   c. $\dfrac{84}{6}$

*Solution:*   **a.** $7\overline{)30}$   $\dfrac{30}{7} = 4\dfrac{2}{7}$
$\quad \dfrac{-28}{2}$

**b.** $15\overline{)16}$   $\dfrac{16}{15} = 1\dfrac{1}{15}$
$\quad \dfrac{-15}{1}$

**c.** $6\overline{)84}$   $\dfrac{84}{6} = 14$   Since the remainder is 0, the result is the whole number 14.
$\quad \dfrac{-6}{24}$
$\quad \dfrac{-24}{0}$

> **Helpful Hint**
>
> When the remainder is 0, the improper fraction is a whole number.
>
> For example, $\dfrac{92}{4} = 23$.
>
> $4\overline{)92}$
> $\dfrac{-8}{12}$
> $\dfrac{-12}{0}$

**Practice Problem 4**

Write each fraction as a mixed number or a whole number.

a. $\dfrac{8}{5}$   b. $\dfrac{17}{6}$

c. $\dfrac{48}{4}$   d. $\dfrac{35}{4}$

e. $\dfrac{51}{7}$   f. $\dfrac{21}{20}$

**Answers**

**4. a.** $1\dfrac{3}{5}$   **b.** $2\dfrac{5}{6}$   **c.** 12   **d.** $8\dfrac{3}{4}$   **e.** $7\dfrac{2}{7}$

**f.** $1\dfrac{1}{20}$

✓ **CONCEPT CHECK**

Which of the following are equivalent to 9?

a. $7\dfrac{6}{3}$    b. $8\dfrac{4}{4}$    c. $8\dfrac{9}{9}$

d. $\dfrac{18}{2}$    e. all of these

**Practice Problem 5**

Multiply: $2\dfrac{1}{2} \cdot \dfrac{8}{15}$

**Practice Problems 6–7**

Multiply.

6. $3\dfrac{1}{5} \cdot 2\dfrac{3}{4}$    7. $\dfrac{2}{3} \cdot 18$

✓ **CONCEPT CHECK**

a. Find the error.

$2\dfrac{1}{4} \cdot \dfrac{1}{2} = 2\dfrac{1 \cdot 1}{4 \cdot 2} = 2\dfrac{1}{8}$

b. How could you estimate the product $3\dfrac{1}{7} \cdot 4\dfrac{5}{6}$?

**Practice Problems 8–10**

Divide.

8. $\dfrac{4}{9} \div 5$    9. $\dfrac{8}{15} \div 3\dfrac{4}{5}$

10. $3\dfrac{2}{5} \div 2\dfrac{2}{15}$

**Answers**

5. $\dfrac{4}{3}$ or $1\dfrac{1}{3}$  6. $\dfrac{44}{5}$ or $8\dfrac{4}{5}$  7. $\dfrac{12}{1}$ or 12  8. $\dfrac{4}{45}$

9. $\dfrac{8}{57}$  10. $1\dfrac{19}{32}$

✓ Concept Check: e

✓ Concept Check: **a.** $2\dfrac{1}{4} \cdot \dfrac{1}{2} = \dfrac{9}{2} \cdot \dfrac{1}{2} = \dfrac{9}{4} = 2\dfrac{1}{4}$

**b.** Answers may vary.

---

TRY THE CONCEPT CHECK IN THE MARGIN.

**D** **MULTIPLYING OR DIVIDING WITH MIXED NUMBERS OR WHOLE NUMBERS**

To multiply or divide with mixed numbers or whole numbers, first write each number as an improper fraction.

**Example 5**    Multiply: $3\dfrac{1}{3} \cdot \dfrac{7}{8}$

*Solution:*    The mixed number $3\dfrac{1}{3}$ can be written as the fraction $\dfrac{10}{3}$. Then,

$$3\dfrac{1}{3} \cdot \dfrac{7}{8} = \dfrac{10}{3} \cdot \dfrac{7}{8} = \dfrac{2 \cdot 5 \cdot 7}{3 \cdot 2 \cdot 4} = \dfrac{35}{12} \text{ or } 2\dfrac{11}{12}$$

Don't forget that a whole number can be written as a fraction by writing the whole number over 1. For example,

$$20 = \dfrac{20}{1} \quad \text{and} \quad 7 = \dfrac{7}{1}$$

**Examples**    Multiply.

6. $1\dfrac{2}{3} \cdot 2\dfrac{1}{4} = \dfrac{5}{3} \cdot \dfrac{9}{4} = \dfrac{5 \cdot 9}{3 \cdot 4} = \dfrac{5 \cdot 3 \cdot 3}{3 \cdot 4} = \dfrac{15}{4}$ or $3\dfrac{3}{4}$

7. $\dfrac{3}{4} \cdot 20 = \dfrac{3}{4} \cdot \dfrac{20}{1} = \dfrac{3 \cdot 20}{4 \cdot 1} = \dfrac{3 \cdot 4 \cdot 5}{4 \cdot 1} = \dfrac{15}{1}$ or 15

TRY THE CONCEPT CHECK IN THE MARGIN.

**Examples**    Divide.

8. $\dfrac{3}{4} \div 5 = \dfrac{3}{4} \cdot \dfrac{1}{5} = \dfrac{3 \cdot 1}{4 \cdot 5} = \dfrac{3}{20}$

9. $\dfrac{11}{18} \div 2\dfrac{5}{6} = \dfrac{11}{18} \div \dfrac{17}{6} = \dfrac{11}{18} \cdot \dfrac{6}{17} = \dfrac{11 \cdot 6}{18 \cdot 17}$

$\qquad = \dfrac{11 \cdot 6}{6 \cdot 3 \cdot 17} = \dfrac{11}{51}$

10. $5\dfrac{2}{3} \div 2\dfrac{5}{9} = \dfrac{17}{3} \div \dfrac{23}{9} = \dfrac{17}{3} \cdot \dfrac{9}{23} = \dfrac{17 \cdot 9}{3 \cdot 23}$

$\qquad = \dfrac{17 \cdot 3 \cdot 3}{3 \cdot 23} = \dfrac{51}{23}$ or $2\dfrac{5}{23}$

**E** **ADDING OR SUBTRACTING MIXED NUMBERS**

We can add or subtract mixed numbers, too, by first writing each mixed number as an improper fraction. But it is often easier to add or subtract the whole-number parts and add or subtract the proper-fraction parts vertically, as shown in the next examples.

**Example 11**  Add: $2\dfrac{1}{3} + 5\dfrac{3}{8}$

*Solution:*  The LCD of 3 and 8 is 24.

$$2\dfrac{1 \cdot 8}{3 \cdot 8} = 2\dfrac{8}{24}$$

$$+5\dfrac{3 \cdot 3}{8 \cdot 3} = 5\dfrac{9}{24}$$

$$7\dfrac{17}{24} \quad \leftarrow \text{Add the fractions.}$$

$$\quad\quad\quad \text{Add the whole numbers.}$$

**Example 12**  Add: $3\dfrac{4}{5} + 1\dfrac{4}{15}$

*Solution:*  The LCD of 5 and 15 is 15.

$$3\dfrac{4}{5} = 3\dfrac{12}{15}$$

$$+1\dfrac{4}{15} = 1\dfrac{4}{15} \quad \text{Add the fractions, then add the whole numbers.}$$

$$4\dfrac{16}{15} \quad \text{Notice that the fractional part is improper.}$$

Since $\dfrac{16}{15}$ is $1\dfrac{1}{15}$, we can write the sum as

$$4\dfrac{16}{15} = 4 + 1\dfrac{1}{15} = 5\dfrac{1}{15}$$

**Example 13**  Add: $1\dfrac{4}{5} + 4 + 2\dfrac{1}{2}$

*Solution:*  The LCD of 5 and 2 is 10.

$$1\dfrac{4}{5} = 1\dfrac{8}{10}$$

$$4 \quad = 4$$

$$+2\dfrac{1}{2} = 2\dfrac{5}{10}$$

$$7\dfrac{13}{10} = 7 + 1\dfrac{3}{10} = 8\dfrac{3}{10}$$

**TRY THE CONCEPT CHECK IN THE MARGIN.**

**Example 14**  Subtract: $9\dfrac{3}{7} - 5\dfrac{4}{21}$

*Solution:*  The LCD of 7 and 21 is 21.

$$9\dfrac{3}{7} = 9\dfrac{9}{21}$$

$$-5\dfrac{4}{21} = -5\dfrac{4}{21}$$

$$4\dfrac{5}{21} \quad \leftarrow \text{Subtract the fractions.}$$

$$\quad\quad\quad \text{Subtract the whole numbers.}$$

When subtracting mixed numbers, borrowing may be needed, as shown in the next example.

**Practice Problem 15**

**Practice Problem 15**

Subtract: $9\dfrac{7}{15} - 5\dfrac{4}{5}$

**Example 15**  Subtract: $7\dfrac{3}{14} - 3\dfrac{6}{7}$

*Solution:*     The LCD of 7 and 14 is 14.

$$7\dfrac{3}{14} = \ 7\dfrac{3}{14}$$

Notice that we cannot subtract $\dfrac{12}{14}$ from $\dfrac{3}{14}$,

$$-3\dfrac{6}{7} = -3\dfrac{12}{14}$$

so we borrow from the whole number 7.

Borrow 1 from 7.

$$7\dfrac{3}{14} = 6 + 1\dfrac{3}{14} = 6 + \dfrac{17}{14} \quad \text{or} \quad 6\dfrac{17}{14}$$

Now subtract.

$$
\begin{array}{ll}
7\dfrac{3}{14} = & 7\dfrac{3}{14} = \ \ 6\dfrac{17}{14} \\
-3\dfrac{6}{7} = & -3\dfrac{12}{14} = -3\dfrac{12}{14} \\
\hline
& \qquad\qquad\ \ 3\dfrac{5}{14} \leftarrow \text{Subtract the fractions.}
\end{array}
$$

$\uparrow$ Subtract the whole numbers.

**✓ CONCEPT CHECK**

In the subtraction problem $5\dfrac{1}{4} - 3\dfrac{3}{4}$, $5\dfrac{1}{4}$ must be rewritten because $\dfrac{3}{4}$ cannot be subtracted from $\dfrac{1}{4}$. Why is it incorrect to rewrite $5\dfrac{1}{4}$ as $5\dfrac{5}{4}$?

**TRY THE CONCEPT CHECK IN THE MARGIN.**

**Example 16**  Subtract: $12 - 8\dfrac{3}{7}$

*Solution:*

$$12 = 11\dfrac{7}{7}$$

Borrow 1 from 12 and write it as $\dfrac{7}{7}$.

$$
\begin{array}{ll}
-8\dfrac{3}{7} = & -8\dfrac{3}{7} \\
\hline
& \ \ 3\dfrac{4}{7} \leftarrow \text{Subtract the fractions.}
\end{array}
$$

$\uparrow$ Subtract the whole numbers.

**Practice Problem 16**

Subtract: $25 - 10\dfrac{2}{9}$

### F SOLVING PROBLEMS CONTAINING MIXED NUMBERS

Now that we know how to perform operations on mixed numbers, we can solve real-life problems.

**Example 17**  Finding Legal Lobster Size

Lobster fishermen must measure the upper body shells of the lobsters they catch. Lobsters that are too small are thrown back into the ocean. Each state has its own size standard for lobsters to help control the breeding stock. In 1988, Massachusetts increased its legal lobster size from $3\dfrac{3}{16}$ inches to $3\dfrac{7}{32}$ inches. How much of an increase was this? (*Source:* Peabody Essex Museum, Salem, MA)

**Answers**

**15.** $3\dfrac{2}{3}$   **16.** $14\dfrac{7}{9}$

**✓ Concept Check:** Rewrite $5\dfrac{1}{4}$ as $4\dfrac{5}{4}$ by borrowing from the 5.

*Solution:*

**1.** UNDERSTAND. Read and reread the problem carefully. The word "increase" found in the problem might make you think that we add to solve the problem. But the phrase "how much of an increase" tells us to subtract to find the increase.

**2.** TRANSLATE.

In words:

| Increase | is | new lobster size | minus | old lobster size |
|----------|-----|------------------|-------|------------------|
| ↓ | ↓ | ↓ | ↓ | ↓ |

Translate:   Increase   =   $3\frac{7}{32}$   −   $3\frac{3}{16}$

**3.** Solve.

$$3\frac{7}{32} = 3\frac{7}{32}$$
$$-3\frac{3}{16} = 3\frac{6}{32}$$
$$\overline{\phantom{-3\frac{3}{16} = }\,\frac{1}{32}}$$

**4.** INTERPRET. *Check* your work. *State* your conclusion: The increase in lobster size is $\frac{1}{32}$ of an inch.

## Example 18   Calculating Manufacturing Materials Needed

In a manufacturing process, a metal-cutting machine cuts strips $1\frac{3}{5}$ inches long from a piece of metal stock. How many such strips can be cut from a 48-inch piece of stock?

*Solution:*

**1.** UNDERSTAND the problem. To do so, read and reread the problem. Then draw a diagram.

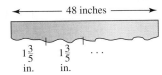

We want to know how many $1\frac{3}{5}$ s there are in 48.

**2.** TRANSLATE.

In words:

| Number of strips | is | 48 | divided by | $1\frac{3}{5}$ |
|------------------|-----|-----|------------|----------------|
| ↓ | ↓ | ↓ | ↓ | ↓ |

Translate:   Number of strips   =   48   ÷   $1\frac{3}{5}$

---

**Practice Problem 17**

The measurement around the trunk of a tree just below shoulder height is called its girth. The largest known American Beech tree in the United States has a girth of $23\frac{1}{4}$ feet. The largest known Sugar Maple tree in the United States has a girth of $19\frac{5}{12}$ feet. How much larger is the girth of the largest known American Beech tree than the girth of the largest known Sugar Maple tree? (*Source: American Forests*)

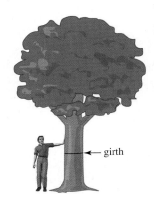

girth

---

**Practice Problem 18**

A designer of women's clothing designs a woman's dress that requires $2\frac{1}{7}$ yards of material. How many dresses can be made from a 30-yard bolt of material?

**Answers**

**17.** $3\frac{5}{6}$ feet   **18.** 14 dresses

**3.** SOLVE.

$$48 \div 1\frac{3}{5} = 48 \div \frac{8}{5} = \frac{48}{1} \cdot \frac{5}{8} = \frac{48 \cdot 5}{1 \cdot 8} = \frac{8 \cdot 6 \cdot 5}{1 \cdot 8} = \frac{30}{1} \quad \text{or} \quad 30$$

**4.** INTERPRET. *Check* your work. *State* your conclusion: Thirty strips can be cut from the 48-inch piece of stock.

## CALCULATOR EXPLORATIONS
### CONVERTING BETWEEN MIXED-NUMBER AND FRACTION NOTATION

If your calculator has a fraction key, such as $\boxed{a^{b\,c}}$, you can use it to convert between mixed-number notation and fraction notation.

To write $13\frac{7}{16}$ as an improper fraction, press

$$\boxed{1}\ \boxed{3}\ \boxed{a^{b/c}}\ \boxed{7}\ \boxed{a^{b/c}}\ \boxed{1}\ \boxed{6}\ \boxed{\text{2ND}}\ \boxed{d/c}$$

The display will read

$$\boxed{215 \lrcorner 16}$$

which represents $\frac{215}{16}$. Thus $13\frac{7}{16} = \frac{215}{16}$

To convert $\frac{190}{13}$ to a mixed number, press

$$\boxed{1}\ \boxed{9}\ \boxed{0}\ \boxed{a^{b/c}}\ \boxed{1}\ \boxed{3}\ \boxed{=}$$

The display will read

$$\boxed{14\_8 \lrcorner 13}$$

which represents $14\frac{8}{13}$. Thus $\frac{190}{13} = 14\frac{8}{13}$.

*Write each mixed number as a fraction and each fraction as a mixed number.*

**1.** $25\frac{5}{11}$ $\quad \frac{280}{11}$  **2.** $67\frac{14}{15}$ $\quad \frac{1019}{15}$  **3.** $107\frac{31}{35}$ $\quad \frac{3776}{35}$

**4.** $186\frac{17}{21}$ $\quad \frac{3923}{21}$  **5.** $\frac{365}{14}$ $\quad 26\frac{1}{14}$  **6.** $\frac{290}{13}$ $\quad 22\frac{4}{13}$

**7.** $\frac{2769}{30}$ $\quad 92\frac{3}{10}$  **8.** $\frac{3941}{17}$ $\quad 231\frac{14}{17}$

**Name** _____ **Section** _____ **Date** _____

## EXERCISE SET 4.8

**A** *Represent the shaded area in each figure group with **(a)** an improper fraction and **(b)** a mixed number. See Examples 1 and 2.*

1.

2.

3.

4.

5.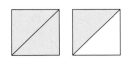

6.

7.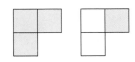

8.

**B** *Write each mixed number as an improper fraction. See Example 3.*

9. $2\frac{1}{3}$

10. $6\frac{3}{4}$

11. $3\frac{3}{8}$

12. $2\frac{5}{8}$

13. $11\frac{6}{7}$

14. $8\frac{9}{10}$

15. $1\frac{6}{7}$

16. $3\frac{3}{13}$

17. $3\frac{2}{15}$

18. $5\frac{5}{12}$

19. $4\frac{5}{8}$

20. $7$

21. $\frac{8}{21}$

22. $2\frac{1}{3}$

23. $4\frac{2}{3}$

24. $1\frac{1}{3}$

25. $5\frac{1}{2}$

26. $7\frac{7}{20}$

27. $18\frac{2}{3}$

28. $\frac{25}{189}$

29. $6\frac{4}{5}$

30. $10\frac{2}{3}$

31. $25\frac{5}{14}$

32. $32\frac{2}{15}$

33. $13\frac{13}{24}$

34. $16\frac{9}{10}$

**Name** _____

**C** *Write each improper fraction as a whole number or a mixed number. See Example 4.*

**15.** $\frac{13}{7}$  **16.** $\frac{42}{13}$  **17.** $\frac{47}{15}$

**18.** $\frac{65}{12}$  **19.** $\frac{37}{8}$  **20.** $\frac{42}{6}$

**D** *Multiply or divide. See Examples 5 through 10.*

**21.** $2\frac{2}{3} \cdot \frac{1}{7}$  **22.** $\frac{5}{9} \cdot 4\frac{1}{5}$  **23.** $8 \div 1\frac{5}{7}$  **24.** $5 \div 3\frac{3}{4}$

**25.** $3\frac{2}{3} \cdot 1\frac{1}{2}$  **26.** $2\frac{4}{5} \cdot 2\frac{5}{8}$  **27.** $2\frac{2}{3} \div \frac{1}{7}$  **28.** $\frac{5}{9} \div 4\frac{1}{5}$

**E** *Add. See Examples 11 through 13.*

**29.** $4\frac{7}{10} + 2\frac{1}{10}$  **30.** $7\frac{4}{9} + 3\frac{2}{9}$  **31.** $15\frac{4}{7}$
$+9\frac{11}{14}$

**32.** $23\frac{3}{5}$
$+8\frac{8}{15}$

**33.** $3\frac{5}{8}$
$2\frac{1}{6}$
$+7\frac{3}{4}$

**34.** $4\frac{1}{3}$
$9\frac{2}{5}$
$+3\frac{1}{6}$

*Subtract. See Examples 14 through 16.*

**35.** $4\dfrac{7}{10}$

$\quad -2\dfrac{1}{10}$

**36.** $7\dfrac{4}{9}$

$\quad -3\dfrac{2}{9}$

**37.** $10\dfrac{13}{14}$

$\quad\ \ -3\dfrac{4}{7}$

**38.** $12\dfrac{5}{12}$

$\quad\ \ -4\dfrac{1}{6}$

**39.** $9\dfrac{1}{5}$

$\quad -8\dfrac{6}{25}$

**40.** $6\dfrac{2}{13}$

$\quad -4\dfrac{7}{26}$

**D** **E** *Perform each indicated operation. See Examples 5 through 16.*

**41.** $2\dfrac{3}{4}$

$\quad +1\dfrac{1}{4}$

**42.** $5\dfrac{5}{8}$

$\quad +2\dfrac{3}{8}$

**43.** $15\dfrac{4}{7}$

$\quad\ \ -9\dfrac{11}{14}$

**44.** $23\dfrac{3}{5}$

$\quad\ \ -8\dfrac{8}{15}$

**45.** $3\dfrac{1}{9}\cdot 2$

**46.** $4\dfrac{1}{2}\cdot 3$

**47.** $1\dfrac{2}{3}\div 2\dfrac{1}{5}$

**48.** $5\dfrac{1}{5}\div 3\dfrac{1}{4}$

**49.** $22\dfrac{4}{9}+13\dfrac{5}{18}$

**50.** $15\dfrac{3}{25}-5\dfrac{2}{5}$

**51.** $5\dfrac{2}{3}-3\dfrac{1}{6}$

**52.** $5\dfrac{3}{8}-2\dfrac{13}{16}$

---

**35.** $2\dfrac{3}{5}$

**36.** $4\dfrac{2}{9}$

**37.** $7\dfrac{5}{14}$

**38.** $8\dfrac{1}{4}$

**39.** $\dfrac{24}{25}$

**40.** $1\dfrac{23}{26}$

**41.** $4$

**42.** $8$

**43.** $5\dfrac{11}{14}$

**44.** $15\dfrac{1}{15}$

**45.** $6\dfrac{2}{9}$

**46.** $13\dfrac{1}{2}$

**47.** $\dfrac{25}{33}$

**48.** $1\dfrac{3}{5}$

**49.** $35\dfrac{13}{18}$

**50.** $9\dfrac{18}{25}$

**51.** $2\dfrac{1}{2}$

**52.** $2\dfrac{9}{16}$

**53.** $72\frac{19}{30}$

**54.** $42\frac{13}{14}$

**55.** $\frac{11}{14}$

**56.** $23\frac{5}{8}$

**57.** $5\frac{4}{7}$

**58.** $18\frac{1}{3}$

**59.** $13\frac{16}{33}$

**60.** $19\frac{5}{16}$

**61.** $10\frac{43}{60}$ min

**62.** $3\frac{4}{15}$ min

**63.** $\frac{49}{60}$ min

**64.** $6\frac{3}{5}$ min

**53.** $15\frac{1}{5}$
  $20\frac{3}{10}$
  $+37\frac{2}{15}$
  _____

**54.** $7\frac{3}{7}$
  $15$
  $+20\frac{1}{2}$
  _____

**55.** $6\frac{4}{7} - 5\frac{11}{14}$

**56.** $47\frac{5}{12} - 23\frac{19}{24}$

**57.** $4\frac{2}{7} \cdot 1\frac{3}{10}$

**58.** $6\frac{2}{3} \cdot 2\frac{3}{4}$

**59.** $6\frac{2}{11}$
  $3$
  $+4\frac{10}{33}$
  _____

**60.** $3\frac{7}{16}$
  $6\frac{1}{2}$
  $+9\frac{3}{8}$
  _____

**F** *Solve. See Examples 17 and 18.*

*During the 21st century, the first three total eclipses of the sun visible from the Atlantic Ocean are listed below. The duration of each eclipse is included in the table. Use this table to answer Exercises 61–64.*

| TOTAL SOLAR ECLIPSES VISIBLE FROM THE ATLANTIC OCEAN | |
| --- | --- |
| Date of Eclipse | Duration (in minutes) |
| June 21, 2001 | $4\frac{14}{15}$ |
| March 29, 2006 | $4\frac{7}{60}$ |
| Nov. 3, 2013 | $1\frac{2}{3}$ |

(*Source: 2000 World Almanac*)

**61.** What is the total duration for the three eclipses?

**62.** How much longer is the June 21, 2001 eclipse than the Nov. 3, 2013 eclipse?

**63.** How much longer is the June 21, 2001 eclipse than the March 29, 2006 eclipse?

**64.** What is the total duration for the two eclipses occurring in odd-numbered years?

**65.** A sidewalk is built six bricks wide by laying each brick side by side. How many inches wide is the sidewalk if each brick measures $3\frac{1}{4}$ inches wide?

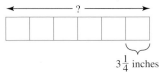

$3\frac{1}{4}$ inches

**66.** Vonshay Bartlet wants to build a $5\frac{1}{2}$-foot-tall bookcase with six equally spaced shelves. How far apart should he space the shelves?

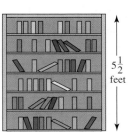

$5\frac{1}{2}$ feet

**67.** On March 19, 1998, the closing price for Kellogg Company stock was $43\frac{3}{4}$ per share. Severo Guiterrez spent $525 buying shares of Kellogg stock at the close of the stock market on that day. How many shares did he purchase? (*Source:* Based on data from Standard & Poor's ComStock)

**68.** The nutrition label on a can of crushed pineapple shows there are 9 grams of carbohydrates for each cup of pineapple. How many grams of carbohydrates are in a $2\frac{1}{2}$-cup can?

△ **69.** To prevent intruding birds, birdhouses built for Eastern Bluebirds should have an entrance hole measuring $1\frac{1}{2}$ inches in diameter. Entrance holes in birdhouses for Mountain Bluebirds should measure $1\frac{9}{16}$ inches in diamater. How much wider should entrance holes for Mountain Bluebirds be than for Eastern Bluebirds? (*Source:* North American Bluebird Society)

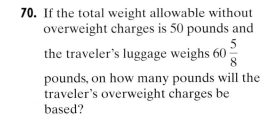

**70.** If the total weight allowable without overweight charges is 50 pounds and the traveler's luggage weighs $60\frac{5}{8}$ pounds, on how many pounds will the traveler's overweight charges be based?

**65.** $\frac{39}{2}$ or $19\frac{1}{2}$ in.

**66.** $\frac{11}{12}$ ft

**67.** 12 shares

**68.** $\frac{45}{2}$ or $22\frac{1}{2}$ g

**69.** $\frac{1}{16}$ in.

**70.** $10\frac{5}{8}$ lb

**321**

**Name** _____

**71.** Charlotte Dowlin has $15\frac{2}{3}$ feet of plastic pipe. She cuts off a $2\frac{1}{2}$-foot length and then a $3\frac{1}{4}$-foot length. If she now needs a 10-foot piece of pipe, will the remaining piece do?

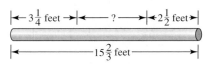

**72.** Shamalika Corning, a trim carpenter, cuts a board $3\frac{3}{8}$ feet long from one 6 feet long. How long is the remaining piece?

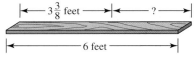

**73.** A small airplane used $58\frac{3}{4}$ gallons of fuel to fly a $7\frac{1}{2}$ hour trip. How many gallons of fuel were used for each hour?

**74.** Kesha Jonston is planning a Memorial Day barbeque. She has $27\frac{3}{4}$ pounds of hamburger. How many quarter-pound hamburgers can she make?

**75.** If Tucson's average rainfall is $11\frac{1}{4}$ inches and Yuma's is $3\frac{3}{5}$ inches, how much more rain, on the average, does Tucson get than Yuma?

**76.** A pair of crutches needs adjustment. One crutch is 43 inches and the other is $41\frac{5}{8}$ inches. Find how much the short crutch should be lengthened to make both crutches the same length.

*For Exercises 77 and 78, find the area of each figure.*

△ **77.**

$1\frac{3}{4}$ yards

2 yards

△ **78.**

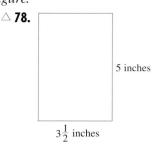

5 inches

$3\frac{1}{2}$ inches

**Name** _____

79. $\frac{121}{4}$ or $30\frac{1}{4}$ sq. in.

△ **79.** A model for a proposed computer chip measures $5\frac{1}{2}$ inches by $5\frac{1}{2}$ inches. Find its area.

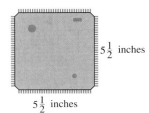

$5\frac{1}{2}$ inches

$5\frac{1}{2}$ inches

△ **80.** The Saltalamachios are planning to build a deck that measures $4\frac{1}{2}$ yards by $6\frac{1}{3}$ yards. Find the area of their proposed deck.

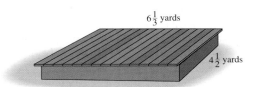

$6\frac{1}{3}$ yards

$4\frac{1}{2}$ yards

80. $\frac{57}{2}$ or $28\frac{1}{2}$ sq. yd

81. 7 mi

*Find the perimeter of each figure.*

△ **81.**

$2\frac{1}{3}$ miles    $2\frac{1}{3}$ miles

$2\frac{1}{3}$ miles

△ **82.**

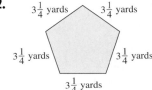

$3\frac{1}{4}$ yards    $3\frac{1}{4}$ yards

$3\frac{1}{4}$ yards    $3\frac{1}{4}$ yards

$3\frac{1}{4}$ yards

82. $16\frac{1}{4}$ yd

83. $21\frac{5}{24}$ m

△ **83.**

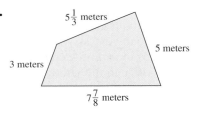

$5\frac{1}{3}$ meters

5 meters

3 meters

$7\frac{7}{8}$ meters

△ **84.**

7 in.    $11\frac{1}{5}$ in.

$12\frac{1}{3}$ in.

84. $30\frac{8}{15}$ in.

**85.** Jessica Callac takes $2\frac{3}{4}$ hours to clean her room. Her brother Matthew takes $1\frac{1}{3}$ hours to clean his room. If they start at the same time, how long does Matthew have to wait for Jessica to finish?

**86.** Jerald Divis, a tax consultant, takes $3\frac{1}{2}$ hours to prepare a personal tax return and $5\frac{7}{8}$ hours to prepare a business return. How much longer does it take him to prepare the business return?

85. $1\frac{5}{12}$ hr

86. $2\frac{3}{8}$ hr

**87.** $306\frac{2}{3}$ ft

**88.** $21\frac{5}{16}$ lb

**89.** $\frac{19}{30}$ in.

**90.** $352\frac{1}{3}$ yd

**91.** $9x - 13$

**92.** $-3y + 6$

**324**

**87.** Located on an island in New York City's harbor, the Statue of Liberty is one of the largest statues in the world. The copper figure is $152\frac{1}{6}$ feet tall from feet to tip of torch. The figure stands on a pedestal that is $154\frac{1}{2}$ feet tall. What is the overall height of the Statue of Liberty from the base of the pedestal to the tip of the torch? (*Source: Microsoft Encarta 97 Encyclopedia*)

**88.** The record for largest rainbow trout ever caught is $42\frac{1}{8}$ pounds and was set in Alaska in 1970. The record for largest tiger trout ever caught is $20\frac{13}{16}$ pounds and was set in Michigan in 1978. How much more did the record-setting rainbow trout weigh than the record-setting tiger trout? (*Source:* International Game Fish Association)

**89.** The record for rainfall during a 24-hour period in Alaska is $15\frac{1}{5}$ inches. This record was set in Angoon, Alaska, in October 1982. How much rain fell per hour on average? (*Source:* National Climatic Data Center)

**90.** The longest floating pontoon bridge in the United States is the Evergreen Point Bridge in Seattle, Washington. It is 2526 yards long. The second-longest pontoon bridge in the United States is the Hood Canal Bridge in Point Gamble, Washington. The second-longest pontoon bridge is $2173\frac{2}{3}$ yards long. How much longer is the Evergreen Point Bridge than the Hood Canal Bridge? (*Source:* Federal Highway Administration)

## REVIEW AND PREVIEW

*Simplify by combining like terms. See Section 1.8.*

**91.** $2x - 5 + 7x - 8$

**92.** $11 + 8y - 5 - 11y$

**93.** $3(y - 2) - 6y$

**94.** $8(3 + 2y) - 20$

*Simplify. See Section 1.7.*

**95.** $2^3 + 3^2$

**96.** $8 \cdot 5 - 5 \div 5$

**97.** $\dfrac{7 - 3}{2^2}$

**98.** $4^2 - 2^4$

### COMBINING CONCEPTS

*Solve.*

**99.** Carmen's Candy Clutch is famous for its "Nutstuff," a special blend of nuts and candy. A Supreme box of Nutstuff has $2\frac{1}{4}$ pounds of nuts and $3\frac{1}{2}$ pounds of candy. A Deluxe box has $1\frac{3}{8}$ pounds of nuts and $4\frac{1}{4}$ pounds of candy. Which box is heavier and by how much?

**100.** Willie Cassidie purchased three Supreme boxes and two Deluxe boxes of Nutstuff from Carmen's Candy Clutch. What is the total weight of his purchase? (See Exercise 99.)

**101.** Explain in your own words why $9\frac{13}{9}$ is equal to $10\frac{4}{9}$.

**102.** In your own words, explain how to borrow when subtracting mixed numbers.

**93.** $-3y - 6$

**94.** $16y + 4$

**95.** $17$

**96.** $41$

**97.** $1$

**98.** $0$

**99.** Supreme is heavier by $\frac{1}{8}$ lb

**100.** $28\frac{1}{2}$ lb

**101.** answers may vary

**102.** answers may vary

Name _____

### Internet Excursions

Go to http://www.prenhall.com/martin-gay

This World Wide Web address will provide you with access to a site called Anita's Tried & True Recipes, or a related site. This is a collection of recipes for reduced-fat dishes. Visit this site and search for two recipes that are interesting to you. Be sure that the ingredient list for each recipe contains at least three measures that are fractions or mixed numbers. Print out each recipe to use with Exercises 103 and 104.

**103.** For your first recipe, rewrite the ingredient list to show the amounts that would be needed to make half of the recipe. How many servings will half the recipe make?

**104.** For your second recipe, rewrite the ingredient list to show the amounts that would be needed to double the recipe. How many servings will a double recipe make?

# CHAPTER 4 ACTIVITY
## INVESTIGATING FRACTIONS

*This activity may be completed by working in groups or individually.*

According to the U.S. National Endowment for the Arts, $\frac{33}{50}$ of all U.S. adults ages 18 and over have gone to at least one movie in the past 12 months. The table shows the participation or attendance rate for other activities.

1. Which of the activities listed in the table was enjoyed by the largest fraction of the U.S. adult population?

2. Which of the activities listed in the table was enjoyed by the smallest fraction of the U.S. adult population?

3. How many people are in your prealgebra class?

4. For each of the activities listed in the table, use the given fraction to find the number of people in your class that you would expect to have participated in the activity in the past 12 months.

5. Choose any five of the activities listed in the table. Survey your class to find the number of people that actually participated in each of your five activities. Compare your results to the numbers you calculated in Question 4. How close are these numbers? Can you explain any large differences?

| PARTICIPATION IN VARIOUS ARTS AND LEISURE ACTIVITIES | |
|---|---|
| Attendance at or Participation in at Least Once During the Last 12 Months | Fraction of U.S. Adult Population |
| Movies | $\frac{33}{50}$ |
| Sports events | $\frac{41}{100}$ |
| Amusement park | $\frac{57}{100}$ |
| Exercise program | $\frac{19}{25}$ |
| Playing sports | $\frac{9}{20}$ |
| Charity work | $\frac{43}{100}$ |
| Home improvement/repair | $\frac{33}{50}$ |
| Computer hobbies | $\frac{2}{5}$ |
| Jazz performance | $\frac{3}{25}$ |
| Classical music performance | $\frac{4}{25}$ |
| Opera | $\frac{1}{20}$ |
| Musical play | $\frac{1}{4}$ |
| Nonmusical play | $\frac{4}{25}$ |
| Ballet | $\frac{3}{50}$ |
| Art museum | $\frac{7}{20}$ |
| Historic park | $\frac{47}{100}$ |
| Reading literature | $\frac{63}{100}$ |

(*Source:* U.S. National Endowment for the Arts, *1997 Survey of Public Participation in the Arts*, Research Division Note #70, July 1998)

# CHAPTER 4 HIGHLIGHTS

| DEFINITIONS AND CONCEPTS | EXAMPLES |
|---|---|

### SECTION 4.1 INTRODUCTION TO FRACTIONS AND EQUIVALENT FRACTIONS

| | |
|---|---|
| A **fraction** is a number of the form $\frac{a}{b}$, where $a$ and $b$ are integers and $b$ is not 0. The number $a$ is the **numerator** of the fraction, and the number $b$ is the **denominator** of the fraction. | |
| A fraction is called a **proper fraction** if its numerator is less than its denominator. | Proper Fractions: $\frac{1}{3}, \frac{2}{5}, \frac{7}{8}, \frac{100}{101}$ |
| A fraction is called an **improper fraction** if its numerator is greater than or equal to its denominator. | Improper Fractions: $\frac{5}{4}, \frac{2}{2}, \frac{9}{7}, \frac{101}{100}$ |
| Fractions that represent the same portion of a whole are called **equivalent fractions**. | |
| FUNDAMENTAL PROPERTY OF FRACTIONS | Write $\frac{3}{7}$ as an equivalent fraction with a denominator of 42. |
| If $a$, $b$, and $c$ are numbers, then $$\frac{a}{b} = \frac{a \cdot c}{b \cdot c} \quad \text{and also} \quad \frac{a}{b} = \frac{a \div c}{b \div c}$$ as long as $b$ and $c$ are not 0. | $$\frac{3}{7} = \frac{3 \cdot 6}{7 \cdot 6} = \frac{18}{42}$$ |

### SECTION 4.2 FACTORS AND SIMPLEST FORM

| | |
|---|---|
| A **prime number** is a whole number greater than 1 whose only divisors are 1 and itself. | Prime Numbers: $2, 3, 5, 7, 11, 13, 17, \ldots$ |
| A **composite number** is a whole number greater than 1 that is not prime. | Composite Numbers: $4, 6, 8, 9, 10, 12, 14, 15, 16, \ldots$ |
| Every whole number greater than 1 has exactly one prime factorization. | Write the prime factorization of 60. $$60 = 6 \cdot 10$$ $$= 2 \cdot 3 \cdot 2 \cdot 5$$ |
| A fraction is in **simplest form** or **lowest terms** when the numerator and the denominator have no common factors other than 1. | Write $\frac{30x}{36x}$ in simplest form. $$\frac{30x}{36x} = \frac{2 \cdot 3 \cdot 5 \cdot x}{2 \cdot 2 \cdot 3 \cdot 3 \cdot x} = \frac{5}{6}$$ |

### SECTION 4.3 MULTIPLYING AND DIVIDING FRACTIONS

| | |
|---|---|
| To **multiply** two fractions, $$\frac{a}{b} \cdot \frac{c}{d} = \frac{a \cdot c}{b \cdot d}, b \neq 0, d \neq 0$$ | Multiply: $\frac{2x}{3} \cdot \frac{5}{7} = \frac{2x \cdot 5}{3 \cdot 7} = \frac{10x}{21}$ |
| Two numbers are **reciprocals** of each other if their product is 1. | The reciprocal of $\frac{3}{5}$ is $\frac{5}{3}$. |
| To **divide** two fractions, $$\frac{a}{b} \div \frac{c}{d} = \frac{a}{b} \cdot \frac{d}{c} = \frac{a \cdot d}{b \cdot c}, b \neq 0, c \neq 0, d \neq 0$$ | Divide: $-\frac{3}{4} \div \frac{3}{8} = -\frac{3}{4} \cdot \frac{8}{3}$ $$= -\frac{3 \cdot 2 \cdot 4}{4 \cdot 3} = -2$$ |

---

### SECTION 4.4 ADDING AND SUBTRACTING LIKE FRACTIONS AND LEAST COMMON DENOMINATOR

Fractions that have a common denominator are called **like fractions**.

To add or subtract like fractions,
$$\frac{a}{b} + \frac{c}{b} = \frac{a+c}{b}, \frac{a}{b} - \frac{c}{b} = \frac{a-c}{b}, b \neq 0$$

The **least common denominator (LCD)** of a list of fractions is the smallest positive number divisible by all the denominators in the list.

Like Fractions:
$$-\frac{1}{3} \text{ and } \frac{2}{3}; \quad \frac{5x}{7} \text{ and } \frac{6}{7}$$

Find $\dfrac{10}{11} + \dfrac{3}{11} - \dfrac{8}{11} = \dfrac{10+3-8}{11} = \dfrac{5}{11}$.

Find the LCD of $\dfrac{7}{30}$ and $\dfrac{1}{12}$.

$$30 = 2 \cdot 3 \cdot 5$$
$$12 = 2 \cdot 2 \cdot 3$$

Notice that the greatest number of times that 2 appears is 2 times.
$$\text{LCD} = 2 \cdot 2 \cdot 3 \cdot 5 = 60$$

---

### SECTION 4.5 ADDING AND SUBTRACTING UNLIKE FRACTIONS

**TO ADD OR SUBTRACT UNLIKE FRACTIONS**

**Step 1.** Find the LCD.

**Step 2.** Write equivalent fractions with the LCD as the denominator.

**Step 3.** Add or subtract the like fractions.

**Step 4.** Write the result in simplest form.

Add: $\dfrac{3}{20} + \dfrac{2}{5}$

**Step 1.** The LCD is 20.

**Step 2.** $\dfrac{2}{5} = \dfrac{2 \cdot 4}{5 \cdot 4} = \dfrac{8}{20}$.

**Step 3.** $\dfrac{3}{20} + \dfrac{2}{5} = \dfrac{3}{20} + \dfrac{8}{20} = \dfrac{11}{20}$.

**Step 4.** $\dfrac{11}{20}$ is in simplest form.

---

### SECTION 4.6 COMPLEX FRACTIONS AND REVIEW OF ORDER OF OPERATIONS

A fraction whose numerator or denominator or both contain fractions is called a **complex fraction**.

One method for simplifying complex fractions is to multiply the numerator and the denominator of the complex fraction by the LCD of all fractions in its numerator and its denominator.

Complex Fractions:
$$\frac{\frac{x}{4}}{\frac{7}{10}}, \quad \frac{\frac{y}{6} - 11}{\frac{4}{3}}$$

$$\frac{\frac{y}{6} - 11}{\frac{4}{3}} = \frac{6\left(\frac{y}{6} - 11\right)}{6\left(\frac{4}{3}\right)} = \frac{6\left(\frac{y}{6}\right) - 6(11)}{6\left(\frac{4}{3}\right)}$$
$$= \frac{y - 66}{8}$$

---

### SECTION 4.7 SOLVING EQUATIONS CONTAINING FRACTIONS

MULTIPLICATION AND DIVISION PROPERTIES OF EQUALITY

Let $a, b,$ and $c$ be numbers, and $c \neq 0$.

If $a = b$, then $a \cdot c = b \cdot c$ and also $\dfrac{a}{c} = \dfrac{b}{c}$.

*(continued)*

---

### SECTION 4.7   SOLVING EQUATIONS CONTAINING FRACTIONS (CONTINUED)

TO SOLVE AN EQUATION IN $x$

**Step 1.** If fractions are present, multiply both sides of the equation by the LCD of the fractions.

**Step 2.** If parentheses are present, use the distributive property.

**Step 3.** Combine any like terms on each side of the equation.

**Step 4.** Use the addition property of equality to rewrite the equation so that variable terms are on one side of the equation and constant terms are on the other side.

**Step 5.** Divide both sides by the numerical coefficient of $x$ to solve.

**Step 6.** Check the answer in the *original equation*.

Solve: $\dfrac{x}{15} + 2 = \dfrac{7}{3}$

$$15\left(\dfrac{x}{15} + 2\right) = 15\left(\dfrac{7}{3}\right)$$

Multiply by the LCD 15.

$$15\left(\dfrac{x}{15}\right) + 15 \cdot 2 = 15\left(\dfrac{7}{3}\right)$$

$$x + 30 = 35$$

$$x + 30 + (-30) = 35 + (-30)$$

$$x = 5$$

Check to see that 5 is the solution.

---

### SECTION 4.8   OPERATIONS ON MIXED NUMBERS

A **mixed number** is the sum of a whole number and a proper fraction.

WRITING A MIXED NUMBER AS AN IMPROPER FRACTION

**Step 1.** Multiply the denominator of the fractional part by the whole-number part. Add the numerator of the fraction to this product.

**Step 2.** Write this sum as the numerator of the improper fraction over the original denominator.

WRITING AN IMPROPER FRACTION AS A MIXED NUMBER OR WHOLE NUMBER

**Step 1.** Divide the denominator into the numerator.

**Step 2.** The whole-number part of the quotient is the whole-number part of the mixed number and the $\dfrac{\text{remainder}}{\text{divisor}}$ is the fractional part.

To perform operations on mixed numbers, first write each mixed number as an improper fraction.

Mixed Numbers: $1\dfrac{2}{5}, 4\dfrac{7}{8}$

$$4\dfrac{7}{8} = \dfrac{8 \cdot 4 + 7}{8} = \dfrac{39}{8}$$

$$\dfrac{29}{7} = 4\dfrac{1}{7}$$

$$
\begin{array}{r}
4\frac{1}{7} \\
7\overline{)29} \\
28 \\
\hline
1
\end{array}
$$

Multiply: $1\dfrac{3}{4} \cdot 2\dfrac{1}{5}$

$$\dfrac{7}{4} \cdot \dfrac{11}{5} = \dfrac{77}{20} = 3\dfrac{17}{20}$$

Addition and subtraction of mixed numbers can be performed using a vertical format.

Add: $2\dfrac{1}{2} + 5\dfrac{7}{8}$

$$2\dfrac{1}{2} = 2\dfrac{4}{8}$$

$$+5\dfrac{7}{8} = 5\dfrac{7}{8}$$

$$\overline{\phantom{+5}7\dfrac{11}{8} = 7 + 1\dfrac{3}{8} = 8\dfrac{3}{8}}$$

# CHAPTER 4 REVIEW

**(4.1)** *Represent the shaded area of each figure with a fraction.*

**1.**   $\dfrac{3}{4}$

**2.** 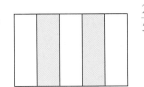  $\dfrac{2}{5}$

*Graph each fraction on a number line.*

**3.** $\dfrac{7}{9}$

**4.** $\dfrac{4}{7}$

**5.** $\dfrac{5}{4}$

**6.** $\dfrac{7}{5}$

**7.** The United States Army offers 242 job specialties to its soldiers. Certain specialties, such as infantry and field artillery, are closed to women. There are 207 army job specialties open to women and 35 army job specialties closed to women. What fraction of army job specialties are closed to women? (*Source:* U.S. Department of the Army)

$\dfrac{35}{242}$ of job specialties

**8.** In the United States Army today, roughly 17 out of every 20 soldiers are men. What fraction of U.S. Army soldiers are men? (*Source:* U.S. Department of the Army)

$\dfrac{17}{20}$ of soldiers

*Write each fraction as an equivalent fraction with the given denominator.*

**9.** $\dfrac{2}{3} = \dfrac{?}{30}$   20

**10.** $\dfrac{5}{8} = \dfrac{?}{56}$   35

**11.** $\dfrac{7a}{6} = \dfrac{?}{42}$   49a

**12.** $\dfrac{9b}{4} = \dfrac{?}{20}$   45b

**13.** $\dfrac{4}{5x} = \dfrac{?}{50x}$   40

**14.** $\dfrac{5}{9y} = \dfrac{?}{18y}$   10

**(4.2)** *Write each fraction in simplest form.*

**15.** $\dfrac{12}{28}$  $\dfrac{3}{7}$

**16.** $\dfrac{15}{27}$  $\dfrac{5}{9}$

**17.** $\dfrac{25x}{75x^2}$  $\dfrac{1}{3x}$

**18.** $\dfrac{36y^3}{72y}$  $\dfrac{y^2}{2}$

**19.** $\dfrac{29ab}{32abc}$  $\dfrac{29}{32c}$

**20.** $\dfrac{18xyz}{23xy}$  $\dfrac{18z}{23}$

**21.** $\dfrac{45x^2y}{27xy^3}$  $\dfrac{5x}{3y^2}$

**22.** $\dfrac{42ab^2c}{30abc^3}$  $\dfrac{7b}{5c^2}$

**23.** There are 12 inches in a foot. What fractional part of a foot does 8 inches represent?

$\dfrac{2}{3}$ of a foot

**24.** Six out of 15 cars are white. What fraction of cars are *not* white?

$\dfrac{3}{5}$ of the cars

**(4.3)** *Multiply.*

**25.** $\dfrac{3}{5} \cdot \dfrac{1}{2}$  $\dfrac{3}{10}$

**26.** $-\dfrac{6}{7} \cdot \dfrac{5}{12}$  $-\dfrac{5}{14}$

**27.** $\dfrac{7}{8x} \cdot -\dfrac{2}{3}$  $-\dfrac{7}{12x}$

**28.** $\dfrac{6}{15} \cdot \dfrac{5y}{8}$  $\dfrac{y}{4}$

**29.** $-\dfrac{24x}{5} \cdot -\dfrac{15}{8x^3}$  $\dfrac{9}{x^2}$

**30.** $\dfrac{27y^3}{21} \cdot \dfrac{7}{18y^2}$  $\dfrac{y}{2}$

**31.** $\left(-\dfrac{1}{3}\right)^3$  $-\dfrac{1}{27}$

**32.** $\left(-\dfrac{5}{12}\right)^2$  $\dfrac{25}{144}$

**33.** $\dfrac{x^3}{y} \cdot \dfrac{y^3}{x}$  $x^2y^2$

**34.** $\dfrac{ac}{b} \cdot \dfrac{b^2}{a^3c}$  $\dfrac{b}{a^2}$

**35.** Evaluate $xy$ if $x = \dfrac{2}{3}$ and $y = \dfrac{1}{5}$.  $\dfrac{2}{15}$

**36.** Evaluate $ab$ if $a = -7$ and $b = \dfrac{9}{10}$.  $-\dfrac{63}{10}$

*Divide.*

**37.** $-\dfrac{3}{4} \div \dfrac{3}{8}$  $-2$

**38.** $\dfrac{21a}{4} \div \dfrac{7a}{5}$  $\dfrac{15}{4}$

**39.** $\dfrac{18x}{5} \div \dfrac{2}{5x}$  $9x^2$

**40.** $-\dfrac{9}{2} \div -\dfrac{1}{3}$  $\dfrac{27}{2}$

**41.** $-\dfrac{5}{3} \div 2y$  $-\dfrac{5}{6y}$

**42.** $\dfrac{5x^2}{y} \div \dfrac{10x^3}{y^3}$  $\dfrac{y^2}{2x}$

**43.** Evaluate $x \div y$ if $x = \dfrac{9}{7}$ and $y = \dfrac{3}{4}$.  $\dfrac{12}{7}$

**44.** Evaluate $a \div b$ if $a = -5$ and $b = \dfrac{2}{3}$.  $-\dfrac{15}{2}$

*Find the area of each figure.*

△ **45.**

$\dfrac{7}{8}$ feet
$\dfrac{11}{6}$ feet
$\dfrac{77}{48}$ ft

△ **46.**

$\dfrac{2}{3}$ meter
$\dfrac{4}{9}$ m

**(4.4)** *Add or subtract as indicated.*

**47.** $\dfrac{7}{11} + \dfrac{3}{11}$  $\dfrac{10}{11}$

**48.** $\dfrac{4}{9} + \dfrac{2}{9}$  $\dfrac{2}{3}$

**49.** $\dfrac{1}{12} - \dfrac{5}{12}$  $-\dfrac{1}{3}$

**50.** $\dfrac{3}{y} - \dfrac{1}{y}$  $\dfrac{2}{y}$

**51.** $\dfrac{11x}{15} + \dfrac{x}{15}$  $\dfrac{4x}{5}$

**52.** $\dfrac{4y}{21} - \dfrac{3}{21}$  $\dfrac{4y - 3}{21}$

**53.** $\dfrac{4}{15} + \dfrac{3}{15} - \dfrac{2}{15}$  $\dfrac{1}{3}$

**54.** $\dfrac{4}{15} - \dfrac{3}{15} - \dfrac{2}{15}$  $-\dfrac{1}{15}$

*Find the LCD of each list of fractions.*

**55.** $\dfrac{2}{3}, \dfrac{5}{x}$  $3x$

**56.** $\dfrac{3}{4}, \dfrac{3}{8}, \dfrac{7}{12}$  $24$

**57.** Determine whether $\dfrac{4}{5}$ is a solution of $z + \dfrac{1}{5} = 1$.  yes

**58.** Determine whether $\dfrac{3}{4}$ is a solution of $x - \dfrac{2}{4} = \dfrac{1}{4}$.  yes

*Solve.*

**59.** If a student studies math for $\frac{3}{8}$ of an hour and geography for $\frac{1}{8}$ of an hour, find how long she studies.
$\frac{1}{2}$ hr

**60.** Beryl Goldstein mixed $\frac{5}{8}$ of a gallon of water with $\frac{1}{8}$ of a gallon of punch concentrate. Then she and her friends drank $\frac{3}{8}$ of a gallon of the punch. Find how much was left.  $\frac{3}{8}$ gal

**61.** One evening Mark Alorenzo did $\frac{3}{8}$ of his homework before supper, another $\frac{2}{8}$ of it while his children did their homework, and $\frac{1}{8}$ of it after his children went to bed. Find what part of his homework he did that day.  $\frac{3}{4}$ of his homework

△ **62.** The Simpsons will be fencing in their land. To do this, they need to find its perimeter. Find the perimeter of their land.  $\frac{3}{2}$ mi

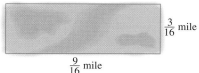

$\frac{3}{16}$ mile

$\frac{9}{16}$ mile

**(4.5)** *Add or subtract as indicated.*

**63.** $\frac{7}{18} + \frac{2}{9}$  $\frac{11}{18}$

**64.** $\frac{4}{13} - \frac{1}{26}$  $\frac{7}{26}$

**65.** $-\frac{1}{3} + \frac{1}{4}$  $-\frac{1}{12}$

**66.** $-\frac{2}{3} + \frac{1}{4}$  $-\frac{5}{12}$

**67.** $\frac{5x}{11} + \frac{2}{55}$  $\frac{25x + 2}{55}$

**68.** $\frac{4}{15} + \frac{b}{5}$  $\frac{4 + 3b}{15}$

**69.** $\frac{5y}{12} - \frac{2y}{9}$  $\frac{7y}{36}$

**70.** $\frac{7x}{18} + \frac{2x}{9}$  $\frac{11x}{18}$

**71.** $\frac{4}{9} + \frac{5}{y}$  $\frac{4y + 45}{9y}$

**72.** $-\frac{9}{14} - \frac{3}{7}$  $-\frac{15}{14}$

**73.** $\frac{4}{25} + \frac{23}{75} + \frac{7}{50}$  $\frac{91}{150}$

**74.** $\frac{2}{3} - \frac{2}{9} - \frac{1}{6}$  $\frac{5}{18}$

*Solve each equation.*

**75.** $a - \frac{2}{3} = \frac{1}{6}$  $\frac{5}{6}$

**76.** $9x + \frac{1}{5} - 8x = -\frac{7}{10}$  $-\frac{9}{10}$

**334**

*Find the perimeter of each figure.*

△ **77.**
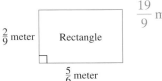

$\frac{2}{9}$ meter | Rectangle | $\frac{19}{9}$ m

$\frac{5}{6}$ meter

△ **78.**
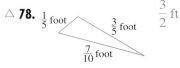

$\frac{1}{5}$ foot   $\frac{3}{5}$ foot   $\frac{3}{2}$ ft

$\frac{7}{10}$ foot

**79.** Determine whether $\frac{8}{11}$ is a solution of $x + \frac{1}{3} = \frac{35}{11}$.

no

**80.** Determine whether $\frac{9}{2}$ is a solution of $\frac{1}{9} y - \frac{1}{4} = \frac{1}{4}$.

yes

**81.** In a group of 100 blood donors, typically $\frac{9}{25}$ have type A Rh-positive blood and $\frac{3}{50}$ have type A Rh-negative blood. What fraction have type A blood?

$\frac{21}{50}$ of the donors

**82.** Find the difference in length of two scarves if one scarf is $\frac{5}{12}$ of a yard long and the other is $\frac{2}{3}$ of a yard long.

$\frac{1}{4}$ yd

$\frac{5}{12}$ of a yard     $\frac{2}{3}$ of a yard

**(4.6)** *Simplify each complex fraction.*

**83.** $\dfrac{\dfrac{2x}{5}}{\dfrac{7}{10}}$   $\dfrac{4x}{7}$

**84.** $\dfrac{\dfrac{3y}{7}}{\dfrac{11}{7}}$   $\dfrac{3y}{11}$

**85.** $\dfrac{2 + \dfrac{3}{4}}{1 - \dfrac{1}{8}}$   $\dfrac{22}{7}$

**86.** $\dfrac{\dfrac{5}{6} + 2}{\dfrac{11}{3} - 1}$   $\dfrac{17}{16}$

**87.** $\dfrac{\dfrac{2}{5} - \dfrac{1}{2}}{\dfrac{3}{4} - \dfrac{7}{10}}$   $-2$

**88.** $\dfrac{\dfrac{5}{6} - \dfrac{1}{4}}{\dfrac{-1}{12y}}$   $-7y$

*Evaluate each expression if $x = \dfrac{1}{2}$, $y = -\dfrac{2}{3}$, and $z = \dfrac{4}{5}$.*

**89.** $2x + y \quad \dfrac{1}{3}$

**90.** $\dfrac{x}{y + z} \quad \dfrac{15}{4}$

**91.** $\dfrac{x + y}{z} \quad -\dfrac{5}{24}$

**92.** $x + y + z \quad \dfrac{19}{30}$

**93.** $y^2 \quad \dfrac{4}{9}$

**94.** $x - z \quad -\dfrac{3}{10}$

**(4.7)** *Solve.*

**95.** $-\dfrac{3}{5}x = 6 \quad -10$

**96.** $\dfrac{2}{9}y = -\dfrac{4}{3} \quad -6$

**97.** $\dfrac{x}{7} - 3 = -\dfrac{6}{7} \quad 15$

**98.** $\dfrac{y}{5} + 2 = \dfrac{11}{5} \quad 1$

**99.** $\dfrac{1}{6} + \dfrac{x}{4} = \dfrac{17}{12} \quad 5$

**100.** $\dfrac{x}{5} - \dfrac{5}{4} = \dfrac{x}{2} - \dfrac{1}{20} \quad -4$

**(4.8)** *Write each improper fraction as a mixed number or a whole number.*

**101.** $\dfrac{15}{4} \quad 3\dfrac{3}{4}$

**102.** $\dfrac{39}{13} \quad 3$

**103.** $\dfrac{7}{7} \quad 1$

**104.** $\dfrac{125}{4} \quad 31\dfrac{1}{4}$

*Write each mixed or whole number as an improper fraction.*

**105.** $2\dfrac{1}{5} \quad \dfrac{11}{5}$

**106.** $5 \quad \dfrac{5}{1}$

**107.** $3\dfrac{8}{9} \quad \dfrac{35}{9}$

**108.** $3 \quad \dfrac{3}{1}$

Perform each indicated operation.

**109.**
$$31\frac{2}{7}$$
$$+14\frac{10}{21}$$
$45\frac{16}{21}$

**110.**
$$24\frac{4}{5}$$
$$+35\frac{1}{5}$$
$60$

**111.**
$$69\frac{5}{22}$$
$$-36\frac{7}{11}$$
$32\frac{13}{22}$

**112.**
$$36\frac{3}{20}$$
$$-32\frac{5}{6}$$
$3\frac{19}{60}$

**113.**
$$29\frac{2}{9}$$
$$27\frac{7}{18}$$
$$+54\frac{2}{3}$$
$111\frac{5}{18}$

**114.**
$$7\frac{3}{8}$$
$$9\frac{5}{6}$$
$$+3\frac{1}{12}$$
$20\frac{7}{24}$

**115.** $1\frac{5}{8} \cdot \frac{2}{3}$  $1\frac{1}{12}$

**116.** $3\frac{6}{11} \cdot \frac{5}{13}$  $1\frac{4}{11}$

**117.** $4\frac{1}{6} \cdot 2\frac{2}{5}$  $10$

**118.** $5\frac{2}{3} \cdot 2\frac{1}{4}$  $12\frac{3}{4}$

**119.** $6\frac{3}{4} \div 1\frac{2}{7}$  $5\frac{1}{4}$

**120.** $5\frac{1}{2} \div 2\frac{1}{11}$  $2\frac{29}{46}$

**121.** $\frac{7}{2} \div 1\frac{1}{2}$  $2\frac{1}{3}$

**122.** $1\frac{3}{5} \div \frac{1}{4}$  $6\frac{2}{5}$

**123.** Two packages of soup bones weigh $3\frac{3}{4}$ pounds and $2\frac{3}{5}$ pounds. Find their combined weight.  $6\frac{7}{20}$ lb

**124.** A ribbon $5\frac{1}{2}$ yards long is cut from a reel of ribbon with 50 yards on it. Find the length of the piece remaining on the reel.  $44\frac{1}{2}$ yd

**125.** The average annual snowfall at a certain ski resort is $62\frac{3}{10}$ inches. Last year it had $54\frac{1}{2}$ inches. Find how many inches below average last year's annual snowfall was.  $7\frac{4}{5}$ in.

△ **126.** Find the area of a rectangular sheet of gift wrap that is $2\frac{1}{4}$ feet by $3\frac{1}{3}$ feet.  $7\frac{1}{2}$ sq. ft

$2\frac{1}{4}$ feet

$3\frac{1}{3}$ feet

△ **127.** Find the perimeter of a sheet of shelf paper needed to fit exactly a square drawer $1\frac{1}{4}$ feet long on each side.   5 ft

$1\frac{1}{4}$ feet

△ **128.** Find the area of the rectangle.

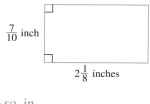

$\frac{7}{10}$ inch

$2\frac{1}{8}$ inches

$\frac{119}{80}$ or $1\frac{39}{80}$ sq. in.

△ **129.** A flower gardener has a square flower bed $\frac{2}{3}$ meter on a side and a rectangular one that is $\frac{7}{8}$ by $\frac{1}{3}$ meters. Find the total perimeter of his flower beds.   $5\frac{1}{12}$ m

**130.** There are 58 calories in 1 ounce of turkey. Find how many calories there are in a $3\frac{1}{2}$-ounce serving of turkey.   203 calories

**131.** There are $3\frac{1}{3}$ grams of fat in each ounce of lean hamburger. Find how many grams of fat are in a 4-ounce hamburger.   $13\frac{1}{3}$ g

**132.** Herman Heltznutt walks 5 days a week for a total distance of $5\frac{1}{4}$ miles per week. If he walks the same distance each day, find the distance he walks each day.   $\frac{21}{20}$ or $1\frac{1}{20}$ mi

# CHAPTER 4 TEST

*Write each mixed number as an improper fraction.*

**1.** $7\dfrac{2}{3}$

**2.** $3\dfrac{6}{11}$

*Write each improper fraction as a mixed number or a whole number.*

**3.** $\dfrac{23}{5}$

**4.** $\dfrac{75}{4}$

*Write each fraction in simplest form.*

**5.** $\dfrac{54}{210}$

**6.** $\dfrac{42}{70}$

*Perform each indicated operation and write the answers in simplest form.*

**7.** $\dfrac{4}{4} \div \dfrac{3}{4}$

**8.** $-\dfrac{4}{3} \cdot \dfrac{4}{4}$

**9.** $\dfrac{7x}{9} + \dfrac{x}{9}$

**10.** $\dfrac{1}{7} - \dfrac{3}{x}$

**11.** $\dfrac{xy^3}{z} \cdot \dfrac{z}{xy}$

**12.** $-\dfrac{2}{3} \cdot -\dfrac{8}{15}$

**13.** $\dfrac{9a}{10} + \dfrac{2}{5}$

**14.** $-\dfrac{8}{15y} - \dfrac{2}{15y}$

**15.** $8y^3 \div \dfrac{y}{3}$

**1.** $\dfrac{23}{3}$

**2.** $\dfrac{39}{11}$

**3.** $4\dfrac{3}{5}$

**4.** $18\dfrac{3}{4}$

**5.** $\dfrac{9}{35}$

**6.** $\dfrac{3}{5}$

**7.** $\dfrac{4}{3}$

**8.** $-\dfrac{4}{3}$

**9.** $\dfrac{8x}{9}$

**10.** $\dfrac{x-21}{7x}$

**11.** $y^2$

**12.** $\dfrac{16}{45}$

**13.** $\dfrac{9a+4}{10}$

**14.** $-\dfrac{2}{3y}$

**15.** $24y^2$

**16.** 9

**17.** $14\frac{1}{40}$

**18.** $\dfrac{1}{a^2}$

**19.** $\dfrac{64}{3}$

**20.** $22\frac{1}{2}$

**21.** $3\frac{3}{5}$

**22.** $\dfrac{1}{3}$

**23.** $\dfrac{3}{4}$

**24.** $\dfrac{3}{4x}$

**25.** $\dfrac{76}{21}$

**26.** $-2$

**27.** $-4$

**28.** 1

**29.** $\dfrac{5}{2}$

**30.** $\dfrac{4}{31}$

**Name** _____

**16.** $5\frac{1}{4} \div \frac{7}{12}$

**17.** $\begin{aligned} &3\frac{7}{8}\\ &7\frac{2}{5}\\ +&2\frac{3}{4}\\ \hline \end{aligned}$

**18.** $\dfrac{3a}{8} \cdot \dfrac{16}{6a^3}$

**19.** $-\dfrac{16}{3} \div -\dfrac{3}{12}$

**20.** $3\frac{1}{3} \cdot 6\frac{3}{4}$

**21.** $12 \div 3\frac{1}{3}$

**22.** $\left(\dfrac{14}{5} \cdot \dfrac{25}{21}\right) \div 10$

**23.** $\dfrac{11}{12} - \dfrac{3}{8} + \dfrac{5}{24}$

*Simplify each complex fraction.*

**24.** $\dfrac{\frac{5x}{7}}{\frac{20x^2}{21}}$

**25.** $\dfrac{5 + \frac{3}{7}}{2 - \frac{1}{2}}$

*Solve.*

**26.** $-\dfrac{3}{8}x = \dfrac{3}{4}$

**27.** $\dfrac{x}{5} + x = -\dfrac{24}{5}$

**28.** $\dfrac{2}{3} + \dfrac{x}{4} = \dfrac{5}{12} + \dfrac{x}{2}$

*Evaluate each expression for the given replacement values.*

**29.** $-5x;\ x = -\dfrac{1}{2}$

**30.** $x \div y;\ x = \dfrac{1}{2},\ y = 3\frac{7}{8}$

**340**

**Name** _____

**31.** $\frac{1}{2}$ of the calories

*Solve.*

**31.** A McDonald's Big Mac® sandwich has 560 calories. There are 280 calories from fat in a Big Mac. What fraction of a Big Mac's calories are from fat? (*Source:* McDonald's Corporation)

**32.** A carpenter cuts a piece $2\frac{3}{4}$ feet long from a cedar plank that is $6\frac{1}{2}$ feet long. How long is the remaining piece?

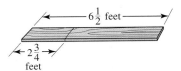

As shown in the circle graph, the market for backpacks is divided among five companies. For instance, Wilderness, Inc.'s backpack accounts for $\frac{1}{4}$ of all backpack sales. Use this graph to answer Questions 33 and 34.

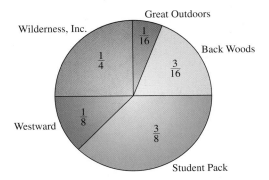

**33.** What fraction of backpack sales goes to Back Woods and Westward combined?

**34.** If a total of 500,000 backpacks are sold each year, how many backpacks does Wilderness, Inc. sell?

△ **35.** How many square yards of artificial turf are necessary to cover a football field, including the end zones and 10 yards beyond the side lines? (*Hint:* A football field measures $100 \times 53\frac{1}{3}$ yards and the end zones are 10 yards deep.)

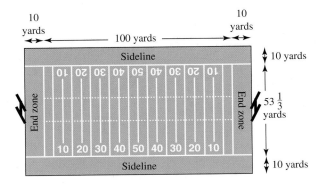

**32.** $3\frac{3}{4}$ ft _____

**33.** $\frac{5}{16}$ of the sales _____

**34.** 125,000 backpacks _____

**35.** 8800 sq. yd _____

**Name** _____

△ **36.** Find the area of the figure.

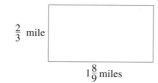

$\frac{2}{3}$ mile

$1\frac{8}{9}$ miles

**37.** During a 258-mile trip, a car used $10\frac{3}{4}$ gallons of gas. How many miles could we expect the car to travel on one gallon of gas?

**38.** Prior to an oil spill, the stock in an oil company sold for $120 per share. As a result of the liability the company incurred from the spill, the price per share fell to $\frac{3}{4}$ of the price before the spill. What did the stock sell for after the spill?

**39.** A suit is on sale for $120, which is $\frac{3}{5}$ of the regular price. What is the regular price of the suit?

**Name** _____ **Section** _____ **Date** _____

# CUMULATIVE REVIEW

ANSWERS

*Write each number in words.*

1. 85

2. 126

3. Add: $23 + 136$

4. Subtract $43 - 29$. Check by adding.

5. Round 278,362 to the nearest thousand.

6. A single-density computer disk can hold 370 thousand bytes of information. How many total bytes can 42 such disks hold?

7. Divide and check: $56,717 \div 8$

*Write using exponential notation.*

8. $4 \cdot 4 \cdot 4$

9. $6 \cdot 6 \cdot 6 \cdot 8 \cdot 8 \cdot 8 \cdot 8 \cdot 8$

10. Evaluate $2(x - y)$ if $x = 8$ and $y = 4$.

11. Jack Mayfield, a miner for the Molly Kathleen Gold Mine, is presently 150 feet below the surface of the Earth. Represent this position using an integer.

12. Add using a number line: $-7 + 3$

13. Simplify: $7 - 8 - (-5) - 1$

14. Evaluate: $(-5)^2$

15. Simplify: $\dfrac{12 - 16}{-1 + 3}$

**ANSWERS**

1. eighty-five (Sec. 1.1, Ex. 4)

2. one hundred twenty-six (Sec. 1.1, Ex. 5)

3. 159 (Sec. 1.2, Ex. 1)

4. 14 (Sec. 1.3, Ex. 3)

5. 278,000 (Sec. 1.4, Ex. 2)

6. 15,540 thousand bytes (Sec. 1.5, Ex. 7)

7. 7089 R 5 (Sec. 1.6, Ex. 7)

8. $4^3$ (Sec. 1.7, Ex. 1)

9. $6^3 \cdot 8^5$ (Sec. 1.7, Ex. 4)

10. 8 (Sec. 1.8, Ex. 2)

11. $-150$ (Sec. 2.1, Ex. 1)

12. $-4$ (Sec. 2.2, Ex. 2)

13. 3 (Sec. 2.3, Ex. 9)

14. 25 (Sec. 2.4, Ex. 6)

15. $-2$ (Sec. 2.5, Ex. 7)

**Name** _____

*Multiply.*

**16.** $5(3y)$

**17.** $-2(4x)$

**18.** Solve: $-8 = x + 1$

**19.** Solve: $8x - 3x = 12 - 17$

**20.** Translate each sentence into an equation.
   **a.** The product of 7 and 6 is 42.
   **b.** Twice the sum of 3 and 5 is equal to 16.
   **c.** The quotient of $-45$ and 5 yields $-9$.

**21.** Twice a number, added to 3, is the same as the number minus 6. Find the number.

**22.** Write $\dfrac{9x}{11}$ as an equivalent fraction whose denominator is 44.

**23.** Write the prime factorization of 45.

*Multiply.*

**24.** $\dfrac{2}{3} \cdot \dfrac{5}{11}$

**25.** $\dfrac{1}{4} \cdot \dfrac{1}{2}$

# Decimals

Decimals are an important part of everyday life. For example, we use decimal numbers in our money system. One penny is 0.01 dollar, and one dime—or ten pennies, is 0.10 dollar. Among many other uses, decimals also express batting averages. A baseball player with a 0.333 batting average is a pretty good batter. Decimal numbers represent parts of a whole, just like fractions. In this chapter, we analyze the relationship between fractions and decimals.

## CONTENTS

The Atlantic hurricane season in the United States lasts from June through November each year. During an average hurricane season, 10 named tropical storms develop in the Atlantic but only 6 of these become hurricanes. As a hurricane reaches land, the damage it causes can range from minimal to catastrophic, depending on such things as sustained winds, central barometric pressure, and storm surge. The Saffir-Simpson Hurricane Intensity Scale, devised in the early 1970s, uses these factors to rate hurricanes for their potential for damage. This five-category scale is shown in the table. In Exercises 81 and 82 on page 359 and Exercises 69 and 70 on page 395, we will use aspects of the Saffir-Simpson Hurricane Intensity Scale to investigate operations on decimal numbers.

| SAFFIR-SIMPSON HURRICANE INTENSITY SCALE | | | | |
|---|---|---|---|---|
| Category | Wind Speed | Barometric Pressure [inches of mercury (Hg)] | Storm Surge | Damage Potential |
| 1 (Weak) | 75–95 mph | ≥ 28.94 in. | 4–5 ft | Minimal damage to vegetation |
| 2 (Moderate) | 96–110 mph | 28.50–28.93 in. | 6–8 ft | Moderate damage to houses |
| 3 (Strong) | 111–130 mph | 27.91–28.49 in. | 9–12 ft | Extensive damage to small buildings |
| 4 (Very Strong) | 131–155 mph | 27.17–27.90 in. | 13–18 ft | Extreme structural damage |
| 5 (Devastating) | >155 mph | <27.17 in. | >18 ft | Catastrophic building failures possible |

**Name** _____  **Section** _____  **Date** _____

# CHAPTER 5 PRETEST

**1.** Write 0.27 as a fraction.

**2.** Insert <, >, or = to form a true statement.
0.205    0.213

**3.** Round 54.651 to the nearest tenth.

*Perform each indicated operation.*

**4.** $38.41 + 14.032 + 7.6$

**5.** $(-3.4)(-2.1)$

**6.** $(2.016)(100)$

**7.** $16.24 \div 0.4$

**8.** $\dfrac{891}{10,000}$

**9.** Evaluate $x - y$ for $x = 12.3$ and $y = 0.61$.

**10.** Simplify: $-9.8 - 6.2x - 7.9 + 1.4x$

**11.** Evaluate $xy$ for $x = 4.2$ and $y = 0.03$.

△ **12.** Find the circumference of a circle whose radius is 6 inches. Then use the approximation 3.14 for $\pi$ to approximate the circumference.

**13.** Perry Sitongia borrowed $576 from his father. He plans to pay it back over the next 20 months with equal payments. How much will each monthly payment be?

**14.** Simplify: $0.2(6.9 - 3.01)$

**15.** Write $\dfrac{3}{8}$ as a decimal.

**16.** Insert <, >, or = to form a true statement.
$\dfrac{9}{11}$    0.8182

**17.** Solve: $4(x + 0.22) = 2x - 3.4$

**18.** Find the square root: $\sqrt{\dfrac{36}{49}}$

**19.** Approximate $\sqrt{46}$ to the nearest hundredth.

△ **20.** Find the length of the hypotenuse of the given right triangle.

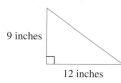

9 inches

12 inches

# 5.1 INTRODUCTION TO DECIMALS

## A DECIMAL NOTATION AND WRITING DECIMALS IN WORDS

Like fractional notation, decimal notation is used to denote a part of a whole. Numbers written in decimal notation are called **decimal numbers**, or simply **decimals**. The decimal 17.758 has three parts.

In Section 1.1, we introduced place value for whole numbers. Place names and place values for the whole-number part of a decimal number are exactly the same, as shown next. Place names and place values for the decimal part are also shown.

**PLACE-VALUE CHART**

| hundreds | tens | ones | ↑ | tenths | hundredths | thousandths | ten-thousandths | hundred-thousandths |
|---|---|---|---|---|---|---|---|---|
| 1 | 7 | . | 7 | 5 | 8 | | | |
| 100 | 10 | 1 | decimal point | $\frac{1}{10}$ | $\frac{1}{100}$ | $\frac{1}{1000}$ | $\frac{1}{10,000}$ | $\frac{1}{100,000}$ |

Notice that the value of each place is $\frac{1}{10}$ of the value of the place to its left. For example,

$$1 \cdot \frac{1}{10} = \frac{1}{10} \quad \text{and} \quad \frac{1}{10} \cdot \frac{1}{10} = \frac{1}{100}$$

ones tenths     tenths   hundredths

For the example above, 17.758, the digit 5 is in the hundredths place, so its value is 5 hundredths or $\frac{5}{100}$.

Writing or reading a decimal in words is similar to writing or reading a whole number. Use the following steps.

---

**WRITING (OR READING) A DECIMAL IN WORDS**

**Step 1.** Write the whole-number part in words.
**Step 2.** Write "and" for the decimal point.
**Step 3.** Write the decimal part in words as though it were a whole number, followed by the place value of the last digit.

---

**Example 1**    Write each decimal in words.

     **a.** 0.3     **b.** 5.82     **c.** 21.093

*Solution:*    **a.** Three tenths
           **b.** Five and eighty-two hundredths
           **c.** Twenty-one and ninety-three thousandths

---

## Objectives

**A** Know the meaning of place value for a decimal number and write decimals in words.
**B** Write decimals in standard form.
**C** Write decimals as fractions.
**D** Compare decimals.
**E** Round decimals to a given place value.

Study Guide    SSM    CD-ROM    Video 5.1

TEACHING TIP

Make sure that students notice the difference in place values which are symmetric to the ones column. For instance, compare "tens" with "tenths."

---

**Practice Problem 1**

Write each decimal in words.
a. 0.08    b. 500.025    c. 0.0329

**Answers**

**1. a.** eight hundredths   **b.** five hundred and twenty-five thousandths   **c.** three hundred twenty-nine ten-thousandths

### Practice Problem 2

Write the decimal 97.28 in words.

### Practice Problem 3

Write the decimal 72.1085 in words.

### Practice Problems 4–5

Write each decimal in standard form.
4. Three hundred and ninety-six hundredths
5. Thirty-nine and forty-two thousandths

TEACHING TIP

After discussing Example 5, ask students to write 6.950 in words using hundredths and using thousandths.

**Answers**

**2.** ninety-seven and twenty-eight hundredths
**3.** seventy-two and one thousand eighty-five ten-thousandths   **4.** 300.96   **5.** 39.042

**Example 2**     Write the decimal in the sentence in words: The Golden Jubilee Diamond is a 545.67 carat cut diamond. (*Source: The Guinness Book of Records*, 1998)

*Solution:*     five hundred forty-five and sixty-seven hundredths ▬▬▬

**Example 3**     Write the decimal in the sentence in words: The oldest known fragments of Earth's crust are Zircon crystals. They were discovered in Australia and are thought to be 4.276 billion years old. (*Source: The Guinness Book of Records*, 1998)

*Solution:*     four and two hundred seventy-six thousandths ▬▬▬

### B WRITING DECIMALS IN STANDARD FORM

A decimal written in words can be written in standard form by reversing the above procedure.

**Examples**     Write each decimal in standard form.

**4.** Forty-eight and twenty-six hundredths

$\downarrow$

48.26

↑ —— hundredths place

**5.** Six and ninety-five thousandths

6.095

↑ —— thousandths place     ▬▬▬

---

**Helpful Hint**

When writing a decimal from words to decimal notation, make sure the last digit is in the correct place by inserting 0s if necessary. For example,

Two and thirty-eight thousandths is 2.038

thousandths place

## **C** WRITING DECIMALS AS FRACTIONS

Once you master reading and writing decimals, writing a decimal as a fraction follows naturally.

| Decimal | In Words | Fraction |
|---------|----------|----------|
| 0.7 | seven tenths | $\dfrac{7}{10}$ |
| 0.51 | fifty-one hundredths | $\dfrac{51}{100}$ |
| 0.009 | nine thousandths | $\dfrac{9}{1000}$ |

Notice that the number of decimal places in a decimal number is the same as the number of zeros in the denominator of the equivalent fraction. We can use this fact to write decimals as fractions.

$$0.51 = \frac{51}{100} \qquad 0.009 = \frac{9}{1000}$$

2 decimal places   2 zeros     3 decimal places   3 zeros

### Example 6

Write 0.43 as a fraction.

*Solution:*  $0.43 = \dfrac{43}{100}$

2 decimal places   2 zeros

### Example 7

Write 5.6 as a mixed number.

*Solution:*  $5.6 = 5\dfrac{6}{10} = 5\dfrac{3}{5}$  in simplest form

1 decimal place   1 zero

### Examples

Write each decimal as a fraction or mixed number. Write your answer in simplest form.

**8.**  $0.125 = \dfrac{125}{1000} = \dfrac{1}{8}$

**9.**  $23.5 = 23\dfrac{5}{10} = 23\dfrac{1}{2}$

**10.**  $105.083 = 105\dfrac{83}{1000}$

---

**Practice Problem 6**

Write 0.037 as a fraction.

**Practice Problem 7**

Write 14.97 as a mixed number.

**Practice Problems 8–10**

Write each decimal as a fraction or mixed number. Write your answer in simplest form.

  8. 0.12
  9. 57.8
10. 209.986

**Answers**

**6.** $\dfrac{37}{1000}$  **7.** $14\dfrac{97}{100}$  **8.** $\dfrac{3}{25}$  **9.** $57\dfrac{4}{5}$

**10.** $209\dfrac{493}{500}$

## D COMPARING DECIMALS

One way to compare decimals is to compare their graphs on a number line. Recall that for any two numbers on a number line, the number to the left is smaller and the number to the right is larger. The decimals 0.5 and 0.8 are graphed as follows:

Comparing decimals by comparing their graphs on a number line can be time-consuming, so we compare the size of decimals by comparing digits in corresponding places.

---

### COMPARING TWO DECIMALS

Compare digits in the same places from left to right. When two digits are not equal, the number with the larger digit is the larger decimal. If necessary, insert 0s after the last digit to the right of the decimal point to continue comparing.

Compare hundredths place digits

28.253          28.263
↑                ↑
5    <    6

so 28.253   <   28.263

---

### Helpful Hint

For any decimal, inserting 0s after the last digit to the right of the decimal point does not change the value of the number.

$$7.6 = 7.60 = 7.600, \text{ and so on}$$

When a whole number is written as a decimal, the decimal point is placed to the right of the ones digit.

$$25 = 25.0 = 25.00, \text{ and so on}$$

---

**Practice Problem 11**

Insert $<$, $>$, or $=$ to form a true statement.

13.208      13.28

**Practice Problem 12**

Insert $<$, $>$, or $=$ to form a true statement.

0.12      0.086

**Example 11**   Insert $<$, $>$, or $=$ to form a true statement.

0.378      0.368

*Solution:*   0. 3 78          0. 3 68    The tenths places are the same.

0.3 7 8          0.3 6 8    The hundredths places are different.

Since $7 > 6$, then $0.378 > 0.368$.    ▬▬▬

**Example 12**   Insert $<$, $>$, or $=$ to form a true statement.

0.052      0.236

*Solution:*   0. 0 52 < 0. 2 36    0 is smaller than 2 in the tenths place.
↑              ↑    ▬▬▬

## E  ROUNDING DECIMALS

We **round the decimal part** of a decimal number in nearly the same way as we round whole numbers. The only difference is that we drop digits to the right of the rounding place, instead of replacing these digits by 0s. For example,

24.954   rounded to the nearest hundredth is   24.95.
↑

---

**ROUNDING DECIMALS TO A PLACE VALUE TO THE RIGHT OF THE DECIMAL POINT**

**Step 1.**  Locate the digit to the right of the given place value.

**Step 2.**  If this digit is 5 or greater, add 1 to the digit in the given place value and drop all digits to its right. If this digit is less than 5, drop all digits to the right of the given place.

---

### Example 13

Round 736.2359 to the nearest tenth.

*Solution:*   **Step 1.** We locate the digit to the right of the tenths place.

┌─── tenths place
↓
736.2 **3** 59
└──────→ digit to the right

**Step 2.**  Since this digit to the right is less than 5, we drop it and all digits to its right.
Thus, 736.2359 rounded to the nearest tenth is 736.2.

Rounding often occurs with money amounts. Since there are 100 cents in a dollar, each cent is $\frac{1}{100}$ of a dollar. This means that if we want to round to the nearest cent, we round to the nearest hundredth of a dollar.

### Example 14   Finding Gasoline Prices

The price of a gallon of gasoline in Aimsville is currently $1.3279. Round this to the nearest cent.

*Solution:*    ┌─── hundredths place
↓
$1.32 **7** 9
└──────→ digit to the right

Since this digit to the right is greater than 5, we add 1 to the hundredths digit and drop all digits to the right of the hundredths digit. Thus, the rounded gasoline price per gallon is $1.33.

**Practice Problem 13**

Round 123.7817 to the nearest thousandth.

**Practice Problem 14**

In Cititown, the price of a gallon of gasoline is $1.2789. Round this to the nearest cent.

**Answers**
**13.** 123.782  **14.** $1.28

✓ **CONCEPT CHECK**

Which digit—1, 2, 3, 4, 5, or 6—is in the tenths place in 123.456? The tens place?

**Practice Problem 15**

Water bills in Gotham City are always rounded to the nearest dime. Lois's water bill was $24.43. Round her bill to the nearest dime (tenth).

**TRY THE CONCEPT CHECK IN THE MARGIN.**

**Example 15**   Determining State Taxable Income

A high school teacher's taxable income is $31,567.72. The tax tables in the teacher's state use amounts to the nearest dollar. Round the teacher's income to the nearest whole dollar.

*Solution:*   Rounding to the nearest whole dollar means rounding to the ones place.

$31,567. 7 2

Since the digit to the right is 5 or greater, we add 1 to the digit in the ones place and drop all digits to the right of the ones place. The teacher's income rounded to the nearest dollar is $31,568.

**Name** _____ **Section** _____ **Date** _____

## MENTAL MATH

*Determine the place value for the 7 in each decimal.*

**1.** 70

**2.** 700

**3.** 0.7

**4.** 0.07

# EXERCISE SET 5.1

**A** *Write each decimal in words. See Examples 1 through 3.*

**1.** 6.52

**2.** 7.59

**3.** 16.23

**4.** 47.65

**5.** 3.205

**6.** 7.495

**7.** 167.009

**8.** 233.056

**B** *Write each decimal number in standard form. See Examples 4 and 5.*

**9.** Six and five tenths

**10.** Three and nine tenths

**11.** Nine and eight hundredths

**12.** Twelve and six hundredths

**13.** Five and six hundred twenty-five thousandths

**14.** Four and three hundred ninety-nine thousandths

**ANSWERS**

**1.** six and fifty-two hundredths

**2.** seven and fifty-nine hundredths

**3.** sixteen and twenty-three hundredths

**4.** forty-seven and sixty-five hundredths

**5.** three and two hundred five thousandths

**6.** seven and four hundred ninety-five thousandths

**7.** one hundred sixty-seven and nine thousandths

**8.** two hundred thirty-three and fifty-six thousandths

**9.** 6.5

**10.** 3.9

**11.** 9.08

**12.** 12.06

**13.** 5.625

**14.** 4.399

**15.** 0.0064

**16.** 0.0038

**17.** 20.33

**18.** 9.62

**19.** 5.9

**20.** 153.2

**21.** $\dfrac{3}{10}$

**22.** $\dfrac{7}{10}$

**23.** $\dfrac{27}{100}$

**24.** $\dfrac{39}{100}$

**25.** $5\dfrac{47}{100}$

**26.** $6\dfrac{3}{10}$

**27.** $\dfrac{6}{125}$

**28.** $\dfrac{41}{500}$

**29.** $7\dfrac{7}{100}$

**30.** $9\dfrac{9}{100}$

**31.** $15\dfrac{401}{500}$

**32.** $11\dfrac{203}{500}$

**354**

---

**15.** Sixty-four ten-thousandths

**16.** Thirty-eight ten-thousandths

**17.** The record rainfall amount for a 24-hour period in Alabama is twenty and thirty-three hundredths inches. This record was set in Axis, Alabama, in 1955. (*Source:* National Climatic Data Center)

**18.** The United States Postal Service vehicle fleet averages nine and sixty-two hundredths miles per gallon of fuel. (*Source:* United States Postal Service)

**19.** In 1997, there was an average of five and nine tenths on-the-job illnesses and injuries for every 100 workers in the mining industry. (*Source:* Bureau of Labor Statistics)

**20.** Kenny Brack won the 1999 Indianapolis 500 with an average speed of one hundred fifty-three and two tenths mph. (*Source:* USA Today, 12/22/99)

**C** *Write each decimal as a fraction or a mixed number. Write your answer in simplest form. See Examples 6 through 10.*

**21.** 0.3

**22.** 0.7

**23.** 0.27

**24.** 0.39

**25.** 5.47

**26.** 6.3

**27.** 0.048

**28.** 0.082

**29.** 7.07

**30.** 9.09

**31.** 15.802

**32.** 11.406

**33.** 0.3005        **34.** 0.2006        **35.** 487.32        **36.** 298.62

**D** *Insert <, >, or = between each pair of numbers to form a true statement. See Examples 11 and 12.*

**37.** 0.15     0.16        **38.** 0.12     0.15        **39.** 0.57     0.54

**40.** 0.59     0.52        **41.** 0.098     0.1        **42.** 0.0756     0.2

**43.** 0.54900     0.549        **44.** 0.98400     0.984        📼 **45.** 167.908     167.980

**46.** 519.3405     519.3054        **47.** 420,000     0.000042        **48.** 0.000987     987,000

**E** *Round each decimal to the given place value. See Examples 13 through 15.*

📼 **49.** 0.57, nearest tenth            **50.** 0.54, nearest tenth

**51.** 0.234, nearest hundredth        **52.** 0.452, nearest hundredth

---

**33.** $\dfrac{601}{2000}$ _____

**34.** $\dfrac{1003}{5000}$ _____

**35.** $487\dfrac{8}{25}$ _____

**36.** $298\dfrac{31}{50}$ _____

**37.** < _____

**38.** < _____

**39.** > _____

**40.** > _____

**41.** < _____

**42.** < _____

**43.** = _____

**44.** = _____

**45.** < _____

**46.** > _____

**47.** > _____

**48.** < _____

**49.** 0.6 _____

**50.** 0.5 _____

**51.** 0.23 _____

**52.** 0.45 _____

**53.** 0.594

**54.** 63.452

**55.** 98,210

**56.** 68,930

**57.** 12.3

**58.** 42.988

**59.** 17.67

**60.** 0.77

**61.** 0.5

**62.** 0.6

**63.** $0.07

**64.** $0.03

**65.** $27

**66.** $14,770

**Name** _____

**53.** 0.5942, nearest thousandth

**54.** 63.4523, nearest thousandth

**55.** 98,207.23, nearest ten

**56.** 68,934.543, nearest ten

**57.** 12.342, nearest tenth

**58.** 42.9878, nearest thousandth

**59.** 17.667, nearest hundredth

**60.** 0.766, nearest hundredth

**61.** 0.501, nearest tenth

**62.** 0.602, nearest tenth

*Round each money amount to the nearest cent or dollar as indicated. See Examples 14 and 15.*

**63.** $0.067, nearest cent

**64.** $0.025, nearest cent

**65.** $26.95, nearest dollar

**66.** $14,769.52, nearest dollar

**Name** _____

**67.** $0.1992, nearest cent

**68.** $0.7633, nearest cent

**69.** Which number(s) rounds to 0.26?
0.26559  0.26499  0.25786  0.25186

**70.** Which number(s) rounds to 0.06?
0.0612  0.066  0.0586  0.0506

*Round each number to the given place value. See Examples 13 through 15.*

**71.** The attendance at a Mets baseball game was reported to be 39,867. Round this number to the nearest thousand.

**72.** A used office desk is advertised at $19.95 by Drawley's Office Furniture. Round this price to the nearest dollar.

**73.** During the 1999 Boston Marathon, Fatuma Roba of Ethiopia was the women's winner for the third time in three years. Her latest time was 2.39027 hours. Round this time to the nearest hundredth. (*Source: 2000 World Almanac*)

**74.** The population density of the state of Ohio is roughly 271.0961 people per square mile. Round this population density to the nearest tenth. (*Source: U.S. Bureau of the Census*)

Ohio

BOSTON MARATHON

**67.** $0.20

**68.** $0.76

**69.** 0.26499; 0.25786

**70.** 0.0612; 0.0586

**71.** 40,000 people

**72.** $20

**73.** 2.39 hr

**74.** 271.1 people per sq. mi

**357**

**Name** _____

**75.** The length of a day on Mars is 24.6229 hours. Round this figure to the nearest thousandth. (*Source:* National Space Science Data Center)

**76.** Venus makes a complete orbit around the sun every 224.695 days. Round this figure to the nearest whole day. (*Source:* National Space Science Data Center)

**77.** The official barefoot water-skiing record is 135.74 mph by Scott Pellaton in November, 1989. Round this figure to the nearest mile per hour. (*Source: The Guinness Book of Records*, 1999)

**78.** Raptor is a roller coaster at Cedar Point, an amusement park in Sandusky, Ohio. It is one of the world's tallest, fastest, and steepest inverted roller coasters. A ride on Raptor lasts about 2.267 minutes. Round this figure to the nearest tenth. (*Source:* Cedar Fair, L.P.)

**79.** The leading NBA scorer for the 1998–1999 regular season was Allen Iverson of the Philadelphia 76ers. The average number of points he scored per game was 26.75. Round this to the nearest whole point. (*Source:* National Basketball Association)

**80.** The leading WNBA scorer for the 1999 regular season was Cynthia Cooper of the Houston Comets. She scored an average 22.12903 points per game. Round this to the nearest hundredth. (*Source:* Women's National Basketball Association)

**81.** Refer to the Saffir-Simpson Hurricane Intensity chart in the chapter opener. In what category would a hurricane be classified if its barometric pressure is 28.21 in. Hg?

**82.** Refer to the Saffir-Simpson Hurricane Intensity chart in the chapter opener. Do barometric pressures associated with various categories increase or decrease with hurricane intensity? Explain.

## REVIEW AND PREVIEW

*Perform each indicated operation. See Sections 1.2 and 1.3.*

**83.** 3452 + 2314

**84.** 8945 + 4536

**85.** 94 − 23

**86.** 82 − 47

**87.** 482 − 239

**88.** 4002 − 3897

## COMBINING CONCEPTS

*The table gives the leading bowling averages for the Professional Bowlers Association Tour for each of the years listed. Use this table to answer Exercises 89–91.*

| LEADING PBA AVERAGES BY YEAR | | |
|---|---|---|
| Year | Bowler | Average Score |
| 1990 | Amleto Monacelli | 215.432 |
| 1991 | Norm Duke | 218.158 |
| 1992 | Dave Ferraro | 219.702 |
| 1993 | Walter Ray Williams, Jr. | 222.980 |
| 1994 | Norm Duke | 222.830 |
| 1995 | Mike Aulby | 225.490 |
| 1996 | Walter Ray Williams, Jr. | 225.370 |
| 1997 | Walter Ray Williams, Jr. | 222.008 |
| 1998 | Walter Ray Williams, Jr. | 226.130 |

(*Source:* Professional Bowlers Association)

**89.** What is the highest average score on the list? Which bowler achieved that average?

**90.** What is the lowest average score on the list? Which bowler achieved that average?

**81.** category 3

**82.** decrease

**83.** 5766

**84.** 13,481

**85.** 71

**86.** 35

**87.** 243

**88.** 105

**89.** 226.130; Walter Ray Williams, Jr.

**90.** 215.432; Amleto Monacelli

**91.** 226.130; 225.490;
225.370; 222.980;
222.830; 222.008;
219.702; 218.158;
215.432

**92.** answers may vary

**93.** answers may vary

**94.** answers may vary

**91.** Make a list of the leading averages in order from greatest to least for the years shown in the table.

**92.** Write a 4-digit number that rounds to 26.3.

**93.** Write a 5-digit number that rounds to 1.7.

**94.** Explain how to identify the value of the 9 in the decimal 486.3297.

# 5.2  ADDING AND SUBTRACTING DECIMALS

## A  ADDING OR SUBTRACTING DECIMALS

Adding or subtracting decimals is similar to adding or subtracting whole numbers. We add or subtract digits in corresponding place values from right to left, carrying or borrowing if necessary. To make sure that digits in corresponding place values are added or subtracted, we line up the decimal points vertically.

---

**ADDING OR SUBTRACTING DECIMALS**

**Step 1.**  Write the decimals so that the decimal points line up vertically.
**Step 2.**  Add or subtract as for whole numbers.
**Step 3.**  Place the decimal point in the sum or difference so that it lines up vertically with the decimal points in the problem.

---

**Example 1**    Add: $23.85 + 1.604$

*Solution:*    Line up the decimal points vertically and add as for whole numbers.

$$
\begin{array}{r}
\overset{1}{2}3.850 \\
+\ 1.604 \\
\hline
25.454
\end{array}
$$

Write one 0.

Place the decimal point in the sum so that all decimal points line up.

---

**Helpful Hint**

Recall that 0s may be inserted to the right of the decimal point after the last digit without changing the value of the decimal. This may be used to help line up place values when adding decimals.

$$
\begin{array}{r}
3.2 \\
15.567 \\
+\ 0.11
\end{array}
\quad \text{becomes} \quad
\begin{array}{r}
3.\mathbf{200} \\
15.567 \\
+\ 0.1\mathbf{10} \\
\hline
18.877
\end{array}
$$

⟵ Two 0s are inserted.

⟵ One 0 is inserted.

---

**Example 2**    Add: $763.7651 + 22.001 + 43.89$

*Solution:*

$$
\begin{array}{r}
763.7651 \\
22.0010 \\
+\ 43.8900 \\
\hline
829.6561
\end{array}
$$

⟵ Write one 0.

⟵ Two 0s are inserted.

Add.

---

**Objectives**

**A**  Add or subtract decimals.
**B**  Evaluate expressions and check solutions with decimal replacement values.
**C**  Simplify expressions containing decimals.
**D**  Solve problems by adding or subtracting decimals.

Study Guide    SSM    CD-ROM    Video 5.2

**Practice Problem 1**

Add.
a. $15.52 + 2.371$
b. $20.06 + 17.612$
c. $0.125 + 122.8$

TEACHING TIP

Remind students that if one number has more decimal places than another, 0s may be inserted as shown in the Helpful Hint.

**Practice Problem 2**

Add.
a. $34.567 + 129.43 + 2.8903$
b. $11.21 + 46.013 + 362.526$

**Answers**

**1. a.** 17.891  **b.** 37.672  **c.** 122.925
**2. a.** 166.8873  **b.** 419.749

## Practice Problem 3

Add: $27 + 0.00043$

## ✓ CONCEPT CHECK

Find and correct the error made in the following addition of 3.05, 2.6, and 1.941.

$$
\begin{array}{r}
3.05 \\
2.6 \\
+1.941 \\
\hline
2.272
\end{array}
$$

## Practice Problem 4

Add: $8.1 + (-99.2)$

## Practice Problem 5

Subtract: $5.8 - 3.92$

**TEACHING TIP**

Remind students that subtraction can be checked by addition.

## Practice Problem 6

Subtract: $53 - 29.31$

**Answers**

**3.** 27.00043   **4.** −91.1   **5.** 1.88   **6.** 23.69
✓ **Concept Check:** Decimal points were not lined up before adding.

$$
\begin{array}{r}
3.050 \\
2.600 \\
+1.941 \\
\hline
7.591
\end{array}
$$

**Example 3**   Add: $39 + 0.0021$

*Solution:*

$$
\begin{array}{r}
39.0000 \\
+\ 0.0021 \\
\hline
39.0021
\end{array}
$$

For a whole number, place the decimal point to the right of the ones digit.

**TRY THE CONCEPT CHECK IN THE MARGIN.**

**Example 4**   Add: $3.62 + (-4.78)$

*Solution:*   Recall from Chapter 2 that to add two numbers with different signs we find the difference of the larger absolute value and the smaller absolute value. The sign of the answer is the same as the sign of the number with the larger absolute value.

$$
\begin{array}{r}
4.78 \\
-3.62 \\
\hline
1.16
\end{array}
$$

Subtract the absolute values.

Thus, $3.62 + (-4.78) = -1.16$

The sign of the number with the larger absolute value.

Just as for whole numbers, borrowing may sometimes be needed when subtracting decimals.

**Example 5**   Subtract: $3.5 - 0.068$

*Solution:*

$$
\begin{array}{r}
\overset{\scriptscriptstyle 4}{\cancel{3}}.\overset{\scriptscriptstyle 9}{\cancel{5}}\overset{\scriptscriptstyle 10}{\cancel{0}}\overset{}{\cancel{0}} \\
-0.068 \\
\hline
3.432
\end{array}
$$

← Write two 0s.

Recall that we can check a subtraction problem by adding.

$$
\begin{array}{r}
3.432 \\
+0.068 \\
\hline
3.500
\end{array}
$$

Difference
Subtrahend
Minuend

**Example 6**   Subtract: $85 - 17.31$

*Solution:*

$$
\begin{array}{r}
\overset{\scriptscriptstyle 7}{\cancel{8}}\overset{\scriptscriptstyle 14}{\cancel{5}}.\overset{\scriptscriptstyle 9}{\cancel{0}}\overset{\scriptscriptstyle 10}{\cancel{0}} \\
-17.31 \\
\hline
67.69
\end{array}
$$

*Check:*

$$
\begin{array}{r}
67.69 \\
17.31 \\
\hline
85.00
\end{array}
$$

## Example 7

Subtract 3 from 6.98

*Solution:*

$$
\begin{array}{r}
6.98 \\
-3.00 \quad \text{Write two 0s.} \\
\hline
3.98
\end{array}
$$

**Practice Problem 7**

Subtract 18 from 26.99.

## Example 8

Subtract: $-5.8 - 1.7$

*Solution:*

Recall from Chapter 2 that to subtract 1.7 we add the opposite of 1.7, or $-1.7$. Thus

$$-5.8 - 1.7 = -5.8 + (-1.7)$$

Then, to add two numbers with the same sign, we add their absolute values. The sign of the answer is the same as their common sign.

$$
\begin{array}{r}
5.8 \\
+1.7 \quad \text{Add the absolute values.} \\
\hline
7.5
\end{array}
$$

Thus, $-5.8 + (-1.7) = -7.5$.

↑

The common sign

**Practice Problem 8**

Subtract: $-3.4 - 9.6$

## B  USING DECIMALS AS REPLACEMENT VALUES

Let's review evaluating expressions with given replacement values. This time the replacement values are decimals.

## Example 9

Evaluate $x - y$ for $x = 2.8$ and $y = 0.92$.

*Solution:*

Replace $x$ with 2.8 and $y$ with 0.92 and simplify.

$$
\begin{aligned}
x - y &= 2.8 - 0.92 \\
&= 1.88
\end{aligned}
\qquad
\begin{array}{r}
2.80 \\
-0.92 \\
\hline
1.88
\end{array}
$$

**Practice Problem 9**

Evaluate $y - z$ for $y = 11.6$ and $z = 10.8$.

## Example 10

Is 2.3 a solution of the equation $6.3 = x + 4$?

*Solution:*

Replace $x$ with 2.3 in the equation $6.3 = x + 4$ to see if the result is a true statement.

$$
\begin{aligned}
6.3 &= x + 4 \\
6.3 &= 2.3 + 4 \quad \text{Replace } x \text{ with 2.3.} \\
6.3 &= 6.3 \quad \text{True.}
\end{aligned}
$$

Since $6.3 = 6.3$ is a true statement, 2.3 is a solution of $6.3 = x + 4$.

**Practice Problem 10**

Is 12.1 a solution of the equation $y - 4.3 = 7.8$?

**Answers**

**7.** 8.99  **8.** $-13$  **9.** 0.8  **10.** yes

## Practice Problem 11

Simplify by combining like terms:
$$-4.3y + 7.8 - 20.1y + 14.6$$

### C  SIMPLIFY EXPRESSIONS CONTAINING DECIMALS

**Example 11**  Simplify by combining like terms:
$$11.1x - 6.3 + 8.9x - 4.6$$

*Solution:*
$$11.1x - 6.3 + 8.9x - 4.6 = 11.1x + 8.9x + (-6.3) + (-4.6)$$
$$= 20x + (-10.9)$$
$$= 20x - 10.9$$

### D  SOLVING PROBLEMS BY ADDING OR SUBTRACTING DECIMALS

Decimals are very common in real-life problems.

## Practice Problem 12

Find the total monthly cost of owning and operating a certain automobile given the expenses shown.

| | |
|---|---|
| Monthly car payment: | $536.50 |
| Monthly insurance cost: | $52.70 |
| Average gasoline bill per month: | $87.50 |

**Example 12**  Calculating the Cost of Owning an Automobile

Find the total monthly cost of owning and operating a certain automobile given the expenses shown.

| | |
|---|---|
| Monthly car payment: | $256.63 |
| Monthly insurance cost: | $ 47.52 |
| Average gasoline bill per month: | $ 95.33 |

*Solution:*

**1.** UNDERSTAND. Read and reread the problem. The phrase "total monthly cost" tells us to add.

**2.** TRANSLATE.

In words:

| Total monthly cost | is | car payment | plus | insurance | plus | gasoline bill |
|---|---|---|---|---|---|---|
| ↓ | ↓ | ↓ | ↓ | ↓ | ↓ | ↓ |

Translate:

| Total monthly cost | = | $256.63 | + | $47.52 | + | $95.33 |
|---|---|---|---|---|---|---|

**3.** SOLVE.
$$\begin{array}{r} {\scriptstyle 1\,1\,1} \\ 256.63 \\ 47.52 \\ +\ 95.33 \\ \hline \$399.48 \end{array}$$

**4.** INTERPRET. *Check* your work. *State* your conclusion: The total monthly cost is $399.48.

## Practice Problem 13

Use the bar graph for Example 13. How much taller is the average height in the Netherlands than the average height in Czechoslovakia?

**Example 13**  Comparing Average Heights

The bar graph shows the current average heights for adults in various countries. How much taller is the average height in Denmark than the average height in the United States?

**Answers**

**11.** $-24.4y + 22.4$  **12.** $676.70  **13.** 1.8 inches

Adding and Subtracting Decimals

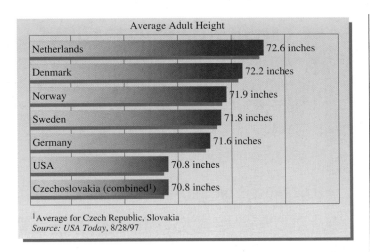

Average Adult Height

| | |
|---|---|
| Netherlands | 72.6 inches |
| Denmark | 72.2 inches |
| Norway | 71.9 inches |
| Sweden | 71.8 inches |
| Germany | 71.6 inches |
| USA | 70.8 inches |
| Czechoslovakia (combined¹) | 70.8 inches |

¹Average for Czech Republic, Slovakia
*Source: USA Today, 8/28/97*

*Solution:*

**1.** UNDERSTAND. Read and reread the problem. Since we want to know "how much taller," we subtract.

**2.** TRANSLATE.

In words:

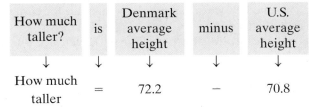

| How much taller? | is | Denmark average height | minus | U.S. average height |
|---|---|---|---|---|
| ↓ | ↓ | ↓ | ↓ | ↓ |

Translate:

| How much taller | = | 72.2 | − | 70.8 |
|---|---|---|---|---|

**3.** SOLVE.

$$
\begin{array}{r}
7\overset{1}{2}.\overset{12}{2} \\
-7\ 0\ .\ 8 \\
\hline
1\ .\ 4
\end{array}
$$

**4.** INTERPRET. *Check* your work. *State* your conclusion: The average height in Denmark is 1.4 inches more than the average height in the U.S.

## CALCULATOR EXPLORATIONS
### ENTERING DECIMAL NUMBERS

To enter a decimal number, find the key marked $\boxed{\cdot}$. To enter the number 2.56, for example, press the keys

$\boxed{2}$ $\boxed{.}$ $\boxed{5}$ $\boxed{6}$

The display will read $\boxed{2.56}$.

### OPERATIONS ON DECIMAL NUMBERS

Operations on decimal numbers are performed in the same way as operations on whole or signed numbers. For example, to find $8.625 - 4.29$, press the keys

$\boxed{8.625}$ $\boxed{-}$ $\boxed{4.29}$ $\boxed{=}$ (or $\boxed{\text{ENTER}}$)

The display will read $\boxed{4.335}$.
(Although entering 8.625, for example, requires pressing more than one key, we group numbers together for easier reading.)

*Use a calculator to perform each indicated operation.*

**1.** $315.782 + 12.96$   328.742

**2.** $29.68 + 85.902$   115.582

**3.** $6.249 - 1.0076$   5.2414

**4.** $5.238 - 0.682$   4.556

**5.**
$$12.555$$
$$224.987$$
$$5.2$$
$$+622.65$$
865.392

**6.**
$$47.006$$
$$0.17$$
$$313.259$$
$$+139.088$$
499.523

**Name** _____ **Section** _____ **Date** _____

## MENTAL MATH

*State each sum or difference.*

**1.**  0.3
  +0.2

**2.**  0.4
  +0.5

**3.**  1.00
  +0.26

**4.**  3.00
  +0.19

**5.**  7.6
  +1.3

**6.**  4.5
  +3.2

**7.**  0.9
  −0.3

**8.**  0.6
  −0.2

## EXERCISE SET 5.2

**A** *Add. See Examples 1 through 4.*

**1.** $1.3 + 2.2$

**2.** $2.5 + 4.1$

**3.** $5.7 + 1.13$

**4.** $2.31 + 6.4$

**5.** $24.6 + 2.39 + 0.0678$

**6.** $32.4 + 1.58 + 0.0934$

**7.**  45.023
   3.006
  + 8.403

**8.**  65.0028
   5.0903
  + 6.9003

**9.** $-2.6 + (-5.97)$

**10.** $-18.2 + (-10.8)$

**11.** $15.78 + (-4.62)$

**12.** $6.91 + (-7.03)$

*Subtract. See Examples 5 through 8.*

**13.** $8.8 - 2.3$

**14.** $7.6 - 2.1$

**15.** $18 - 2.7$

**16.** $28 - 3.3$

**17.**  654.9
  −56.67

**18.**  863.23
  −39.453

**MENTAL MATH ANSWERS**

**1.**  0.5

**2.**  0.9

**3.**  1.26

**4.**  3.19

**5.**  8.9

**6.**  7.7

**7.**  0.6

**8.**  0.4

**ANSWERS**

**1.**  3.5

**2.**  6.6

**3.**  6.83

**4.**  8.71

**5.**  27.0578

**6.**  34.0734

**7.**  56.432

**8.**  76.9934

**9.**  −8.57

**10.**  −29

**11.**  11.16

**12.**  −0.12

**13.**  6.5

**14.**  5.5

**15.**  15.3

**16.**  24.7

**17.**  598.23

**18.**  823.777

**Name** _____

**19.** Subtract 6.7 from 23   **20.** Subtract 9.2 from 45   ▭ **21.** −1.12 − 5.2

**22.** −8.63 − 5.6   **23.** 7.7 − 14.1   **24.** 10.25 − 21.76

*Perform each indicated operation. See Examples 1 through 8.*

**25.** 0.9 + 2.2   **26.** 0.7 + 3.4   **27.** 362.3 + 1.098 + 87.34

**28.** 462.8 + 1.164 + 23.845   **29.** −5.9 − 4   **30.** −6.4 − 3.4

**31.** 45.67 − 20   **32.** 56.89 − 30   **33.** −6.06 + 0.44

**34.** −5.05 + 0.88   **35.** 900.34 − 123.45   **36.** 800.74 − 463.98

**37.** 3490.23 + 8493.09   **38.** 600.004 + 7983.0062   **39.**  234.89
                                                              +230.67

**40.**  734.89   **41.** 50.2 − 600   **42.** 40.3 − 700
      +640.56

▭ **43.** Subtract 61.9 from 923.5   **44.** Subtract 45.8 from 845.93

**45.**  100.009   **46.**  200.89   **47.**  1000
          6.08             7.49          − 123.4
       +  9.034          + 62.83

**368**

**48.**   2000
      − 327.47

**49.** −0.003 + 0.091

**50.** −0.004 + 0.085

**51.** 500 − 34.098

**52.** 300 − 98.345

**B** *Evaluate each expression for x = 3.6, y = 5, and z = 0.21. See Example 9.*

**53.** $x + z$

**54.** $y + x$

**55.** $x - z$

**56.** $y - z$

**57.** $y - x + z$

**58.** $x + y + z$

*Determine whether the given values are solutions to the given equations. See Example 10.*

**59.** Is 7 a solution to $x + 2.7 = 9.3$?

**60.** Is 3.7 a solution to $x + 5.9 = 8.6$?

**61.** Is 11.4 a solution to $27.4 - y = 16$?

**62.** Is 22.9 a solution to $45.9 - z = 23$?

**63.** Is 1 a solution to $2.3 + x = 5.3 - x$?

**64.** Is 0.9 a solution to $1.9 - x = x + 0.1$?

**C** *Simplify by combining like terms. See Example 11.*

**65.** $30.7x + 17.6 - 23.8x - 10.7$

**66.** $14.2z + 11.9 - 9.6z - 15.2$

**67.** $-8.61 + 4.23y - 2.36 - 0.76y$

**68.** $-8.96x - 2.31 - 4.08x + 9.68$

**48.** 1672.53

**49.** 0.088

**50.** 0.081

**51.** 465.902

**52.** 201.655

**53.** 3.81

**54.** 8.6

**55.** 3.39

**56.** 4.79

**57.** 1.61

**58.** 8.81

**59.** no

**60.** no

**61.** yes

**62.** yes

**63.** no

**64.** yes

**65.** $6.9x + 6.9$

**66.** $4.6z - 3.3$

**67.** $-10.97 + 3.47y$

**68.** $-13.04x + 7.37$

**Name** _____

**D** *Solve. See Examples 12 and 13.*

**69.** Find the total monthly cost of owning and maintaining a car given the information shown.

| | |
|---|---|
| Monthly car payment: | $275.36 |
| Monthly insurance cost: | $83.00 |
| Average cost of gasoline per month: | $81.60 |
| Average maintenance cost per month: | $14.75 |

**70.** Find the total monthly cost of owning and maintaining a car given the information shown.

| | |
|---|---|
| Monthly car payment: | $306.42 |
| Monthly insurance cost: | $53.50 |
| Average cost of gasoline per month: | $123.00 |
| Average maintenance cost per month: | $23.50 |

**71.** Gasoline was $1.039 per gallon on one day and $0.979 per gallon the next day. By how much did the price change?

**72.** A pair of eyeglasses costs a total of $347.89. The frames of the glasses are $97.23. How much do the lenses of the eyeglasses cost?

**73.** Ann-Margaret Tober bought a book for $32.48. If she paid with two $20 bills, what was her change?

**74.** Tom Mackey bought a car part for $18.26. If he paid with two $10 bills, what was his change?

**75.** Swaziland and Singapore are the number one and number two sugar-consuming countries in the world, respectively. Swaziland's citizens consume an annual average of 373.7 pounds of sugar per person. Singapore's citizens consume an annual average of 175.9 pounds of sugar per person. How much more sugar does the average person in Swaziland consume than the average person in Singapore in a year? (*Source: The Top 10 of Everything, 1997,* by Russell Ash)

**76.** In 1997, the average wage for U.S. production workers was $12.28 per hour. By 1998, this average wage had climbed to $12.77 per hour. How much of an increase was this? (*Source:* Bureau of Labor Statistics)

**77.** The average annual rainfall in Houston, Texas, is 46.07 inches. The average annual rainfall in New Orleans, Louisiana, is 61.88 inches. On average, how much more rain does New Orleans receive annually than Houston? (*Source:* National Climatic Data Center)

**78.** The average wind speed at the weather station on Mt. Washington in New Hampshire is 35.3 miles per hour. The average wind speed in Chicago, Illinois, is 10.4 miles per hour. How much faster is the average wind speed on Mt. Washington than in Chicago? (*Source:* National Climatic Data Center)

**79.** In October 1997, Andy Green set a new one-mile land speed record. This record was 129.567 miles per hour faster than a previous record of 633.468 set in 1983. What was Green's record-setting speed? (*Source:* United States Auto Club)

**80.** It costs $2.80 to send a 2-pound package locally via parcel post at a U.S. Post Office. A 5-pound package costs $3.45 to send locally via parcel post. What is the total cost of sendng a 2-pound and a 5-pound package locally via parcel post? (*Source:* United States Postal Service)

**81.** The three North America concert tours that have earned the most money are the Rolling Stones (1994) $121.2 million, Pink Floyd (1994) $103.5 million, and the Rolling Stones (1989) $98 million. What was the total amount of money earned from these three concerts? (*Source:* Pollstar, Fresno, CA)

**82.** In 1995, the average credit-card late fee was $12.53. In 1998, the average late fee had increased to $21.82. By how much did the average credit-card late fee increase from 1995 to 1998? (*Source:* Consumer Action)

**83.** The snowiest city in the United States is Blue Canyon, California, which receives an average of 111.6 more inches of snow than the second snowiest city. The second snowiest city in the United States is Marquette, Michigan. Marquette receives an average of 129.2 inches of snow annually. How much snow does Blue Canyon receive on average each year? (*Source:* National Climatic Data Center)

**84.** The driest city in the world is Aswan, Egypt, which receives an average of only 0.02 inches of rain per year. Yuma, Arizona, is the driest city in the United States. Yuma receives an average of 2.63 more inches of rain each year than Aswan. What is the average annual rainfall in Yuma? (*Source:* National Climatic Data Center)

**Name** _____

**85.** A landscape architect is planning a border for a flower garden shaped like a triangle. The sides of the garden measure 12.4 feet, 29.34 feet, and 25.7 feet. Find the amount of border material needed.

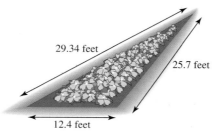

29.34 feet

25.7 feet

12.4 feet

△ **86.** A contractor needs to buy railing to completely enclose a newly built deck. Find the amount of railing needed.

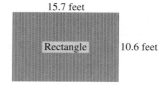

15.7 feet

Rectangle     10.6 feet

*The table shows the average speeds for the Indianapolis 500 winners for three years. Use this table to answer Exercises 87–88.*

| INDIANAPOLIS 500 WINNERS | | |
|---|---|---|
| Year | Winner | Average Speed |
| 1995 | Jacques Villeneuve | 153.616 mph |
| 1950 | Johnnie Parsons | 124.002 mph |
| 1911 | Ray Harroun | 74.602 mph |

(*Source:* Indianapolis Motor Speedway)

**87.** How much faster was the average Indianapolis 500 winning speed in 1995 than in 1950?

**88.** How much faster was the average Indianapolis 500 winning speed in 1995 than in 1911?

*The table shows spaceflight information for astronaut James A. Lovell. Use this table to answer Exercises 89–90.*

| SPACEFLIGHTS OF JAMES A. LOVELL | | |
|---|---|---|
| Year | Mission | Duration (in hours) |
| 1965 | Gemini 6 | 330.583 |
| 1966 | Gemini 12 | 94.567 |
| 1968 | Apollo 8 | 147.0 |
| 1970 | Apollo 13 | 142.9 |

(*Source:* NASA)

**89.** Find the total time spent in spaceflight by astronaut James A. Lovell.

**90.** Find the total time James A. Lovell spent in spaceflight on all Apollo missions.

*The table shows the five top chocolate-consuming nations in the world. Use this table to answer Exercises 91–95.*

**91.** Which country in the table has the greatest chocolate consumption per person?

| THE WORLD'S TOP CHOCOLATE-CONSUMING COUNTRIES | |
|---|---|
| Country | Pounds of Chocolate Per Person |
| Belgium | 13.9 |
| Germany | 15.8 |
| Norway | 16.0 |
| Switzerland | 22.0 |
| United Kingdom | 14.5 |

(*Source:* Hershey Foods Corporation)

**92.** Which country in the table has the least chocolate consumption per person?

**93.** How much more is the greatest chocolate consumption than the least chocolate consumption shown in the table?

**94.** How much more chocolate does the average German consume than the average citizen of the United Kingdom?

**95.** Make a new chart listing the countries and their corresponding chocolate consumption in order from greatest to least.

**REVIEW AND PREVIEW**

*Multiply. See Section 4.3.*

**96.** $\left(\dfrac{2}{3}\right)^2$   **97.** $\left(\dfrac{1}{5}\right)^3$   **98.** $\dfrac{12}{7} \cdot \dfrac{14}{3}$   **99.** $\dfrac{25}{36} \cdot \dfrac{24}{40}$

**91.** Switzerland

**92.** Belgium

**93.** 8.1 lb

**94.** 1.3 lb

| Country | Pounds of Chocolate Per Person |
|---|---|
| Switzerland | 22.0 |
| Norway | 16.0 |
| Germany | 15.8 |
| United Kingdom | 14.5 |
| Belgium | 13.9 |

**95.**

**96.** $\dfrac{4}{9}$

**97.** $\dfrac{1}{125}$

**98.** 8

**99.** $\dfrac{5}{12}$

## COMBINING CONCEPTS

**100.** Laser beams can be used to measure the distance to the moon. One measurement showed the distance to the moon to be 256,435.235 miles. A later measurement showed that the distance is 256,436.012 miles. Find how much farther away the moon is in the second measurement compared to the first.

**101.** Explain how adding or subtracting decimals is similar to adding or subtracting whole numbers.

*Combine like terms and simplify.*

**102.** $-8.689 + 4.286x - 14.295 - 12.966x + 30.861x$

**103.** $14.271 - 8.968x + 1.333 - 201.815x + 101.239x$

**104.** Can the sum of two negative decimals ever be a positive decimal? Why or why not?

---

**100.** 0.777 mi

**101.** answers may vary

**102.** $22.181x - 22.984$

**103.** $-109.544x + 15.604$

**104.** no; answers may vary

# 5.3   MULTIPLYING DECIMALS AND CIRCUMFERENCE OF A CIRCLE

## A   MULTIPLYING DECIMALS

Multiplying decimals is similar to multiplying whole numbers. The only difference is that we place a decimal point in the product. To discover where a decimal point is placed in the product, let's multiply $0.6 \times 0.03$. We first write each decimal as an equivalent fraction and then multiply.

$$0.6 \quad \times \quad 0.03 \quad = \frac{6}{10} \times \frac{3}{100} = \frac{18}{1000} = 0.018$$

↑ 1 decimal place   ↑ 2 decimal places   ↑ 3 decimal places

Now let's multiply $0.03 \times 0.002$.

$$0.03 \quad \times \quad 0.002 \quad = \frac{3}{100} \times \frac{2}{1000} = \frac{6}{100,000} = 0.00006$$

↑ 2 decimal places   ↑ 3 decimal places   ↑ 5 decimal places

Instead of writing decimals as fractions each time we want to multiply, we notice a pattern from these examples and state a rule that we can use.

---

**MULTIPLYING DECIMALS**

**Step 1.**   Multiply the decimals as though they were whole numbers.

**Step 2.**   The decimal point in the product is placed so the number of decimal places in the product is equal to the *sum* of the number of decimal places in the factors.

---

### Example 1   Multiply: $23.6 \times 0.78$

*Solution:*
$$
\begin{array}{r}
23.6 \\
\times\ 0.78 \\
\hline
1888 \\
16520 \\
\hline
18.408
\end{array}
$$

23.6   1 decimal place
× 0.78   2 decimal places
18.408   3 decimal places

### Example 2   Multiply: $0.283 \times 0.3$

*Solution:*
$$
\begin{array}{r}
0.283 \\
\times\quad 0.3 \\
\hline
0.0849
\end{array}
$$

0.283   3 decimal places
×   0.3   1 decimal place
0.0849   4 decimal places

↑ Insert one 0 since the product must have 4 decimal places.

### Example 3   Multiply: $0.0531 \times 16$

*Solution:*
$$
\begin{array}{r}
0.0531 \\
\times\quad 16 \\
\hline
3186 \\
05310 \\
\hline
0.8496
\end{array}
$$

0.0531   4 decimal places
×   16   0 decimal places
0.8496   4 decimal places

---

**Objectives**

**A**  Multiply decimals.

**B**  Multiply decimals by powers of 10.

**C**  Evaluate expressions and check solutions with decimal replacement values.

**D**  Find the circumference of a circle.

**E**  Solve problems by multiplying decimals.

Study   SSM   CD-ROM   Video
Guide                    5.3

---

TEACHING TIP

Before discussing the answer to $0.6 \times 0.03$, point out to students that if they know what $6 \times 3$ is, they should know what digits will be in the product. The only other thing they need to find the product is to determine where the decimal point is located.

---

**Practice Problem 1**

Multiply: $45.9 \times 0.42$

---

**Practice Problem 2**

Multiply: $0.112 \times 0.6$

---

**Practice Problem 3**

Multiply: $0.0721 \times 48$

---

**Answers**

**1.** 19.278   **2.** 0.0672   **3.** 3.4608

TRY THE CONCEPT CHECK IN THE MARGIN.

## ✓ CONCEPT CHECK

True or false? The number of decimal places in the product of 0.261 and 0.78 is 6. Explain.

**Example 4**     Multiply: $(-2.6)(0.8)$

*Solution:*     Recall that the product of a negative number and a positive number is a negative number.

$$(-2.6)(0.8) = -2.08$$

## Practice Problem 4

Multiply: $(5.4)(-1.3)$

## TEACHING TIP

Tell your students that multiplying by powers of 10 will be used later when we work with the metric system.

### **B** MULTIPLYING DECIMALS BY POWERS OF 10

There are some patterns that occur when we multiply a number by a power of 10, such as 10, 100, 1000, 10,000, and so on.

$$23.6951 \times 10 = 236.951$$     Move the decimal point *1 place* to the *right*.
1 zero

$$23.6951 \times 100 = 2369.51$$     Move the decimal point *2 places* to the *right*.
2 zeros

$$23.6951 \times 100,000 = 2,369,510.$$     Move the decimal point *5 places* to the *right* (insert a 0).
5 zeros

Notice that we move the decimal point the same number of places as there are zeros in the power of 10.

> **MULTIPLYING DECIMALS BY POWERS OF 10 SUCH AS 10, 100, 1000, 10,000, …**
>
> Move the decimal point to the *right* the same number of places as there are *zeros* in the power of 10.

## Practice Problems 5–7

Multiply.
5. $23.7 \times 10$
6. $203.004 \times 100$
7. $1.15 \times 1000$

**Examples**     Multiply.

**5.** $7.68 \times 10 = 76.8$     $7.68$

**6.** $23.702 \times 100 = 2370.2$     $23.702$

**7.** $(-76.3)(1000) = -76,300$     $-76.300$  Recall that the product of a negative number and a positive number is a negative number.

There are also powers of 10 that are less than 1. The decimals 0.1, 0.01, 0.001, 0.0001, and so on, are examples of powers of 10 less than 1. Notice the pattern when we multiply by these powers of 10.

$$569.2 \times 0.1 = 56.92$$     Move the decimal point *1 place* to the *left*.
1 decimal place

$$569.2 \times 0.01 = 5.692$$     Move the decimal point *2 places* to the *left*.
2 decimal places

$$569.2 \times 0.0001 = 0.05692$$     Move the decimal point *4 places* to the *left* (insert one 0).
4 decimal places

## Answers

**4.** $-7.02$   **5.** 237   **6.** 20,300.4   **7.** 1150
✓ **Concept Check:**  False; 3 decimal places + 2 decimal places is 5 decimal places in the product.

MULTIPLYING DECIMALS BY POWERS OF 10 SUCH AS 0.1, 0.01, 0.001, 0.0001, ...

Move the decimal point to the *left* the same number of places as there are *decimal places* in the power of 10.

## Examples

Multiply.

**8.** $42.1 \times 0.1 = 4.21$    42.1

**9.** $76{,}805 \times 0.01 = 768.05$    76,805.

**10.** $(-9.2)(-0.001) = 0.0092$    0009.2   Recall that the product of a negative number and a negative number is a positive number.

**TRY THE CONCEPT CHECK IN THE MARGIN.**

Many times we see large numbers written, for example, in the form 270.3 million rather than in the longer standard notation. The next example shows how to interpret these numbers.

**Example 11**   The population of the United States is 270.3 million. Write this number in standard notation. (*Source: The World Almanac*, 2000)

*Solution:*   270.3 million $= 270.3 \times 1$ million

$= 270.3 \times 1{,}000{,}000 = 270{,}300{,}000$

## **C** USING DECIMALS AS REPLACEMENT VALUES

Now let's practice working with variables.

**Example 12**   Evaluate $xy$ for $x = 2.3$ and $y = 0.44$.

*Solution:*   Recall that $xy$ means $x \cdot y$.

$xy = (2.3)(0.44)$      2.3
               $\times$ 0.44
                92
             920
$= 1.012$ ⟵    1.012

**Example 13**   Is 9 a solution of the equation $3.7y = 3.33$?

*Solution:*   Replace $y$ with 9 in the equation $3.7y = 3.33$ to see if a true equation results.

$3.7y = 3.33$

$3.7(9) = 3.33$    Replace $y$ with 9.

$33.3 = 3.33$    False.

Since $33.3 = 3.33$ is a false statement, 9 is **not** a solution of $3.7y = 3.33$.

---

**Practice Problems 8–10**

Multiply.

8. $7.62 \times 0.1$

9. $1.9 \times 0.01$

10. $7682 \times 0.001$

**✓ CONCEPT CHECK**

True or false? 372.511 multiplied by 100 is 3.72511.

**Practice Problem 11**

There are 2192 thousand farms in the United States. Write this number in standard notation. (*Source: The World Almanac*, 2000)

**Practice Problem 12**

Evaluate $7y$ for $y = 0.028$.

**Practice Problem 13**

Is 5.5 a solution of the equation $4x = 22$?

**Answers**

**8.** 0.762   **9.** 0.019   **10.** 7.682   **11.** 2,192,000
**12.** 0.196   **13.** yes
**✓ Concept Check:** False.

## △ D  Finding the Circumference of a Circle

Recall that the distance around a polygon is called its perimeter. The distance around a circle is given a special name called the **circumference**, and this distance depends on the radius or the diameter of the circle.

---

**Circumference of a Circle**

Circumference = $2 \cdot \pi \cdot$ **r**adius   or   Circumference = $\pi \cdot$ **d**iameter

$C = 2\pi r$        or        $C = \pi d$

---

The symbol $\pi$ is the Greek letter pi, pronounced "pie." It is a constant between 3 and 4. A decimal approximation for $\pi$ is 3.14. Also, a fraction approximation for $\pi$ is $\frac{22}{7}$.

### Practice Problem 14

Find the circumference of a circle whose radius is 11 meters. Then use the approximation 3.14 for $\pi$ to approximate this circumference.

**Example 14**   Find the circumference of a circle whose radius is 5 inches. Then use the approximation 3.14 for $\pi$ to approximate the circumference.

*Solution:*   Let $r = 5$ in the formula $C = 2\pi r$.

$$C = 2\pi r$$
$$= 2\pi \cdot 5$$
$$= 10\pi$$

Next, replace $\pi$ with the approximation 3.14.

$$C = 10\pi$$
(is approximately) ⟶ $\approx 10(3.14)$
$$= 31.4$$

The **exact** circumference or distance around the circle is $10\pi$ inches, which is **approximately** 31.4 inches. ▬▬▬

### E  Solving Problems by Multiplying Decimals

The solutions to many real-life problems are found by multiplying decimals. We continue using our four problem-solving steps to solve such problems.

TEACHING TIP   Classroom Activity

Ask students to work in groups to make up their own word problems involving the multiplication of decimals. Then have them give their problems to another group to solve.

**Answer**

**14.** $22\pi$ meters $\approx 69.08$ meters

**Example 15**    Finding Total Cost of Materials for a Job

A college student is hired to paint a billboard with paint costing $2.49 per quart. If the job requires 3 quarts of paint, what is the total cost of the paint?

*Solution:*    **1.** UNDERSTAND. Read and reread the problem. The phrase "total cost" might make us think addition, but since this is repeated addition, let's multiply.

**2.** TRANSLATE.

In words:

| Total cost | is | cost per quart of paint | times | number of quarts |
|------------|----|-----|------|------|
| ↓ | ↓ | ↓ | ↓ | ↓ |
| Translate:  Total cost | = | 2.49 | × | 3 |

**3.** SOLVE.

$$\begin{array}{r} 2.49 \\ \times \quad 3 \\ \hline 7.47 \end{array}$$

**4.** INTERPRET. *Check* your work. *State* your conclusion: The total cost of the paint is $7.47.    ■■■■■

**Practice Problem 15**

Elaine Rehmann is fertilizing her garden. She uses 5.6 ounces of fertilizer per square yard. The garden measures 60.5 square yards. How much fertilizer does she need?

Answer

**15.** 338.8 ounces

# Focus on Business and Career

## TIME SHEETS

Many jobs require employees to fill out a weekly time sheet. A time sheet allows employees to show how much time they worked on various projects and to find how much time they worked overall. For ease of calculation, it is standard to round times on a time sheet to the nearest quarter hour. The following is an example of a time sheet.

| | | | TIME SHEET<br>FOR THE WEEK OF: 5/9 | | | | | |
|---|---|---|---|---|---|---|---|---|
| Employee: Darla Williams | | | | | | | Employee Number: 630087 | |
| | Project Number | Monday | Tuesday | Wednesday | Thursday | Friday | Saturday/Sunday | TOTALS |
| 1 | 315 | 5.5 | 1 | | 3.25 | 1.5 | | |
| 2 | 629 | 2.5 | | 4 | 1.25 | 1.5 | 3 | |
| 3 | 471 | | 3.25 | 2 | | 1.75 | | |
| 4 | 511 | | 3.75 | 1.5 | 2.75 | | | |
| 5 | 512 | | | | 0.75 | 2.5 | 3.25 | |
| 6 | | | | | | | | |
| 7 | | | | | | | | |
| 8 | | | | | | | | |
| 9 | | | | | | | | |
| 10 | | | | | | | | |
| | TOTALS | | | | | | | |

## CRITICAL THINKING

1. Find the column totals for columns Monday through Saturday/Sunday on the time sheet. What do these column totals represent?

2. Find the row totals for rows 1 through 5 on the time sheet. What do these row totals represent?

3. Find the sum of the row totals and the sum of the column totals. Are these sums the same? If so, explain why. Otherwise, explain why having different row sums and column sums should prompt you to recheck your math.

4. Darla Williams is a full-time employee who is expected to work 40 hours per week. Any time over 40 hours is considered overtime. Did Darla work any overtime this week? If so, how much?

5. Darla is paid an hourly rate of $16.60 per hour. For overtime, she receives "time and a quarter," or 1.25 times her normal hourly pay rate. How much will Darla be paid for this week of work?

6. Describe how Darla's time was distributed among the five projects on which she worked during this week.

# Exercise Set 5.3

**A** *Multiply. See Examples 1 through 4.*

**1.**  0.2
× 0.6

**2.**  0.7
× 0.9

**3.**  1.2
× 0.5

**4.**  6.8
× 0.3

**5.** $(-2.3)(7.65)$

**6.** $(4.7)(-9.02)$

**7.** $(-6.89)(-5.7)$

**8.** $(-6.45)(-2.8)$

**B** *Multiply. See Examples 5 through 10.*

**9.** $6.5 \times 10$

**10.** $7.2 \times 10$

**11.** $6.5 \times 0.1$

**12.** $7.2 \times 0.1$

**13.** $(-7.093)(1000)$

**14.** $(-1.123)(1000)$

**15.** $(-9.83)(-0.01)$

**16.** $(-4.72)(-0.01)$

**A** **B** *Multiply. See Examples 1 through 10.*

**17.**  5.62
× 7.7

**18.**  8.03
× 5.5

**19.**  1.0047
× 8.2

**20.**  2.0005
× 5.5

**21.** $(147.9)(100)$

**22.** $(345.2)(10)$

**23.** $(937.62)(-0.01)$

**24.** $(-0.001)(562.01)$

**25.** $49.02 \times 0.023$

**26.** $30.09 \times 0.0032$

**27.** $(-0.023)(6.28)$

**28.** $(0.071)(-5.19)$

**1.** 0.12

**2.** 0.63

**3.** 0.6

**4.** 2.04

**5.** −17.595

**6.** −42.394

**7.** 39.273

**8.** 18.06

**9.** 65

**10.** 72

**11.** 0.65

**12.** 0.72

**13.** −7093

**14.** −1123

**15.** 0.0983

**16.** 0.0472

**17.** 43.274

**18.** 44.165

**19.** 8.23854

**20.** 11.00275

**21.** 14,790

**22.** 3452

**23.** −9.3762

**24.** −0.56201

**25.** 1.12746

**26.** 0.096288

**27.** −0.14444

**28.** −0.36849

**Name** _____

*Write each number in standard notation. See Example 11.*

**29.** The storage silos at the main Hershey chocolate factory in Hershey, Pennsylvania, can hold enough cocoa beans to make 5.5 billion Hershey's milk chocolate bars. (*Source:* Hershey Foods Corporation)

**30.** In 1998, the restaurant industry in the United States employed 9.5 million people. (*Source:* National Restaurant Association)

**31.** About 36.4 million American households own at least one dog. (*Source:* American Pet Products Manufacturers Association)

**32.** The most-visited national park in the United States is the Blue Ridge Parkway in Virginia. An estimated 17.17 million people visit the park each year. (*Source:* National Park Service)

**33.** Americans lose more than 1.6 million hours each day stuck in traffic. (*Source:* United States Department of Transportation—Federal Highway Administration)

**34.** The top advertiser in the United States in 1998 was General Motors, who spent $2.12 billion on advertising. (*Source:* Competitive Media Reporting and Publishers Information Bureau)

**C** *Evaluate each expression for $x = 3$, $y = −0.2$, and $z = 5.7$. See Example 12.*

**35.** $xy$

**36.** $yz$

**37.** $xz$

**38.** $−5y$

**39.** $20z$

**40.** $−0.45x$

*Determine whether the given value is a solution of each given equation. See Example 13.*

**41.** Is 14.2 a solution of $0.6x = 4.92$?

**42.** Is 1414 a solution of $100z = 14.14$?

**43.** Is 0.08 a solution of $−3x = −2.4$?

**44.** Is 17.8 a solution of $2x = 3.56$?

**45.** Is −4 a solution of $3.5y = −14$?

**46.** Is −3.6 a solution of $0.7x = −2.52$?

**Name** _____

**D** *Find the circumference of each circle. Then use the approximation 3.14 for $\pi$ and approximate each circumference. See Example 14.*

△ **47.**

4 meters

△ **48.**

8 feet

▭ **49.**
△

10 centimeters

△ **50.**

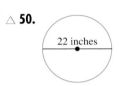

22 inches

△ **51.**

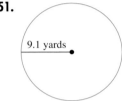

9.1 yards

△ **52.**

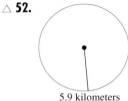

5.9 kilometers

△ **53.** In 1893, the first ride called a Ferris wheel was constructed by Washington Gale Ferris. Its diameter was 250 feet. Find its circumference. Give an exact answer and an approximation using 3.14 for $\pi$. (*Source: The Handy Science Answer Book*, Visible Ink Press, 1994)

△ **54.** The radius of Earth is approximately 3950 miles. Find the distance around Earth at the equator. Give an exact answer and an approximation using 3.14 for $\pi$. (*Hint:* Find the circumference of a circle with radius 3950 miles.)

3950 miles

62.8 m
**55. a.** 125.6 m

**b.** yes

50.24 in. and
**56. a.** 100.48 in.

**b.** yes

**57.** 24.8 g

**58.** 0.3 g

**59.** $2700

**60.** $53,500

**61.** 64.9605 in.

**62.** 84.6455 in.

**384**

**Name** _____

△ **55. a.** Approximate the circumference of each circle.

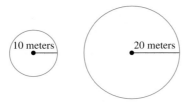

**b.** If the radius of a circle is doubled, is its corresponding circumference doubled?

△ **56. a.** Approximate the circumference of each circle.

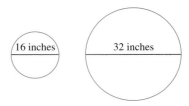

**b.** If the diameter of a circle is doubled, is its corresponding circumference doubled?

**E** *Solve. See Example 15.*

**57.** A 1-ounce serving of cream cheese contains 6.2 grams of saturated fat. How much saturated fat is in 4 ounces of cream cheese? (*Source: Home and Garden Bulletin No. 72*; U.S. Department of Agriculture)

**58.** A 3.5-ounce serving of lobster meat contains 0.1 gram of saturated fat. How much saturated fat is in 3 servings of lobster meat? (*Source:* The National Institute of Health)

**59.** In 1998, American farmers received an average of $2.70 per bushel of wheat. How much did a farmer receive for selling 1000 bushels of wheat? (*Source:* National Agricultural Statistics Service)

**60.** In 1998, American farmers received an average of $5.35 per bushel of soybeans. How much did a farmer receive for selling 10,000 bushels of soybeans? (*Source:* National Agricultural Statistics Service)

**61.** A meter is a unit of length in the metric system that is approximately equal to 39.37 inches. Sophia Wagner is 1.65 meters tall. Find her approximate height in inches.

**62.** The doorway to a room is 2.15 meters tall. Approximate this height in inches. (*Hint:* See Exercise 61.)

**63.** Jose Severos, an electrician for Central Power and Light, worked 40 hours last week. Calculate his pay before taxes for last week if his hourly wage is $13.88.

**64.** Maribel Chin, an assembly line worker, worked 20 hours last week. Her hourly rate is $8.52 per hour. Calculate her pay before taxes.

*The table shows currency exchange rates for various countries on September 10, 1999. To find the amount of foreign currency equivalent to an amount of U.S. money, multiply the U.S. dollar amount by the exchange rate listed in the table. Use this table to answer Exercises 65–68.*

**65.** How many Japanese yen are equivalent to 675 U.S. dollars?

| FOREIGN CURRENCY EXCHANGE RATES | |
|---|---|
| Country | Exchange Rate |
| Australian dollar | 1.5337 |
| British pounds | 0.6123 |
| Canadian dollars | 1.4800 |
| French francs | 6.2217 |
| German marks | 1.8551 |
| Japanese yen | 108.0000 |

(*Source:* New York Federal Reserve Bank)

**66.** Suppose you wish to exchange 500 U.S. dollars into French francs. How much money, in French francs, would you receive?

**67.** The Goldthwaite family is taking a vacation to Canada. How much Canadian currency can they "buy" with 350 U.S. dollars?

**68.** A British tourist bought a $30 sweatshirt in New York City. What did he pay for the sweatshirt in British pounds?

## REVIEW AND PREVIEW

*Divide. See Sections 1.6 and 4.3.*

**69.** $130 \div 5$

**70.** $495 \div 27$

**71.** $2016 \div 56$

**72.** $1863 \div 69$

**73.** $2920 \div 365$

**74.** $2916 \div 6$

**75.** $\dfrac{24}{7} \div \dfrac{8}{21}$

**76.** $\dfrac{162}{25} \div \dfrac{9}{75}$

Answers (right margin):

**63.** $555.20

**64.** $170.40

**65.** 72,900 yen

**66.** 3110.85 francs

**67.** 518 Canadian dollars

**68.** 18.369 pounds

**69.** 26

**70.** 18 R 9

**71.** 36

**72.** 27

**73.** 8

**74.** 486

**75.** 9

**76.** 54

**77.** 3,831,600 mi

**78.** 92,628,000 mi

**79.** answers may vary

**80.** answers may vary

**81.** answers may vary

**82.** answers may vary

 **COMBINING CONCEPTS**

 **77.** Find how far radio waves travel in 20.6 seconds. (Radio waves travel at a speed of 1.86 × 100,000 miles per second.)

**78.** If it takes radio waves approximately 8.3 minutes to travel from the sun to Earth, find approximately how far it is from the sun to Earth. (*Hint:* See Exercise 77.)

**79.** In your own words, explain how to find the number of decimal places in a product of decimal numbers.

**80.** In your own words, explain how to multiply by a power of 10.

**Internet Excursions**

Go to http://www.prenhall.com/martin-gay

This World Wide Web address will provide you with access to a listing of current foreign currency exchange rates for the U.S. dollar, or a related site. The exchange rates in the "To United States Dollar" column are used to convert U.S. dollars to other currencies. The exchange rates in the "In United States Dollar" column are used to convert other currencies into U.S. dollars. Visit this site and use today's exchange rates to answer the following questions.

**81.** Convert $650 U.S. dollars to each given currancy.
  **a.** Chinese renminbi
  **b.** Italian lira
  **c.** Swiss francs

**82.** Determine the cost in U.S. dollars for each item.
  **a.** A silk scarf costing 1400 Japanese yen
  **b.** A paperback book costing 9.95 Canadian dollars
  **c.** A sweater costing 50 British pounds

# 5.4 DIVIDING DECIMALS

## A DIVIDING DECIMALS

Division of decimal numbers is similar to division of whole numbers. The only difference is that we place a decimal point in the quotient. If the divisor is a whole number, divide as for whole numbers; then place the decimal point in the quotient directly above the decimal point in the dividend. Recall that division can be checked by multiplication.

$$
\begin{array}{r}
0.26 \leftarrow \text{quotient} \\
\text{divisor} \rightarrow 32\overline{)8.32} \leftarrow \text{dividend} \\
\underline{-6\,4} \\
192 \\
\underline{-192} \\
0
\end{array}
$$

Check:

$$
\begin{array}{r}
0.26 \quad \text{quotient} \\
\times \quad 32 \quad \text{divisor} \\
\hline
52 \\
7\,80 \\
\hline
8.32 \quad \text{dividend}
\end{array}
$$

If the divisor is not a whole number, we need to move the decimal point to the right until the divisor is a whole number before we divide.

$$
1.5\overline{)64.85}
$$

divisor ⟶　　　⟵ dividend

To understand how this works, let's rewrite

$$1.5\overline{)64.85} \quad \text{as} \quad \frac{64.85}{1.5}$$

and then multiply the numerator and the denominator by 10.

$$\frac{64.85}{1.5} = \frac{64.85 \times 10}{1.5 \times 10} = \frac{648.5}{15}$$

which can be written as $15\overline{)648.5}$. Notice that

$$1.5\overline{)64.85} \quad \text{is equivalent to} \quad 15.\overline{)648.5}$$

The decimal points in the dividend and the divisor were both moved one place to the right, and the divisor is now a whole number. This procedure is summarized next.

---

**DIVIDING BY A DECIMAL**

**Step 1.** Move the decimal point in the divisor to the right until the divisor is a whole number.

**Step 2.** Move the decimal point in the dividend to the right the *same number of places* as the decimal point was moved in Step 1.

**Step 3.** Divide. Place the decimal point in the quotient directly over the moved decimal point in the dividend.

---

**Objectives**

**A** Divide decimals.

**B** Divide decimals by powers of 10.

**C** Evaluate expressions and check solutions with decimal replacement values.

**D** Solve problems by dividing decimals.

Study　　SSM　CD-ROM　Video
Guide　　　　　　　　　5.4

## Practice Problem 1

Divide: $5.6\overline{)166.88}$

## Practice Problem 2

Divide: $-2.808 \div (-104)$

## Practice Problem 3

Divide: $23.4 \div 0.57$. Round the quotient to the nearest hundredth.

TEACHING TIP

Remind students not to forget to move the decimal point in the dividend the same number of places as it was moved in the divisor. Then place the decimal point in the quotient directly above the moved decimal point in the dividend.

**Answers**

**1.** 29.8   **2.** 0.027   **3.** 41.05

**Example 1**   Divide: $2.3\overline{)10.764}$

*Solution:*   Move the decimal points in the divisor and the dividend one place to the right so that the divisor is a whole number.

$$2.3\overline{)10.764} \quad \text{becomes} \quad 23.\overline{)107.64}$$

$$
\begin{array}{r}
4.68 \\
23.\overline{)107.64} \\
-92\phantom{.64} \\
\hline
15\,6\phantom{4} \\
-13\,8\phantom{4} \\
\hline
1\,84 \\
-1\,84 \\
\hline
0
\end{array}
$$

To check, see that $4.68 \times 2.3 = 10.764$.

**Example 2**   Divide: $-5.98 \div 115$

*Solution:*   Recall that a negative number divided by a positive number gives a negative quotient. The divisor is a whole number, so the decimal point in the divisor is not moved.

$$
\begin{array}{r}
0.052 \\
115\overline{)5.980} \\
-5\,75\phantom{0} \\
\hline
230 \\
-230 \\
\hline
0
\end{array}
$$
$\leftarrow$ Insert one 0.

Thus $-5.98 \div 115 = -0.052$.

**Example 3**   Divide: $17.5 \div 0.48$. Round the quotient to the nearest hundredth.

*Solution:*   Move the decimal points in the divisor and the dividend 2 places.

hundredths place

$$36.458 \approx 36.46$$
$$
\begin{array}{r}
48.\overline{)1750.000} \\
-144\phantom{0.000} \\
\hline
310\phantom{0.00} \\
-288\phantom{0.00} \\
\hline
220\phantom{0.0} \\
-192\phantom{0.0} \\
\hline
280\phantom{0} \\
-240\phantom{0} \\
\hline
400 \\
-384 \\
\hline
16
\end{array}
$$

approximately

If rounding to the nearest hundredth, carry the division process out to one more decimal place, the thousandths place.

TRY THE CONCEPT CHECK IN THE MARGIN.

**B  DIVIDING DECIMALS BY POWERS OF 10**

There are patterns that occur when we divide decimals by powers of 10 such as 10, 100, 1000, and so on.

$$\frac{569.2}{10} = 56.92$$   Move the decimal point *1 place* to the *left*.

└── 1 zero

$$\frac{569.2}{10,000} = 0.05692$$   Move the decimal point *4 places* to the *left*.

└── 4 zeros

This pattern suggests the following rule.

---

**DIVIDING DECIMALS BY POWERS OF 10 SUCH AS 10, 100, 1000, ...**

Move the decimal point of the dividend to the *left* the same number of places as there are *zeros* in the power of 10.

---

Notice that this is the same pattern as *multiplying* by powers of 10 such as 0.1, 0,01, or 0.001. This is because dividing by a power of 10 such as 100 is the same as multiplying by its reciprocal $\frac{1}{100}$, or 0.01. For example,

$$\frac{569.2}{10} = 569.2 \times \frac{1}{10} = 569.2 \times 0.1 = 56.92$$

**Examples**   Divide.

**4.** $\dfrac{786}{10,000} = 0.0786$   Move the decimal point *4 places* to the *left*.

└── 4 zeros

**5.** $\dfrac{0.12}{10} = 0.012$   Move the decimal point *1 place* to the *left*.

└── 1 zero

TRY THE CONCEPT CHECK IN THE MARGIN.

**C  USING DECIMALS AS REPLACEMENT VALUES**

**Example 6**   Evaluate $x \div y$ for $x = 2.5$ and $y = 0.05$.

*Solution:*   Replace $x$ with 2.5 and $y$ with 0.05.

$$x \div y = 2.5 \div 0.05$$
$$= 50$$

$0.05\overline{)2.5}$   becomes   $5\overline{)250}$ ... 50

✓ **CONCEPT CHECK**

If a quotient is to be rounded to the nearest thousandth, to what place should the division be carried out?

**Practice Problems 4–5**

Divide.

**4.** $\dfrac{28}{1000}$   **5.** $\dfrac{8.56}{100}$

✓ **CONCEPT CHECK**

Describe how to check the answer in Example 4.

**Practice Problem 6**

Evaluate $x \div y$ for $x = 0.035$ and $y = 0.02$.

**Answers**

**4.** 0.028   **5.** 0.0856   **6.** 1.75
✓ **Concept Check:** The ten-thousandths place.
✓ **Concept Check:** Answers may vary.

## Practice Problem 7

Is 39 a solution of the equation

$\dfrac{x}{100} = 3.9$?

**Example 7** Is 720 a solution of the equation $\dfrac{y}{100} = 7.2$?

*Solution:* Replace $y$ with 720 to see if a true statement results.

$\dfrac{y}{100} = 7.2$    Original equation

$\dfrac{720}{100} = 7.2$    Replace $y$ with 720.

$7.2 = 7.2$    True.

Since $7.2 = 7.2$ is a true statement, 720 is a solution of the equation. ▬▬▬

### D SOLVING PROBLEMS BY DIVIDING DECIMALS

Many real-life problems involve dividing decimals.

**Example 8** Calculating Materials Needed for a Job

A gallon of paint covers a 250-square-foot area. If Betty Adkins wishes to paint a wall that measures 1450 square feet, how many gallons of paint does she need? If she can buy gallon containers of paint only, how many gallon containers does she need?

## Practice Problem 8

A bag of fertilizer covers 1250 square feet of lawn. Tim Parker's lawn measures 14,800 square feet. How many bags of fertilizer does he need? If he can buy whole bags of fertilizer only, how many whole bags does he need?

*Solution:*

**1. UNDERSTAND.** Read and reread the problem. To find the number of gallons, divide 1450 by 250.

**2. TRANSLATE.**

| In words: | Number of gallons | is | square feet | divided by | square feet per gallon |
|---|---|---|---|---|---|
| | ↓ | ↓ | ↓ | ↓ | ↓ |
| Translate: | Number of gallons | = | 1450 | ÷ | 250 |

**3. SOLVE.**

$$\begin{array}{r} 5.8 \\ 250\overline{)1450.0} \\ -1250 \phantom{.0}\\ \hline 200\,0 \\ -200\,0 \\ \hline 0 \end{array}$$

**4. INTERPRET.** *Check* your work. *State* your conclusion: Betty needs 5.8 gallons of paint. If she can buy gallon containers of paint only, she needs 6 gallon containers of paint to complete the job. ▬▬▬

TEACHING TIP   Classroom Activity

Have each person bring two circular objects with different diameters. Provide each group with a centimeter and inch measuring tape and a worksheet to record their results. Have the groups measure and record the diameter and circumference of each object in centimeters and inches. Then have them use a calculator to find the circumference/diameter for each object and ask them to make a conclusion from their data. After they have reached a conclusion, have them share their insights with the class. Be sure they see that no matter how large or small a circle is, the circumference ÷ diameter = $\pi$, which is about 3.14.

**Answers**

**7.** no   **8.** 11.84 bags; 12 bags

## EXERCISE SET 5.4

**A** *Divide. See Examples 1 and 2.*

**1.** $0.47 \div 5$          **2.** $11.8 \div 2$          **3.** $-18 \div 0.06$

**4.** $20 \div (-0.04)$     **5.** $4.756 \div 0.82$     **6.** $3.312 \div 0.92$

**7.** $-36.3 \div 5.5$      **8.** $-21.78 \div 2.2$

*Divide, and round each quotient as indicated. See Example 3.*

**9.** Divide 429.34 by 2.4 and round the quotient to the nearest hundred.

**10.** Divide 54.8 by 2.6 and round the quotient to the nearest ten.

**11.** Divide 0.549 by 0.023 and round the quotient to the nearest hundredth.

**12.** Divide 0.0453 by 0.98 and round the quotient to the nearest thousandth.

**13.** Divide 45.23 by 0.4 and round the quotient to the nearest ten.

**14.** Divide 983.32 by 0.061 and round the quotient to the nearest thousand.

**B** *Divide. See Examples 4 and 5.*

**15.** $54.982 \div 100$     **16.** $342.54 \div 100$     **17.** $12.9 \div (-1000)$

**18.** $13.49 \div (-10,000)$   **19.** $87 \div 10$     **20.** $0.27 \div 10$

**391**

**21.** 0.413

**22.** 0.045

**23.** −7

**24.** 4

**25.** 4.8

**26.** −6.4

**27.** 2100

**28.** 800

**29.** 30

**30.** 50

**31.** 7000

**32.** 50,000

**33.** −0.69

**34.** −2.2

**35.** 0.024

**36.** 0.08679

**37.** 65

**38.** −680

**39.** 0.0003

**40.** 4000

**41.** 120,000

**42.** 0.49

**Name** _____

**A** **B** *Divide. See Examples 1 through 5.*

**21.** $1.239 \div 3$      **22.** $0.54 \div 12$      **23.** Divide $-4.2$ by $0.6$

**24.** Divide $3.6$ by $0.9$      **25.** $1.296 \div 0.27$      **26.** $-2.176 \div 0.34$

**27.** Divide $42$ by $0.02$      **28.** Divide $24$ by $0.03$      **29.** Divide $-18$ by $-0.6$

**30.** Divide $20$ by $0.4$      **31.** Divide $35$ by $0.005$      **32.** Divide $-35$ by $-0.0007$

**33.** $-1.104 \div 1.6$      **34.** $-2.156 \div 0.98$      **35.** $-2.4 \div (-100)$

**36.** $-86.79 \div (-1000)$      **37.** $\dfrac{4.615}{0.071}$      **38.** $\dfrac{23.8}{-0.035}$

**39.** Divide $0.00263$ by $8.9$ and round the quotient to the nearest ten-thousandth.

**40.** Divide $200$ by $0.054$ and round the quotient to the nearest thousand.

**41.** Divide $500$ by $0.0043$ and round the quotient to the nearest ten-thousand.

**42.** Divide $4.56$ by $9.34$ and round the quotient to the nearest hundredth.

**C** *Evaluate each expression for x = 5.65, y = 0.8, and z = 4.52. See Example 6.*

**43.** $z \div y$                **44.** $z \div x$                **45.** $x \div y$

**46.** $z \div 2$                **47.** $y \div 2$                **48.** $x \div 5$

*Determine whether the given values are solutions of the given equations. See Example 7.*

**49.** $\dfrac{x}{4} = 3.04$; $x = 12.16$       **50.** $\dfrac{y}{8} = 0.89$; $y = 7.12$       **51.** $\dfrac{x}{4.3} = 2$; $x = 0.86$

**52.** $\dfrac{z}{3.7} = 5$; $z = 185$       **53.** $\dfrac{z}{10} = 0.8$; $z = 8$       **54.** $\dfrac{x}{100} = 0.23$; $x = 23$

**D** *Solve. See Example 8.*

**55.** Dorren Schmidt pays $73.86 per month to pay back a loan of $1772.64. In how many months will the loan be paid off?

**56.** Josef Jones is painting the walls of a room. The walls have a total area of 546 square feet. If a quart of paint covers 52 square feet, how many quarts of paint does he need to buy?

**57.** The leading money winner in men's professional golf in 1998 was David Duvall. He earned $2,591,031. Suppose he had earned this working 40 hours per week for a year. Determine his hourly wage to the nearest cent. (*Source: The World Almanac*, 2000)

**58.** Juanita Gomez bought unleaded gasoline for her car at $1.169 per gallon. She wanted to keep a record of how many gallons her car is using but forgot to write down how many gallons she purchased. She wrote a check for $27.71 to pay for it. How many gallons, to the nearest tenth of a gallon, did she buy?

**43.** 5.65

**44.** 0.8

**45.** 7.0625

**46.** 2.26

**47.** 0.4

**48.** 1.13

**49.** yes

**50.** yes

**51.** no

**52.** no

**53.** yes

**54.** yes

**55.** 24 mo

**56.** 11 qt

**57.** $1245.69

**58.** 23.7 gal

**59.** 202.1 lb

**60.** 493.3 pages

**61.** 5.1 m

**62.** 19.7 in.

**63.** $0.596 per lb

**64.** $0.723 per lb

**65.** 128.6 mph

**66.** 7.9 m/sec

**59.** A pound of fertilizer covers 39 square feet of lawn. Vivian Bulgakov's lawn measures 7883.5 square feet. How much fertilizer, to the nearest tenth of a pound, does she need to buy?

**60.** A page of a book contains about 1.5 kilobytes of information. If a computer disk can hold 740 kilobytes of information, how many pages of a book can be stored on one computer disk? Round to the nearest tenth of a page.

**61.** There are approximately 39.37 inches in 1 meter. How many meters, to the nearest tenth of a meter, are there in 200 inches?

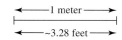

**62.** There are approximately 2.54 centimeters in 1 inch. How many inches are there in 50 centimeters? Round to the nearest tenth.

**63.** In 1998, American farmers received an average of $59.60 per 100 pounds of beef cattle. What was the average price per pound for beef cattle? (*Source:* National Agricultural Statistics Service)

**64.** In 1998, American farmers received an average of $72.30 per 100 pounds of lamb. What was the average price per pound for lambs? (*Source:* National Agricultural Statistics Service)

**65.** During the 24 Hours of Le Mans endurance auto race in 1999, the winning team of Yannick Dalmas, Joachim Winkelhock, and Pierluigi Martini drove a total of 3087.12 miles in 24 hours. What was their average speed in miles per hour? Round to the nearest tenth. (*Source: The World Almanac, 2000*)

**66.** In 1997, Danish runner Wilson Kipketer set a new world record for the 800-meter event. His time for the event was 101.11 seconds. What was his average speed in meters per second? Round to the nearest tenth. (*Source:* International Amateur Athletic Federation)

**67.** The highest WNBA scorer for the 1999 season was Cynthia Cooper of the Houston Comets. She scored a total of 686 points during 31 basketball games. Find her average number of points scored per game. Round to the nearest tenth. (*Source:* Women's National Basketball Association)

**68.** During the 1999 Major League Soccer season, DC United was the top scoring team with a total of 65 goals throughout the season. DC United played 32 games. What was the average number of goals scored by DC United per game? Round to the nearest hundredth. (*Source:* Major League Soccer)

*To convert wind speeds in miles per hour to knots, divide by 1.15. Use this information and the Saffir-Simpson Hurricane Intensity chart in the chapter opener to answer Exercises 69–70. Round to the nearest tenth.*

**69.** The chart gives wind speeds in miles per hour. What is the range of wind speeds for a Category 1 hurricane in knots?

**70.** What is the range of wind speeds for a Category 4 hurricane in knots?

## REVIEW AND PREVIEW

*Round each decimal to the given place value. See Section 5.1.*

**71.** 345.219, nearest hundredth

**72.** 902.155, nearest hundredth

**73.** 1000.994, nearest tenth

**74.** 234.1029, nearest thousandth

*Use the order of operations to simplify each expression. See Section 1.8.*

**75.** $2 + 3 \cdot 6$

**76.** $9 \cdot 2 + 8 \cdot 3$

**77.** $20 - 10 \div 5$

**78.** $(20 - 10) \div 5$

---

**67.** 22.1 points/game

**68.** 2.03 goals/game

**69.** 65.2–82.6 knots

**70.** 113.9–134.8 knots

**71.** 345.22

**72.** 902.16

**73.** 1001.0

**74.** 234.103

**75.** 20

**76.** 42

**77.** 18

**78.** 2

**Name** _____

◆ **COMBINING CONCEPTS**

*Recall from Section 1.6 that the average of a list of numbers is their total divided by how many numbers there are in the list. Use this to find the average of the test scores listed. If necessary, round to the nearest tenth.*

**79.** 86, 78, 91, 85

**80.** 56, 75, 80

△ **81.** The area of the rectangle is 38.7 square feet. If its width is 4.5 feet, find its length.

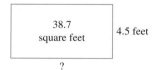

△ **82.** The perimeter of the square is 180.8 centimeters. Find the length of a side.

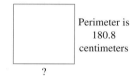

✎ **83.** Explain how to decide where to place the decimal point in a quotient of decimals.

✎ **84.** Describe how to round a decimal number to the nearest hundredth.

**Name** _____ **Section** _____ **Date** _____

# INTEGRATED REVIEW—OPERATIONS ON DECIMALS

*Perform each indicated operation.*

**1.** $1.6 + 0.97$

**2.** $3.2 + 0.85$

**3.** $9.8 - 0.9$

**4.** $10.2 - 6.7$

**5.**
$$\begin{array}{r} 0.8 \\ \times\ 0.2 \\ \hline \end{array}$$

**6.**
$$\begin{array}{r} 0.6 \\ \times\ 0.4 \\ \hline \end{array}$$

**7.** $8\overline{)2.16}$

**8.** $6\overline{)3.12}$

**9.** $(9.6)(-0.5)$

**10.** $(-8.7)(-0.7)$

**11.**
$$\begin{array}{r} 123.6 \\ -\ 48.04 \\ \hline \end{array}$$

**12.**
$$\begin{array}{r} 325.2 \\ -\ 36.08 \\ \hline \end{array}$$

**13.** $-25 + 0.026$

**14.** $0.125 + (-44)$

**15.** $29.24 \div (-3.4)$

**16.** $-10.26 \div (-1.9)$

**17.** $-2.8 \times 100$

**18.** $1.6 \times 1000$

**19.**
$$\begin{array}{r} 96.21 \\ 7.028 \\ +121.7 \\ \hline \end{array}$$

**20.**
$$\begin{array}{r} 0.268 \\ 1.93 \\ +142.881 \\ \hline \end{array}$$

**21.** $-25.76 \div -46$

**22.** $-27.09 \div 43$

**23.**
$$\begin{array}{r} 12.004 \\ \times\ \ \ \ 2.3 \\ \hline \end{array}$$

**24.**
$$\begin{array}{r} 28.006 \\ \times\ \ \ \ 5.2 \\ \hline \end{array}$$

**25.** Subtract 4.6 from 10

**26.** Subtract 18 from 0.26

**1.** 2.57

**2.** 4.05

**3.** 8.9

**4.** 3.5

**5.** 0.16

**6.** 0.24

**7.** 0.27

**8.** 0.52

**9.** −4.8

**10.** 6.09

**11.** 75.56

**12.** 289.12

**13.** −24.974

**14.** −43.875

**15.** −8.6

**16.** 5.4

**17.** −280

**18.** 1600

**19.** 224.938

**20.** 145.079

**21.** 0.56

**22.** −0.63

**23.** 27.6092

**24.** 145.6312

**25.** 5.4

**26.** −17.74

**27.** $-414.44$

**28.** $-1295.03$

**29.** $-34$

**30.** $-28$

**31.** $116.81$

**32.** $18.79$

**33.** yes

**34.** 7640.25 points

**35.** 26.3 points

**27.** $-268.19 - 146.25$

**28.** $-860.18 - 434.85$

**29.** $\dfrac{2.958}{-0.087}$

**30.** $\dfrac{-1.708}{0.061}$

**31.** $160 - 43.19$

**32.** $120 - 101.21$

**33.** According to the federal Nutrition Labeling and Education Act of 1990, a food manufacturer may label a food product "Fat Free" if it contains less than 0.5 grams of fat per serving. A new type of cookie contains 0.38 grams of fat per cookie. Can the packaging for this cookie be labeled "Fat Free"? Explain.

**34.** One of the greatest one-day point gains on the Dow Jones Industrial Average (DJIA) occurred on September 8, 1998. On that day, the DJIA changed by +380.53 points and its value at the close of the stock market was 8020.78 points. What was the value of the DJIA when the stock market opened on September 8, 1998? (*Source:* Dow Jones & Co.)

**35.** The second-highest NBA scorer for the 1998–1999 regular season was Shaquille O'Neal of the Los Angeles Lakers. He scored a total of 1289 points in 49 games. Find his average number of points per game to the nearest tenth of a point. (*Source:* National Basketball Association)

# 5.5 ESTIMATING AND ORDER OF OPERATIONS

## A ESTIMATING OPERATIONS ON DECIMALS

Estimating sums, differences, products, and quotients of decimal numbers is an important skill whether you use a calculator or perform decimal operations by hand. When you can estimate results as well as calculate them, you can judge whether the calculations are reasonable. If they aren't, you know you've made an error somewhere along the way.

**Example 1**    Subtract: $78.62 − $16.85. Then estimate the difference to see whether the proposed result is reasonable by rounding each decimal to the nearest dollar and then subtracting.

*Solution:*

| Given | | Estimate |
|---|---|---|
| $78.62 | rounds to | $79 |
| −$16.85 | rounds to | −$17 |
| $61.77 | | $62 |

The estimated difference is $62, so $61.77 is reasonable.

**TRY THE CONCEPT CHECK IN THE MARGIN.**

**Example 2**    Multiply: 28.06 × 1.95. Then estimate the product to see whether the proposed result is reasonable.

*Solution:*

| Given | Estimate 1 | Estimate 2 |
|---|---|---|
| 28.06 | 28 | 30 |
| × 1.95 | × 2 | × 2 |
| 14030 | 56 | 60 |
| 252540 | | |
| 280600 | | |
| 54.7170 | | |

The answer 54.7170 is reasonable.

As shown in Example 2, estimated results will vary depending on what estimates are used. Notice that estimating results is a good way to see whether the decimal point has been correctly placed.

**Example 3**    Divide: 272.356 ÷ 28.4. Then estimate the quotient to see whether the proposed result is reasonable.

*Solution:*

Given

$$
\begin{array}{r}
9.59 \\
284.\overline{)2723.56} \\
-2556 \\
\hline
1675 \\
-1420 \\
\hline
2556 \\
-2556 \\
\hline
0
\end{array}
$$

Estimate

$$
30\overline{)270} \quad \text{or} \quad 300\overline{)2700}
$$

with 9 above each.

The estimate is 9, so 9.59 is reasonable.

### Objectives

**A** Estimate operations on decimals.
**B** Simplify expressions containing decimals using the order of operations.
**C** Evaluate expressions and check solutions given decimal replacement values.

Study Guide    SSM    CD-ROM    Video 5.5

### Practice Problem 1

Subtract: $65.34 − $14.68. Then estimate the difference to see whether the proposed result is reasonable.

### ✓ CONCEPT CHECK

Should you estimate the sum 21.98 + 42.36 as 30 + 50 = 80? Why or why not?

### Practice Problem 2

Multiply: 30.26 × 2.98. Then estimate the product to see whether the proposed solution is reasonable.

### TEACHING TIP

Point out that when estimating, estimates could be larger or smaller than the actual answer. For instance, for the second estimate in Example 2, both numbers were rounded up so the estimate will be larger than the actual answer.

### Practice Problem 3

Divide: 713.7 ÷ 91.5. Then estimate the quotient to see whether the proposed answer is reasonable.

**Answers**

**1.** $50.66    **2.** 90.1748    **3.** 7.8
✓ **Concept Check:** No; each number is rounded incorrectly. The estimate is too high.

## Practice Problem 4

Using the figure below, estimate how much farther it is between Dodge City and Pratt than it is between Garden City and Dodge City by rounding each given distance to the nearest mile.

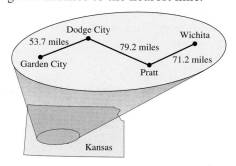

## Example 4    Estimating Distance

Use the figure in the margin to estimate the distance in miles between Garden City, Kansas, and Wichita, Kansas, by rounding each given distance to the nearest ten.

*Solution:*

| Calculated Distance | | Estimate |
|---|---|---|
| 53.7 | rounds to | 50 |
| 79.2 | rounds to | 80 |
| +71.2 | rounds to | +70 |
| | | 200 |

The distance between Garden City and Wichita is approximately 200 miles. (The calculated distance is 204.1 miles.)

## B    SIMPLIFYING EXPRESSIONS CONTAINING DECIMALS

In the remaining examples, we will review the order of operations by simplifying expressions that contain decimals.

> ### ORDER OF OPERATIONS
>
> 1. Do all operations within grouping symbols such as parentheses or brackets.
> 2. Evaluate any expressions with exponents.
> 3. Multiply or divide in order from left to right.
> 4. Add or subtract in order from left to right.

## Practice Problem 5

Simplify: $-8.6(3.2 - 1.8)$

## Example 5    Simplify: $-0.5(8.6 - 1.2)$

*Solution:*    According to the order of operations, simplify inside the parentheses first.

$$-0.5(8.6 - 1.2) = -0.5(7.4) \quad \text{Subtract.}$$
$$= -3.7 \quad \text{Multiply. The product of a negative number and a positive number is a negative number.}$$

## Practice Problem 6

Simplify: $(0.7)^2$

## Example 6    Simplify: $(-1.3)^2$

*Solution:*    Recall the meaning of an exponent.

$$(-1.3)^2 = (-1.3)(-1.3) \quad \text{Use the definition of an exponent.}$$
$$= 1.69 \quad \text{Multiply. The product of two negative numbers is a positive number.}$$

## Practice Problem 7

Simplify: $\dfrac{8.78 - 2.8}{20}$

## Example 7    Simplify: $\dfrac{0.7 + 1.84}{0.4}$

*Solution:*    First simplify the numerator of the fraction. Then divide.

$$\frac{0.7 + 1.84}{0.4} = \frac{2.54}{0.4} \quad \text{Simplify the numerator.}$$
$$= 6.35 \quad \text{Divide.}$$

**Answers**

**4.** 25 miles    **5.** −12.04    **6.** 0.49    **7.** 0.299

## C USING DECIMALS AS REPLACEMENT VALUES

**Example 8**    Evaluate $-2x + 5$ for $x = 3.8$.

*Solution:*    Replace $x$ with 3.8 in the expression $-2x + 5$ and simplify.

$$-2x + 5 = -2(3.8) + 5 \quad \text{Replace } x \text{ with 3.8.}$$
$$= -7.6 + 5 \quad \text{Multiply.}$$
$$= -2.6 \quad \text{Add.}$$

**Example 9**    Determine whether $-3.3$ is a solution of $2x - 1.2 = -7.8$.

*Solution:*    Replace $x$ with $-3.3$ in the equation $2x - 1.2 = -7.8$ to see if a true statement results.

$$2x - 1.2 = -7.8$$
$$2(-3.3) - 1.2 = -7.8 \quad \text{Replace } x \text{ with } -3.3.$$
$$-6.6 - 1.2 = -7.8 \quad \text{Multiply.}$$
$$-7.8 = -7.8 \quad \text{True.}$$

Since $-7.8 = -7.8$ is a true statement, $-3.3$ is a solution of $2x - 1.2 = -7.8$.

**Practice Problem 8**

Evaluate $1.7y - 2$ for $y = 2.3$.

**Practice Problem 9**

Is $-2.1$ a solution of $3x + 7.5 = 1.2$?

Answers

**8.** 1.91    **9.** yes

# Focus on the Real World

## McDonald's Menu

Today's worldwide chain of McDonald's fast-food restaurants started as a single drive-in restaurant run by the McDonald brothers, Dick and Mac, in San Bernardino, California. In the late 1940s, they decided to revamp their business, eliminating car hops in favor of counter-only service and reducing their 25-item menu to just 9 items. They focused on quick service, large volume, and low prices. The newly revamped business was hugely popular and attracted the attention of a milkshake machine salesman named Ray Kroc. Seeing the franchising potential of the McDonald brothers' fast-food formula, Ray Kroc became the exclusive McDonald's franchising agent for the entire United States in 1955. By 1956, there were 14 McDonald's restaurants, and that number had grown to 228 by 1960. Today, there are over 23,000 McDonald's restaurants worldwide—with a presence on every continent except Antarctica.

Ray Kroc's franchise McDonald's menu from 1955 is shown at the right. Local franchise owners have frequently been responsible for additions to the basic menu. The Filet-O-Fish sandwich was the first item to be added to the original menu in 1963 after it was developed by a franchise owner in Cincinnati, Ohio. The Big

| ORIGINAL 1955 McDONALD'S MENU | |
| --- | --- |
| Hamburger | $0.15 |
| Cheeseburger | $0.19 |
| Fries | $0.10 |
| Small soft drink | $0.10 |
| Large soft drink | $0.15 |
| Coffee | $0.10 |
| Shake | $0.20 |

Mac sandwich, the invention of a Pittsburgh franchise owner, followed in 1968. And the Egg McMuffin was introduced nationally in 1973 after a Santa Barbara, California, franchisee searched for a breakfast product to offer at his restaurant. Other items that are standards on today's menu followed: the Quarter Pounder in 1972, Chicken McNuggets in 1983, and fresh tossed salads in 1987.

International McDonald's restaurants include local specialties on the menu. McDonald's restaurants in Norway include a grilled salmon sandwich called McLaks. McDonald's customers will find McSpaghetti on the menu in the Philippines, a Kiwi burger in New Zealand, and the Samurai Pork burger in Thailand.

## Group Activity

Visit a McDonald's restaurant near you and record the present-day prices for each of the seven items on the original 1955 menu. (Unless otherwise noted, assume that sizes on the original menu are "small" for the sake of comparison with today's menu.) For each menu item, find the number of times greater that today's price is than the original price.

**Name** _____  **Section** _____ **Date** _____

# EXERCISE SET 5.5

**A** *Perform each indicated operation. Then estimate to see whether each proposed result is reasonable. See Examples 1 through 3.*

**1.** 4.9 − 2.1

**2.** 7.3 − 3.8

**3.** 6 × 483.11

**4.** 98.2 × 405.8

**5.** 62.16 ÷ 14.8

**6.** 186.75 ÷ 24.9

**7.** 69.2 + 32.1 + 48.57

**8.** 79.52 + 21.47 + 97.2

**9.** 34.92 − 12.03

**10.** 349.87 − 251.32

*Solve. See Example 4.*

△ **11.** Estimate the perimeter of the rectangle by rounding each measurement to the nearest whole inch.

5.9 inches

12.2 inches

△ **12.** Joe Armani wants to put plastic edging around his rectangular flower garden to help control the weeds. The length and width are 16.2 meters and 12.9 meters, respectively. Find how much edging he needs by rounding each measurement to the nearest whole meter.

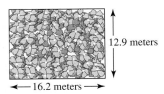

12.9 meters

16.2 meters

**1.** 2.8

**2.** 3.5

**3.** 2898.66

**4.** 39,849.56

**5.** 4.2

**6.** 7.5

**7.** 149.87

**8.** 198.19

**9.** 22.89

**10.** 98.55

**11.** 36 in.

**12.** 58 m

**Name** _____

△ **13.** Estimate the perimeter of the triangle by rounding each measurement to the nearest whole foot.

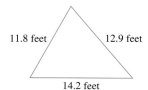

11.8 feet    12.9 feet

14.2 feet

**14.** The area of a triangle is

△    $$\text{area} = \frac{1}{2} \cdot \text{base} \cdot \text{height}.$$

Approximate the area of a triangle whose base is 21.9 centimeters and whose height is 9.9 centimeters. Round each measurement to the nearest whole centimeter.

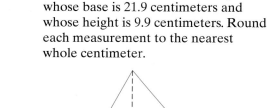

9.9 centimeters

21.9 centimeters

△ **15.** Use 3.14 for $\pi$ to approximate the circumference of a circle whose radius is 7 meters.

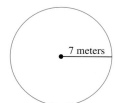

7 meters

△ **16.** Use 3.14 for $\pi$ to approximate the circumference of a circle whose radius is 6 centimeters.

6 centimeters

**17.** Mike and Sandra Hallahan are taking a trip and plan to travel about 1550 miles. Their car gets 32.8 miles to the gallon. Approximate how many gallons of gasoline their car will use by rounding 32.8 to the nearest ten. Round the result to the nearest gallon.

**18.** Refer to Exercise 17. Suppose gasoline costs $1.059 per gallon. Approximate how much money Mike and Sandra need for gasoline for their 1550-mile trip by rounding $1.059 to the nearest cent.

**19.** Ken and Wendy Holladay purchased a new car. They pay $198.79 per month on their car loan for 5 years. Approximate how much they will pay for the car altogether by rounding the monthly payment to the nearest hundred.

**20.** Yoshikazu Sumo is purchasing the following groceries. He has only a ten dollar bill in his pocket. Estimate the cost of each item to see whether he has enough money. Bread, $1.49; milk, $2.09; carrots, $0.97; corn, $0.89; salt, $0.53; and butter, $2.89.

**Name** _____

**21.** Estimate the total distance to the nearest mile between Grove City and Jerome by rounding each distance to the nearest mile.

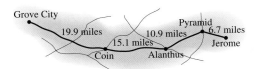

△ **22.** Estimate the perimeter of the figure shown by rounding each measurement to the nearest whole foot.

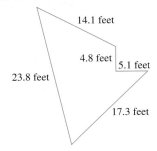

**23.** The all-time top six movies (that is, those that have earned the most money in the United States) along with the approximate amount of money they have earned are listed in the table. Estimate the total amount of money that these movies have earned by rounding each earning to the nearest hundred million.

| ALL-TIME TOP SIX AMERICAN MOVIES | |
|---|---|
| Movie | Gross Domestic Earnings |
| *Titanic* (1997) | $600.8 million |
| *Star Wars* (1977) | $461.0 million |
| *The Phantom Menace* (1999) | $421.4 million |
| *E.T.* (1982) | $399.8 million |
| *Jurassic Park* (1993) | $357.1 million |
| *Forrest Gump* (1994) | $329.7 million |

(*Source: Variety* magazine)

**24.** San Marino is one of the smallest countries in the world. In 1999, it had a population density of 1253 people per square mile. The area of San Marino is about 20 square miles. Estimate the population of San Marino in 1999 by rounding 1253 to the nearest ten. (*Source: 2000 World Almanac*)

**25.** Micronesia is a string of islands in the western Pacific Ocean. In 1999, it had a population density of 485 people per square mile. The area of Micronesia is about 271 square miles. Estimate the population of Micronesia in 1999 by rounding both 485 and 271 to the nearest ten. (*Source: 2000 World Almanac*)

**26.** In 1998, American manufacturers shipped approximately 56 million music CD singles to retailers. The value of these shipments was about $213.36 million. Estimate the value of an individual CD single by rounding $213.36 to the nearest ten. (*Source:* Recording Industry Association of America)

**27.** 0.16 _____

**28.** −260 _____

**29.** −3 _____

**30.** 0.0081 _____

**31.** 0.28 _____

**32.** 0.66 _____

**33.** 2.3 _____

**34.** 23 _____

**35.** 5.29 _____

**36.** 738 _____

**37.** 7.6 _____

**38.** −0.191 _____

**39.** 0.2025 _____

**40.** −0.81 _____

**B** *Simplify each expression. See Examples 5 through 7.*

**27.** $(0.4)^2$

**28.** $(-100)(2.3) - 30$

**29.** $\dfrac{1 + 0.8}{-0.6}$

**30.** $(-0.09)^2$

**31.** $1.4(2 - 1.8)$

**32.** $\dfrac{0.29 + 1.69}{3}$

**33.** $4.83 \div 2.1$

**34.** $62.1 \div 2.7$

**35.** $(-2.3)^2$

**36.** $(8.2)(100) - (8.2)(10)$

**37.** $(3.1 + 0.7)(2.9 - 0.9)$

**38.** $\dfrac{0.707 - 3.19}{13}$

**39.** $\dfrac{(4.5)^2}{100}$

**40.** $0.9(5.6 - 6.5)$

**Name** _____

**41.** $\dfrac{7 + 0.74}{-6}$

**42.** $(1.5)^2 + 0.5$

**C** *Evaluate each expression for $x = 6$, $y = 0.3$, and $z = -2.4$. See Example 8.*

**43.** $z^2$

**44.** $y^2$

**45.** $x - y$

**46.** $x - z$

**47.** $4y - z$

**48.** $\dfrac{x}{y}$

*Determine whether the given value is a solution of the given equation. See Example 9.*

**49.** Is $-1.3$ a solution to $7x + 2.1 = -7$?

**50.** Is $6.2$ a solution to $5y - 8.6 = 23.9$?

**51.** Is $-4.7$ a solution to $x - 6.5 = 2x + 1.8$?

**52.** Is $-0.9$ a solution to $2x - 3.7 = x - 4.6$?

_Answer column_

**41.** $-1.29$

**42.** $2.75$

**43.** $5.76$

**44.** $0.09$

**45.** $5.7$

**46.** $8.4$

**47.** $3.6$

**48.** $20$

**49.** yes

**50.** no

**51.** no

**52.** yes

**Name** _____

## REVIEW AND PREVIEW

*Perform each indicated operation. See Sections 4.3, 4.5, and 4.8.*

**53.** $\dfrac{3}{4} \cdot \dfrac{5}{12}$

**54.** $\dfrac{56}{23} \cdot \dfrac{46}{8}$

**55.** $\dfrac{36}{56} \div \dfrac{30}{35}$

**56.** $\dfrac{5}{7} \div \dfrac{14}{10}$

**57.** $\dfrac{5}{12} - \dfrac{1}{3}$

**58.** $\dfrac{12}{15} + \dfrac{5}{21}$

## COMBINING CONCEPTS

*Simplify each expression. Then estimate to see whether the proposed result is reasonable.*

**59.** $1.96(7.852 - 3.147)^2$

**60.** $(6.02)^2 + (2.06)^2$

**61.** Suppose that your approximation of 12,743 is 13,000 and your friend's approximation is 12,700. Who is correct? Explain your answer.

# 5.6  FRACTIONS AND DECIMALS

## A  WRITING FRACTIONS AS DECIMALS

To write a fraction as a decimal, we interpret the fraction bar as division and find the quotient.

> **WRITING FRACTIONS AS DECIMALS**
>
> To write a fraction as a decimal, divide the numerator by the denominator.

**Example 1**  Write $\frac{1}{4}$ as a decimal.

*Solution:*    $\frac{1}{4} = 1 \div 4$

$$
\begin{array}{r}
0.25 \\
4\overline{)1.00} \\
-8 \phantom{0} \\
\hline
20 \\
-20 \\
\hline
0
\end{array}
$$

Thus, $\frac{1}{4}$ written as a decimal is 0.25.

**Example 2**  Write $\frac{2}{3}$ as a decimal.

*Solution:*

$$
\begin{array}{r}
0.666\ldots \\
3\overline{)2.000} \\
-1\,8 \phantom{0} \\
\hline
20 \\
-18 \\
\hline
20 \\
-18 \\
\hline
2
\end{array}
$$

This pattern will continue so that $\frac{2}{3} = 0.6666\ldots$ .

We can place a bar over the digit 6 to indicate that it repeats.

$$\frac{2}{3} = 0.666\ldots = 0.\overline{6}$$

We can also write a decimal approximation for $\frac{2}{3}$. For example, $\frac{2}{3}$ rounded to the nearest hundredth is 0.67. This can be written as $\frac{2}{3} \approx 0.67$.

**Practice Problem 1**

a. Write $\frac{2}{5}$ as a decimal.

b. Write $\frac{9}{40}$ as a decimal.

**Practice Problem 2**

a. Write $\frac{5}{6}$ as a decimal.

b. Write $\frac{2}{9}$ as a decimal.

**TEACHING TIP**

Point out to students that the long division symbol resembles a small multiplication chart where you know one factor and the result and need to find the other factor. Using division of two whole numbers, $20 \div 4$ is the same as asking, 4 times what number is 20? Using a fraction, $\frac{3}{4} = 3 \div 4$ is the same as asking, 4 times what number is 3?

$$
\begin{array}{c|c}
\times & ? \\
\hline
4 & 20
\end{array}
\quad \rightarrow \quad 4\overline{)20}
$$

$$
\begin{array}{c|c}
\times & ? \\
\hline
4 & 3
\end{array}
\quad \rightarrow \quad 4\overline{)3}
$$

**Answers**

**1. a.** 0.4  **b.** 0.225
**2. a.** $0.8\overline{3} \approx 0.83$  **b.** $0.\overline{2} \approx 0.22$

## Practice Problem 3

Write $\frac{1}{9}$ as a decimal. Round to the nearest thousandth.

## ✓ CONCEPT CHECK

Suppose you are rewriting the fraction $\frac{9}{16}$ as a decimal. How do you know you have made a mistake if you come up with 1.735?

## Practice Problem 4

Insert $<$, $>$, or $=$ to form a true statement.

$$\frac{1}{5} \qquad 0.25$$

**Answers**

**3.** $\approx 0.11$ **4.** $<$
✓ **Concept Check:** Answers may vary.

**Example 3** Write $\frac{22}{7}$ as a decimal. Round to the nearest hundredth. (The fraction $\frac{22}{7}$ is an approximation for $\pi$.)

*Solution:*

$$\begin{array}{r} 3.142 \approx 3.14 \\ 7\overline{)22.000} \\ -21 \phantom{.000} \\ \hline 1\,0 \phantom{00} \\ -\,7 \phantom{00} \\ \hline 30 \phantom{0} \\ -28 \phantom{0} \\ \hline 20 \\ -14 \\ \hline 6 \end{array}$$

Carry the division out to the thousandths place.

The fraction $\frac{22}{7}$ in decimal form is approximately 3.14.

**TRY THE CONCEPT CHECK IN THE MARGIN.**

## B COMPARING FRACTIONS AND DECIMALS

Now we can compare decimals and fractions by writing fractions as equivalent decimals.

**Example 4** Insert $<$, $>$, or $=$ to form a true statement.

$$\frac{1}{8} \qquad 0.12$$

*Solution:* First we write $\frac{1}{8}$ as an equivalent decimal. Then we compare decimal places.

$$\begin{array}{r} 0.125 \\ 8\overline{)1.000} \\ -8 \phantom{.00} \\ \hline 20 \phantom{0} \\ -16 \phantom{0} \\ \hline 40 \\ -40 \\ \hline 0 \end{array}$$

| Original numbers | $\frac{1}{8}$ | 0.12 |
|---|---|---|
| Decimals | 0.125 | 0.120 |
| Compare | 0.125 > 0.12 | |
| Thus, | $\frac{1}{8} > 0.12$ | |

**Example 5**   Insert $<$, $>$, or $=$ to form a true statement.

$0.\overline{7}$      $\dfrac{7}{9}$

*Solution:*   We write $\dfrac{7}{9}$ as a decimal and then compare.

$$
\begin{array}{r}
0.77\ldots = 0.\overline{7} \\
9\overline{)7.00} \\
-6\,3 \phantom{0} \\
\hline
70 \\
-63 \\
\hline
7
\end{array}
$$

| Original numbers | $0.\overline{7}$ | $\dfrac{7}{9}$ |
|---|---|---|
| Decimals | $0.\overline{7}$ | $0.\overline{7}$ |
| Compare | $0.\overline{7} = 0.\overline{7}$ | |
| Thus, | $0.\overline{7} = \dfrac{7}{9}$ | |

**Example 6**   Write the numbers in order from smallest to largest.

$\dfrac{9}{20}$,   $\dfrac{4}{9}$,   $0.456$

*Solution:*

| Original numbers | $\dfrac{9}{20}$ | $\dfrac{4}{9}$ | $0.456$ |
|---|---|---|---|
| Decimals | $0.450$ | $0.444\ldots$ | $0.456$ |
| Compare in order | 2nd | 1st | 3rd |

The numbers written in order are

$\dfrac{4}{9}$,   $\dfrac{9}{20}$,   $0.456$

△ **C** **SOLVING AREA PROBLEMS CONTAINING FRACTIONS AND DECIMALS**

Sometimes real-life problems contain both fractions and decimals. In this section, we will solve such problems about area.

**Example 7**   **Finding the Area of a Triangle**

The area of a triangle is Area $= \dfrac{1}{2} \cdot$ base $\cdot$ height. Find the area of the triangle shown.

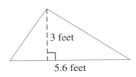
3 feet
5.6 feet

**Practice Problem 5**

Insert $<$, $>$, or $=$ to form a true statement.

a. $\dfrac{1}{2}$      $0.54$        b. $0.\overline{2}$      $\dfrac{2}{9}$

c. $\dfrac{5}{7}$      $0.72$

**Practice Problem 6**

Write the numbers in order from smallest to largest.

a. $\dfrac{1}{3}$,   $0.302$,   $\dfrac{3}{8}$

b. $1.26$,   $1\dfrac{1}{4}$,   $1\dfrac{2}{5}$

c. $0.4$,   $0.41$,   $\dfrac{5}{7}$

**Practice Problem 7**

Find the area of the triangle.

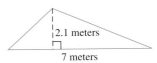
2.1 meters
7 meters

**Answers**

**5. a.** $<$   **b.** $=$   **c.** $<$

**6. a.** $0.302, \dfrac{1}{3}, \dfrac{3}{8}$  **b.** $1\dfrac{1}{4}, 1.26, 1\dfrac{2}{5}$  **c.** $0.4, 0.41, \dfrac{5}{7}$

**7.** 7.35 square meters

*Solution:*    area $= \dfrac{1}{2} \cdot$ base $\cdot$ height

$\phantom{area} = \dfrac{1}{2} \cdot 5.6 \cdot 3$          Let base $= 5.6$ and height $= 3$.

$\phantom{area} = 0.5 \cdot 5.6 \cdot 3$          Write $\dfrac{1}{2}$ as the decimal $0.5$.

$\phantom{area} = 8.4$

The area of the triangle is 8.4 square feet.

## Name _____ Section _____ Date _____

# EXERCISE SET 5.6

**A** *Write each fraction as a decimal. See Examples 1 through 3.*

**1.** $\dfrac{1}{5}$

**2.** $\dfrac{2}{5}$

**3.** $\dfrac{4}{8}$

**4.** $\dfrac{6}{8}$

**5.** $\dfrac{3}{4}$

**6.** $\dfrac{6}{5}$

**7.** $\dfrac{2}{25}$

**8.** $\dfrac{3}{25}$

**9.** $\dfrac{3}{8}$

**10.** $\dfrac{1}{4}$

**11.** $\dfrac{11}{12}$

**12.** $\dfrac{19}{25}$

**13.** $\dfrac{17}{40}$

**14.** $\dfrac{5}{12}$

**15.** $\dfrac{9}{20}$

**16.** $\dfrac{31}{40}$

**17.** $\dfrac{1}{3}$

**18.** $\dfrac{7}{9}$

**19.** $\dfrac{7}{16}$

**20.** $\dfrac{27}{16}$

**21.** $\dfrac{2}{9}$

**22.** $\dfrac{9}{11}$

**23.** $\dfrac{5}{3}$

**24.** $\dfrac{4}{11}$

*Round each number as indicated.*

**25.** Round your decimal answer to Exercise 17 to the nearest hundredth.

**26.** Round your decimal answer to Exercise 18 to the nearest hundredth.

| ANSWER | |
|---|---|
| **1.** | 0.2 |
| **2.** | 0.4 |
| **3.** | 0.5 |
| **4.** | 0.75 |
| **5.** | 0.75 |
| **6.** | 1.2 |
| **7.** | 0.08 |
| **8.** | 0.12 |
| **9.** | 0.375 |
| **10.** | 0.25 |
| **11.** | $0.91\overline{6}$ |
| **12.** | 0.76 |
| **13.** | 0.425 |
| **14.** | $0.41\overline{6}$ |
| **15.** | 0.45 |
| **16.** | 0.775 |
| **17.** | $0.\overline{3}$ |
| **18.** | $0.\overline{7}$ |
| **19.** | 0.4375 |
| **20.** | 1.6875 |
| **21.** | $0.\overline{2}$ |
| **22.** | $0.\overline{81}$ |
| **23.** | $1.\overline{6}$ |
| **24.** | $0.\overline{36}$ |
| **25.** | 0.33 |
| **26.** | 0.78 |

**413**

**27.** Round your decimal answer to Exercise 19 to the nearest hundredth.

**28.** Round your decimal answer to Exercise 20 to the nearest hundredth.

**29.** Round your decimal answer to Exercise 21 to the nearest tenth.

**30.** Round your decimal answer to Exercise 22 to the nearest tenth.

**31.** Round your decimal answer to Exercise 23 to the nearest tenth.

**32.** Round your decimal answer to Exercise 24 to the nearest tenth.

*Write each fraction as a decimal. See Examples 1 through 3.*

**33.** In the United States' National Park System, $\frac{73}{376}$ of all the park units are classified as national monuments. Round your decimal answer to the nearest thousandth. (*Source:* National Park System)

**34.** About $\frac{21}{50}$ of all blood donors have type A blood. (*Source:* American Red Cross Biomedical Services)

**35.** The National Athletic Trainers' Association is an organization dedicated to advancing the athletic training profession. Women make up approximately $\frac{11}{25}$ of its membership. (*Source:* National Athletic Trainers' Association)

**36.** In 1999, approximately $\frac{16}{25}$ of all individuals who had flown in space were citizens of the United States. (*Source:* Congressional Research Service)

**37.** Approximately $\frac{4}{5}$ of the funding for public television comes from sources other than the federal government, such as individuals, businesses, and state governments. (*Source:* Corporation for Public Broadcasting)

**38.** In 1998, approximately $\frac{73}{250}$ of the civilian labor force had at least a bachelor's degree. (*Source:* U.S. Bureau of the Census)

**B** *Insert* $<$, $>$, *or* $=$ *between each pair of numbers to form a true statement. See Examples 4 and 5.*

**39.** 0.562    0.569          **40.** 0.983    0.988          **41.** 0.823    0.813

**42.** 0.824    0.821          **43.** 0.0923    0.0932          **44.** 0.00536    0.00563

**45.** $\dfrac{2}{3}$    $\dfrac{5}{6}$     **46.** $\dfrac{1}{9}$    $\dfrac{2}{17}$     **47.** $\dfrac{5}{9}$    $\dfrac{51}{91}$     **48.** $\dfrac{7}{12}$    $\dfrac{6}{11}$

**49.** $\dfrac{4}{7}$    0.14     **50.** $\dfrac{5}{9}$    0.557     **51.** 1.38    $\dfrac{18}{13}$     **52.** 0.372    $\dfrac{22}{59}$

**53.** 7.123    $\dfrac{456}{64}$     **54.** 12.713    $\dfrac{89}{7}$

*Write each list of numbers in order from smallest to largest. See Example 6.*

**55.** 0.34,    0.35,    0.32          **56.** 0.47,    0.42,    0.40          **57.** 0.49,    0.491,    0.498

**58.** 0.72,    0.727,    0.728          **59.** $\dfrac{3}{4}$,    0.78,    0.73          **60.** $\dfrac{2}{5}$,    0.49,    0.42

**61.** $\dfrac{4}{7}$,    0.453,    0.412          **62.** $\dfrac{6}{9}$,    0.663,    0.668          **63.** 5.23,    $\dfrac{42}{8}$,    5.34

**39.** $<$ _____

**40.** $<$ _____

**41.** $>$ _____

**42.** $>$ _____

**43.** $<$ _____

**44.** $<$ _____

**45.** $<$ _____

**46.** $<$ _____

**47.** $<$ _____

**48.** $>$ _____

**49.** $>$ _____

**50.** $<$ _____

**51.** $<$ _____

**52.** $<$ _____

**53.** $<$ _____

**54.** $<$ _____

**55.** 0.32, 0.34, 0.35 _____

**56.** 0.40, 0.42, 0.47 _____

**57.** 0.49, 0.491, 0.498 _____

**58.** 0.72, 0.727, 0.728 _____

**59.** 0.73, $\dfrac{3}{4}$, 0.78 _____

**60.** $\dfrac{2}{5}$, 0.42, 0.49 _____

**61.** 0.412, 0.453, $\dfrac{4}{7}$ _____

**62.** 0.663, $\dfrac{6}{9}$, 0.668 _____

**63.** 5.23, $\dfrac{42}{8}$, 5.34 _____

**416**

**64.** 7.56,  $\frac{67}{9}$,  7.562  **65.** $\frac{12}{5}$,  2.37,  $\frac{17}{8}$  **66.** $\frac{29}{16}$,  1.75,  $\frac{58}{32}$

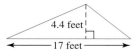 *Find the area of each triangle or rectangle. See Example 7.*

△ **67.**

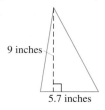

9 inches

5.7 inches

△ **68.**

4.4 feet

17 feet

△ **69.**

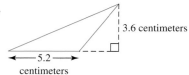

3.6 centimeters

5.2 centimeters

△ **70.**

10 meters

25.6 meters

△ **71.**

0.62 yard

$\frac{2}{5}$ yard

△ **72.**

1.2 miles

$\frac{7}{8}$ mile

## REVIEW AND PREVIEW

*Simplify. See Sections 1.7 and 4.3.*

**73.** $2^3$  **74.** $5^4$  **75.** $6^2 \cdot 2$  **76.** $4 \cdot 3^4$

**77.** $\left(\frac{1}{3}\right)^4$  **78.** $\left(\frac{4}{5}\right)^3$  **79.** $\left(\frac{3}{5}\right)^2$  **80.** $\left(\frac{7}{2}\right)^2$

**81.** $\left(\frac{2}{5}\right)\left(\frac{5}{2}\right)^2$  **82.** $\left(\frac{2}{3}\right)\left(\frac{3}{2}\right)^2$

**Name** _____

## COMBINING CONCEPTS

_In 1999, there were 10,506 commercial radio stations in the United States. The most popular formats are listed in the table along with their counts for 1999. Use this table to answer Exercises 83–86._

| TOP COMMERCIAL RADIO STATION FORMATS IN 1999 | |
| --- | --- |
| Format | Number of Stations |
| Country | 2321 |
| Adult Contemporary | 1576 |
| News/Talk/Business/Sports | 1396 |
| Oldies/Classic Hits/R&B | |
| Oldies | 1109 |
| Religious | 1088 |
| Rock | 803 |

(_Source:_ M Street Corporation)

**83.** Write the fraction of radio stations with a country music format as a decimal. Round to the nearest thousandth.

**84.** Write the fraction of radio stations with a talk format as a decimal. Round to the nearest hundredth.

**85.** Estimate to the nearest hundred, the total number of stations with the top six formats in 1999.

**86.** Use your estimate from Exercise 85 to write the fraction of radio stations accounted for by the top five formats as a decimal. Round to the nearest hundredth.

**87.** Suppose that you are writing a paper on a word processor. You have been instructed to use $\frac{5}{8}$-inch margins. To change margins in the word processing software, you must enter margin widths in inches as a decimal. What number should you enter?

**88.** Suppose that you are using a word processor to format a table for a presentation. The first column of the table must be $3\frac{7}{16}$ inches wide. To adjust the width of the column in the word processing software, you must enter the column width in inches as a decimal. What number should you enter?

**89.** Describe how to determine the larger of two fractions.

**Name** _____

*Find the value of each expression. Give the result as a decimal.*

**90.** $\dfrac{1}{5} - 2(7.8)$

**91.** $\dfrac{3}{4} - (9.6)(5)$

**92.** $8.25 - \left(\dfrac{1}{2}\right)^2$

**93.** $\left(\dfrac{1}{10}\right)^2 + (1.6)(2.1)$

**94.** $\dfrac{1}{4}(-9.6 - 5.2)$

**95.** $\dfrac{3}{8}(4.7 - 5.9)$

# 5.7  EQUATIONS CONTAINING DECIMALS

## A  SOLVING EQUATIONS CONTAINING DECIMALS

In this section, we continue our work with decimals and algebra by solving equations containing decimals. First, we review the steps given earlier for solving an equation.

---

**STEPS FOR SOLVING AN EQUATION IN $x$**

**Step 1.**  If fractions are present, multiply both sides of the equation by the LCD of the fractions.

**Step 2.**  If parentheses are present, use the distributive property.

**Step 3.**  Combine any like terms on each side of the equation.

**Step 4.**  Use the addition property of equality to rewrite the equation so that variable terms are on one side of the equation and constant terms are on the other side.

**Step 5.**  Divide both sides by the numerical coefficient of $x$ to solve.

**Step 6.**  Check the answer in the **original equation**.

---

**Example 1**    Solve for $x$:  $x - 1.5 = 8$

*Solution:*    To get $x$ alone on one side of the equation, add 1.5 to both sides.

$$x - 1.5 = 8 \qquad \text{Original equation}$$
$$x - 1.5 + 1.5 = 8 + 1.5 \qquad \text{Add 1.5 to both sides.}$$
$$x = 9.5 \qquad \text{Simplify.}$$

*Check:*    To check, replace $x$ with 9.5 in the *original equation*.

$$x - 1.5 = 8 \qquad \text{Original equation}$$
$$9.5 - 1.5 = 8 \qquad \text{Replace } x \text{ with 9.5.}$$
$$8 = 8 \qquad \text{True.}$$

Since $8 = 8$ is a true statement, 9.5 is a solution of the equation.

**Example 2**    Solve for $y$:  $-2y = 6.7$

*Solution:*    To solve for $y$, divide both sides by the coefficient of $y$, which is $-2$.

$$-2y = 6.7 \qquad \text{Original equation}$$
$$\frac{-2y}{-2} = \frac{6.7}{-2} \qquad \text{Divide both sides by } -2.$$
$$y = -3.35 \qquad \text{Simplify.}$$

*Check:*    To check, replace $y$ with $-3.35$ in the original equation.

$$-2y = 6.7 \qquad \text{Original equation}$$
$$-2(-3.35) = 6.7 \qquad \text{Replace } y \text{ with } -3.35.$$
$$6.7 = 6.7 \qquad \text{True.}$$

Thus $-3.35$ is a solution of the equation $-2y = 6.7$.

---

**Objective**

**A** Solve equations containing decimals.

Study Guide    SSM    CD-ROM    Video 5.7

---

**Practice Problem 1**

Solve for $z$:  $z + 0.9 = 1.3$

---

**Practice Problem 2**

Solve for $x$:  $0.17x = -0.34$

---

**Answers**

**1.** 0.4   **2.** $-2$

## Practice Problem 3

Solve for $x$:
$6.3 - 5x = 3(x + 2.9)$

**Example 3**     Solve for $x$: $5(x - 0.36) = -x + 2.4$

*Solution:*     First use the distributive property to distribute the factor 5.

$$5(x - 0.36) = -x + 2.4 \quad \text{Original equation}$$
$$5x - 1.8 = -x + 2.4 \quad \text{Apply the distributive property.}$$

Next, get $x$ alone on the left side of the equation by adding 1.8 to both sides of the equation and then adding $x$ to both sides of the equation.

$$5x - 1.8 + 1.8 = -x + 2.4 + 1.8 \quad \text{Add 1.8 to both sides.}$$
$$5x = -x + 4.2 \quad \text{Simplify.}$$
$$5x + x = -x + 4.2 + x \quad \text{Add } x \text{ to both sides.}$$
$$6x = 4.2 \quad \text{Simplify.}$$
$$\frac{6x}{6} = \frac{4.2}{6} \quad \text{Divide both sides by 6.}$$
$$x = 0.7 \quad \text{Simplify.}$$

To verify that 0.7 is the solution, replace $x$ with 0.7 in the original equation. ▬▬▬

Instead of solving equations with decimals, sometimes it may be easier to first rewrite the equation so it contains integers only. Recall that multiplying a decimal by a power of 10, such as 10, 100, or 1000, has the effect of moving the decimal point to the right. We can use the multiplication property of equality to multiply both sides of the equation through by an appropriate power of 10. The resulting equivalent equation will contain integers only.

## Practice Problem 4

Solve for $y$: $0.2y + 2.6 = 4$

**Example 4**     Solve for $y$: $0.5y + 2.3 = 1.65$

*Solution:*     Multiply the equation through by 100. This will move the decimal point in each term two places to the right.

$$0.5y + 2.3 = 1.65 \quad \text{Original equation}$$
$$100(0.5y + 2.3) = 100(1.65) \quad \text{Multiply both sides by 100.}$$
$$100(0.5y) + 100(2.3) = 100(1.65) \quad \text{Apply the distributive property.}$$
$$50y + 230 = 165 \quad \text{Simplify.}$$

Now the equation contains integers only. Finish solving by subtracting 230 from both sides.

$$50y + 230 = 165$$
$$50y + 230 - 230 = 165 - 230 \quad \text{Subtract 230 from both sides.}$$
$$50y = -65 \quad \text{Simplify.}$$
$$\frac{50y}{50} = \frac{-65}{50} \quad \text{Divide both sides by 50.}$$
$$y = -1.3 \quad \text{Simplify.}$$

Check to see that $-1.3$ is the solution by replacing $y$ with $-1.3$ in the original equation. ▬▬▬

## ✓ CONCEPT CHECK

By what number would you multiply both sides of the following equation to make calculations easier? Explain your choice.
$1.7x + 3.655 = -14.2$

**Answers**

**3.** $-0.3$  **4.** 7

✓ **Concept Check:** Multiply by 1000.

**TRY THE CONCEPT CHECK IN THE MARGIN.**

## EXERCISE SET 5.7

**A** *Solve each equation. See Examples 1 through 3.*

**1.** $x + 1.2 = 7.1$

**2.** $y - 0.5 = 9$

**3.** $5y = 2.15$

**4.** $0.4x = 50$

**5.** $6x + 8.65 = 3x + 10$

**6.** $7x - 9.64 = 5x + 2.32$

**7.** $2(x - 1.3) = 5.8$

**8.** $5(x + 2.3) = 19.5$

*Solve each equation by first multiplying both sides through by an appropriate power of 10 so that the equation contains integers only. See Example 4.*

**9.** $0.4x + 0.7 = -0.9$

**10.** $0.7x + 0.1 = 1.5$

**11.** $7x - 10.8 = x$

**12.** $3y = 7y + 24.4$

**13.** $2.1x + 5 - 1.6x = 10$

**14.** $1.5x + 2 - 1.2x = 12.2$

*Solve.*

**15.** $y - 3.6 = 4$

**16.** $x + 5.7 = 8.4$

**17.** $-0.02x = -1.2$

**18.** $-9y = -0.162$

**19.** $6.5 = 10x + 7.2$

**20.** $2x - 4.2 = 8.6$

**21.** $200x - 0.67 = 100x + 0.81$

**22.** $7(x + 8.6) = 6x$

ANSWERS

**1.** 5.9

**2.** 9.5

**3.** 0.43

**4.** 125

**5.** 0.45

**6.** 5.98

**7.** 4.2

**8.** 1.6

**9.** −4

**10.** 2

**11.** 1.8

**12.** −6.1

**13.** 10

**14.** 34

**15.** 7.6

**16.** 2.7

**17.** 60

**18.** 0.018

**19.** −0.07

**20.** 6.4

**21.** 0.0148

**22.** −60.2

**23.** $3(x + 2.71) = 2x$

**24.** $1.5 = 0.4x + 0.5$

**25.** $1.2 + 0.3x = 0.9$

**26.** $50x + 0.81 = 40x - 0.48$

**27.** $0.9x + 2.65 = 0.5x + 5.45$

**28.** $4(2x - 1.6) = 5x - 6.4$

**29.** $4x + 7.6 = 2(3x - 3.2)$

**30.** $0.7x + 13.8 = x - 2.16$

## REVIEW AND PREVIEW

*Simplify each expression by combining like terms. If parentheses are present, first use the distributive property. See Section 3.1.*

**31.** $2x - 6 + 4x - 10$

**32.** $x - 4 - x - 4$

**33.** $3(x - 5) + 10$

**34.** $9x + 4(x + 20)$

**35.** $5y - 1.2 - 7y + 8$

**36.** $7.76 + 8z - 12z + 8.91$

## COMBINING CONCEPTS

**37.** Explain in your own words the property of equality that allows us to multiply an equation through by a power of 10.

**38.** Construct an equation whose solution is 1.4.

*Solve.*

**39.** $-5.25x = -40.33575$

**40.** $7.68y = -114.98496$

**41.** $1.95y + 6.834 = 7.65y - 19.8591$

**42.** $6.11x + 4.683 = 7.51x + 18.235$

**23.** $-8.13$

**24.** $2.5$

**25.** $-1$

**26.** $-0.129$

**27.** $7$

**28.** $0$

**29.** $7$

**30.** $53.2$

**31.** $6x - 16$

**32.** $-8$

**33.** $3x - 5$

**34.** $13x + 80$

**35.** $-2y + 6.8$

**36.** $-4z + 16.67$

**37.** answers may vary

**38.** answers may vary

**39.** $7.683$

**40.** $-14.972$

**41.** $4.683$

**42.** $-9.68$

# 5.8    SQUARE ROOTS AND THE PYTHAGOREAN THEOREM

## A    FINDING SQUARE ROOTS

The square of a number is the number times itself. For example,

> The square of 5 is 25 because $5^2$ or $5 \cdot 5 = 25$.
> The square of $-5$ is $(-5)^2$ or $(-5)(-5) = 25$.
> The square of 3 is 9 because $3^2$ or $3 \cdot 3 = 9$.
> The square of 10 is $10^2$ or $10 \cdot 10 = 100$.

The reverse process of squaring is finding a **square root**. For example,

> A square root of 9 is 3 because $3^2 = 9$.
> A square root of 25 is 5 because $5^2 = 25$.
> A square root of 25 is also $-5$ because $(-5)^2 = 25$.
> A square root of 100 is 10 because $10^2 = 100$.

We use the symbol $\sqrt{\ }$, called a **radical sign**, to indicate square roots. For example,

$$\sqrt{9} = 3 \quad \text{because } 3^2 = 9 \text{ and 3 is positive.}$$
$$\sqrt{25} = 5 \quad \text{because } 5^2 = 25 \text{ and 5 is positive.}$$

---

> **SQUARE ROOT OF A NUMBER**
>
> The square root of a positive number $a$ is the positive number $b$ whose square is $a$. In symbols,
>
> $$\sqrt{a} = b, \quad \text{if } b^2 = a$$
>
> Also,
>
> $$\sqrt{0} = 0$$

---

**Example 1**    Find each square root.

    **a.** $\sqrt{49}$     **b.** $\sqrt{36}$     **c.** $\sqrt{1}$     **d.** $\sqrt{81}$

*Solution:*    **a.** $\sqrt{49} = 7$   because   $7^2 = 49$.
           **b.** $\sqrt{36} = 6$   because   $6^2 = 36$.
           **c.** $\sqrt{1} = 1$   because   $1^2 = 1$.
           **d.** $\sqrt{81} = 9$   because   $9^2 = 81$.

**Example 2**    Find: $\sqrt{\dfrac{1}{36}}$

*Solution:*   $\sqrt{\dfrac{1}{36}} = \dfrac{1}{6}$   because   $\dfrac{1}{6} \cdot \dfrac{1}{6} = \dfrac{1}{36}$.

**Example 3**    Find: $\sqrt{\dfrac{4}{25}}$

*Solution:*   $\sqrt{\dfrac{4}{25}} = \dfrac{2}{5}$   because   $\dfrac{2}{5} \cdot \dfrac{2}{5} = \dfrac{4}{25}$.

---

### Objectives

**A** Find the square root of a number.
**B** Approximate square roots.
**C** Use the Pythagorean theorem.

 Study Guide     SSM     CD-ROM     Video 5.8

---

### Practice Problem 1

Find each square root.
a. $\sqrt{100}$
b. $\sqrt{64}$
c. $\sqrt{121}$
d. $\sqrt{0}$

### Practice Problem 2

Find: $\sqrt{\dfrac{1}{4}}$

### Practice Problem 3

Find: $\sqrt{\dfrac{9}{16}}$

**Answers**

**1. a.** 10   **b.** 8   **c.** 11   **d.** 0   **2.** $\dfrac{1}{2}$   **3.** $\dfrac{3}{4}$

## B APPROXIMATING SQUARE ROOTS

Thus far, we have found square roots of perfect squares. Numbers like $\frac{1}{4}$, 36, $\frac{4}{25}$, and 1 are called **perfect squares** because their square root is a whole number or a fraction. A square root such as $\sqrt{5}$ cannot be written as a whole number or a fraction since 5 is not a perfect square.

Although $\sqrt{5}$ cannot be written as a whole number or a fraction, it can be approximated by estimating, by using a table (as in Appendix E), or by using a calculator.

### Practice Problem 4

Use Appendix E or a calculator to approximate the square root of 11 to the nearest thousandth.

**Example 4**    Use Appendix E or a calculator to approximate the square root of 43 to the nearest thousandth.

*Solution:*    $\sqrt{43} \approx 6.557$

---

**Helpful Hint**

$\sqrt{43}$ is *approximately* 6.557. This means that if we multiply 6.557 by 6.557, the product is *close* to 43.

$$6.557 \times 6.557 = 42.994249$$

### Practice Problem 5

Approximate $\sqrt{29}$ to the nearest thousandth.

**Example 5**    Approximate $\sqrt{32}$ to the nearest thousandth.

*Solution:*    $\sqrt{32} \approx 5.657$

## C USING THE PYTHAGOREAN THEOREM

One important application of square roots has to do with right triangles. Recall that a **right triangle** is a triangle in which one of the angles is a right angle, or measures 90° (degrees). The **hypotenuse** of a right triangle is the side opposite the right angle. The **legs** of a right triangle are the other two sides. These are shown in the following figure. The right angle in the triangle is indicated by the small square drawn in that angle.

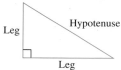

The following theorem is true for all right triangles.

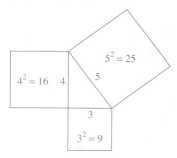
---

**PYTHAGOREAN THEOREM**

If $a$ and $b$ are the lengths of the legs of a right triangle and $c$ is the length of the hypotenuse, then

$$a^2 + b^2 = c^2$$

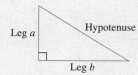

In other words, $(\text{leg})^2 + (\text{other leg})^2 = (\text{hypotenuse})^2$.

**Example 6**    Find the length of the hypotenuse of the given right triangle.

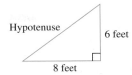

*Solution:*    Let $a = 6$ and $b = 8$. According to the Pythagorean theorem,

$$a^2 + b^2 = c^2$$
$$6^2 + 8^2 = c^2 \quad \text{Let } a = 6 \text{ and } b = 8.$$
$$36 + 64 = c^2 \quad \text{Evaluate } 6^2 \text{ and } 8^2.$$
$$100 = c^2 \quad \text{Add.}$$

In the equation $c^2 = 100$, the solutions of $c$ are the square roots of 100. This means that the solutions are 10 and $-10$. Since $c$ represents a length, we are interested in the positive square root of $c^2$ only.

$$c = \sqrt{100}$$
$$= 10$$

The hypotenuse is 10 feet long.    ▬▬▬

**Example 7**    Approximate the length of the hypotenuse of the given right triangle. Round the length to the nearest whole unit.

*Solution:*    Let $a = 17$ and $b = 10$.

$$a^2 + b^2 = c^2$$
$$17^2 + 10^2 = c^2$$
$$289 + 100 = c^2$$
$$389 = c^2$$
$$\sqrt{389} = c \text{ or } c \approx 20 \quad \text{From Appendix E or a calculator.}$$

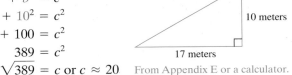

The hypotenuse is exactly $\sqrt{389}$ meters, which is approximately 20 meters.    ▬▬▬

**Example 8**    Find the length of the leg in the given right triangle. Give the exact length and a two-decimal-place approximation.

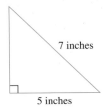

*Solution:*    Notice that the hypotenuse measures 7 inches and that the length of one leg measures 5 inches. Thus, let $c = 7$ and $a$ or $b$ be 5. We will let $a = 5$.

$$a^2 + b^2 = c^2$$
$$5^2 + b^2 = 7^2 \quad \text{Let } a = 5 \text{ and } c = 7.$$
$$25 + b^2 = 49 \quad \text{Evaluate } 5^2 \text{ and } 7^2.$$
$$b^2 = 24 \quad \text{Subtract 25 from both sides.}$$
$$b = \sqrt{24} \approx 4.90$$

The length of the leg is exactly $\sqrt{24}$ inches and approximately 4.90 inches.    ▬▬▬

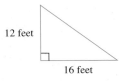
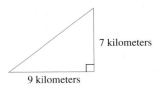

✓ **CONCEPT CHECK**

The following lists are the lengths of the sides of two triangles. Which set forms a right triangle?

a. 8, 15, 17

b. 24, 30, 40

## Practice Problem 9

A football field is a rectangle measuring 100 yards by 53 yards. Draw a diagram and find the length of the diagonal of a football field to the nearest yard.

**Example 9**     Finding the Dimensions of a Park

An inner-city park is in the shape of a square that measures 300 feet on a side. A sidewalk is to be constructed along the diagonal of the park. Find the length of the sidewalk rounded to the nearest whole foot.

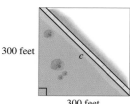

300 feet

300 feet

*Solution:*     The diagonal is the hypotenuse of a right triangle, which we label $c$.

$$a^2 + b^2 = c^2$$

$$300^2 + 300^2 = c^2 \qquad \text{Let } a = 300 \text{ and } b = 300.$$

$$90{,}000 + 90{,}000 = c^2 \qquad \text{Evaluate } (300)^2.$$

$$180{,}000 = c^2 \qquad \text{Add.}$$

$$\sqrt{180{,}000} = c \text{ or } c \approx 424$$

The length of the sidewalk is approximately 424 feet.

---

## CALCULATOR EXPLORATIONS
### FINDING SQUARE ROOTS

To simplify or approximate square roots using a calculator, locate the key marked $\boxed{\sqrt{\phantom{x}}}$.

To simplify $\sqrt{64}$, for example, press the keys

$\boxed{64}\ \boxed{\sqrt{\phantom{x}}}$  or  $\boxed{\sqrt{\phantom{x}}}\ \boxed{64}$

The display should read $\boxed{\phantom{xxxxxxx}8}$. Then

$$\sqrt{64} = 8$$

To *approximate* $\sqrt{10}$, press the keys

$\boxed{10}\ \boxed{\sqrt{\phantom{x}}}$  or  $\boxed{\sqrt{\phantom{x}}}\ \boxed{10}$

The display should read $\boxed{3.16227766}$. This is an *approximation* for $\sqrt{10}$. A three-decimal-place approximation is

$$\sqrt{10} \approx 3.162$$

Is this answer reasonable? Since 10 is between the perfect squares 9 and 16, $\sqrt{10}$ is between $\sqrt{9} = 3$ and $\sqrt{16} = 4$. Our answer is reasonable since 3.162 is between 3 and 4.

*Simplify.*

**1.** $\sqrt{1024}$  32     **2.** $\sqrt{676}$  26

*Approximate each square root. Round each answer to the nearest thousandth.*

**3.** $\sqrt{15}$  3.873     **4.** $\sqrt{19}$  4.359

**5.** $\sqrt{97}$  9.849     **6.** $\sqrt{56}$  7.483

**Answers**

**9.** 113 yards

✓ Concept Check:  a

Name _____ Section _____ Date _____

# EXERCISE SET 5.8

**A** *Find each square root. See Examples 1 through 3.*

**1.** $\sqrt{4}$    **2.** $\sqrt{9}$    **3.** $\sqrt{625}$    **4.** $\sqrt{16}$

**5.** $\sqrt{\dfrac{1}{81}}$    **6.** $\sqrt{\dfrac{1}{64}}$    **7.** $\sqrt{\dfrac{144}{64}}$    **8.** $\sqrt{\dfrac{36}{81}}$

**9.** $\sqrt{256}$    **10.** $\sqrt{144}$    **11.** $\sqrt{\dfrac{9}{4}}$    **12.** $\sqrt{\dfrac{121}{169}}$

**B** *Use Appendix E or a calculator to approximate each square root. Round the square root to the nearest thousandth. See Examples 4 and 5.*

**13.** $\sqrt{3}$    **14.** $\sqrt{5}$    **15.** $\sqrt{15}$    **16.** $\sqrt{17}$

**17.** $\sqrt{14}$    **18.** $\sqrt{18}$    **19.** $\sqrt{47}$    **20.** $\sqrt{85}$

**21.** $\sqrt{8}$    **22.** $\sqrt{10}$    **23.** $\sqrt{26}$    **24.** $\sqrt{35}$

**25.** $\sqrt{71}$    **26.** $\sqrt{62}$    **27.** $\sqrt{7}$    **28.** $\sqrt{2}$

**C** *Find the unknown length in each right triangle. If necessary, approximate the length to the nearest thousandth. See Examples 6 through 8.*

△ **29.**

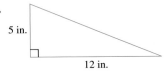

5 in.    12 in.

△ **30.**

24 ft    25 ft

**31.** 6.633 cm

**32.** 8.485 yd

**33.** 5

**34.** 15

**35.** 8

**36.** 22.472

**37.** 17.205

**38.** 37.216

**39.** 16.125

**40.** 45

**41.** 12

**42.** 108

**43.** 44.822

**44.** 33.541

**45.** 42.426

**46.** 171.825

**47.** 1.732

**48.** 3.606

**Name** _____

△ **31.**

10 cm   12 cm

△ **32.**

3 yd   9 yd

*Sketch each right triangle and find the length of the side not given. If necessary, approximate the length to the nearest thousandth. See Examples 6 through 8.*

△ **33.** leg = 3, leg = 4          △ **34.** leg = 9, leg = 12

△ **35.** leg = 6, hypotenuse = 10          △ **36.** leg = 48, hypotenuse = 53

▭ **37.** leg = 10, leg = 14          △ **38.** leg = 32, leg = 19
△

△ **39.** leg = 2, leg 16          △ **40.** leg = 27, leg = 36

▭ **41.** leg = 5, hypotenuse = 13          △ **42.** leg = 45, hypotenuse = 117
△

△ **43.** leg = 35, leg = 28          △ **44.** leg = 30, leg = 15

△ **45.** leg = 30, leg = 30          △ **46.** leg = 110, leg = 132

△ **47.** hypotenuse = 2, leg = 1          △ **48.** hypotenuse = 7, leg = 6

**428**

*Solve. See Example 9*

△ **49.** A standard city block is a square with each side measuring 100 yards. Find the length of the diagonal of a city block to the nearest hundredth yard.

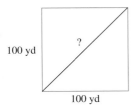

100 yd

? 

100 yd

△ **50.** A section of land is a square with each side measuring 1 mile. Find the length of the diagonal of a section of land to the nearest thousandth mile.

? 1 mi

△ **51.** Find the height of the tree. Round the height to one decimal place.

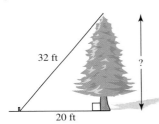

32 ft

?

20 ft

△ **52.** Find the height of the antenna. Round the height to one decimal place.

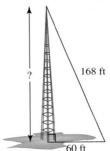

? 168 ft

60 ft

△ **53.** A football field is a rectangle that is 300 feet long by 160 feet wide. Find, to the nearest foot, the length of a straight-line run that started at one corner and went diagonally to end at the opposite corner.

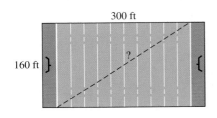

300 ft

160 ft { ? {

△ **54.** A baseball diamond is the shape of a square with 90-foot sides. Find the distance across the diamond from third base to first base to the nearest tenth of a foot.

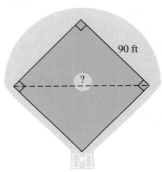

90 ft

?

**55.** $\dfrac{5}{6}$

**56.** $\dfrac{2}{3}$

**57.** $\dfrac{2}{5}$

**58.** $\dfrac{7}{15}$

**59.** $\dfrac{5}{12}$

**60.** $\dfrac{3}{5}$

**61.** 6, 7

**62.** 5, 6

**63.** 10, 11

**64.** 9, 10

**65.** answers may vary

**Name** _____

## REVIEW AND PREVIEW

*Write each fraction in simplest form. See Section 4.2.*

**55.** $\dfrac{10}{12}$        **56.** $\dfrac{10}{15}$        **57.** $\dfrac{24}{60}$

**58.** $\dfrac{35}{75}$        **59.** $\dfrac{30}{72}$        **60.** $\dfrac{18}{30}$

## COMBINING CONCEPTS

*Determine what two whole numbers each square root is between without using a calculator or table. Then use a calculator or table to check.*

**61.** $\sqrt{38}$      **62.** $\sqrt{27}$      **63.** $\sqrt{101}$      **64.** $\sqrt{85}$

**65.** Without using a calculator, explain why you know that $\sqrt{11}$ is between 3 and 4.

# CHAPTER 5 ACTIVITY
## ORDERING FROM A CATALOG

*This activity may be completed by working in groups or individually.*

Almost any financial transaction involves operations on decimals. Ordering items from a catalog is no exception. Complete the following order form for the following quantities of mail-order plants. Then answer the questions below.

▲ Catalog # 4277; 3 burning bushes, $12.95 each

▲ Catalog # 1796; 6 cushion mums, $3.99 each

▲ Catalog # 4284; 2 spring beauty rose bushes, $5.99 each

▲ Catalog # 3160; 12 variegated vinca, $1.49 each

| FAITH MEADOW NURSERIES | | | | |
|---|---|---|---|---|
| Mail-Order Plants and Shrubs Since 1970 | | | | |
| Catalog Number | Description | Quantity | Item Price | Total Charge |
|  |  |  |  |  |
|  |  |  |  |  |
|  |  |  |  |  |
|  |  |  |  |  |
|  |  |  |  |  |
|  |  |  |  |  |
|  |  |  |  |  |
|  |  |  | Merchandise Total |  |
|  |  |  | Shipping & Handling* |  |
|  |  |  | Total Amount |  |

| *Shipping and Handling Charges: | |
|---|---|
| For Orders Totaling | ADD |
| $20.00 | $4.95 |
| $20.01 to $40.00 | $6.95 |
| $40.01 to $60.00 | $9.95 |
| $60.01 to $80.00 | $11.95 |
| Over $80.00 | $11.95 plus $0.10 per $1 of merchandise over $80.00 |

1. Which operations are needed to complete this order form?

2. What would be the shipping costs for an order of merchandise costing $100?

# CHAPTER 5 HIGHLIGHTS

| DEFINITIONS AND CONCEPTS | EXAMPLES |
|---|---|

## SECTION 5.1   INTRODUCTION TO DECIMALS

PLACE-VALUE CHART

| | 1 | 7 | . | 7 | 5 | 8 | | |
|---|---|---|---|---|---|---|---|---|
| hundreds | tens | ones | decimal point | tenths | hundredths | thousandths | ten-thousandths | hundred-thousandths |
| 100 | 10 | 1 | | $\frac{1}{10}$ | $\frac{1}{100}$ | $\frac{1}{1000}$ | $\frac{1}{10,000}$ | $\frac{1}{100,000}$ |

**Rounding Decimals to a Place Value to the Right of the Decimal Point**

**Step 1.** Locate the digit to the right of the given place value.

**Step 2.** If this digit is 5 or greater, add 1 to the digit in the given place value and drop all digits to its right. If this digit is less than 5, drop all digits to the right of the given place value.

Write 3.08 in words.

Three and eight hundredths

Round 86.1256 to the nearest hundredth.

```
              ┌──── hundredths place
              ↓
86.1256
         ↑
         └──── 5 or greater
```

rounds to 86.13

## SECTION 5.2   ADDING AND SUBTRACTING DECIMALS

**Adding or Subtracting Decimals**

**Step 1.** Write the decimals so that the decimal points line up vertically.

**Step 2.** Add or subtract as for whole numbers.

**Step 3.** Place the decimal point in the sum or difference so that it lines up vertically with the decimal points in the problem.

Subtract: 2.8 − 1.04

$$\begin{array}{r} 2.8\overset{7\ 10}{0} \\ -1.04 \\ \hline 1.76 \end{array}$$

## SECTION 5.3   MULTIPLYING DECIMALS AND CIRCUMFERENCE OF A CIRCLE

**Multiplying Decimals**

**Step 1.** Multiply the decimals as though they were whole numbers.

**Step 2.** The decimal point in the product is placed so that the number of decimal places in the product is equal to the *sum* of the number of decimal places in the factors.

The circumference of a circle is the distance around the circle.

$C = \pi d$ or $C = 2\pi r$

where $\pi \approx 3.14$ or $\pi \approx \frac{22}{7}$.

Multiply: $1.48 \times 5.9$

$$\begin{array}{r} 1.48 \quad \leftarrow\text{2 decimal places.} \\ \times\ 5.9 \quad \leftarrow\text{1 decimal place.} \\ \hline 1332 \\ 7400 \\ \hline 8.732 \quad \leftarrow\text{3 decimal places.} \end{array}$$

Find the exact circumference and a two-decimal-place approximation.

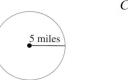

5 miles

$$\begin{aligned} C &= 2\pi r \\ &= 2\pi(5) \\ &= 10\pi \\ &\approx 10(3.14) \\ &= 31.4 \end{aligned}$$

The circumference is exactly $10\pi$ miles and approximately 31.4 miles.

## SECTION 5.4   DIVIDING DECIMALS

**DIVIDING DECIMALS**

**Step 1.** Move the decimal point in the divisor to the right until the divisor is a whole number.

**Step 2.** Move the decimal point in the dividend to the right the *same number of places* as the decimal point was moved in Step 1.

**Step 3.** Divide. The decimal point in the quotient is directly over the moved decimal point in the dividend.

Divide: $1.118 \div 2.6$

$$
\begin{array}{r}
0.43 \\
2.6\overline{\smash{)}1.118} \\
\end{array}
$$

$$
\begin{array}{r}
-1\,04 \\
\hline
78 \\
-78 \\
\hline
0
\end{array}
$$

## SECTION 5.5   ESTIMATING AND ORDER OF OPERATIONS

**ORDER OF OPERATIONS**

1. Do all operations within grouping symbols such as parentheses or brackets.
2. Evaluate any expressions with exponents.
3. Multiply or divide in order from left to right.
4. Add or subtract in order from left to right.

Simplify.

$$-1.9(12.8 - 4.1) = -1.9(8.7) \quad \text{Subtract.}$$
$$= -16.53 \quad \text{Multiply.}$$

## SECTION 5.6   FRACTIONS AND DECIMALS

**To write fractions as decimals**, divide the numerator by the denominator.

Write $\dfrac{3}{8}$ as a decimal.

$$
\begin{array}{r}
0.375 \\
8\overline{\smash{)}3.000} \\
\end{array}
$$

$$
\begin{array}{r}
-2\,4 \\
\hline
60 \\
-56 \\
\hline
40 \\
-40 \\
\hline
0
\end{array}
$$

## SECTION 5.7   EQUATIONS CONTAINING DECIMALS

**STEPS FOR SOLVING AN EQUATION IN** $x$

**Step 1.** If fractions are present, multiply both sides of the equation by the LCD of the fractions.

**Step 2.** If parentheses are present, use the distributive property.

**Step 3.** Combine any like terms on each side of the equation.

**Step 4.** Use the addition property of equality to rewrite the equation so that variable terms are on one side of the equation and constant terms are on the other side.

**Step 5.** Divide both sides by the numerical coefficient of $x$ to solve.

**Step 6.** Check the answer in the *original equation*.

Solve: $3(x + 2.6) = 10.92$

$$3x + 7.8 = 10.92 \qquad \text{Apply the distributive property.}$$
$$3x + 7.8 - 7.8 = 10.92 - 7.8 \qquad \text{Subtract 7.8 from both sides.}$$
$$3x = 3.12 \qquad \text{Simplify.}$$
$$\frac{3x}{3} = \frac{3.12}{3} \qquad \text{Divide by 3.}$$
$$x = 1.04 \qquad \text{Simplify.}$$

Check 1.04 in the original equation.

## SECTION 5.8    SQUARE ROOTS AND THE PYTHAGOREAN THEOREM

**SQUARE ROOT OF A NUMBER**

The square root of a positive number $a$ is the positive number $b$ whose square is $a$. In symbols,

$$\sqrt{a} = b, \quad \text{if} \quad b^2 = a$$

Also,

$$\sqrt{0} = 0.$$

**PYTHAGOREAN THEOREM**

If $a$ and $b$ are the lengths of the legs of a right triangle and $c$ is the length of the hypotenuse, then

$$a^2 + b^2 = c^2$$

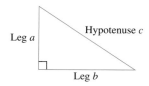

Leg $a$    Hypotenuse $c$    Leg $b$

$$\sqrt{9} = 3, \quad \sqrt{100} = 10, \quad \sqrt{1} = 1$$

Find $c$.    $a = 3$ inches    $c$    $b = 8$ inches

$$a^2 + b^2 = c^2$$

$$3^2 + 8^2 = c^2 \qquad \text{Let } a = 3 \text{ and } b = 8.$$

$$9 + 64 = c^2 \qquad \text{Multiply.}$$

$$73 = c^2 \qquad \text{Simplify.}$$

$$\sqrt{73} = c \text{ or } c \approx 8.5$$

**Name** _____ **Section** _____ **Date** _____

## CHAPTER 5 REVIEW

**(5.1)** *Determine the place value of the number 4 in each decimal.*

**1.** 23.45   tenths

**2.** 0.000345   hundred-thousandths

*Write each decimal in words.*

**3.** 23.45   twenty-three and forty-five hundredths

**4.** 0.00345
three hundred forty-five hundred-thousandths

**5.** 109.23   one hundred nine and twenty-three hundredths

**6.** 200.000032   two hundred and thirty-two millionths

*Write each decimal in standard form.*

**7.** Two and fifteen hundredths   2.15

**8.** Five hundred three and one hundred two thousandths   503.102

**9.** Sixteen thousand twenty-five and four-teen ten-thousandths   16,025.0014

*Write each decimal as a fraction or a mixed number.*

**10.** 0.16   $\dfrac{4}{25}$

**11.** 12.023   $12\dfrac{23}{1000}$

**12.** 1.0045   $1\dfrac{9}{2000}$

**13.** 0.00231   $\dfrac{231}{100,000}$

**14.** 25.25   $25\dfrac{1}{4}$

*Insert $<$, $>$, or $=$ between each pair of numbers to make a true statement.*

**15.** 0.49   0.43   $>$

**16.** 0.973   0.9730   $=$

**17.** 402.00032   402.000032   $>$

**18.** 0.230505   0.23505   $<$

*Round each decimal to the given place value.*

**19.** 0.623, nearest tenth   0.6

**20.** 0.9384, nearest hundredth   0.94

**21.** 42.895, nearest hundredth   42.90

**22.** 16.34925, nearest thousandth   16.349

**23.** Every day in America an average of 13,490.5 people get married. Round this number to the nearest hundred.   13,500 people

**24.** A certain kind of chocolate candy bar contains 10.75 teaspoons of sugar. Convert this number to a mixed number.

$$10\frac{3}{4} \text{ tsp}$$

**(5.2)** *Add.*

**25.** 2.4 + 7.1   9.5

**26.** 3.9 + 1.2   5.1

**27.** −6.4 + (−0.88)   −7.28

**28.** −19.02 + 6.98   −12.04

**29.** 200.49 + 16.82 + 103.002   320.312

**30.** 0.00236 + 100.45 + 48.29   148.74236

*Subtract.*

**31.** 4.9 − 3.2   1.7

**32.** 5.23 − 2.74   2.49

**33.** −892.1 − 432.4   −1324.5

**34.** 0.064 − 10.2   −10.136

**35.** 100 − 34.98   65.02

**36.** 200 − 0.00198   199.99802

**37.** One of the greatest one-day point losses on the Dow Jones Industrial Average (DJIA) occurred on April 14, 2000. On that day, the DJIA changed by −617.78 points and its value at the close of the stock market was 10,305.77. What was the value of the DJIA when the stock market opened on April 14, 2000? (*Source:* Dow Jones & Co.)   10,923.55 points

**38.** Evaluate $x - y$ for $x = 1.2$ and $y = 6.9$.   −5.7

*Simplify by combining like terms.*

**39.** $2.3x + 6.5 + 1.9x + 6.3$   $4.2x + 12.8$

**40.** $8.6y - 7.61 + 1.29y + 3.44$   $9.89y - 4.17$

**(5.3)** *Multiply.*

**41.** $7.2 \times 10$   $72$

**42.** $9.345 \times 1000$   $9345$

**43.** $-34.02 \times 2.3$   $-78.246$

**44.** $-839.02 \times (-87.3)$
   $73{,}246.446$

△ **45.** Find the exact circumference of the circle. Then use the approximation 3.14 for $\pi$ and approximate the circumference.   $14\pi$ m, $\approx 43.96$ m

7 meters

**46.** A kilometer is approximately 0.625 mile. It is 102 kilometers from Hays to Colby. Write 102 kilometers in miles to the nearest tenth mile.   63.8 mi

**(5.4)** *Divide. Round the quotient to the nearest thousandth if necessary.*

**47.** $21 \div 0.3$   $70$

**48.** $-0.0063 \div 0.03$   $-0.21$

**49.** $24.5 \div (-0.005)$   $-4900$

**50.** $54.98 \div 2.3$   $23.904$

**51.** $274 \div 34$   $8.059$

**52.** $-3165 \div (-20)$   $158.25$

**53.** There are approximately 3.28 feet in 1 meter. Find how many meters there are in 24 feet to the nearest tenth of a meter.   7.3 m

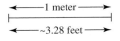

← 1 meter →

← ~3.28 feet →

**54.** George Strait pays $69.71 per month to pay back a loan of $3136.95. In how many months will the loan be paid off?   45 mo

**(5.5)** *Perform each indicated operation. Then estimate to see whether each proposed result is reasonable.*

**55.** $2.4 + 6.7 + 9.1$   $18.2$

**56.** $15.9 + 34.1$   $50$

**57.** $340.03 - 240.98$   $99.05$

**58.** $100 - 45.9$   54.1

**59.** $6.02 \times 5.91$   35.5782

**60.** $0.205 \times 1.72$   0.3526

**61.** $62.13 \div 1.9$   32.7

**62.** $601.92 \div 19.8$   30.4

**63.** The American Red Cross spent the following amounts for disaster services during fiscal years 1994 through 1998:

| Fiscal Year | Amount |
|-------------|--------|
| 1994 | $220.9 million |
| 1995 | $233.3 million |
| 1996 | $216.5 million |
| 1997 | $214.5 million |
| 1998 | $192.6 million |

Estimate the total expenditures for disaster services by the Red Cross during these years by rounding each amount to the nearest ten million. (*Source:* The American National Red Cross)   $1070 million

△ **64.** Julio Tomaso is going to fertilize his lawn, a rectangle measuring 77.3 feet by 115.9 feet. Approximate the area of the lawn by rounding each measurement to the nearest foot.   8932 sq. ft

115.9 feet

77.3 feet

**65.** Estimate the cost of the items to see whether the groceries can be purchased with a $5 bill.   Yes, $5 is enough.

$1.07     $1.89     3 cans for $0.99

*Simplify each expression.*

**66.** $(-7.6)(1.9) + 2.5$   $-11.94$

**67.** $2.3^2 - 1.4$   3.89

**68.** $\dfrac{(-3.2)^2}{100}$   0.1024

**69.** $(2.6 + 1.4)(4.5 - 3.6)$   3.6

**(5.6)** *Write each fraction as a decimal. Round to the nearest thousandth if necessary.*

**70.** $\dfrac{4}{5}$  0.8

**71.** $\dfrac{12}{13}$  0.923

**72.** $\dfrac{3}{7}$  0.429

**73.** $\dfrac{13}{60}$  0.217

**74.** $\dfrac{9}{80}$  0.113

**75.** $\dfrac{8935}{175}$  51.057

*Insert* <, >, *or* = *between each pair of numbers to form a true statement.*

**76.** 0.3920    0.392  =

**77.** $\dfrac{4}{7}$    $\dfrac{5}{8}$  <

**78.** 0.293    $\dfrac{5}{17}$  <

**79.** $\dfrac{6}{11}$    0.55  <

*Write each list of numbers in order from smallest to largest.*

**80.** 0.837, 0.839, 0.832  0.832, 0.837, 0.839

**81.** $\dfrac{3}{7}$, 0.42, 0.43  0.42, $\dfrac{3}{7}$, 0.43

**82.** $\dfrac{18}{11}$, 1.63, $\dfrac{19}{12}$  $\dfrac{19}{12}$, 1.63, $\dfrac{18}{11}$

**83.** $\dfrac{6}{7}$, $\dfrac{8}{9}$, $\dfrac{3}{4}$  $\dfrac{3}{4}$, $\dfrac{6}{7}$, $\dfrac{8}{9}$

*Find each area.*

△ **84.**  6.9 sq. ft

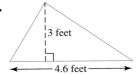

3 feet

4.6 feet

△ **85.**  5.46 sq. in.

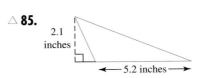

2.1 inches

5.2 inches

**(5.7)** *Solve.*

**86.** $x + 3.9 = 4.2$  0.3

**87.** $70 = y + 22.81$  47.19

**88.** $2x = 17.2$  8.6

**89.** $1.1y = 88$  80

**90.** $\dfrac{x}{4} = 0.12$  0.48

**91.** $6.8 = \dfrac{y}{5}$  34

**92.** $x + 0.78 = 1.2$    0.42

**93.** $0.56 = 2x$    0.28

**94.** $1.3x - 9.4 = 0.4x + 8.6$    20

**95.** $3(x - 1.1) = 5x - 5.3$    1

**(5.8)** *Simplify.*

**96.** $\sqrt{64}$    8

**97.** $\sqrt{144}$    12

**98.** $\sqrt{36}$    6

**99.** $\sqrt{1}$    1

**100.** $\sqrt{\dfrac{4}{25}}$    $\dfrac{2}{5}$

**101.** $\sqrt{\dfrac{1}{100}}$    $\dfrac{1}{10}$

*Find the unknown length in each given right triangle. If necessary, round to the nearest tenth.*

△ **102.** leg = 12, leg = 5    13

△ **103.** leg = 20, leg = 21    29

△ **104.** leg = 9, hypotenuse = 14    10.7

△ **105.** leg = 124, hypotenuse = 155    93

△ **106.** leg = 66, leg = 56    86.6

△ **107.** Find the length to the nearest hundredth of the diagonal of a square that has a side of length 20 centimeters.    28.28 cm

△ **108.** Find the height of the building rounded to the nearest tenth.    88.2 ft

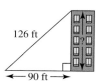

126 ft

← 90 ft →

**Name** _____

# Chapter 5 Test

_Write each decimal as indicated._

**1.** 45.092, in words

**2.** Three thousand and fifty-nine thousandths, in standard form

_Perform each indicated operation. Round the result to the nearest thousandth if necessary._

**3.** $2.893 + 4.21 + 10.492$

**4.** $-47.92 - 3.28$

**5.** $9.83 - 30.25$

**6.** $10.2 \times 4.01$

**7.** $(-0.00843) \div (-0.23)$

_Round each decimal to the indicated place value._

**8.** 34.8923, nearest tenth

**9.** 0.8623, nearest thousandth

_Insert <, >, or = between each pair of numbers to form a true statement._

**10.** 25.0909     25.9090

**11.** $\dfrac{4}{9}$     0.445

_Write each decimal as a fraction or a mixed number._

**12.** 0.345

**13.** 24.73

_Write each fraction as a decimal. If necessary, round to the nearest thousandth._

**14.** $\dfrac{13}{26}$

**15.** $\dfrac{16}{17}$

_Simplify._

**16.** $(-0.6)^2 + 1.57$

**17.** $\dfrac{0.23 + 1.63}{-0.3}$

**18.** $2.4x - 3.6 - 1.9x - 9.8$

_Find each square root and simplify. Round to the nearest thousandth if necessary._

**19.** $\sqrt{49}$

**20.** $\sqrt{157}$

**21.** $\sqrt{\dfrac{64}{100}}$

_Solve._

**22.** $0.2x + 1.3 = 0.7$

**23.** $2(x + 5.7) = 6x - 3.4$

**1.** forty-five and ninety-two thousandths

**2.** 3000.059

**3.** 17.595

**4.** −51.20

**5.** −20.42

**6.** 40.902

**7.** 0.037

**8.** 34.9

**9.** 0.862

**10.** <

**11.** <

**12.** $\dfrac{69}{200}$

**13.** $24\dfrac{73}{100}$

**14.** 0.5

**15.** 0.941

**16.** 1.93

**17.** −6.2

**18.** $0.5x - 13.4$

**19.** 7

**20.** 12.530

**21.** $\dfrac{8}{10} = \dfrac{4}{5}$

**22.** −3

**23.** 3.7

**Name** _____

△ **24.** Approximate to the nearest hundredth of a centimeter the length of the missing side of a right triangle with legs of 4 centimeters each.

△ **25.** Find the area.

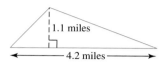
1.1 miles
4.2 miles

△ **26.** Vivian Thomas is going to put insecticide on her lawn to control grubworms. The lawn is a rectangle measuring 123.8 feet by 80 feet. The amount of insecticide required is 0.02 ounces per square foot. Find how much insecticide Vivian needs to purchase.

△ **27.** Find the exact circumference of the circle. Then use the approximation 3.14 for $\pi$ and approximate the circumference.

9 miles

**28.** A Superdisk diskette holds 120 megabytes of data. A high-density diskette holds 1.4 megabytes of data. How many high-density diskettes does it take to hold as much data as a Superdisk diskette? Round to the nearest whole diskette.

120 MB    1.4 MB

**29.** Estimate the total distance from Bayette to Center City by rounding each distance to the nearest mile.

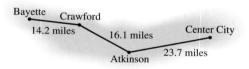
Bayette    Crawford
14.2 miles    16.1 miles    Center City
23.7 miles
Atkinson

**30.** Due to recently recognized health benefits and greater convenience, broccoli has enjoyed a surge in popularity. Use the data given in the table to complete the table. Did frozen broccoli consumption increase or decrease from 1992 to 1998? How much more total broccoli did the average American eat in 1998 than in 1992?

| AVERAGE AMERICAN BROCCOLI CONSUMPTION (POUNDS PER PERSON) | | |
|---|---|---|
| | 1992 | 1998 |
| Fresh broccoli | 3.4 | 5.6 |
| Frozen broccoli | 2.4 | 2.1 |
| Total | 5.8 | 7.7 |

(*Source:* U.S. Department of Agriculture)

# CUMULATIVE REVIEW

1. Add: $34,285 + 149,761$

2. Round each number to the nearest hundred to find an estimated sum:
$$294$$
$$625$$
$$1071$$
$$+ \ 349$$

3. Divide: $6819 \div 17$

4. Evaluate $\dfrac{x - 5y}{y}$ for $x = 21$ and $y = 3$.

5. Find the opposite of each number.

   **a.** 11    **b.** $-2$    **c.** 0

6. Add: $-2 + (-21)$

*Find the value of each expression.*

7. $2 \cdot 5^2$

8. $(-4)^3$

9. $-2^4$

10. Simplify by combining like terms.
    **a.** $3x + 2x$
    **b.** $y - 7y$
    **c.** $3x^2 + 5x^2 - 2$

11. Solve: $5x + 2 - 4x = 7 - 9$

12. Solve for $x$: $-5x = 15$

13. Solve: $7(x - 2) = 9x - 6$

14. In an election, the incumbent city council member, Laura Hartley, received 275 *more* votes than her challenger. If a total of 7857 votes were cast in the election, find how many votes the incumbent, Laura Hartley, received.

**ANSWERS**

1. 184,046 (Sec. 1.2, Ex. 2)

2. 2300 (Sec. 1.4, Ex. 4)

3. 401 R 2 (Sec. 1.6, Ex. 8)

4. 2 (Sec. 1.8, Ex. 3)

5. **a.** $-11$

   **b.** 2

   **c.** 0 (Sec. 2.1, Ex. 5)

6. $-23$ (Sec. 2.2, Ex. 3)

7. 50 (Sec. 2.5, Ex. 3)

8. $-64$ (Sec. 2.5, Ex. 4)

9. $-16$ (Sec. 2.5, Ex. 5)

10. **a.** $5x$

    **b.** $-6y$

    **c.** $8x^2 - 2$ (Sec. 3.1, Ex. 1)

11. $-4$ (Sec. 3.2, Ex. 7)

12. $-3$ (Sec. 3.3, Ex. 1)

13. $-4$ (Sec. 3.4, Ex. 4)

14. 4066 votes (Sec. 3.5, Ex. 3)

**Name** _____

*Identify the numerator and the denominator of each fraction.*

**15.** $\dfrac{3}{7}$

**16.** $\dfrac{13x}{5}$

**17.** Simplify: $\dfrac{12}{20}$

**18.** Multiply: $-\dfrac{1}{4} \cdot \dfrac{1}{2}$

*Add and simplify.*

**19.** $\dfrac{2}{7} + \dfrac{3}{7}$

**20.** $\dfrac{7}{8} + \dfrac{6}{8} + \dfrac{3}{8}$

**21.** Add: $\dfrac{2}{5} + \dfrac{4}{15}$

**22.** Simplify: $\dfrac{\frac{x}{4}}{\frac{3}{2}}$

**23.** Solve for $a$: $\dfrac{3}{5}a = 9$

**24.** Add: $763.7651 + 22.001 + 43.89$

**25.** Multiply: $23.6 \times 0.78$

# Ratio and Proportion

Having studied equations in Chapter 3 and quotients in fraction form in Chapter 4, we are ready to explore the useful notations of ratio and proportion. Ratio is another name for quotient and is usually written in fraction form. A proportion is an equation of ratios.

## CONTENTS

"CHAMPION" IMPROVED ICE CREAM FREEZERS.
Geared, With Fly Wheel and Frame. For Hand or Power.
Patented July 9, 1872. Improved, 1877.

Ice cream was probably invented in ancient China. It first made its appearance in America in the late 1700s. In fact, George Washington spent an unheard-of sum of $200 on ice cream during the summer of 1770. But making ice cream at that time was a tedious process, involving continuous stirring of ice cream mixture in bowls nestled in pots filled with salted ice. Without the stirring, the mixture would freeze into unpleasantly large ice crystals. In 1846, an American housewife named Nancy Johnson saved the world from monotonous stirring when she invented the hand-cranked churn ice cream freezer. The paddles in the freezer did the stirring, turning the mixture into the undetectably small ice crystals of a truly creamy ice cream more easily and faster than ever before. In Exercises 37–38 on page 481, we will see how a ratio can be used to describe the proper mix of rock salt and crushed ice needed in a hand-cranked ice cream freezer.

**Name** _____ **Section** _____ **Date** _____

# Chapter 6 Pretest

*Write each ratio using fractional notation.*

1. 15 to 19

2. $3\dfrac{4}{5}$ to $9\dfrac{1}{8}$

*Write each ratio as a fraction in simplest form.*

3. 5.1 to 7.9

4. $30 to $20

5. Write the rate as a fraction in simplest form: $6 for 3 pounds

6. Write as a unit rate: 292.5 miles every 15 gallons of gas

7. A store charges $3.27 for 3 boxes of cat food. What is the unit price in dollars per box?

*Write each sentence as a proportion.*

8. 18 credits is to 6 classes as 15 credits is to 5 classes.

9. 49 days is to 7 weeks as 35 days is to 5 weeks.

10. Is $\dfrac{3}{5} = \dfrac{21}{34}$ a true proportion?

11. Is $\dfrac{72}{9} = \dfrac{216}{27}$ a true proportion?

*Solve.*

12. $\dfrac{22}{x} = \dfrac{66}{12}$

13. $\dfrac{x}{8} = \dfrac{0.6}{2.4}$

14. $\dfrac{\frac{1}{2}}{8} = \dfrac{3}{y}$

15. $\dfrac{1\frac{1}{3}}{\frac{6}{5}} = \dfrac{n}{9}$

16. On Jill Tilbert's map, 1 inch represents 20 miles. Find the distance represented by $5\dfrac{3}{8}$ inches on her map.

△ 17. Find the ratio of the corresponding sides of the given similar triangles.

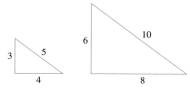

*Given that the pairs of triangles are similar, find the length of the side labeled n.*

△ 18.

△ 19.

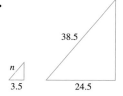

20. If a 40-foot tree casts a 22-foot shadow, find the length of the shadow cast by a 36-foot tree.

# 6.1  RATIOS

## A  WRITING RATIOS AS FRACTIONS

A **ratio** is the quotient of two quantities. A ratio, in fact, is no different than a fraction, except that a ratio is sometimes written using notation other than fractional notation. For example, the ratio of 1 to 2 can be written as

$$1 \text{ to } 2 \quad \text{or} \quad \frac{1}{2} \quad \text{or} \quad 1:2$$

fractional notation                     colon notation

These ratios are all read as "the ratio of 1 to 2."

**TRY THE CONCEPT CHECK IN THE MARGIN.**

In this section, we will write ratios using fractional notation.

> ### WRITING A RATIO AS A FRACTION
>
> The order of the quantities is important when writing ratios. To write a ratio as a fraction, write the *first number* of the ratio as the *numerator* of the fraction and the *second number* as the *denominator*.

For example, the ratio of 6 to 11 is $\frac{6}{11}$, *not* $\frac{11}{6}$.

### Example 1

Write the ratio of 12 to 17 using fractional notation.

*Solution:*    The ratio is $\frac{12}{17}$.

> **HELPFUL HINT**
> Don't forget that order is important when writing ratios. The ratio $\frac{17}{12}$ is *not* the same as the ratio $\frac{12}{17}$.

### Examples

Write each ratio using fractional notation.

**2.** The ratio of 2.6 to 3.1 is $\frac{2.6}{3.1}$.

**3.** The ratio of $1\frac{1}{2}$ to $7\frac{3}{4}$ is $\dfrac{1\frac{1}{2}}{7\frac{3}{4}}$.

## B  SIMPLIFYING RATIOS

To simplify a ratio, we just write the fraction in simplest form. Common factors can be divided out as well as common units.

### Example 4

Write the ratio of $15 to $10 as a fraction in simplest form.

*Solution:*    $\dfrac{\$15}{\$10} = \dfrac{15}{10} = \dfrac{3 \cdot 5}{2 \cdot 5} = \dfrac{3}{2}$

---

**Objectives**

**A** Write ratios as fractions.
**B** Write ratios in simplest form.

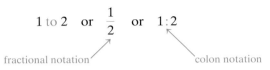

Study Guide    SSM    CD-ROM    Video 6.1

## ✓ CONCEPT CHECK

How should each ratio be read aloud?

a. $\dfrac{8}{5}$          b. $\dfrac{5}{8}$

**TEACHING TIP**
While discussing Example 1, have students come up with several situations that would have a ratio of 12 to 17. For example, this is the ratio of female students to male students in a class having 12 female students and 17 male students. Then ask, "Are there any other numbers of female students and male students in a class that would have the ratio 12 to 17?"

## Practice Problem 1

Write the ratio of 20 to 23 using fractional notation.

**TEACHING TIP**
While discussing Example 2, try giving students a real-life situation. For example, miles are often measured in tenths. This ratio could represent the ratio of miles run to miles walked if I ran 2.6 miles and walked 3.1 miles. What else could it represent?

## Practice Problems 2–3

Write each ratio using fractional notation.

2. The ratio of 10.3 to 15.1

3. The ratio of $3\frac{1}{3}$ to $12\frac{1}{5}$

## Practice Problem 4

Write the ratio of $8 to $6 as a fraction in simplest form.

**Answers**

**1.** $\dfrac{20}{23}$  **2.** $\dfrac{10.3}{15.1}$  **3.** $\dfrac{3\frac{1}{3}}{12\frac{1}{5}}$  **4.** $\dfrac{4}{3}$

✓ **Concept Check:**
**a.** Eight to five.  **b.** Five to eight.

While discussing Example 3, have students work in groups to come up with several situations that would have this ratio. Then have them share their situations with the other groups in the class. To help them get started, ask them what is commonly measured in mixed numbers with fractional parts in halves or fourths.

## Practice Problem 5

Write the ratio of 1.71 to 4.56 as a fraction in simplest form.

## Practice Problem 6

Use the circle graph below to write the ratio of work miles to total miles as a fraction in simplest form.

△ **Practice Problem 7**

Given the triangle shown:

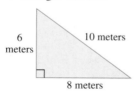

a. Find the ratio of the length of the shortest side to the length of the longest side.
b. Find the ratio of the length of the longest side to the perimeter of the triangle.

## ✓ CONCEPT CHECK

Explain why the answer $\frac{7}{5}$ would be incorrect for part (a) of Example 7.

**Answers**

5. $\frac{3}{8}$   6. $\frac{8}{25}$   7. a. $\frac{3}{5}$   b. $\frac{5}{12}$

✓ **Concept Check:** $\frac{7}{5}$ would be the ratio of the rectangle's length to its width.

---

In Example 4, although the fraction $\frac{3}{2}$ is equal to the mixed number $1\frac{1}{2}$, a ratio is a quotient of *two* quantities. For that reason, ratios are not written as mixed numbers.

If a ratio compares two decimal numbers, we will write the ratio as a ratio of whole numbers.

### Example 5

Write the ratio of 2.6 to 3.1 as a fraction in simplest form.

*Solution:*    The ratio is

$$\frac{2.6}{3.1}$$

Now let's clear the ratio of decimals.

$$\frac{2.6}{3.1} = \frac{2.6 \cdot 10}{3.1 \cdot 10} = \frac{26}{31} \quad \text{Simplest form}$$

### Example 6    Writing a Ratio from a Circle Graph

The circle graph in the margin shows the part of a car's total mileage that falls into a particular category. Write the ratio of medical miles to total miles as a fraction in simplest form.

*Solution:*

$$\frac{\text{medical miles}}{\text{total miles}} = \frac{150 \text{ miles}}{15,000 \text{ miles}} = \frac{150}{15,000} = \frac{150}{150 \cdot 100} = \frac{1}{100}$$

### △ Example 7    Given the rectangle shown:

a. Find the ratio of its width to its length.
b. Find the ratio of its length to its perimeter.

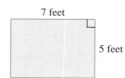

*Solution:*    a. The ratio of its width to its length is

$$\frac{\text{width}}{\text{length}} = \frac{5 \text{ feet}}{7 \text{ feet}} = \frac{5}{7}$$

b. Recall that the perimeter of the rectangle is the distance around the rectangle: $7 + 5 + 7 + 5 = 24$ feet. The ratio of its length to its perimeter is

$$\frac{\text{length}}{\text{perimeter}} = \frac{7 \text{ feet}}{24 \text{ feet}} = \frac{7}{24}$$

**TRY THE CONCEPT CHECK IN THE MARGIN.**

# EXERCISE SET 6.1

**A** *Write each ratio using fractional notation. Do not simplify. See Examples 1 through 3.*

**1.** 11 to 14    **2.** 7 to 12    ▭ **3.** 23 to 10    **4.** 8 to 5

**5.** 151 to 201    **6.** 673 to 1000    **7.** 2.8 to 7.6    **8.** 3.9 to 4.2

**9.** 5 to $7\frac{1}{2}$    **10.** $5\frac{3}{4}$ to 3    **11.** $3\frac{3}{4}$ to $1\frac{2}{3}$    **12.** $2\frac{2}{5}$ to $6\frac{1}{2}$

**B** *Write each ratio as a ratio of whole numbers using fractional notation. Write the fraction in simplest form. See Examples 4 and 5.*

▭ **13.** 16 to 24    **14.** 25 to 150    ▭ **15.** 7.7 to 10

**16.** 8.1 to 10    **17.** 4.63 to 8.21    **18.** 9.61 to 7.62

**19.** 9 inches to 12 inches    **20.** 14 centimeters to 20 centimeters

**21.** 10 hours to 24 hours    **22.** 18 quarts to 30 quarts

**23.** $32 to $100    **24.** $46 to $102

**ANSWERS**

**1.** $\frac{11}{14}$

**2.** $\frac{7}{12}$

**3.** $\frac{23}{10}$

**4.** $\frac{8}{5}$

**5.** $\frac{151}{201}$

**6.** $\frac{673}{1000}$

**7.** $\frac{2.8}{7.6}$

**8.** $\frac{3.9}{4.2}$

**9.** $\frac{5}{7\frac{1}{2}}$

**10.** $\frac{5\frac{3}{4}}{3}$

**11.** $\frac{3\frac{3}{4}}{1\frac{2}{3}}$

**12.** $\frac{2\frac{2}{5}}{6\frac{1}{2}}$

**13.** $\frac{2}{3}$

**14.** $\frac{1}{6}$

**15.** $\frac{77}{100}$

**16.** $\frac{81}{100}$

**17.** $\frac{463}{821}$

**18.** $\frac{961}{762}$

**19.** $\frac{3}{4}$

**20.** $\frac{7}{10}$

**21.** $\frac{5}{12}$

**22.** $\frac{3}{5}$

**23.** $\frac{8}{25}$

**24.** $\frac{23}{51}$

**449**

**25.** $\frac{12}{7}$

**26.** $\frac{2}{3}$

**27.** $\frac{16}{23}$

**28.** $\frac{4}{1}$

**29.** $\frac{3}{50}$

**30.** $\frac{3}{25}$

**31.** $\frac{5}{3}$

**32.** $\frac{1}{3}$

**33.** $\frac{17}{40}$

**34.** $\frac{2}{13}$

**35.** $\frac{4}{9}$

**36.** $\frac{6}{1}$

**37.** $\frac{5}{4}$

**38.** $\frac{4}{9}$

**450**

■ **25.** 24 days to 14 days

**26.** 80 miles to 120 miles

**27.** 32,000 bytes to 46,000 bytes

**28.** 600 copies to 150 copies

*Find the ratio described in each problem. For Exercises 29 and 30, use the circle graph by Practice Problem 6. See Examples 6 and 7.*

**29.** Write the ratio of vacation/other miles to total miles as a fraction in simplest form.

**30.** Write the ratio of shopping miles to total miles as a fraction in simplest form.

△ **31.** Find the ratio of the length to the width of the swimming pool.

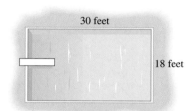
30 feet
18 feet

△ **32.** Find the ratio of the base to the height of the triangular mainsail.

18 feet (height)
6 feet (base)

△ **33.** Find the ratio of the longest side to the perimeter of the right-triangular-shaped billboard.

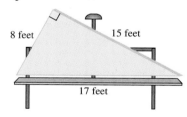
8 feet
15 feet
17 feet

△ **34.** Find the ratio of the width to the perimeter of the rectangular vegetable garden.

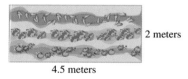
2 meters
4.5 meters

**35.** A large order of McDonald's french fries has 450 calories. Of this total, 200 calories are from fat. Find the ratio of calories from fat to total calories in a large order of McDonald's french fries. (*Source:* McDonald's Corporation)

**36.** A McDonald's Quarter Pounder® with Cheese contains 30 grams of fat. A plain McDonald's Grilled Chicken Deluxe™ sandwich contains 5 grams of fat. Find the ratio of the amount of fat in a Quarter Pounder to the amount of fat in a Grilled Chicken Deluxe. (*Source:* McDonald's Corporation)

At the Honey Island College Glee Club meeting one night, there were 125 women and 100 men present.

■ **37.** Find the ratio of women to men.

**38.** Find the ratio of men to the total number of people present.

A poll at State University revealed that 4500 students out of 6000 students are single, and the rest are married.

**39.** Find the ratio of single students to married students.

**40.** Find the ratio of married students to the total student population.

Blood contains three types of cells: red blood cells, white blood cells, and platelets. For approximately every 600 red blood cells in healthy humans, there are 40 platelets and 1 white blood cell. (*Source:* American Red Cross Biomedical Services)

**41.** Write the ratio of red blood cells to platelet cells.

**42.** Write the ratio of white blood cells to red blood cells.

**43.** When the first national minimum wage was enacted in 1938, minimum wage was $0.25 per hour. The Senate recently passed a law that would raise the minimum wage to $6.15 in March, 2002. Find the ratio of the 2002 proposed minimum wage to the 1938 minimum wage. (*Source:* U.S. Department of Labor)

**44.** Citizens of the United States eat an average of 25 pints of ice cream per year. Residents of the New England states eat an average of 39 pints of ice cream per year. Find the ratio of the amount of ice cream eaten by New Englanders to the amount eaten by the average U.S. citizen. (*Source:* International Dairy Foods Association)

**45.** At the Winter Olympic Games in Nagano, Japan, a total of 205 medals were awarded. Norwegian athletes won a total of 25 medals. Find the ratio of medals won by Norway to the total medals awarded. (*Source:* Organizing Committee for the XVIII Olympic Winter Games, Nagano)

**46.** At the Winter Olympic Games in Nagano, Japan, a total of 205 medals were awarded. Japanese athletes won a total of 10 medals. Find the ratio of medals won by Japan to the total medals awarded. (*Source:* Organizing Committee for the XVIII Olympic Winter Games, Nagano)

## REVIEW AND PREVIEW

*Divide. See Section 5.4.*

**47.** $9\overline{)20.7}$  **48.** $7\overline{)60.2}$  **49.** $3.7\overline{)0.555}$  **50.** $4.6\overline{)1.15}$

**39.** $\dfrac{3}{1}$

**40.** $\dfrac{1}{4}$

**41.** $\dfrac{15}{1}$

**42.** $\dfrac{1}{600}$

**43.** $\dfrac{103}{5}$

**44.** $\dfrac{39}{25}$

**45.** $\dfrac{5}{41}$

**46.** $\dfrac{2}{41}$

**47.** 2.3

**48.** 8.6

**49.** 0.15

**50.** 0.25

**451**

**Name** _____

 **COMBINING CONCEPTS**

**51.** As of April 2000, statewide mandatory bicycle helmet laws had been adopted by 15 states. Find the ratio of states with bicycle helmet laws to states with no such laws. (*Source:* Bicycle Helmet Safety Institute)

**52.** A total of 17 states have 200 or more public libraries. Find the ratio of states with 200 or more public libraries to states with fewer than 200 public libraries. (*Source:* U.S. Department of Education)

**53.** Suppose that the ratio of water to soap concentrate is 5 to 1. Why is it then incorrect to say that the ratio is 1 to 5.

**54.** Write the ratio $2\dfrac{1}{2}$ to $5\dfrac{3}{4}$ as a fraction in simplest form.

**55.** A grocer will refuse a shipment of tomatoes if the ratio of bruised tomatoes to the total batch is at least 1 to 10. A sample is found to contain 3 bruised tomatoes and 30 good tomatoes. Determine whether the shipment should be refused.

**56.** A pantyhose manufacturing machine will be repaired if the ratio of defective pantyhose to good pantyhose is at least 1 to 20. A quality-control engineer found 10 defective pantyhose in a batch of 200. Determine whether the machine should be repaired.

 **Internet Excursions**

Go to http://www.prenhall.com/martin-gay

One way to evaluate the performance of a business is to use ratio analysis on its financial statements. This World Wide Web address will provide you with access to information about ratio analysis and ratio calculators on the Deloitte Touche Tohmatsu Website, or a related site.

**57. a.** What, in general, do liquidity ratios measure?
   **b.** What specifically does the current ratio measure?
   **c.** Calculate the current ratio for the Greylock Company, which has $1,000,000 in current assets and $525,000 in current liabilities. Interpret Greylock Company's current ratio.

**58. a.** What, in general, do leverage ratios measure?
   **b.** What specifically does the debt/equity ratio measure?
   **c.** Calculate the debt/equity ratio for
      **i.** Carlton, Inc., which has $1,200,000 in current and long-term debt and $1,600,000 in equity
      **ii.** Suzuka Ltd., which has $2,500,000 in current and long-term debt and $2,000,000 in equity
   **d.** Which company is more reliant on external funding?

# 6.2 RATES

## A WRITING RATES AS FRACTIONS

A special type of ratio is a rate. **Rates** are used to compare *different* kinds of quantities. For example, suppose that a recreational runner can run 3 miles in 33 minutes. If we write this rate as a fraction, we have

$$\frac{3 \text{ miles}}{33 \text{ minutes}} = \frac{1 \text{ mile}}{11 \text{ minutes}} \quad \text{In simplest form}$$

### Helpful Hint

When comparing quantities with different units, write the units as part of the comparison. They do not divide out.

**Same Units:** $\dfrac{3 \text{ inches}}{12 \text{ inches}} = \dfrac{1}{4}$

**Different Units:** $\dfrac{2 \text{ miles}}{20 \text{ minutes}} = \dfrac{1 \text{ mile}}{10 \text{ minutes}}$   Units are still written.

### Example 1

Write the rate as a fraction in simplest form: 10 nails every 6 feet

*Solution:*   $\dfrac{10 \text{ nails}}{6 \text{ feet}} = \dfrac{5 \text{ nails}}{3 \text{ feet}}$

### Examples

Write each rate as a fraction in simplest form.

**2.** $2160 for 12 weeks is

$$\frac{2160 \text{ dollars}}{12 \text{ weeks}} = \frac{180 \text{ dollars}}{1 \text{ week}}$$

**3.** 360 miles on 16 gallons of gasoline is

$$\frac{360 \text{ miles}}{16 \text{ gallons}} = \frac{45 \text{ miles}}{2 \text{ gallons}}$$

## TRY THE CONCEPT CHECK IN THE MARGIN.

## B FINDING UNIT RATES

A **unit rate** is a rate with a denominator of 1. A familiar example of a unit rate is 55 mph, read as "55 **miles per hour.**" This means 55 miles per 1 hour or

$$\frac{55 \text{ miles}}{1 \text{ hour}} \quad \text{Denominator of 1}$$

### WRITING A RATE AS A UNIT RATE

To write a rate as a unit rate, divide the numerator of the rate by the denominator.

---

### Objectives

**A** Write rates as fractions.
**B** Find unit rates.
**C** Find unit prices.

Study     SSM     CD-ROM     Video
Guide                                    6.2

**TEACHING TIP**
Before beginning this lesson, ask students to estimate each of the following quantities.
  cost of long distance phone call
  cost of movie ticket
  mileage per gallon of gas
  speed of a car on the highway
Then help students see that each estimate is described using two units—an amount per an amount. These are examples of rates.

### Practice Problem 1

Write the rate as a fraction in simplest form: 12 commercials every 45 minutes

### Practice Problems 2–3

Write each rate as a fraction in simplest form.
2. $1680 for 8 weeks
3. 236 miles on 12 gallons of gasoline

### ✓ CONCEPT CHECK

True of false: $\dfrac{16 \text{ gallons}}{4 \text{ gallons}}$ is a rate.

Explain.

**Answers**

**1.** $\dfrac{4 \text{ commercials}}{15 \text{ minutes}}$  **2.** $\dfrac{210 \text{ dollars}}{1 \text{ week}}$  **3.** $\dfrac{59 \text{ miles}}{3 \text{ gallons}}$

✓ **Concept Check:** False; a rate compares different kinds of quantities.

454 CHAPTER 6 Ratio and Proportion

## Practice Problem 4

Write as a unit rate: 3600 feet every 12 seconds

TEACHING TIP

After discussing Example 4, point out that the unit rate was given in dollars per month. Then ask students if they could write it as a unit rate in dollars per year.

## Practice Problem 5

Write as a unit rate: 52 bushels of fruit from 8 trees

TEACHING TIP

Tell students that grocery stores often label the shelves with the unit prices so shoppers can easily compare prices. You may want to have students make a price comparison for different sizes of a given product. Is there any relationship between the amount of the item being purchased and the unit price?

## Practice Problem 6

An automobile rental agency charges $170 for 5 days for a certain model car. What is the unit price in dollars per day?

## Practice Problem 7

Approximate each unit price to decide which is the better buy for a bag of nacho chips: 11 ounces for $2.32 or 16 ounces for $3.59.

11 oz.    16 oz.

**Answers**

**4.** $\frac{300 \text{ feet}}{1 \text{ second}}$ or 300 feet/second

**5.** $\frac{6.5 \text{ bushels}}{1 \text{ tree}}$ or 6.5 bushels/tree

**6.** $34 per day   **7.** 11-ounce bag

---

**Example 4**   Write as a unit rate: $27,000 every 6 months

*Solution:*   $\frac{27,000 \text{ dollars}}{6 \text{ months}}$     $6\overline{)27,000}$  = 4500

The unit rate is

$\frac{4500 \text{ dollars}}{1 \text{ month}}$   or   4500 dollars/month   Read as "4500 dollars per month."

**Example 5**   Write as a unit rate: 318.5 miles every 13 gallons of gas

*Solution:*   $\frac{318.5 \text{ miles}}{13 \text{ gallons}}$     $13\overline{)318.5}$ = 24.5

The unit rate is

$\frac{24.5 \text{ miles}}{1 \text{ gallon}}$   or   24.5 miles/gallon   Read as "24.5 miles per gallon."

or   24.5 mpg

### C FINDING UNIT PRICES

Rates are used extensively in sports, business, medicine, and science applications. One of the most common uses of rates is in consumer economics. When a unit rate is "money per item" or "price per unit," it is also called a **unit price**.

**Example 6**   Finding Unit Price

A store charges $3.36 for a 16-ounce jar of picante sauce. What is the unit price in dollars per ounce?

*Solution:*   unit price $= \frac{\$3.36}{16 \text{ ounces}} = \frac{\$0.21}{1 \text{ ounce}}$   or   $0.21 per ounce

**Example 7**   Finding the Better Buy

Approximate each unit price to decide which is the better buy: $0.99 for 4 bars of soap or $1.19 for 5 bars of soap. Assume each bar is the same size.

*Solution:*

unit price $= \frac{\$0.99}{4 \text{ bars}} \approx$ $0.25 per bar of soap     $4\overline{)0.990}$ = 0.247 ≈ 0.25 ↑__(is approximately)

unit price $= \frac{\$1.19}{5 \text{ bars}} \approx$ $0.24 per bar of soap     $5\overline{)1.190}$ = 0.238 ≈ 0.24

Thus, the 5-bar package is the better buy.

# EXERCISE SET 6.2

**A** *Write each rate as a fraction in simplest form. See Examples 1 through 3.*

1. 5 shrubs every 15 feet

2. 14 lab tables for 28 students

3. 15 returns for 100 sales

4. 150 graduate students for 8 advisors

5. 8 phone lines for 36 employees

6. 6 laser printers for 28 computers

7. 18 gallons of pesticide for 4 acres of crops

8. 4 inches of rain in 18 hours

9. 6 flight attendants for 200 passengers

10. 240 pounds of grass seed for 9 lawns

11. 355 calories in a 10-fluid ounce chocolate milkshake (*Source: Home and Garden Bulletin No. 72*; U.S. Department of Agriculture)

12. 160 calories in an 8-fluid ounce serving of cream of tomato soup (*Source: Home and Garden Bulletin No. 72*; U.S. Department of Agriculture)

**B** *Write each rate as a unit rate. See Examples 4 and 5.*

13. 375 riders in 5 subway cars

14. 18 campaign yard signs in 6 blocks

15. 330 calories in a 3-ounce serving

16. 275 miles in 11 hours

17. 144 diapers for 24 babies

18. 420 feet in 3 seconds

19. $1,000,000 lottery paid over 20 years

20. 400,000 library books for 8000 students

## Answer column (left margin)

**21.** $\approx 6.67$ km/min

**22.** 2.5 vehicles/space

**23.** 1,225,000 voters/senator

**24.** 68,300 residents/parish

**25.** 300 good/defective

**26.** 1,250,000 tickets/winner

**27.** 0.48 tons/acre

**28.** 250 checks/American

**29.** $50,000/species

**30.** $6284/hr

## Problems

**B** *Write each rate as a unit rate. See Examples 4 and 5.*

**13.** 375 riders in 5 subway cars

**14.** 18 campaign yard signs in 6 blocks

**23.** The state of Tennessee has 2,450,000 registered voters for 2 senators. (*Source:* U.S. Bureau of the Census)

**24.** The state of Louisiana has 4,371,200 residents for 64 parishes. (*Source:* U.S. Bureau of the Census)

**25.** 12,000 good assemblyline products to 40 defective products

**26.** 5,000,000 lottery tickets for 4 lottery winners

**27.** 12 million tons of dust and dirt are trapped by the 25 million acres of lawns in the United States each year. (*Source:* Professional Lawn Care Association of America)

**28.** Approximately 65,000,000,000 checks are written each year by a total of approximately 260,000,000 Americans. (*Source:* Board of Governors of the Federal Reserve System)

**29.** The National Zoo in Washington, D.C., has an annual budget of $20,000,000 for its approximately 400 different species. (*Source: World Almanac 2000*)

**30.** On average, it costs $25,136 for 4 hours of flight in a B747-100 aircraft. (*Source:* Air Transport Association of America)

**31.** In Maryland, there is $123,219,000 of funding for 24 public libraries. (*Source:* U.S. Department of Education)

**32.** The top-grossing concert tour in North America was the 1994 Rolling Stones tour, which grossed $121,200,000 for 60 shows. (*Source:* Pollstar)

**33.** Greer Krieger can assemble 250 computer boards in an 8-hour shift, while Lamont Williams can assemble 400 computer boards in a 12-hour shift.
  **a** Find the unit rate of Greer.
  **b.** Find the unit rate of Lamont.
  **c.** Who can assemble computer boards faster, Greer or Lamont?

**34.** Jerry Stein laid 713 bricks in 46 minutes, while his associate, Bobby Burns, laid 396 bricks in 30 minutes.
  **a.** Find the unit rate of Jerry.
  **b.** Find the unit rate of Bobby.
  **c.** Who is the faster bricklayer?

**C** *Find each unit price. See Example 6.*

**35.** $57.50 for 5 compact disks

**36.** $0.87 for 3 apples

**37.** $1.19 for 7 bananas

**38.** $73.50 for 6 lawn chairs

*Find each unit price and decide which is the better buy. Round to three decimal places. Assume that we are comparing different sizes of the same brand. See Example 7.*

**39.** Crackers:
    $1.19 for 8 ounces

**40** Pickles:
    $1.89 for 32 ounces

$1.59 for 12 ounces

$0.89 for 18 ounces

---

**31.** $5,134,125/library

**32.** $2,020,000/show

**33. a.** 31.25 computer boards/hr
  **b.** ≈33.3 computer boards/hr
  **c.** Lamont

**34. a.** 15.5 bricks/min
  **b.** 13.2 bricks/min
  **c.** Jerry

**35.** $11.50 per compact disk

**36.** $0.29 per apple

**37.** $0.17 per banana

**38.** $12.25 per lawn chair

**39.** 8 oz: $0.149 per oz; 12 oz: $0.133 per oz; 12-oz size

**40.** 32 oz: $0.059 per oz; 18 oz: $0.049 per oz; 18-oz size

**Name** _____

**41.** Frozen orange juice:
$1.69 for 16 ounces

$0.69 for 6 ounces

**42.** Eggs:
$0.69 per dozen

$2.10 for a flat $\left(2\frac{1}{2}\text{ dozen}\right)$

**43.** Soy sauce:
$2.29 for 12 ounces

$1.49 for 8 ounces

**44.** Shampoo:
$1.89 for 20 ounces

$3.19 for 32 ounces

**45.** Napkins:
100 for $0.59

180 for $0.93

**46.** Crackers:
$2.39 for 20 ounces

$0.99 for 8 ounces

**Name** _____

## REVIEW AND PREVIEW

*Multiply or divide as indicated. See Sections 5.3 and 5.4.*

**47.**   $\begin{array}{r} 1.7 \\ \times\ 6 \\ \hline \end{array}$

**48.**   $\begin{array}{r} 2.3 \\ \times\ 9 \\ \hline \end{array}$

**49.**   $\begin{array}{r} 3.7 \\ \times\ 1.2 \\ \hline \end{array}$

**50.**   $\begin{array}{r} 6.6 \\ \times\ 2.5 \\ \hline \end{array}$

**51.** $2.3\overline{)4.37}$

**52.** $3.5\overline{)22.75}$

## COMBINING CONCEPTS

**53.** Fill in the table to calculate miles per gallon.

| Beginning Odometer Reading | Ending Odometer Reading | Miles Driven | Gallons of Gas Used | Miles Per Gallon (round to the nearest tenth) |
|---|---|---|---|---|
| 79,286 | 79,543 | | 13.4 | |
| 79,543 | 79,895 | | 15.8 | |
| 79,895 | 80,242 | | 16.1 | |

*Find each unit rate.*

**54.** Sammy Joe Wingfield from Arlington, Texas, is the fastest bricklayer on record. On May 20, 1994, he laid 1048 bricks in 60 minutes. Find his unit rate of bricks per minute rounded to the nearest tenth of a brick. (*Source: The Guinness Book of Records*, 1996)

**55.** The longest stairway is the service stairway for the Niesenbahn Cable railway near Spiez, Switzerland. It has 11,674 steps and rises to a height of 7759 feet. Find the unit rate of steps per foot rounded to the nearest tenth of a step. (*Source: The Guinness Book of Records*, 1996)

**47.** 10.2 _____

**48.** 20.7 _____

**49.** 4.44 _____

**50.** 16.5 _____

**51.** 1.9 _____

**52.** 6.5 _____

**53.** miles driven: 257, 352, 347; miles per gallon: 19.2, 22.3, 21.6 _____

**54.** 17.5 bricks/min _____

**55.** 1.5 steps/ft _____

**Name** _____

**56.** In New Mexico, the total enrollment in public schools was 329,640 in a recent year. At the same time, there were 19,368 public school teachers. Write a unit rate in students per teacher. Round to the nearest whole student. (*Source:* National Center for Education Statistics)

**57.** In Utah, the total number of students enrolled in public schools during a recent year was 477,121. At the same time, there were 20,039 public school teachers. Write a unit rate in students per teacher. Round to the nearest whole. (*Source:* National Center for Education Statistics)

**58.** Suppose that the amount of a product decreases, say from an 80-ounce container to a 70-ounce container, but the price of the container remains the same. Does the unit price increase or decrease? Explain why.

**59.** In your own words, define rate.

# 6.3 PROPORTIONS

## A WRITING PROPORTIONS

A **proportion** is a statement that two ratios or rates are equal.

> **PROPORTION**
>
> A proportion states that two ratios are equal. If $\dfrac{a}{b}$ and $\dfrac{c}{d}$ are two ratios, then
>
> $\dfrac{a}{b} = \dfrac{c}{d}$ is a proportion.

For example, $\dfrac{5}{6} = \dfrac{10}{12}$ is a proportion. When we want to emphasize this equation as a proportion, we read it as "5 is to 6 as 10 is to 12."

**Example 1**    Write each sentence as a proportion.

    **a.** 12 diamonds is to 15 rubies as 4 diamonds is to 5 rubies.

    **b.** 5 hits is to 9 at bats as 20 hits is to 36 at bats.

*Solution:*   **a.** $\begin{array}{c} \text{diamonds} \rightarrow \\ \text{rubies} \rightarrow \end{array} \dfrac{12}{15} = \dfrac{4}{5} \begin{array}{c} \leftarrow \text{diamonds} \\ \leftarrow \text{rubies} \end{array}$

       **b.** $\begin{array}{c} \text{hits} \rightarrow \\ \text{at bats} \rightarrow \end{array} \dfrac{5}{9} = \dfrac{20}{36} \begin{array}{c} \leftarrow \text{hits} \\ \leftarrow \text{at bats} \end{array}$

> **Helpful Hint**
>
> Notice in the above examples of proportions that numerators contain the same units and denominators contain the same units. In this text, proportions will be written in this way.

## B DETERMINING WHETHER PROPORTIONS ARE TRUE

Like other mathematical statements, a proportion may be either true or false. A proportion is true if its ratios are equal. Since ratios are fractions, one way to determine whether a proportion is true is to write each fraction in simplest form and compare them. Another way to see if a proportion is true is by comparing cross products.

> **USING CROSS PRODUCTS TO DETERMINE WHETHER PROPORTIONS ARE TRUE OR FALSE**
>
> $$\text{cross product} \rightarrow \overbrace{b \cdot c} \quad \dfrac{a}{b} \diagup\!\!\!\!= \dfrac{c}{d} \quad \overbrace{a \cdot d} \leftarrow \text{cross product}$$
>
> If cross products are *equal*, the proportion is *true*.
>
> If cross products are *not equal*, the proportion is *false*.

**Practice Problem 1**

Write each sentence as a proportion.

a. 24 right is to 6 wrong as 4 right is to 1 wrong.

b. 32 Cubs fans is to 18 Mets fans as 16 Cubs fans is to 9 Mets fans.

TEACHING TIP

Before beginning this lesson, ask students if they ever use the idea of proportions in everyday life. If someone volunteers, ask them how they use proportions. If nobody thinks they use proportions, point out that they do any time they double a recipe, or reduce or enlarge a picture or diagram.

**Answers**

**1. a.** $\dfrac{24}{6} = \dfrac{4}{1}$    **b.** $\dfrac{32}{18} = \dfrac{16}{9}$

**Practice Problem 2**

Is $\dfrac{3}{6} = \dfrac{4}{8}$ a true proportion?

**Practice Problem 3**

Is $\dfrac{3.6}{6} = \dfrac{5.4}{8}$ a true proportion?

**Practice Problem 4**

Is $\dfrac{4\frac{1}{5}}{2\frac{1}{3}} = \dfrac{3\frac{3}{10}}{1\frac{5}{6}}$ a true proportion?

✓ **CONCEPT CHECK**

Using the numbers in the proportion $\dfrac{21}{27} = \dfrac{7}{9}$, write two other true proportions.

**Answers**

**2.** yes  **3.** no  **4.** yes

✓ **Concept Check:** One way is $\dfrac{27}{21} = \dfrac{9}{7}$.

**Example 2**  Is $\dfrac{2}{3} = \dfrac{4}{6}$ a true proportion?

*Solution:*

$$\dfrac{2}{3} \,\,\, \diagdown\!\!\!\!\diagup \,\,\, \dfrac{4}{6}$$

$$3 \cdot 4 \overset{?}{=} 2 \cdot 6$$

$$12 = 12 \quad \text{Equal}$$

Since the cross products are equal, the proportion is true.

**Example 3**  Is $\dfrac{4.1}{7} = \dfrac{2.9}{5}$ a true proportion?

*Solution:*

$$\dfrac{4.1}{7} \,\,\, \diagdown\!\!\!\!\diagup \,\,\, \dfrac{2.9}{5}$$

$$7 \cdot 2.9 \overset{?}{=} 4.1 \cdot 5$$

$$20.3 \neq 20.5 \quad \text{Not equal}$$

Since the cross products are not equal, $\dfrac{4.1}{7} \neq \dfrac{2.9}{5}$. The proportion is not a true proportion.

**Example 4**  Is $\dfrac{1\frac{1}{6}}{10\frac{1}{2}} = \dfrac{\frac{1}{2}}{4\frac{1}{2}}$ a true proportion?

*Solution:*

$$\dfrac{1\frac{1}{6}}{10\frac{1}{2}} \,\,\, \diagdown\!\!\!\!\diagup \,\,\, \dfrac{\frac{1}{2}}{4\frac{1}{2}}$$

$$10\frac{1}{2} \cdot \frac{1}{2} \overset{?}{=} 1\frac{1}{6} \cdot 4\frac{1}{2}$$

$$\frac{21}{2} \cdot \frac{1}{2} \overset{?}{=} \frac{7}{6} \cdot \frac{9}{2}$$

$$\frac{21}{4} = \frac{21}{4} \quad \text{Equal}$$

Since the cross products are equal, the proportion is true.

**TRY THE CONCEPT CHECK IN THE MARGIN.**

**C  FINDING UNKNOWN NUMBERS IN PROPORTIONS**

When one number of a proportion is unknown, we can use cross products to find the unknown number. For example, to find the unknown number $x$ in the proportion $\dfrac{2}{3} = \dfrac{x}{30}$, we use cross products.

**Example 5**    Solve $\frac{2}{3} = \frac{x}{30}$ for $x$.

*Solution:*    If the cross products are equal, then the proportion is true. We begin, then, by setting cross products equal to each other.

$$\frac{2}{3} \diagdown\!\!\!\diagup \frac{x}{30}$$

$3 \cdot x = 2 \cdot 30$    Set cross products equal.

$3x = 60$    Multiply.

Recall that to find $x$, we divide both sides of the equation by 3.

$\frac{3x}{3} = \frac{60}{3}$    Divide both sides by 3.

$x = 20$    Simplify.

To check, we replace $x$ with 20 in the original proportion to see if the result is a true statement.

$\frac{2}{3} = \frac{x}{30}$    Original proportion

$\frac{2}{3} = \frac{20}{30}$    Replace $x$ with 20.

$\frac{2}{3} = \frac{2}{3}$    True.

Since $\frac{2}{3} = \frac{2}{3}$ is a true statement, 20 is the solution. ▬▬

**Example 6**    Solve $\frac{25}{x} = \frac{20}{4}$ for $x$ and then check.

*Solution:*

$$\frac{25}{x} \diagdown\!\!\!\diagup \frac{20}{4}$$

$20x = 25 \cdot 4$    Set cross products equal.

$20x = 100$    Multiply.

$\frac{20x}{20} = \frac{100}{20}$    Divide both sides by 20.

$x = 5$    Simplify.

*Check:*    $\frac{25}{x} = \frac{20}{4}$    Original proportion

$\frac{25}{5} = \frac{20}{4}$    Replace $x$ with 5.

$5 = 5$    Simplify.

Since 5 makes the original proportion true, 5 is the solution. ▬▬

**Practice Problem 5**

Solve $\frac{2}{7} = \frac{x}{35}$ for $x$.

**Practice Problem 6**

Solve $\frac{2}{15} = \frac{y}{60}$ for $y$ and then check.

**Answers**

**5.** 10   **6.** 8

**Practice Problem 7**

Solve for $z$: $\dfrac{8}{z} = \dfrac{1}{9}$

**Example 7**    Solve for $y$: $\dfrac{6}{1} = \dfrac{7}{y}$

*Solution:*

$$\dfrac{6}{1} = \dfrac{7}{y}$$

$7 \cdot 1 = 6y$    Set cross products equal.

$7 = 6y$    Multiply.

$\dfrac{7}{6} = \dfrac{6y}{6}$    Divide both sides by 6.

$\dfrac{7}{6} = y$

Verify that $\dfrac{7}{6}$ is the solution.

*Check:*    To check, replace $y$ with $\dfrac{7}{6}$ in the original proportion and see that a true statement results. For this example, let's check by cross products.

$$\dfrac{6}{1} = \dfrac{7}{y}$$

$$\dfrac{6}{1} = \dfrac{7}{\frac{7}{6}}$$

$7 \cdot 1 = 6 \cdot \dfrac{7}{6}$

$7 = 7$    True.

True, so the solution is $\dfrac{7}{6}$.

**Practice Problem 8**

Solve for $y$: $\dfrac{y}{6} = \dfrac{0.7}{1.2}$

**Example 8**    Solve for $x$: $\dfrac{x}{3} = \dfrac{0.8}{1.5}$

*Solution:*

$$\dfrac{x}{3} = \dfrac{0.8}{1.5}$$

$3(0.8) = 1.5x$    Set cross products equal.

$2.4 = 1.5x$    Multiply.

$\dfrac{2.4}{1.5} = \dfrac{1.5x}{1.5}$    Divide both sides by 1.5.

$1.6 = x$    Simplify.

*Check:*    To check, replace $x$ with 1.6 in the original proportion to see if the result is a true statement.

$$\dfrac{x}{3} = \dfrac{0.8}{1.5}$$

$$\dfrac{1.6}{3} = \dfrac{0.8}{1.5}$$

$3(0.8) = (1.6)(1.5)$    Set cross products equal.

$2.4 = 2.4$    True.

True, so the solution is 1.6.

**Answers**

**7.** 72    **8.** 3.5

**Example 9**  Solve for $y$: $\dfrac{14}{y} = \dfrac{12}{16}$

*Solution:*

$$\dfrac{14}{y} = \dfrac{12}{16}$$

$12y = 224$  Set cross products equal.

$\dfrac{12y}{12} = \dfrac{224}{12}$  Divide both sides by 12.

$y = \dfrac{56}{3}$  Simplify.

Check to see that the solution is $\dfrac{56}{3}$.

---

**Helpful Hint**

In Example 9, the fraction $\dfrac{12}{16}$ may be simplified to $\dfrac{3}{4}$ before solving the equation. The solution will remain the same.

---

**Example 10**  Solve for $x$: $\dfrac{1.6}{1.1} = \dfrac{x}{0.3}$. Round the solution to the nearest hundredth.

*Solution:*

$$\dfrac{1.6}{1.1} = \dfrac{x}{0.3}$$

$1.1x = 0.48$  Set cross products equal.

$\dfrac{1.1x}{1.1} = \dfrac{0.48}{1.1}$  Divide both sides by 1.1.

$x \approx 0.44$  Round to the nearest hundredth.

**TRY THE CONCEPT CHECK IN THE MARGIN.**

**Practice Problem 9**

Solve for $z$: $\dfrac{15}{z} = \dfrac{8}{10}$

**Practice Problem 10**

Solve for $y$: $\dfrac{3.4}{1.8} = \dfrac{y}{3}$. Round the solution to the nearest tenth.

**✓ CONCEPT CHECK**

True or false: the first step in solving the proportion $\dfrac{4}{z} = \dfrac{12}{15}$ yields the equation $4z = 180$.

**Answers**

**9.** $\dfrac{75}{4}$ or 18.75  **10.** 5.7

**✓ Concept Check:** False.

# Focus on Business and Career

### COST OF LIVING

In the working world, you may find it necessary to relocate to get the job you want. When considering job offers in different cities, you should keep in mind that the cost of living in one city is not necessarily the same as the cost of living in another. For example, an annual salary of $25,000 might be plenty to live on in one city but might barely cover housing expenses in another city.

A cost-of-living index helps us gauge how much more or less expensive it is to live in one city when compared to another. With it, we can find the salary level needed in one city to be equivalent to a given salary in another city using the following proportion:

| COST-OF-LIVING INDEX | |
|---|---|
| City | Index Value |
| Atlanta, GA | 99.5 |
| Boston, MA | 142.5 |
| Chicago, IL | 105.3 |
| Cleveland, OH | 104.4 |
| Houston, TX | 93.0 |
| Los Angeles, CA | 119.7 |
| Miami, FL | 107.7 |
| New Orleans, LA | 94.9 |
| New York, NY | 234.5 |
| Oklahoma City, OK | 90.0 |
| Pittsburgh, PA | 109.5 |

(*Source:* Virtual Relocation.com, Inc.)

$$\frac{\text{equivalent salary in City A}}{\text{index value for City A}} = \frac{\text{salary in City B}}{\text{index value for City B}}$$

For example, we can use the given cost-of-living index to find the salary that a person would need to earn in Chicago to be equivalent to his or her current annual salary of $25,000 in New Orleans. The index value is 105.3 for Chicago and 94.9 for New Orleans. We will let $x$ represent the equivalent salary in Chicago,

$$\frac{\text{equivalent salary in Chicago}}{\text{index value for Chicago}} = \frac{\text{salary in New Orleans}}{\text{index value for New Orleans}}$$

$$\frac{x}{105.3} = \frac{25,000}{94.9}$$

$$94.9 \cdot x = 105.3(25,000)$$

$$94.9x = 2,632,500$$

$$\frac{94.9x}{94.9} = \frac{2,632,500}{94.9}$$

$$x \approx \$27,740$$

Thus, the person would need to earn $27,740 in Chicago to be equivalent to his or her current salary of $25,000 in New Orleans. Another way to look at this situation is that what this person can afford with $25,000 in New Orleans would require $27,740 in Chicago.

### CRITICAL THINKING

1. In which city on the Cost-of-Living Index list is it most expensive to live? In which city is it least expensive to live? Explain your reasoning.

2. Suppose you currently live in Cleveland, Ohio, and earn an annual salary of $28,500. You are considering moving to Los Angeles, California. How much would you have to earn in Los Angeles to maintain the same standard of living you have in Cleveland?

3. Suppose you have just graduated from college and have been offered two comparable jobs in different cities. Job A is in Miami, Florida, and pays $32,000 per year. Job B is in Boston, Massachusetts, and pays $39,000 per year. Which job offer would you choose? Explain your reasoning.

**Name** _____ **Section** _____ **Date** _____

## MENTAL MATH

*State whether each proportion is true or false.*

**1.** $\dfrac{2}{1} = \dfrac{6}{3}$

**2.** $\dfrac{3}{1} = \dfrac{15}{5}$

**3.** $\dfrac{1}{2} = \dfrac{3}{5}$

**4.** $\dfrac{2}{11} = \dfrac{1}{5}$

**5.** $\dfrac{2}{3} = \dfrac{4}{6}$

**6.** $\dfrac{3}{4} = \dfrac{6}{8}$

## EXERCISE SET 6.3

**A** *Write each sentence as a proportion. See Example 1.*

**1.** 10 diamonds is to 6 opals as 5 diamonds is to 3 opals.

**2.** 1 raisin is to 5 cornflakes as 8 raisins is to 40 cornflakes.

**3.** 3 printers is to 12 computers as 1 printer is to 4 computers.

**4.** 4 hit songs is to 16 releases as 1 hit song is to 4 releases.

**5.** 6 eagles is to 58 sparrows as 3 eagles is to 29 sparrows.

**6.** 12 errors is to 8 pages as 1.5 errors is to 1 page.

**7.** $2\dfrac{1}{4}$ cups of flour is to 24 cookies as $6\dfrac{3}{4}$ cups of flour is to 72 cookies.

**8.** $1\dfrac{1}{2}$ cups milk is to 10 bagels as $\dfrac{3}{4}$ cup milk is to 5 bagels.

**9.** 22 vanilla wafers is to 1 cup of cookie crumbs as 55 vanilla wafers is to 2.5 cups of cookie crumbs. (*Source:* Based on data from *Family Circle* magazine)

**10.** 1 cup of instant rice is to 1.5 cups cooked rice as 1.5 cups of instant rice is to 2.25 cups of cooked rice. (*Source:* Based on data from *Family Circle* magazine)

**B** *Determine whether each proportion is true or false. See Examples 2 through 4.*

**11.** $\dfrac{15}{9} = \dfrac{5}{3}$    **12.** $\dfrac{8}{6} = \dfrac{20}{15}$    **13.** $\dfrac{8}{6} = \dfrac{9}{7}$    **14.** $\dfrac{7}{12} = \dfrac{4}{7}$

**15.** $\dfrac{9}{36} = \dfrac{2}{8}$    **16.** $\dfrac{8}{24} = \dfrac{3}{9}$    **17.** $\dfrac{5}{8} = \dfrac{625}{1000}$    **18.** $\dfrac{30}{50} = \dfrac{600}{1000}$

**19.** $\dfrac{0.8}{0.3} = \dfrac{0.2}{0.6}$    **20.** $\dfrac{0.7}{0.4} = \dfrac{0.3}{0.1}$    **21.** $\dfrac{4.2}{8.4} = \dfrac{5}{10}$    **22.** $\dfrac{8}{10} = \dfrac{5.6}{0.7}$

**23.** $\dfrac{\frac{3}{4}}{\frac{4}{3}} = \dfrac{\frac{1}{2}}{\frac{8}{9}}$    **24.** $\dfrac{\frac{2}{5}}{\frac{2}{7}} = \dfrac{\frac{1}{10}}{\frac{1}{3}}$    **25.** $\dfrac{2\frac{2}{5}}{\frac{2}{3}} = \dfrac{\frac{10}{9}}{\frac{1}{4}}$    **26.** $\dfrac{5\frac{5}{8}}{\frac{5}{3}} = \dfrac{4\frac{1}{2}}{1\frac{1}{5}}$

**C** *Solve each proportion for the given variable. Round the solution where indicated. See Examples 5 through 10.*

**27.** $\dfrac{x}{5} = \dfrac{6}{10}$    **28.** $\dfrac{x}{3} = \dfrac{12}{9}$    **29.** $\dfrac{30}{10} = \dfrac{15}{y}$    **30.** $\dfrac{25}{100} = \dfrac{7}{y}$

**31.** $\dfrac{z}{8} = \dfrac{50}{100}$    **32.** $\dfrac{12}{18} = \dfrac{z}{21}$    **33.** $\dfrac{n}{6} = \dfrac{8}{15}$    **34.** $\dfrac{24}{n} = \dfrac{60}{96}$

**35.** $\dfrac{12}{10} = \dfrac{x}{16}$

**36.** $\dfrac{18}{54} = \dfrac{3}{x}$

**37.** $\dfrac{\frac{n}{6}}{\frac{6}{5}} = \dfrac{4\frac{1}{6}}{6\frac{2}{3}}$

**38.** $\dfrac{8}{\frac{1}{3}} = \dfrac{24}{n}$

**39.** $\dfrac{\frac{3}{4}}{12} = \dfrac{y}{48}$

**40.** $\dfrac{\frac{11}{4}}{\frac{25}{8}} = \dfrac{7\frac{3}{5}}{y}$

**41.** $\dfrac{\frac{2}{3}}{\frac{6}{9}} = \dfrac{12}{z}$

**42.** $\dfrac{z}{24} = \dfrac{\frac{5}{8}}{3}$

**43.** $\dfrac{n}{0.6} = \dfrac{0.05}{12}$

**44.** $\dfrac{0.2}{0.7} = \dfrac{8}{n}$

**45.** $\dfrac{3.5}{12.5} = \dfrac{7}{z}$

**46.** $\dfrac{7.8}{13} = \dfrac{z}{2.6}$

**47.** $\dfrac{3.2}{0.3} = \dfrac{x}{1.4}$. Round to the nearest tenth.

**48.** $\dfrac{1.8}{n} = \dfrac{2.5}{8.4}$. Round to the nearest tenth.

**49.** $\dfrac{z}{5.2} = \dfrac{0.08}{6}$. Round to the nearest hundredth.

**50.** $\dfrac{4.25}{6.03} = \dfrac{5}{y}$. Round to the nearest hundredth.

**51.** $\dfrac{9}{11} = \dfrac{x}{4}$. Round to the nearest tenth.

**52.** $\dfrac{24}{x} = \dfrac{7}{3}$. Round to the nearest thousandth.

**53.** $\dfrac{43}{17} = \dfrac{8}{z}$. Round to the nearest thousandth.

**54.** $\dfrac{n}{12} = \dfrac{18}{7}$. Round to the nearest hundredth.

**Name** _____

## REVIEW AND PREVIEW

*Insert $<$ or $>$ between each pair of numbers to form a true statement. See Sections 2.1 and 4.8.*

**55.** $-8$   $8$     **56.** $7$   $-7$     **57.** $-2$   $-3$     **58.** $0$   $-2$

**59.** $-5$   $0$     **60.** $-5\dfrac{1}{3}$   $-6\dfrac{2}{3}$     **61.** $-1\dfrac{1}{2}$   $-2\dfrac{1}{2}$     **62.** $-4$   $-1$

## COMBINING CONCEPTS

*For each proportion, find the unknown number, n.*

**63.** $\dfrac{n}{7} = \dfrac{0}{8}$

**64.** $\dfrac{0}{2} = \dfrac{n}{3.5}$

**65.** $\dfrac{n}{1150} = \dfrac{588}{483}$

**66.** $\dfrac{585}{n} = \dfrac{117}{474}$

**67.** $\dfrac{222}{1515} = \dfrac{37}{n}$

**68.** $\dfrac{1425}{1062} = \dfrac{n}{177}$

**69.** Explain the difference between a ratio and a proportion.

**70.** Explain how to find the unknown number in a proportion such as $\dfrac{n}{18} = \dfrac{12}{8}$.

# INTEGRATED REVIEW—RATIO, RATE, AND PROPORTION

*Write each ratio as a ratio of whole numbers using fractional notation. Write the fraction in simplest form.*

**1.** 18 to 20

**2.** 36 to 100

**3.** 8.6 to 10

**4.** 1.6 to 4.6

**5.** 8.65 to 6.95

**6.** 7.2 to 8.4

**7.** $3\frac{1}{2}$ to 13

**8.** $1\frac{2}{3}$ to $2\frac{3}{4}$

**9.** 8 inches to 12 inches

**10.** 3 hours to 24 hours

*Find the ratio described in each problem.*

**11.** American women between the ages of 25 and 54 watch an average of 3 hours of television between 4:30 P.M. and 7:30 P.M. Monday through Friday. American men between the ages of 25 and 54 watch an average of $2\frac{7}{20}$ hours of television in the same time slot. Find the ratio of the time spent by women watching television to the time spent by men watching television during this time slot. (*Source:* Nielsen Media Research)

**12.** At the end of 1998, Marriott International had $6232 million in assets and $944 million in long-term debts. Find the ratio of assets to long-term debt. (*Source:* Marriott International, Inc.)

*Write each rate as a fraction in simplest form.*

**13.** 5 offices for every 20 graduate assistants

**14.** 6 lights every 15 feet

**15.** 100 U.S. Senators for 50 states

**16.** 5 teachers for every 140 students

**17.** 55 mph

**18.** 140 ft/sec

**19.** 21 employees/ fax line

**20.** 17 phone calls/ teenager

**21.** 8 lb: $0.27 per lb; 18 lb: $0.28 per lb; 8-lb size

**22.** 100:$0.020 per plate; 500: $0.018 per plate; 500 paper plates

**23.** 3 packs: $0.80 per pack; 8 packs: $0.75 per pack; 8 packs

**24.** 4: $0.92 per battery; 10: $0.99 per battery; 4 batteries

**Name** _____

*Write each rate as a unit rate.*

**17.** 165 miles in 3 hours

**18.** 560 feet in 4 seconds

**19.** 63 employees per 3 fax lines

**20.** 85 phone calls for 5 teenagers

*Write each unit price and decide which is the better buy.*

**21.** Dog food:
8 pounds for $2.16
18 pounds for $4.99

**22.** Paper plates:
100 for $1.98
500 for $8.99

**23.** Microwave popcorn:
3 packs for $2.39
8 packs for $5.99

**24.** AA batteries:
4 for $3.69
10 for $9.89

# 6.4    PROPORTIONS AND PROBLEM SOLVING

## A    SOLVING PROBLEMS BY WRITING PROPORTIONS

Writing proportions is a powerful tool for solving problems in almost every field, including business, chemistry, biology, health sciences, and engineering, as well as in daily life, too. Given a specified ratio (or rate) of two quantities, a proportion can be used to determine an unknown quantity.

**Example 1**    Determining Distances from a Map

On a Chamber of Commerce map of Abita Springs, 5 miles corresponds to 2 inches. How many miles correspond to 7 inches?

*Solution:*

1. UNDERSTAND. Read and reread the problem. You may want to draw a diagram.

| 5 miles | 5 miles | 5 miles | ? | = a little over 15 miles |
|---------|---------|---------|-----|--------------------------|
| 2 inches | 2 inches | 2 inches | 1 inch | = 7 inches |

From the diagram we can see that our solution should be a little over 15 miles.

2. TRANSLATE. We will let $n$ represent our unknown number. Since 5 miles corresponds to 2 inches as $n$ miles corresponds to 7 inches, we have the proportion:

$$\text{miles} \rightarrow \frac{5}{2} = \frac{n}{7} \leftarrow \text{miles}$$
$$\text{inches} \rightarrow \phantom{\frac{5}{2}} \phantom{=} \phantom{\frac{n}{7}} \leftarrow \text{inches}$$

3. SOLVE.

$$\frac{5}{2} = \frac{n}{7}$$

$$2n = 35 \qquad \text{Set cross products equal.}$$

$$\frac{2n}{2} = \frac{35}{2} \qquad \text{Divide both sides by 2.}$$

$$n = 17.5 \qquad \text{Simplify.}$$

4. INTERPRET. *Check* your work. This result is reasonable since it is a little over 15 miles. *State* your conclusion: 7 inches corresponds to 17.5 miles.

---

**Objective**

**A** Solve problems by writing proportions.

Study Guide    SSM    CD-ROM    Video 6.4

**Practice Problem 1**

On an architect's blueprint, 1 inch corresponds to 12 feet. How long is a wall represented by a $3\frac{1}{2}$-inch line on the blueprint?

TEACHING TIP

When students are writing their own proportion, remind then to check and make sure that units in the numerators are the same and that units in the denominators are the same. (Although other proportions are correct in this text, we will only show these types.)

**Answer**

**1.** 42 feet

> **Helpful Hint**
>
> We can also solve Example 1 by writing the proportion
>
> $$\frac{2 \text{ inches}}{5 \text{ miles}} = \frac{7 \text{ inches}}{n \text{ miles}}$$
>
> Although other proportions may be used to solve Example 1, we will solve by writing proportions so that the numerators have the same unit measures and the denominators have the same unit measures.

## Practice Problem 2

An auto mechanic recommends that 3 ounces of isopropyl alcohol be mixed with a tankful of gas (14 gallons) to increase the octane of the gasoline for better engine performance. At this rate, how many gallons of gas can be treated with a 16-ounce bottle of alcohol?

### Example 2     Finding Medicine Dosage

The standard dose of an antibiotic is 4 cc (cubic centimeters) for every 25 pounds (lb) of body weight. At this rate, find the standard dose for a 140-lb woman.

*Solution:*     **1.** UNDERSTAND. Read and reread the problem. You may want to draw a diagram to estimate a reasonable solution.

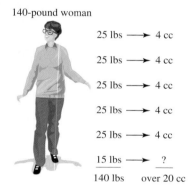

140-pound woman

| 25 lbs | → | 4 cc |
| 25 lbs | → | 4 cc |
| 25 lbs | → | 4 cc |
| 25 lbs | → | 4 cc |
| 25 lbs | → | 4 cc |
| 15 lbs | → | ? |
| 140 lbs | | over 20 cc |

From the diagram, we can see that a reasonable solution is a little over 20 cc.

**2.** TRANSLATE. We will let $n$ represent the unknown number. From the problem, we know that 4 cc is to 25 lb as $n$ cc is to 140 lb, or

$$\begin{array}{ll} \text{cc} \rightarrow \\ \text{lb} \rightarrow \end{array} \frac{4}{25} = \frac{n}{140} \begin{array}{ll} \leftarrow \text{cc} \\ \leftarrow \text{lb} \end{array}$$

**3.** SOLVE.

$$\frac{4}{25} \diagdown \frac{n}{140}$$

$$25n = 560 \quad \text{Set cross products equal.}$$

$$\frac{25n}{25} = \frac{560}{25} \quad \text{Divide both sides by 25.}$$

$$n = 22.4 \quad \text{Simplify.}$$

**Answer**

**2.** $74\frac{2}{3}$ gallons

**4.** INTERPRET. *Check* your work. This result is reasonable since it is a little over 20 cc. *State* your conclusion: The standard dose for a 140-lb woman is 22.4 cc.

---

**Example 3**   Calculating Supplies Needed to Fertilize a Lawn

A 50-pound bag of fertilizer covers 2400 square feet of lawn. How many bags of fertilizer are needed to cover a town square containing 15,360 square feet of lawn? Round the answer up to the nearest whole bag.

*Solution:*   **1.** UNDERSTAND. Read and reread the problem. Draw a picture.

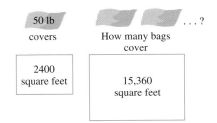

**2.** TRANSLATE. We'll let *n* represent the unknown number. From the problem, we know that 1 bag is to 2400 square feet as *n* bags is to 15,360 square feet.

$$\text{bags} \rightarrow \frac{1}{2400} = \frac{n}{15{,}360} \leftarrow \text{bags} \atop \leftarrow \text{square feet}$$

**3.** SOLVE.

$$\frac{1}{2400} \diagup\!\!\!\!\!\times \diagdown \frac{n}{15{,}360}$$

$$2400n = 15{,}360 \qquad \text{Set cross products equal.}$$

$$\frac{2400n}{2400} = \frac{15{,}360}{2400} \qquad \text{Divide both sides by 2400.}$$

$$n = 6.4 \qquad \text{Simplify.}$$

---

**Practice Problem 3**

If a gallon of paint covers 400 square feet, how many gallons must be bought to paint a retaining wall 260 feet long and 4 feet high? Round the answer up to the nearest whole gallon.

TEACHING TIP    Classroom Activity

Have students work in groups to create a word problem involving proportions. Then have them give their problem to another group to solve.

**Answer**
**3.** 3 gallons

**4.** INTERPRET. *Check* that replacing $n$ with 6.4 makes the proportion true. Is the answer reasonable?

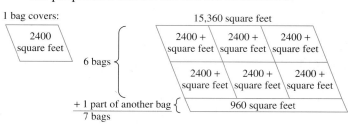

Yes. Since we must buy whole bags of fertilizer, 7 bags are needed. *State* your conclusion: To cover 15,360 square feet of lawn, 7 bags are needed. ▬▬

**TRY THE CONCEPT CHECK IN THE MARGIN.**

✓ **CONCEPT CHECK**

You are told that 12 ounces of ground coffee will brew enough coffee to serve 20 people. How could you estimate how much ground coffee will be needed to serve 95 people?

**Answer**

✓ **Concept Check:** Find how much will be needed for 100 people ($20 \times 5$) by multiplying 12 ounces by 5, which is 60 ounces.

Name _____ Section _____ Date _____

## EXERCISE SET 6.4

**A** Solve. See Examples 1 through 3.

The ratio of a quarterback's completed passes to attempted passes is 4 to 9.

**1.** If he attempted 27 passes, find how many passes he completed.

**2.** If he completed 20 passes, find how many passes he attempted.

It takes Sandra Hallahan 30 minutes to word process and spell check 4 pages.

**3.** Find how long it takes her to word process and spell check 22 pages.

**4.** Find how many pages she can word process and spell check in 4.5 hours.

University Law School accepts 2 out of every 7 applicants.

**5.** If the school received 630 applications, find how many students were accepted.

**6.** If the school accepted 150 students, find how many applications were received.

On an architect's blueprint, 1 inch corresponds to 8 feet.

**7.** Find the length of a wall represented by a line $2\frac{7}{8}$ inches long on the blueprint.

**8.** If an exterior wall is 42 feet long, find how long the blueprint measurement should be.

A human factors expert recommends that there be at least 9 square feet of floor space in a college classroom for every student in the class.

**9.** Find the minimum floor space that 30 students require.

**10.** Due to a space crunch, a university converts a 21′ by 15′ conference room into a classroom. Find the maximum number of students the room can accommodate.

**1.** 12 passes

**2.** 45 passes

**3.** 165 min

**4.** 36 pages

**5.** 180 students

**6.** 525 applications

**7.** 23 ft

**8.** $5\frac{1}{4}$ in.

**9.** 270 sq. ft

**10.** 35 students

**Name** _____

A Honda Civic averages 450 miles on a 12-gallon tank of gas.

**11.** If Dave Smythe runs out of gas in a Honda Civic and AAA comes to his rescue with $1\frac{1}{2}$ gallons of gas, determine how far he can go. Round to the nearest mile.

**12.** Find how many gallons of gas Denise Wolcott can expect to burn on a 2000-mile vacation trip in a Honda Civic. Round to the nearest gallon.

The scale on an Italian map states that 1 centimeter corresponds to 30 kilometers (a unit of length in the metric system).

**13.** Find how far apart Milan and Rome are if their corresponding points on the map are 15 centimeters apart.

**14.** On the map, a small Italian village is located 0.4 centimeters from the Mediterranean Sea. Find the actual distance.

Milan

Rome

A drink called Sea Breeze punch is made by mixing 3 parts of grapefruit juice with 4 parts of cranberry juice.

**15.** Find how much grapefruit juice should be mixed with 32 ounces of cranberry juice.

**16.** For a party, 6 quarts of grapefruit juice have been purchased to make Sea Breeze punch. Find how much cranberry juice should be purchased.

**Name** _____

A bag of Scott fertilizer covers 3000 square feet of lawn.

**17.** Find how many bags of fertilizer should be purchased to cover a rectangular lawn 260 feet by 180 feet.

**18.** Find how many bags of fertilizer should be purchased to cover a square lawn measuring 160 feet on each side.

Yearly homeowner property taxes are figured at a rate of $1.45 tax for every $100 of house value.

**19.** If Janet Blossom, a homeowner, pays $2349 in property taxes, find the value of her home.

**20.** Find the property taxes on a condominium valued at $72,000.

A Cubs baseball player makes 3 hits in every 8 times at bat.

**21.** If this Cubs player comes up to bat 40 times in a World Series playoff series, find how many hits he would be expected to make.

**22.** At this rate, if he made 12 hits, find how many times he batted.

A survey reveals that 2 out of 3 people prefer Coke to Pepsi.

**23.** In a room of 40 people, how many people are likely to prefer Coke? Round the answer to the nearest person.

**24.** In a college class of 36 students, find how many students are likely to prefer Pepsi.

An office uses 5 boxes of envelopes every 3 weeks.

**25.** Find how long a gross of envelope boxes is likely to last. (A gross of boxes is 144 boxes). Round to the nearest week.

**26.** Find how many boxes should be purchased to last a month. Round to the nearest box.

**17.** 16 bags

**18.** 9 bags

**19.** $162,000

**20.** $1044

**21.** 15 hits

**22.** 32 times at bat

**23.** 27 people

**24.** 12 students

**25.** 86 weeks

**26.** 7 boxes

**479**

Name _____

**27.** The daily supply of oxygen for one person is provided by 625 square feet of lawn. A total of 3750 square feet of lawn would provide the daily supplies of oxygen for how many people? (*Source:* Professional Lawn Care Association of America)

**28.** In the United States, approximately 71 million of the 200 million cars and light trucks in service have driver air bags. In a parking lot containing 800 cars and light trucks, how many would be expected to have driver air bags? (*Source:* Insurance Institute for Highway Safety)

**29.** A student would like to estimate the height of the Statue of Liberty in New York City's harbor. According to the *2000 World Almanac*, the length of the Statue of Liberty's right arm is 42 feet. The student's right arm is 2 feet long and her height is $5\frac{1}{3}$ feet. Use this information to estimate the height of the Statue of Liberty. (The actual height of the Statue of Liberty, from heel to top of the head, is 111 feet, 1 inch. How close is the estimated height to the actual height of the statue?)

**30.** There are 72 milligrams of cholesterol in a 3.5 ounce serving of lobster. How much cholesterol is in 5 ounces of lobster? (*Source:* The National Institute of Health)

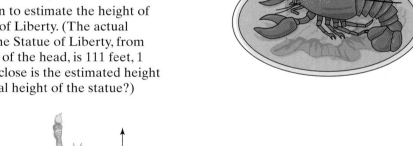

42 feet

$5\frac{1}{3}$ feet   2 feet

**31.** One pound of firmly packed brown sugar yields $2\frac{1}{4}$ cups. How many pounds of brown sugar will be required by a recipe that calls for 6 cups of firmly packed brown sugar? (*Source:* Based on data from *Family Circle* magazine)

**32.** Eleven out of every 25 greeting cards sold in the United States are Hallmark brand cards. If a consumer purchased 75 greeting cards in the past year, how many do you expect would have been Hallmark brand cards? (*Source:* Hallmark Cards, Inc.)

**33.** Medication is prescribed in 7 out of every 10 hospital emergency room visits that involve an injury. If a large urban hospital had 620 emergency room visits involving an injury in the past month, how many of these visits would you expect included a prescription for medication? (*Source:* National Center for Health Statistics)

**34.** Currently in the American population of people aged 65 years old and older, there are 145 women for every 100 men. In a nursing home with 280 male residents over the age of 65, how many female residents over the age of 65 would be expected? (*Source:* U.S. Bureau of the Census)

**35.** McDonald's four-piece Chicken McNuggets® has 190 calories. How many calories are in a nine-piece Chicken McNuggets? (*Source:* McDonald's Corporation)

**36.** A small order of McDonald's french fries weighs 68 grams and contains 10 grams of fat. McDonald's Super Size® french fries weighs 176 grams. How many grams of fat are in McDonald's SuperSize french fries? Round to the nearest tenth. (*Source:* McDonald's Corporation)

When making homemade ice cream in a hand-cranked freezer, the tub containing the ice cream mix is surrounded by a brine solution. To freeze the ice cream mix rapidly so that smooth and creamy ice cream results, the brine solution should combine crushed ice and rock salt in a ratio of 5 to 1. (*Source:* White Mountain Freezers, The Rival Company)

**37.** A small ice cream freezer requires 12 cups of crushed ice. How much rock salt should be mixed with the ice to create the necessary brine solution?

**38.** A large ice cream freezer requires $18\frac{3}{4}$ cups of crushed ice. How much rock salt will be needed?

## REVIEW AND PREVIEW

*Find the prime factorization of each number. See Section 4.2.*

**39.** 15        **40.** 21        **41.** 20        **42.** 24

**43.** 200       **44.** 300       **45.** 32        **46.** 81

**33.** 434 emergency room visits

**34.** 406 female residents

**35.** 427.5 cal

**36.** 25.9 g

**37.** 2.4 cups

**38.** $3\frac{3}{4}$ cups

**39.** $3 \cdot 5$

**40.** $3 \cdot 7$

**41.** $2^2 \cdot 5$

**42.** $2^3 \cdot 3$

**43.** $2^3 \cdot 5^2$

**44.** $2^2 \cdot 3 \cdot 5^2$

**45.** $2^5$

**46.** $3^4$

**Name** _____

### ◤ COMBINING CONCEPTS

_A board such as the one pictured below will balance if the following is true:_

$$\frac{\text{first weight}}{\text{second distance}} = \frac{\text{second weight}}{\text{first distance}}$$

_or, by using cross products,_

first weight · first distance = second weight · second distance

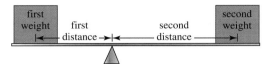

_Use either equation to solve Exercises 47–48._

**47.** Find the distance _n_ that will allow the board to balance.

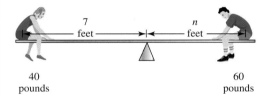

**48.** Find the length _n_ needed to lift the weight below.

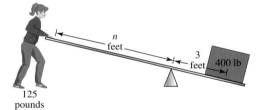

**49.** Describe a situation in which writing a proportion might solve a problem related to driving a car.

△ **6.5**  SIMILAR TRIANGLES AND PROBLEM SOLVING

**A**  FINDING THE RATIOS OF CORRESPONDING SIDES IN SIMILAR TRIANGLES

Two triangles are **similar** if they have the same shape but not necessarily the same size. In similar triangles, the measures of corresponding angles are equal and corresponding sides are in proportion. The following triangles are similar.

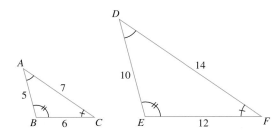

Objectives

 **A** Find the ratio of corresponding sides in similar triangles.

 **B** Find unknown lengths of sides in similar triangles.

 **C** Solve problems containing similar triangles.

Study Guide    SSM    CD-ROM    Video 6.5

Since these triangles are similar, the measures of corresponding angles are equal. Corresponding angles in the two triangles are marked with the same arc notation.

Angles with equal measure: $\angle A$ and $\angle D$, $\angle B$ and $\angle E$, $\angle C$ and $\angle F$

Also, the lengths of corresponding sides are in proportion.

Sides in proportion: $\dfrac{AB}{DE} = \dfrac{5}{10} = \dfrac{1}{2}$, $\dfrac{BC}{EF} = \dfrac{6}{12} = \dfrac{1}{2}$, $\dfrac{CA}{FD} = \dfrac{7}{14} = \dfrac{1}{2}$

The ratio of corresponding sides is $\dfrac{1}{2}$.

TEACHING TIP

Ask students if they can find any examples of similar triangles in any local architecture.

**Example 1**  Find the ratio of corresponding sides for the similar triangles $ABC$ and $DEF$.

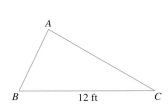  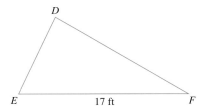

*Solution:*  We are given the lengths of two corresponding sides. Their ratio is

$$\frac{12 \text{ feet}}{17 \text{ feet}} = \frac{12}{17}$$

**Practice Problem 1**

Find the ratio of corresponding sides for the similar triangles $QRS$ and $XYZ$.

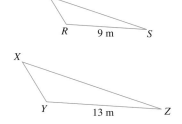

**Answer**

1. $\dfrac{9}{13}$

## B FINDING UNKNOWN LENGTHS OF SIDES IN SIMILAR TRIANGLES

Because the ratios of lengths of corresponding sides are equal, we can use proportions to find unknown lengths in similar triangles.

**Practice Problem 2**

Given that the triangles are similar, find the unknown length $n$.

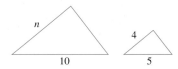

**Example 2**     Given that the triangles are similar, find the unknown length $n$.

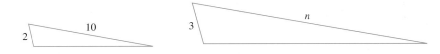

*Solution:*     Since the triangles are similar, corresponding sides are in proportion. Thus, the ratio of 2 to 3 is the same as the ratio of 10 to $n$, or

$$\frac{2}{3} = \frac{10}{n}$$

To find the unknown length $n$, we set cross products equal.

$$30 = 2n \qquad \text{Set cross products equal.}$$

$$\frac{30}{2} = \frac{2n}{2} \qquad \text{Divide both sides by 2.}$$

$$15 = n$$

The unknown length $n$ is 15 units.     ▬▬▬

**TRY THE CONCEPT CHECK IN THE MARGIN.**

### ✓ CONCEPT CHECK

The following two triangles are similar. Which vertices of the first triangle appear to correspond to which vertices of the second triangle?

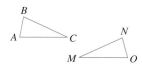

**Answer**

**2.** $n = 8$

✓ **Concept Check:** *A* corresponds to *O*; *B* corresponds to *N*; *C* corresponds to *M* (or *A* corresponds to *N*; *B* corresponds to *O*).

## C SOLVING PROBLEMS CONTAINING SIMILAR TRIANGLES

Many applications involve a diagram containing similar triangles. Surveyors, astronomers, and many other professionals use ratios of similar triangles continually in their work.

**Example 3**   **Finding the Height of a Tree**

Mel Rose is a 6-foot-tall park ranger who needs to know the height of a particular tree. He notices that when the shadow of the tree is 69 feet long, his own shadow is 9 feet long. Find the height of the tree.

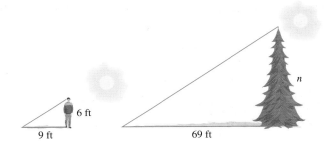

*Solution:*   **1.** UNDERSTAND. Read and reread the problem. Notice that the triangle formed by the sun's rays, Mel, and his shadow is similar to the triangle formed by the sun's rays, the tree, and its shadow.

**2.** TRANSLATE. Write a proportion from the similar triangles formed.

Mel's height $\rightarrow \dfrac{6}{n} = \dfrac{9}{69} \leftarrow$ length of Mel's shadow
height of tree $\rightarrow$           $\leftarrow$ length of tree's shadow

or $\dfrac{6}{n} = \dfrac{3}{23}$   Write in simplest form.

**Practice Problem 3**

Tammy Shultz, a firefighter, needs to estimate the height of a building. She estimates the length of her shadow to be 8 feet long and the length of the building's shadow to be 200 feet long. Find the height of the building if she is 5 feet tall.

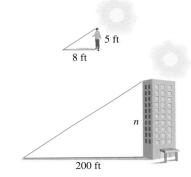

**Answer**

**3.** 125 feet

**3.** SOLVE for $n$.

$$\frac{6}{n} = \frac{3}{23}$$

$3n = 138$    Set cross products equal.

$$\frac{3n}{3} = \frac{138}{3}$$    Divide both sides by 3.

$n = 46$

**4.** INTERPRET. *Check* to see that replacing $n$ with 46 in the proportion makes the proportion true. *State* your conclusion: The height of the tree is 46 feet.

## EXERCISE SET 6.5

**A** *Find each ratio of the corresponding sides of the given similar triangles. See Example 1.*

**1.**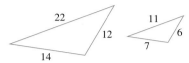

**2.**

**3.**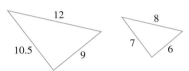

**4.**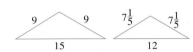

**B** *Given that the pairs of triangles are similar, find the unknown length n. See Example 2.*

**5.**

**6.**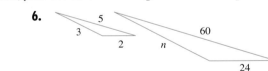

**7.**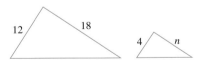

**8.**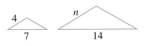

**9.**

**10.**

**1.** $\dfrac{2}{1}$

**2.** $\dfrac{1}{4}$

**3.** $\dfrac{3}{2}$

**4.** $\dfrac{5}{4}$

**5.** 4.5

**6.** 36

**7.** 6

**8.** 8

**9.** 5

**10.** 13.5

**Name** _____

**11.**

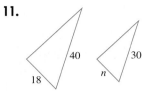

**12.**

**13.**

**14.**

21.6  $n$     7.2  9.6

**15.**

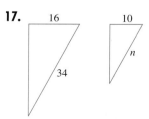

**16.**

9     $n$     6     9

**17.**

16     10

34     $n$

**18.**

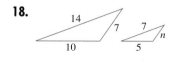

**Name** _____

**19.** 50 ft

**C** *Solve. See Example 3.*

**19.** Lloyd White, a firefighter, needs to estimate the height of a burning building. He estimates the length of his shadow to be 9 feet long and the length of the building's shadow to be 75 feet long. Find the height of the building if he is 6 feet tall.

**20.** Samantha Black, a 5-foot-tall park ranger, needs to know the height of a tree. She notices that when the shadow of the tree is 48 feet long her shadow is 4 feet long. Find the height of the tree.

**20.** 60 ft

**21.** $x = 4.4$ ft; $y = 5.6$ ft

**21.** Gepetto the toy maker wants to make a triangular mainsail for a toy sailboat in the same shape as a full-size sailboat's mainsail. Use the diagram to find the lengths of the toy mainsail's sides.

**22.** A parent wants to make a doll's triangular diaper in the same shape as a full-size diaper. Use the diagram to find the lengths of the doll diaper's sides.

**22.** $x = 9.6$ in.; $y = 8$ in.

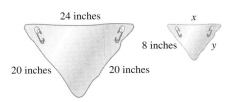

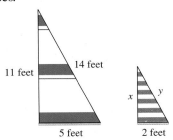

**23.** 14.4 ft

**24.** 58.7 ft

**25.** 0.25

**23.** If a 30-foot tree casts an 18-foot shadow, find the length of the shadow cast by a 24-foot tree.

**24.** If a 24-foot flagpole casts a 32-foot shadow, find the length of the shadow cast by a 44-foot antenna. Round to the nearest tenth.

**26.** 0.6

**REVIEW AND PREVIEW**

*Write each fraction as a decimal. See Section 5.6.*

**25.** $\dfrac{1}{4}$

**26.** $\dfrac{3}{5}$

**27.** $\dfrac{13}{20}$

**27.** 0.65

**28.** 0.275

**28.** $\dfrac{11}{40}$

**29.** $\dfrac{9}{10}$

**30.** $\dfrac{7}{10}$

**29.** 0.9

**30.** 0.7

**Name** _____

### ◆ COMBINING CONCEPTS

*Given that the pairs of triangles are similar, find the unknown length n. Round your result to one decimal place.*

▣ **31.**

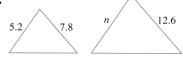

▣ **32.**

✎ **33.** Congruent triangles are triangles that have exactly the same shape and size. In your own words, describe any differences between similar triangles and congruent triangles.

# CHAPTER 6 ACTIVITY
## USING ASPECT RATIOS

*This activity may be completed by working in groups or individually.*

In the movie industry, a film's **aspect ratio** describes its image format as a ratio of the width to the height of the picture (width:height). For example, an aspect ratio of 2.35:1 means that the width of the picture is 2.35 times the height of the picture. The following table lists various cinematographic processes (some no longer used in filming new movies), along with an example of a movie filmed with each process, the movie's aspect ratio, and the year it was originally released. In many cases, a cinematographic process allows more than one aspect ratio. CinemaScope, for instance, offers aspect ratios of 2.66:1, 2.55:1, and 2.35:1. Vista Vision offers 1.96:1, 1.85:1, and 1.66:1 aspect ratios.

| Cinematographic Process | Movie Filmed with This Process | Aspect Ratio of Film | Release Date |
| --- | --- | --- | --- |
| CinemaScope | *How to Marry a Millionaire* | 2.55:1 | 1953 |
| Cinerama | *South Seas Adventure* | 2.77:1 | 1958 |
| Panavision | *Star Wars* | 2.35:1 | 1977 |
| Spherical | *All Quiet on the Western Front* | 1.33:1 | 1930 |
| | *Batman* | 1.85:1 | 1989 |
| Super 35 | *Titanic* | 2.35:1 | 1997 |
| Super-Cinerama | *How the West Was Won* | 2.68:1 | 1962 |
| Todd-AO | *The Sound of Music* | 2.20:1 | 1965 |
| Ultra Panavision 70 | *Mutiny on the Bounty* | 2.75:1 | 1962 |
| Vista Vision | *Vertigo* | 1.85:1 | 1958 |
| | *North by Northwest* | 1.66:1 | 1959 |

(*Source:* The Internet Movie Database Ltd)

1. If a movie screen is 12 feet tall, how wide would it need to be to show the following movies?
   a. *North by Northwest*   b. *Batman*
   c. *The Sound of Music*   d. *Titanic*
   e. *Mutiny on the Bounty*

2. The smallest film aspect ratio is 1.33:1. If a movie screen is to be 30 feet wide, how tall should it be to show a movie with this aspect ratio?

3. The largest film aspect ratio, offered by the Cinerama process, was 3.0:1. If a movie screen is to be 30 feet wide, how tall should it be to show a movie with this aspect ratio?

4. a. A standard television set has an aspect ratio of 4:3. How tall is the screen of a standard television with screen width of 20 inches?
   b. Suppose a television station wishes to broadcast *Star Wars* in its original image format without cropping the picture. Because the aspect ratio of *Star Wars* is different from that of a standard television set, black bands (called *letterboxing*) are added above and below the picture to fill the excess space so that the movie's aspect ratio can be maintained. What will be the dimensions (in inches) of this movie when shown on the standard television set described in part (a)?
   c. If the letterboxing added to *Star Wars* for this broadcast is the same height at the top and the bottom of the screen, how tall is each band of

letterboxing on the standard television set described in part (a)?

5. a. By 2006, U.S. broadcasters will be required to broadcast only a digital television signal that can be received by high-definition televisions (HDTV). HDTV sets have an aspect ratio of 16:9. How tall is the screen of an HDTV whose screen width is 20 inches?
   b. Suppose it is 2006 and a television station wishes to digitally broadcast *Vertigo* in its original image format without cropping the picture. Because the aspect ratio of *Vertigo* is different from that of HDTV, letterboxing will be added above and below the picture to fill the excess space to maintain the movie's aspect ratio. What will be the dimensions (in inches) of this movie when shown on the HDTV set described in part (a)?

Letterboxing

*Star wars*

   c. If the letterboxing added to *Vertigo* for this broadcast is the same height at the top and the bottom of the screen, how tall is each band of letterboxing on the HDTV set described in part (a)?

# CHAPTER 6 HIGHLIGHTS

| DEFINITIONS AND CONCEPTS | EXAMPLES |
|---|---|

## SECTION 6.1 RATIOS

| | |
|---|---|
| A **ratio** is the quotient of two quantities. | The ratio of 3 to 4 can be written as $$\frac{3}{4} \quad \text{or} \quad 3:4$$ $\underset{\text{fraction notation}}{\uparrow} \qquad \underset{\text{colon notation}}{\uparrow}$ |

## SECTION 6.2 RATES

| | |
|---|---|
| **Rates** are used to compare different kinds of quantities. | Write the rate 12 spikes every 8 inches as a fraction in simplest form. $$\frac{12 \text{ spikes}}{8 \text{ inches}} = \frac{3 \text{ spikes}}{2 \text{ inches}}$$ |
| A **unit rate** is a rate with a denominator of 1. | Write as a unit rate: 117 miles on 5 gallons of gas $$\frac{117 \text{ miles}}{5 \text{ gallons}} = \frac{23.4 \text{ miles}}{1 \text{ gallon}} \quad \text{or } 23.4 \text{ miles per gallon} \\ \text{or } 23.4 \text{ mpg}$$ |
| A **unit price** is a "money per item" unit rate. | Write as a unit price: $5.88 for 42 ounces of detergent $$\frac{\$5.88}{42 \text{ ounces}} = \frac{\$0.14}{1 \text{ ounce}} = \$0.14 \text{ per ounce}$$ |

## SECTION 6.3 PROPORTIONS

| | |
|---|---|
| A **proportion** is a statement that two ratios or rates are equal. $$\frac{a}{b} \diagdown\!\!\!\diagup \frac{c}{d}$$ cross product $\rightarrow \overset{\frown}{b \cdot c} = \overset{\frown}{a \cdot d}$ ← cross product  If cross products are equal, the proportion is true. If cross products are not equal, the proportion is false. | Find the value of $x$ that makes the proportion true. $$\frac{9}{45} \diagdown\!\!\!\diagup \frac{21}{x}$$ $\overline{45 \cdot 21} = \overline{9 \cdot x}$    Set cross products equal. $945 = 9x$    Multiply. $\dfrac{945}{9} = \dfrac{9x}{9}$    Divide both sides by 9. $105 = x$    Simplify. |

---

### SECTION 6.4 PROPORTIONS AND PROBLEM SOLVING

Given a specified ratio (or rate) of two quantities, a proportion can be used to determine an unknown quantity.

On a map, 50 miles corresponds to 3 inches. How many miles correspond to 10 inches?

1. UNDERSTAND. Read and reread the problem.

2. TRANSLATE. We let $n$ represent the unknown number. We are given that 50 miles is to 3 inches as $n$ miles is to 10 inches.

$$\text{miles} \rightarrow \quad \frac{50}{3} = \frac{n}{10} \quad \leftarrow \text{miles}$$
$$\text{inches} \rightarrow \qquad\qquad\qquad \leftarrow \text{inches}$$

3. SOLVE.

$$\frac{50}{3} = \frac{n}{10}$$

$3n = 500$     Set cross products equal.

$$\frac{3n}{3} = \frac{500}{3}$$     Divide both sides by 3.

$$n = 166\frac{2}{3}$$

4. INTERPRET. *Check* your work. *State* your conclusion: On the map, $166\frac{2}{3}$ miles corresponds to 10 inches.

---

### SECTION 6.5 SIMILAR TRIANGLES AND PROBLEM SOLVING

**Similar triangles** have exactly the same shape but not necessarily the same size.

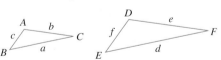

Corresponding angles are equal and the ratios of the lengths of corresponding sides are equal.

$$\frac{a}{d} = \frac{b}{e} = \frac{c}{f}$$

If the two triangles are similar, find the unknown length $x$.

$$\frac{6}{4} = \frac{9}{x}$$     Set ratios of the lengths of corresponding sides equal.

$4 \cdot 9 = 6 \cdot x$     Set cross products equal.

$36 = 6x$     Multiply.

$$\frac{36}{6} = \frac{6x}{6}$$     Divide both sides by 6.

$6 = x$     Simplify.

The unknown length $x$ is 6 units.

# CHAPTER 6 REVIEW

**(6.1)** *Write each ratio as a fraction in simplest form.*

**1.** 6000 people to 4800 people $\dfrac{5}{4}$

**2.** 121 births to 143 births $\dfrac{11}{13}$

**3.** $2\dfrac{1}{4}$ days to 10 days $\dfrac{9}{40}$

**4.** 14 quarters to 5 quarters $\dfrac{14}{5}$

**5.** 4 weeks to 15 weeks $\dfrac{4}{15}$

**6.** 4 yards to 8 yards $\dfrac{1}{2}$

**7.** $3\dfrac{1}{2}$ dollars to 7 dollars $\dfrac{1}{2}$

**8.** 3.5 centimeters to 75 centimeters $\dfrac{7}{150}$

**(6.2)** *Write each rate as a fraction in simplest form.*

**9.** 8 stillborn births to 1000 live births $\dfrac{1 \text{ stillborn birth}}{125 \text{ live births}}$

**10.** 6 professors for 20 graduate research assistants $\dfrac{3 \text{ professors}}{10 \text{ assistants}}$

**11.** 15 word-processing pages printed in 6 minutes $\dfrac{5 \text{ pages}}{2 \text{ min}}$

**12.** 8 computers assembled in 6 hours $\dfrac{4 \text{ computers}}{3 \text{ hr}}$

*Find each unit rate.*

**13.** 468 miles in 9 hours  52 mph

**14.** 180 feet in 12 seconds  15 ft/sec

**15.** $0.93 for 3 pears  $0.31/pear

**16.** $6.96 for 4 diskettes  $1.74/diskette

**17.** 260 kilometers in 4 hours   65 km/hr

**18.** 8 gallons of pesticide for 6 acres of crops
$1\frac{1}{3}$ gal/acre

**19.** $184 for books for 5 college courses   $36.80/course

**20.** 52 bushels of fruit from 4 trees   13 bushels/tree

*Find each unit price and decide which is the better buy. Assume that we are comparing different sizes of the same brand.*

**21.** Taco sauce: 8 ounces for $0.99 or 12 ounces for $1.69

8 oz: $0.124 per oz;
12 oz: $0.141 per oz;
8-oz size

**22.** Peanut butter: 18 ounces for $1.49 or 28 ounces for $2.39

18 oz: $0.828; per oz;
28 oz: $0.854; per oz;
18-oz size

**23.** 2% milk: 16 ounces for $0.59, $\frac{1}{2}$ gallon for $1.69, or 1 gallon for $2.29 (1 gallon = 128 fluid ounces)

16 oz: $0.037 per oz;
1 gal: $0.018 per oz;
$\frac{1}{2}$ gal: $0.026 per oz;
1-gal size

**24.** Coca-Cola: 12 ounces for $0.59, 16 ounces for $0.79, or 32 ounces for $1.19
12 oz: $0.0492 per oz;
16 oz: $0.0494 per oz;
32 oz: $0.0372 per oz;
32-oz size

**(6.3)** *Write each sentence as a proportion.*

**25.** 20 men is to 14 women as 10 men is to 7 women.

$$\frac{20\text{ men}}{14\text{ women}} = \frac{10\text{ men}}{7\text{ women}}$$

**26.** 50 tries is to 4 successes as 25 tries is to 2 successes.

$$\frac{50\text{ tries}}{4\text{ successes}} = \frac{25\text{ tries}}{2\text{ successes}}$$

**27.** 16 sandwiches is to 8 players as 2 sandwiches is to 1 player.

$$\frac{16\text{ sandwiches}}{8\text{ players}} = \frac{2\text{ sandwiches}}{1\text{ player}}$$

**28.** 12 tires is to 3 cars as 4 tires is to 1 car.

$$\frac{12\text{ tires}}{3\text{ cars}} = \frac{4\text{ tires}}{1\text{ car}}$$

*Determine whether each proportion is true or false.*

**29.** $\dfrac{21}{8} = \dfrac{14}{6}$   false

**30.** $\dfrac{3}{5} = \dfrac{60}{100}$   true

**31.** $\dfrac{3.1}{6.2} = \dfrac{0.8}{0.16}$   false

**32.** $\dfrac{3.75}{3} = \dfrac{7.5}{6}$   true

*Solve each proportion for the given variable.*

**33.** $\dfrac{x}{6} = \dfrac{15}{18}$   5

**34.** $\dfrac{y}{9} = \dfrac{5}{3}$   15

**35.** $\dfrac{4}{13} = \dfrac{10}{x}$   32.5

**36.** $\dfrac{8}{5} = \dfrac{9}{z}$   5.625

**37.** $\dfrac{16}{3} = \dfrac{y}{6}$   32

**38.** $\dfrac{x}{3} = \dfrac{9}{2}$   13.5

**39.** $\dfrac{x}{5} = \dfrac{27}{2\frac{1}{4}}$   60

**40.** $\dfrac{2\frac{1}{2}}{6} = \dfrac{3}{z}$   $7\frac{1}{5}$

**41.** $\dfrac{x}{0.4} = \dfrac{4.7}{3}$. Round to the nearest hundredth.   0.63

**42.** $\dfrac{0.07}{0.3} = \dfrac{7.2}{n}$. Round to the nearest tenth.   30.9

**(6.4)** *Solve.*

The ratio of a quarterback's completed passes to attempted passes is 3 to 7.

**43.** If he attempts 32 passes, find how many passes he completed. Round to the nearest whole pass.
14 passes

**44.** If he completed 15 passes, find how many passes he attempted.   35 passes

One bag of pesticide covers 4000 square feet of crops.

**45.** Find how many bags of pesticide should be purchased to cover a rectangular garden 180 feet by 175 feet.   8 bags

**46.** Find how many bags of pesticide should be purchased to cover a square garden 250 feet on each side.   16 bags

An owner of a Ford Escort can drive 420 miles on 11 gallons of gas.

**47.** If Tom Aloiso ran out of gas in an Escort and AAA comes to his rescue with $1\frac{1}{2}$ gallons of gas, determine whether Tom can then drive to a gas station 65 miles away.   no

**48.** Find how many gallons of gas Tom can expect to burn on a 3000-mile trip. Round to the nearest gallon.   79 gal

Yearly homeowner property taxes are figured at a rate of $1.15 tax for every $100 of house value.

**49.** If a homeowner pays $627.90 in property taxes, find the value of the house.   $54,600

**50.** Find the property taxes on a townhouse valued at $89,000.   $1023.50

On an architect's blueprint, 1 inch = 12 feet.

**51.** Find the length of a wall represented by a $3\frac{3}{8}$ -inch line on the blueprint.   $40\frac{1}{2}$ ft

**52.** If an exterior wall is 99 feet long, find how long the blueprint measurement should be.

$8\frac{1}{4}$ in.

**(6.5)** *Given that the pairs of triangles are similar, find the unknown length x.*

△ **53.**

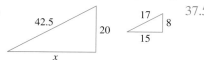

37.5

△ **54.**

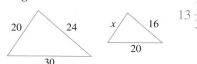

$13\frac{1}{3}$

△ **55.**

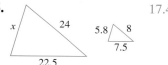

17.4

△ **56.**

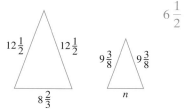

$6\frac{1}{2}$

*Solve*

**57.** A housepainter needs to estimate the height of a condominium. He estimates the length of his shadow to be 7 feet long and the length of the building's shadow to be 42 feet long. Find the height of the building if the housepainter is $5\frac{1}{2}$ feet tall.   33 ft

△ **58.** Santa's elves are making a triangular sail for a toy sailboat. The toy sail is to be the same shape as a real sailboat's sail. Use the following diagram to find the unknown lengths *x* and *y*.

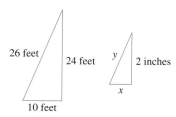

$x = \frac{5}{6}$ in., $y = 2\frac{1}{6}$ in.

# Chapter 6 Test

*Write each ratio as a fraction in simplest form.*

**1.** 4500 trees to 6500 trees

**2.** $75 to $10

*Write each rate as a fraction in simplest form.*

**3.** 28 men to every 4 women

**4.** 9 inches of rain in 30 days

*Find each unit rate.*

**5.** 650 kilometers in 8 hours

**6.** 8 inches of rain in 12 hours

**7.** 140 students for 5 teachers

*Find each unit price and decide which is the better buy.*

**8.** Steak sauce:
8 ounces for $1.19
12 ounces for $1.89

**9.** Jelly:
16 ounces for $1.49
24 ounces for $2.39

*Determine whether each proportion is true or false.*

**10.** $\dfrac{28}{16} = \dfrac{14}{8}$

**11.** $\dfrac{3.6}{2.2} = \dfrac{1.9}{1.2}$

*Solve each proportion for the given variable.*

**12.** $\dfrac{n}{3} = \dfrac{15}{9}$

**13.** $\dfrac{8}{x} = \dfrac{11}{6}$

**14.** $\dfrac{4}{3} = \dfrac{y}{\frac{1}{4}}$ over $\dfrac{3}{7}$

**15.** $\dfrac{1.5}{5} = \dfrac{2.4}{n}$

**Answers**

**1.** $\dfrac{9}{13}$

**2.** $\dfrac{15}{2}$

**3.** $\dfrac{7\ \text{men}}{1\ \text{woman}}$

**4.** $\dfrac{3\ \text{in.}}{10\ \text{days}}$

**5.** 81.25 km/hr

**6.** $\dfrac{2}{3}$ in./hr

**7.** 28 students/teacher

**8.** 8 oz: $0.149 per oz;
12 oz: $0.158 per oz;
8-oz size

**9.** 16 oz: $.0931 per oz;
24 oz: $.0996 per oz;
16-oz size

**10.** true

**11.** false

**12.** 5

**13.** $4\dfrac{4}{11}$

**14.** $\dfrac{7}{3}$

**15.** 8

*Solve.*

**16.** On an architect's drawing, 2 inches corresponds to 9 feet. Find the length of a home represented by a line that is 11 inches long.

**17.** If a car can be driven 80 miles in 3 hours, how long will it take to travel 100 miles?

**18.** The standard dose of medicine for a dog is 10 grams for every 15 pounds of body weight. What is the standard dose for a dog that weighs 80 pounds?

**19.** Jerome Grant worked 6 hours and packed 86 cartons of books. At this rate, how many cartons can he pack in 8 hours?

**20.** In the 1998 American adult population, 12 out of 25 people drank coffee. In a town with a population of 31,000 adults, how many of these adults would you expect to be coffee drinkers? (*Source: Chicago Tribune,* 1/24/98.)

△ **21.** Given that the following triangles are similar, find the unknown length *n*.

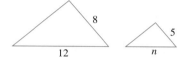

△ **22.** Tamara Watford, a surveyor, needs to estimate the height of a tower. She estimates the length of her shadow to be 4 feet long and the length of the tower's shadow to be 48 feet long. Find the height of the tower if she is $5\frac{3}{4}$ feet tall.

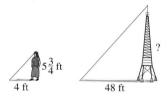

---

**16.** $49\frac{1}{2}$ ft

**17.** $3\frac{3}{4}$ hr

**18.** $53\frac{1}{3}$ g

**19.** $144\frac{2}{3}$ cartons

**20.** 14,880 adults

**21.** 7.5

**22.** 69 ft

| | |
|---|---|
| **Name** _____ | **Section** _____ **Date** _____ |

## CUMULATIVE REVIEW

**1.** Divide: $51,600 \div 403$

**2.** Write as an algebraic expression. Use $x$ to represent "a number."
  **a.** 7 more than a number
  **b.** 15 decreased by a number
  **c.** The product of 2 and a number
  **d.** The quotient of a number and 5
  **e.** 2 subtracted from a number

**3.** Simplify.
  **a.** $-(-4)$  **b.** $-|-5|$  **c.** $-|6|$

**4.** Subtract.
  $-10 - 5$

**5.** Simplify.
  $7 + (-12) - 3 - (-8)$

*Simplify each expression by combining like terms.*

**6.** $6x + 2x - 5$

**7.** $2x - 5 + 3y + 4x - 10y + 11$

**8.** Determine whether 6 is a solution of the equation $4(x - 3) = 12$.

**9.** Solve for $y$: $-8 = 2y$

**10.** Solve: $20 - x = 21$

**11.** Write $\dfrac{2}{5}$ as an equivalent fraction whose denominator is 15.

**12.** Write the prime factorization of 80.

**13.** Multiply: $\dfrac{3x}{4} \cdot \dfrac{8}{5x}$

*Subtract and simplify.*

**14.** $\dfrac{8}{9} - \dfrac{1}{9}$

**15.** $\dfrac{7}{8} - \dfrac{5}{8}$

**Name** _____

**16.** Subtract: $2 - \dfrac{x}{3}$

**17.** Simplify: $\dfrac{\dfrac{y}{5} - 2}{\dfrac{3}{10}}$

**18.** Solve for $x$: $\dfrac{1}{3}x = 7$

**19.** Multiply: $3\dfrac{1}{3} \cdot \dfrac{7}{8}$

**20.** Write 0.43 as a fraction.

**21.** Divide: $2.3\overline{)10.764}$

**22.** Simplify: $-0.5(8.6 - 1.2)$

**23.** Write the numbers in order from smallest to largest.
$\dfrac{9}{20}, \dfrac{4}{9}, 0.456$

**24.** Solve for $x$: $x - 1.5 = 8$

**25.** Solve for $y$: $\dfrac{6}{1} = \dfrac{7}{y}$

# Percent

This chapter is devoted to percent, a concept used virtually every day in ordinary and business life. Understanding percent and using it efficiently depends on understanding ratios, because a percent is a ratio whose denominator is 100. We present techniques to write percents as fractions and as decimals and then solve problems relating to interest rates, sales tax, discounts, and other real-life situations by writing percent equations.

## CONTENTS

Three species of bluebirds are native to North America: the Eastern, the Western, and the Mountain Bluebird. These once-abundant songbirds make their nests in hollow cavities, such as the holes in trees made by woodpeckers. However, years of land-clearing for farms, logging, housing developments, and urban sprawl has greatly reduced the number of naturally available bluebird nesting cavities. This, along with other factors, has contributed to a sharp decline in the bluebird population since the early 1900s. Since the 1970s, human efforts to bolster the bluebird population have been paying off. Backyard and roadside bluebird nesting boxes have helped to slowly reverse the decline. In Exercise 40 on page 544, we will see how a percent can be used to describe the increase in the Eastern Bluebird population.

**503**

1. 12%; (7.1A)

2. 0.57; (7.1B)

3. 275%; (7.1C)

4. $\frac{3}{40}$; (7.1D)

5. 15%; (7.1E)

6. $18\% \cdot 50 = x$; (7.2A)

7. $4\% \cdot x = 89$; (7.2A)

8. $\frac{82}{b} = \frac{90}{100}$; (7.3A)

9. $\frac{48}{112} = \frac{p}{100}$; (7.3A)

10. 20%; (7.2B, 7.3B)

11. 3.3; (7.2B, 7.3B)

12. 200; (7.2B, 7.3B)

13. 3 lightbulbs; (7.4A)

14. decrease of 84 students; current enrollment: 4116 students; (7.4B)

15. 7%; (7.5A)

16. $22,400; (7.5B)

17. discount: $78; sales price: $572; (7.5C)

18. $144; (7.6A)

19. $7147.50; (7.6B)

20. $37.33; (7.6C)

**Name** _____  **Section** _____ **Date** _____

# CHAPTER 7 PRETEST

**1.** In a group of 100 people, 12 are female. What percent of the group is female?

**2.** Write 57% as a decimal.

**3.** Write 2.75 as a percent.

**4.** Write 7.5% as a fraction in simplest form.

**5.** Write $\frac{3}{20}$ as a percent.

*Translate each question to an equation.*

**6.** 18% of 50 is what number?

**7.** 4% of what number is 89?

*Translate each question to a proportion.*

**8.** 90% of what number is 82?

**9.** 48 is what percent of 112?

*Solve.*

**10.** What percent of 80 is 16?

**11.** What number is 1.5% of 220?

**12.** 32 is 16% of what number?

**13.** In a box of 250 lightbulbs, 1.2% were found to be defective. How many lightbulbs were defective?

**14.** The enrollment at a local high school decreased 2% over last year's enrollment of 4200. Find the decrease in enrollment and the current enrollment.

**15.** The sales tax on a $499 printer is $34.93. Find the sales tax rate.

**16.** Jerry Williams receives 2% commission on his sales of computer equipment. Last week his commission check was $448. Find the amount of his sales last week.

**17.** A television that normally sells for $650 is on sale at 12% off. What is the discount and what is the sales price?

**18.** Find the simple interest after 3 years on $600 at an interest rate of 8%.

**19.** $5000 is invested at 6% compounded quarterly for 6 years. Find the total amount at the end of 6 years. Use Appendix F.

**20.** Find the monthly payment on a $700 loan for 2 years if the interest on the 2-year loan is $196.

# 7.1 PERCENTS, DECIMALS, AND FRACTIONS

## A UNDERSTANDING PERCENT

The word **percent** comes from the Latin phrase *per centum*, which means "**per 100**." For example, 53 percent (%) means 53 per 100. In the square below, 53 of the 100 squares are shaded. Thus 53% of the figure is shaded.

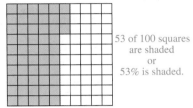

53 of 100 squares
are shaded
or
53% is shaded.

Since 53% means 53 per 100, 53% is the ratio of 53 to 100, or $\frac{53}{100}$.

$$53\% = \frac{53}{100}$$

---

**PERCENT**

**Percent** means **per one hundred**. The "%" symbol is used to denote percent.

---

Also,

$$7\% = \frac{7}{100} \quad \text{7 parts per 100 parts}$$

$$73\% = \frac{73}{100} \quad \text{73 parts per 100 parts}$$

$$109\% = \frac{109}{100} \quad \text{109 parts per 100 parts}$$

Percent is used in a variety of everyday situations. For example:

The interest rate is 5.7%.

28% of U.S. homes have Internet access.

The store is having a 25% off sale.

78% of us trust our local fire department.

The federal debt has increased 112% in the last 10 years.

**Example 1**     In a survey of 100 people, 17 people drive blue cars. What percent of people drive blue cars?

*Solution:*     Since 17 people out of 100 drive blue cars, the fraction is $\frac{17}{100}$. Then

$$\frac{17}{100} = 17\%$$

---

**Objectives**

**A** Know the meaning of percent.
**B** Write percents as decimals.
**C** Write decimals as percents.
**D** Write percents as fractions.
**E** Write fractions as percents.
**F** Convert percents, decimals, and fractions.

Study     SSM     CD-ROM     Video
Guide                              7.1

TEACHING TIP

Ask students to find the percent of the figure that is unshaded. Then ask how the two percents are related.

**Practice Problem 1**

Of 100 students in a club, 23 are freshmen. What percent of the students are freshmen?

**Answer**

**1.** 23%

## Practice Problem 2

29 out of 100 executives are in their forties. What percent of executives are in their forties?

**Example 2**    46 out of every 100 college students live at home. What percent of students live at home? (*Source:* Independent Insurance Agents of America)

*Solution:*    $\dfrac{46}{100} = 46\%$

### B  WRITING PERCENTS AS DECIMALS

To write a percent as a decimal, we can first write the percent as a fraction.

$$53\% = \dfrac{53}{100}$$

Now we can write the fraction as a decimal as we did in Section 5.6.

$$\dfrac{53}{100} = 0.53 \qquad \text{(53 hundredths)}$$

Notice that the result is

$53.\% = 0.53$    Drop the percent symbol and move the decimal point two places to the left.

---

**WRITING A PERCENT AS A DECIMAL**

Drop the percent symbol and move the decimal point two places to the left.

$$43\% = 0.43$$

---

## Practice Problem 3

Write 89% as a decimal.

**Example 3**    Write 23% as a decimal.

*Solution:*    $23\% = 23.\%$    Drop the percent symbol and move the decimal point two places to the left.
$= 0.23$

## Practice Problems 4–6

Write each percent as a decimal.
4. 2.7%          5. 150%          6. 0.69%

**Examples**    Write each percent as a decimal.

**4.** $4.6\% = 0.046$    Drop the percent symbol and move the decimal point two places to the left.

**5.** $190\% = 190.\% = 1.90 \quad \text{or} \quad 1.9$

**6.** $0.74\% = 0.0074$

**TRY THE CONCEPT CHECK IN THE MARGIN.**

### C  WRITING DECIMALS AS PERCENTS

## ✓ CONCEPT CHECK

Why is it incorrect to write the percent 0.033% as 3.3 in decimal form?

To write a decimal as a percent, we can first write the decimal as a fraction.

$$0.38 = \dfrac{38}{100}$$

Now we can write the fraction as a percent.

$$\dfrac{38}{100} = 38\%$$

Notice that the result is

$0.38 = 38.\%$    Move the decimal point two places to the right and attach a percent symbol.

---

### WRITING A DECIMAL AS A PERCENT

Move the decimal point two places to the right and attach the percent symbol, %.

$$0.27 = 27.\%$$

---

**Example 7**    Write 0.65 as a percent.

*Solution:*    $0.65 = 65.\%$    Move the decimal point two places to the right and attach a percent symbol.

$$= 65\%$$

**Examples**    Write each decimal as a percent.

**8.** $1.25 = 125.\%$ or $125\%$

**9.** $0.012 = 001.2\%$ or $1.2\%$

**10.** $0.6 = 060.\%$ or $60\%$

---

TRY THE CONCEPT CHECK IN THE MARGIN.

### D  WRITING PERCENTS AS FRACTIONS

When we write a percent as a fraction, we usually then write the fraction in simplest form. For example, $50\% = \dfrac{50}{100}$. Then, in simplest form, $\dfrac{50}{100} = \dfrac{1}{2}$.

$$50\% = \frac{50}{100} = \frac{1}{2}$$

---

### WRITING A PERCENT AS A FRACTION

Drop the percent symbol and write the number over 100. Then simplify the fraction if possible.

$$7\% = \frac{7}{100}$$

---

**Examples**    Write each percent as a fraction or mixed number in simplest form.

**11.** $40\% = \dfrac{40}{100} = \dfrac{2 \cdot 20}{5 \cdot 20} = \dfrac{2}{5}$

**12.** $1.9\% = \dfrac{1.9}{100}$. Next, multiply the numerator and denominator by 10 so that the numerator 1.9 becomes a whole number.

$$\frac{1.9}{100} = \frac{1.9 \cdot 10}{100 \cdot 10} = \frac{19}{1000}$$

---

Write 0.19 as a percent.

---

**Practice Problems 8–10**

Write each decimal as a percent.
8. 1.75        9. 0.044        10. 0.7

---

**✓ CONCEPT CHECK**

Why is it incorrect to write the decimal 0.0345 as 34.5% in percent form?

---

**Practice Problems 11–15**

Write each percent as a fraction in simplest form.
11. 25%        12. 2.3%        13. 150%

14. $66\dfrac{2}{3}\%$        15. 8%

**Answers**

**7.** 19%   **8.** 175%   **9.** 4.4%   **10.** 70%

**11.** $\dfrac{1}{4}$   **12.** $\dfrac{23}{1000}$   **13.** $\dfrac{3}{2}$   **14.** $\dfrac{2}{3}$   **15.** $\dfrac{2}{25}$

**✓ Concept Check:** To change a decimal to a percent, the decimal point should be moved *only two* places to the right. So the correct answer is 3.45%.

**13.** $125\% = \dfrac{125}{100} = \dfrac{5 \cdot \boxed{25}}{4 \cdot \boxed{25}} = \dfrac{5}{4}$ or $1\dfrac{1}{4}$

**14.** $33\dfrac{1}{3}\% = \dfrac{33\dfrac{1}{3}}{100} = \dfrac{\dfrac{100}{3}}{100}$. Recall that the fraction bar

means divison. Thus

$$\dfrac{100}{3} \div 100 = \dfrac{100}{3} \cdot \dfrac{1}{100} = \dfrac{1}{3}$$

**15.** $100\% = \dfrac{100}{100} = 1$

**TRY THE CONCEPT CHECK IN THE MARGIN.**

## E WRITING FRACTIONS AS PERCENTS

Recall that to write a percent as a fraction, we drop the percent symbol and divide by 100. We can reverse these steps to write a fraction as a percent.

> **WRITING A FRACTION AS A PERCENT**
>
> Multiply by 100 and attach a percent symbol. This is the same as multiplying by 100%.
>
> $$\dfrac{1}{8} = \dfrac{1}{8} \cdot 100\% = \dfrac{1}{8} \cdot \dfrac{100}{1}\% = \dfrac{100}{8}\% = 12\dfrac{1}{2}\% \text{ or } 12.5\%$$

> **Helpful Hint**
>
> From Example 15, we know that
>
> $$100\% = 1$$
>
> Recall that when we multiply a number by 1, we are not changing the value of that number. This means that when we multiply a number by 100%, we are not changing its value but rather writing the number as an equivalent percent.

**Examples**    Write each fraction or mixed number as a percent.

**16.** $\dfrac{9}{20} = \dfrac{9}{20} \cdot 100\% = \dfrac{9}{20} \cdot \dfrac{100}{1}\% = \dfrac{900}{20}\% = 45\%$

**17.** $\dfrac{2}{3} = \dfrac{2}{3} \cdot 100\% = \dfrac{2}{3} \cdot \dfrac{100}{1}\% = \dfrac{200}{3}\% = 66\dfrac{2}{3}\%$

**18.** $1\dfrac{1}{2} = \dfrac{3}{2} \cdot 100\% = \dfrac{3}{2} \cdot \dfrac{100}{1}\% = \dfrac{300}{2}\% = 150\%$

**TRY THE CONCEPT CHECK IN THE MARGIN.**

---

### ✓ CONCEPT CHECK

True or false? $16.5\% = \dfrac{16.5}{1000}$

**TEACHING TIP**

For Examples 16 and 18, you may want to point out that students could also write the fraction as a percent by rewriting the fraction with a denominator of 100.

### Practice Problems 16–18

Write each fraction or mixed number as a percent.

**16.** $\dfrac{1}{2}$      **17.** $\dfrac{7}{40}$      **18.** $2\dfrac{1}{4}$

### ✓ CONCEPT CHECK

Which digit in the percent 76.4582% represents
a. A tenth percent?
b. A thousandth percent?
c. A hundredth percent?
d. A whole percent?

**Answers**

**16.** 50%   **17.** $17\dfrac{1}{2}\%$   **18.** 225%

✓ Concept Check: False.
✓ Concept Check: **a.** 4  **b.** 8  **c.** 5  **d.** 6

**Example 19**    Write $\frac{1}{12}$ as a percent. Round to the nearest hundredth percent.

*Solution:*

$$\frac{1}{12} = \frac{1}{12} \cdot 100\% = \frac{1}{12} \cdot \frac{100\%}{1} = \frac{100}{12}\% \overset{\text{approximately}}{\approx} 8.33\%$$

$$\begin{array}{r} 8.333 \\ 12\overline{)100.000} \approx 8.33 \\ \underline{-96} \\ 4\,0 \\ \underline{-3\,6} \\ 40 \\ \underline{-36} \\ 40 \\ \underline{-36} \\ 4 \end{array}$$

Thus, $\frac{1}{12}$ is approximately 8.33%.

**Practice Problem 19**

Write $\frac{3}{17}$ as a percent. Round to the nearest hundredth percent.

## F    CONVERTING PERCENTS, DECIMALS, AND FRACTIONS

Let's summarize what we have learned so far about percents, decimals, and fractions.

TEACHING TIP

Before going over the summary, you may want to have students work in groups and write their own summaries of conversion strategies.

---

**SUMMARY OF CONVERTING PERCENTS, DECIMALS, AND FRACTIONS**

▲  *To write a percent as a decimal*, drop the % symbol and move the decimal point two places to the left.

▲  *To write a decimal as a percent*, move the decimal point two places to the right and attach the % symbol.

▲  *To write a percent as a fraction*, drop the % symbol and write the number over 100.

▲  *To write a fraction as a percent*, multiply the fraction by 100%.

---

**Example 20**    17.8% of automobile thefts in the continental United States occur in the Midwest. Write this percent as a decimal. (*Source:* The American Automobile Manufacturers Association)

*Solution:*    $17.8\% = 0.178$. Thus, 17.8% written as a decimal is 0.178.

**Practice Problem 20**

A family decides to spend no more than 25% of its monthly income on rent. Write 25% as a decimal.

**Example 21**    An advertisement for a stereo system reads "$\frac{1}{4}$ off." What percent off is this?

*Solution:*    Write $\frac{1}{4}$ as a percent.

$$\frac{1}{4} = \frac{1}{4} \cdot 100\% = \frac{1}{4} \cdot \frac{100\%}{1} = \frac{100}{4}\% = 25\%$$

Then "$\frac{1}{4}$ off" is the same as "25% off."

**Practice Problem 21**

Provincetown's budget for waste disposal increased by $1\frac{1}{4}$ times over the budget from last year. What percent increase is this?

It is helpful to know a few basic percent conversions. Appendix D contains a handy reference of percent, decimal, and fraction equivalences.

**Answers**

**19.** 17.65%  **20.** 0.25  **21.** 125%

# Focus on the Real World

### M&M's®

In the 1930s, many American stores did not stock chocolates during the summer because, without widespread air conditioning, the chocolate tended to melt and sales declined. In 1940, Forrest E. Mars, Sr., set out to develop a chocolate candy that could be sold year-round without the melting problem. So he formed a company in Newark, New Jersey, to make bite-sized melt-proof chocolate candies encased in a thin sugar shell. Thus M&M's® Plain Chocolate Candies were born!

The first Plain M&M's colors were brown, green, orange, red, violet, and yellow. Violet was replaced by tan in 1950. When the safety of a particular type of red food coloring was publicly questioned in 1976, red was completely eliminated from the M&M's color mix to avoid alarming consumers, even though the coloring in question was never used in M&Ms. After an 11-year hiatus, red returned to the color mix in 1987. In 1995, over 10 million Americans responded to a marketing campaign asking for help in choosing a new M&M's color. Given the choices of blue, pink, purple, or no change, 54% of the respondents chose blue, and it replaced tan. The new color mix of Plain M&M's became blue, brown, green, orange, red, and yellow in the percentages shown in the circle graph.

Color Mix for
M&M's® Plain Chocolate Candies

Yellow 20%
Brown 30%
Red 20%
Orange 10%
Green 10%
Blue 10%

*Source:* Mars, Incorporated

M&M's Peanut Chocolate Candies debuted in 1954. At first, Peanut M&M's were all brown. However, in 1960, red, green, and yellow were added. Orange joined the mix in 1976, and, finally, blue was introduced in 1995. The current color mix for Peanut M&M's is shown in the circle graph.

Color Mix for
M&M's® Peanut Chocolate Candies

Yellow 20%
Brown 20%
Orange 10%
Red 20%
Blue 20%
Green 10%

*Source:* Mars, Incorporated

### GROUP ACTIVITY

1. Use a single-serving pouch of M&M's Plain Chocolate Candies and find the percentage of each color in the bag. How does it compare to the overall percentages officially reported?

2. Repeat Question 2 with a single-serving pouch of M&M's Peanut Chocolate Candies.

3. For each type of M&M's Candies, combine your group's color counts with those of the other groups in your class. Find the percentage of each color for the combined figures. How do these percentages compare to the official color mix?

**Name** _____ **Section** _____ **Date** _____

## MENTAL MATH

*Write each fraction as a percent.*

1. $\dfrac{13}{100}$

2. $\dfrac{92}{100}$

3. $\dfrac{87}{100}$

4. $\dfrac{71}{100}$

5. $\dfrac{1}{100}$

6. $\dfrac{2}{100}$

## EXERCISE SET 7.1

**A** *Solve. See Examples 1 and 2.*

1. A basketball player makes 81 out of 100 attempted free throws. What percent of free throws was made?

2. In a survey of 100 people, 54 preferred chocolate syrup on their ice cream. What percent preferred chocolate syrup?

*Adults were asked what type of cookie was their favorite. The circle graph shows the results for every 100 people. Use this graph to answer Exercises 3–6. See Examples 1 and 2.*

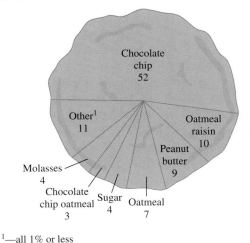

Chocolate chip 52
Other[1] 11
Oatmeal raisin 10
Peanut butter 9
Molasses 4
Chocolate chip oatmeal 3
Sugar 4
Oatmeal 7

[1]—all 1% or less
*Source: USA Today, 2/23/96*

3. What percent preferred peanut butter cookies?

4. What percent preferred oatmeal raisin cookies?

5. What type of cookie was preferred by most adults? What percent preferred this type of cookie?

6. What two types of cookies were preferred by the same number of adults? What percent preferred each type?

**7.** 12 out of 100 adults have watched an entire infomercial. What percent is this? (*Source:* Aragon Consulting Group)

**8.** In 1999, 99 out of 100 elementary schools had computers. What percent is this? (*Source: The World Almanac, 2000*)

**B** *Write each percent as a decimal. See Examples 3 through 6.*

**9.** 48%

**10.** 64%

**11.** 6%

**12.** 9%

**13.** 100%

**14.** 136%

**15.** 61.3%

**16.** 52.7%

**17.** 2.8%

**18.** 1.7%

**19.** 0.6%

**20.** 0.9%

**21.** 300%

**22.** 500%

**23.** 32.58%

**24.** 72.18%

*Write each percent as a decimal. See Examples 3 through 6.*

**25.** 73.7% of the work force in the United States works 35 hours or more per week. (*Source:* U.S. Bureau of Labor)

**26.** Approximately 23.6% of new pickup trucks and vans are white, making white the most popular new vehicle color for that class. (*Source:* American Automobile Manufacturers Association)

**27.** The United States is the largest consumer of energy in the world, using 25% of all energy produced worldwide each year. (*Source:* Energy Information Administration)

**28.** A health insurance company pays 80% of a person's medical costs.

**7.** 12%

**8.** 99%

**9.** 0.48

**10.** 0.64

**11.** 0.06

**12.** 0.09

**13.** 1

**14.** 1.36

**15.** 0.613

**16.** 0.527

**17.** 0.028

**18.** 0.017

**19.** 0.006

**20.** 0.009

**21.** 3

**22.** 5

**23.** 0.3258

**24.** 0.7218

**25.** 0.737

**26.** 0.236

**27.** 0.25

**28.** 0.80

**29.** In the first quarter of 1999, 80.2% of all Southwest Airlines' flights were on time, making Southwest the leading airline for on-time arrivals that period. (*Source:* U.S. Department of Transportation)

**30.** In 1999, 98.1% of all public high schools in the United States owned at least one computer. (*Source:* Quality Education Data, Inc.)

**31.** 46.2% of registered dental hygienists in the United States have earned an associate's degree or higher. (*Source:* The American Dental Hygienists' Association)

**32.** 27.3% of all households in the United States own a pet cat. (*Source:* American Veterinary Medical Association)

**C** *Write each decimal as a percent. See Examples 7 through 10.*

**33.** 3.1  **34.** 4.8  **35.** 29  **36.** 56

**37.** 0.003  **38.** 0.006  **39.** 0.22  **40.** 0.45

**41.** 0.056  **42.** 0.027  **43.** 0.3328  **44.** 0.1115

**45.** 3.00  **46.** 5.00  **47.** 0.7  **48.** 0.8

*Write each decimal as a percent. See Examples 7 through 10.*

**49.** The Munoz family saves 0.10 of their take-home pay.

**50.** The cost of an item for sale is 0.7 of the sales price.

| | |
|---|---|
| **29.** | 0.802 |
| **30.** | 0.981 |
| **31.** | 0.462 |
| **32.** | 0.273 |
| **33.** | 310% |
| **34.** | 480% |
| **35.** | 2900% |
| **36.** | 5600% |
| **37.** | 0.3% |
| **38.** | 0.6% |
| **39.** | 22% |
| **40.** | 45% |
| **41.** | 5.6% |
| **42.** | 2.7% |
| **43.** | 33.28% |
| **44.** | 11.15% |
| **45.** | 300% |
| **46.** | 500% |
| **47.** | 70% |
| **48.** | 80% |
| **49.** | 10% |
| **50.** | 70% |

51. 75.3%

52. 52.2%

53. 38%

54. 8.98%

55. $\dfrac{1}{25}$

56. $\dfrac{1}{50}$

57. $\dfrac{9}{200}$

58. $\dfrac{3}{40}$

59. $1\dfrac{3}{4}$

60. $2\dfrac{1}{2}$

61. $\dfrac{73}{100}$

62. $\dfrac{43}{50}$

63. $\dfrac{1}{8}$

64. $\dfrac{5}{8}$

65. $\dfrac{1}{16}$

66. $\dfrac{3}{8}$

67. $\dfrac{31}{300}$

68. $\dfrac{31}{400}$

69. $\dfrac{179}{800}$

70. $\dfrac{5}{32}$

71. 75%

72. 50%

73. 70%

74. 30%

75. 40%

76. 80%

77. 59%

78. 73%

79. 34%

80. 94%

81. $37\dfrac{1}{2}\%$

82. $62\dfrac{1}{2}\%$

**514**

**51.** In 1998, 0.753 of all recorded music sales were in the full-length CD format. (*Source: Recording Industry Association of America*)

**52.** In a recent year, 0.522 of all mail delivered by the United States Postal Service was first-class mail. (*Source: United States Postal Service*)

**53.** People take aspirin for a variety of reasons. The most common use of aspirin is to prevent heart disease, accounting for 0.38 of all aspirin use. (*Source:* Bayer Market Research)

**54.** The highest state income tax rate in the state of Iowa is 0.0898 on taxable income over $51,660. (*Source: CCH State Tax Guide*)

**D** *Write each percent as a fraction or mixed number in simplest form. See Examples 11 through 15.*

**55.** 4%    **56.** 2%    **57.** 4.5%    **58.** 7.5%

**59.** 175%    **60.** 250%    **61.** 73%    **62.** 86%

**63.** 12.5%    **64.** 62.5%    **65.** 6.25%    **66.** 37.5%

**67.** $10\dfrac{1}{3}\%$    **68.** $7\dfrac{3}{4}\%$    **69.** $22\dfrac{3}{8}\%$    **70.** $15\dfrac{5}{8}\%$

**E** *Write each fraction or mixed number as a percent. See Examples 16 through 19.*

**71.** $\dfrac{3}{4}$    **72.** $\dfrac{1}{2}$    **73.** $\dfrac{7}{10}$    **74.** $\dfrac{3}{10}$

**75.** $\dfrac{2}{5}$    **76.** $\dfrac{4}{5}$    **77.** $\dfrac{59}{100}$    **78.** $\dfrac{73}{100}$

**79.** $\dfrac{17}{50}$    **80.** $\dfrac{47}{50}$    **81.** $\dfrac{3}{8}$    **82.** $\dfrac{5}{8}$

**83.** $\dfrac{5}{16}$　　　　**84.** $\dfrac{7}{16}$　　　　**85.** $\dfrac{2}{3}$　　　　**86.** $\dfrac{1}{3}$

**87.** $2\dfrac{1}{2}$　　　　**88.** $2\dfrac{1}{5}$　　　　**89.** $1\dfrac{9}{10}$　　　　**90.** $2\dfrac{7}{10}$

*Write each fraction as a percent. Round to the nearest hundredth percent. See Example 19.*

**91.** $\dfrac{7}{11}$　　　**92.** $\dfrac{5}{12}$　　　📼 **93.** $\dfrac{4}{15}$　　　**94.** $\dfrac{10}{11}$

**95.** $\dfrac{1}{7}$　　　**96.** $\dfrac{1}{9}$　　　**97.** $\dfrac{11}{12}$　　　**98.** $\dfrac{5}{6}$

**F** *Complete each table. See Examples 20 and 21.*

**99.**

| Percent | Decimal | Fraction |
|---|---|---|
| 35% | 0.35 | $\dfrac{7}{20}$ |
| 20% | 0.2 | $\dfrac{1}{5}$ |
| 50% | 0.5 | $\dfrac{1}{2}$ |
| 70% | 0.7 | $\dfrac{7}{10}$ |
| 37.5% | 0.375 | $\dfrac{3}{8}$ |

**100.**

| Percent | Decimal | Fraction |
|---|---|---|
| 52.5% | 0.525 | $\dfrac{21}{40}$ |
| 75% | 0.75 | $\dfrac{3}{4}$ |
| $66\dfrac{2}{3}\%$ | $0.666\overline{6}$ | $\dfrac{2}{3}$ |
| $83\dfrac{1}{3}\%$ | $0.833\overline{3}$ | $\dfrac{5}{6}$ |
| 100% | 1 | 1 |

**101.**

| Percent | Decimal | Fraction |
|---|---|---|
| 40% | 0.4 | $\dfrac{2}{5}$ |
| $23\dfrac{1}{2}\%$ | 0.235 | $\dfrac{47}{200}$ |
| 80% | 0.8 | $\dfrac{4}{5}$ |
| $33\dfrac{1}{3}\%$ | $0.333\overline{3}$ | $\dfrac{1}{3}$ |
| 87.5% | 0.875 | $\dfrac{7}{8}$ |
| 7.5% | 0.075 | $\dfrac{3}{40}$ |

**102.**

| Percent | Decimal | Fraction |
|---|---|---|
| 50% | 0.5 | $\dfrac{1}{2}$ |
| 40% | 0.4 | $\dfrac{2}{5}$ |
| 25% | 0.25 | $\dfrac{1}{4}$ |
| 12.5% | 0.125 | $\dfrac{1}{8}$ |
| 62.5% | 0.625 | $\dfrac{5}{8}$ |
| 14% | 0.14 | $\dfrac{7}{50}$ |

**83.** $31\dfrac{1}{4}\%$

**84.** $43\dfrac{3}{4}\%$

**85.** $66\dfrac{2}{3}\%$

**86.** $33\dfrac{1}{3}\%$

**87.** 250%

**88.** 220%

**89.** 190%

**90.** 270%

**91.** 63.64%

**92.** 41.67%

**93.** 26.67%

**94.** 90.91%

**95.** 14.29%

**96.** 11.11%

**97.** 91.67%

**98.** 83.33%

**99.** see table

**100.** see table

**101.** see table

**102.** see table

**103.** $\frac{47}{250}$

**104.** 0.757

**105.** 8.5%

**106.** 21.2%

**107.** 0.674

**108.** $\frac{59}{500}$

**109.** 30%

**110.** 74%

*Solve. See Examples 20 and 21.*

**103.** Approximately 18.8% of new full-size cars are dark green, making dark green the most popular new vehicle color for that class. Write this percent as a fraction. (*Source:* American Automobile Manufacturers Association)

**104.** In 1950, the United States produced 75.7% of all motor vehicles made worldwide. Write this percent as a decimal. (*Source:* American Automobile Manufacturers Association)

**105.** In 1998, $\frac{17}{200}$ of all new cars sold in the United States were imported from Japan. Write this fraction as a percent. (*Source:* Ward's Communications)

**106.** In 1980, $\frac{53}{250}$ of all new cars sold in the United States were imported from Japan. Write this fraction as a percent. (*Source:* American Automobile Manufacturers Association)

**107.** In 1998, 67.4% of all households with televisions subscribed to a cable television service. Write this percent as a decimal. (*Source:* Nielsen Media Research)

**108.** In 1998, 11.8% of Wisconsin residents were not covered by any type of health insurance. Write this percent as a fraction. (*Source:* U.S. Bureau of the Census)

**109.** In the first half of 1999, the top-rated prime-time television program was *E.R.*, which had an average audience share of approximately $\frac{3}{10}$ of all those watching television during that time slot. Write this fraction as a percent. (*Source:* Nielsen Media Research)

**110.** Approximately $\frac{37}{50}$ of all structure fires take place in personal residences. Write this fraction as a percent. (*Source:* National Fire Protection Association)

**Name** _____

## REVIEW AND PREVIEW

*Find the value of n. See Section 3.3.*

**111.** $3n = 45$

**112.** $7n = 48$

**113.** $-8n = 80$

**114.** $-2n = 16$

**115.** $6n = -72$

**116.** $5n = -35$

## COMBINING CONCEPTS

*Write each fraction as a decimal and then write each decimal as a percent. Round the decimal to three decimal places and the percent to the nearest tenth of a percent.*

**117.** $\dfrac{850}{736}$

**118.** $\dfrac{506}{248}$

*Fill in the blanks.*

**119.** A fraction written as a percent is greater than 100% when the numerator is _____ than the greater/less denominator.

**120.** A decimal written as a percent is less than 100% when the decimal is _____ than 1. greater/less

**121.** In your own words, explain how to write a percent as a fraction.

**122.** In your own words, explain how to write a fraction as a decimal.

---

**111.** 15

**112.** $\dfrac{48}{7}$

**113.** $-10$

**114.** $-8$

**115.** $-12$

**116.** $-7$

**117.** 1.155; 115.5%

**118.** 2.040; 204.0%

**119.** greater

**120.** less

**121.** answers may vary

**122.** answers may vary

**517**

Database
administrators,
computer
support specialists,
and other computer
**123.** scientists

**Name** _____

*The bar graph shows the predicted fastest-growing occupations. Use this graph to answer Exercises 123–126.*

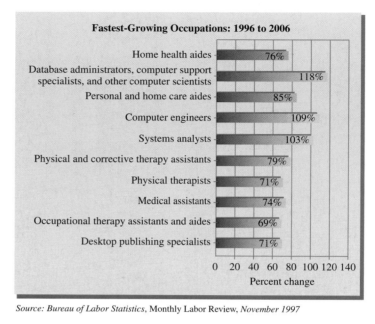

**Fastest-Growing Occupations: 1996 to 2006**

| Occupation | Percent change |
|---|---|
| Home health aides | 76% |
| Database administrators, computer support specialists, and other computer scientists | 118% |
| Personal and home care aides | 85% |
| Computer engineers | 109% |
| Systems analysts | 103% |
| Physical and corrective therapy assistants | 79% |
| Physical therapists | 71% |
| Medical assistants | 74% |
| Occupational therapy assistants and aides | 69% |
| Desktop publishing specialists | 71% |

0  20  40  60  80  100 120 140
Percent change

*Source: Bureau of Labor Statistics,* Monthly Labor Review, *November 1997*

**123.** What occupation is predicted to be the fastest growing?

**124.** What occupation is predicted to be the second fastest growing?

**125.** Write the percent change for Personal and home care aides as a decimal.

**126.** Write the percent change for Systems analysts as a decimal.

**127.** In your own words, explain how to write a percent as a decimal.

**128.** In your own words, explain how to write a decimal as a percent.

# 7.2 SOLVING PERCENT PROBLEMS WITH EQUATIONS

Sections 7.2 and 7.3 introduce two methods for solving percent problems. It is not necessary that you cover both sections. You may want to check with your instructor for further advice.

   To solve percent problems in this section, we will translate the problems into mathematical statements, or equations.

## A WRITING PERCENT PROBLEMS AS EQUATIONS

Recognizing key words in a percent problem is helpful in writing the problem as an equation. Three key words in the statement of a percent problem and their meanings are as follows:

   **of** means **multiplication** ( $\cdot$ )
   **is** means **equals** ( = )
   **what** (or some equivalent) means **the unknown number**.

In our examples, we will let the letter $x$ stand for the unknown number.

**Example 1**    Translate to an equation:

   5 is what percent of 20?

*Solution:*    5 is what percent of 20?

   $\downarrow\downarrow$        $\downarrow$        $\downarrow$ $\downarrow$
   $5 =$        $x$        $\cdot$ 20

---

**Helpful Hint**

Remember that an equation is simply a mathematical statement that contains an equal sign (=).

   $5 = 20x$

   $\uparrow$
   equal sign

---

**Example 2**    Translate to an equation:

   1.2 is 30% of what number?

*Solution:*    1.2 is 30% of what number?

   $\downarrow$ $\downarrow$ $\downarrow$ $\downarrow$        $\downarrow$
   $1.2 = 30\% \cdot$        $x$

**Example 3**    Translate to an equation:

   What number is 25% of 0.008?

*Solution:*    What number is 25% of 0.008?

   $\downarrow$        $\downarrow$ $\downarrow$ $\downarrow$ $\downarrow$
   $x$        $= 25\% \cdot 0.008$

---

### Objectives

**A** Write percent problems as equations.

**B** Solve percent problems.

Study Guide    SSM    CD-ROM    Video 7.2

TEACHING TIP

You may want to begin this section by asking students if they recall how to find the unknown number $x$ in an equation such as $7 \cdot x = 105$.

### Practice Problem 1

Translate: 6 is what percent of 24?

TEACHING TIP

Remember that it is not necessary to cover both Section 7.2 and Section 7.3. You may want to read both sections and then choose one for your students and omit the other from your syllabus.

### Practice Problem 2

Translate: 1.8 is 20% of what number?

### Practice Problem 3

Translate: What number is 40% of 3.6?

**Answers**

**1.** $6 = x \cdot 24$    **2.** $1.8 = 20\% \cdot x$
**3.** $x = 40\% \cdot 3.6$

### Practice Problems 4–6

Translate each question to an equation.

4. 42% of 50 is what number?
5. 15% of what number is 9?
6. What percent of 150 is 90?

### ✓ CONCEPT CHECK

In the equation $2x = 10$, what step is taken to solve the equation?

### Practice Problem 7

What number is 20% of 85?

### Answers

4. $42\% \cdot 50 = x$    5. $15\% \cdot x = 9$
6. $x \cdot 150 = 90$    7. 17
✓ **Concept Check:** Divide both sides of the equation by 2.

---

### Examples

Translate each question to an equation.

4. 38% of 200 is what number?

$$38\% \cdot 200 = x$$

5. 40% of what number is 80?

$$40\% \cdot x = 80$$

6. What percent of 85 is 34?

$$x \cdot 85 = 34$$

**TRY THE CONCEPT CHECK IN THE MARGIN.**

## B SOLVING PERCENT PROBLEMS

You may have noticed by now that each percent problem has contained three numbers—in our examples, two are known and one is unknown. Each of these numbers is given a special name.

| 15% | of | 60 | is | 9 |
|---|---|---|---|---|

$$\underset{\text{percent}}{15\%} \cdot \underset{\text{base}}{60} = \underset{\text{amount}}{9}$$

We call this equation the **percent equation**.

> **PERCENT EQUATION**
>
> $\text{percent} \cdot \text{base} = \text{amount}$

Once a percent problem has been written as a percent equation, we can use the equation to find the unknown number.

### Solving Percent Equations for the Amount

### Example 7

What number is 35% of 60?

*Solution:*

$$x = 35\% \cdot 60 \quad \text{Translate to an equation.}$$
$$x = 0.35 \cdot 60 \quad \text{Write 35\% as 0.35.}$$
$$x = 21$$

Multiply:
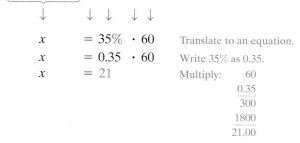

Then 21 is 35% of 40. Is this reasonable? Since 0.35 is close to 0.40, then 35% of 60 should be close to 40% of 60, or 24. Our result is reasonable since 21 is close to 24.

> **Helpful Hint**
>
> When solving a percent equation, write the percent as a decimal or fraction.

**Example 8**   85% of 300 is what number?

$$\downarrow \quad \downarrow \quad \downarrow \quad \downarrow \qquad \downarrow$$

*Solution:*   $85\% \cdot 300 = \quad x$   Translate to an equation.

$0.85 \cdot 300 = \quad x$   Write 85% as 0.85.

$225 = \quad x$   Multiply: $0.85 \cdot 300 = 255$.

Then 85% of 300 is 255. Is this result reasonable? To check, round 85% to 90%. Then 90% of 300 is 270, which is close to 255.

**Practice Problem 8**

90% of 150 is what number?

## Solving Percent Equations for the Base

**Example 9**   12% of what number is 0.6?

$$\downarrow \quad \downarrow \qquad \downarrow \qquad \downarrow \quad \downarrow$$

*Solution:*   $12\% \cdot \qquad x \quad = 0.6$   Translate to an equation.

$0.12 \cdot \qquad x \quad = 0.6$   Write 12% as 0.12.

$$\frac{0.12 \cdot x}{0.12} = \frac{0.6}{0.12}$$   Divide both sides by 0.12.

$$x \quad = 5$$

$$\begin{array}{r} 5. \\ 0.12\overline{)0.60} \\ \phantom{0.12)}60 \\ \hline \phantom{0.12)}0 \end{array}$$

Then 12% of 5 is 0.6.

**Practice Problem 9**

15% of what number is 1.2?

**Example 10**   13   is   $6\frac{1}{2}\%$   of   what number?

$$\downarrow \quad \downarrow \quad \downarrow \quad \downarrow \qquad \downarrow$$

*Solution:*   $13 \quad = \quad 6\frac{1}{2}\% \quad \cdot \qquad x$   Translate to an equation.

$13 \quad = 0.065 \cdot x$        $6\frac{1}{2}\% = 6.5\% = 0.065$.

$$\frac{13}{0.065} = \frac{0.065 \cdot x}{0.065}$$        Divide both sides by 0.065.

$$200 \quad = \quad x$$

$$\begin{array}{r} 200. \\ 0.065\overline{)13.000} \\ \phantom{0.065)}130 \\ \hline \phantom{0.065)}0 \end{array}$$

Then 13 is $6\frac{1}{2}\%$ of 200.

**Practice Problem 10**

27 is $4\frac{1}{2}\%$ of what number?

**Answers**

**8.** 135   **9.** 8   **10.** 600

## Practice Problem 11

What percent of 80 is 8?

## Solving Percent Equations for the Percent

**Example 11**

What percent of 12 is 9?

*Solution:*

$$x \cdot 12 = 9$$    Translate to an equation.

$$\frac{x \cdot 12}{12} = \frac{9}{12}$$    Divide both sides by 12.

$$x = 0.75$$

Next, since we are looking for percent, we write 0.75 as a percent.

$$x = 75\%$$

Then $75\%$ of 12 is 9. To check, see that $75\% \cdot 12 = 9$.

---

### Helpful Hint

If your unknown in the percent equation is percent, don't forget to convert your answer to a percent.

---

## Practice Problem 12

35 is what percent of 25?

**Example 12**

78 is what percent of 65?

*Solution:*

$$78 = x \cdot 65$$    Translate to an equation.

$$\frac{78}{65} = \frac{x \cdot 65}{65}$$    Divide both sides by 65.

$$1.2 = x$$

$$120\% = x$$    Write 1.2 as a percent.

Then 78 is $120\%$ of 65.

**TRY THE CONCEPT CHECK IN THE MARGIN.**

## ✓ CONCEPT CHECK

Consider the problem

10 is 40% of what number?

Will the solution be greater than or less than 10? How do you know without solving the problem?

**Answers**

**11.** 10%  **12.** 140%

✓ Concept Check: Greater.

## MENTAL MATH

*Identify the percent, the base, and the amount in each equation. Recall that* percent · base = amount.

**1.** $42\% \cdot 50 = 21$

**2.** $30\% \cdot 65 = 19.5$

**3.** $107.5 = 125\% \cdot 86$

**4.** $99 = 110\% \cdot 90$

# EXERCISE SET 7.2

**A** *Translate each question to an equation. Do not solve. See Examples 1 through 6.*

**1.** 15% of 72 is what number?

**2.** What number is 25% of 55?

**3.** 30% of what number is 80?

**4.** 0.5 is 20% of what number?

**5.** What percent of 90 is 20?

**6.** 8 is 50% of what number?

**7.** 1.9 is 40% of what number?

**8.** 72% of 63 is what number?

**9.** What number is 9% of 43?

**10.** 4.5 is what percent of 45?

**B** *Solve. See Examples 7 and 8.*

**11.** 10% of 35 is what number?

**12.** 25% of 60 is what number?

**13.** What number is 14% of 52?

**14.** What number is 30% of 17?

---

### MENTAL MATH ANSWERS

**1.** percent: 42; base: 50; amount: 21

**2.** percent: 30; base: 65; amount: 19.5

**3.** percent: 125; base: 86; amount: 107.5

**4.** percent: 110; base: 90; amount: 99

### ANSWERS

**1.** $15\% \cdot 72 = x$

**2.** $x = 25\% \cdot 55$

**3.** $30\% \cdot x = 80$

**4.** $0.5 = 20\% \cdot x$

**5.** $x \cdot 90 = 20$

**6.** $8 = 50\% \cdot x$

**7.** $1.9 = 40\% \cdot x$

**8.** $72\% \cdot 63 = x$

**9.** $x = 9\% \cdot 43$

**10.** $4.5 = x \cdot 45$

**11.** 3.5

**12.** 15

**13.** 7.28

**14.** 5.1

**Name** _____

*Solve. See Examples 9 and 10.*

**15.** 30 is 5% of what number?

**16.** 25 is 25% of what number?

**17.** 1.2 is 12% of what number?

**18.** 0.22 is 44% of what number?

*Solve. See Examples 11 and 12.*

**19.** 66 is what percent of 60?

**20.** 30 is what percent of 20?

**21.** 16 is what percent of 50?

**22.** 27 is what percent of 50?

*Solve. See Examples 7 through 12.*

**23.** 0.1 is 10% of what number?

**24.** 0.5 is 5% of what number?

**25.** 125% of 36 is what number?

**26.** 200% of 13.5 is what number?

**27.** 82.5 is $16\frac{1}{2}$% of what number?

**28.** 7.2 is $6\frac{1}{4}$% of what number?

**29.** 2.58 is what percent of 50?

**30.** 264 is what percent of 33?

**31.** What number is 42% of 60?

**32.** What number is 36% of 80?

**33.** What percent of 150 is 67.5?

**34.** What percent of 105 is 88.2?

**35.** 120% of what number is 42?

**36.** 160% of what number is 40?

**Name** _____

## REVIEW AND PREVIEW

*Find the value of n in each proportion. See Section 6.3.*

**37.** $\dfrac{27}{n} = \dfrac{9}{10}$ **38.** $\dfrac{35}{n} = \dfrac{7}{5}$ **39.** $\dfrac{n}{5} = \dfrac{8}{11}$ **40.** $\dfrac{n}{3} = \dfrac{6}{13}$

*Write each sentence as a proportion.*

**41.** 17 is to 12 as *n* is to 20.

**42.** 20 is to 25 as *n* is to 10.

**43.** 8 is to 9 as 14 is to *n*.

**44.** 5 is to 6 as 15 is to *n*.

## COMBINING CONCEPTS

*Solve.*

 **45.** 1.5% of 45,775 is what number?

 **46.** What percent of 75,528 is 27,945.36?

 **47.** 22,113 is 180% of what number?

**48.** In your own words, explain how to solve a percent equation.

**37.** 30

**38.** 25

**39.** $3\dfrac{7}{11}$

**40.** $1\dfrac{5}{13}$

**41.** $\dfrac{17}{12} = \dfrac{n}{20}$

**42.** $\dfrac{20}{25} = \dfrac{n}{10}$

**43.** $\dfrac{8}{9} = \dfrac{14}{n}$

**44.** $\dfrac{5}{6} = \dfrac{15}{n}$

**45.** 686.625

**46.** 37%

**47.** 12,285

**48.** answers may vary

# Focus on Mathematical Connections

### PROBABILITY

The probability of an event is a measure of the chance or likelihood of the event occurring. If a fair coin is tossed, the probability of it landing tails up is $\frac{1}{2}$ because tails is one of two equally likely outcomes (heads or tails) when the coin is tossed. Probabilities are always between 0 and 1. A probability of 0 means an event is certain *not* to occur, and a probability of 1 means an event is certain to occur.

We know that fractions can be reported as decimals or percents. In fact, probabilities are frequently given as percents in the news media. A weather forecaster might say that there is a 30% chance of rain. A medical report might say that the probability of successful treatment with a certain drug is 90%.

In Section 8.6, you will learn how to calculate probabilities in detail. Here we will discuss estimating probabilities based on data. Suppose, for example, you wished to know the probability of a tossed coin landing heads up but you didn't know how or didn't want to calculate it. You could estimate the probability by tossing the coin many times and recording the results. The percentage of coin tosses that landed heads up is an estimate of the probability of the coin landing heads up. Weather forecasters take this kind of approach to estimate the probability of rain. They look back in their records for days having the same weather conditions, such as temperature and barometric pressure, as today. The percentage of these days on which it rained is an estimate of the probability that it will rain today. For example, if there were 2000 days in the past with weather conditions matching today's conditions and if it rained on 500 of those days, forecasters would say there is a 25% chance of rain today.

### GROUP ACTIVITY

1. Find several instances of probabilities used in media. Explain what each means.

2. Try to determine whether each probability used in Question 1 was calculated or estimated based on data. Explain your reasoning.

# 7.3  SOLVING PERCENT PROBLEMS WITH PROPORTIONS

There is more than one method that can be used to solve percent problems. (See the note at the beginning of Section 7.2) In the last section, we used the percent equation. In this section, we will use proportions.

## A  WRITING PERCENT PROBLEMS AS PROPORTIONS

To understand the proportion method, recall that 70% means the ratio of 70 to 100, or $\frac{70}{100}$.

$$70\% = \frac{70}{100} = \frac{7}{10}$$

$\frac{7}{10}$ shaded

70% or $\frac{70}{100}$ shaded

Since the ratio $\frac{70}{100}$ is equal to the ratio $\frac{7}{10}$, we have the proportion

$$\frac{7}{10} = \frac{70}{100}$$

We call this proportion the **percent proportion**. In general, we can name the parts of this proportion as follows.

---

**PERCENT PROPORTION**

$$\frac{\text{amount}}{\text{base}} = \frac{\text{percent}}{100} \quad \leftarrow \text{ always 100}$$

or

$$\text{amount} \rightarrow \frac{a}{b} = \frac{p}{100} \leftarrow \text{ percent} \\ \text{base} \rightarrow$$

---

When we translate percent problems to proportions, the **percent** can be identified by looking for the symbol % or the word *percent*. The **base** usually follows the word *of*. The **amount** is the part compared to the whole.

---

**Helpful Hint**

This table may be useful when identifying the parts of a proportion.

| Part of Proportion | How It's Identified |
|---|---|
| Percent | % or percent |
| Base | Appears after *of* |
| Amount | Part compared to whole |

---

Objectives

**A** Write percent problems as proportions.

**B** Solve percent problems.

Study Guide   SSM   CD-ROM   Video 7.3

## Practice Problem 1

Translate to a proportion: 15% of what number is 55?

## Practice Problem 2

Translate to a proportion: 35 is what percent of 70?

## Practice Problem 3

Translate to a proportion: What number is 25% of 68?

TEACHING TIP

Ask students how they are identifying the amount, the base, and the percent. It often helps students to hear how fellow students are accomplishing a task.

## Practice Problem 4

Translate to a proportion:
520 is 65% of what number?

**Answers**

1. $\dfrac{15}{100} = \dfrac{55}{b}$   2. $\dfrac{35}{70} = \dfrac{p}{100}$   3. $\dfrac{a}{68} = \dfrac{25}{100}$

4. $\dfrac{520}{b} = \dfrac{65}{100}$

**Example 1**   Translate to a proportion:

12% of what number is 47?

$\downarrow$     $\downarrow$     $\downarrow$

| percent | base<br>It appears after the word *of.* | amount<br>It is the part compared to the whole. |

$\text{amount} \rightarrow \dfrac{47}{b} = \dfrac{12}{100} \leftarrow \text{percent}$
$\text{base} \rightarrow$

**Example 2**   Translate to a proportion:

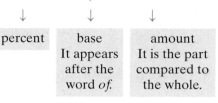

101   is   what percent   of   200?

$\downarrow$     $\downarrow$     $\downarrow$

| amount<br>It is the part compared to the whole. | percent | base<br>It appears after the word *of.* |

$\text{amount} \rightarrow \dfrac{101}{200} = \dfrac{p}{100} \leftarrow \text{percent}$
$\text{base} \rightarrow$

**Example 3**   Translate to a proportion:

What number   is 90%   of   45?

$\downarrow$     $\downarrow$     $\downarrow$

| amount<br>It is the part compared to the whole. | percent | base<br>It appears after the word *of.* |

$\text{amount} \rightarrow \dfrac{a}{45} = \dfrac{90}{100} \leftarrow \text{percent}$
$\text{base} \rightarrow$

**Example 4**   Translate to a proportion:

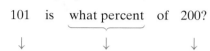

238   is   40%   of what number?

$\downarrow$     $\downarrow$     $\downarrow$

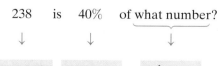

| amount | percent | base |

$\dfrac{238}{b} = \dfrac{40}{100}$

**Example 5**　Translate to a proportion:

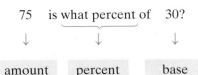

75　is what percent of　30?
↓　　↓　　↓

*Solution:*　[amount]　[percent]　[base]

$$\frac{75}{30} = \frac{p}{100}$$

**Practice Problem 5**

Translate to a proportion:
65 is what percent of 50?

**Example 6**　Translate to a proportion:

45%　of　105　is　what number?
↓　·　↓　　↓

*Solution:*　[percent]　[base]　[amount]

$$\frac{a}{105} = \frac{45}{100}$$

**Practice Problem 6**

Translate to a proportion:
36% of 80 is what number?

**TRY THE CONCEPT CHECK IN THE MARGIN.**

**B　SOLVING PERCENT PROBLEMS**

The proportions that we have written in this section contain three values that can change: the percent, the base, and the amount. If any two of these values are known, we can find the third (unknown value). To do this, we write a percent proportion and find the unknown value as we did in Section 6.3.

**Solving Percent Proportions for the Amount**

**Example 7**　What number is　30%　of　9?

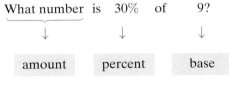

↓　　↓　　↓

*Solution:*　[amount]　[percent]　[base]

$$\frac{a}{9} = \frac{30}{100}$$

To solve, we set cross products equal to each other.

$$\frac{a}{9} = \frac{30}{100}$$

$a \cdot 100 = 9 \cdot 30$　　Set cross products equal.

$a \cdot 100 = 270$　　Multiply.

**✓ CONCEPT CHECK**

When solving a percent problem with a proportion, describe how you can check the result.

**Practice Problem 7**

What number is 8% of 120?

TEACHING TIP

After Example 6, you may want to have students translate each of the following proportions into a question of the form "_____ % of _____ is _____?"

a. $\dfrac{a}{480} = \dfrac{26}{100}$

b. $\dfrac{87}{b} = \dfrac{38}{100}$

c. $\dfrac{360}{135} = \dfrac{p}{100}$

**Answers**

**5.** $\dfrac{65}{50} = \dfrac{p}{100}$　**6.** $\dfrac{a}{80} = \dfrac{36}{100}$　**7.** 9.6

**✓ Concept Check:** Put the result into the proportion and check that the proportion is true.

$$\frac{a \cdot 100}{100} = \frac{270}{100} \qquad \text{Divide both sides by 100.}$$

$$a = 2.7 \qquad \text{Simplify.}$$

Then 2.7 is 30% of 9.

### ✓ CONCEPT CHECK

Consider the problem
      78 is what percent of 350?
Which part of the percent proportion is unknown?

a. the amount
b. the base
c. the percent

**TRY THE CONCEPT CHECK IN THE MARGIN.**

**Helpful Hint**

The proportion in Example 7 contained the ratio $\frac{30}{100}$. A ratio in a proportion may be simplified before solving the proportion. The unknown number in both

$$\frac{a}{9} = \frac{30}{100} \qquad \text{and} \qquad \frac{a}{9} = \frac{3}{10}$$

is 2.7

### Solving Percent Proportions for the Base

**Practice Problem 8**

75% of what number is 60?

**Example 8**   150% of what number is 30?

*Solution:*
   ↓      ↓      ↓
  percent    base    amount

$$\frac{30}{b} = \frac{150}{100} \qquad \text{Write the proportion.}$$

$$\frac{30}{b} = \frac{3}{2} \qquad \text{Simplify } \frac{150}{100}.$$

$$30 \cdot 2 = b \cdot 3 \qquad \text{Set cross products equal.}$$

$$60 = b \cdot 3 \qquad \text{Write } 30 \cdot 2 \text{ as } 60.$$

$$\frac{60}{3} = \frac{b \cdot 3}{3} \qquad \text{Divide both sides by 3.}$$

$$20 = b \qquad \text{Simplify.}$$

Then 150% of 20 is 30.

**Answers**

**8.** 80
✓ Concept Check: c

**Example 9**   20.8   is   40%   of   what number?

↓     ↓     ↓

*Solution:*   amount    percent      base

$$\frac{20.8}{b} = \frac{40}{100} \quad \text{or} \quad \frac{20.8}{b} = \frac{2}{5}$$   Write the proportion and simplify $\frac{40}{100}$.

$20.8 \cdot 5 = b \cdot 2$   Set cross products equal.

$104 = b \cdot 2$   Multiply.

$\frac{104}{2} = \frac{b \cdot 2}{2}$   Divide both sides by 2.

$52 = b$   Simplify.

Then 20.8 is 40% of 52.

## Solving Percent Proportions for the Percent

**Example 10**   What percent   of   50   is   8?

↓     ↓     ↓

*Solution:*   percent      base      amount

$$\frac{8}{50} = \frac{p}{100} \quad \text{or} \quad \frac{4}{25} = \frac{p}{100}$$   Write the proportion and simplify $\frac{8}{50}$.

$4 \cdot 100 = 25 \cdot p$   Set cross products equal.

$400 = 25 \cdot p$   Multiply.

$\frac{400}{25} = \frac{25 \cdot p}{25}$   Divide both sides by 25.

$16 = p$   Simplify.

Then 16% of 50 is 8.

**Example 11**   504   is   what percent   of 360?

↓     ↓     ↓

*Solution:*   amount      percent      base

$$\frac{504}{360} = \frac{p}{100}$$

**Practice Problem 9**

15 is 5% of what number?

**Practice Problem 10**

What percent of 40 is 5?

**HELPFUL HINT**

Recall from our percent proportion that this number already is a percent. Just keep the number as is and attach a % symbol.

**Practice Problem 11**

What percent of 160 is 336?

**Answers**

**9.** 300  **10.** 12.5%  **11.** 210%

TEACHING TIP

You may want to ask students to compare the amount and the base when the percent > 100. Then suggest that they check that
amount > base when percent > 100,
amount < base when percent < 100,
and amount = base when percent = 100.

Let's choose not to simplify the ratio $\frac{504}{360}$.

$$504 \cdot 100 = 360 \cdot p \qquad \text{Set cross products equal.}$$

$$50{,}400 = 360 \cdot p \qquad \text{Multiply.}$$

$$\frac{50{,}400}{360} = \frac{360 \cdot p}{360} \qquad \text{Divide both sides by 360.}$$

$$140 = p \qquad \text{Simplify.}$$

Notice by choosing not to simplify $\frac{504}{360}$, we had larger

numbers in our equation. Either way, we find that 504 is 140% of 360.

Name _____   Section _____   Date _____

## MENTAL MATH

Identify the amount, the base, and the percent in each equation. Recall that $\dfrac{\text{amount}}{\text{base}} = \dfrac{\text{percent}}{100}$.

1. $\dfrac{12.6}{42} = \dfrac{30}{100}$    2. $\dfrac{201}{300} = \dfrac{67}{100}$    3. $\dfrac{20}{100} = \dfrac{102}{510}$    4. $\dfrac{40}{100} = \dfrac{248}{620}$

# EXERCISE SET 7.3

**A** *Translate each question to a proportion. Do not solve. See Examples 1 through 6.*

1. 32% of 65 is what number?

2. What number is 5% of 125?

3. 40% of what number is 75?

4. 1.2 is 47% of what number?

5. What percent of 200 is 70?

6. 520 is 85% of what number?

7. 2.3 is 58% of what number?

8. 92% of 30 is what number?

9. What number is 19% of 130?

10. 8.2 is what percent of 82?

**B** *Solve. See Example 7.*

11. 10% of 55 is what number?

12. 25% of 84 is what number?

🔲 13. What number is 18% of 105?

14. What number is 40% of 29?

*Solve. See Examples 8 and 9.*

15. 60 is 15% of what number?

16. 75 is 75% of what number?

🔲 17. 7.8 is 78% of what number?

18. 1.1 is 44% of what number?

*Solve. See Examples 10 and 11.*

19. 105 is what percent of 84?

20. 77 is what percent of 44?

🔲 21. 14 is what percent of 50?

22. 37 is what percent of 50?

MENTAL MATH ANSWERS

1. amount: 12.6; base: 42; percent: 30

2. amount: 201; base: 300; percent: 67

3. amount: 102; base: 510; percent: 20

4. amount: 248; base: 620; percent: 40

ANSWERS

1. $\dfrac{a}{65} = \dfrac{32}{100}$

2. $\dfrac{a}{125} = \dfrac{5}{100}$

3. $\dfrac{75}{b} = \dfrac{40}{100}$

4. $\dfrac{1.2}{b} = \dfrac{47}{100}$

5. $\dfrac{70}{200} = \dfrac{p}{100}$

6. $\dfrac{520}{b} = \dfrac{85}{100}$

7. $\dfrac{2.3}{b} = \dfrac{58}{100}$

8. $\dfrac{a}{30} = \dfrac{92}{100}$

9. $\dfrac{a}{130} = \dfrac{19}{100}$

10. $\dfrac{8.2}{82} = \dfrac{p}{100}$

11. 5.5

12. 21

13. 18.9

14. 11.6

15. 400

16. 100

17. 10

18. 2.5

19. 125%

20. 175%

21. 28%

22. 74%

**533**

**23.** 29

**24.** 124

**25.** 1.92

**26.** 7.8

**27.** 1000

**28.** 500

**29.** 210%

**30.** 320%

**31.** 55.18

**32.** 68.9

**33.** 45%

**34.** 32%

**35.** 85

**36.** 130

**37.** $\dfrac{7}{8}$

**38.** $\dfrac{1}{24}$

**39.** $3\dfrac{2}{15}$

**40.** $7\dfrac{1}{6}$

**41.** 0.7

**42.** 15.29

**43.** 2.19

**44.** 13.76

**45.** 12,011.2

**46.** 80.0%

**47.** 7270.6

**48.** answers may vary

**Name** _____

*Solve. See Examples 7 through 11.*

**23.** 2.9 is 10% of what number?

**24.** 6.2 is 5% of what number?

**25.** 2.4% of 80 is what number?

**26.** 6.5% of 120 is what number?

**27.** 160 is 16% of what number?

**28.** 30 is 6% of what number?

**29.** 348.6 is what percent of 166?

**30.** 262.4 is what percent of 82?

**31.** What number is 89% of 62?

**32.** What number is 53% of 130?

**33.** What percent of 8 is 3.6?

**34.** What percent of 5 is 1.6?

**35.** 140% of what number is 119?

**36.** 170% of what number is 221?

## REVIEW AND PREVIEW

*Add or subtract as indicated. See Sections 4.4, 4.5, and 4.8.*

**37.** $\dfrac{11}{16} + \dfrac{3}{16}$  **38.** $\dfrac{5}{8} - \dfrac{7}{12}$  **39.** $3\dfrac{1}{2} - \dfrac{11}{30}$  **40.** $2\dfrac{2}{3} + 4\dfrac{1}{2}$

*Add or subtract as indicated. See Section 5.2.*

**41.**    0.41
      $+0.29$

**42.**   10.78
        4.3
      $+\ 0.21$

**43.**    2.38
      $-0.19$

**44.**   16.37
      $-\ 2.61$

## COMBINING CONCEPTS

*Solve. Round to the nearest tenth, if necessary.*

**45.** What number is 22.3% of 53,862?  **46.** What percent of 110,736 is 88,542?

**47.** 8652 is 119% of what number?

**48.** In your own words, describe how to identify the percent, the base, and the amount in a percent problem.

**534**

# Name _____ Section _____ Date _____

## INTEGRATED REVIEW—PERCENT AND PERCENT PROBLEMS

*Write each number as a percent.*

**1.** 0.12  **2.** 0.68  **3.** $\dfrac{1}{4}$  **4.** $\dfrac{1}{2}$

**5.** 5.2  **6.** 7.8  **7.** $\dfrac{3}{50}$  **8.** $\dfrac{11}{25}$

**9.** $2\dfrac{1}{2}$  **10.** $3\dfrac{1}{4}$  **11.** 0.03  **12.** 0.05

*Write each percent as a decimal.*

**13.** 65%  **14.** 31%  **15.** 8%  **16.** 7%

**17.** 142%  **18.** 538%  **19.** 2.9%  **20.** 6.6%

*Write each percent as a fraction or mixed number in simplest form.*

**21.** 3%  **22.** 8%  **23.** 5.25%  **24.** 12.75%

**25.** 38%  **26.** 45%  **27.** $12\dfrac{1}{3}\%$  **28.** $16\dfrac{2}{3}\%$

*Solve each percent problem.*

**29.** 12% of 70 is what number?  **30.** 36 is 36% of what number?

**31.** 212.5 is 85% of what number?  **32.** 66 is what percent of 55?

ANSWERS

**1.** 12%
**2.** 68%
**3.** 25%
**4.** 50%
**5.** 520%
**6.** 780%
**7.** 6%
**8.** 44%
**9.** 250%
**10.** 325%
**11.** 3%
**12.** 5%
**13.** 0.65
**14.** 0.31
**15.** 0.08
**16.** 0.07
**17.** 1.42
**18.** 5.38
**19.** 0.029
**20.** 0.066
**21.** $\dfrac{3}{100}$
**22.** $\dfrac{2}{25}$
**23.** $\dfrac{21}{400}$
**24.** $\dfrac{51}{400}$
**25.** $\dfrac{19}{50}$
**26.** $\dfrac{9}{20}$
**27.** $\dfrac{37}{300}$
**28.** $\dfrac{1}{6}$
**29.** 8.4
**30.** 100
**31.** 250
**32.** 120%

**Name** _____

**33.** 23.8 is what percent of 85?

**34.** 38% of 200 is what number?

**35.** What number is 25% of 44?

**36.** What percent of 99 is 128.7?

**37.** What percent of 250 is 215?

**38.** What number is 45% of 84?

**39.** 63 is 42% of what number?

**40.** 58.9 is 95% of what number?

# 7.4   APPLICATIONS OF PERCENT

## A   SOLVING APPLICATIONS INVOLVING PERCENT

The next few examples show just a few ways that percent occurs in real-life settings. (Each of these examples shows two ways of solving these problems. If you studied Section 7.2 only, see *Method 1*. If you studied Section 7.3 only, see *Method 2*.)

**Example 1**   Finding Totals Using Percents

Mr. Buccaran, the principal at Slidell High School, counted 31 freshmen absent during a particular day. If this is 4% of the total number of freshmen, how many freshmen are there at Slidell High School?

*Solution:*   *Method 1.* First we state the problem in words, then we translate.

In words:  31  is  4%  of  what number?

Translate:  $31 = 4\% \cdot x$

Next, we solve for $x$.

$31 = 0.04 \cdot x$   Write 4% as a decimal.

$\dfrac{31}{0.04} = \dfrac{0.04 \cdot x}{0.04}$   Divide both sides by 0.04.

$775 = x$   Simplify.

There are 775 freshmen at Slidell High School.

*Method 2.* First we state the problem in words, then we translate.

In words:   31   is   4% of what number?

amount   percent   base

Translate:   $\underset{\text{base} \to}{\overset{\text{amount} \to}{}} \dfrac{31}{b} = \dfrac{4}{100} \leftarrow \text{percent}$

Next we solve for $b$.

$31 \cdot 100 = b \cdot 4$   Set cross products equal.

$3100 = b \cdot 4$   Multiply.

$\dfrac{3100}{4} = \dfrac{b \cdot 4}{4}$   Divide both sides by 4.

$775 = b$   Simplify.

There are 775 freshmen at Slidell High School.   ■■■■

---

### Objectives

**A** Solve applications involving percent.

**B** Find percent increase and percent decrease.

Study   SSM   CD-ROM   Video
Guide                          7.4

---

**Practice Problem 1**

The freshmen class of 775 students is 31% of all students at Euclid University. How many students go to Euclid University?

TEACHING TIP

Method 1 corresponds to Section 7.2 and Method 2 corresponds to Section 7.3.

**Answer**

**1.** 2500 students

## Practice Problem 2

The nutrition label below is from a can of cashews. Find what percent of total calories are from fat. Round to the nearest tenth of a percent.

**Nutrition Facts**

Serving Size ¼ cup (33g)
Servings Per Container About 9

Amount Per Serving

**Calories** 190  Calories from Fat 130

| | % Daily Value |
|---|---|
| **Total Fat** 16g | **24%** |
| Saturated Fat 3g | **16%** |
| **Cholesterol** 0mg | **0%** |
| **Sodium** 135mg | **6%** |
| **Total Carbohydrate** 9g | **3%** |
| Dietary Fiber 1g | **5%** |
| Sugars 2g | |
| **Protein** 5g | |

Vitamin A 0% • Vitamin C 0%
Calcium 0%  •  Iron 8%

## Example 2     Finding Percents

Standardized nutrition labeling like the one shown has been on foods since 1994. It is recommended that no more than 30% of your calorie intake be from fat. Find what percent of the total calories shown are fat.

**Nutrition Facts**

Serving Size 1 pouch (20g)
Servings Per Container 6

Amount Per Serving

| Calories | 80 |
|---|---|
| Calories from fat | 10 |

| | % Daily Value* |
|---|---|
| **Total Fat** 1g | **2%** |
| **Sodium** 45mg | **2%** |
| **Total Carbohydrate** 17g | **6%** |
| Sugars 9g | |
| **Protein** 0g | |

| Vitamin C | 25% |
|---|---|

Not a significant source of saturated fat, cholesterol, dietary fiber, vitamin A, calcium and iron.

*Percent Daily Values are based on a 2,000 calorie diet.

Fruit snacks nutrition label

*Solution:*     *Method 1.*

In words:   10   is   what percent   of   80?

Translate:   $10 = x \cdot 80$

Next we solve for $x$.

$$\frac{10}{80} = \frac{x \cdot 80}{80} \qquad \text{Divide both sides by 80.}$$

$$0.125 = x \qquad \text{Simplify.}$$

$$12.5\% = x \qquad \text{Write 0.125 as a percent.}$$

This food contains $12.5\%$ of its total calories from fat.

*Method 2.*

In words:     10  is  what percent  of  80?

| amount | percent | base |
|---|---|---|

Translate:   $\text{amount} \rightarrow \dfrac{10}{80} = \dfrac{p}{100} \leftarrow \text{percent}$
             $\text{base} \rightarrow$

Next we solve for $p$.

$$10 \cdot 100 = 80 \cdot p \qquad \text{Set cross products equal.}$$

$$1000 = 80 \cdot p \qquad \text{Multiply.}$$

$$\frac{1000}{80} = \frac{80 \cdot p}{80} \qquad \text{Divide both sides by 80.}$$

$$12.5 = p \qquad \text{Simplify.}$$

This food contains $12.5\%$ of its total calories from fat.

**Answer**

**2.** 68.4%

## B Finding Percent Increase and Percent Decrease

We often use percents to show how much an amount has increased or decreased.

**Example 3** Finding an Increase

From 1970 to 1998, the number of U.S. drivers on the road increased by 65%. If the number of drivers on the road in 1970 was 112 million, find the number of drivers on the road in 1998. (*Source:* Federal Highway Administration)

*Solution:* **Method 1.** First we find the increase in drivers.

In words:  What number  is  65%  of  112?

Translate:   $x$   $=$   65%   $\cdot$   112

$x = 0.65 \cdot 112$    Write 65% as a decimal.

$x = 72.8$    Multiply.

The increase in drivers is 72.8 million. This means that the number of drivers in 1998 was

112 million + 72.8 million = 184.8 million

**Method 2.** First we find the increase in drivers.

In words:  What number  is  65%  of  112?

   amount    percent    base

Translate:   $\dfrac{\text{amount} \rightarrow a}{\text{base} \rightarrow 112} = \dfrac{65}{100} \leftarrow \text{percent}$

$a \cdot 100 = 112 \cdot 65$    Set cross products equal.

$a \cdot 100 = 7280$    Multiply.

$\dfrac{a \cdot 100}{100} = \dfrac{7280}{100}$    Divide both sides by 100.

$a = 72.8$    Simplify.

The increase in drivers is 72.8 million. This means that the number of drivers in 1998 was

112 million + 72.8 million = 184.8 million    ▬▬▬

Suppose that the population of a town is 10,000 people and then it increases by 2000 people. The **percent increase** is

$$\dfrac{\text{amount of increase} \rightarrow 2000}{\text{original amount} \rightarrow 10,000} = 0.2 = 20\%$$

---

### PERCENT INCREASE

$$\text{percent increase} = \dfrac{\text{amount of increase}}{\text{original amount}}$$

**Practice Problem 3**

From 1970 to 1998, the number of vehicles on the road has increased by 94%. If the number of vehicles on the road in 1970 was 109 million, find the number of vehicles on the road in 1998. (*Source:* Federal Highway Administration)

**Answer**

**3.** 211.46 million

## Practice Problem 4

Saturday's attendance at the play *Peter Pan* increased to 333 people over Friday's attendance of 285 people. What was the percent increase in attendance? Round to the nearest tenth of a percent.

## ✓ CONCEPT CHECK

A student is calculating the percent increase in enrollment from 180 students one year to 200 students the next year. Explain what is wrong with the following calculations.

$$\frac{\text{amount}}{\text{of increase}} = 200 - 180 = 20$$

$$\frac{\text{percent}}{\text{increase}} = \frac{20}{200} = 0.1 = 10\%$$

## Practice Problem 5

A town with a population of 20,145 decreased to 18,430 over a 10-year period. What was the percent decrease? Round to the nearest tenth of a percent.

## ✓ CONCEPT CHECK

An ice cream stand sold 6000 ice cream cones last summer. This year the same stand sold 5400 cones. Was there a 10% increase, a 10% decrease, or neither? Explain.

**Answers**

**4.** 16.8%  **5.** 8.5%

✓ **Concept Check:** To find the percent increase, you have to divide the amount of increase by the original amount $\left(\frac{20}{180}\right)$.

✓ **Concept Check:** 10% decrease.

## Example 4    Finding Percent Increase

The number of applications for a mathematics scholarship at Yale increased from 34 to 45 in one year. What is the percent increase? Round to the nearest whole percent.

*Solution:*    First we find the amount of increase by subtracting the original number of applicants from the new number of applicants.

amount of increase = 45 − 34 = 11

The amount of increase is 11 applicants. To find the percent increase,

$$\text{percent increase} = \frac{\text{amount of increase}}{\text{original amount}} = \frac{11}{34} \approx 0.32 = 32\%$$

> **HELPFUL HINT**
> Make sure that this number is the original number and not the new number.

The number of applications increased by about 32%.

**TRY THE CONCEPT CHECK IN THE MARGIN.**

Suppose that your income was $300 a week and then it decreased by $30. The percent decrease is

$$\begin{array}{l}\text{amount of decrease} \rightarrow \\ \text{original amount} \rightarrow\end{array} \frac{\$30}{\$300} = 0.1 = 10\%$$

> **PERCENT DECREASE**
> $$\text{percent decrease} = \frac{\text{amount of decrease}}{\text{original amount}}$$

## Example 5    Finding Percent Decrease

In response to a decrease in sales, a company with 1500 employees reduces the number of employees to 1230. What is the percent decrease?

*Solution:*    First we find the amount of decrease by subtracting 1230 from 1500.

amount of decrease = 1500 − 1230 = 270

The amount of decrease is 270. To find the percent decrease,

$$\frac{\text{percent}}{\text{decrease}} = \frac{\text{amount of decrease}}{\text{original amount}} = \frac{270}{1500} = 0.18 = 18\%$$

The number of employees decreased by 18%.

**TRY THE CONCEPT CHECK IN THE MARGIN.**

# EXERCISE SET 7.4

**A** *Solve. See Examples 1 and 2. If necessary, round percents to the nearest tenth and all other answers to the nearest whole.*

**1.** An inspector found 24 defective bolts during an inspection. If this is 1.5% of the total number of bolts inspected, how many bolts were inspected?

**2.** A daycare worker found 28 children absent one day during an epidemic of chicken pox. If this was 35% of the total number of children attending the daycare, how many children attend this daycare?

**3.** An owner of a repair service company estimates that, for every 40 hours a repairperson is on the job, he can only bill for 75% of the hours. The remaining hours, the repairperson is idle or driving to or from a job. Determine the number of hours per 40-hour week the owner can bill for a repairperson.

**4.** The Hodder family paid 20% of the purchase price of a $75,000 home as a down payment. Determine the amount of the down payment.

**5.** Vera Faciane earns $2000 per month and budgets $300 per month for food. What percent of her monthly income is spent on food?

**6.** Last year, Mai Toberlan bought a share of stock for $83. She was paid a dividend of $4.15. Determine what percent of the stock price is the dividend.

**7.** A manufacturer of electronic components expects 1.04% of its product to be defective. Determine the number of defective components expected in a batch of 28,350 components. Round to the nearest whole component.

**8.** 18% of Marvin Frank's wages are withheld for income tax. Find the amount withheld from his wages of $3680 per month.

**9.** Of the 535 members of the 105th U.S. Congress, 82 have either attended, taught at, or had a family member attend a community college. What percent of the members of the 105th Congress have had some direct connection with community colleges? (*Source:* American Association of Community Colleges)

**10.** 31.6% of all households in the United States own at least one pet dog. There are 11,250 households in Anytown. How many of these households would you expect own a dog? (*Source:* American Veterinary Medical Association)

1. 1600 bolts

2. 80 children

3. 30 hr

4. $15,000

5. 15%

6. 5%

7. 295 components

8. $662.40

9. 15.3%

10. 3555 households

**11.** There are about 98,400 female dental hygienists registered in the United States. If this represents about 98.3% of the nation's dental hygienists, find the number of dental hygienists in the United States. (*Source:* The American Dental Hygienists' Association)

**12.** The Los Angeles County courts excused 775,130 prospective jurors from jury duty in a recent year. This represented 28% of all juror qualification affidavits sent out that year. How many juror qualification affidavits were sent out that year? (*Source:* Los Angeles Superior Court)

*For each food described, find what percent of total calories is from fat. If necessary, round to the nearest tenth of a percent. See Example 2.*

**13.**

**Nutrition Facts**

Serving Size 18 crackers (29g)
Servings Per Container About 9

Amount Per Serving

**Calories** 120 Calories from Fat 35

| | % Daily Value* |
|---|---|
| **Total Fat** 4g | **6%** |
| Saturated Fat 0.5g | **3%** |
| Polyunsaturated Fat 0g | |
| Monounsaturated Fat 1.5g | |
| **Cholesterol** 0mg | **0%** |
| **Sodium** 220mg | **9%** |
| **Total Carbohydrate** 21g | **7%** |
| Dietary Fiber 2g | **7%** |
| Sugars 3g | |
| **Protein** 2g | |

Vitamin A 0% • Vitamin C 0%
Calcium 2% • Iron 4%
Phosphorus 10%

**14.**

**Nutrition Facts**

Serving Size 28 crackers (31g)
Servings Per Container About 6

Amount Per Serving

**Calories** 130 Calories from Fat 35

| | % Daily Value* |
|---|---|
| **Total Fat** 4g | **6%** |
| Saturated Fat 2g | **10%** |
| Polyunsaturated Fat 1g | |
| Monounsaturated Fat 1g | |
| **Cholesterol** 0mg | **0%** |
| **Sodium** 470mg | **20%** |
| **Total Carbohydrate** 23g | **8%** |
| Dietary Fiber 1g | **4%** |
| Sugars 4g | |
| **Protein** 2g | |

Vitamin A 0% • Vitamin C 0%
Calcium 0% • Iron 2%

**B** *Solve. Round money amounts to the nearest cent and all other amounts to the nearest tenth. See Examples 3 through 5.*

**15.** Sport utility vehicle sales are increasing by 19% each year. If sales were 3,340,000 in 1999, determine the sales in 2000 and in 2001.

**16.** The enrollment at a local college increased 5% over last year's enrollment of 7640. Find the increase in enrollment and the current enrollment.

**17.** By carefully planning their meals, a family was able to decrease their weekly grocery bill by 20%. Their weekly grocery bill used to be $170. What is their new weekly grocery bill?

**18.** The profit of Ramone Company last year was $175,000. This year's profit decreased by 11%. Find this year's profit.

**19.** A car manufacturer announced that next year the price of a certain model car would increase 4.5%. This year the price is $19,286. Find the increase and the new price.

**20.** A union contract calls for a 6.5% salary increase for all employees. Determine the increase and the new salary that a worker currently making $28,500 under this contract can expect.

**21.** The nation's Hispanic population is projected to increase 33% during the 10-year period from 1995 to 2005. The number of Hispanics in 1995 was 26.8 million. Find the increase and the projected 2005 population. (Do not round.) (*Source:* U.S. Bureau of the Census, *U.S. Census of Population*)

**22.** From 1995 to 2004, the number of doctorates awarded to women is projected to increase 15%. The number of women who received doctorates in 1995 was 17,000. Find the predicted number of women to be awarded doctorates in 2004. (*Source:* U.S. National Center for Education Statistics)

*Find the amount of increase and the percent increase. See Example 4.*

| | Original Amount | New Amount | Amount of Increase | Percent Increase |
|---|---|---|---|---|
| **23.** | 40 | 50 | _____ | _____ |
| **24.** | 10 | 15 | _____ | _____ |
| **25.** | 85 | 187 | _____ | _____ |
| **26.** | 78 | 351 | _____ | _____ |

*Find the amount of decrease and the percent decrease. See Example 5.*

| | Original Amount | New Amount | Amount of Decrease | Percent Decrease |
|---|---|---|---|---|
| **27.** | 8 | 6 | _____ | _____ |
| **28.** | 25 | 20 | _____ | _____ |
| **29.** | 160 | 40 | _____ | _____ |
| **30.** | 200 | 162 | _____ | _____ |

*Solve. Round percents to the nearest tenth, if necessary. See Examples 3 through 5.*

**31.** There are 150 calories in a cup of whole milk and only 84 in a cup of skimmed milk. In switching to skimmed milk, find the percent decrease in number of calories.

**32.** In reaction to a slow economy, the number of employees at a soup company decreased from 530 to 477. What was the percent decrease in employees?

**33.** 21.5%

**34.** 35.4%

**35.** 1.3%

**36.** 137.5%

**37.** 86.2%

**38.** 24.8%

**39.** 300%

**40.** 3.0%

**33.** By changing his driving routines, Alan Miller increased his car's rate of miles per gallon from 19.5 to 23.7. Find the percent increase.

**34.** John Smith decided to decrease the number of calories in his diet from 3250 to 2100. Find the percent decrease.

**35.** The number of cable TV systems recently decreased from 10,845 to 10,700. Find the percent decrease.

**36.** Before taking a typing course, Geoffry Landers could type 32 words per minute. By the end of the course, he was able to type 76 words per minute. Find the percent increase.

**37.** In 1940, there were approximately 52.1 million sheep on farms in the United States. In 1999, this number had decreased to 7.2 million sheep. What was the percent decrease? (*Source:* National Agricultural Statistics Service)

**38.** In 1970, there were approximately 12.1 million milk cows on farms in the United States. In 1999, this number had decreased to 9.1 million milk cows. What was the percent decrease? (*Source:* National Agricultural Statistics Service)

**39.** In 1999, discarded electronics, including obsolete computer equipment, accounted for 75,000 tons of solid waste per year in Massachusetts. By 2006, discarded electronic waste is expected to increase to 300,000 tons of waste per year in the state. Find the percent increase. (*Source:* Massachusetts Department of Environmental Protection)

**40.** In 1966, the population of Eastern Bluebirds in the United States was 6974. By 1996, this number had increased to 7183. Find the percent increase. (*Source:* Patuxent Wildlife Research Center)

**Name** _____

**41.** 81.3%

 **41.** In 1994, approximately 16,000 occupational therapy assistants and aides were employed in the United States. By 2005, this number is expected to increase to 29,000 assistants and aides. What is the percent increase? (*Source:* Bureau of Labor Statistics)

**42.** In 1994, approximately 206,000 medical assistants were employed in the United States. By 2005, this number is expected to increase to 327,000 medical assistants. What is the percent increase? (*Source:* Bureau of Labor Statistics)

**42.** 58.7%

**43.** 4.56

**44.** 29.4

## REVIEW AND PREVIEW

*Perform each indicated operation. See Sections 5.2 and 5.3.*

**45.** 11.18

**43.**
$$\begin{array}{r} 0.12 \\ \times\ 38 \end{array}$$

**44.**
$$\begin{array}{r} 42 \\ \times\ 0.7 \end{array}$$

**45.** $9.20 + 1.98$

**46.** 53.89

**46.** $46 + 7.89$

**47.** $78 - 19.46$

**48.** $64.80 - 10.72$

**47.** 58.54

## COMBINING CONCEPTS

*Solve. Round percents to the nearest tenth.*

**49.** The population of Tokyo is expected to increase from 26,518 thousand in 1994 to 28,700 thousand in 2015. Find the percent increase. (*Source:* United Nations, Dept. for Economic and Social Information and Policy Analysis)

**50.** In 1994, approximately 179,000 personal and home care aides were employed in the United States. By 2005, this number is expected to increase to 391,000 aides. What is the percent increase? (*Source:* Bureau of Labor Statistics)

**48.** 54.08

**49.** 8.2%

**51.** If a number is increased by 100%, how does the increased number compare with the original number? Explain your answer.

**50.** 118.4%

The increased number is double the
**51.** original number.

**Name** _____

## Internet Excursions

Go to http://www.prenhall.com/martin-gay

This World Wide Web address will provide you with access to an American Savings Education Council worksheet for figuring a ballpark estimate of the savings needed for retirement, or a related site. Print out two copies of the worksheet to use with Exercises 52–53.

**52.** Fill out a copy of the ballpark worksheet for a person who currently is 25 years old, earns $32,000 a year, plans to retire at age 60, expects no traditional employer pension, expects $5000 per year in part-time income in retirement, and has $2000 in retirement savings. How much does this person need to save each year toward retirement?

**53.** Fill out a copy of the ballpark worksheet for your own situation. How much do you need to save each year toward retirement?

# 7.5 PERCENT AND PROBLEM SOLVING: SALES TAX, COMMISSION, AND DISCOUNT

Objectives

**A** Calculate sales tax and total price.
**B** Calculate commissions.
**C** Calculate discount and sale price.

Study     SSM   CD-ROM   Video
Guide                      7.5

## A CALCULATING SALES TAX AND TOTAL PRICE

Percents are frequently used in the retail trade. For example, most states charge a tax on certain items when purchased. This tax is called a **sales tax**, and retail stores collect it for the state. Sales tax is almost always stated as a percent of the purchase price.

A 6% sales tax rate on a purchase of a $10.00 item gives a sales tax of

$$\text{sales tax} = 6\% \text{ of } \$10 = 0.06 \cdot \$10.00 = \$0.60$$

The total price to the customer would be

| purchase price | plus | sales tax |
|:---:|:---:|:---:|
| ↓ | ↓ | ↓ |
| $10.00 | + | $0.60 = $10.60 |

This example suggests the following equations.

---

**SALES TAX AND TOTAL PRICE**

sales tax = tax rate · purchase price

total price = purchase price + sales tax

---

In this section, we will round dollar amounts to the nearest cent.

## Example 1    Finding Sales Tax and Purchase Price

Find the sales tax and the total price on a purchase of an $85.50 trench coat in a city where the sales tax rate is 7.5%.

$85.50
+ 7.5% tax

TEACHING TIP

You may want to begin this section by showing students how easy it is to find 1% and 10% of a price. Then show how they can use these percents to calculate other percents.

TEACHING TIP

You may want to show students how to quickly calculate tips. Typically people leave a 15% or 20% tip at a restaurant where they received good service. You can quickly estimate a tip by using an estimate of 10% of the bill. A 15% tip can be found by adding on half of the 10% estimate. A 20% tip can be found by doubling the 10% estimate.

### Practice Problem 1

If the sales tax rate is 6%, what is the sales tax and the total amount due on a $29.90 Goodgrip tire?

**Answer**

**1.** tax: $1.79; total: $31.69

*Solution:* The purchase price is $85.50 and the tax rate is 7.5%.

$$\underset{\downarrow}{\text{sales tax}} \quad = \quad \underset{\downarrow}{\text{tax rate}} \quad \cdot \quad \underset{\downarrow}{\text{purchase price}}$$

$$\text{sales tax} \quad = \quad 7.5\% \quad \cdot \quad \$85.50$$

$$= \quad 0.075 \quad \cdot \quad \$85.5 \qquad \text{Write 7.5\% as a decimal.}$$

$$\approx \quad \$6.41 \qquad \qquad \text{Round to the nearest cent.}$$

Thus

$$\underset{\downarrow}{\text{total price}} \quad = \quad \underset{\downarrow}{\text{purchase price}} \quad + \quad \underset{\downarrow}{\text{sales tax}}$$

$$\text{total price} \quad = \quad \$85.50 \quad + \quad \$6.41$$

$$= \quad \$91.91$$

The sales tax on $85.50 is $6.41 and the total price is $91.91.

**TRY THE CONCEPT CHECK IN THE MARGIN.**

✓ **CONCEPT CHECK**

The purchase price of a textbook is $50 and sales tax is 10%. If you are told by the cashier that the total price is $75, how can you tell that a mistake has been made?

**Example 2** Finding a Sales Tax Rate

The sales tax on a $300 printer is $22.50. Find the sales tax rate.

$300 + $22.50 sales tax

**Practice Problem 2**

The sales tax on a $13,500 automobile is $1080.00. Find the sales tax rate.

*Solution:* Let $r$ be the unknown sales tax rate. Then

$$\underset{\downarrow}{\text{sales tax}} \quad = \quad \underset{\downarrow}{\text{tax rate}} \quad \cdot \quad \underset{\downarrow}{\text{purchase price}}$$

$$22.50 \quad = \quad r \quad \cdot \quad 300$$

$$\frac{22.50}{300} \quad = \quad \frac{r \cdot 300}{300} \qquad \text{Divide both sides by 300.}$$

$$0.075 \quad = \quad r \qquad \text{Simplify.}$$

$$7.5\% \quad = \quad r \qquad \text{Write 0.075 as a percent.}$$

The sales tax rate is 7.5%.

## B CALCULATING COMMISSIONS

A **wage** is payment for performing work. Hourly wage, commissions, and salary are some of the ways wages can be paid. Many people who work in sales are paid a commission. An employee who is paid a **commission** is paid a percent of his or her total sales.

**Answers**

**2.** 8%

✓ **Concept Check:** Since $10\% = \frac{1}{10}$, the sales tax is $\frac{\$50}{10} = \$5$. The total price should have been $55.

---

**COMMISSION**

commission = commission rate · sales

---

## Example 3   Finding a Commission

Sherry Souter, a real estate broker for Wealth Invest-
ments, sold a house for $114,000 last week. If her commis-
sion is 1.5% of the selling price of the home, find the
amount of her commission.

*Solution:*

| commission | = | commission rate | · | sales |
|---|---|---|---|---|
| ↓ | | ↓ | | ↓ |

| commission | = | 1.5% | · | 114,000 | Write 1.5% as 0.015. |
| | = | 0.015 | · | 114,000 | |
| | = | 1710 | | | Multiply. |

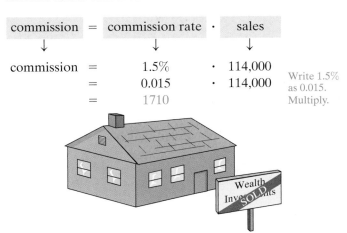

Her commission on the house is $1710.

## Example 4   Finding a Commission Rate

A salesperson earned $1560 for selling $13,000 worth of
television and stereo systems. Find the commission rate.

*Solution:*   Let $r$ stand for the unknown commission rate. Then

| commission | = | commission rate | · | sales |
|---|---|---|---|---|
| ↓ | | ↓ | | ↓ |

$$1560 = r \cdot 13,000$$

$$\frac{1560}{13,000} = \frac{r \cdot 13,000}{13,000} \quad \text{Divide both sides by 13,000.}$$

$$0.12 = r \quad \text{Simplify.}$$

$$12\% = r \quad \text{Write 0.12 as a percent.}$$

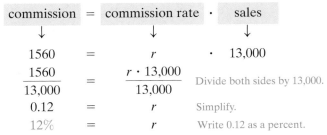

Commission: $1560

The commission rate is 12%.

## C CALCULATING DISCOUNT AND SALE PRICE

Suppose that an item that normally sells for $40 is on sale for 25% off. This means that the **original price** of $40 is reduced, or **discounted** by 25% of $40, or $10. The **discount rate** is 25%, the **amount of discount** is $10, and the **sale price** is $40 − $10 or $30.

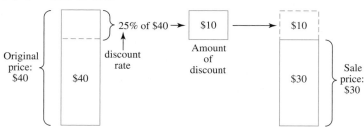

To calculate discounts and sale prices, we can use the following equations.

> **DISCOUNT AND SALE PRICE**
>
> amount of discount = discount rate · original price
> sale price = original price − amount of discount

### Practice Problem 5

A Panasonic TV is advertised on sale for 15% off the regular price of $700. Find the discount and the sale price.

### Example 5    Finding a Discount and a Sale Price

A speaker that normally sells for $65.00 is on sale at 25% off. What is the discount and what is the sale price?

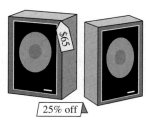

25% off

*Solution:*   First we find the discount.

| amount of discount | = | discount rate | · | original price |
|---|---|---|---|---|
| ↓ | | ↓ | | ↓ |

amount of discount = 25% · 65
= 0.25 · 65   Write 25% as 0.25.
= 16.25   Multiply.

The discount is $16.25. Next we find the sale price.

| sale price | = | original price | − | discount |
|---|---|---|---|---|
| ↓ | | ↓ | | ↓ |

sale price = $65 − $16.25
= $48.75   Subtract.

The sale price is $48.75.

# EXERCISE SET 7.5

**A** *Solve. See Examples 1 and 2.*

1. What is the sales tax on a suit priced at $150.00 if the sales tax rate is 5%?

2. If the sales tax rate is 6%, find the sales tax on a microwave oven priced at $188.

3. The purchase price of a camcorder is $799. What is the total price if the sales tax rate is 7.5%?

4. A stereo system has a purchase price of $426. What is the total price if the sales tax rate is 8%?

5. A chair and ottoman have a purchase price of $600. If the sales tax on this purchase is $54, find the sales tax rate.

6. The sales tax on the purchase of a $2500 computer is $162.50. Find the sales tax rate.

7. A computer desk sells for $220. With a sales tax rate of 8.5%, find the total price.

$220
+ 8.5%
tax

8. A one-half carat diamond ring is priced at $800. The sales tax rate is 6.5%. Find the total price.

$800
+ 6.5% tax

9. A gold and diamond bracelet sells for $1800. Find the total price if the sales tax rate is 6.5%.

10. The purchase price of a personal computer is $1890.00. If the sales tax rate is 8%, what is the total price?

11. The sales tax on the purchase of a truck is $920. If the tax rate is 8%, find the purchase price of the truck.

12. The sales tax on the purchase of a desk is $27.50. If the tax rate is 5%, find the purchase price of the desk.

1. $7.50

2. $11.28

3. $858.93

4. $460.08

5. 9%

6. 6.5%

7. $238.70

8. $852

9. $1917

10. $2041.20

11. $11,500

12. $550

**Name** _____

**13.** A cordless phone costs $90 and a battery recharger costs $15. What is the total price for purchasing these items if the sales tax rate is 7%?

**14.** Ms. Warner bought a blouse for $35, a skirt for $55, and a blazer for $95. Find the total price she paid, given a sales tax rate of 6.5%.

**15.** The sales tax is $98.70 on a stereo sound system purchase of $1645. Find the sales tax rate.

**16.** The sales tax is $103.50 on a necklace purchase of $1150. Find the sales tax rate.

**B** *Solve. See Examples 3 and 4.*

**17.** Jane Moreschi, a sales representative for a large furniture warehouse, is paid a commission rate of 4%. Find her commission if she sold $1,236,856 worth of furniture last month.

**18.** Rosie Davis-Smith is a beauty consultant for a home cosmetic business. She is paid a commission rate of 4.8%. Find her commission if she sold $1638 in cosmetics last month.

**19.** A salesperson earned a commission of $1380.40 for selling $9860.00 worth of paper products. Find the commission rate.

**20.** A salesperson earned a commission of $3575 for selling $32,500 worth of books to various book stores. Find the commission rate.

**21.** How much commission will Jack Pruet make on the sale of a $125,900 house if he receives 1.5% of the selling price?

**22.** Frankie Lopez sold $9638.00 of jewelry this week. Find her commission for the week if she is paid a commission rate of 5.6%.

**23.** A house sold for $85,500 and the real estate agent earned a commission of $2565. Find the rate of commission.

**24.** A salesperson earned $1750 for selling $25,000 worth of fertilizer. Find the commission rate.

**C** *Find the amount of discount and the sale price. See Example 5.*

| | Original Price | Discount Rate | Amount of Discount | Sale Price |
|---|---|---|---|---|
| **25.** | $68.00 | 10% | _____ | _____ |
| **26.** | $47.00 | 20% | _____ | _____ |
| **27.** | $96.50 | 50% | _____ | _____ |
| **28.** | $110.60 | 40% | _____ | _____ |
| **29.** | $215.00 | 35% | _____ | _____ |
| **30.** | $370.00 | 25% | _____ | _____ |
| **31.** | $21,700.00 | 15% | _____ | _____ |
| **32.** | $17,800.00 | 12% | _____ | _____ |

**33.** A $300 fax machine is on sale at 15% off. Find the discount and the sale price.

**34.** A $2000 designer dress is on sale at 30% off. Find the discount and the sale price.

**23.** 3%

**24.** 7%

**25.** $6.80; $61.20

**26.** $9.40; $37.60

**27.** $48.25; $48.25

**28.** $44.24; $66.36

**29.** $75.25; $139.75

**30.** $92.50; $277.50

**31.** $3255.00; $18,445.00

**32.** $2136.00; $15,664.00

**33.** $45; $255

**34.** $600; $1400

**Name** _____

## REVIEW AND PREVIEW

*Multiply. See Sections 5.3 and 5.6.*

**35.** $2000 \cdot 0.3 \cdot 2$

**36.** $500 \cdot 0.08 \cdot 3$

**37.** $400 \cdot 0.03 \cdot 11$

**38.** $1000 \cdot 0.05 \cdot 5$

**39.** $600 \cdot 0.04 \cdot \dfrac{2}{3}$

**40.** $6000 \cdot 0.06 \cdot \dfrac{3}{4}$

## COMBINING CONCEPTS

**41.** A diamond necklace sells for $24,966. If the tax rate is 7.5%, find the total price.

**42.** A house recently sold for $562,560. The commission rate on the sale is 5.5%. If a real estate agent is to receive 60% of the commission, find the amount received by the agent.

**43.** Suppose that the original price of a shirt is $50. Which is better, a 60% discount or a discount of 30% followed by a discount of 35% of the reduced price. Explain your answer.

# 7.6   PERCENT AND PROBLEM SOLVING: INTEREST

## **A** CALCULATING SIMPLE INTEREST

**Interest** is money charged for using other people's money. When you borrow money, you pay interest. When you loan or invest money, you earn interest. The money borrowed, loaned, or invested is called the **principal amount**, or simply **principal**. Interest is normally stated in terms of a percent of the principal for a given period of time. The **interest rate** is the percent used in computing the interest. Unless stated otherwise, *the rate is understood to be per year*. When the interest is computed on the original principal, it is called **simple interest**. Simple interest is calculated using the following equation.

> **SIMPLE INTEREST**
>
> simple interest = principal · rate · time
>
> or   $I = P \cdot R \cdot T$
>
> where the rate is understood to be per year and time is in years.

**Objectives**

**A** Calculate simple interest.

**B** Use a compound interest table to calculate compound interest.

**C** Calculate monthly payments on loans.

Study Guide   SSM   CD-ROM   Video 7.6

### Example 1   Finding Simple Interest

Find the simple interest after 2 years on $500 at an interest rate of 12%.

*Solution:*   In this example, $P = \$500$, $R = 12\%$, and $T = 2$ years. Replace the variables with values in the formula $I = PRT$.

$I = P \cdot R \cdot T$

$I = \$500 \cdot 12\% \cdot 2$   Let $P = \$500$, $R = 12\%$, and $T = 2$.

$= \$500 \cdot (0.12) \cdot 2$   Write 12% as a decimal.

$= \$120$   Multiply.

The simple interest is $120.

**TRY THE CONCEPT CHECK IN THE MARGIN.**

If time is not given in years, we need to convert the given time to years.

### Example 2   Finding Simple Interest

Ivan Borski borrowed $2400 at 10% simple interest for 8 months to buy a used Chevy S-10. Find the simple interest he paid.

*Solution:*   Since there are 12 months in a year, we first find what part of a year 8 months is.

$8 \text{ months} = \dfrac{8}{12} \text{ year} = \dfrac{2}{3} \text{ year}$

Now we find the simple interest.

$I = P \cdot R \cdot T$

$= \$2400 \cdot (0.10) \cdot \dfrac{2}{3}$   Let $P = \$2400$, $R = 10\%$ or $0.10$, and $T = \dfrac{2}{3}$ year.

$= \$160$

The interest on Ivan's loan is $160.

**Practice Problem 1**

Find the simple interest after 3 years on $750 at an interest rate of 8%.

TEACHING TIP

For Example 1, you may want to have students also calculate the interest after 1, 3, and 4 years. Then ask them to predict the amount of simple interest after 5 years and 10 years.

**✓ CONCEPT CHECK**

If $20 is borrowed at 12% simple interest for 14 months, is 20 the variable $T$, $P$, or $I$ in the simple interest formula

$I = P \cdot R \cdot T$

**Practice Problem 2**

Juanita Lopez borrowed $800 for 9 months at a simple interest rate of 20%. How much interest did she pay?

**Answers**

**1.** $180   **2.** $120

✓ **Concept Check:** 20 is $P$, principal.

## ✓ CONCEPT CHECK

Suppose in Example 2 you had obtained an answer of $16,000. How would you know that you had made a mistake in this problem?

**Practice Problem 3**

If $500 is borrowed at a simple interest rate of 12% for 6 months, find the total amount paid.

## ✓ CONCEPT CHECK

Which investment would earn more interest: an amount of money invested at 8% interest for 2 years or the same amount of money invested at 8% for 3 years? Explain.

**Answers**

**3.** $530

✓ **Concept Check:** Answers may vary.

✓ **Concept Check:** 8% for 3 years. Since the interest rate is the same, the longer you keep the money invested, the more interest you earn.

## TRY THE CONCEPT CHECK IN THE MARGIN.

When money is borrowed, the borrower pays the original amount borrowed, or the principal, as well as the interest. When money is invested, the investor receives the original amount invested, or the principal, as well as the interest. In either case, the **total amount** is the sum of the principal and the interest.

> **FINDING THE TOTAL AMOUNT OF A LOAN OR INVESTMENT**
>
> total amount (paid or received) = principal + interest

### Example 3   Finding the Total Amount of an Investment

An accountant invested $2000 at a simple interest rate of 10% for 2 years. What total amount of money will she have from her investment in 2 years?

*Solution:*    First we find her interest.

$$I = P \cdot R \cdot T$$
$$= \$2000 \cdot (0.10) \cdot 2 \qquad \text{Let } P = \$2000, R = 10\% \text{ or } 0.10, \text{ and}$$
$$= \$400 \qquad\qquad\qquad\quad T = 2 \text{ years.}$$

The interest is $400.

Next we add the interest to the principal.

total amount  =  principal  +  interest
↓            ↓         ↓

total amount  =  $2000  +  $400

             =   $2400

After 2 years, she will have a total amount of $2400.

## TRY THE CONCEPT CHECK IN THE MARGIN.

### B   CALCULATING COMPOUND INTEREST

Recall that simple interest depends on the original principal only. Another type of interest is compound interest. **Compound interest** is computed on not only the principal, but also on the interest already earned in previous compounding periods. Compound interest is used more often than simple interest.

Let's see how compound interest differs from simple interest. Suppose that $2000 is invested at 7% interest **compounded annually** for 3 years. This means that interest is added to the principal at the end of each year and next year's interest is computed on this new amount. In this section, we will round dollar amounts to the nearest cent.

| | Amount at Beginning of Year | Principal | · Rate · Time = Interest | Amount at End of Year |
|---|---|---|---|---|
| 1st year | $2000 | $2000 | · 0.07 · 1 = $140 | $2000 + 140 = $2140 |
| 2nd year | $2140 | $2140 | · 0.07 · 1 = $149.80 | $2140 + 149.80 = $2289.80 |
| 3rd year | $2289.80 | $2289.80 | · 0.07 · 1 = $160.29 | $2289.80 + 160.29 = $2450.09 |

The compound interest earned can be found by

total amount − original principal = compound interest
↓ ↓ ↓
$2450.09 − $2000 = $450.09

The simple interest earned would have been

principal · rate · time = interest
↓ ↓ ↓ ↓
$2000 · 0.07 · 3 = $420

*Since compound interest earns "interest on interest," compound interest is more than simple interest.*

Computing compound interest using the method above can be tedious. We can use a **compound interest table** to compute interest more quickly. The compound interest table is found in Appendix F. This table gives the total compound interest and principal paid on $1 for given rates and number of years. Then we can use the following equation to find the total amount of interest and principal.

> **FINDING TOTAL AMOUNTS WITH COMPOUND INTEREST**
>
> total amount = original principal · compound interest factor
> (from table)

## Example 4   Finding Total Amount Received on an Investment

$4000 is invested at 8% compounded semiannually for 10 years. Find the total amount at the end of 10 years.

*Solution:*  Look in Appendix F. The compound interest factor for 10 years at 8% in the compounded semiannually section is 2.19112.

total amount = original principal · compound interest factor
↓ ↓ ↓
total amount = $4000 · 2.19112
= $8764.48

Therefore, the total amount at the end of 10 years is $8764.48.

## Example 5   Finding Compound Interest Earned

In Example 4, we found that the total amount for $4000 invested at 8% compounded semiannually for 10 years is $8764.48. Find the compound interest earned.

*Solution:*

interest earned = total amount − original principal
↓ ↓ ↓
interest earned = $8764.48 − $4000
= $4764.48

The compound interest earned is $4764.48.

## C   CALCULATING A MONTHLY PAYMENT

We conclude this section with a method to find the monthly payment on a loan.

TEACHING TIP   Classroom Activity

Have students add the 4th and 5th year to the table of 7% interest compounded annually. Then have them use the table to calculate how much more money was earned each year as compared to the previous year for years 2–5.

TEACHING TIP

You may want to point out that when interest is compounded semiannually it is added to the principal every six months so you will earn "interest on your interest" sooner than you would if the interest was compounded annually. Then have students look in Appendix F to compare the compound interest factor for 10 years at 8% compounded semiannually to the compound interest factor for 10 years at 8% compounded annually.

**Practice Problem 4**

$5500 is invested at 7% compounded daily for 5 years. Find the total amount at the end of 5 years.

**Practice Problem 5**

If the total amount is $9933.14 when $5500 is invested, find the compound interest earned.

Answers
**4.** $7804.61   **5.** $4433.14

> **FINDING THE MONTHLY PAYMENT OF A LOAN**
>
> $$\text{monthly payment} = \frac{\text{principal} + \text{interest}}{\text{total number of payments}}$$

**Practice Problem 6**

Find the monthly payment on a $3000 3-year loan if the interest on the loan is $1123.58.

**Example 6**    Finding a Monthly Payment

Find the monthly payment on a $2000 loan for 2 years. The interest on the 2-year loan is $435.88.

*Solution:*    First we determine the total number of monthly payments. The loan is for 2 years. Since there are 12 months per year, the number of payments is 2 · 12, or 24. Now we can calculate the monthly payment.

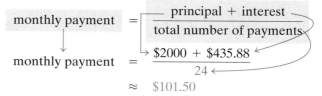

$$\text{monthly payment} = \frac{\text{principal} + \text{interest}}{\text{total number of payments}}$$

$$\text{monthly payment} = \frac{\$2000 + \$435.88}{24}$$

$$\approx \$101.50$$

The monthly payment is $101.50.

## CALCULATOR EXPLORATIONS
### COMPOUND INTEREST FACTOR

A compound interest factor may be found by using your calculator and evaluating the formula

$$\textbf{compound interest factor} = \left(1 + \frac{r}{n}\right)^{nt}$$

where $r$ is the interest rate, $t$ is the time in years, and $n$ is the number of times compounded per year. For example, we stated earlier that the compound interest factor for 10 years at 8% compounded semiannually is 2.19112. Let's find this factor by evaluating the compound interest factor formula when $r = 8\%$ or 0.08, $t = 10$, and $n = 2$ (compounded semiannually means 2 times per year). Thus,

$$\text{compound interest factor} = \left(1 + \frac{0.08}{2}\right)^{2 \cdot 10} \text{ or } \left(1 + \frac{0.08}{2}\right)^{20}$$

To evaluate, press the keys

( 1 + 0.08 ÷ 2 ) $y^x$ 20 = (or ENTER )

The display will read ⟨ 2.1911231 ⟩. Rounded to five decimal places, this is 2.19112.

*Find the compound interest factor. Use the table in Appendix F to check your answer.*

**1.** 5 years, 9%, compounded quarterly    1.56051

**2.** 15 years, 14%, compounded daily    8.16288

**3.** 20 years, 11%, compounded annually    8.06231

**4.** 1 year, 7%, compounded semiannually    1.07123

**5.** Find the total amount after 4 years if $500 is invested at 6% compounded quarterly.    $634.50

**6.** Find the total amount for 19 years if $2500 is invested at 5% compounded daily.    $6463.85

**Answers**

**6.** $114.54

## EXERCISE SET 7.6

**A** *Find the simple interest. See Examples 1 and 2.*

| | Principal | Rate | Time | | Principal | Rate | Time |
|---|---|---|---|---|---|---|---|
| **1.** | $200 | 8% | 2 years | **2.** | $800 | 9% | 3 years |
| **3.** | $160 | 11.5% | 4 years | **4.** | $950 | 12.5% | 5 years |
| **5.** | $5000 | 10% | $1\frac{1}{2}$ years | **6.** | $1500 | 14% | $2\frac{1}{4}$ years |
| **7.** | $375 | 18% | 6 months | **8.** | $1000 | 10% | 18 months |
| **9.** | $2500 | 16% | 21 months | **10.** | $775 | 15% | 8 months |

*Solve. See Examples 1 through 3.*

**11.** A company borrows $62,500 for 2 years at a simple interest rate of 12.5% to buy an airplane. Find the total amount paid on the loan.

**12.** $65,000 is borrowed to buy a house. If the simple interest rate on the 30-year loan is 10.25%, find the total amount paid on the loan.

**13.** A money market fund advertises a simple interest rate of 9%. Find the total amount received on an investment of $5000 for 15 months.

**14.** The Real Service Company takes out a 270-day (9-month) short-term, simple interest loan of $4500 to finance the purchase of some new equipment. If the interest rate is 14%, find the total amount that the company pays back.

**ANSWERS**

1. $32
2. $216
3. $73.60
4. $593.75
5. $750
6. $472.50
7. $33.75
8. $150
9. $700
10. $77.50
11. $78,125
12. $264,875
13. $5562.50
14. $4972.50

**559**

**15.** $12,580

**16.** $2800

**17.** $46,815.40

**18.** $8333.85

**19.** $2327.15

**20.** $6379.70

**21.** $58,163.60

**22.** $7788.73

**23.** $240.75

**24.** $653.44

**25.** $938.66

**26.** $960.48

**27.** $971.90

**28.** $983.52

**Name** _____

**15.** Marsha Waide borrows $8500 and agrees to pay it back in 4 years. If the simple interest rate is 12%, find the total amount she pays back.

**16.** Ms. Lapchinski gives her 18-year-old daughter a graduation gift of $2000. If this money is invested at 8% simple interest for 5 years, find the total amount after 5 years.

**B** *Find the total amount in each compound interest account. See Example 4.*

**17.** $6150 is compounded semiannually at a rate of 14% for 15 years.

**18.** $2060 is compounded annually at a rate of 15% for 10 years.

**19.** $1560 is compounded daily at a rate of 8% for 5 years.

**20.** $1450 is compounded quarterly at a rate of 10% for 15 years.

**21.** $10,000 is compounded semiannually at a rate of 9% for 20 years.

**22.** $3500 is compounded daily at a rate of 8% for 10 years.

*Find the total amount of compound interest earned. See Example 5.*

**23.** $2675 is compounded annually at a rate of 9% for 1 year.

**24.** $6375 is compounded semiannually at a rate of 10% for 1 year.

**25.** $2000 is compounded annually at a rate of 8% for 5 years.

**26.** $2000 is compounded semiannually at a rate of 8% for 5 years.

**27.** $2000 is compounded quarterly at a rate of 8% for 5 years.

**28.** $2000 is compounded daily at a rate of 8% for 5 years.

**560**

**Name** _____

**C** *Solve. See Example 6.*

**29.** A college student borrows $1500 for 6 months to pay for a semester of school. If the interest is $61.88, find the monthly payment.

**30.** Jim Tillman borrows $1800 for 9 months. If the interest is $148.90, find his monthly payment.

**31.** $20,000 is borrowed for 4 years. If the interest on the loan is $10,588.70, find the monthly payment.

**32.** $105,000 is borrowed for 15 years. If the interest on the loan is $181,125.00, find the monthly payment.

## REVIEW AND PREVIEW

*Perform each indicated operation. See Sections 2.2 to 2.4.*

**33.** $-5 + (-24)$

**34.** $7 - (-31)$

**35.** $(-5)(-30)$

**36.** $-30 \div (-5)$

**37.** $\dfrac{7 - 10}{3}$

**38.** $\dfrac{22 + (-4)}{-4 - 5}$

## COMBINING CONCEPTS

**39.** Explain how to look up the compound interest factor in the compound interest table.

**40.** Explain how to find the amount of interest on a compounded account.

**41.** Compare the following accounts: Account 1: $1000 is invested for 10 years at a simple interest rate of 6%. Account 2: $1000 is compounded semiannually at a rate of 6% for 10 years. Discuss how the interest is computed for each account. Determine which account earns more interest. Why?

# Focus on Business and Career

## FASTEST-GROWING OCCUPATIONS

According to U.S. Bureau of Labor Statistics projections, the careers listed below are the top ten fastest-growing jobs, ranked by expected percent increase through the year 2006.

| Occupation | Employment in 1996 | Percent Increase from 1996 to 2006 |
|---|---|---|
| Database administrators, computer support specialists, and all other computer scientists | 212,000 | 118% |
| Computer engineers | 216,000 | 109% |
| Systems analysts | 506,000 | 103% |
| Personal and home care aides | 202,000 | 85% |
| Physical and corrective therapy assistants and aides | 84,000 | 79% |
| Home health aides | 495,000 | 76% |
| Medical assistants | 225,000 | 74% |
| Desktop publishing specialists | 30,000 | 74% |
| Physical therapists | 115,000 | 71% |
| Occupational therapy assistants and aides | 16,000 | 69% |

(*Source:* Bureau of Labor Statistics, *Monthly Labor Review*, November 1997)

What do all of these fast-growing occupations have in common? They all require a knowledge of math! For some careers, such as desktop publishing specialists, medical assistants, and computer engineers, the ways math is used on the job may be obvious. For other occupations, the use of math may not be quite as apparent. However, tasks common to many jobs—filling in a time sheet, writing up an expense or mileage report, planning a budget, figuring a bill, ordering supplies, and even making a work schedule—all require math.

## GROUP ACTIVITY

1. List the top five occupations by order of employment figures for 1996.
2. Using the 1996 employment figures and the percent increase from 1996 to 2006, find the expected 2006 employment figures for each occupation listed in the table.
3. List the top five occupations by order of employment figures for 2006. Did the order change at all from 1996? Explain.
4. How many of the occupations on the list are expected to have more than double the number of positions in 2006 than in 1996? Explain how you identified these occupations.

# CHAPTER 7 ACTIVITY
## INVESTIGATING PROBABILITY

*This activity may be completed by working in groups or individually.*

The measure of the chance of an event occurring is its probability. In this activity, you will investigate one way that probabilities can be estimated. You will need a cup and 30 thumbtacks.

1. Place the thumbtacks in the cup. Shake the cup and toss out the thumbtacks onto a flat surface. Count the number of tacks that land point up, and record this number in the table. Repeat for 60, 90, 120, and 150 tacks. Record your results in the second column of the table. (*Hint:* For 60 thumbtacks, count the number of tacks landing point up in two tosses of the 30 thumbtacks, etc.)

2. For each row of the table, find the fraction of tacks that landed point up. Express this fraction as a percent. Complete the third and fourth columns of the table.

3. Each of the percents you computed in Question 2 is an *estimate* of the probability that a single thumbtack will land point up when tossed. With an estimate like this, the larger the number of tack

tosses used, the better the estimate of the actual probability. What do you suppose is the value of the actual probability? Explain your reasoning.

4. Combine your results for all 450 of your tack tosses recorded in the table with the results of other students or groups in your class. Of this total number of tack tosses, compute the percent of tacks that landed point up. This is your best estimate of the probability that a tack will land point up when tossed.

5. If you tossed 200 thumbtacks, what percent would you expect to land point up? How many tacks would you expect to land point up? Use the percent (probability) you computed in Question 4 to make this calculation. What if you tossed 300 thumbtacks?

| Number of Tacks | Number of Tacks Landing Point Up | Fraction of Point-Up Tacks | Percent of Point-Up Tacks |
|---|---|---|---|
| 30 | | | |
| 60 | | | |
| 90 | | | |
| 120 | | | |
| 150 | | | |

# CHAPTER 7 HIGHLIGHTS

| DEFINITIONS AND CONCEPTS | EXAMPLES |
|---|---|

## SECTION 7.1    PERCENTS, DECIMALS, AND FRACTIONS

| | |
|---|---|
| **Percent** means per hundred. The % symbol denotes percent. | $51\% = \dfrac{51}{100}$   51 per 100 <br><br> $7\% = \dfrac{7}{100}$   7 per 100 |
| **To write a percent as a decimal**, drop the % symbol and move the decimal point two places to the left. | $32\% = 0.32$ |
| **To write a decimal as a percent**, move the decimal point two places to the right and attach the % symbol. | $0.08 = 08\% = 8\%$ |
| **To write a percent as a fraction**, drop the % symbol and write the number over 100. | $25\% = \dfrac{25}{100} = \dfrac{25}{4 \cdot 25} = \dfrac{1}{4}$ |
| **To write a fraction as a percent**, multiply the fraction by 100%. | $\dfrac{1}{6} = \dfrac{1}{6} \cdot 100\% = \dfrac{1}{6} \cdot \dfrac{100}{1}\% = \dfrac{100}{6}\% = 16\dfrac{2}{3}\%$ |

## SECTION 7.2    SOLVING PERCENT PROBLEMS WITH EQUATIONS

| | |
|---|---|
| Three key words in the statement of a percent problem are <br><br> **of**, which means multiplication ( $\cdot$ ) <br> **is**, which means equals ( $=$ ) <br><br> **what** (or some equivalent word or phrase), which stands for the unknown | Solve. <br><br> $\begin{array}{ccccc} 6 & \text{is} & 12\% & \text{of} & \text{what number?} \\ \downarrow & \downarrow & \downarrow & \downarrow & \downarrow \\ 6 & = & 12\% & \cdot & x \end{array}$ <br><br> $6 = 0.12 \cdot x$    Write 12% as a decimal. <br><br> $\dfrac{6}{0.12} = \dfrac{0.12 \cdot x}{0.12}$    Divide both sides by 0.12. <br><br> $50 = x$ <br><br> Thus, 6 is 12% of 50. |

## SECTION 7.3    SOLVING PERCENT PROBLEMS WITH PROPORTIONS

| | |
|---|---|
| PERCENT PROPORTION <br><br> $\dfrac{\text{amount}}{\text{base}} = \dfrac{\text{percent}}{100}$   ← always 100 <br><br> or <br><br> amount → $\dfrac{a}{b} = \dfrac{p}{100}$   ← percent <br> base → | Solve. <br><br> 20.4 is what percent of 85? <br> $\downarrow$       $\downarrow$       $\downarrow$ <br> 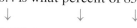 <br> amount → $\dfrac{20.4}{85} = \dfrac{p}{100}$ ← percent <br> base → <br><br> $20.4 \cdot 100 = 85 \cdot p$    Set cross products equal. <br><br> $2040 = 85 \cdot p$    Multiply. <br><br> $\dfrac{2040}{85} = \dfrac{85 \cdot p}{85}$    Divide both sides by 85. <br><br> $24 = p$    Simplify. <br><br> Thus, 20.4 is 24% of 85. |

| SECTION 7.4 | APPLICATIONS OF PERCENT |
|---|---|

PERCENT INCREASE

$$\text{percent increase} = \frac{\text{amount of increase}}{\text{original amount}}$$

PERCENT DECREASE

$$\text{percent decrease} = \frac{\text{amount of decrease}}{\text{original amount}}$$

A town with a population of 16,480 decreased to 13,870 over a 12-year period. Find the percent decrease. Round to the nearest whole percent.

$$\text{amount of decrease} = 16{,}480 - 13{,}870$$
$$= 2610$$

$$\text{percent decrease} = \frac{\text{amount of decrease}}{\text{original amount}}$$

$$= \frac{2610}{16{,}480} \approx 0.16$$

$$= 16\%$$

The town's population decreased by 16%.

| SECTION 7.5 | PERCENT AND PROBLEM SOLVING: SALES TAX, COMMISSION, AND DISCOUNT |
|---|---|

SALES TAX

$$\text{sales tax} = \text{sales tax rate} \cdot \text{purchase price}$$
$$\text{total price} = \text{purchase price} + \text{sales tax}$$

Find the sales tax and the total price of a purchase of $42.00 if the sales tax rate is 9%.

$$\text{sales tax} = \text{sales tax rate} \cdot \text{purchase price}$$
$$\downarrow \qquad \downarrow \qquad \downarrow$$
$$\text{sales tax} = \quad 9\% \quad \cdot \quad \$42$$
$$= 0.09 \cdot \$42$$
$$= \$3.78$$

The total price is

$$\text{total price} = \text{purchase price} + \text{sales tax}$$
$$\downarrow \qquad \downarrow \qquad \downarrow$$
$$\text{total price} = \quad \$42.00 \quad + \quad \$3.78$$
$$= \$45.78$$

COMMISSION

$$\text{commission} = \text{commission rate} \cdot \text{sales}$$

A salesperson earns a commission of 3%. Find the commission from sales of $12,500 worth of appliances.

$$\text{commission} = \text{commission rate} \cdot \text{sales}$$
$$\downarrow \qquad \downarrow \qquad \downarrow$$
$$\text{commission} = \quad 3\% \quad \cdot \$12{,}500$$
$$= 0.03 \cdot 12{,}500$$
$$= \$375$$

## SECTION 7.5 (CONTINUED)

DISCOUNT AND SALE PRICE

amount of discount = discount rate · original price

sale price = original price − amount of discount

A suit is priced at $320 and is on sale today for 25% off. What is the sale price?

$$\text{amount of discount} = \text{discount rate} \cdot \text{original price}$$

$$\text{amount of discount} = 25\% \cdot \$320$$

$$= 0.25 \cdot 320$$
$$= \$80$$

$$\text{sale price} = \text{original price} - \text{amount of discount}$$

$$\text{sale price} = \$320 - \$80$$
$$= \$240$$

The sale price is $240.

## SECTION 7.6 PERCENT AND PROBLEM SOLVING: INTEREST

SIMPLE INTEREST

interest = principal · rate · time

where the rate is understood to be per year.

COMPOUND INTEREST

total amount = original principal · compound interest factor

Compound interest is computed not only on the principal, but also on interest already earned in previous compounding periods. (See Appendix F.)

Find the simple interest after 3 years on $800 at an interest rate of 5%.

$$\text{interest} = \text{principal} \cdot \text{rate} \cdot \text{time}$$

$$\text{interest} = \$800 \cdot 5\% \cdot 3$$

$$= \$800 \cdot 0.05 \cdot 3 \qquad \text{Write 5\% as 0.05.}$$
$$= \$120 \qquad \text{Multiply.}$$

The interest is $120.

$800 is invested at 5% compounded quarterly for 10 years. Find the total amount at the end of 10 years.

$$\text{total amount} = \text{original principal} \cdot \text{compound interest factor}$$

$$\text{total amount} = \$800 \cdot 1.64362$$

$$\approx \$1314.90$$

# Chapter 7 Review

**(7.1)** *Solve*

1. In a survey of 100 adults, 37 preferred pepperoni on their pizza. What percent preferred pepperoni?
   37%

2. A basketball player made 77 out 100 attempted free throws. What percent of free throws was made?
   77%

*Write each percent as a decimal.*

**3.** 83%   0.83

**4.** 75%   0.75

**5.** 73.5%   0.735

**6.** 1.5%   0.015

**7.** 125%   1.25

**8.** 145%   1.45

**9.** 0.5%   0.005

**10.** 0.7%   0.007

**11.** 200%   2.00 or 2

**12.** 400%   4.00 or 4

**13.** 26.25%   0.2625

**14.** 85.34%   0.8534

*Write each decimal as a percent.*

**15.** 2.6   260%

**16.** 0.055   5.5%

**17.** 0.35   35%

**18.** 1.02   102%

**19.** 0.725   72.5%

**20.** 0.25   25%

**21.** 0.076   7.6%

**22.** 0.085   8.5%

**23.** 0.75   75%

**24.** 0.65   65%

**25.** 4.00   400%

**26.** 9.00   900%

*Write each percent as a fraction or a mixed number in simplest form.*

**27.** 1%   $\dfrac{1}{100}$

**28.** 10%   $\dfrac{1}{10}$

**29.** 25%   $\dfrac{1}{4}$

**30.** 8.5%   $\dfrac{17}{200}$

**31.** 10.2%   $\dfrac{51}{500}$

**32.** $16\dfrac{2}{3}\%$   $\dfrac{1}{6}$

**33.** $33\dfrac{1}{3}\%$   $\dfrac{1}{3}$

**34.** 110%   $1\dfrac{1}{10}$

*Write each fraction or mixed number as a percent.*

**35.** $\dfrac{1}{5}$   20%

**36.** $\dfrac{7}{10}$   70%

**37.** $\dfrac{5}{6}$   $83\dfrac{1}{3}\%$

**38.** $\dfrac{5}{8}$   62.5%

**39.** $1\dfrac{2}{3}$   $166\dfrac{2}{3}\%$

**40.** $1\dfrac{1}{4}$   125%

**41.** $\dfrac{3}{5}$   60%

**42.** $\dfrac{1}{16}$   6.25%

**43.** More and more consumers are following the "90-10 rule," that is, eating healthy foods 90% of the time, but indulging in junk food the other 10% of the time. Write 90% and 10% as fractions. (*Source:* Grocery Manufacturers of America.)   $\dfrac{9}{10}, \dfrac{1}{10}$

**44.** In medium to large firms in the United States, 96% of full-time employees receive paid vacation benefits. Write 96% as a fraction. (*Source:* U.S. Bureau of Labor Statistics)   $\dfrac{24}{25}$

**45.** Precut broccoli florets were purchased by $\dfrac{21}{25}$ of all broccoli consumers during 1998. Write $\dfrac{21}{25}$ as a percent. (*Source:* U.S. Department of Agriculture)   84%

**46.** The number of U.S. coffee drinkers drinking cappuccino and other espresso drinks has grown 150% since the early 1990s. Write 150% as a decimal. (*Source:* National Coffee Association)   1.5

**(7.2)** *Translate each question into an equation and solve.*

**47.** 1250 is 1.25% of what number?   100,000

**48.** What number is $33\dfrac{1}{3}\%$ of 24,000?   8000

**49.** 124.2 is what percent of 540?   23%

**50.** 22.9 is 20% of what number?   114.5

**51.** What number is 40% of 7500?   3000

**52.** 693 is what percent of 462?   150%

**(7.3)** *Translate each question into a proportion and solve.*

**53.** 104.5 is 25% of what number?   418

**54.** 16.5 is 5.5% of what number?   300

**55.** What number is 36% of 180?   64.8

**56.** 63 is what percent of 35?   180%

**57.** 93.5 is what percent of 85?   110%

**58.** What number is 33% of 500?   165

**(7.4)** *Solve.*

**59.** In a survey of 2000 people, it was found that 1320 have a microwave oven. Find the percent of people who own microwaves.   66%

**60.** Of the 12,360 freshmen entering County College, 2000 are enrolled in Basic College Mathematics. Find the percent of entering freshmen who are enrolled in Basic College Mathematics. Round to the nearest whole percent.   16%

**61.** The current charge for dumping waste in a local landfill is $16 per cubic foot. To cover new environmental costs, the charge will increase to $33 per cubic foot. Find the percent increase.   106.25%

**62.** The number of violent crimes in a city decreased from 675 to 534. Find the percent decrease. Round to the nearest tenth.   20.9%

**63.** This year the fund drive for a charity collected $215,000. Next year, a 4% decrease is expected. Find how much is expected to be collected in next year's drive.   $206,400

**64.** A local union negotiated a new contract that increases the hourly pay 15% over last year's pay. The old hourly rate was $11.50. Find the new hourly rate rounded to the nearest cent.   $13.23

**(7.5)** *Solve.*

**65.** If the sales tax rate is 5.5%, what is the total amount charged for a $250 coat?   $263.75

**66.** Find the sales tax paid on a $25.50 purchase if the sales tax rate is 4.5%.   $1.15

**67.** Russ James is a sales representative for a chemical company and is paid a commission rate of 5% on all sales. Find his commission if he sold $100,000 worth of chemicals last month.   $5000

**68.** Carol Sell is a sales clerk in a clothing store. She receives a commission of 7.5% on all sales. Find her commission for the week if her sales for the week were $4005. Round to the nearest cent.   $300.38

**69.** A $3000 mink coat is on sale for 30% off. Find the discount and the sale price.   discount: $900; sale price: $2100

**70.** A $90 calculator is on sale for 10% off. Find the discount and the sale price.   discount: $9; sale price: $81

**569**

**(7.6)** *Solve.*

**71.** Find the simple interest due on $4000 loaned for 3 months at 12% interest.   $120

**72.** Find the total amount due on an 8-month loan of $1200 at a simple interest rate of 15%.   $1320

**73.** Find the total amount in an account if $5500 is compounded annually at 12% for 15 years.   $30,104.64

**74.** Find the total amount in an account if $6000 is compounded semiannually at 11% for 10 years. $17,506.56

**75.** Find the compound interest earned if $100 is compounded quarterly at 12% for 5 years.   $80.61

**76.** Find the compound interest earned if $1000 is compounded quarterly at 18% for 20 years.   $32,830.10

# Chapter 7 Test

*Write each percent as a decimal.*

**1.** 85%

**2.** 500%

**3.** 0.6%

*Write each decimal as a percent.*

**4.** 0.056

**5.** 6.1

**6.** 0.35

*Write each percent as a fraction or a mixed number in simplest form.*

**7.** 120%

**8.** 38.5%

**9.** 0.2%

*Write each fraction or mixed number as a percent.*

**10.** $\dfrac{11}{20}$

**11.** $\dfrac{3}{8}$

**12.** $1\dfrac{3}{4}$

**13.** Sales of bottled water have recently surged. Bottled water accounts for $\dfrac{1}{5}$ of the total noncarbonated beverage category in convenience stores. Write $\dfrac{1}{5}$ as a percent. (*Source:* Grocery Manufacturers of America)

**14.** In small firms in the United States, 64% of full-time employees receive medical insurance benefits. Write 64% as a fraction. (*Source:* U.S. Bureau of Labor Statistics)

*Solve.*

**15.** What number is 42% of 80?

**16.** 0.6% of what number is 7.5?

**17.** 567 is what percent of 756?

**Answers**

**1.** 0.85

**2.** 5

**3.** 0.006

**4.** 5.6%

**5.** 610%

**6.** 35%

**7.** $1\dfrac{1}{5}$

**8.** $\dfrac{77}{200}$

**9.** $\dfrac{1}{500}$

**10.** 55%

**11.** 37.5%

**12.** 175%

**13.** 20%

**14.** $\dfrac{16}{25}$

**15.** 33.6

**16.** 1250

**17.** 75%

18. 38.4 lb

19. $56,750

20. $358.43

21. 5%

22. discount: $18;
    sale price: $102

23. $395

24. 1%

25. $647.50

26. $2005.64

27. $427

28. 16.9%

*Solve. If necessary, round percents to the nearest tenth, dollar amounts to the nearest cent, and all other numbers to the nearest whole.*

**18.** An alloy is 12% copper. How much copper is contained in 320 pounds of this alloy?

**19.** A farmer in Nebraska estimates that 20% of his potential crop, or $11,350, has been lost to a hard freeze. Find the total value of his potential crop.

**20.** If the local sales tax rate is 1.25%, find the total amount charged for a stereo system priced at $354.

**21.** A town's population increased from 25,200 to 26,460. Find the percent increase.

**22.** A $120 framed picture is on sale for 15% off. Find the discount and the sale price.

**23.** Randy Nguyen is paid a commission rate of 4% on all sales. Find Randy's commission if his sales were $9875.

**24.** A sales tax of $1.53 is added to an item's price of $152.99. Find the sales tax rate. Round to the nearest whole percent.

**25.** Find the simple interest earned on $2000 saved for $3\frac{1}{2}$ years at an interest rate of 9.25%.

**26.** $1365 is compounded annually at 8%. Find the total amount in the account after 5 years.

**27.** A couple borrowed $400 from a bank at 13.5% for 6 months for car repairs. Find the total amount due the bank at the end of the 6-month period.

**28.** The number of crimes reported in El Paso, Texas, was 21,212 in the first half of 1997 and 17,637 in the first half of 1998. Find the percent decrease in the number of reported crimes in El Paso from 1997 to 1998. (*Source:* Federal Bureau of Investigation, *Uniform Crime Reports,* January–June 1998)

**Name** _____  **Section** _____  **Date** _____

# CUMULATIVE REVIEW

1. Multiply: $236 \times 86$

2. Subtract 7 from $-3$

3. Solve for $x$: $x - 2 = 1$

4. Solve: $3(2x - 6) + 6 = 0$

5. Write 3 as an equivalent fraction whose denominator is 7.

6. Simplify: $\dfrac{84x}{90}$

7. Divide: $-\dfrac{5}{16} \div -\dfrac{3}{4}$

8. Evaluate $x - y$ if $x = -\dfrac{3}{10}$ and $y = \dfrac{2}{10}$.

9. Find: $-\dfrac{3}{4} - \dfrac{1}{14} + \dfrac{6}{7}$

10. Simplify: $\dfrac{\frac{1}{2} + \frac{3}{8}}{\frac{3}{4} - \frac{1}{6}}$

11. Solve: $\dfrac{x}{2} = \dfrac{x}{3} + \dfrac{1}{2}$

12. Write each mixed number as an improper fraction.

    a. $4\dfrac{2}{9}$    b. $1\dfrac{8}{11}$

*Write each decimal as a fraction or mixed number. Write your answer in simplest form.*

13. 0.125

14. 105.083

15. Subtract: $85 - 17.31$

**573**

**16.** 76.8 (Sec. 5.3, Ex. 5)

**17.** −76,300 (Sec. 5.3, Ex. 7)

**18.** yes (Sec. 5.4, Ex. 7)

**19.** $\frac{2}{5}$ (Sec. 5.8, Ex. 3)

**20.** $\frac{26}{31}$ (Sec. 6.1, Ex. 5)

**21.** $0.21/oz (Sec. 6.2, Ex. 6)

**22.** no (Sec. 6.3, Ex. 3)

**23.** 17.5 mi (Sec. 6.4, Ex. 1)

**24.** $\frac{19}{1000}$ (Sec. 7.1, Ex. 12)

**25.** $\frac{1}{3}$ (Sec. 7.1, Ex. 14)

**Name** _____

*Multiply.*

**16.** $7.68 \times 10$

**17.** $(-76.3)(1000)$

**18.** Is 720 a solution of the equation $\dfrac{y}{100} = 7.2$?

**19.** Find: $\sqrt{\dfrac{4}{25}}$

**20.** Write the ratio of 2.6 to 3.1 as a fraction in simplest form.

**21.** A store charges $3.36 for a 16-ounce jar of picante sauce. What is the unit price in dollars per ounce?

**22.** Is $\dfrac{4.1}{7} = \dfrac{2.9}{5}$ a true proportion?

**23.** On a Chamber of Commerce map of Abita Springs, 5 miles corresponds to 2 inches. How many miles correspond to 7 inches?

*Write each percent as a fraction or mixed number in simplest form.*

**24.** 1.9%

**25.** $33\dfrac{1}{3}\%$

# Graphing and Introduction to Statistics

**8**

Numerical data is all around us—in newspaper and magazine articles and in television news reports. Frequently, numerical data is summarized in graphs and basic statistics. To understand this data, we must first know how to read graphs and interpret basic statistics. We learn to do just that in this chapter.

## CONTENTS

The game of basketball was invented during the winter of 1891. Its inventor, James Naismith, was a physical education instructor at the YMCA International Training School, now known as Springfield College, in Springfield, Massachusetts. Dr. Naismith was looking for a sport that could be played indoors during the cold New England winters and would keep his students in shape between the football season played outdoors in the fall and the baseball and track seasons played outdoors in the spring. After deciding on 13 basic rules to the game, such as no player could run with the ball, Dr. Naismith asked a janitor to nail a wooden peach basket to the balcony at opposite ends of the gymnasium and gave his students a soccer ball to play with. One of his students named the game "basket ball," which stuck, and the sport quickly spread. Backboards became mandatory in 1904 to prevent spectators in the balconies from interfering with the ball. In 1912, the bottomless basket came into widespread use.

Three years before his death, Dr. Naismith saw the sport he invented played as an official Olympic sport at the 1936 Olympic Games in Berlin, Germany. Soon afterward, several competing professional basketball leagues appeared in the United States. In 1949, two of these leagues combined to form the National Basketball Association (NBA). In Exercise 31 on page 594, we will see how a graph can be used to display the number of regular season basketball games won by a modern NBA team.

# CHAPTER 8 PRETEST

1.  The following table shows a break-down of an average day for Dawn Miller.

    | | |
    |---|---|
    | Attending college classes | 4 hours |
    | Studying | 3 hours |
    | Working | 5 hours |
    | Sleeping | 8 hours |
    | Driving | 1 hour |
    | Other | 3 hours |

    *Draw a circle graph showing this data.*

*The line graph below shows the number of burglaries in a town during the months of March through September.*

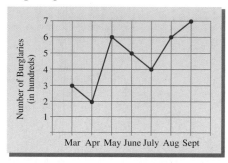

2.  During which month, between March and September, did the least number of burglaries occur?

3.  During which month, between March and September, were there 400 burglaries?

4.  How many burglaries were there in September?

5.  Is $(-2, 7)$ a solution of the equation $3x + y = -1$?

6.  Complete the ordered-pair solutions of the equation $y = 2x - 6$.
    a. $(0, \ )$   b. $(-3, \ )$   c. $( \ , 0)$

7.  Graph $y = 4x - 1$

    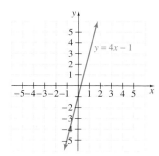

8.  Find the mean, the median, and the mode for the given set of numbers. If necessary round the mean to one decimal place.
    28, 36, 81, 73, 28, 74, 31, 74, 64, 25, 74

9.  Find the grade point average if the following grades were earned in one semester. Round to two decimal places.

    | Grade | Credit Hours |
    |---|---|
    | B | 4 |
    | B | 3 |
    | A | 3 |
    | C | 5 |
    | D | 2 |

*A single die is tossed. Find the probability that the die is each of the following.*

10. a 4

11. a number greater than 3

12. a 3 or a 5

# 8.1    Reading Circle Graphs

## A    Reading Circle Graphs

In Section 7.1, the following circle graph was shown. This particular graph shows the favorite cookie for every 100 people.

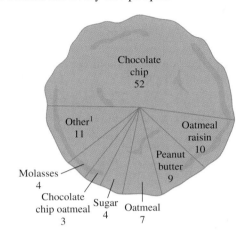

Chocolate chip 52

Other[1] 11

Oatmeal raisin 10

Peanut butter 9

Molasses 4

Chocolate chip oatmeal 3    Sugar 4    Oatmeal 7

[1]–all 1% or less
*Source: USA Today, 2/23/96*

Each sector of the graph (shaped like a piece of pie) shows a category and the relative size of the category. In other words, the most popular cookie is the chocolate chip cookie, and it is represented by the largest sector.

**Example 1**    Find the ratio of people preferring chocolate chip cookies to total people. Write the ratio as a fraction in simplest form.

*Solution:*    The ratio is

$$\frac{52 \text{ people preferring chocolate chip}}{100 \text{ people}} = \frac{52}{100} = \frac{13}{25}$$ ▬

A circle graph is often used to show percents in different categories, with the whole circle representing 100%. For example, in 1999, 65,800 public schools had CD-ROM technology. The next circle graph shows the percent of these schools that are elementary, junior high, or senior high. Notice that the percents in each category sum to 100%.

Type of U.S. Public Schools with
CD-ROM Technology in 1999

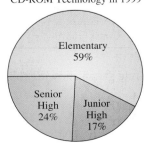

Elementary 59%

Senior High 24%    Junior High 17%

*Source:* Quality Education Data, Inc.

**Objectives**

**A**  Read circle graphs.
**B**  Draw circle graphs.

Study Guide    SSM    CD-ROM    Video 8.1

TEACHING TIP

Have students add the numbers in each sector. Ask what the sum represents.

**Practice Problem 1**

Find the ratio of people preferring oatmeal raisin cookies to total people. Write the ratio as a fraction in simplest form.

**Answer**

1. $\frac{1}{10}$

### Practice Problem 2

Determine the percent of schools with CD-ROM technology that are either elementary or junior high schools.

TEACHING TIP

Ask students to bring a ruler, compass, and protractor to class to use for this chapter.

### Practice Problem 3

Find the number of junior high schools with CD-ROM technology in 1999.

### ✓ CONCEPT CHECK

Can the following data be represented by a circle graph? Why or why not?

| Responses to the Question "In which activities are you involved?" | |
|---|---|
| Intramural sports | 60% |
| On-campus job | 42% |
| Fraternity/sorority | 27% |
| Academic clubs | 21% |
| Music programs | 14% |

### Practice Problem 4

Use the data shown to draw a circle graph.

| Freshmen | 30% |
|---|---|
| Sophomores | 27% |
| Juniors | 25% |
| Seniors | 18% |

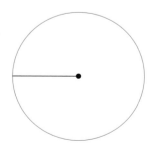

### Answers

**2.** 76%   **3.** 11,186 junior high schools
**4.** see next page

✓ **Concept Check:** No, since the percents add up to more than 100%.

---

**Example 2**    Determine the percent of schools with CD-ROM technology that are either junior or senior high schools.

*Solution:*    To find this percent, we add the junior high and senior high percents. The percent of schools with CD-ROM technology that are either junior or senior high schools is

$$24\% + 17\% = 41\%$$    ▬▬▬

**Example 3**    Find the number of elementary schools with CD-ROM technology in 1999.

*Solution:*    We use the percent equation:

amount $=$ percent $\cdot$ base
$\downarrow$        $\downarrow$        $\downarrow$
amount $=$    59%    $\cdot$ 65,800 $= 0.59(65,800) = 38,822$

Thus, 38,822 elementary schools had CD-ROM technology in 1999.    ▬▬▬

**TRY THE CONCEPT CHECK IN THE MARGIN.**

### B DRAWING CIRCLE GRAPHS

To draw a circle graph, we will use the fact that a whole circle contains 360° (degrees).

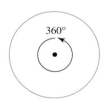

**Example 4**    The following table shows the percent of U.S. armed forces personnel that are in each branch of the service. (*Source:* U.S. Department of Defense)

| Branch of Service | Percent |
|---|---|
| Army | 33% |
| Navy | 27% |
| Marine Corps | 12% |
| Air Force | 26% |
| Coast Guard | 2% |

Draw a circle graph showing this data.

*Solution:*    First we find the number of degrees in each sector representing each branch of service. Remember that the whole circle contains 360°. (We will round degrees to the nearest whole.)

| Sector | Degrees in Each Sector |
|--------|------------------------|
| Army | 33% × 360° = 118.8° ≈ 119° |
| Navy | 27% × 360° = 97.2° ≈ 97° |
| Marine Corps | 12% × 360° = 43.2° ≈ 43° |
| Air Force | 26% × 360° = 93.6° ≈ 94° |
| Coast Guard | 2% × 360° = 7.2° ≈ 7° |

TEACHING TIP

Before making an accurate circle graph of the U.S. Armed Forces Personnel, consider having students make a quick rough sketch. When they complete the accurate graph, have them compare their accurate graph with the sketch to see where their estimates may have been off.

Next we draw a circle and mark its center. Then we draw a line from the center of the circle to the circle itself.

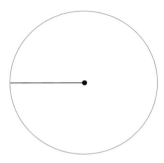

To construct the sectors, we will use a **protractor**. We place the hole in the protractor over the center of the circle. Then we adjust the protractor so that 0° on the protractor is aligned with the line that we drew.

It makes no difference which sector we draw first. To construct the "Army" sector, we find 119° on the protractor and mark our circle. Then we remove the protractor and use this mark to draw a second line from the center to the circle itself.

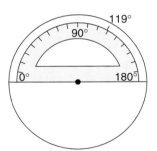
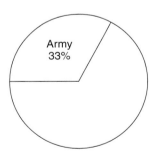

To construct the "Navy" sector next, we follow the same procedure as above except that we line 0° up with the second line we drew and mark the protractor this time at 97°.

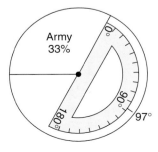
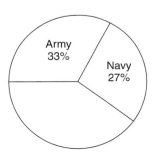

Answer
4.

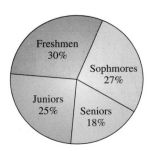

We continue in this manner until the circle graph is complete.

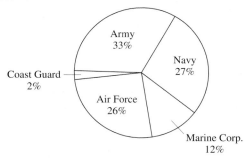

TRY THE CONCEPT CHECK IN THE MARGIN.

✓ **CONCEPT CHECK**

True or false? The larger a sector in a circle graph, the larger the percent of the total it represents. Explain your answer.

Answer
✓ Concept Check:  True.

**Name** _____ **Section** _____ **Date** _____

## EXERCISE SET 8.1

**A** *The circle graph is a result of surveying 700 college students. They were asked where they live while attending college. Use this graph to answer Exercises 1–6. Write all ratios as fractions in simplest form. See Example 1.*

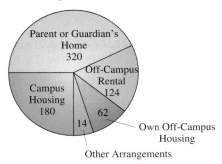

**1.** Where do most of these college students live?

**2.** Besides the category "Other Arrangements," where do least of these college students live?

**3.** Find the ratio of students living in campus housing to total students.

**4.** Find the ratio of students living in off-campus rentals to total students.

**5.** Find the ratio of students living in campus housing to students living at home.

**6.** Find the ratio of students living in off-campus rentals to students living at home.

*The circle graph shows how much money teenagers spend on a backpack. Use this graph for Exercises 7–14. See Examples 2 and 3.*

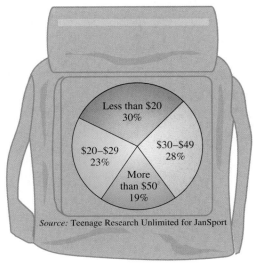

*Source:* Teenage Research Unlimited for JanSport

**7.** Backpacks in which price range are purchased most often?

**8.** Backpacks in which price range are purchased least often?

**Name** _____

**9.** What percent of backpacks purchased cost less than $30?

**10.** What percent of backpacks purchased cost $30 or more?

*If Central High School has 4700 teenagers, use the backpack graph to find how many Central High teenagers spent the dollar amounts given in Exercises 11–14 on backpacks.*

**11.** Less than $20

**12.** $20–$29

**13.** $30–$49

**14.** More than $50

*The circle graph shows the percent of types of books available at Midway Memorial Library. Use this graph to answer Exercises 15–24. See Examples 2 and 3.*

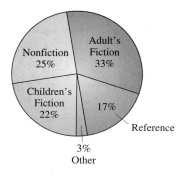

**15.** What percent of books are classified as some type of fiction?

**16.** What percent of books are nonfiction or reference?

**17.** What is the second-largest category of books?

**18.** What is the third-largest category of books?

**Name** _____

*If this library has 125,600 books, find how many books are in each category given in Exercises 19–24.*

**19.** Nonfiction

**20.** Reference

**21.** Children's fiction

**22.** Adult's fiction

**23.** Reference or other

**24.** Nonfiction or other

**B** *Draw a circle graph to represent the information given in each table. See Example 4.*

**25.**

| Absence from Work Due to Carpal Tunnel Syndrome | |
|---|---|
| Under 3 days | 7% |
| 3–10 days | 18% |
| 11–20 days | 14% |
| 21 or more days | 61% |

(*Note:* Carpal tunnel syndrome is a nerve disorder causing pain mostly in the wrist and hand.)
(*Source:* Bureau of Labor Statistics)

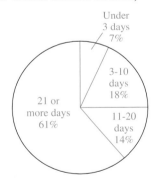

**26.**

| Most Important Reason for Picking a Favorite Restaurant Among Wendy's, Burger King, and McDonald's | |
|---|---|
| Convenience | 13% |
| Taste of food | 70% |
| Price | 9% |
| Atmosphere | 4% |
| Others | 4% |

(*Source:* TeleNation)

## REVIEW AND PREVIEW

*Write the prime factorization of each number. See Section 4.2.*

**27.** 20

**28.** 25

**29.** 40

**30.** 16

**31.** 85

**32.** 105

**19.** 31,400 books

**20.** 21,352 books

**21.** 27,632 books

**22.** 41,448 books

**23.** 25,120 books

**24.** 35,168 books

**25.** see graph

**26.** see graph

**27.** $2^2 \cdot 5$

**28.** $5^2$

**29.** $2^3 \cdot 5$

**30.** $2^4$

**31.** $5 \cdot 17$

**32.** $3 \cdot 5 \cdot 7$

**Name** _____

*The circle graph shows the relative sizes of the great oceans. These oceans together make up 264,489,800 square kilometers of Earth's surface. Find the square kilometers for each ocean.*

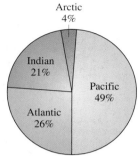

Source: *World Almanac, 2000*

 **33.** Pacific Ocean

**34.** Atlantic Ocean

**35.** Indian Ocean

**36.** Arctic Ocean

**37.** In your own words, explain why the percents in a circle graph should have a sum of 100%.

# 8.2   READING PICTOGRAPHS, BAR GRAPHS, AND LINE GRAPHS

Often data is presented visually in a graph. In this section, we practice reading several kinds of graphs including pictographs, bar graphs, and line graphs.

## A   READING PICTOGRAPHS

A **pictograph** such as the one below is a graph in which pictures or symbols are used. This type of graph will contain a key that explains the meaning of the symbol used. An advantage of using a pictograph to display information is that comparisons can easily be made. A disadvantage of using a pictograph is that it is often hard to tell what fractional part of a symbol is shown. For example, in the pictograph below, Ukraine shows a part of a symbol, but it's hard to determine what fractional part of a symbol is shown with any accuracy.

**Example 1**   The pictograph shows the approximate nuclear energy generated by selected countries. Use this pictograph to answer the questions.

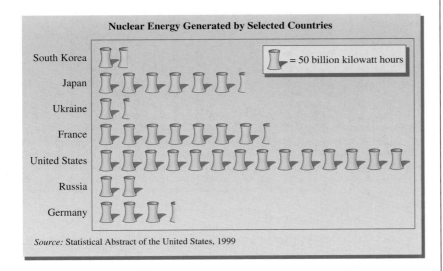

Source: Statistical Abstract of the United States, 1999

**a.** Approximate the amount of nuclear energy that is generated in Japan.

**b.** Approximate how much more nuclear energy is generated in Japan than in Russia.

*Solution:*   **a.** Japan corresponds to 6.5 symbols and each symbol represents 50 billion kilowatt hours of energy. This means that Japan generates approximately 6.5 · (50 billion) or 325 billion kilowatt hours of energy.

**b.** Japan shows 4.5 more symbols than Russia. This means that Japan generates 4.5 · (50 billion) or 225 billion more kilowatt hours of nuclear energy than Russia.

### Objectives

**A** Read pictographs.
**B** Read bar graphs.
**C** Construct bar graphs.
**D** Read line graphs.

Study    SSM    CD-ROM    Video
Guide                              8.2

### Practice Problem 1

Use the pictograph shown in Example 1 to answer the questions.

a. Approximate the amount of nuclear energy that is generated in the United States.

b. Approximate the total nuclear energy generated in Japan and Ukraine.

**Answers**

**1. a.** 700 billion kilowatt hours   **b.** 400 billion kilowatt hours

## B READING BAR GRAPHS

Another way to present data by a graph is with a **bar graph**. Bar graphs can appear with vertical bars or horizontal bars. Although we have studied bar graphs in previous sections, we now practice reading the height of the bars contained in a bar graph. An advantage to using bar graphs is that a scale is usually included for greater accuracy. Care must be taken when reading bar graphs, as well as other types of graphs—they may be misleading, as shown later in this section.

### Practice Problem 2

Use the bar graph in Example 2 to answer the questions.

a. Approximate the number of endangered species that are insects.

b. Which category shows the most endangered species?

**Example 2**    The bar graph shows the number of endangered species in different categories. Use this graph to answer the questions.

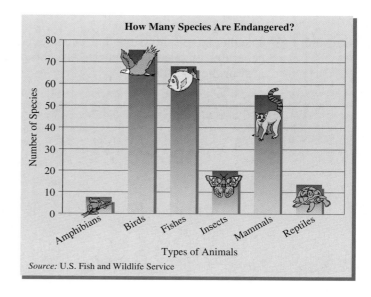

*Source:* U.S. Fish and Wildlife Service

**a.** Approximate the number of endangered species that are mammals.

**b.** Which category shows the fewest endangered species?

*Solution:*

**a.** To approximate the number of endangered species that are mammals, we go to the top of the bar that represents mammals. From the top of this bar, we move horizontally to the left until the scale is reached. We read the height of the bar on the scale as approximately 55. There are approximately 55 mammal species that are endangered, as shown in the figure on the following page.

**b.** The fewest endangered species is represented by the shortest bar. The shortest bar corresponds to amphibians.

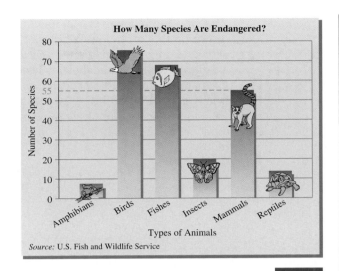

How Many Species Are Endangered?

*Source:* U.S. Fish and Wildlife Service

## TRY THE CONCEPT CHECK IN THE MARGIN.

As mentioned previously, graphs can be misleading. Both graphs below show the same information, but with different scales. Special care should be taken when forming conclusions from the appearance of a graph.

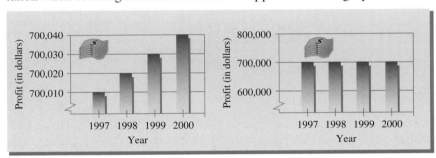

Are profits shown in the graphs above greatly increasing or are they remaining about the same?

## C CONSTRUCTING BAR GRAPHS

Next, we practice constructing a bar graph.

**Example 3**    Draw a vertical bar graph using the information in the table about the caffeine content of selected foods.

| AVERAGE CAFFEINE CONTENT OF SELECTED FOODS | | | |
|---|---|---|---|
| Food | Milligrams | Food | Milligrams |
| Brewed coffee (percolator, 8 ounces) | 124 | Instant coffee (8 ounces) | 104 |
| Brewed decaffeinated coffee (8 ounces) | 3 | Brewed tea (U.S. brands, 8 ounces) | 64 |
| Coca-Cola classic (8 ounces) | 31 | Mr. Pibb (8 ounces) | 27 |
| Dark chocolate (semi-sweet, $1\frac{1}{2}$ ounces) | 30 | Milk chocolate (8 ounces) | 9 |

(*Source:* International Food Information Council and the Coca-Cola Company)

## ✓ CONCEPT CHECK

True or false? With a bar graph, the width of the bar is just as important as the height of the bar. Explain your answer.

## Practice Problem 3

Draw a vertical bar graph using the information in the table about electoral votes for selected states.

| ELECTORAL VOTES FOR PRESIDENT BY SELECTED STATES | |
|---|---|
| State | Electoral Votes |
| Texas | 31 |
| California | 54 |
| Florida | 25 |
| Nebraska | 5 |
| Indiana | 12 |
| Georgia | 13 |

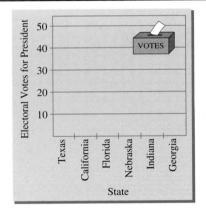

**Answers**

**3.**

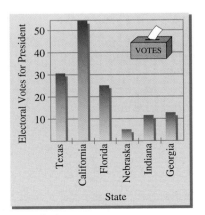

✓ **Concept Check:** False.

*Solution:* We draw and label a vertical line and a horizontal line as shown on the left below. We place the different food categories along the horizontal line. Along the vertical line, we place a scale. The scale shown starts at 0 and then shows multiples of 20 placed at equally spaced intervals. It may also be helpful to draw horizontal lines along the scale markings to help draw the vertical bars at the correct height. The finished bar graph is shown below on the right.

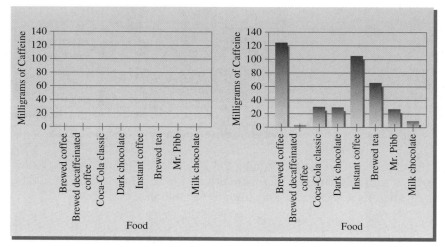

### D   READING LINE GRAPHS

Another common way to display information with a graph is by using a **line graph**. An advantage of a line graph is that it can be used to visualize relationships between two quantities. A line graph can also be very useful in showing change over time.

**Practice Problem 4**

Use the temperature graph in Example 4 to answer the questions.

a. During what month is the average daily temperature the lowest?

b. During what month is the average daily temperature 25°F?

c. During what months is the average daily temperature greater than 70°F?

**Example 4**    The line graph shows the average daily temperature for each month in Omaha, Nebraska. Use this graph to answer the questions.

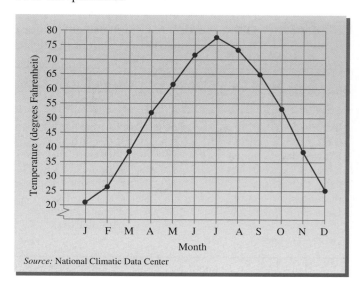

*Source:* National Climatic Data Center

**Answers**

**4. a.** January   **b.** December   **c.** June, July, and August

**a.** During what month is the average daily temperature the highest?

**b.** During what month is the average daily temperature 65°F?

**c.** During what months is the average daily temperature less than 30°F?

**d.** Write a sentence using the information from the point above the month April.

*Solution:*    **a.** The month with the highest temperature corresponds to the highest point. We follow the highest point downward to the horizontal month scale and see that this point corresponds to July.

**b.** We find the 65°F mark on the vertical scale and move to the right until a darkened point on the graph is reached. From that point, we move downward to the month scale and read the corresponding month. During the month of September, the average daily temperature is 65°F.

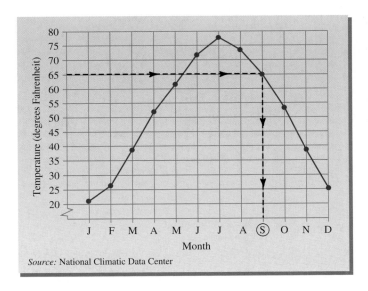

*Source:* National Climatic Data Center

**c.** To see what months the temperature is less than 30°F, we find what months correspond to darkened points that fall below the 30°F mark on the vertical scale. These months are January, February, and December.

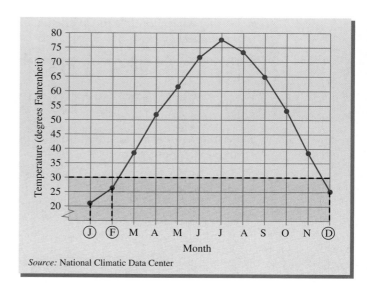

*Source:* National Climatic Data Center

**d.** The dot above April appears to correspond to the temperature 52°F. Here is a sample sentence: "In April, the average daily temperature is 52°F."

**TRY THE CONCEPT CHECK IN THE MARGIN.**

✓ **CONCEPT CHECK**

On the line graph in Example 4, what two quantities does each point on the graph represent?

**Answer**
✓ **Concept Check:** month and average daily temperature

Name ______________________ Section __________ Date __________

ANSWERS

# EXERCISE SET 8.2

**A** *The pictograph shows the annual automobile production by one plant for the years 1993–1999. Use this graph to answer Exercises 1–8. See Example 1.*

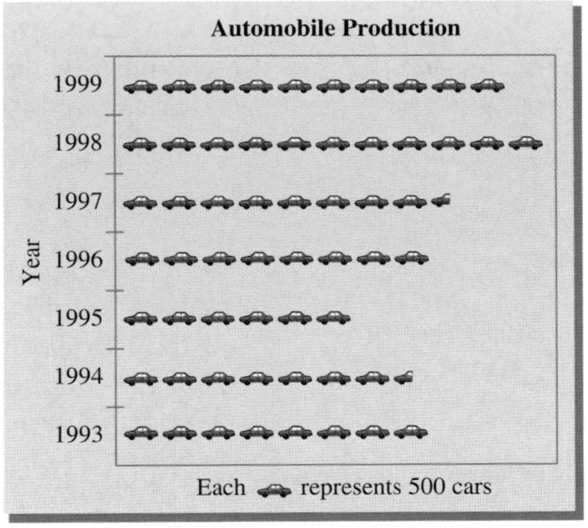


1. In what year was the greatest number of cars manufactured?

2. In what year was the least number of cars manufactured?

3. Approximate the number of cars manufactured in the year 1996.

4. Approximate the number of cars manufactured in the year 1997.

5. In what year(s) did the production of cars decrease from the previous year?

6. In what year(s) did the production of cars increase from the previous year?

7. In what year(s) were 4000 cars manufactured?

8. In what year(s) were 5500 cars manufactured?

*The pictograph shows the average number of ounces of chicken consumed per person per week in the United States. Use this graph to answer Exercises 9–16. See Example 1.*

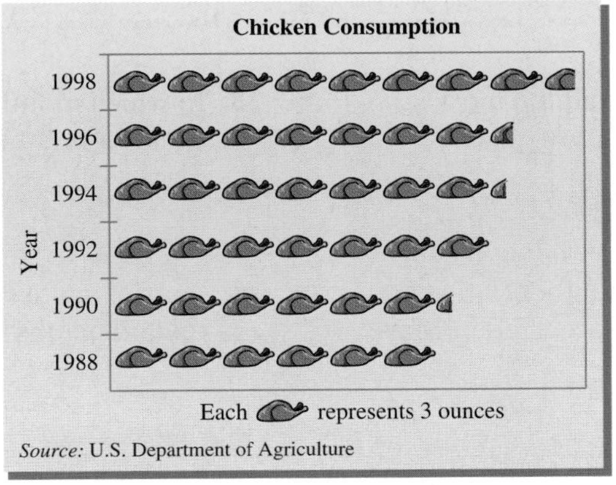


9. Approximate the number of ounces of chicken consumed per week in 1992.

10. Approximate the number of ounces of chicken consumed per week in 1996.

**Answers**

1. 1998
2. 1995
3. 4000 cars
4. 4250 cars
5. 1994, 1995, 1999
6. 1996, 1997, 1998
7. 1993, 1996
8. 1998
9. 21 oz
10. 22.5 oz

**Name** _____

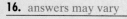 **11.** In what year(s) was the number of ounces of chicken consumed per week greater than 21 ounces?

**12.** In what year(s) was the number of ounces of chicken consumed per week 21 ounces or less?

**13.** What was the increase in average chicken consumption from 1988 to 1992?

**14.** What was the increase in average chicken consumption from 1992 to 1998?

**15.** Describe a trend in eating habits shown by this graph.

**16.** In 1998, did you eat less than, greater than, or about the same as the U.S. average number of ounces of chicken consumed per week?

**B** *The bar graph shows the average number of people killed by tornadoes during the months of the year. Use this graph to answer Exercises 17–22. See Example 2.*

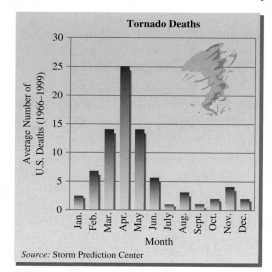

*Source:* Storm Prediction Center

**17.** In which month(s) did the most tornado-related deaths occur?

**18.** In which month(s) did the fewest tornado-related deaths occur?

**19.** Approximate the number of tornado-related deaths that occurred in May.

**20.** Approximate the number of tornado-related deaths that occurred in April.

**21.** In which month(s) did over 5 tornado-related deaths occur?

**22.** In which month(s) did over 10 tornado-related deaths occur?

*The horizontal bar graph shows the population (in millions) of the world's largest cities. Use this graph to answer Exercises 23–28. See Example 2.*

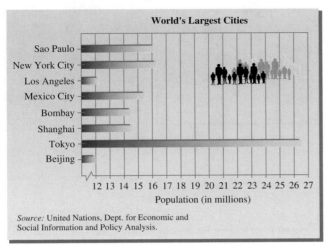

**World's Largest Cities**

Population (in millions)

*Source:* United Nations, Dept. for Economic and Social Information and Policy Analysis.

**23.** Name the city with the largest population and estimate its population.

**24.** Name the city whose population is between 15 and 16 million and estimate its population.

**25.** Name the United States city with the largest population and estimate its population.

**26.** Name the city with the smallest population and estimate its population.

**27.** How much larger is Tokyo than Bombay?

**28.** How much larger is Mexico City than Beijing?

**C** *Use the information given in the table to draw a vertical bar graph. Clearly label the bars in each graph. See Example 3.*

**29.**

| BEST-SELLING MAGAZINES (IN MILLIONS) | |
|---|---|
| Magazine | Circulation |
| *Modern Maturity* | 21 |
| *NRTA/AARP* Bulletin | 20 |
| *Reader's Digest* | 14 |
| *TV Guide* | 13 |
| *National Geographic* | 9 |
| *Better Homes and Gardens* | 8 |

(*Source:* Audit Bureau of Circulations 1998)

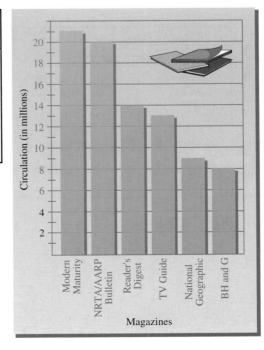

Circulation (in millions)

Magazines

**23.** Tokyo, 26.5 million or 26,500,000

**24.** Mexico City, 15.5 million or 15,500,000

**25.** New York City, 16.2 million or 16,200,000

**26.** Beijing, 12 million or 12,000,000

**27.** 12 million or 12,000,000

**28.** 3.5 million or 3,500,000

**29.** see graph

**Name** _____

**30.**

| LEADING U.S. PASSENGER AIRLINES (IN MILLIONS) | |
|---|---|
| Airline | Passengers |
| Delta | 105 |
| United | 87 |
| American | 81 |
| Southwest | 59 |
| US Airways | 58 |
| Northwest | 50 |

(*Source:* Air Transport Association of America 1998)

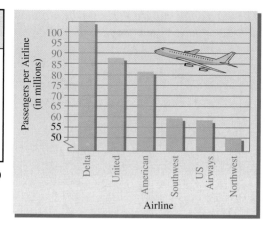

**31.**

| SAN ANTONIO REGULAR SEASON WINS | |
|---|---|
| Season | Games won |
| 93–94 | 55 |
| 94–95 | 62 |
| 95–96 | 59 |
| 96–97 | 20 |
| 97–98 | 56 |
| 98–99 | 37 |

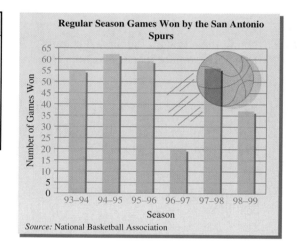

**32.**

| MICHAEL JORDAN'S SCORING STATISTICS | |
|---|---|
| Season | Average Points Per Game |
| 91–92 | 30.1 |
| 92–93 | 32.6 |
| 94–95 | 26.9 |
| 95–96 | 30.4 |
| 96–97 | 29.6 |
| 97–98 | 28.7 |

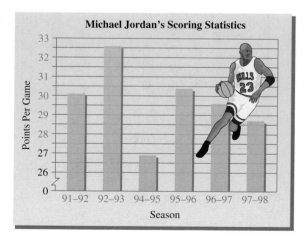

**D** *The line graph shows NCAA Division I average number of field goals per game. Use this graph to answer Exercises 39–46. See Example 4.*

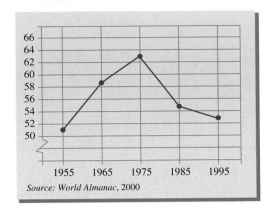

Source: World Almanac, 2000

**33.** Approximate the average number of field goals per game in 1985.

**34.** Approximate the average number of field goals per game in 1975.

**35.** During what year shown was the average number of field goals per game the highest?

**36.** During what year shown was the average number of field goals per game the lowest?

**37.** Between 1955 and 1965, did the average number of field goals per game increase or decrease?

**38.** Between 1985 and 1995, did the average number of field goals per game increase or decrease?

Star Wars Episode I: The Phantom Menace *opened in movie theaters on May 19, 1999. The line graph shows the weekly gross box office receipts for the first 8 weeks of its release. Use this graph to answer Exercises 39–46. See Example 4.*

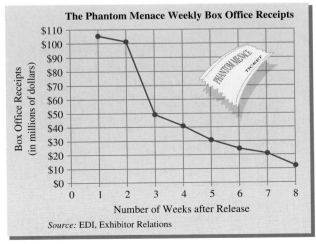

Source: EDI, Exhibitor Relations

**39.** Approximate the weekly box office receipts 2 weeks after release.

**40.** Approximate the weekly box office receipts 3 weeks after release.

**Name** _____

**41.** In what week after release were the
receipts the greatest?

**42.** In what week after release were the
receipts the least?

**43.** Approximate the decrease in receipts
from week 1 to week 8.

**44.** Approximate the decrease in receipts
from week 2 to week 6.

**45.** Write a sentence using the information
from the point above the number 8.

**46.** Write a sentence using the information
from the point above the number 1.

## REVIEW AND PREVIEW

*Find each percent. See Sections 7.2 and 7.3.*

**47.** Find 30% of 12.　**48.** Find 45% of 120.　**49.** Find 10% of 62.　**50.** Find 95% of 50.

*Write each fraction as a percent. See Section 7.1.*

**51.** $\dfrac{1}{4}$　　**52.** $\dfrac{2}{5}$　　**53.** $\dfrac{17}{50}$　　**54.** $\dfrac{9}{10}$

## COMBINING CONCEPTS

*The double line graph shows temperature highs and lows for a week. Use this graph to answer
Exercises 55–60.*

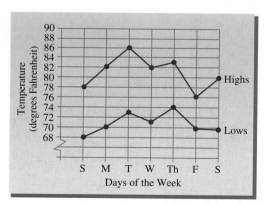

**55.** What was the high temperature reading on Thursday?

**56.** What was the low temperature reading
on Thursday?

**596**

**57.** On what day of the week was the low temperature the lowest? What was this low temperature?

**58.** On what day of the week was the high temperature the highest? What was this high temperature?

**59.** On what day of the week was the difference between the high temperature and the low temperature the greatest? What was this difference in temperature?

**60.** On what day of the week was the difference between the high temperature and the low temperature the least? What was this difference in temperature?

*The following graph is called a double bar graph. Study this graph and use it to answer Exercises 61–64.*

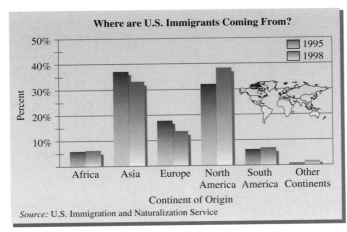

**61.** Which continent shows the greatest change in percent of U.S. immigrants?

**62.** Which continent accounted for the greatest percent of U.S. immigrants in 1995?

**63.** Which continent accounted for the greatest percent of U.S. immigrants in 1998?

**64.** How might a graph of this type be helpful?

**57.** Sunday, 68°F _____

**58.** Tuesday, 86°F _____

**59.** Tuesday, 13°F _____

**60.** Friday, 6°F _____

**61.** North America _____

**62.** Asia _____

**63.** North America _____

**64.** answers may vary _____

**597**

## Internet Excursions

Go to http://www.prenhall.com/martin-gay

Ultraviolet (UV) radiation from the sun can damage the skin and eyes, causing skin cancer, premature aging, and cataracts. The National Oceanic and Atmospheric Administration (NOAA) and Environmental Protection Agency (EPA) provide a daily forecast, called the UV Index, of the amount of UV radiation expected to reach Earth's surface. The values range from 0 to 15 and are grouped into five different UV exposure categories.

| UV INDEX VALUES | EXPOSURE CATEGORY |
| --- | --- |
| 0–2 | Minimal |
| 3–4 | Low |
| 5–6 | Moderate |
| 7–9 | High |
| 10+ | Very High |

   This World Wide Web address will provide you with access to graphs of the NOAA/EPA UV Index for an entire year in various cities, or a related site. Select the most recent graph for the city closest to your college, and use it to answer the following questions.

**65.** Use the graph to estimate when the UV Index Forecast was the highest for your area. Approximate the UV Index value and corresponding exposure category for that point in time.

**66.** Use the graph to estimate when the UV Index Forecast was the lowest for your area. Approximate the UV Index value and corresponding exposure category for that point in time.

# 8.3 THE RECTANGULAR COORDINATE SYSTEM

## A PLOTTING POINTS

In the last section, we saw how bar and line graphs can be used to show relationships between items listed on the horizontal and vertical axes. We can use this same horizontal and vertical axis idea to describe the location of points in a plane.

The system that we use to describe the location of points in a plane is called the **rectangular coordinate system**. It consists of two number lines, one horizontal and one vertical, intersecting at the point 0 on each number line. This point of intersection is called the **origin**. We call the horizontal number line the *x*-axis and the vertical number line the *y*-axis. Notice that the axes divide the plane into four regions, called **quadrants**. They are numbered as shown.

Objectives

**A** Plot points on a rectangular coordinate system.

**B** Determine whether ordered pairs are solutions of equations.

**C** Complete ordered-pair solutions of equations.

Study Guide    SSM    CD-ROM    Video 8.3

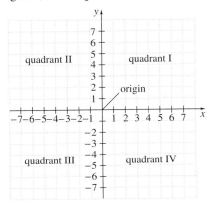

Every point in the rectangular coordinate system corresponds to an **ordered pair of numbers**, such as (3, 4). The first number, 3, of an ordered pair is associated with the *x*-axis and is called the ***x*-coordinate** or ***x*-value**. The second number, 4, is associated with the *y*-axis and is called the ***y*-coordinate** or ***y*-value**. To find the **single point** on the rectangular coordinate system corresponding to the ordered pair (3, 4), start at the origin. Move 3 units in the positive direction along the *x*-axis. From there, move 4 units in the positive direction parallel to the *y*-axis. This process of locating a point on the rectangular coordinate system is called **plotting the point**. Since the origin is located at 0 on the *x*-axis and 0 on the *y*-axis, the origin corresponds to the ordered pair (0, 0).

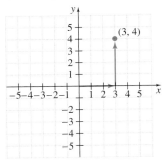

In general, to plot the ordered pair $(x, y)$, start at the origin. Next,

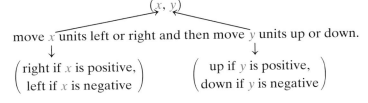

TEACHING TIP

Remind students that the graph of **1** ordered pair is **1** point.

## Helpful Hint

Since the first number, or *x*-coordinate, of an ordered pair is associated with the *x*-axis, it tells how many units to move left or right. Similarly, the second number, or *y*-coordinate, tells how many units to move up or down. For example, to plot $(-1, 5)$, start at the origin. Move 1 unit left (because the *x*-coordinate is negative) and then 5 units up and draw a dot at that point. This dot is the graph of the point that corresponds to the ordered pair $(-1, 5)$.

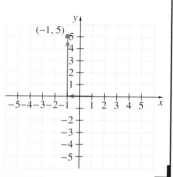

Here are some more plotted points with their corresponding ordered pairs.

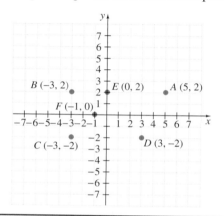

## Helpful Hint

Remember that **each point** in the rectangular coordinate system corresponds to exactly **one ordered pair** and that **each ordered pair** corresponds to exactly **one point**.

**Example 1**    Plot each point corresponding to the ordered pairs on the same set of axes.

$$(5, 4), (-2, 3), (-1, -2), (6, -3), (0, 2), (-4, 0), \left(3, \frac{1}{2}\right)$$

*Solution:*

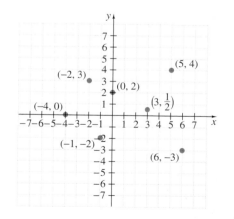

---

TEACHING TIP

For those students having trouble graphing ordered pairs, tell them to think of an ordered pair as directions to a location. They must perform both directions in order to find the correct location. Ordered pair directions are given as (left or right, up or down).

## Practice Problem 1

Plot each point corresponding to the ordered pairs on the same set of axes.
$(1, 3), (-3, 2), (-6, -5), (2, -2), (5, 0), (0, -3)$

**Answer**

**1.**

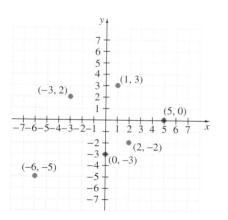

TRY THE CONCEPT CHECK IN THE MARGIN.

**Example 2**     Find the ordered pair corresponding to each point plotted on the rectangular coordinate system.

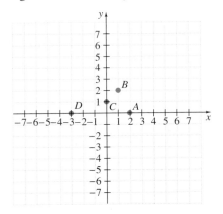

*Solution:*     Point $A$ has coordinates $(2, 0)$.
Point $B$ has coordinates $(1, 2)$.
Point $C$ has coordinates $(0, 1)$.
Point $D$ has coordinates $(-3, 0)$.

Have you noticed a pattern from the preceding examples? **If an ordered pair has a $y$-coordinate of 0, its graph lies on the $x$-axis. If an ordered pair has an $x$-coordinate of 0, its graph lies on the $y$-axis.**

## B   DETERMINING WHETHER AN ORDERED PAIR IS A SOLUTION

Let's see how we can use ordered pairs of numbers to record solutions of equations containing two variables.

Recall that an equation with one variable, such as $3 + x = 5$, has one solution—in this case, 2 because 2 is the only value that can be substituted for $x$ so that the resulting equation $3 + 2 = 5$ is a true statement. We can graph this solution on a number line.

An equation with two variables such as $x + y = 7$ has many solutions. Each solution is a pair of numbers, one for each variable that makes the equation a true statement. For example, $x = 4$ and $y = 3$ is a solution of the equation $x + y = 7$ because, when $x$ is replaced with 4 and $y$ is replaced with 3, $x + y = 7$ becomes $4 + 3 = 7$, which is a true statement. We can write the solution $x = 4$ and $y = 3$ as the ordered pair $(4, 3)$. Also, $x = 0$ and $y = 7$ is a solution because $0 + 7 = 7$ is a true statement. We can write the solution $x = 0$ and $y = 7$ as the ordered pair of numbers $(0, 7)$. In general, we say that an ordered pair of numbers is a solution of an equation if the equation is a true statement when the variables are replaced by the coordinates of the ordered pair.

Unlike equations with one variable, equations with two variables, such as $x + y = 7$, have so many solutions we cannot simply list them. Instead, to "see" all the solutions of a two-variable equation, we draw a graph of its solutions.

First, let's practice deciding whether an ordered pair of numbers is a solution of an equation in two variables.

✓ **CONCEPT CHECK**

a. Is the ordered pair $(4, -3)$ plotted to the left or right of the $y$-axis?

b. Is the ordered pair $(6, -2)$ plotted above or below the $x$-axis?

**Practice Problem 2**

Find the ordered pair corresponding to each point plotted on the rectangular coordinate system.

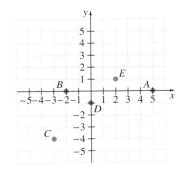

**Answers**
**2.** $A(5, 0)$; $B(-2, 0)$; $C(-3, -4)$; $D(0, -1)$; $E(2, 1)$
✓ **Concept Check: a.** Right. **b.** Below.

## Practice Problem 3

Is $(0, -4)$ a solution of the equation $x + 3y = -12$?

## Practice Problem 4

Each ordered pair listed is a solution of the equation $x - y = 5$. Plot them on the same set of axes.

a. $(6, 1)$      b. $(5, 0)$

c. $(0, -5)$      d. $(7, 2)$

e. $(-1, -6)$      f. $(2, -3)$

**Example 3**    Is $(-1, 6)$ a solution of the equation $2x + y = 4$?

*Solution:*    Replace $x$ with $-1$ and $y$ with 6 in the equation $2x + y = 4$ to see if the result is a true statement.

$$2x + y = 4 \quad \text{Original equation}$$
$$2(-1) + 6 = 4 \quad \text{Replace } x \text{ with } -1 \text{ and } y \text{ with 6.}$$
$$-2 + 6 = 4 \quad \text{Multiply.}$$
$$4 = 4 \quad \text{True.}$$

Since $4 = 4$ is true, $(-1, 6)$ is a solution of the equation $2x + y = 4$.

**Example 4**    Each ordered pair listed is a solution of the equation $x + y = 7$. Plot them on the same set of axes.

a. $(3, 4)$      b. $(0, 7)$      c. $(-1, 8)$

d. $(7, 0)$      e. $(5, 2)$      f. $(4, 3)$

*Solution:*

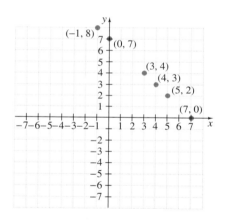

Notice that the points in Example 4 all seem to lie on the same line. We will discuss this more in the next section.

An equation such as $x + y = 7$ is called a **linear equation in two variables**. It is called a **linear equation** because the exponent on $x$ and $y$ is an understood 1. The equation is called an equation **in two variables** because it contains two different variables, $x$ and $y$.

### ⓒ COMPLETING ORDERED-PAIR SOLUTIONS

In the next section, we will graph linear equations in two variables. Before that, we need to practice finding ordered-pair solutions of these equations. If one coordinate of an ordered-pair solution is known, the other coordinate can be determined. To find the unknown coordinate, replace the appropriate variable with the known coordinate in the equation. Doing so results in an equation with one variable that we can solve.

**Answers**

**3.** yes   **4.**

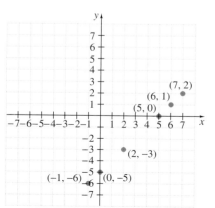

**Example 5**    Complete each ordered-pair solution of the equation $y = 2x$.

a. $(3, \quad)$        b. $(\quad, 0)$        c. $(-2, \quad)$

*Solution:*    **a.** In the ordered pair $(3, \quad)$, the $x$-value is 3. To find the corresponding $y$-value, let $x = 3$ in the equation $y = 2x$ and calculate the value for $y$.

$y = 2x$        Original equation

$y = 2(3)$      Replace $x$ with 3.

$y = 6$

*Check:*    To check, replace $x$ with 3 and $y$ with 6 in the original equation to see that a true statement results.

$2x = y$        Original equation

$2(3) \stackrel{?}{=} 6$    Let $x = 3$ and $y = 6$.

$6 = 6$        True.

The ordered-pair solution is $(3, 6)$.

**b.** Replace $y$ with 0 in the equation and solve for $x$.

$y = 2x$        Original equation

$0 = 2x$        Replace $y$ with 0.

$\dfrac{0}{2} = \dfrac{2x}{2}$    Solve for $x$.

$0 = x$        Simplify.

The ordered-pair solution is $(0, 0)$, the origin.

**c.** Replace $x$ with $-2$ in the equation and solve for $y$.

$y = 2x$        Original equation

$y = 2(-2)$     Replace $x$ with $-2$.

$y = -4$        Solve for $y$.

The ordered-pair solution is $(-2, -4)$.    ▬▬▬

**Practice Problem 5**

Complete each ordered-pair solution of the equation $x + y = 10$.

a. $(5, \quad)$    b. $(0, \quad)$    c. $(\quad, -2)$

Answers

**5. a.** $(5, 5)$    **b.** $(0, 10)$    **c.** $(12, -2)$

# Focus on Mathematical Connections

### SCATTER DIAGRAMS

In Section 8.2, we learned about presenting data visually in a graph, such as a pictograph, bar graph, or line graph. Now we learn about another type of graph, based on ordered pairs, that can be used to present data.

Data that can be represented as an ordered pair is called paired data. Many types of data collected from the real world are paired data. For instance, the total amount of rainfall a location receives each year can be written as an ordered pair of the form (year, total rainfall in inches) and is paired data. The graph of paired data as points in the rectangular coordinate system is called a **scatter diagram,** or scatter plot. Scatter diagrams can be used to look for patterns and trends in paired data.

For example, the data shown in the table is paired data that can be written as a set of ordered pairs of the form (year, number of countries), such as (1994, 79) and (1998, 114). A scatter diagram of the paired data is shown below. Notice that the horizontal axis is labeled "Year" to describe the *x*-coordinates in the ordered pairs. The vertical axis is labeled "Number of Countries" to describe the *y*-coordinates in the ordered pairs.

| NUMBER OF COUNTRIES WITH McDONALD'S RESTAURANTS | |
|---|---|
| Year | Number of Countries |
| 1994 | 79 |
| 1995 | 89 |
| 1996 | 101 |
| 1997 | 109 |
| 1998 | 114 |

(*Source:* McDonald's Corporation)

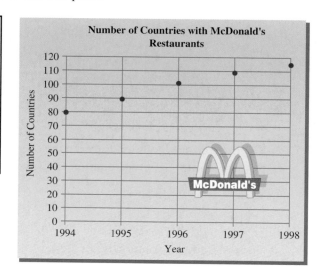

### CRITICAL THINKING

*The table gives the number of McDonald's restaurants in operation worldwide each year. Use the table to answer the following questions.*

1.  Write this paired data as a set of ordered pairs of the form (year, number of restaurants in thousands).
2.  Create a scatter diagram of the paired data.
3.  What trend in the paired data does the scatter diagram show?

### GROUP ACTIVITY

4.  Find or collect your own paired data and present it in a scatter diagram. What does the graph show?

| NUMBER OF McDONALD'S RESTAURANTS WORLDWIDE | |
|---|---|
| Year | Number of Restaurants (in thousands) |
| 1993 | 14 |
| 1994 | 16 |
| 1995 | 18 |
| 1996 | 21 |
| 1997 | 23 |
| 1998 | 25 |

(*Source:* McDonald's Corporation)

# EXERCISE SET 8.3

**A** *Plot the points corresponding to the ordered pairs on the same set of axes. See Example 1.*

**1.** $(1, 3)$, $(-2, 4)$, $(0, 7)$,
$(-5, 0)$, $(-6, -3)$, $(5, -5)$

**2.** $(5, 2)$, $(3, -4)$, $(-1, -1)$,
$(0, -6)$, $(4, 0)$, $(-2, 4)$

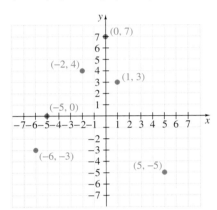

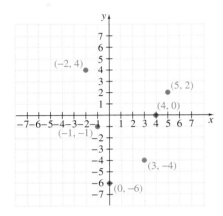

**3.** $\left(2\frac{1}{2}, 3\right)$, $(0, -3)$, $(-4, -6)$,
$\left(-1, 5\frac{1}{2}\right)$, $(1, 0)$, $(3, -5)$

**4.** $\left(5, \frac{1}{2}\right)$, $\left(-3\frac{1}{2}, 0\right)$, $(-1, 4)$,
$(4, -1)$, $(0, 2)$, $(-5, -5)$

$A(0, 0); B\left(3\frac{1}{2}, 0\right);$
$C(3, 2); D(-1, 3);$
$E(-2, -2); F(0, -1);$

**5.** $G(2, -1)$

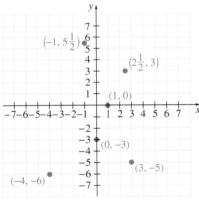

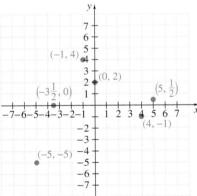

*Find the x- and y-coordinates of each labeled point. See Example 2.*

**5.**

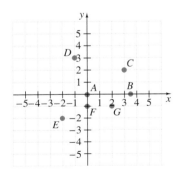

**6.**

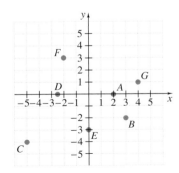

$A(2, 0); B(3, -2);$
$C(-5, -4);$
$D\left(-2\frac{1}{2}, 0\right);$
$E(0, -3); F(-2, 3);$

**6.** $G(4, 1)$

**7.** yes

**8.** no

**9.** no

**10.** yes

**11.** yes

**12.** no

**13.** yes

**14.** yes

**15.** yes

**16.** yes

**17.** no

**18.** no

**19.–22.** see graphs

**606**

**Name** _____

**B** *Determine whether each ordered pair is a solution of the given linear equation. See Example 3.*

**7.** $(0, 0)$; $y = -10x$

**8.** $(1, 7)$; $x = 7y$

**9.** $(1, 2)$; $x - y = 3$

**10.** $(-1, 9)$; $x + y = 8$

**11.** $(-2, -3)$; $y = 2x + 1$

**12.** $(1, 1)$; $y = -x$

**13.** $(2, -8)$; $y = -4x$

**14.** $(9, 1)$; $x = 9y$

**15.** $(5, 0)$; $2x + 3y = 10$

**16.** $(1, 1)$; $4x - 5y = -1$

**17.** $(3, 1)$; $x - 5y = -1$

**18.** $(0, 2)$; $x - 7y = -15$

*Plot the three ordered-pair solutions of the given equation. See Example 4.*

**19.** $2x + y = 5$; $(1, 3), (0, 5), (3, -1)$

**20.** $x + y = 5$; $(-1, 6), (0, 5), (4, 1)$

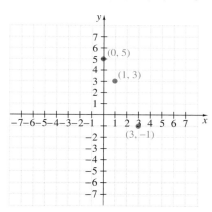

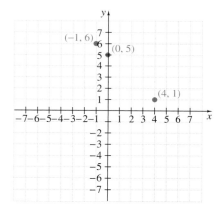

**21.** $x = 5y$; $(5, 1), (0, 0), (-5, -1)$

**22.** $y = -3x$; $(1, -3), (2, -6), (-1, 3)$

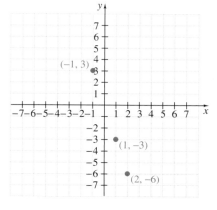

**Name** _____

**23.** $x - y = 7$; $(8, 1)$, $(2, -5)$, $(0, -7)$

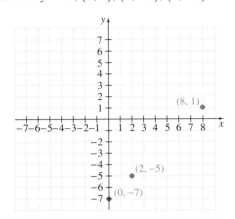

**24.** $y = 3x + 1$; $(0, 1)$, $(1, 4)$, $(-1, -2)$

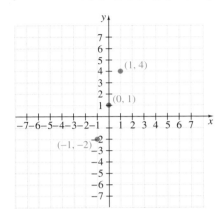

**C** *Complete each ordered-pair solution of the given equations. See Example 5.*

**25.** $y = 8x$; $(1, \quad)$, $(0, \quad)$, $(\quad, -16)$

**26.** $x = -7y$; $(\quad, 2)$, $(14, \quad)$, $(\quad, -1)$

**27.** $x + y = 14$; $(2, \quad)$, $(\quad, -8)$, $(0, \quad)$

**28.** $x - y = 8$; $(0, \quad)$, $(\quad, 0)$, $(5, \quad)$

**29.** $y = x + 5$; $(1, \quad)$, $(\quad, 7)$, $(3, \quad)$

**30.** $x = y$; $(-8, \quad)$, $(\quad, 3)$, $(100, \quad)$

**31.** $y = 3x - 5$; $(1, \quad)$, $(2, \quad)$, $(3, \quad)$

**32.** $x = -12y$; $(\quad, -1)$, $(\quad, 1)$, $(36, \quad)$

**33.** $y = -x$; $(\quad, 0)$, $(2, \quad)$, $(\quad, 2)$

**34.** $y = 5x + 1$; $(0, \quad)$, $(-1, \quad)$, $(2, \quad)$

**35.** $x + y = -2$; $(-2, \quad)$, $(1, \quad)$, $(\quad, 5)$

**36.** $x - y = -3$; $(\quad, 0)$, $(0, \quad)$, $(4, \quad)$

**Name** _____

## REVIEW AND PREVIEW

*Perform each indicated operation on decimals. See Sections 5.2 to 5.4.*

**37.** $5.6 - 3.9$　　　　　**38.** $5 + 2.54 + 8.7$　　　　　**39.** $5.6 \times 3.9$

**40.** $0.56 \div 0.8$　　　　　**41.** $(0.236)(-100)$　　　　　**42.** $44.72 \div 100$

## COMBINING CONCEPTS

*Recall that the axes divide the plane into four quadrants as shown. If a and b are both positive numbers, determine whether each statement is true or false.*

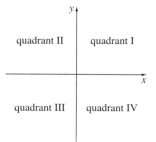

**43.** $(a, b)$ lies in quadrant I.

**44.** $(-a, -b)$ lies in quadrant IV.

**45.** $(0, b)$ lies on the $y$-axis.

**46.** $(a, 0)$ lies on the $x$-axis.

**47.** $(0, -b)$ lies on the $x$-axis.

**48.** $(-a, 0)$ lies on the $y$-axis.

**49.** $(-a, b)$ lies in quadrant III.

**50.** $(a, -b)$ lies in quadrant IV.

**51.** In your own words, describe how to plot a point.

Name _____  Section _____ Date _____

# INTEGRATED REVIEW—READING GRAPHS AND THE RECTANGULAR COORDINATE SYSTEM

*The following pictograph shows the average number of pounds of pork consumed per person per year in the United States. Use this graph to answer Exercises 1–4.*

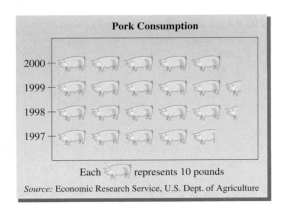

Each 🐖 represents 10 pounds

*Source:* Economic Research Service, U.S. Dept. of Agriculture

1. Approximate the number of pounds of pork consumed per person for the year 2000.

2. Approximate the number of pounds of pork consumed per person for the year 1998.

3. In what year(s) was the number of pounds consumed the greatest?

4. In what year(s) was the number of pounds consumed the least?

*The bar graph shows the tallest United States dams. Use this graph to answer Exercises 5–8.*

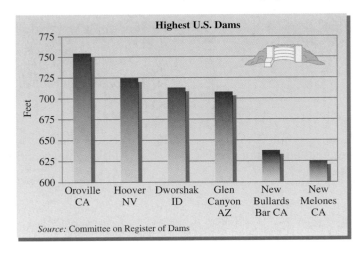

*Source:* Committee on Register of Dams

5. Name the United States dam with the greatest height and estimate this height.

6. Name the United States dam whose height is between 625 and 650 feet and estimate its height.

7. Estimate how much higher the Hoover Dam is than the Glen Canyon Dam.

8. How many United States dams have heights over 700 feet?

**610**

**Name** _____

*The line graph shows the daily high temperatures for one week in Annapolis, Maryland. Use this graph to answer Exercises 13–16.*

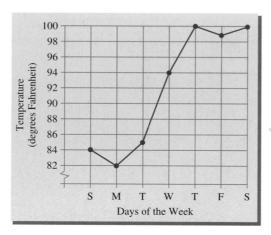

**9.** Name the day(s) of the week with the highest temperature and give that high temperature.

**10.** Name the day(s) of the week with the lowest temperature and give that low temperature.

**11.** On what days of the week was the temperature less than 90°F?

**12.** On what days of the week was the temperature greater than 90°F?

*The circle graph shows the type of beverage milk consumed in the United States. If a store in Kerrville, Texas, sells 200 quart containers of milk per week, estimate how many quart containers are sold in each category. See Exercises 13–16.*

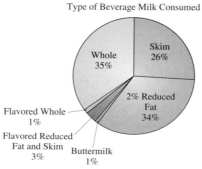

Type of Beverage Milk Consumed

Whole 35%
Skim 26%
2% Reduced Fat 34%
Flavored Whole 1%
Flavored Reduced Fat and Skim 3%
Buttermilk 1%

*Source:* U.S. Department of Agriculture

**13.** Whole milk

**14.** Skim milk

**15.** Buttermilk

**16.** Flavored Reduced Fat and Skim milk.

**17.** Plot the points corresponding to the ordered pairs on the same set of axes.
$(0, 2), (-1, 4)\ (2, 1)\ (-3, 0)\ (-3, -5)$
$(4, -1)$

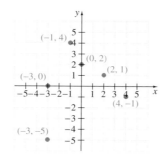

**18.** Determine whether the ordered pair $(1, 3)$ is a solution of the linear equation $x = 3y$.

**19.** Determine whether the ordered pair $(-2, -4)$ is a solution of the linear equation $x + y = -6$.

**20.** Complete each ordered pair solution of the equation $x - y = 6$.
$(0,\ \ ), (\ \ , 0), (2,\ \ )$

# 8.4 GRAPHING LINEAR EQUATIONS

Now that we know how to plot points in a rectangular coordinate system and how to find ordered-pair solutions, we are ready to graph linear equations in two variables. First, we give a formal definition.

---

**LINEAR EQUATION IN TWO VARIABLES**

A *linear equation in two variables* is an equation that can be written in the form

$$ax + by = c$$      Examples:   $\begin{cases} 2x + 3y = 6 \\ 2x = 5 \\ y = 3 \end{cases}$

where $a$, $b$, and $c$ are numbers, and $a$ and $b$ are not both 0.

---

In the last section, we realized that a linear equation in two variables has many solutions. For the linear equation $x + y = 7$, we listed, for example, the solutions $(3, 4), (0, 7), (-1, 8), (7, 0), (5, 2)$, and $(4, 3)$. Are these all the solutions? No. There are infinitely many solutions of the equation $x + y = 7$ since there are infinitely many pairs of numbers whose sum is 7. Every linear equation in two variables has infinitely many ordered-pair solutions. Since it is impossible to list every solution, we graph the solutions instead.

## A GRAPHING LINEAR EQUATIONS BY PLOTTING POINTS

Fortunately, the pattern described by the solutions of a linear equation makes "seeing" the solutions possible by graphing. This is so because **all the solutions of a linear equation in two variables correspond to points on a single straight line.** If we plot just a few of these points and draw the straight line connecting them, we have a complete graph of all the solutions.

To graph the equation $x + y = 7$, then, we plot a few ordered-pair solutions, say $(3, 4), (0, 7)$, and $(-1, 8)$.

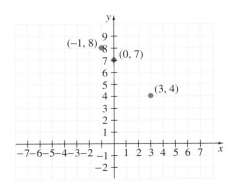

Now we connect these three points by drawing a straight line through them. The arrows at both ends of the line indicate that the line goes on forever in both directions.

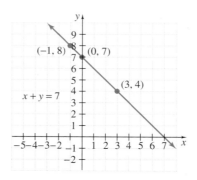

Every point on this line corresponds to an ordered-pair solution of the equation $x + y = 7$. Also, every ordered-pair solution of the equation $x + y = 7$ corresponds to a point on this line. In other words, this line is the graph of the equation $x + y = 7$. Although a line can be drawn using just two points, we will graph a third solution to check our work.

### Practice Problem 1

Graph the equation $x - y = 1$ by plotting the following points that satisfy the equation and drawing a line through the points.
$$(3, 2), (0, -1), (1, 0)$$

### Example 1

Graph the equation $y = 3x$ by plotting the following points that satisfy the equation and drawing a line through the points.
$$(1, 3), (0, 0), (-1, -3)$$

*Solution:*    Plot the points and draw a line through them. The line is the graph of the linear equation $y = 3x$.

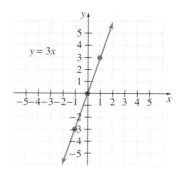

**Answer**

**1.**

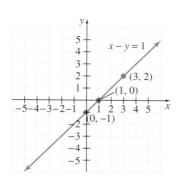

To graph a linear equation in two variables, find three ordered-pair solutions, graph the solutions, and draw the line through the plotted points. To find an ordered-pair solution of an equation, choose either an $x$-value or $y$-value of the ordered pair and complete the ordered pair as we did in the previous section.

## Example 2

Graph: $x - y = 6$

*Solution:*   Find any three ordered-pair solutions. For each solution, we choose a value for $x$ or $y$, and replace $x$ or $y$ by its chosen value in the equation. Then we solve the resulting equation for the other variable.

For example, let $x = 3$.

| | |
|---|---|
| $x - y = 6$ | Original equation |
| $3 - y = 6$ | Let $x = 3$. |
| $-y = 6 - 3$ | Subtract 3 from both sides. |
| $-y = 3$ | Simplify. |
| $\dfrac{-y}{-1} = \dfrac{3}{-1}$ | Divide both sides by $-1$. |
| $y = -3$ | Simplify. |

**The ordered pair is $(3, -3)$.**

Also, let $y = 0$.

| | |
|---|---|
| $x - y = 6$ | Original equation |
| $x - 0 = 6$ | Let $y = 0$. |
| $x = 6$ | Simplify. |

**The ordered pair is $(6, 0)$.**

If we let $x = 7$, and solve the resulting equation for $y$, we will see that $y = 1$. **A third ordered-pair solution is, then, $(7, 1)$.**

Plot the three ordered-pair solutions on a rectangular coordinate system and then draw a line through the three points. This line is the graph of $x - y = 6$.

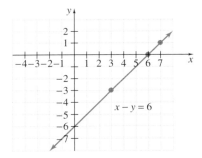

---

**Helpful Hint**

All three points should fall on the same straight line. If not, check your ordered-pair solutions for a mistake.

---

## TRY THE CONCEPT CHECK IN THE MARGIN.

## Example 3

Graph: $y = 5x + 1$

*Solution:*   For this equation, we will choose three $x$-values and find the corresponding $y$-values. The table shown will be used to list our ordered-pair solutions. We choose $x = 0$, $x = 1$, and $x = 2$.

| If $x = 0$, then | If $x = 1$, then | If $x = 2$, then |
|---|---|---|
| $y = 5x + 1$ becomes | $y = 5x + 1$ becomes | $y = 5x + 1$ becomes |
| $y = 5 \cdot 0 + 1$ or | $y = 5 \cdot 1 + 1$ or | $y = 5 \cdot 2 + 1$ or |
| $y = 1$ | $y = 6$ | $y = 11$ |

**Practice Problem 2**

Graph: $y - x = 6$

## ✓ CONCEPT CHECK

In Example 2, is the point $(9, -3)$ on the graph of $x - y = 6$?

**Practice Problem 3**

Graph: $y = -3x + 2$

Answers

**2.**

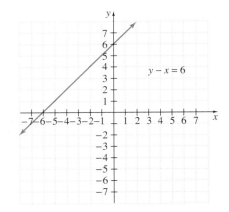

**3.**

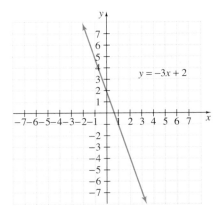

✓ Concept Check: No.

Now we complete the table of values, and graph the equation $y = 5x + 1$.

| x | y |
|---|---|
| 0 | |
| 1 | |
| 2 | |

| x | y |
|---|---|
| 0 | 1 |
| 1 | 6 |
| 2 | 11 |

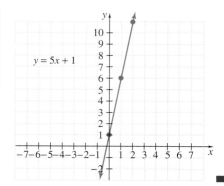

Next, we will graph a few special linear equations.

**Example 4** Graph: $y = 4$

*Solution:* The equation $y = 4$ can be written as $0x + y = 4$. When the equation is written in this form, notice that no matter what value we choose for $x$, $y$ is always 4.

$$0 \cdot x + y = 4$$
$$0 \cdot \text{(any number)} + y = 4$$
$$0 + y = 4$$
$$y = 4$$

Fill in a table listing ordered-pair solutions of $y = 4$. Choose any three $x$-values. The $y$-values must be 4. Plot the ordered-pair solutions and graph $y = 4$.

| x | y |
|---|---|
| 0 | 4 |
| 1 | 4 |
| −2 | 4 |

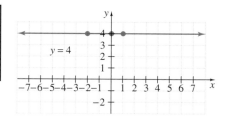

The graph is a horizontal line that crosses the $y$-axis at 4.

**Practice Problem 4**

Graph: $y = -1$

**Answer**

**4.**

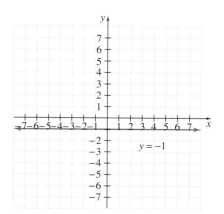

### HORIZONTAL LINES

If $a$ is a number, then the graph of $y = a$ is a *horizontal line* that crosses the $y$-axis at $a$. For example, the graph of $y = 2$ is a horizontal line that crosses the $y$-axis at 2.

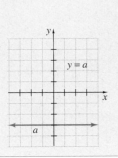

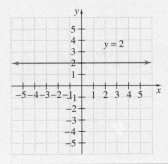

**Example 5**    Graph: $x = -2$

*Solution:*    The equation $x = -2$ can be written as $x + 0y = -2$. No matter what $y$-value we choose, $x$ is always $-2$. Fill in a table listing ordered-pair solutions of $x = -2$. Choose any three $y$-values. The $x$-values must be $-2$. Plot the ordered-pair solutions and graph $x = -2$.

| $x$ | $y$ |
|-----|-----|
| $-2$ | $3$ |
| $-2$ | $0$ |
| $-2$ | $-2$ |

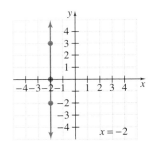

The graph is a vertical line that crosses the $x$-axis at $-2$.

**Practice Problem 5**

Graph: $x = 5$

TEACHING TIP

Remind students that they don't need to memorize that $x = a$ is a vertical line and $y = a$ is a horizontal line. When graphing $x = 2$, for example, they just need to form and graph ordered pairs whose $x$-value is 2. The line they graph will be vertical.

**Answer**

**5.**

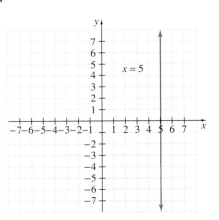

**VERTICAL LINES**

If *a* is a number, then the graph of **x = a** is a *vertical line* that crosses the *x*-axis at *a*. For example, the graph of $x = -3$ is a vertical line that crosses the *x*-axis at $-3$.

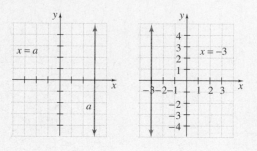

**✓ CONCEPT CHECK**

Determine whether the following equations represent vertical lines, horizontal lines, or neither.

a. $y = -6$

b. $x + y = 7$

c. $x = \dfrac{1}{2}$

d. $x = \dfrac{1}{2} y$

**TRY THE CONCEPT CHECK IN THE MARGIN.**

**Answer**

✓ Concept Check: **a.** Horizontal line.  **b.** Neither.  **c.** Vertical line.  **d.** Neither.

# EXERCISE SET 8.4

**A** *Graph each equation. See Examples 1 through 3.*

**1.** $x + y = 6$

**2.** $x + y = 7$

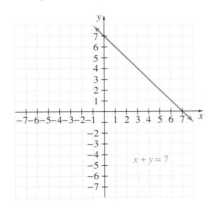

**3.** $x - y = -2$

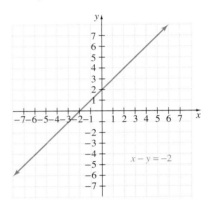

**4.** $y - x = 6$

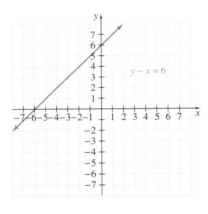

**5.** $y = 4x$

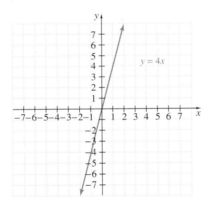

**6.** $x = 2y$

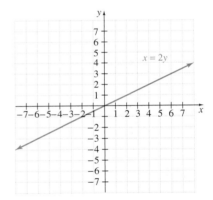

**7.** $y = 2x - 1$

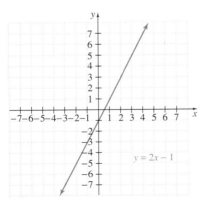

**8.** $y = x + 5$

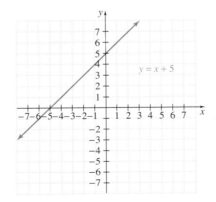

*Graph each equation. See Examples 4 and 5.*

**9.** $x = 5$

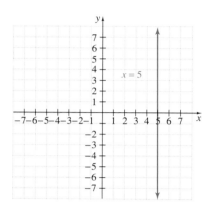

**10.** $y = 1$

**11.** $y = -3$

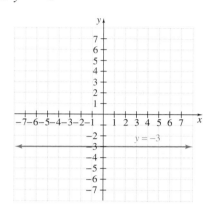

**12.** $x = -7$

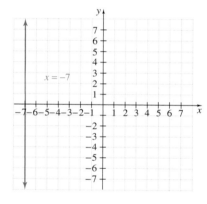

**13.** $x = 0$

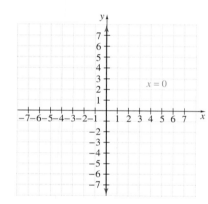

**14.** $y = 0$

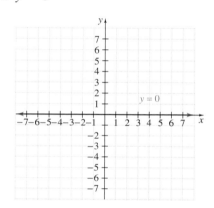

*Graph each equation. See Examples 1 through 5.*

**15.** $y = -2x$

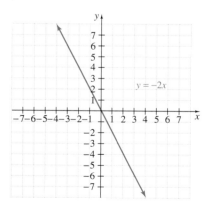

**16.** $x = y$

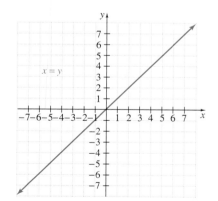

**17.** $y = -2$

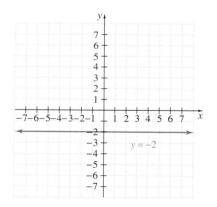

**18.** $x = 1$

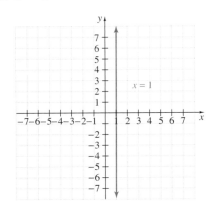

**19.** $x + 2y = 12$

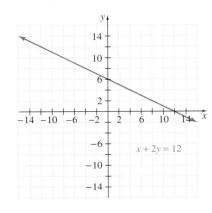

**20.** $3x - y = 3$

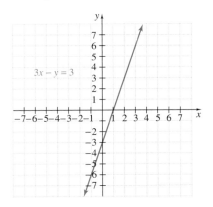

**21.** $x = 6$

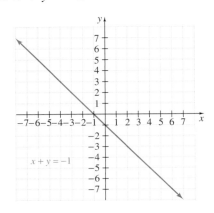

**22.** $y = x + 7$

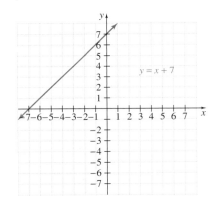

**23.** $y = x - 3$

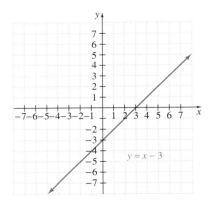

**24.** $x + y = -1$

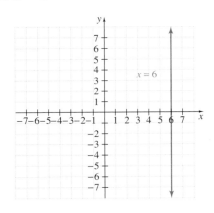

**25.** $x = y - 4$

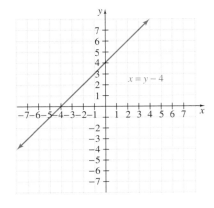

**26.** $x + y = 5$

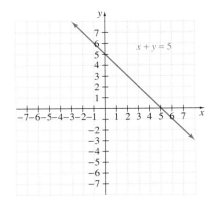

**27.** $x + 3 = 0$

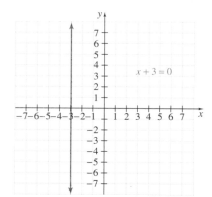

**28.** $y = -4$

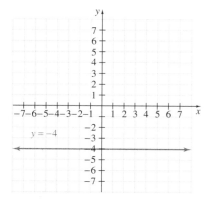

**29.** $x = 4y$

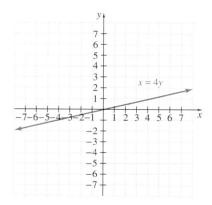

**30.** $y = -6x$

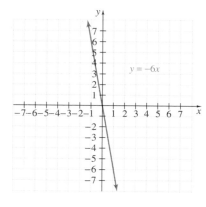

**31.** $y = \frac{1}{3}x$

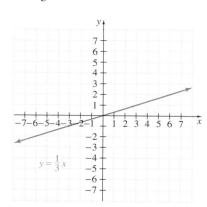

**32.** $y = -\frac{1}{3}x$

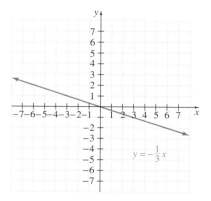

**33.** $y = 4x + 2$

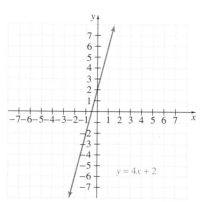

**34.** $y = -2x + 3$

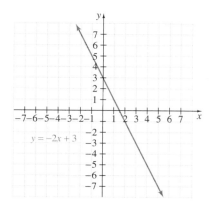

**35.** $2x + 3y = 6$

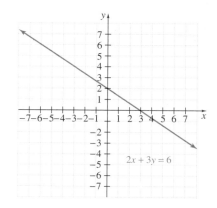

**36.** $5x - 2y = 10$

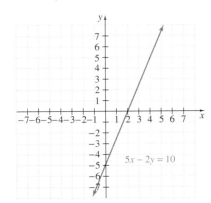

**37.** $x = -3.5$

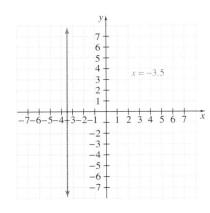

**38.** $y = 5.5$

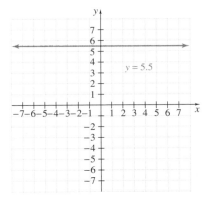

**39.** $3x - 4y = 24$

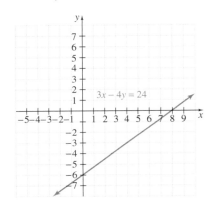

**40.** $4x + 2y = 16$

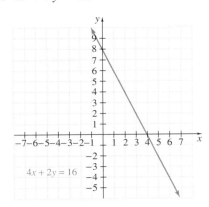

## Review and Preview

*Find the average of each list of numbers. See Section 1.6. (Recall that the average of a list of numbers is the sum of the numbers divided by the number of numbers.)*

**41.** $86, 94$

**42.** $75, 87$

**43.** $12, 28, 20$

**44.** $19, 10, 22$

**45.** $30, 22, 23, 33$

**46.** $39, 25, 31, 37$

**ANSWERS**

**41.** 90

**42.** 81

**43.** 20

**44.** 17

**45.** 27

**46.** 33

**COMBINING CONCEPTS**

**47.** see graph

**47.** Fill in the table and use it to graph $y = |x|$.

| $x$ | $y$ |
|-----|-----|
| $-3$ | 3 |
| $-2$ | 2 |
| $-1$ | 1 |
| 0 | 0 |
| 1 | 1 |
| 2 | 2 |
| 3 | 3 |

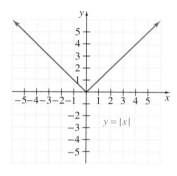

$y = |x|$

**48.** Graph: $15x - 18y = 270$

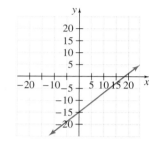

**48.** see graph

**49.** answers may vary

**49.** In your own words, explain how to graph a linear equation in two variables.

# 8.5  MEAN, MEDIAN, AND MODE

## A  FINDING THE MEAN

Sometimes we want to summarize data by displaying them in a graph, but sometimes it is also desirable to be able to describe a set of data, or a set of numbers, by a single "middle" number. Three such **measures of central tendency** are the mean, the median, and the mode.

The most common measure of central tendency is the mean (sometimes called the arithmetic mean or the average). Recall that we first introduced finding the average of a list of numbers in Section 1.6.

> ### MEAN (AVERAGE)
>
> The **mean (average)** of a set of number items is the sum of the items divided by the number of items.

**Example 1**  Seven students in a psychology class conducted an experiment on mazes. Each student was given a pencil and asked to successfully complete the same maze. The timed results are below.

| Student | Ann | Thanh | Carlos | Jesse | Melinda | Ramzi | Dayni |
|---|---|---|---|---|---|---|---|
| Time (seconds) | 13.2 | 11.8 | 10.7 | 16.2 | 15.9 | 13.8 | 18.5 |

**a.** Who completed the maze in the shortest time? Who completed the maze in the longest time?

**b.** Find the mean time.

**c.** How many students took longer than the mean time? How many students took shorter than the mean time?

*Solution:*

**a.** Carlos completed the maze in 10.7 seconds, the shortest time. Dayni completed the maze in 18.5 seconds, the longest time.

**b.** To find the mean (or average), we find the sum of the number items and divide by 7, the number of items.

$$\text{mean} = \frac{13.2 + 11.8 + 10.7 + 16.2 + 15.9 + 13.8 + 18.5}{7}$$

$$= \frac{100.1}{7} = 14.3$$

**c.** Three students, Jesse, Melinda, and Dayni, had times longer than the mean time. Four students, Ann, Thanh, Carlos, and Ramzi, had times shorter than the mean time.

**TRY THE CONCEPT CHECK IN THE MARGIN.**

Often in college, the calculation of a **grade point average** (GPA) is a **weighted mean** and is calculated as shown in Example 2.

---

**Practice Problem 1**

Find the mean of the following test scores: 77, 85, 86, 91, and 88.

✓ **CONCEPT CHECK**

Estimate the mean of the following set of data:
5, 10, 10, 10, 10, 15

Answers

**1.** 85.4
✓ Concept Check: 10

## Practice Problem 2

Find the grade point average if the following grades were earned in one semester.

| Grade | Credit Hours |
|-------|--------------|
| A | 2 |
| C | 4 |
| B | 5 |
| D | 2 |
| A | 2 |

**TEACHING TIP**

Point out that in the weighted mean the grade is weighted so a grade for a 3 credit hour class counts three times as much as a grade for a 1 credit hour class. An A in a 3 credit hour class is the same as As in three 1 credit hour classes.

## Example 2

The following grades were earned by a student during one semester. Find the student's grade point average.

| Course | Grade | Credit Hours |
|--------|-------|--------------|
| College Mathematics | A | 3 |
| Biology | B | 3 |
| English | A | 3 |
| PE | C | 1 |
| Social Studies | D | 2 |

*Solution:*   To calculate the grade point average, we need to know the point values for the different possible grades. The point values of grades commonly used in colleges and universities are given below.

A: 4,    B: 3,    C: 2,    D: 1,    F: 0

Now, to find the grade point average, we multiply the number of credit hours for each course by the point value of each grade. The grade point average is the sum of these products divided by the sum of the credit hours.

| Course | Grade | Point Value of Grade | Credit Hours | (Point Value) · (Credit Hours) |
|--------|-------|---------------------|--------------|-------------------------------|
| College Mathematics | A | 4 | 3 | 12 |
| Biology | B | 3 | 3 | 9 |
| English | A | 4 | 3 | 12 |
| PE | C | 2 | 1 | 2 |
| Social Studies | D | 1 | 2 | 2 |
| | | | Totals: 12 | 37 |

$$\text{grade point average} = \frac{37}{12} \approx 3.08 \text{ rounded to two decimal places}$$

The student earned a grade point average of 3.08.

### B   FINDING THE MEDIAN

You may have noticed that a very low number or a very high number can affect the mean of a list of numbers. Because of this, you may sometimes want to use another measure of central tendency, called the **median**. The median of a list of numbers is not affected by a low or high number in the list.

> **MEDIAN**
>
> The **median** of an ordered set of numbers is the middle number. If the number of items is even, the median is the mean of the two middle numbers.

## Practice Problem 3

Find the median of the list of numbers:
7, 9, 13, 23, 24, 35, 38, 41, 43

## Example 3

Find the median of the list of numbers:

25, 54, 56, 57, 60, 71, 98

*Solution:*   Because this list is in numerical order, the median is the middle number, 57.

25, 54, 56, 57, 60, 71, 78
                ↑

**Answers**
**2.** 2.73   **3.** 24

## Example 4

Find the median of the list of scores: 67, 91, 75, 86, 55, 91

*Solution:*   First we list the scores in numerical order and then find the middle number.

55, 67, 75, 86, 91, 91

Since there is an even number of scores, there are two middle numbers. The median is the mean of the two middle numbers.

$$\text{median} = \frac{75 + 86}{2} = 80.5$$

The median is 80.5.

**TRY THE CONCEPT CHECK IN THE MARGIN.**

## C  FINDING THE MODE

The last common measure of central tendency is called the **mode**.

> **MODE**
>
> The **mode** of a set of numbers is the number that occurs most often. (It is possible for a set of numbers to have more than one mode or to have no mode.)

## Example 5

Find the mode of the list of numbers:

11, 14, 14, 16, 31, 56, 65, 77, 77, 78, 79

*Solution:*   There are two numbers that occur the most often. They are 14 and 77. This list of numbers has two modes, 14 and 77.

---

**Practice Problem 4**

Find the median of the list of scores:
43, 89, 78, 65, 95, 95, 88, 71

TEACHING TIP    Classroom Activity

In groups, have students discuss the following:
  If you managed a shoe store, which measure of central tendency would be more useful to you? Explain.
  If you coached baseball, which measure of central tendency would be most useful to you? Explain.
Then have the groups share their conclusions with the class.

### ✓ CONCEPT CHECK

Without making any computations, decide whether the median of the following list of numbers will be a whole number. Explain your reasoning.

36, 77, 29, 58, 43

**Practice Problem 5**

Find the mode of the list of numbers:
9, 10, 10, 13, 15, 15, 15, 17, 18, 18, 20

Answers

**4.** 83   **5.** 15

✓ Concept Check:  Yes. Answers may vary.

## Practice Problem 6

Find the median and the mode of the list of numbers:

26, 31, 15, 15, 26, 30, 16, 18, 15, 35

---

## ✓ CONCEPT CHECK

True or False? Every set of numbers *must* have a mean, median, and mode. Explain your answer.

---

**Example 6**    Find the median and the mode of the following set of numbers. These numbers were high temperatures for fourteen consecutive days in a city in Montana.

76, 80, 85, 86, 89, 87, 82, 77, 76, 79, 82, 89, 89, 92

*Solution:*    First we write the numbers in numerical order.

76, 76, 77, 79, 80, 82, 82, 85, 86, 87, 89, 89, 89, 92

Since there is an even number of items, the median is the mean of the two middle numbers.

$$\text{median} = \frac{82 + 85}{2} = 83.5$$

The mode is 89, since 89 occurs most often.    ▬▬

**TRY THE CONCEPT CHECK IN THE MARGIN.**

---

**Helpful Hint**

Don't forget that it is possible for a list of numbers to have no mode. For example, the list

2, 4, 5, 6, 8, 9

has no mode. There is no number or numbers that occur more often than the others.

---

**Answers**

**6.** median: 22; mode: 15

✓ **Concept Check:**  False; a set of numbers may have no mode.

Name _____ Section _____ Date _____

## Mental Math

*State the mean for each list of numbers.*

**1.** 3, 5     **2.** 10, 20     **3.** 1, 3, 5     **4.** 7, 7, 7

# Exercise Set 8.5

**A B C** *For each set of numbers, find the mean, the median, and the mode. If necessary, round the mean to one decimal place. See Examples 1 and 3 through 6.*

**1.** 21, 28, 16, 42, 38

**2.** 42, 35, 36, 40, 50

**3.** 7.6, 8.2, 8.2, 9.6, 5.7, 9.1

**4.** 4.9, 7.1, 6.8, 6.8, 5.3, 4.9

**5.** 0.2, 0.3, 0.5, 0.6, 0.6, 0.9, 0.2, 0.7, 1.1

**6.** 0.6, 0.6, 0.8, 0.4, 0.5, 0.3, 0.7, 0.8, 0.1

**7.** 231, 543, 601, 293, 588, 109, 334, 268

**8.** 451, 356, 478, 776, 892, 500, 467, 780

*The ten tallest buildings in the United States are listed in the table. Use this table to answer Exercises 9–12. If necessary, round results to one decimal place. See Examples 1 and 3 through 6.*

| Building | Height (in feet) |
|---|---|
| Sears Tower, Chicago | 1450 |
| One World Trade Center (1972), New York | 1368 |
| Two World Trade Center (1973), New York | 1362 |
| Empire State Building, New York | 1250 |
| Amoco, Chicago | 1136 |
| John Hancock Center, Chicago | 1127 |
| Stratosphere Tower, Las Vegas | 1049 |
| Chrysler Building, New York | 1046 |
| NationsBank Tower, Atlanta | 1023 |
| First Interstate World Center, Los Angeles | 1018 |

(*Source: World Almanac, 1998*)

**9.** Find the mean height for the five tallest buildings.

**10.** Find the median height for the five tallest buildings.

**11.** Find the median height for the ten tallest buildings.

**12.** Find the mean height for the ten tallest buildings.

**627**

13. 2.79

14. 2.18

15. 3.46

16. 3.0

17. 6.8

18. 6.95

19. 6.9

20. 84.67

21. 85.5

22. no mode

23. 73

24. 71

25. 70 and 71

26. 6 rates

27. 9 rates

*For Exercises 13–16, the grades are given for a student for a particular semester. Find the grade point average. If necessary, round the grade point average to the nearest hundredth. See Example 2.*

**13.**

| Grade | Credit Hours |
|-------|--------------|
| B | 3 |
| C | 3 |
| A | 4 |
| C | 4 |

**14.**

| Grade | Credit Hours |
|-------|--------------|
| D | 1 |
| F | 1 |
| C | 4 |
| B | 5 |

**15.**

| Grade | Credit Hours |
|-------|--------------|
| A | 3 |
| A | 3 |
| B | 4 |
| B | 1 |
| B | 2 |

**16.**

| Grade | Credit Hours |
|-------|--------------|
| B | 2 |
| B | 2 |
| A | 3 |
| C | 3 |
| B | 3 |

*During an experiment, the following times (in seconds) were recorded: 7.8, 6.9, 7.5, 4.7, 6.9, 7.0.*

**17.** Find the mean. Round to the nearest tenth.

**18.** Find the median.

**19.** Find the mode.

*In a mathematics class, the following test scores were recorded for a student: 86, 95, 91, 74, 77, 85.*

**20.** Find the mean. Round to the nearest hundredth.

**21.** Find the median.

**22.** Find the mode.

*The following pulse rates were recorded for a group of 15 students: 78, 80, 66, 68, 71, 64, 82, 71, 70, 65, 70, 75, 77, 86, 72.*

**23.** Find the mean.

**24.** Find the median.

**25.** Find the mode.

**26.** How many rates were higher than the mean?

**27.** How many rates were lower than the mean?

**Name** _____

## REVIEW AND PREVIEW

*Write each fraction in simplest form. See Section 4.2.*

**28.** $\dfrac{12}{20}$

**29.** $\dfrac{6}{18}$

**30.** $\dfrac{4}{36}$

**31.** $\dfrac{18}{30}$

**32.** $\dfrac{35}{100}$

**33.** $\dfrac{55}{75}$

## ◢ COMBINING CONCEPTS

*Find the missing numbers in each set of numbers.*

**34.** 16, 18, _____, _____, _____.
The mode is 21.
The median is 20.

**35.** _____, _____, _____, 40, _____.
The mode is 35.
The median is 37.
The mean is 38.

**36.** Write a list of numbers for which you feel the median would be a better measure of central tendency than the mean. Explain why.

**28.** $\dfrac{3}{5}$ _____

**29.** $\dfrac{1}{3}$ _____

**30.** $\dfrac{1}{9}$ _____

**31.** $\dfrac{3}{5}$ _____

**32.** $\dfrac{7}{20}$ _____

**33.** $\dfrac{11}{15}$ _____

**34.** 20, 21, 21 _____

**35.** 35, 35, 37, 43 _____

**36.** answers may vary _____

# Focus on Mathematical Connections

### RANGE

In addition to measures of central tendency, *measures of dispersion* can also help to describe or summarize a set of data. These measures indicate how "spread out" the numbers are in a set of data. One measure of dispersion is the **range** of a set of data. The range is computed as the difference between the largest and the smallest number in the data set.

Take, for example, the data set 39, 27, 30, 66, 63, and 57. The smallest number in the set is 27 and the largest number in the set is 66. The difference between these two numbers is $66 - 27 = 39$. Thus, the range of the data is 39.

What can the range tell us about a set of data? Suppose we have two different sets of data. The median of the first set is 18 and the median of the second set is also 18. So far, the data sets seem to be similar. But if we are then told that the range of the first set is 2 and the range of the second set is 28, we know that the two sets of data are likely to be quite different. In the first set, the numbers are clustered very closely to the median, 18. In the second set, the numbers are probably spread out much farther, covering a wider slice of values. In fact, the first set are ages of the students in a high school math class and the second set are ages of the students in a college math class (see below). We would expect the age makeup of these two classes to be quite different, and comparing the ranges of the ages confirms this.

**Ages in High School Math Class**

17  17  17  17  17  17  17  18  18  18  18  18  18  18  18  18  18  18  18  19

Median:  18     Range:  2

**Ages in College Math Class:**

18  18  18  18  18  18  18  18  18  18  18  18  19  19  20  21  23  27  33  41  46

Median:  18     Range:  28

### GROUP ACTIVITY

Collect class grade information for two different tests or quizzes in this class or another class; both sets of grades should be for the same class. Find the mean, median, mode, and range for each set of grades. Compare statistics for each set of grades. What can you conclude?

# 8.6   COUNTING AND INTRODUCTION TO PROBABILITY

## A   USING A TREE DIAGRAM

In our daily conversations, we often talk about the likelihood or the probability of a given result occurring. For example:

The *chance* of thundershowers is 70 percent.

What are the *odds* that the Saints will go to the Super Bowl?

What is the *probability* that you will finish cleaning your room today?

Each of these chance happenings—thundershowers, the Saints playing in the Super Bowl, and cleaning your room today—is called an **experiment**. The possible results of an experiment are called **outcomes**. For example, flipping a coin is an experiment and the possible outcomes are heads (H) or tails (T).

One way to picture the outcomes of an experiment is to draw a *tree diagram*. Each outcome is shown on a separate branch. For example, the outcomes of flipping a coin are

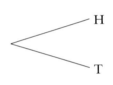          

Head                    Tail

**Example 1**   Draw a tree diagram for tossing a coin twice. Then use the diagram to find the number of possible outcomes.

*Solution:*   There are 4 possible outcomes when tossing a coin twice.

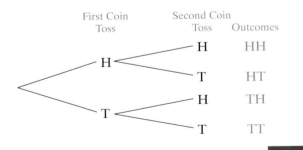

### Objectives

**A** Use a tree diagram to count outcomes.

**B** Find the probability of an event.

Study Guide    SSM    CD-ROM    Video 8.6

**TEACHING TIP**

Have students make a row at the bottom of their tree diagram with the following information:

|  | First Coin Toss | Second Coin Toss | Both Coin Tosses Together |
|---|---|---|---|
| Number of Possible Outcomes | 2 | 2 | 4 |

**Practice Problem 1**

Draw a tree diagram for tossing a coin three times. Then use the diagram to find the number of possible outcomes.

**Answer**

**1.**

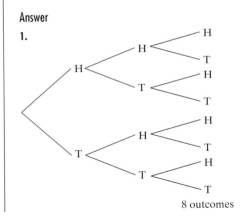

8 outcomes

## Practice Problem 2

Draw a tree diagram for an experiment consisting of tossing a coin and then rolling a die. Then use the diagram to find the number of possible outcomes.

TEACHING TIP

Have students make a row at the bottom of their tree diagram with the following information:

|  | Die Roll | Coin Toss | Die Roll & Coin Toss Together |
|---|---|---|---|
| Number of Possible Outcomes | 6 | 2 | 12 |

**Answer**

**3.**

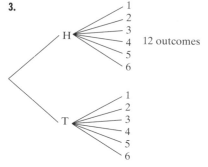

## Example 2

Draw a tree diagram for an experiment consisting of rolling a die and then tossing a coin. Then use the diagram to find the number of possible outcomes.

Die

*Solution:*    Recall that a die has six sides and each side represents a number, 1 through 6.

| Roll a Die | Toss a coin | Outcomes |
|---|---|---|
| 1 | H | 1, H |
|  | T | 1, T |
| 2 | H | 2, H |
|  | T | 2, T |
| 3 | H | 3, H |
|  | T | 3, T |
| 4 | H | 4, H |
|  | T | 4, T |
| 5 | H | 5, H |
|  | T | 5, T |
| 6 | H | 6, H |
|  | T | 6, T |

There are 12 possible outcomes for rolling a die and then tossing a coin.

Any number of outcomes considered together is called an **event**. For example, when tossing a coin twice, HH is an event. The event is tossing a head first and then tossing a head second. Another event would be tossing a tail first and then a head (TH), and so on.

### B   FINDING THE PROBABILITY OF AN EVENT

As we mentioned earlier, the *probability* of an event is a measure of the chance or likelihood of it occuring. For example, if a coin is tossed, what is the probability that a head occurs? Since 1 of 2 equally likely possible outcomes is heads, the probability is $\frac{1}{2}$.

---

**THE PROBABILITY OF AN EVENT**

$$\text{probability of an event} = \frac{\text{number of ways that the event can occur}}{\text{number of possible outcomes}}$$

---

**Helpful Hint**

Note from the definition of probability that the probability of an event is always between 0 and 1, inclusive (including 0 and 1). A probability of 0 means an event won't occur and a probability of 1 means that an event is certain to occur.

**TEACHING TIP**

Ask students to describe an event having a probability of 0 using the roll of a die. If they are stumped, give them a hint such as, "Is there any number you are certain you will not roll?" Then ask students to describe an event using the roll of a die with a probability of 1.

---

**Example 3**   If a coin is tossed twice, find the probability of tossing a head and then a head (HH).

*Solution:*

1 way the event can occur

$$\text{HT, HH, TH, TT}$$

4 possible outcomes

$$\text{probability} = \frac{1}{4} \quad \begin{array}{l}\text{Number of ways the event can occur}\\ \text{Number of possible outcomes}\end{array}$$

The probability of tossing a head and then a head is $\frac{1}{4}$.

**Practice Problem 3**

If a coin is tossed three times, find the probability of tossing a head, then a tail, then a tail (HTT).

---

**Example 4**   If a die is rolled one time, find the probability of rolling a 3 or a 4.

*Solution:*   Recall that there are 6 possible outcomes when rolling a die.

2 ways that the event can occur

$$\text{possible outcomes:} \quad 1, \quad 2, \quad 3, \quad 4, \quad 5, \quad 6$$

6 possible outcomes

$$\text{probability of a 3 or a 4} = \frac{2}{6} \quad \begin{array}{l}\text{Number of ways the event}\\ \text{can occur}\\ \text{Number possible outcomes}\end{array}$$

$$= \frac{1}{3} \quad \text{Simplest form}$$

**Practice Problem 4**

If a die is rolled one time, find the probability of rolling a 1 or a 2.

**Answers**

**3.** $\frac{1}{8}$   **4.** $\frac{1}{3}$

✓ **CONCEPT CHECK**

Suppose you have calculated a probability of $\frac{11}{9}$. How do you know that you have made an error in your calculation?

**Practice Problem 5**

Use the diagram from Example 5 and find the probability of choosing a blue marble from the box.

**TRY THE CONCEPT CHECK IN THE MARGIN.**

**Example 5**    Find the probability of choosing a red marble from a box containing 1 red, 1 yellow, and 2 blue marbles.

*Solution:*

1 way that event can occur

yellow    blue    blue    red

4 possible outcomes

$$\text{probability} = \frac{1}{4}$$

✓ **Concept Check:** The number of ways an event can occur can't be larger than the number of possible outcomes.

**Name** _____ **Section** _____ **Date** _____

## MENTAL MATH

*If a coin is tossed once, find the probability of each event.*

**1.** The coin lands heads up.

**2.** The coin lands tails up.

*If the spinner shown is spun once, find the probability of each event.*

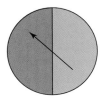

**3.** The spinner stops on red.

**4.** The spinner stops on blue.

## EXERCISE SET 8.6

**A** *Draw a tree diagram for each experiment. Then use the diagram to find the number of possible outcomes. See Examples 1 and 2.*

**1.** Choose a vowel, a, e, i, o, u, and then a number, 1, 2, or 3.

**2.** Choose a number 1 or 2 and then a vowel, a, e, i, o, u.

**3.** Spin Spinner A once.

**4.** Spin Spinner B once.

**5.** Spin Spinner B twice.

**6.** Spin Spinner A twice.

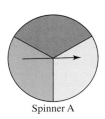
Spinner A

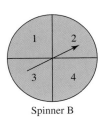
Spinner B

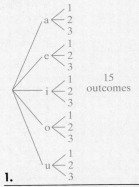

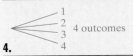
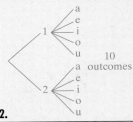
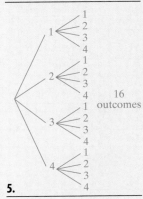
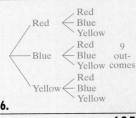
**635**

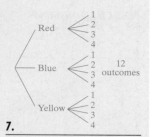

**7.**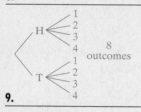

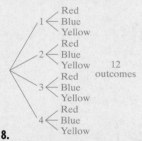

**8.** 

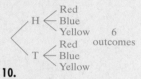

**9.** 

**10.** 

**11.** $\dfrac{1}{6}$

**12.** 0

**13.** $\dfrac{1}{3}$

**14.** $\dfrac{1}{3}$

**15.** $\dfrac{1}{2}$

**16.** $\dfrac{1}{2}$

**17.** $\dfrac{1}{3}$

**18.** $\dfrac{1}{3}$

**19.** $\dfrac{2}{3}$

**20.** $\dfrac{1}{3}$

**7.** Spin Spinner A then Spinner B.

**8.** Spin Spinner B then Spinner A.

**9.** Toss a coin and then spin Spinner B.

**10.** Toss a coin and then spin Spinner A.

**B** *If a single die is tossed once, find the probability of each event. See Examples 3 through 5.*

**11.** A 5

**12.** A 7

**13.** A 1 or a 4

**14.** A 2 or a 3

**15.** An even number

**16.** An odd number

*Suppose the spinner shown is spun once. Find the probability of each event. See Examples 3 through 5.*

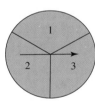

**17.** The result of the spin is 2.

**18.** The result of the spin is 3.

**19.** The result of the spin is an odd number.

**20.** The result of the spin is an even number.

**Name** _____

21. $\dfrac{1}{7}$ _____

*If a single choice is made from the bag of marbles shown, find the probability of each event. See Examples 3 through 5.*

22. $\dfrac{1}{7}$ _____

**21.** A red marble is chosen.

**22.** A blue marble is chosen.

23. $\dfrac{2}{7}$ _____

24. $\dfrac{3}{7}$ _____

**23.** A yellow marble is chosen.

**24.** A green marble is chosen.

25. $\dfrac{5}{6}$ _____

## REVIEW AND PREVIEW

*Perform each indicated operation. See Sections 4.3 and 4.5.*

26. $\dfrac{3}{10}$ _____

**25.** $\dfrac{1}{2} + \dfrac{1}{3}$

**26.** $\dfrac{7}{10} - \dfrac{2}{5}$

27. $\dfrac{1}{6}$ _____

**27.** $\dfrac{1}{2} \cdot \dfrac{1}{3}$

**28.** $\dfrac{7}{10} \div \dfrac{2}{5}$

28. $\dfrac{7}{4}$ or $1\dfrac{3}{4}$ _____

29. $\dfrac{20}{3}$ or $6\dfrac{2}{3}$ _____

**29.** $5 \div \dfrac{3}{4}$

**30.** $\dfrac{3}{5} \cdot 10$

30. 6 _____

**Name** _____

 **COMBINING CONCEPTS**

*Recall that a deck of cards contains 52 cards. These cards consist of four suits (hearts, spades, clubs, and diamonds) containing each of the following: 2, 3, 4, 5, 6, 7, 8, 9, 10, Jack, Queen, and King, and Ace. If a card is chosen from a deck of cards, find the probability of each event.*

**31.** The King of hearts

**32.** The 10 of spades

**33.** A King

**34.** A 10

**35.** A heart

**36.** A club

*Two dice are tossed. Find the probability of each sum of the dice. (Hint: Draw a tree diagram of the possibilities of two tosses of a die and then find the sum of the numbers on each branch.)*

**37.** A sum of 4

**38.** A sum of 11

**39.** A sum of 13

**40.** A sum of 2

**41.** Can the probability of an event be $\frac{3}{2}$? Can the probability of an event ever be greater than 1? Explain why.

# CHAPTER 8 ACTIVITY
## CONDUCTING A SURVEY

*This activity may be completed by working in groups or individually.*

How often have you read an article in a newspaper or in a magazine that included results from a survey or poll? Surveys seem to have become very popular ways of getting feedback on anything from a political candidate, to a new product, to services offered by a health club. In this activity, you will conduct a survey and analyze the results.

**1.** Conduct a survey of 30 students in one of your classes. Ask each student to report his or her age.

**2.** Classify each age according to the following categories: under 20, 20 to 24, 25 to 29, 30 to 39, 40 to 49, and 50 or over. Tally the number of your survey respondents that fall into each category. Make a bar graph of your results. What does this graph tell you about the ages of your survey respondents?

**3.** Find the average age of your survey respondents.

**4.** Find the median age of your survey respondents.

**5.** Find the mode of the ages of your survey respondents.

**6.** Compare the mean, median, and mode of your age data. Are these measures similar? Which is largest? Which is smallest? If there is a noticeable difference between any of these measures, can you explain why?

# CHAPTER 8 HIGHLIGHTS

| DEFINITIONS AND CONCEPTS | EXAMPLES |
|---|---|

### SECTION 8.1   READING CIRCLE GRAPHS

| | |
|---|---|
| In a **circle graph**, each section (shaped like a piece of pie) shows a category and the relative size of the category. | The circle graph classifies tornadoes by wind speed. 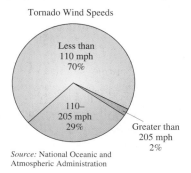 Tornado Wind Speeds<br><br>*Source:* National Oceanic and Atmospheric Administration<br><br>**1.** What percent of tornadoes have wind speeds of 110 mph or greater?<br><br>$29\% + 2\% = 31\%$<br><br>**2.** If there were 1235 tornadoes in the United States in 1995, how many of these might we expect to have had wind speeds less than 110 mph? Find 70% of 1235.<br><br>$70\% (1235) = 0.70(1235) = 864.5 \approx 865$<br><br>Around 865 tornadoes would be expected to have had wind speeds of less than 110 mph. |

### SECTION 8.2   READING PICTOGRAPHS, BAR GRAPHS, AND LINE GRAPHS

| | |
|---|---|
| A **pictograph** is a graph in which pictures or symbols are used to visually present data.<br>A **bar graph** presents data using vertical or horizontal bars.<br>A **line graph** displays information with a line that connects data points. | The bar graph shows the number of acres of wheat harvested in 1998 for leading states. 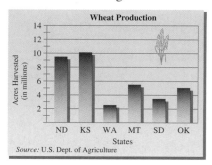 *Source:* U.S. Dept. of Agriculture<br><br>**1.** Approximately how many acres of wheat were harvested in Kansas?<br><br>10,000,000 acres<br><br>**2.** About how many more acres of wheat were harvested in North Dakota than South Dakota?<br>$\phantom{-}9.5 \text{ million}$<br>$\underline{-3.5 \text{ million}}$<br>$\phantom{-}6 \text{ million} \quad \text{or} \quad 6{,}000{,}000 \text{ acres}$ |

## SECTION 8.3    THE RECTANGULAR COORDINATE SYSTEM

The **rectangular coordinate system** consists of two number lines intersecting at the point 0 on each number line. The horizontal number line is called the *x*-axis and the vertical number line is called the *y*-axis.

Every point in the rectangular coordinate system corresponds to an **ordered pair of numbers** such as

$$(2, -2)$$
$$\nearrow \quad \nwarrow$$

*x*-coordinate    *y*-coordinate
or *x*-value      or *y*-value

An ordered pair is a **solution** of an equation if the equation is a true statement when the variables are replaced by the coordinates of the ordered pair.

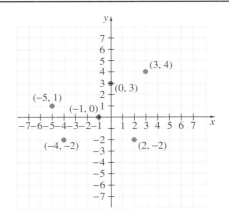

Is $(2, -1)$ a solution of $5x - y = 11$?

$$5x - y = 11$$

$5(2) - (-1) = 11$    Replace *x* with 2 and *y* with −1.

$10 + 1 = 11$    Multiply.

$11 = 11$    True.

Yes, $(2, -1)$ is a solution of $5x - y = 11$.

## SECTION 8.4    GRAPHING LINEAR EQUATIONS

LINEAR EQUATIONS IN TWO VARIABLES

A linear equation in two variables is an equation that can be written in the form

$$ax + by = c$$

where $a, b,$ and $c$ are numbers, and $a$ and $b$ are not both 0.

**To graph a linear equation in two variables**, find three ordered-pair solutions and draw the line through the plotted points.

Graph: $y = 4x$

Let $x = 1$.         Let $x = -1$.        Let $x = 0$.
$y = 4(1)$          $y = 4(-1)$          $y = 4(0)$
$y = 4$             $y = -4$             $y = 0$

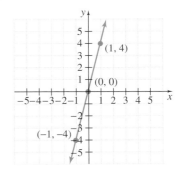

| x  | y  |
|----|----|
| 1  | 4  |
| −1 | −4 |
| 0  | 0  |

The graph of $y = a$ is a **horizontal line** that crosses the *y*-axis at *a*.
The graph of $x = a$ is a **vertical line** that crosses the *x*-axis at *a*.

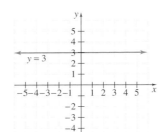

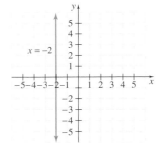

## SECTION 8.5    MEAN, MEDIAN, AND MODE

The **mean** (or **average**) of a set of number items is

$$\text{mean} = \frac{\text{sum of items}}{\text{number of items}}$$

The **median** of an ordered set of numbers is the middle number. If the number of items is even, the median is the mean of the two middle numbers.

The **mode** of a set of numbers is the number that occurs most often. (A set of numbers may have no mode or more than one mode.)

Find the mean, median, and mode of the set of numbers: 33, 35, 35, 43, 68, 68

$$\text{mean} = \frac{33 + 35 + 35 + 43 + 68 + 68}{6} = 47$$

The median is the mean of the two middle numbers:

$$\text{median} = \frac{35 + 43}{2} = 39$$

There are two modes because there are two numbers that both occur twice:

modes: 35 and 68

## SECTION 8.6    COUNTING AND INTRODUCTION TO PROBABILITY

An **experiment** is an activity being considered, such as tossing a coin or rolling a die. The possible results of an experiment are the **outcomes**. A **tree diagram** is one way to picture and count outcomes.

Draw a tree diagram for tossing a coin and then choosing a number from 1 to 4.

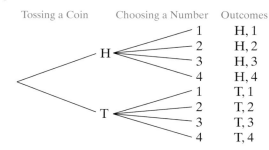

Any number of outcomes considered together is called an **event**. The **probability** of an event is a measure of the chance or likelihood of it occurring.

$$\text{probability of an event} = \frac{\text{number of ways that the event can occur}}{\text{number of possible outcomes}}$$

Find the probability of tossing a coin twice and a tail occurring each time.

1 way the event can occur

$$\underbrace{\text{HH} \quad \text{HT} \quad \text{TH} \quad \text{TT}}_{\text{4 possible outcomes}}$$

$$\text{probability} = \frac{1}{4}$$

# CHAPTER 8 REVIEW

**(8.1)** *The circle graph shows a family's $4000 monthly budget. Use this graph to answer Exercises 1–6. Write all ratios as fractions in simplest form.*

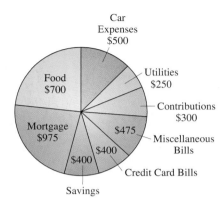

**1.** What is the largest budget item?   Mortgage

**2.** What is the smallest budget item?   Utilities

**3.** How much money is budgeted for either the mortgage or utilities?   $1225

**4.** How much money is budgeted for either savings or contributions?   $700

**5.** Find the ratio of mortgage to the total monthly budget.
$\frac{39}{160}$

**6.** Find the ratio of food budget to the total monthly budget.   $\frac{7}{40}$

*The circle graph shows the percent of states with various interstate highway speed limits. Use this graph to determine the number of states with each speed limit in Exercises 7–10.*

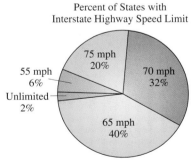

*Source:* National Motorist Association

**7.** How many states have an interstate highway speed limit of 65 mph?   20 states

**8.** How many states have an interstate highway speed limit of 75 mph?   10 states

**9.** How many states have an interstate highway speed limit of either 70 mph or 75 mph?   26 states

**10.** How many states have an interstate highway speed limit of either 55 mph or 65 mph?   23 states

**(8.2)** *The pictograph shows the number of new homes constructed, by state. Use this graph to answer Exercises 11–16.*

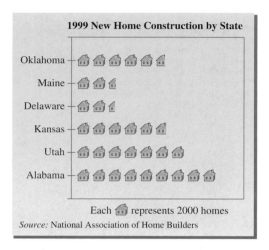

**1999 New Home Construction by State**

Each 🏠 represents 2000 homes

*Source:* National Association of Home Builders

**11.** How many new homes were constructed in Oklahoma?   11,500 homes

**12.** How many new homes were constructed in Kansas?   11,500 homes

**13.** Which state shown had the most new homes constructed?   Alabama

**14.** Which state shown had the fewest new homes constructed?   Delaware

**15.** Which state(s) shown had more than 13,000 new homes constructed?   Utah and Alabama

**16.** Which state(s) shown had fewer than 8000 new homes constructed?   Maine and Delaware

*The bar graph shows percent of persons age 25 or more who completed four or more years of college. Use this graph to answer Exercises 17–20.*

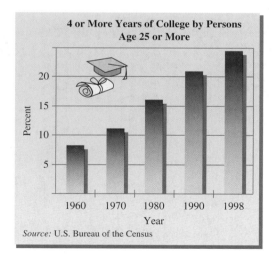

**4 or More Years of College by Persons Age 25 or More**

Percent

Year

*Source:* U.S. Bureau of the Census

**17.** Approximate the percent of persons who completed four or more years of college in 1960.   8%

**18.** What year shown had the greatest percent of persons completing four or more years of college?   1998

**19.** What years shown had 15% or more of persons completing four or more years of college?   1980, 1990, 1998

**20.** Describe any patterns you notice in this graph.   answers may vary

*The double line graph shows the U.S. meat consumption for red meats and poultry. Use this graph to answer Exercises 21–24.*

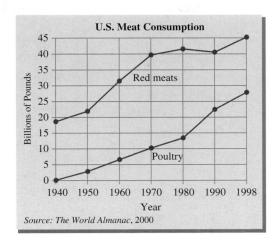

**21.** Approximate the number of pounds of poultry eaten in 1970.   10 billion lb

**22.** Approximate the number of pounds of red meat eaten in 1970.   39.5 billion lb

**23.** In what 10-year period does the graph show the greatest increase in the number of pounds of poultry eaten?   1980–1990

**24.** In what 10-year period does the graph show the smallest increase in the number of pounds of poultry eaten?   1940–1950

**(8.3)** *Complete and graph the ordered-pair solutions of each given equation.*

**25.** $x = -7y$; $(0, \ )$, $( \ , -1)$, $(-7, \ )$

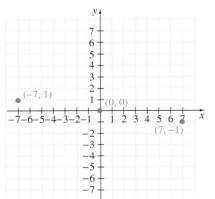

$(0, 0)$,
$(7, -1)$,
$(-7, 1)$

**26.** $y = 3x - 2$; $(0, \ )$, $(1, \ )$, $(-2, \ )$

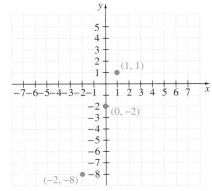

$(0, -2)$,
$(1, 1)$,
$(-2, -8)$

**Name** _____

**27.** $x + y = -9$; $(-1, \ )$, $( \ , 0)$, $(-5, \ )$

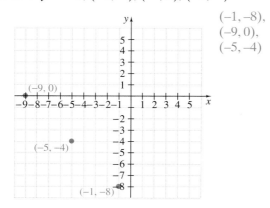

$(-1, -8)$,
$(-9, 0)$,
$(-5, -4)$

**28.** $x - y = 3$; $(4, \ )$, $(0, \ )$, $( \ , 3)$

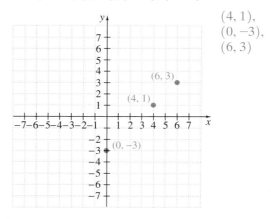

$(4, 1)$,
$(0, -3)$,
$(6, 3)$

**29.** $y = 3x$; $(1, \ )$, $(-2, \ )$, $( \ , 0)$    $(1, 3)$, $(-2, -6)$, $(0, 0)$

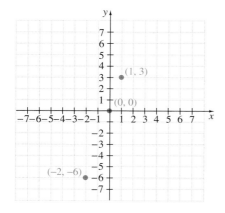

**30.** $x = y + 6$; $(1, \ )$, $(6, \ )$, $(-1, \ )$

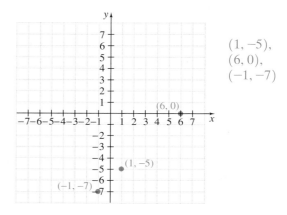

$(1, -5)$,
$(6, 0)$,
$(-1, -7)$

**(8.4)** *Graph each linear equation.*

**31.** $x = -6$

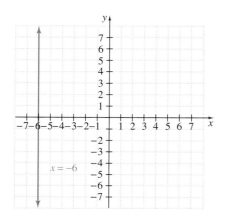

**32.** $y = 0$

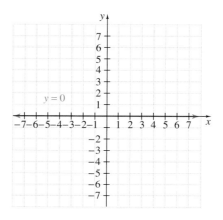

**33.** $x + y = 11$

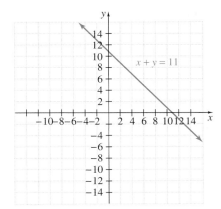

**34.** $x - y = 11$

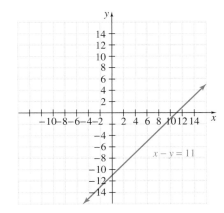

**35.** $y = 4x - 2$

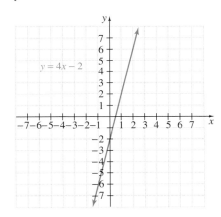

**36.** $y = 5x$

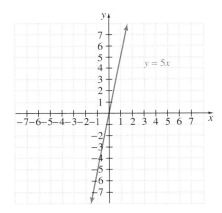

**37.** $x = -2y$

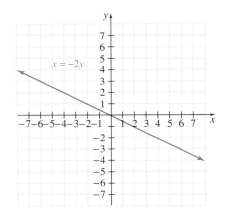

**38.** $x + y = -1$

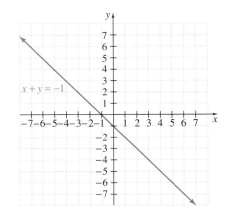

**39.** $2x - 3y = 12$

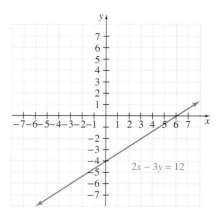

**40.** $x = \dfrac{1}{2} y$

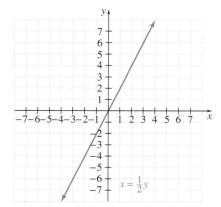

**(8.5)** *Find the mean, the median, and any mode(s) for each list of numbers.*

**41.** $13, 23, 33, 14, 6$   mean: 17.8; median: 14; no mode

**42.** $45, 21, 60, 86, 64$   mean: 55.2; median: 60; no mode

**43.** $14,000, $20,000, $12,000, $20,000, $36,000, $45,000
mean: $24,500; median: $20,000; mode: $20,000

**44.** $560, 620, 123, 400, 410, 300, 400, 780, 430, 450$   mean: 447.3; median: 420; mode: 400

*For Exercises 45 and 46, the grades are given for a student for a particular semester. Find each grade point average. If necessary, round the grade point average to the nearest hundredth.*

**45.**   3.25

| Grade | Credit Hours |
|-------|--------------|
| A | 3 |
| A | 3 |
| C | 2 |
| B | 3 |
| C | 1 |

**46.**   2.57

| Grade | Credit Hours |
|-------|--------------|
| B | 3 |
| B | 4 |
| C | 2 |
| D | 2 |
| B | 3 |

**Name** _____

**(8.6)** *Draw a tree diagram for each experiment. Then use the diagram to determine the number of outcomes.*

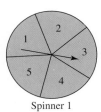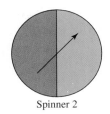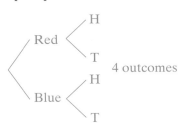

Spinner 1          Spinner 2

**47.** Toss a coin and then spin Spinner 1.

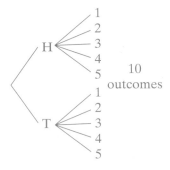

H
  1
  2
  3
  4
  5
  10 outcomes
T
  1
  2
  3
  4
  5

**48.** Spin Spinner 2 and then toss a coin.

Red
  H
  T
  4 outcomes
Blue
  H
  T

**49.** Spin Spinner 1 twice.

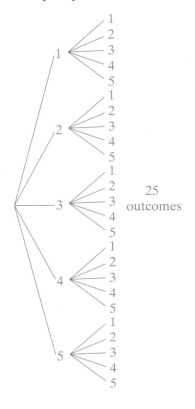

1
  1
  2
  3
  4
  5
2
  1
  2
  3
  4
  5
3
  1
  2
  3
  4
  5
  25 outcomes
4
  1
  2
  3
  4
  5
5
  1
  2
  3
  4
  5

**50.** Spin Spinner 2 twice.

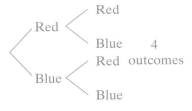

Red
  Red
  Blue
  4 outcomes
Blue
  Red
  Blue

**51.** Spin Spinner 1 and then Spinner 2.

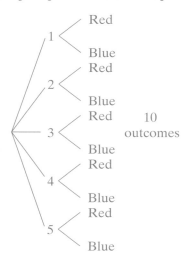

Red
1
Blue
Red
2
Blue
Red
3          10
outcomes
Blue
Red
4
Blue
Red
5
Blue

*Find the probability of each event. For Exercises 54–57, see spinner 1 on previous page.*

Die

**52.** Roll a 4 on a die.  $\dfrac{1}{6}$

**53.** Roll a 3 on a die.  $\dfrac{1}{6}$

**54.** Spin a 4 on Spinner 1.  $\dfrac{1}{5}$

**55.** Spin a 3 on Spinner 1.  $\dfrac{1}{5}$

**56.** Spin either a 1, 3, or 5 on Spinner 1.  $\dfrac{3}{5}$

**57.** Spin either a 2 or 4 on Spinner 1.  $\dfrac{2}{5}$

# CHAPTER 8 TEST

*The pictograph shows the money collected each week from a wrapping paper fundraiser. Use this graph to answer Exercises 1–3.*

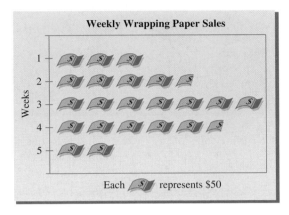

**Weekly Wrapping Paper Sales**

Each 💵 represents $50

1. How much money was collected during the second week?

2. During which week was the most money collected? How much money was collected during that week?

3. What was the total money collected for the fund-raiser?

*The bar graph shows the normal monthly precipitation in centimeters for Chicago, Illinois. Use this graph to answer Exercises 4–6.*

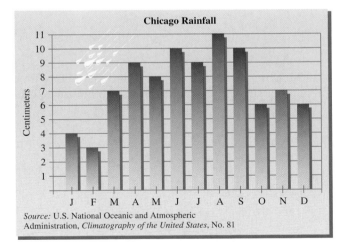

**Chicago Rainfall**

*Source:* U.S. National Oceanic and Atmospheric Administration, *Climatography of the United States*, No. 81

4. During which month(s) does Chicago normally have greater than 9 centimeters of rainfall?

5. During which month does Chicago normally have the least amount of rainfall? How much rain falls during that month?

6. During which month(s) does 7 centimeters of rain normally fall?

**1.** $225

**2.** 3rd week, $350

**3.** $1100

**4.** June, August, September

**5.** February, 3 cm

**6.** March and November

**Name** _____

*The line graph shows online sales of recorded music (CDs, albums, and cassettes). Use this graph to answer Exercises 7–9.*

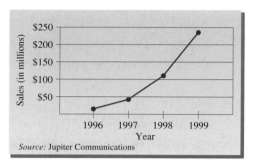

Source: Jupiter Communications

**7.** Approximate the 1998 online sales of recorded music.

**8.** What year has shown the greatest increase in online sales?

**9.** During what year(s) shown were sales less than $50 million?

*The result of a survey of 200 people is shown in the circle graph. Each person was asked to tell his or her favorite type of music. Use this graph to answer Exercises 10–11.*

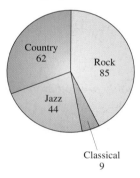

**10.** Find the ratio of those who prefer rock music to the total number surveyed.

**11.** Find the ratio of those who prefer country music to those who prefer jazz.

*The circle graph shows U.S. car sales by type in 1998. If there were approximately 8,139,000 cars sold in the U.S in 1998, find how many cars of the type given were sold.*

Source: Ward's Automotive Reports

**12.** Small cars

**13.** Luxury cars

*Find the coordinates of each point.*

**14.** *A*          **15.** *B*          **16.** *C*          **17.** *D*

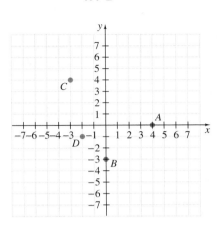

**14.** $(4, 0)$ _____

**15.** $(0, -3)$ _____

**16.** $(-3, 4)$ _____

**17.** $(-2, -1)$ _____

*Complete and graph the ordered-pair solutions of each given equation.*

**18.** $x = -6y$; $(0, \phantom{x})$, $(\phantom{x}, 1)$, $(12, \phantom{x})$

**19.** $y = 7x - 4$; $(2, \phantom{x})$, $(-1, \phantom{x})$, $(0, \phantom{x})$

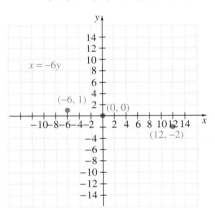

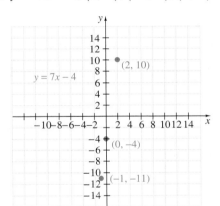

**18.** $(0, 0)$, $(-6, 1)$, $(12, -2)$ _____

**19.** $(2, 10)$, $(-1, -11)$, $(0, -4)$ _____

*Graph each linear equation.*

**20.** $y + x = -4$          **21.** $y = -4$          **22.** $y = 3x - 5$

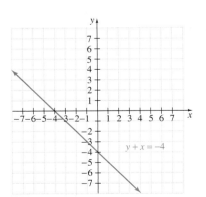

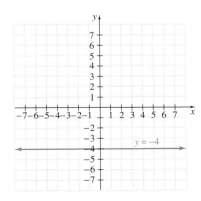

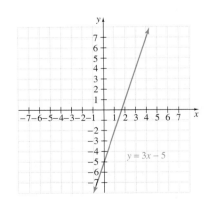

**23.** $x = 5$

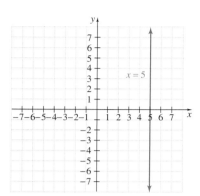

**24.** $y = -\dfrac{1}{2}x$

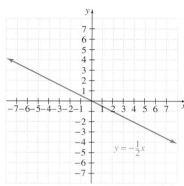

**25.** $3x - 2y = 12$

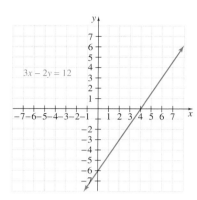

*Find the mean, median, and mode of each list of numbers.*

**26.** 26, 32, 42, 43, 49

**27.** 8, 10, 16, 16, 14, 12, 12, 13

*Find the grade point average. If necessary, round to the nearest hundredth.*

**28.**

| Grade | Credit Hours |
|-------|--------------|
| A | 3 |
| B | 3 |
| C | 3 |
| B | 4 |
| A | 1 |

**29.** Draw a tree diagram for the experiment of spinning the spinner twice.

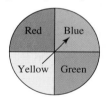

**30.** Draw a tree diagram for the experiment of tossing a coin twice.

*Suppose that the numbers 1 through 10 are each written on a scrap of paper and placed in a bag. You then select one number from the bag.*

**31.** What is the probability of choosing a 6 from the bag?

**32.** What is the probability of choosing a 3 or a 4 from the bag?

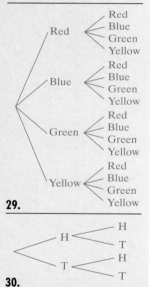

**26.** mean: 38.4; median: 42; no mode

**27.** mean: 12.625; median: 12.5; mode: 16 and 12

**28.** 3.07

**29.**

**30.**

**31.** $\dfrac{1}{10}$

**32.** $\dfrac{1}{5}$

**654**

# CUMULATIVE REVIEW

**1.** Simplify: $4^3 + [3^2 - (10 \div 2)] - 7 \cdot 3$

**2.** Evaluate $x - y$ for $x = -3$ and $y = 9$.

**3.** Solve for $y$: $3y - 7y = 12$

**4.** Solve for $x$: $\dfrac{x}{6} + 1 = \dfrac{4}{3}$

**5.** Add: $2\dfrac{1}{3} + 5\dfrac{3}{8}$

**6.** Write 5.6 as a mixed number.

**7.** Subtract: $3.5 - 0.068$

**8.** Multiply: $0.283 \times 0.3$

**9.** Divide: $-5.98 \div 115$

**10.** Simplify: $(-1.3)^2$

**11.** Write $\dfrac{1}{4}$ as a decimal.

**12.** Solve for $x$: $5(x - 0.36) = -x + 2.4$

**13.** Approximate $\sqrt{32}$ to the nearest thousandth.

**14.** Write the ratio of 12 to 17 using fractional notation.

**Name** _____

**15.** Write as a unit rate: 318.5 miles every 13 gallons of gas.

**16.** Solve $\dfrac{25}{x} = \dfrac{20}{4}$ for $x$ and then check.

△ **17.** Find the ratio of corresponding sides for the similar triangles $ABC$ and $DEF$.

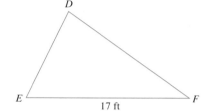

*Write each percent as a decimal.*

**18.** 4.6%

**19.** 0.74%

**20.** What number is 35% of 60?

**21.** 20.8 is 40% of what number?

**22.** The sales tax on a $300 printer is $22.50. Find the sales tax rate.

**23.** Find the monthly payment on a $2000 loan for 2 years. The interest on the 2-year loan is $435.88.

**24.** Find the median of the list of scores: 67, 91, 75, 86, 55, 91

**25.** If a die is rolled one time, find the probability of rolling a 3 or a 4.

# Geometry and Measurement

T he word *geometry* is formed from the Greek words *geo*, meaning earth, and *metron*, meaning measure. Geometry literally means to measure the earth. In this chapter, we learn about various geometric figures and their properties, such as perimeter, area, and volume. Knowledge of geometry can help us solve practical problems in real-life situations. For instance, knowing certain measures of a circular swimming pool allows us to calculate how much water it can hold.

We also learn about some of the measures frequently used in geometry, as well as other types of measurements. In the United States, two systems of measurement are commonly used. They are the United States (U.S.), or English, system and the metric system. The U.S. system is familiar to most Americans. Units such as feet, miles, ounces, and gallons are used. However, the metric system is also commonly used in fields such as medicine, sports, international marketing, and certain physical sciences. We are used to buying a 2-liter bottle of Dr. Pepper, watching televised coverage of the 500-meter dash in the Olympic Games, or taking a 200-milligram dose of pain reliever.

## CONTENTS

T he Vietnam Veterans Memorial, also known simply as The Wall, is the most visited memorial in Washington, D.C. It was dedicated in November 1982 at Constitution Gardens on the Mall. Although the memorial also includes several other elements, the most visible and well-known portion of the memorial is a pair of black granite walls, each $246\frac{2}{3}$ feet long, arranged in the shape of a V and inscribed with the names of the 58,209 U.S. casualties of the Vietnam War between 1959 and 1975. The design for this contemplative memorial was created by Maya Lin of Athens, Ohio, then a 21-year-old senior at Yale University's School of Architecture. She created sketches of the memorial for a class project and then entered her idea in the national design competition for the memorial. Her design was unanimously selected by a panel of judges from a field of 1421 entries. In Exercise 67 on page 671 we will learn about the measure of the angle between the two walls of the Vietnam Veterans Memorial.

**Name** _____ **Section** _____ **Date** _____

# CHAPTER 9 PRETEST

*Classify each angle as acute, right, obtuse or straight.*

△ **1.**

A

△ **2.**
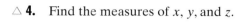
B

△ **3.** Find the supplement of a 92° angle.

△ **4.** Find the measures of $x$, $y$, and $z$.

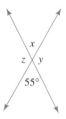

△ **5.** Find the measure of $\angle x$.

40°    30°

**6.** Convert 11 feet to yards.

**7.** Convert 6,250,000 cm to kilometers.

△ **8.** Find the perimeter of a rectangle with a length of 24 inches and a width of 6 inches.

△ **9.** Find the circumference of the given circle. Use $\pi \approx 3.14$.

9 yd

△ **10.** Find the area of the given triangle.

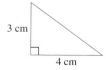

3 cm

4 cm

**11.** Divide 15 lb 8 oz by 2.

**12.** Subtract 2 qt from 9 gal 1 qt.

**13.** Convert 25 L to milliters.

**14.** Convert 118°F to Celsius. If necessary, round to the nearest tenth of a degree.

## △ **9.1** LINES AND ANGLES

### **A** IDENTIFYING LINES, LINE SEGMENTS, RAYS, AND ANGLES

Let's begin with a review of two important concepts—plane and space.

A **plane** is a flat surface that extends indefinitely. Surfaces like a plane are a classroom floor or a blackboard or whiteboard.

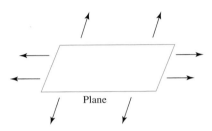

Plane

**Space** extends in all directions indefinitely. Examples of *objects* in space are houses, grains of salt, bushes, your Prealgebra text-book, and you.

The most basic concept of geometry is the idea of a point in space. A **point** has no length, no width, and no height, but it does have location. We will represent a point by a dot, and we will label points with letters.

Point *P*

A **line** is a set of points extending indefinitely in two directions. A line has no width or height, but it does have length. We can name a line by any two of its points. A **line segment** is a piece of a line with two endpoints.

Line *AB* or $\overleftrightarrow{AB}$          Line segment *AB* or $\overline{AB}$

A **ray** is a part of a line with one endpoint. A ray extends indefinitely in one direction. An **angle** is made up of two rays that share the same endpoint. The common endpoint is called the **vertex**.

Ray *AB* or $\overrightarrow{AB}$          Vertex

## Objectives

**A** Identify lines, line segments, rays, and angles.

**B** Classify angles as acute, right, obtuse, or straight.

**C** Identify complementary and supplementary angles.

**D** Find measures of angles.

Study Guide    SSM    CD-ROM    Video 9.1

TEACHING TIP

When discussing a concept such as a point, remind students that we draw a dot and therefore give the point dimensions so we can see it.

The angle in the figure above can be named

$$\angle ABC \qquad \angle CBA \qquad \angle B \qquad \text{or} \qquad \angle x$$

The vertex is the middle point.

Rays $BA$ and $BC$ are **sides** of the angle.

## Practice Problem 1

Identify each figure as a line, a ray, a line segment, or an angle. Then name the figure using the given points.

a.

b.

c.
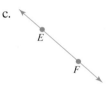

d.

## Example 1

Identify each figure as a line, a ray, a line segment, or an angle. Then name the figure using the given points.

a.
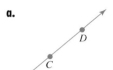

b.

c.
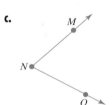

d.
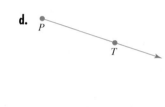

*Solution:* Figure (a) extends indefinitely in two directions. It is line $CD$ or $\overleftrightarrow{CD}$.
Figure (b) has two endpoints. It is line segment $EF$ or $\overline{EF}$.
Figure (c) has two rays with a common endpoint. It is $\angle MNO, \angle ONM,$ or $\angle N$.
Figure (d) is part of a line with one endpoint. It is ray $PT$ or $\overrightarrow{PT}$.

## Practice Problem 2

Use the figure for Example 2 to list other ways to name $\angle z$.

## Example 2

List other ways to name $\angle y$.

*Solution:* Two other ways to name $\angle y$ are $\angle QTR$ and $\angle RTQ$. We may not use the vertex alone to name this angle because three different angles have $T$ as their vertex.

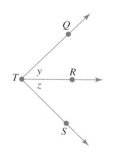

**Answers**

**1. a.** line segment; line segment $RS$ or $\overline{RS}$
**b.** ray; ray $AB$ or $\overrightarrow{AB}$   **c.** line; line $EF$ or $\overleftrightarrow{EF}$
**d.** angle; $\angle TVH, \angle HVT,$ or $\angle V$
**2.** $\angle RTS, \angle STR$

## B CLASSIFYING ANGLES

An angle can be measured in **degrees**. The symbol for degrees is a small, raised circle, °. There are 360° in a full revolution, or full circle.

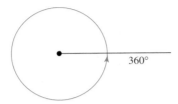

$\frac{1}{2}$ of a revolution measures $\frac{1}{2}(360°) = 180°$. An angle that measures 180° is called a **straight angle**.

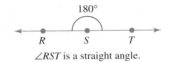

∠RST is a straight angle.

$\frac{1}{4}$ of a revolution measures $\frac{1}{4}(360°) = 90°$. An angle that measures 90° is called a **right angle**. The symbol ⌐ is used to denote a right angle.

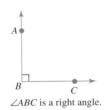

∠ABC is a right angle.

An angle whose measure is between 0° and 90° is called an **acute angle**.

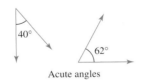

Acute angles

An angle whose measure is between 90° and 180° is called an **obtuse angle**.

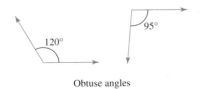

Obtuse angles

## Practice Problem 3

Classify each angle as acute, right, obtuse, or straight.

a.

b.

c.

d.

## Example 3

Classify each angle as acute, right, obtuse, or straight.

a.

b.

c.

d.

*Solution:*

**a.** $\angle R$ is a right angle, denoted by ⌐.

**b.** $\angle S$ is a straight angle.

**c.** $\angle T$ is an acute angle. It measures between 0° and 90°.

**d.** $\angle Q$ is an obtuse angle. It measures between 90° and 180°.

### C   IDENTIFYING COMPLEMENTARY AND SUPPLEMENTARY ANGLES

Two angles that have a sum of 90° are called **complementary angles**. We say that each angle is the **complement** of the other.

$\angle R$ and $\angle S$ are complementary angles because

$$60° + 30° = 90°$$

Complementary angles
$60° + 30° = 90°$

Two angles that have a sum of 180° are called **supplementary angles**. We say that each angle is the **supplement** of the other.

$\angle M$ and $\angle N$ are supplementary angles because

$$125° + 55° = 180°$$

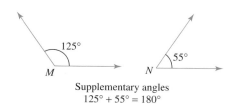

Supplementary angles
$125° + 55° = 180°$

## Practice Problem 4

Find the complement of a 36° angle.

## Practice Problem 5

Find the supplement of an 88° angle.

**Answers**

**3. a.** acute   **b.** straight   **c.** obtuse   **d.** right
**4.** 54°   **5.** 92°

## Example 4

Find the complement of a 48° angle.

*Solution:*    The complement of an angle that measures 48° is an angle that measures 90° − 48° = 42°.

## Example 5

Find the supplement of a 107° angle.

*Solution:*    The supplement of an angle that measures 107° is an angle that measures 180° − 107° = 73°.

**TRY THE CONCEPT CHECK IN THE MARGIN.**

## D FINDING MEASURES OF ANGLES

Measures of angles can be added or subtracted to find measures of related angles.

**Example 6**   Find the measure of ∠x.

*Solution:*   ∠x = 87° − 52° = 35°

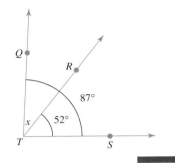

Two lines in a plane can be either parallel or intersecting. **Parallel lines** never meet. **Intersecting lines** meet at a point. The symbol ‖ is used to indicate "is parallel to." For example, in the figure p‖q.

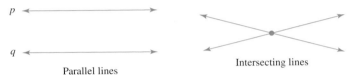

Parallel lines          Intersecting lines

Some intersecting lines are perpendicular. Two lines are **perpendicular** if they form right angles when they intersect. The symbol ⊥ is used to denote "is perpendicular to." For example, in the figure n ⊥ m.

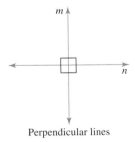

Perpendicular lines

When two lines intersect, four angles are formed. Two of these angles that are opposite each other are called **vertical angles**. Vertical angles have the same measure. Two angles that share a common side are called **adjacent angles**. Adjacent angles formed by intersecting lines are supplementary. That is, they have a sum of 180°.

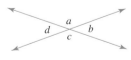

Vertical angles:
∠a and ∠c
∠d and ∠b

Adjacent angles:
∠a and ∠b
∠b and ∠c
∠c and ∠d
∠d and ∠a

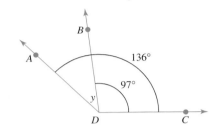

## Practice Problem 7

Find the measure of $\angle a$, $\angle b$, and $\angle c$.

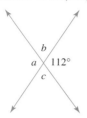

## Example 7

Find the measure of $\angle x$, $\angle y$, and $\angle z$ if the measure of $\angle t$ is 42°.

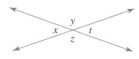

*Solution:*    Since $\angle t$ and $\angle x$ are vertical angles, they have the same measure, so $\angle x$ measures 42°.

Since $\angle t$ and $\angle y$ are adjacent angles, their measures have a sum of 180°. So $\angle y$ measures 180° − 42° = 138°.

Since $\angle y$ and $\angle z$ are vertical angles, they have the same measure. So $\angle z$ measures 138°.

A line that intersects two or more lines at different points is called a **transversal**. Line $l$ is a transversal that intersects lines $m$ and $n$. The eight angles formed have special names. Some of these names are:

Corresponding Angles: $\angle a$ and $\angle e$, $\angle c$ and $\angle g$, $\angle b$ and $\angle f$, $\angle d$ and $\angle h$

Alternate Interior Angles: $\angle c$ and $\angle f$, $\angle d$ and $\angle e$

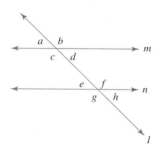

When two lines cut by a transversal are *parallel*, the following are true.

---

**PARALLEL LINES CUT BY A TRANSVERSAL**

If two parallel lines are cut by a transversal, then the measures of **corresponding angles are equal** and **alternate interior angles are equal**.

---

## Practice Problem 8

Given $m \| n$ and that the measure of $\angle w = 40°$, find the measures of $\angle x$, $\angle y$, and $\angle z$.

## Example 8

Given that $m \| n$ and that the measure of $\angle w$ is 100°, find the measures of $\angle x$, $\angle y$, and $\angle z$.

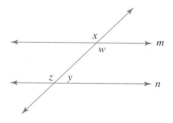

*Solution:*

The measure of $\angle x = 100°$.     $\angle x$ and $\angle w$ are vertical angles.

The measure of $\angle z = 100°$.     $\angle x$ and $\angle z$ are corresponding angles.

The measure of $\angle y = 180° - 100° = 80°$.     $\angle z$ and $\angle y$ are supplementary angles.

**Answers**

**7.** $\angle a = 112°$; $\angle b = 68°$; $\angle c = 68°$

**8.** $\angle x = 40°$; $\angle y = 40°$; $\angle z = 140°$

The sum of the measures of the angles of a triangle is 180°. We can use this fact to find measures of unknown angles in a triangle.

**Example 9**    Find the measure of ∠a.

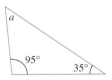

*Solution:*    Since the sum of the measures of the three angles is 180°, we have

measure of ∠a = 180° − 95° − 35° = 50°

To check, see that 95° + 35° + 50° = 180°.    ▬▬▬

An important type of triangle is a right triangle. A **right triangle** is a triangle with a right angle. The side opposite the right angle is called the **hypotenuse** and the other two sides are called **legs**.

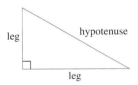

**Example 10**    Find the measure of ∠b.

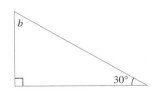

**Practice Problem 9**

Find the measure of ∠x.

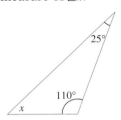

**Practice Problem 10**

Find the measure of ∠y.

**Answers**

**9.** 45°  **10.** 65°

*Solution:*     We know that the measure of the right angle, ⌐, is 90°. Since the sum of the measures of the angles is 180°, we have

measure of $\angle b = 180° - 90° - 30° = 60°$

---

**Helpful Hint**

From the previous example, can you see that in a right triangle, the sum of the other two acute angles is 90°? This is because

$$90° + 90° = 180°$$

| ↑ | ↑ | ↑ |
|---|---|---|
| right angle's measure | sum of other two angles' measures | sum of angles' measures |

---

**Name** _____ **Section** _____ **Date** _____

# EXERCISE SET 9.1

**A** *Identify each figure as a line, a ray, a line segment, or an angle. Then name the figure using the given points. See Example 1.*

**1.**

**2.**

**3.**

**4.**

**5.**

**6.**

**7.**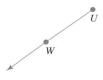

**8.**

**B** *Find the measure of each angle in the figure.*

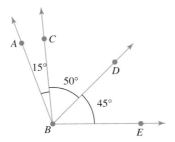

**9.** ∠ABC   **10.** ∠EBD   **11.** ∠CBD   **12.** ∠CBA

**13.** ∠DBA   **14.** ∠EBC   **15.** ∠CBE   **16.** ∠ABE

**667**

**17.** 90°

**18.** 180°

**19.** 0°; 90°

**20.** 90°; 180°

**21.** straight

**22.** acute

**23.** right

**24.** obtuse

**25.** obtuse

**26.** straight

**27.** right

**28.** acute

**29.** 73°

**30.** 3°

**31.** 163°

**32.** 93°

**33.** 45°

**34.** 68°

**Name** _____

*Fill in each blank. See Example 3.*

**17.** A right angle has a measure of _____.

**18.** A straight angle has a measure of _____.

**19.** An acute angle measures between _____ and _____.

**20.** An obtuse angle measures between _____ and _____.

*Classify each angle as acute, right, obtuse, or straight. See Example 3.*

**21.**

**22.**

**23.**

**24.**

**25.**

**26.**

**27.**

**28.**

**C** *Find each complementary or supplementary angle as indicated. See Examples 4 and 5.*

**29.** Find the complement of a 17° angle.

**30.** Find the complement of an 87° angle.

**31.** Find the supplement of a 17° angle.

**32.** Find the supplement of an 87° angle.

**33.** Find the complement of a 45° angle.

**34.** Find the complement of a 22° angle.

**35.** Find the supplement of a 125° angle.

**36.** Find the supplement of a 155° angle.

**37.** Identify the pairs of complementary angles.

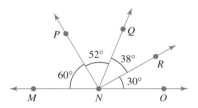

**38.** Identify the pairs of complementary angles.

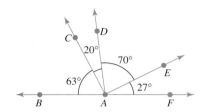

**39.** Identify the pairs of supplementary angles.

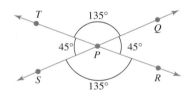

**40.** Identify the pairs of supplementary angles.

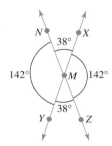

**D** *Find the measure of ∠x in each figure. See Example 6.*

**41.**

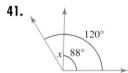

**42.**

**43.**

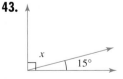

**44.**

**Name** _____

*Find the measures of angles x, y, and z in each figure. See Examples 7 and 8.*

**45.**

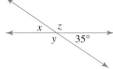

**46.**

 **47.**

**48.**

**49.** $m \parallel n$

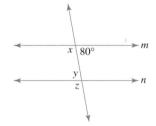

**50.** $m \parallel n$

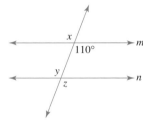

**51.** $m \parallel n$

**52.** $m \parallel n$

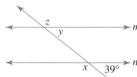

*Find the measure of ∠x in each figure. See Examples 9 and 10.*

**53.**

**54.**

**55.**

**Name** _____

**56.**  35° _____

**57.**  40° _____

**58.**  70° _____

**59.**  $\frac{9}{8}$ or $1\frac{1}{8}$ _____

**60.**  $\frac{5}{8}$ _____

**61.**  $\frac{7}{32}$ _____

**62.**  $\frac{7}{2}$ or $3\frac{1}{2}$ _____

**63.**  $\frac{5}{6}$ _____

**64.**  $\frac{35}{6}$ or $5\frac{5}{6}$ _____

**65.**  $\frac{4}{3}$ or $1\frac{1}{3}$ _____

**66.**  $\frac{25}{3}$ or $8\frac{1}{3}$ _____

**67.**  54.8° _____

**68.**  36.87° _____

**56.**

**57.**

**58.**

## REVIEW AND PREVIEW

*Perform each indicated operation. See Sections 4.3, 4.5, and 4.8.*

**59.** $\frac{7}{8} + \frac{1}{4}$  **60.** $\frac{7}{8} - \frac{1}{4}$  **61.** $\frac{7}{8} \cdot \frac{1}{4}$  **62.** $\frac{7}{8} \div \frac{1}{4}$

**63.** $3\frac{1}{3} - 2\frac{1}{2}$  **64.** $3\frac{1}{3} + 2\frac{1}{2}$  **65.** $3\frac{1}{3} \div 2\frac{1}{2}$  **66.** $3\frac{1}{3} \cdot 2\frac{1}{2}$

## COMBINING CONCEPTS

**67.** The angle between the two walls of the Vietnam Veterans Memorial in Washington, D.C., is 125.2°. Find the supplement of this angle. (*Source:* National Park Service)

**68.** The faces of Khafre's Pyramid at Giza, Egypt, are inclined at an angle of 53.13°. Find the complement of this angle. (*Source:* PBS *NOVA* Online)

**Name** _____

**69.** If lines *m* and *n* are parallel, find the measures of angles *a* through *e*.

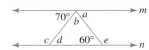

**70.** In your own words, describe how to find the complement and the supplement of a given angle.

**71.** The following demonstration is credited to the mathematician Pascal, who is said to have developed it as a young boy.

Cut a triangle from a piece of paper. The length of the sides and the size of the angles is unimportant. Tear the points off the triangle.

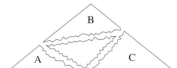

Place the points of the triangle together. Notice that a straight line is formed. What was Pascal trying to show?

# 9.2  LINEAR MEASUREMENT

## A  CONVERTING U.S. SYSTEM UNITS OF LENGTH

In the United States, two systems of measurement are commonly used. They are the **United States (U.S.) or English system** and the **metric system**. The U.S. system is familiar to most Americans. Units such as feet, miles, ounces, and gallons are used. However, the metric system is also commonly used in fields such as medicine, sports, international marketing, and certain physical sciences. We are used to buying 2-liter bottles of soft drinks, watching televised coverage of the 100-meter dash at the Olympic Games, or taking a 200-milligram dose of pain reliever.

The U.S. system of measurement uses the **inch, foot, yard**, and **mile** to measure **length**. The following is a summary of equivalencies between units of length.

| U.S. UNITS OF LENGTH | UNIT FRACTIONS |
|---|---|
| 12 inches (in.) = 1 foot (ft) | $\dfrac{12 \text{ in.}}{1 \text{ ft}} = \dfrac{1 \text{ ft}}{12 \text{ in.}} = 1$ |
| 3 feet = 1 yard (yd) | $\dfrac{3 \text{ ft}}{1 \text{ yd}} = \dfrac{1 \text{ yd}}{3 \text{ ft}} = 1$ |
| 5280 feet = 1 mile (mi) | $\dfrac{5280 \text{ ft}}{1 \text{ mi}} = \dfrac{1 \text{ mi}}{5280 \text{ ft}} = 1$ |

To convert from one unit of length to another, **unit fractions** may be used. A unit fraction is a fraction that equals 1. For example, since 12 in. = 1 ft, we have the unit fractions

$$\frac{12 \text{ in.}}{1 \text{ ft}} = \frac{1 \text{ ft}}{12 \text{ in.}} = 1$$

To convert 48 inches to feet, we *multiply by a unit fraction* that relates feet to inches. The unit fraction should be written so that *the units we are converting to*, feet, *are in the numerator and the original units*, inches, *are in the denominator*. We do this so that like units will divide out, as shown next.

$$
\begin{aligned}
48 \text{ in.} &= \frac{48 \ \cancel{\text{in.}}}{1} \cdot \frac{1 \text{ ft}}{12 \ \cancel{\text{in.}}} && \leftarrow \text{Units converting to} \\
&&& \leftarrow \text{Original units} \\
&= \frac{48 \cdot 1 \text{ ft}}{1 \cdot 12} \\
&= \frac{48 \text{ ft}}{12} \\
&= 4 \text{ ft}
\end{aligned}
$$

Unit fraction

Therefore, 48 inches equals 4 feet, as seen in the diagram.

| 12 in. | 12 in. | 12 in. | 12 in. | |
|---|---|---|---|---|
| | | | | 48 in. = 4 ft |
| 1 ft | 1 ft | 1 ft | 1 ft | |

## Objectives

**A** Define U.S. units of length and convert from one unit to another.

**B** Use mixed units of length.

**C** Perform arithmetic operations on U.S. units of length.

**D** Define the metric units of length and convert from one unit to another.

**E** Perform arithmetic operations on metric units of length.

Study Guide    SSM    CD-ROM    Video 9.2

TEACHING TIP

When discussing unit fractions, make sure that students understand that the units must be included in the numerator and denominator: $\dfrac{1}{12}$ is not a unit fraction but $\dfrac{1 \text{ ft}}{12 \text{ in.}}$ is a unit fraction.

TEACHING TIP

While discussing unit fractions, take a moment and have students write unit fractions on their own. You may want to write down other equivalencies for them to use. For example, 36 in. = 1 yd.

## Practice Problem 1

Convert 5 feet to inches.

## Practice Problem 2

Convert 7 yards to feet.

**Example 1**     Convert 8 feet to inches.

*Solution:*     We multiply 8 feet by a unit fraction that compares 12 inches to 1 foot. The unit fraction should be $\dfrac{\text{units converting to}}{\text{original units}}$ or $\dfrac{12 \text{ inches}}{1 \text{ foot}}$.

$$8 \text{ ft} = \frac{8 \not{\text{ft}}}{1} \cdot \overbrace{\frac{12 \text{ in.}}{1 \not{\text{ft}}}}^{\text{Unit fraction}}$$

$$= 8 \cdot 12 \text{ in.}$$

$$= 96 \text{ in.} \qquad \text{Multiply.}$$

Then 8 ft = 96 in., as shown in the diagram.

| 1 ft | 1 ft | 1 ft | 1 ft | 1 ft | 1 ft | 1 ft | 1 ft | |
|---|---|---|---|---|---|---|---|---|
| 12 in. | 12 in. | 12 in. | 12 in. | 12 in. | 12 in. | 12 in. | 12 in. | 8 ft = 96 in. |

**Example 2**     Convert 7 feet to yards.

*Solution:*     We multiply by a unit fraction that compares 1 yard to 3 feet.

$$7 \text{ ft} = \frac{7 \not{\text{ft}}}{1} \cdot \frac{1 \text{ yd}}{3 \not{\text{ft}}} \quad \leftarrow \text{Units converting to}$$
$$\qquad\qquad\qquad\quad \leftarrow \text{Original units}$$

$$= \frac{7 \text{ yd}}{3}$$

$$= 2\frac{1}{3} \text{ yd} \qquad \text{Divide.}$$

Thus 7 ft = $2\frac{1}{3}$ yd.

### B USING MIXED U.S. SYSTEM UNITS OF LENGTH

Sometimes it is more meaningful to express a measurement of length with mixed units, such as 1 ft and 5 in. We usually condense this and write 1 ft 5 in.

In Example 2, we found that 7 feet was the same as $2\frac{1}{3}$ yards. The measurement can also be written as a mixture of yards and feet. That is,

7 ft = _____ yd _____ ft

Because 3 ft = 1 yd, we divide 3 into 7 to see how many whole yards are in 7 feet. The quotient is the number of yards, and the remainder is the number of feet.

$$
\begin{array}{r}
2 \text{ yd } 1 \text{ ft} \\
3\overline{)7} \\
-6 \\
\hline
1
\end{array}
$$

Thus 7 ft = 2 yd 1 ft, as seen in the diagram.

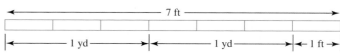

**Example 3**    Convert: 134 in. = _____ ft _____ in.

*Solution:*    Because 12 in. = 1 ft, we divide 12 into 134. The quotient is the number of feet. The remainder is the number of inches. To see why we divide 12 into 134, notice that

$$134 \text{ in.} = \frac{134 \text{ in.}}{1} \cdot \frac{1 \text{ ft}}{12 \text{ in.}} = \frac{134}{12} \text{ ft}$$

$$
\begin{array}{r}
11 \text{ ft } 2 \text{ in.} \\
12\overline{)134} \\
-12 \\
\hline
14 \\
-12 \\
\hline
2
\end{array}
$$

Thus 134 in. = 11 ft 2 in.

**Practice Problem 3**

Convert: 68 in. = _____ ft _____in.

**Example 4**    Convert 3 feet 7 inches to inches.

*Solution:*    First, we will convert 3 feet to inches. Then we will add 7 inches.

$$3 \text{ ft} = \frac{3 \text{ ft}}{1} \cdot \frac{12 \text{ in.}}{1 \text{ ft}} = 36 \text{ in.}$$

Then

$$3 \text{ ft } 7 \text{ in.} = 36 \text{ in.} + 7 \text{ in.} = 43 \text{ in.}$$

**Practice Problem 4**

Convert 5 yards 2 feet to feet.

TEACHING TIP

Point out to students that mixed units are just a way of writing mixed numbers. For example, 3 ft 2 in. = $3\frac{2}{12}$ ft. You may want to have them add the measurements in Example 5 as mixed numbers.

## C  Operations on U.S. System Units of Length

Finding sums or differences of measurements often involves converting units as shown in the next example. Just remember that, as usual, only like units can be added or subtracted.

**Example 5**    Add 3 ft 2 in. and 5 ft 11 in.

*Solution:*    To add, we line up the similar units.

$$
\begin{array}{r}
3 \text{ ft } \phantom{0}2 \text{ in.} \\
+5 \text{ ft } 11 \text{ in.} \\
\hline
8 \text{ ft } 13 \text{ in.}
\end{array}
$$

Since 13 inches is the same as 1 ft 1 in., we have

8 ft 13 in. = 8 ft + 1 ft 1 in.

= 9 ft 1 in.

Try the Concept Check in the margin.

**Practice Problem 5**

Add 4 ft 8 in. and 8 ft 11 in.

✓ **Concept Check**

How could you estimate the following sum?

$$
\begin{array}{r}
7 \text{ yd } \phantom{2}4 \text{ in.} \\
+3 \text{ yd } 27 \text{ in.}
\end{array}
$$

**Example 6**    Multiply 8 ft 9 in. by 3.

*Solution:*    By the distributive property, we multiply 8 ft by 3 and 9 in. by 3.

$$
\begin{array}{r}
8 \text{ ft } \phantom{2}9 \text{ in.} \\
\times \phantom{8 \text{ ft } 9 \text{ in}} 3 \\
\hline
24 \text{ ft } 27 \text{ in.}
\end{array}
$$

**Practice Problem 6**

Multiply 4 ft 7 in. by 4.

**Answers**

**3.** 5 ft 8 in.    **4.** 17 ft    **5.** 13 ft 7 in.    **6.** 18 ft 4 in.
✓ **Concept Check:** Round each to the nearest yard: 7 yd + 4 yd = 11 yd.

Since 27 in. is the same as 2 ft 3 in., we simplify the product as

$$24 \text{ ft } 27 \text{ in.} = 24 \text{ ft } + 2 \text{ ft } 3 \text{ in.}$$
$$= 26 \text{ ft } 3 \text{ in.}$$

## Practice Problem 7

Divide 18 ft 6 in. by 2.

**TEACHING TIP**

After going over Example 7, have students try the problem again after first writing 24 yd 6 in. as either yards or inches.

**Example 7**    Divide 24 yd 6 in. by 3.

*Solution:*    We divide each of the units by 3.

$$
\begin{array}{r}
8 \text{ yd } 2 \text{ in.} \\
3\overline{)24 \text{ yd } 6 \text{ in.}} \\
-24 \text{ yd} \\
\hline
6 \text{ in.} \\
-6 \text{ in.} \\
\hline
0
\end{array}
$$

The quotient is 8 yd 2 in. To check, see that 8 yd 2 in. multiplied by 3 is 24 yd 6 in.

## Practice Problem 8

A carpenter cuts 1 ft 9 in. from a board of length 5 ft 8 in. Find the remaining length of the board.

**Example 8**    Finding the Length of a Piece of Rope

A rope of length 6 yd 1 ft has 2 yd 2 ft cut from one end. Find the length of the remaining rope.

*Solution:*    Subtract 2 yd 2 ft from 6 yd 1 ft.

$$
\begin{array}{ll}
\text{beginning length} \rightarrow & 6 \text{ yd } 1 \text{ ft} \\
- \quad \text{amount cut} \rightarrow & -2 \text{ yd } 2 \text{ ft} \\
\hline
\text{remaining length}
\end{array}
$$

We cannot subtract 2 ft from 1 ft, so we borrow 1 yd from the 6 yd. One yard is converted to 3 ft and combined with the 1 ft already there.

Borrow 1 yd = 3 ft      The problem now reads:

$$5 \text{ yd } + 1 \text{ yd } 3 \text{ ft}$$

$$
\begin{array}{r}
\cancel{6 \text{ yd }} 1 \text{ ft} \\
-2 \text{ yd } 2 \text{ ft}
\end{array}
\qquad
\begin{array}{r}
5 \text{ yd } 4 \text{ ft} \\
-2 \text{ yd } 2 \text{ ft} \\
\hline
3 \text{ yd } 2 \text{ ft}
\end{array}
$$

The remaining rope is 3 yd 2 ft long.

**TRY THE CONCEPT CHECK IN THE MARGIN.**

## ✓ CONCEPT CHECK

If you were describing the distance between houses in the suburb of a large city, which unit of measurement would you use: inch, foot, yard, or mile? Why? Would you want to use a different unit to describe the distances between houses in the country? Explain.

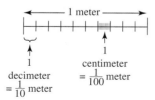

1 decimeter $= \frac{1}{10}$ meter

1 centimeter $= \frac{1}{100}$ meter

## **D** CONVERTING METRIC SYSTEM UNITS OF LENGTH

The basic unit of length in the metric system is the **meter**. A meter is slightly longer than a yard. It is approximately 39.37 inches long, and recall that a yard is 36 inches long.

1 yard — 36 inches

1 meter — 39.37 inches

All units of length in the metric system are based on the meter. The following is a summary of the prefixes used in the metric system. Also shown

**Answers**

**7.** 9 ft 3 in.    **8.** 3 ft 11 in.

✓ **Concept Check:** Answers may vary.

are equivalencies between units of length. Like the decimal system, the metric system uses powers of 10 to define units.

| Prefix | Meaning | Metric Unit of Length |
|--------|---------|------------------------|
| kilo | 1000 | 1 **kilo**meter (km) = 1000 meters (m) |
| hecto | 100 | 1 **hecto**meter (hm) = 100 m |
| deka | 10 | 1 **deka**meter (dam) = 10 m |
| | | **1 meter (m) = 1 m** |
| deci | 1/10 | 1 **deci**meter (dm) = 1/10 m or 0.1 m |
| centi | 1/100 | 1 **centi**meter (cm) = 1/100 m or 0.01 m |
| milli | 1/1000 | 1 **milli**meter (mm) = 1/1000 m or 0.001 m |

These same prefixes are used in the metric system for mass and capacity. The most commonly used measurements of length in the metric system are the **meter, millimeter, centimeter**, and **kilometer**.

Being comfortable with the metric units of length means gaining a "feeling" for metric lengths, just as you have a "feeling" for the length of an inch, a foot, and a mile. To help you accomplish this, study the following examples.

A millimeter is about the thickness of a large paper clip wire.

A centimeter is about the width of a large paper clip.

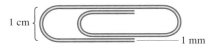

A meter is slightly longer than a yard.

A kilometer is about two-thirds of a mile.

The length of this workbook is approximately 27.5 centimeters.

The width of this workbook is approximately 21.5 centimeters.

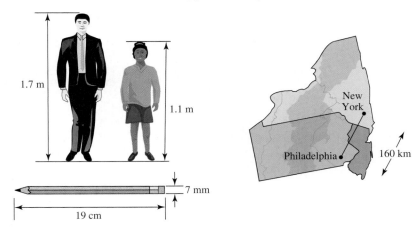

## TRY THE CONCEPT CHECK IN THE MARGIN.

Just as for the U.S. system of measurement, unit fractions may be used to convert from one unit of length to another. The metric system does, however, have a distinct advantage over the U.S. system: the ease of converting from one unit of length to another. Since all units of length are powers of 10 of the meter, converting from one unit of length to another is as simple as moving the decimal point. Listing units of length in order from largest to smallest helps to keep track of how many places to move the decimal point when converting.

## ✓ CONCEPT CHECK

Which unit of measurement would you use for the length of a soccer field: millimeter, centimeter, meter, or kilometer? Why?

**Answer**

✓ **Concept Check:** Answers may vary.

For example, let's convert 1200 meters to kilometers. To convert from meters to kilometers, we move along the chart shown from meters 3 units to the left to kilometers. This means that we move the decimal point 3 places to the left to convert from meters to kilometers.

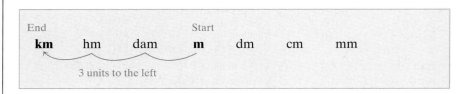

1200 m = 1.200 km

| | 1000 m | 200 m |
|---|---|---|

3 places to the left ◄——————— 1 km ———————► ◄0.2 km►

The same conversion can be made using unit fractions.

Unit fraction

$$1200 \text{ m} = \frac{1200 \text{ m}}{1} \cdot \frac{1 \text{ km}}{1000 \text{ m}} = \frac{1200 \text{ km}}{1000} = 1.2 \text{ km}$$

**Practice Problem 9**

Convert 3.5 m to kilometers.

**Example 9**   Convert 2.3 m to centimeters.

*Solution:*   First we will convert by using a unit fraction.

Unit fraction

$$2.3 \text{ m} = \frac{2.3 \text{ m}}{1} \cdot \frac{100 \text{ cm}}{1 \text{ m}} = 230 \text{ cm}$$

Now we will convert by listing the units of length in a chart and moving from meters to centimeters.

| km | hm | dam | **Start** m | dm | **End** cm | mm |
|---|---|---|---|---|---|---|

2 units to the right

2.30 m = 230 cm

2 places to the right

With either method, we get 230 cm.   ▬▬▬

**Practice Problem 10**

Convert 2.5 m to millimeters.

**Example 10**   Convert 450,000 mm to meters.

*Solution:*   We list the units of length in a chart and move from millimeters to meters.

| km | hm | dam | **End** m | dm | cm | **Start** mm |
|---|---|---|---|---|---|---|

3 units to the left

450,000 mm = 450.000 m   or   450 m   ▬▬▬

**Answers**

**9.** 0.0035 km   **10.** 2500 mm

TRY THE CONCEPT CHECK IN THE MARGIN.

### E   OPERATIONS ON METRIC SYSTEM UNITS OF LENGTH

To add, subtract, multiply, or divide with metric measurements of length, we write all numbers using the same unit of length and then add, subtract, multiply, or divide as with decimals.

**Example 11**   Subtract 430 m from 1.3 km.

*Solution:*   First we convert both measurements to kilometers or both to meters.

430 m = 0.43 km ⎤   or   1.3 km = 1300 m ⎤

$$\begin{array}{r} 1.30 \text{ km} \\ -0.43 \text{ km} \\ \hline 0.87 \text{ km} \end{array} \qquad \begin{array}{r} 1300 \text{ m} \\ -\phantom{0}430 \text{ m} \\ \hline 870 \text{ m} \end{array}$$

The difference is 0.87 km or 870 m.

**Example 12**   Multiply 5.7 mm by 4.

*Solution:*   Here we simply multiply the two numbers. Note that the unit of measurement remains the same.

$$\begin{array}{r} 5.7 \text{ mm} \\ \times\phantom{0} 4 \\ \hline 22.8 \text{ mm} \end{array}$$

**Example 13**   Finding a Person's Height

Fritz Martinson was 1.2 meters tall on his last birthday. Since then he has grown 14 centimeters. Find his current height in meters.

*Solution:*
$$\begin{array}{ll} \text{original height} \rightarrow & 1.20 \text{ m} \\ +\text{ height grown} \rightarrow & +0.14 \text{ m} \\ \hline \text{current height} & 1.34 \text{ m} \end{array}$$   (since 14 cm = 0.14 m)

Fritz is now 1.34 meters tall.

---

✓ **CONCEPT CHECK**

What is wrong with the following conversion? Convert 150 cm to meters.

150.00 cm = 15,000 m

**Practice Problem 11**

Subtract 640 m from 2.1 km.

**Practice Problem 12**

Multiply 18.3 hm by 5.

**Practice Problem 13**

Doris Blackwell is knitting a scarf that is currently 0.8 meter long. If she knits an additional 45 centimeters, how long will the scarf be?

**Answers**

**11.** 1.46 km or 1460 m   **12.** 91.5 hm
**13.** 125 cm or 1.25 m
✓ **Concept Check:** The decimal should be moved to the left: 1.5 m.

TEACHING TIP

Ask students to use the approximations to determine which unit is longer:
    meter or foot
    meter or yard
    centimeter or inch
    kilometer or mile

# CALCULATOR EXPLORATIONS
## METRIC TO U.S. SYSTEM CONVERSIONS IN LENGTH

To convert between the two systems of measurement in length, the following **approximations** may be used.

| | |
|---|---|
| meters × 3.28 ≈ feet | feet × 0.305 ≈ meters |
| meters × 1.09 ≈ yards | yards × 0.914 ≈ meters |
| centimeters × 0.39 ≈ inches | inches × 2.54 ≈ centimeters |
| kilometers × 0.62 ≈ miles | miles × 1.609 ≈ kilometers |

**Example**    The distance from New Orleans, Louisiana, to Pensacola, Florida, is about 400 miles. How many kilometers is this?

*Solution:*    From the above approximations,

miles × 1.609 ≈ kilometers
   ↓
400 × 1.609 ≈ kilometers

To multiply on your calculator, press the keys

| 400 | × | 1.609 | = | (or ENTER ).

The display will read | 643.6 |. Then
400 miles ≈ 643.6 kilometers.
   ↑
(is approximately)

*Convert as indicated.*

**1.** 7 m to feet  ≈22.96 ft

**2.** 11.5 yd to meters  ≈10.511 m

**3.** 8.5 in. to centimeters  ≈21.59 cm

**4.** 15 km to miles  ≈9.3 mi

**5.** A 5-kilometer race is being held today. How many miles is this?  ≈3.1 mi

**6.** A 100-meter dash is being held today. How many yards is this? ≈109 yd

**Name** _____ **Section** _____ **Date** _____

## MENTAL MATH

*Convert as indicated.*

**1.** 12 in. to feet

**2.** 6 ft to yards

**3.** 24 in. to feet

**4.** 36 in. to feet

**5.** 36 in. to yards

**6.** 2 yd to inches

*Determine whether the measurement in each statement is reasonable.*

**7.** The screen of a home television set has a 30-meter diagonal.

**8.** A window measures 1 meter by 0.5 meter.

**9.** A drinking glass is made of glass 2 millimeters thick.

**10.** A paper clip is 4 kilometers long.

**11.** The distance across the Colorado River is 50 kilometers.

**12.** A model's hair is 30 centimeters long.

## EXERCISE SET 9.2

**A** *Convert each measurement as indicated. See Examples 1 and 2.*

**1.** 60 in. to feet

**2.** 84 in. to feet

**3.** 12 yd to feet

**4.** 18 yd to feet

**5.** 42,240 ft to miles

**6.** 36,960 ft to miles

**7.** 102 in. to feet

**8.** 150 in. to feet

**9.** 10 ft to yards

**10.** $8\frac{1}{3}$ yd _____

**11.** 33,792 ft _____

**12.** 20,064 ft _____

**13.** 13 yd 1 ft _____

**14.** 33 yd 1 ft _____

**15.** 3 ft 5 in. _____

**16.** 6 ft 3 in. _____

**17.** 1 mi 4720 ft _____

**18.** 4 mi 3880 ft _____

**19.** 62 in. _____

**20.** 59 in. _____

**21.** 17 ft _____

**22.** 22 ft _____

**23.** 84 in. _____

**24.** 60 in. _____

**25.** 12 ft 3 in. _____

**26.** 18 ft 2 in. _____

**Name** _____

**10.** 25 ft to yards          **11.** 6.4 mi to feet          **12.** 3.8 mi to feet

**B** *Convert each measurement as indicated. See Examples 3 and 4.*

**13.** 40 ft = _____ yd _____ ft          **14.** 100 ft = _____ yd _____ ft

**15.** 41 in. = _____ ft _____ in.          **16.** 75 in. = _____ ft _____ in.

**17.** 10,000 ft = _____ mi _____ ft          **18.** 25,000 ft = _____ mi _____ ft

**19.** 5 ft 2 in. = _____ in.          **20.** 4 ft 11 in. = _____ in.

**21.** 5 yd 2 ft = _____ ft          **22.** 7 yd 1 ft = _____ ft

**23.** 2 yd 1 ft = _____ in.          **24.** 1 yd 2 ft = _____ in.

**C** *Perform each indicated operation. Simplify the result if possible. See Examples 5 through 7.*

**25.** 5 ft 8 in. + 6 ft 7 in.          **26.** 9 ft 10 in. + 8 ft 4 in.

**27.** 12 yd 2 ft + 9 yd 2 ft

**28.** 16 yd 2 ft + 8 yd 1 ft

**29.** 24 ft 8 in. − 16 ft 3 in.

**30.** 15 ft 5 in. − 8 ft 2 in.

**31.** 16 ft 3 in. − 10 ft 9 in.

**32.** 14 ft 8 in. − 3 ft 11 in.

**33.** 6 ft 8 in. ÷ 2

**34.** 26 ft 10 in. ÷ 2

**35.** 12 yd 2 ft × 4

**36.** 15 yd 1 ft × 8

*Solve. Remember to insert units when writing your answers. See Example 8.*

**37.** The National Zoo maintains a small patch of bamboo, which it grows as a food supply for its pandas. Two weeks ago, the bamboo was 6 ft 10 in. tall. Since then, the bamboo has grown 3 ft 8 in. taller. How tall is the bamboo now?

**38.** While exploring in the Marianas Trench, a submarine probe was lowered to a point 1 mile 1400 feet below the ocean's surface. Later it was lowered an additional 1 mile 4000 feet below this point. How far is the probe below the surface of the Pacific?

**39.** The right arm of the Statue of Liberty in New York harbor is 42 ft long. The tablet held in the Statue's left arm is 23 ft 7 in. long. How much longer is the right arm than the tablet? (*Source: 2000 World Almanac*)

**40.** The length of one of the Statue of Liberty's hands is 16 ft 5 in. The nose of the Statue of Liberty is 4 ft 6 in. long. How much longer is a hand than the nose? (*Source: 2000 World Almanac*)

**27.** 22 yd 1 ft

**28.** 25 yd

**29.** 8 ft 5 in.

**30.** 7 ft 3 in.

**31.** 5 ft 6 in.

**32.** 10 ft 9 in.

**33.** 3 ft 4 in.

**34.** 13 ft 5 in.

**35.** 50 yd 2 ft

**36.** 122 yd 2 ft

**37.** 10 ft 6 in.

**38.** 3 mi 120 ft

**39.** 18 ft 5 in.

**40.** 11 ft 11 in.

**41.** The Amana Corporation stacks up its microwave ovens in a distribution warehouse. Each stack is 1 ft 9 in. wide. How far from the wall would 9 of these stacks extend?

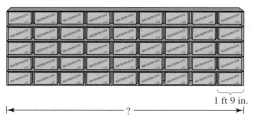

1 ft 9 in.

?

**42.** The highway commission is installing concrete barriers along a highway. Each barrier is 1 yd 2 ft long. How far will 25 barriers in a row reach?

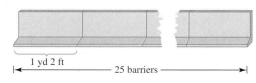

1 yd 2 ft — 25 barriers —

**43.** A carpenter needs to cut a board into thirds. If the board is 9 ft 3 in. long originally, how long will each cut piece be?

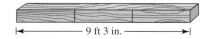

— 9 ft 3 in. —

**44.** A wall is erected exactly halfway between two buildings that are 192 ft 8 in. apart. If the wall is 8 in. wide, how far is it from the wall to either of the buildings?

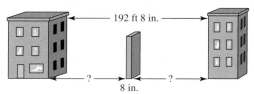

— 192 ft 8 in. —

? ? 8 in.

△ **45.** Evelyn Pittman plans to erect a fence around her garden to keep the rabbits out. If the garden is a rectangle 24 ft 9 in. long by 18 ft 6 in. wide, what is the length of the fencing material she must purchase?

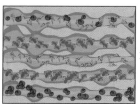

18 ft 6 in.

24 ft 9 in.

△ **46.** Ronnie Hall needs new gutters for the front and *both sides* of his home. The front of the house is 50 ft 8 in., and each side is 22 ft 9 in. wide. What length of gutter must he buy?

50 ft 8 in.

22 ft 9 in.

**47.** The world's longest Coca-Cola truck is in Sweden and is 79 feet long. How many *yards* long are 4 of these trucks? (*Source: Coca-Cola Today*)

△ **48.** The world's largest Coca-Cola sign is in Arica, Chile. It is in the shape of a rectangle whose length is 400 feet and whose width is 131 feet. Find the area of the sign. (*Source: Coca-Cola Today*) (*Hint:* Recall that area of a rectangle is the product: length · width.)

**D** *Convert as indicated. See Examples 9 and 10.*

**49.** 40 m to centimeters

**50.** 18 m to centimeters

**51.** 40 mm to centimeters

**52.** 18 mm to centimeters

**53.** 300 m to kilometers

**54.** 400 m to kilometers

**55.** 1400 mm to meters

**56.** 6400 mm to meters

📼 **57.** 1500 cm to meters

**58.** 6400 cm to meters

**59.** 8.3 cm to millimeters

**60.** 4.6 cm to millimeters

**61.** 20.1 mm to decimeters

**62.** 140.2 mm to decimeters

📼 **63.** 0.04 m to millimeters

**64.** 0.2 m to millimeters

**E** *Perform each indicated operation. See Examples 11 and 12.*

**65.** 8.6 m + 0.34 m

**66.** 14.1 cm + 3.96 cm

**67.** 2.9 m + 40 mm

---

**47.** $105\frac{1}{3}$ yd

**48.** 52,400 sq. ft

**49.** 4000 cm

**50.** 1800 cm

**51.** 4 cm

**52.** 1.8 cm

**53.** 0.3 km

**54.** 0.4 km

**55.** 1.4 m

**56.** 6.4 m

**57.** 15 m

**58.** 64 m

**59.** 83 mm

**60.** 46 mm

**61.** 0.201 dm

**62.** 1.402 dm

**63.** 40 mm

**64.** 200 mm

**65.** 8.94 m

**66.** 18.06 cm

**67.** 2.94 m or 2940 mm

Name _____

**68.** 30 cm + 8.9 m

**69.** 24.8 mm − 1.19 cm

**70.** 45.3 m − 2.16 dam

**71.** 15 km − 2360 m

**72.** 14 cm − 15 mm

**73.** 18.3 m × 3

**74.** 14.1 m × 4

**75.** 6.2 km ÷ 4

**76.** 9.6 m ÷ 5

*Solve. Remember to insert units when writing your answers. See Example 13.*

**77.** A 3.4-m rope is attached to a 5.8-m rope. However, when the ropes are tied, 8 cm of length is lost to form the knot. What is the length of the tied ropes?

**78.** A 2.15-m-long sash cord has become frayed at both ends, so 1 cm is trimmed from each end. How long is the remaining cord?

**79.** The ice on Doc Miller's pond is 5.33 cm thick. For safe skating, Doc insists that it must be 80 mm thick. How much thicker must the ice be before Doc goes skating?

**80.** The sediment on the bottom of the Towamencin Creek is normally 14 cm thick, but the recent flood washed away 22 mm of sediment. How thick is it now?

**81.** An art class is learning how to make kites. The two sticks used for each kite have lengths of 1 m and 65 cm. What total length of wood must be ordered for the sticks if 25 kites are to be built?

**82.** The total pages of a hard-bound economics text are 3.1 cm thick. The front and back covers are each 2 mm thick. How high would a stack of 10 of these texts be?

**Name** _____

**83.** A logging firm needs to cut a 67-m-long redwood log into 20 equal pieces before loading it onto a truck for shipment. How long will each piece be?

**84.** An 18.3-m-tall flagpole is mounted on a 65-cm-high pedestal. How far is the top of the flagpole above the ground?

**85.** At one time it was believed that the fort of Basasi, on the Indian-Tibetan border, was the highest located structure since it was at an elevation of 5.988 km above sea level. However, a settlement has been located that is 21 m higher than the fort. What is the elevation of this settlement?

**86.** The average American male at age 35 is 1.75 m tall. The average 65-year-old male is 48 mm shorter. How tall is the average 65-year-old male?

**87.** A floor tile is 22.86 cm wide. How many tiles in a row are needed to cross a room 3.429 m wide?

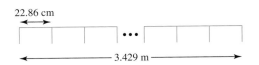

**88.** A standard postcard is 1.6 times longer than it is wide. If it is 9.9 cm wide, what is its length?

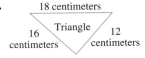

## REVIEW AND PREVIEW

*Recall that the perimeter of a figure is the distance around the figure. Find the perimeter of each figure. See Section 1.2.*

△ **89.**

△ **90.**
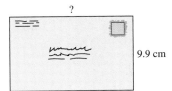

△ **91.**

△ **92.**

**83.** 3.35 m

**84.** 18.95 m or 1895 cm

**85.** 6.009 km or 6009 m

**86.** 1.702 m or 1702 mm

**87.** 15 tiles

**88.** 15.84 cm

**89.** 32 yd

**90.** 46 cm

**91.** 35 m

**92.** 84 mi

**Name** _____

 **COMBINING CONCEPTS**

**93.** To convert from meters to centimeters, the decimal point is moved two places to the right. Explain how this relates to the fact that the prefix centi- means $\frac{1}{100}$.

**94.** Explain why conversions in the metric system are easier to make than conversions in the U.S. system.

△ **95.** Anoa Longway plans to use 26.3 meters of leftover fencing material to enclose a square garden plot for her daughter. How long will each side of the garden be?

**96.** A marathon is a running race over a distance of 26 mi 385 yd. If a runner runs five marathons in a year, what is the total distance he or she has run in marathons? (*Source: Microsoft Encarta Encyclopedia*)

# △ 9.3 PERIMETER

## A USING FORMULAS TO FIND PERIMETERS

Recall from Section 1.2 that the perimeter of a polygon is the distance around the polygon. This means that the perimeter of a polygon is the sum of the lengths of its sides.

**Example 1**    Find the perimeter of the rectangle below.

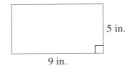

*Solution:*    perimeter = 9 inches + 9 inches + 5 inches + 5 inches
            = 28 inches

Notice that the perimeter of the rectangle in Example 1 can be written as
2 · (9 inches) + 2 · (5 inches).
        ↑                    ↑
    length              width

In general, we can say that the perimeter of a rectangle is always

    2 · length + 2 · width

As we have just seen, the perimeter of some special figures such as rectangles form patterns. These patterns are given as **formulas**. The formula for the perimeter of a rectangle is shown next.

---

**PERIMETER OF A RECTANGLE**

    perimeter = 2 · length + 2 · width

In symbols, this can be written as

    $P = 2l + 2w$

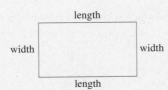

---

**Example 2**    Find the perimeter of a rectangle with a length of 11 inches and a width of 3 inches.

*Solution:*    We will use the formula for perimeter and replace the letters by their known lengths.

    $P = 2l + 2w$
        $= 2 · 11$ in. $+ 2 · 3$ in.    Replace *l* with 11 in. and *w* with 3 in.
        $= 22$ in. $+ 6$ in.
        $= 28$ in.

The perimeter is 28 inches.

---

## Objectives

**A** Use formulas to find the perimeter of special geometric figures.

**B** Use formulas to find the circumference of a circle.

Study     SSM    CD-ROM    Video
Guide                        9.3

---

### Practice Problem 1

Find the perimeter of the rectangular lot below.

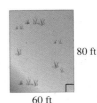

TEACHING TIP

Point out the word "rim" embedded in pe*rim*eter to help students remember its meaning.

---

### Practice Problem 2

Find the perimeter of a rectangle with a length of 22 cm and a width of 10 cm.

**Answers**

**1.** 280 ft    **2.** 64 cm

TEACHING TIP

Consider reminding students that measurement problems require units in the answer. Then point out that in Example 2, 28 is meaningless without inches.

---

**Helpful Hint**

Perimeter is always measured in units.

---

Recall that a square is a special rectangle with all four sides the same length. The formula for the perimeter of a square is shown next.

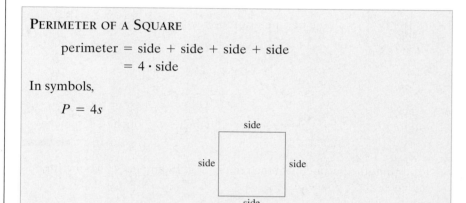

**PERIMETER OF A SQUARE**

$$\text{perimeter} = \text{side} + \text{side} + \text{side} + \text{side}$$
$$= 4 \cdot \text{side}$$

In symbols,

$$P = 4s$$

## Practice Problem 3

How much fencing is needed to enclose a square field 50 yd on a side?

50 yd

## Example 3    Finding the Perimeter of a Table Top

Find the perimeter of a square table top if each side is 5 feet long.

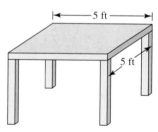

5 ft

5 ft

*Solution:*    The formula for the perimeter of a square is $P = 4 \cdot s$. We will use this formula and replace $s$ with 5 feet.

$$P = 4s$$
$$= 4 \cdot 5 \text{ ft} \qquad \text{Replace } s \text{ with 5 ft.}$$
$$= 20 \text{ ft}$$

The perimeter of the square table top is 20 feet. ■

The formula for the perimeter of a triangle with sides of length $a$, $b$, and $c$ is given next.

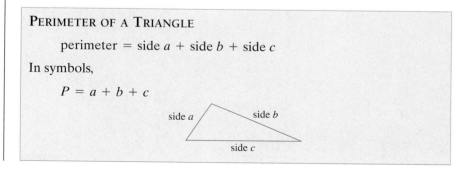

**PERIMETER OF A TRIANGLE**

$$\text{perimeter} = \text{side } a + \text{side } b + \text{side } c$$

In symbols,

$$P = a + b + c$$

side $a$    side $b$

side $c$

**Answer**

**3.** 200 yd

**Example 4**    Find the perimeter of a triangle if the sides are 3 inches, 7 inches, and 6 inches in length.

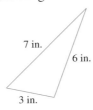

7 in.  6 in.  3 in.

*Solution:*    The formula is $P = a + b + c$, where $a, b,$ and $c$ are the lengths of the sides. Thus,

$P = a + b + c$
$\quad = 3 \text{ in.} + 7 \text{ in.} + 6 \text{ in.}$
$\quad = 16 \text{ in.}$

The perimeter of the triangle is 16 inches. ▬▬▬

Recall that to find the perimeter of other polygons, we find the sum of the lengths of their sides.

**Example 5**    Find the perimeter of the trapezoid shown.

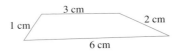

3 cm  2 cm  1 cm  6 cm

*Solution:*    To find the perimeter, we find the sum of the lengths of its sides.

perimeter $= 3 \text{ cm} + 2 \text{ cm} + 6 \text{ cm} + 1 \text{ cm} = 12 \text{ cm}$

The perimeter is 12 centimeters. ▬▬▬

**Example 6**    **Finding the Perimeter of a Room**

Find the perimeter of the room shown.

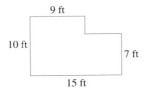

9 ft  10 ft  7 ft  15 ft

*Solution:*    To find the perimeter of the room, we first need to find the lengths of all sides of the room.

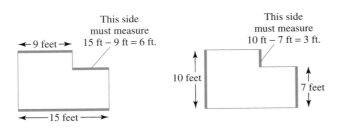

This side must measure 15 ft − 9 ft = 6 ft.
This side must measure 10 ft − 7 ft = 3 ft.
9 feet    10 feet    7 feet    15 feet

Find the perimeter of a triangle if the sides are 5 cm, 9 cm, and 7 cm in length.

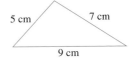

5 cm  7 cm  9 cm

Find the perimeter of the trapezoid shown.

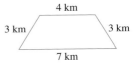

4 km  3 km  3 km  7 km

Find the perimeter of the room shown.

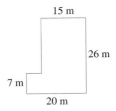

15 m  26 m  7 m  20 m

**Answers**

**4.** 21 cm    **5.** 17 km    **6.** 92 m

Now that we know the measures of all sides of the room, we can add the measures to find the perimeter.

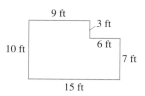

$$\text{perimeter} = 10\text{ ft} + 9\text{ ft} + 3\text{ ft} + 6\text{ ft} + 7\text{ ft} + 15\text{ ft}$$
$$= 50\text{ ft}$$

The perimeter of the room is 50 feet.

## Practice Problem 7

A rectangular lot measures 60 ft by 120 ft. Find the cost to install fencing around the lot if the cost of fencing is $1.90 per foot.

**Example 7   Calculating the Cost of Baseboard**

A rectangular room measures 10 feet by 12 feet. Find the cost to install new baseboard around the room if the cost of the baseboard is $0.66 per foot.

*Solution:*   First we find the perimeter of the room.

$$P = 2l + 2w$$
$$= 2 \cdot 12\text{ ft} + 2 \cdot 10\text{ ft}$$
$$= 24\text{ ft} + 20\text{ ft}$$
$$= 44\text{ ft}$$

Replace *l* with 12 ft and *w* with 10 ft.

The cost for the baseboard is

$$\text{cost} = 0.66 \cdot 44 = 29.04$$

The cost of the baseboard is $29.04.

TEACHING TIP

Ask students to suggest ways they can remember that circumference is the perimeter of a circle.

## B   USING FORMULAS TO FIND CIRCUMFERENCES

Recall from Section 5.3 that the distance around a circle is given a special name called the **circumference**. This distance depends on the radius or the diameter of the circle.

The formulas for circumference are reviewed next.

> ### CIRCUMFERENCE OF A CIRCLE
>
>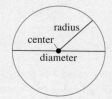
>
> circumference = $2 \cdot \pi \cdot$ radius   or   circumference = $\pi \cdot$ diameter
>
> In symbols,
>
> $$C = 2\pi r \quad \text{or} \quad C = \pi d$$
>
> where $\pi \approx 3.14$   or   $\pi \approx \dfrac{22}{7}$.

**Answer**

**7.** $684

To better understand circumference and $\pi$ (pi), try the following experiment. Take any can and measure its circumference and its diameter.

The can in the figure above has a circumference of 23.5 centimeters and a diameter of 7.5 centimeters. Now divide the circumference by the diameter.

$$\frac{\text{circumference}}{\text{diameter}} = \frac{23.5 \text{ cm}}{7.5 \text{ cm}} \approx 3.1$$

Try this with other sizes of cylinders and circles—you should always get a number close to 3.1. The exact ratio of circumference to diameter is $\pi$.

$\left(\text{Recall that } \pi \approx 3.14 \text{ or } \pi \approx \frac{22}{7}\right).$

## Example 8    Finding the Circumference of a Spa

Mary Catherine Dooley plans to install a border of new tiling around the circumference of her circular spa. If her spa has a diameter of 14 feet, find its circumference.

*Solution:*    Because we are given the diameter, we use the formula $C = \pi d$.

$C = \pi d$

$\quad = \pi \cdot 14 \text{ ft}$    Replace $d$ with 14 ft.

$\quad = 14\pi \text{ ft}$

The circumference of the spa is *exactly* $14\pi$ feet. By replacing $\pi$ with the *approximation* 3.14, we find that the circumference is *approximately* 14 feet $\cdot$ 3.14 $= 43.96$ feet.

**TRY THE CONCEPT CHECK IN THE MARGIN.**

---

**Practice Problem 8**

An irrigation device waters a circular region with a diameter of 20 yd. What is the circumference of the watered region?

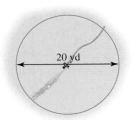

TEACHING TIP

After doing Example 8, point out to students that they could quickly check the reasonableness of their answer by multiplying the diameter by 3. The circumference will always be slightly more than 3 times the diameter or slightly more than 6 times the radius. Then have them use this check on Example 8.

**✓ CONCEPT CHECK**

The distance around which figure is greater: a square with side length 5 in. or a circle with radius 3 in.?

**Answers**

**8.** 62.8 yd

**✓ Concept Check:** The distance around a square with side length 5 in. is greater.

# Focus On the Real World

### SPEEDS

A speed measures how far something travels in a given unit of time. You already learned in Section 6.2 that the speed 55 miles per hour is a unit rate that can be written as $\dfrac{55 \text{ miles}}{1 \text{ hour}}$. Just as there are different units of measurement for length or distance, there are different units of measurement for speed as well. It is also possible to perform unit conversions on speeds. Before we learn about converting speeds, we will review units of time. The following is a summary of equivalencies between various units of time.

| Units of Time | Unit Fractions |
|---|---|
| 60 seconds (s) = 1 minute (min) | $\dfrac{60 \text{ s}}{1 \text{ min}} = \dfrac{1 \text{ min}}{60 \text{ s}} = 1$ |
| 60 minutes = 1 hour (h) | $\dfrac{60 \text{ min}}{1 \text{ h}} = \dfrac{1 \text{ h}}{60 \text{ min}} = 1$ |
| 3600 seconds = 1 hour | $\dfrac{3600 \text{ s}}{1 \text{ h}} = \dfrac{1 \text{ h}}{3600 \text{ s}}$ |

Here are some common speeds.

**Speeds**

Miles per hour (mph)
Miles per minute (mi/min)
Miles per second (mi/s)
Feet per second (ft/s)
Feet per minute (ft/min)
Kilometers per hour (kmph or km/h)
Kilometers per second (kmps or km/s)
Meters per second (m/s)
Knots

> **HELPFUL HINT**
>
> A **knot** is 1 nautical mile per hour and is a measure of speed used for ships.
>
> 1 nautical mile (nmi) ≈ 1.15 miles (mi)
>
> 1 nautical mile (nmi) ≈ 6076.12 feet (ft)

To convert from one speed to another, unit fractions may be used. To convert from miles per hour to feet per second, first write the original speed as a unit rate. Then multiply by a unit fraction that relates miles to feet and by a unit fraction that relates hours to seconds. The unit fractions should be written so that like units will divide out. For example, to convert 55 mph to feet per second:

$$55 \text{ mph} = \frac{55 \text{ miles}}{1 \text{ hour}} = \frac{55 \text{ } \cancel{\text{miles}}}{1 \text{ } \cancel{\text{hour}}} \cdot \frac{5280 \text{ ft}}{1 \text{ } \cancel{\text{mile}}} \cdot \frac{1 \text{ } \cancel{\text{hour}}}{3600 \text{ s}}$$

$$= \frac{55 \cdot 5280 \text{ ft}}{3600 \text{ s}}$$

$$= \frac{290,400 \text{ ft}}{3600 \text{ s}}$$

$$= 80\frac{2}{3} \text{ ft/s}$$

### GROUP ACTIVITY

1. Research the current world land speed record. Convert the speed from mph to feet per second.
2. Research the current world water speed record. Convert from mph to knots.
3. Research and then describe the Beaufort Wind Scale, its origins, and how it is used. Give the scale keyed to both miles per hour and knots. Why would both measures be useful?

**Name** _____

## EXERCISE SET 9.3

**A** *Find the perimeter of each figure. See Examples 1 through 6.*

**1.**

15 ft | Rectangle

17 ft

**2.**

Rectangle | 10 m

4 m

**3.**

Square

9 cm

**4.**

Square

46 mi

**5.**

5 in.   7 in.

9 in.

**6.**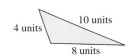

4 units   10 units

8 units

**7.**

Parallelogram / 25 cm

35 cm

**8.**

Parallelogram

3 yd

2 yd

**9.**

10 ft   8 ft

7 ft   8 ft

15 ft

**10.**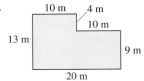

10 m   4 m

10 m

13 m   9 m

20 m

**11.** 66 in.

**12.** 183 cm

**13.** 21 ft

**14.** 30 in.

**15.** 60 ft

**16.** 200 ft

**17.** 346 yd

**18.** 96 in.

**19.** 22 ft

**20.** 182 ft

**21.** $66

**22.** $364

**11.**

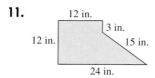

**12.**

*Solve. See Example 7.*

**13.** A polygon has sides of length 5 ft, 3 ft, 2 ft, 7 ft, and 4 ft. Find its perimeter.

**14.** A triangle has sides of length 8 in., 12 in., and 10 in. Find its perimeter.

**15.** Baseboard is to be installed in a square room that measures 15 ft on one side. Find how much baseboard is needed.

**16.** Find how much fencing is needed to enclose a rectangular rose garden 85 ft by 15 ft.

**17.** If a football field is 53 yd wide and 120 yd long, what is the perimeter?

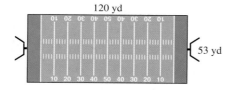

**18.** A stop sign has eight equal sides of length 12 in. Find its perimeter.

**19.** A metal strip is being installed around a workbench that is 8 ft long and 3 ft wide. Find how much stripping is needed.

**20.** Find how much fencing is needed to enclose a rectangular garden 70 ft by 21 ft.

**21.** If the stripping in Exercise 19 costs $3 per foot, find the total cost of the stripping.

**22.** If the fencing in Exercise 20 costs $2 per foot, find the total cost of the fencing.

**23.** A regular hexagon has a side length of 6 in. Find its perimeter. (A regular hexagon is a polygon with six sides of equal length.)

**24.** A regular pentagon has a side length of 14 m. Find its perimeter. (A regular pentagon is a polygon with five sides of equal length.)

**25.** Find the perimeter of the top of a square compact disc case if the length of one side is 7 in.

**26.** Find the perimeter of a square ceramic tile with a side of length 5 in.

**27.** A rectangular room measures 6 ft by 8 ft. Find the cost of installing a strip of wallpaper around the room if the wallpaper costs $0.86 per foot.

**28.** A rectangular house measures 75 ft by 60 ft. Find the cost of installing gutters around the house if the cost is $2.36 per foot.

*Find the perimeter of each figure. See Example 6.*

**29.**
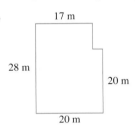
17 m
28 m
20 m
20 m

**30.**
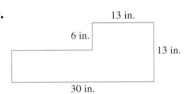
13 in.
6 in.
13 in.
30 in.

**31.**
3 ft   4 ft
5 ft
6 ft
15 ft

**32.**
16 cm
2 cm
11 cm
4 cm
3 cm
9 cm

**33.**
12 mi
34 mi
10 mi
8 mi

**34.**
22 km
12 km
5 km
6 km

**23.** 36 in.

**24.** 70 m

**25.** 28 in.

**26.** 20 in.

**27.** $24.08

**28.** $637.20

**29.** 96 m

**30.** 86 in.

**31.** 66 ft

**32.** 58 cm

**33.** 128 mi

**34.** 90 km

**Name** _____

**B** *Find the circumference of each circle. Give the exact circumference and then an approximation. Use π ≈ 3.14. See Example 8.*

**35.**

17 cm

**36.**

6 in.

**37.**

8 mi

**38.**

50 ft

**39.**

26 m

**40.**

10 yd

**41.** A circular fountain has a radius of 5 ft. Approximate the distance around the fountain. Use $\frac{22}{7}$ for π.

**42.** A circular walkway has a radius of 40 m. Approximate the distance around the walkway. Use 3.14 for π.

**43.** Meteor Crater, near Winslow, Arizona is 4000 ft in diameter. Approximate the distance around the crater. Use 3.14 for π. (*Source: The Handy Science Answer Book*, 1994)

**44.** The largest pearl, the *Pearl of Lao-tze*, has a diameter of $5\frac{1}{2}$ in. Approximate the distance around the pearl. Use $\frac{22}{7}$ for π. (*Source: The Guinness Book of Records*)

**Name** _____

**45.** 23

**46.** 4

**47.** 1

**48.** 60

**49.** 10

**50.** 6

**51.** 216

**52.** 0

**53.** perimeter

**54.** area

**55.** area

**56.** perimeter

**57.** area

**58.** perimeter

**59.** perimeter

**60.** area

## REVIEW AND PREVIEW

*Simplify. See Section 1.7.*

**45.** $5 + 6 \cdot 3$

**46.** $25 - 3 \cdot 7$

**47.** $(20 - 16) \div 4$

**48.** $6 \cdot (8 + 2)$

**49.** $(18 + 8) - (12 + 4)$

**50.** $72 \div (2 \cdot 6)$

**51.** $(72 \div 2) \cdot 6$

**52.** $4^1 \cdot (2^3 - 8)$

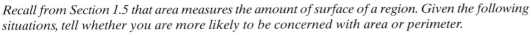

## COMBINING CONCEPTS

*Recall from Section 1.5 that area measures the amount of surface of a region. Given the following situations, tell whether you are more likely to be concerned with area or perimeter.*

**53.** Ordering fencing to fence a yard

**54.** Ordering grass seed to plant in a yard

**55.** Buying carpet to install in a room

**56.** Buying gutters to install on a house

**57.** Ordering paint to paint a wall

**58.** Ordering baseboards to install in a room

**59.** Buying a wallpaper border to go on the walls around a room

**60.** Buying fertilizer for your yard

**Name** _____

**61. a.** Find the circumference of each circle. Approximate the circumference by using 3.14 for $\pi$.

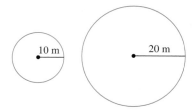

**b.** If the radius of a circle is doubled, is its corresponding circumference doubled?

**62. a.** Find the circumference of each circle. Approximate the circumference by using 3.14 for $\pi$.

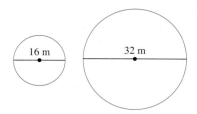

**b.** If the diameter of a circle is doubled, is its corresponding circumference doubled?

**63.** Find the perimeter of the skating rink.

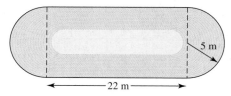

**64.** In your own words, explain how to find the perimeter of any polygon.

# △ **9.4** AREA AND VOLUME

## **A** FINDING AREA

Recall that area measures the amount of surface of a region. Thus far, we know how to find the area of a rectangle and a square. These formulas, as well as formulas for finding the areas of other common geometic figures, are given next.

**AREA FORMULAS OF COMMON GEOMETRIC FIGURES**

| Geometric Figure | Area Formula |
|---|---|
| RECTANGLE <br> 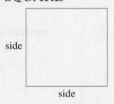 | Area of a rectangle: <br> **area** = **length · width** <br> $A = lw$ |
| SQUARE <br> 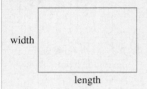 | Area of a square: <br> **area** = **side · side** <br> $A = s \cdot s = s^2$ |
| TRIANGLE <br> 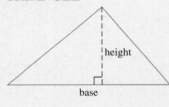 | Area of a triangle: <br> **area** = $\dfrac{1}{2} \cdot$ **base · height** <br> $A = \dfrac{1}{2}bh$ |
| PARALLELOGRAM <br> 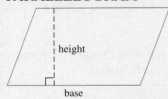 | Area of a parallelogram: <br> **area** = **base · height** <br> $A = bh$ |
| TRAPEZOID <br> 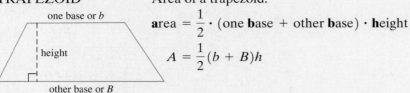 | Area of a trapezoid: <br> **area** = $\dfrac{1}{2} \cdot$ (one **base** + other **base**) · **height** <br> $A = \dfrac{1}{2}(b + B)h$ |

**Objectives**

**A** Find the area of geometric figures.
**B** Find the volume of geometric figures.

Study    SSM    CD-ROM    Video
Guide                            9.4

TEACHING TIP

Begin this lesson by asking students to list some everyday situations in which they would need to calculate the area (painting a room, carpeting a room, tiling a patio). Remind them also of when perimeter would be appropriate (installing base-board around room, etc.).

Use these formulas for the following examples.

> **Helpful Hint**
>
> Area is always measured in square units.

## Practice Problem 1

Find the area of the triangle.

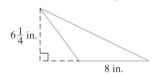

## Example 1

Find the area of the triangle.

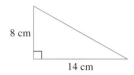

*Solution:*    $A = \dfrac{1}{2}bh$

$\qquad = \dfrac{1}{2} \cdot 14 \text{ cm} \cdot 8 \text{ cm}$     Replace $b$, base, with 14 cm and $h$, height, with 8 cm.

$\qquad = \dfrac{\overset{1}{\cancel{2}} \cdot 7 \cdot 8}{\underset{1}{\cancel{2}}}$ square centimeters     Wrote 14 as $2 \cdot 7$.

$\qquad = 56$ square centimeters

The area is 56 square centimeters.

## Practice Problem 2

Find the area of the trapezoid.

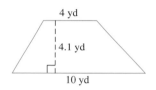

## Example 2

Find the area of the parallelogram.

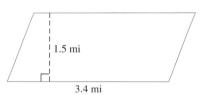

*Solution:*    $A = bh$

$\qquad = 3.4 \text{ miles} \cdot 1.5 \text{ miles}$     Replace $b$, base, with 3.4 miles and $h$, height, with 1.5 miles.

$\qquad = 5.1$ square miles

The area is 5.1 square miles.

> **Helpful Hint**
>
> When finding the area of figures, check to make sure that all measurements are in the same units before calculations are made.

## Practice Problem 3

Find the area of the figure.

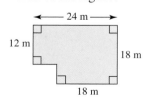

## Example 3

Find the area of the figure.

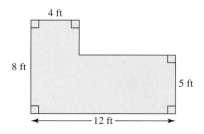

**Answers**

**1.** 25 sq. in.    **2.** 28.7 sq. yd    **3.** 396 sq. m

*Solution:*   Split the figure into two rectangles. To find the area of the figure, we find the sum of the areas of the two rectangles.

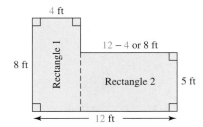

area of Rectangle 1 = $lw$

   = 8 feet · 4 feet

   = 32 square feet

Notice that the length of Rectangle 2 is 12 feet − 4 feet or 8 feet.

area of Rectangle 2 = $lw$

   = 8 feet · 5 feet

   = 40 square feet

area of the figure = area of Rectangle 1 + area of Rectangle 2

   = 32 square feet + 40 square feet

   = 72 square feet

To better understand the formula for area of a circle, try the following. Cut a circle into many pieces, as shown.

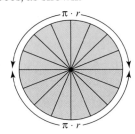

The circumference of a circle is $2\pi r$. This means that the circumference of half a circle is half of $2\pi r$, or $\pi r$.

Then unfold the two halves of the circle and place them together, as shown.

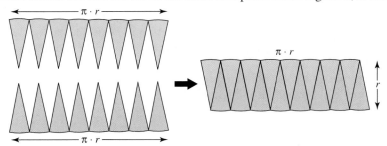

The figure on the right is almost a parallelogram with a base of $\pi r$ and a height of $r$. The area is

$$A = \text{base} \cdot \text{height}$$

$$= (\pi r) \cdot r$$

$$= \pi r^2$$

This is the formula for area of a circle.

**HELPFUL HINT**

The figure in Example 3 could also be split into two rectangles as shown.

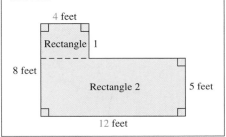

TEACHING TIP   Classroom Activity

Before discussing Example 4, have students work in groups to estimate the number of square feet in a circle with radius 3 feet. You may want to hand out a scaled drawing of the circle on 1/4-inch graph paper where 1 foot is scaled to 1/4 inch. Have them first use the drawing to give an upper bound by finding the area of the 6′ × 6′ square that circumscribes the circle. Then have them give a better estimate of the number of square feet contained in the circle. Have each group share its estimate and the way it arrived at its estimate. Then compare their results with the answer.

> ### AREA FORMULA OF A CIRCLE
>
> Circle
>
>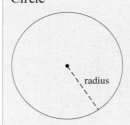
>
> Area of a circle
> **a**rea = $\pi \cdot$ (radius)$^2$
> $\quad A = \pi r^2$
>
> (A fraction approximation for $\pi$ is $\dfrac{22}{7}$.)
>
> (A decimal approximation for $\pi$ is 3.14.)

## Practice Problem 4

Find the area of the given circle. Find the exact area and an approximation. Use 3.14 as an approximation for $\pi$.

7 cm

## ✓ CONCEPT CHECK

Use estimation to decide which figure would have a larger area: a circle of diameter 10 in. or a square 10 in. long on each side.

**Example 4**  Find the area of a circle with a radius of 3 feet. Find the exact area and an approximation. Use 3.14 as an approximation for $\pi$.

3 ft

*Solution:*  We let $r = 3$ feet and use the formula

$A = \pi r^2$
$\quad = \pi \cdot (3 \text{ feet})^2$  Replace $r$ with 3 feet.
$\quad = 9 \cdot \pi \text{ square feet}$  Replace $(3 \text{ feet})^2$ with 9 sq. ft.

To approximate this area, we substitute 3.14 for $\pi$.

$9 \cdot \pi \text{ square feet} \approx 9 \cdot 3.14 \text{ square feet}$
$\qquad\qquad\qquad\quad = 28.26 \text{ square feet}$

The *exact* area of the circle is $9\pi$ square feet, which is *approximately* 28.26 square feet.

──────

**TRY THE CONCEPT CHECK IN THE MARGIN.**

## B FINDING VOLUME

**Volume** is a measure of the space of a region. The volume of a box or can, for example, is the amount of space inside. Volume can be used to describe the amount of juice in a pitcher or the amount of concrete needed to pour a foundation for a house.

The volume of a solid is the number of **cubic units** in the solid. A cubic centimeter and a cubic inch are illustrated.

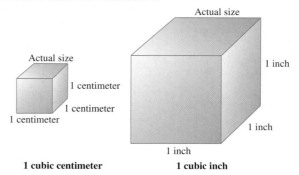

**1 cubic centimeter**          **1 cubic inch**

### Answers

**4.** $49\pi$ sq. cm $\approx$ 153.86 sq. cm

✓ **Concept Check:** A square 10 in. long on each side would have a larger area.

Formulas for finding the volumes of some common solids are given next.

## VOLUME FORMULAS OF COMMON SOLIDS

| Solid | Volume Formulas |
|---|---|

**RECTANGLULAR SOLID**

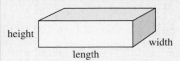

Volume of a rectangular solid:
**volume = length · width · height**
$$V = lwh$$

**CUBE**

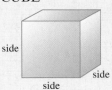

Volume of a cube:
**volume = side · side · side**
$$V = s \cdot s \cdot s = s^3$$

**SPHERE**

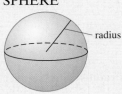

Volume of a sphere:
**volume** $= \dfrac{4}{3} \cdot \pi \cdot (\textbf{radius})^3$
$$V = \frac{4}{3}\pi r^3$$

**CIRCULAR CYLINDER**

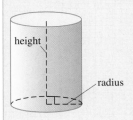

Volume of a circular cylinder:
**volume** $= \pi \cdot (\textbf{radius})^2 \cdot \textbf{height}$
$$V = \pi r^2 h$$

**CONE**

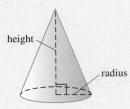

Volume of a cone:
**volume** $= \dfrac{1}{3} \cdot \pi (\textbf{radius})^2 \cdot \textbf{height}$
$$V = \frac{1}{3}\pi r^2 h$$

**SQUARE-BASED PYRAMID**

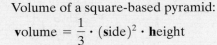

Volume of a square-based pyramid:
**volume** $= \dfrac{1}{3} \cdot (\textbf{side})^2 \cdot \textbf{height}$
$$V = \frac{1}{3}s^2 h$$

**TEACHING TIP**

Point out that the volume of a rectangular solid, a cube, and a circular cylinder are just the product of the area of their bases and their height.

> **Helpful Hint**
>
> Volume is always measured in cubic units.

## Practice Problem 5

Draw a diagram and find the volume of a rectangular box that is 5 ft long, 2 ft wide, and 4 ft deep.

### ✓ CONCEPT CHECK

Juan Lopez is calculating the volume of the rectangular solid shown. Find the error in his calculation.

$$\text{volume} = l + w + h$$
$$= 14 + 8 + 5$$
$$= 27 \text{ cubic cm}$$

## Practice Problem 6

Draw a diagram and approximate the volume of a ball of radius $\frac{1}{2}$ cm. Use $\frac{22}{7}$ for $\pi$.

### Answers

**5.** 40 cu. ft    **6.** $\frac{11}{21}$ cu. cm

✓ **Concept Check:** volume = $l \cdot w \cdot h$
                = 14 · 8 · 5
                = 560 cu. cm

**Example 5**    Find the volume of a rectangular cardboard box that is 12 inches long, 6 inches wide, and 3 inches high.

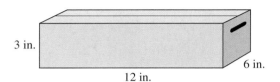

*Solution:*    $V = lwh$
             $V = 12 \text{ inches} \cdot 6 \text{ inches} \cdot 3 \text{ inches} = 216 \text{ cubic inches}$

The volume of the rectangular box is 216 cubic inches.

### TRY THE CONCEPT CHECK IN THE MARGIN.

**Example 6**    Approximate the volume of a ball of radius 3 inches. Use the approximation $\frac{22}{7}$ for $\pi$.

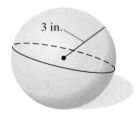

*Solution:*    $V = \frac{4}{3}\pi r^3$

             $\approx \frac{4}{3} \cdot \frac{22}{7} \cdot (3 \text{ inches})^3$

             $= \frac{4}{3} \cdot \frac{22}{7} \cdot 27 \text{ cubic inches}$

             $= \frac{4 \cdot 22 \cdot \overset{1}{\cancel{3}} \cdot 9}{\underset{1}{\cancel{3}} \cdot 7}$ cubic inches    Multiply and write 27 as 3 · 9.

             $= \frac{792}{7}$   or   $113\frac{1}{7}$ cubic inches

The volume is *approximately* $113\frac{1}{7}$ cubic inches.

**Example 7**    Approximate the volume of a can that has a $3\frac{1}{2}$-inch radius and a height of 6 inches. Use $\frac{22}{7}$ for $\pi$.

3½ in.

6 in.

*Solution:*    Using the formula for volume of a circular cylinder, we have

$$V = \pi r^2 h$$

$3\frac{1}{2} = \frac{7}{2}$

$$= \pi \cdot \left(\frac{7}{2} \text{ inches}\right)^2 \cdot 6 \text{ inches}$$

or approximately

$$\approx \frac{22}{7} \cdot \frac{49}{4} \cdot 6 \text{ cubic inches}$$

$$= 231 \text{ cubic inches}$$

The volume is approximately 231 cubic inches.    ▬▬▬

**Example 8**    Approximate the volume of a cone that has a height of 14 centimeters and radius of 3 centimeters. Use $\frac{22}{7}$ for $\pi$.

14 cm

3 cm

*Solution:*    Using the formula for volume of a cone, we have

$$V = \frac{1}{3}\pi r^2 h$$

$$= \frac{1}{3} \cdot \pi \cdot (3 \text{ centimeters})^2 \cdot 14 \text{ centimeters}$$    Replace $r$ with 3 cm and $h$ with 14 cm.

$$= 42\pi$$

or approximately

$$\approx 42 \cdot \frac{22}{7} \text{ cubic centimeters}$$

$$= 132 \text{ cubic centimeters}$$

The volume is approximately 132 cubic centimeters.    ▬▬▬

**Practice Problem 7**

Approximate the volume of a cylinder of radius 5 in. and height 7 in. Use 3.14 for $\pi$.

TEACHING TIP    Classroom Activity

Have students work in groups to create a sketch of a solid which combines 2 or more solids. Then have them find the volume of their solid. Finally have them exchange the sketch of their solid with another group to find the volume.

**Practice Problem 8**

Find the volume of a square-based pyramid that has a 3-meter side and height of 5.1 meters.

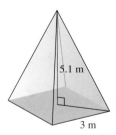

5.1 m

3 m

**Answers**

**7.** 549.5 cu. in.    **8.** 15.3 cu. m

# Focus On Mathematical Connections

POLYGONS

A **plane figure** is a figure that lies on a plane. Plane figures, like planes, have length and width but no thickness or depth.

A **polygon** is a closed plane figure that basically consists of three or more line segments that meet at their endpoints. A polygon is named according to the number of its sides. The table summarizes information about some basic polygons.

| POLYGONS | | | |
|---|---|---|---|
| Number of Sides | Name | Sum of Interior Angles | Example |
| 3 | Triangle | 180° | |
| 4 | Quadrilateral | 360° | |
| 5 | Pentagon | 540° | |
| 6 | Hexagon | 720° | |
| 7 | Heptagon | | |
| 8 | Octagon | | |
| 9 | Nonagon | | |
| 10 | Decagon | | |

CRITICAL THINKING

Study the sums of interior angles shown in the table for the triangle, quadrilateral, pentagon, and hexagon. What pattern do you notice. Use this pattern to complete the Sum of Interior Angles column in the table.

## EXERCISE SET 9.4

 **A** *Find the area of each geometric figure. If the figure is a circle, give an exact area and then use the given* **approximation** *for* π *to approximate the area. See Examples 1 through 4.*

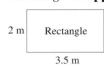 **1.**

2 m | Rectangle

3.5 m

**2.**

2.75 ft | Rectangle

7 ft

 **3.**

3 yd

$6\frac{1}{2}$ yd

**4.**

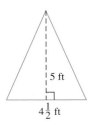

5 ft

$4\frac{1}{2}$ ft

**5.**

6 yd

5 yd

**6.**

5 ft       7 ft

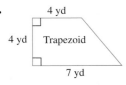 **7.** Use 3.14 for π.

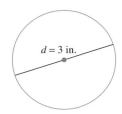

d = 3 in.

**8.** Use $\frac{22}{7}$ for π.

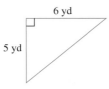

r = 2 cm

**9.**

Parallelogram

5.25 ft

7 ft

**10.**

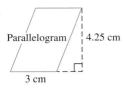

Parallelogram | 4.25 cm

3 cm

**11.**

5 m

Trapezoid

4 m

9 m

**12.**

Trapezoid

5 in.    6 in.    $8\frac{1}{2}$ in.

**13.**

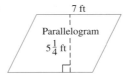

4 yd

4 yd | Trapezoid

7 yd

**14.**

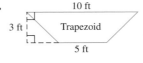

10 ft

3 ft | Trapezoid

5 ft

**15.**

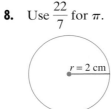

7 ft

Parallelogram

$5\frac{1}{4}$ ft

**16.**

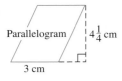

Parallelogram | $4\frac{1}{4}$ cm

3 cm

**17.**

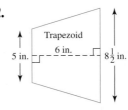

$4\frac{1}{2}$ in. | Parallelogram

5 in.

**18.**

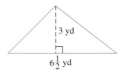

4 m

6 m

Parallelogram

**ANSWERS**

1. ___7 sq. m___

2. ___19.25 sq. ft___

3. ___$9\frac{3}{4}$ sq. yd___

4. ___$11\frac{1}{4}$ sq. ft___

5. ___15 sq. yd___

6. ___17.5 sq. ft___

7. ___2.25π sq. in. ≈ 7.065 sq. in___

8. ___4π sq. cm ≈ $12\frac{4}{7}$ sq. cm___

9. ___36.75 sq. ft___

10. ___12.75 sq. cm___

11. ___28 sq. m___

12. ___$40\frac{1}{2}$ sq. in.___

13. ___22 sq. yd___

14. ___22.5 sq. ft___

15. ___$36\frac{3}{4}$ sq. ft___

16. ___$12\frac{3}{4}$ sq. cm___

17. ___$22\frac{1}{2}$ sq. in.___

18. ___24 sq. m___

**Name** _____

**19.**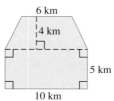
$1\frac{1}{2}$ cm  2 cm  $1\frac{1}{2}$ cm
3 cm
7 cm

**20.**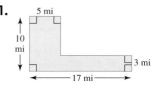
6 km
4 km
5 km
10 km

 **21.**
5 mi
10 mi
3 mi
17 mi

**22.**
25 cm
15 cm
12 cm
5 cm

**23.**
5 cm
3 cm

**24.**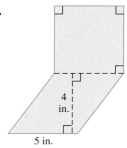
4 in.
5 in.

**25.** Use $\frac{22}{7}$ for $\pi$.

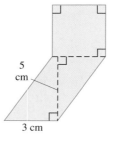
$r = 6$ in.

**26.** Use 3.14 for $\pi$.

$d = 5$ m

**B** *Find the volume of each solid. Use $\frac{22}{7}$ for $\pi$. See Examples 5 through 8.*

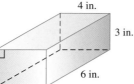 **27.**
4 in.
3 in.
6 in.

**28.**
3 mi

**29.**
8 cm
8 cm
8 cm

**30.**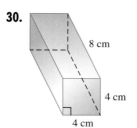
8 cm
4 cm
4 cm

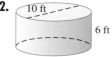 **31.**
3 yd
2 yd

**32.**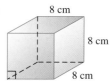
10 ft
6 ft

**33.**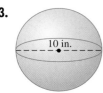
10 in.

**34.**
$1\frac{3}{4}$ in.
9 in.

**35.**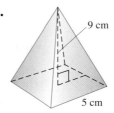
9 cm
5 cm

**36.**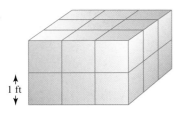
1 ft

**A** **B** *Solve. See Examples 1 through 8.*

**37.** Find the volume of a cube with edges of $1\frac{1}{3}$ inches.

$1\frac{1}{3}$ in.

**38.** A water storage tank is in the shape of a cone with the pointed end down. If the radius is 14 ft and the depth of the tank is 15 ft, approximate the volume of the tank in cubic feet. Use $\frac{22}{7}$ for $\pi$.

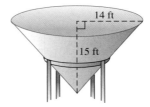
14 ft
15 ft

**39.** Find the volume of a rectangular box 2 ft by 1.4 ft by 3 ft.

**40.** Find the volume of a box in the shape of a cube that is 5 ft on each side.

**41.** Reebok International Ltd. of Massachusetts flew the largest banner from a single-seater plane in 1990. Use the diagram to find its area. (*Source: The Guinness Book of Records*, 1998)

100 ft
Reebok Totally Beachin'
50 ft

**42.** The longest illuminated sign is in Ramat Gan, Israel, and measures 197 ft by 66 ft. Find its area. (*Source: The Guinness Book of Records*, 1998)

66 ft
197 ft

**33.** $523\frac{17}{21}$ cu. in.

**34.** $28\frac{7}{8}$ cu. in.

**35.** 75 cu. cm

**36.** 18 cu. ft

**37.** $2\frac{10}{27}$ cu. in.

**38.** 3080 cu. ft

**39.** 8.4 cu. ft

**40.** 125 cu. ft

**41.** 5000 sq. ft

**42.** 13,002 sq. ft

**43.** 168 sq. ft

**44.** 599.5 sq. cm

**45.** 960 cu. cm

**46.** $2095\frac{5}{21}$ cu. in.

**47.** 9200 sq. ft

**48.** 4320 sq. ft

**49.** 381 sq. ft

**50. a.** 240 sq. ft

    **b.** 1440 bricks

**43.** A drapery panel measures 6 ft by 7 ft. Find how many square feet of material are needed for *four* panels.

**44.** A page in this book measures 27.5 cm by 21.8 cm. Find its area.

**45.** A paperweight is in the shape of a square-based pyramid 20 cm tall. If an edge of the base is 12 cm, find the volume of the paperweight.

**46.** A bird bath is made in the shape of a hemisphere (half-sphere). If its radius is 10 in., approximate the volume. Use $\frac{22}{7}$ for $\pi$.

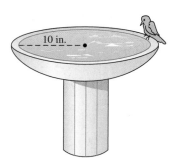

10 in.

**47.** Find how many square feet of land are in the plot shown.

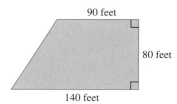

90 feet
80 feet
140 feet

**48.** For Gerald Gomez to determine how much grass seed he needs to buy, he must know the size of his yard. Use the drawing to determine how many square feet are in his yard.

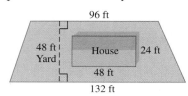

96 ft
48 ft Yard
House
24 ft
48 ft
132 ft

**49.** The shaded part of the roof shown is in the shape of a trapezoid and needs to be shingled. The number of packages of shingles to buy depends on the area. Use the dimensions given to find the area of the shaded part of the roof to the nearest whole square foot.

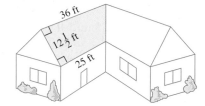

36 ft
$12\frac{1}{2}$ ft
25 ft

**50.** The end of the building shaded in the drawing is to be bricked. The number of bricks to buy depends on the area.
**a.** Find the area.

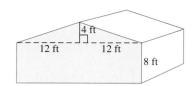

4 ft
12 ft
12 ft
8 ft

**b.** If the side area of each brick (including mortar room) is $\frac{1}{6}$ sq. ft, find the number of bricks needed to buy.

$\frac{1}{4}$ ft
$\frac{2}{3}$ ft

**Name** _____

## REVIEW AND PREVIEW

*Evaluate. See Section 1.7.*

**51.** $5^2$        **52.** $7^2$        **53.** $3^2$        **54.** $20^2$

**55.** $1^2 + 2^2$      **56.** $5^2 + 3^2$      **57.** $4^2 + 2^2$      **58.** $1^2 + 6^2$

## COMBINING CONCEPTS

**59.** A pizza restaurant recently advertised two specials. The first special was a 12-inch pizza for $10. The second special was two 8-inch pizzas for $9. Determine the better buy. (*Hint:* First compare the areas of the two specials and then find a price per square inch for both specials.)

**60.** Find the approximate area of the state of Utah.

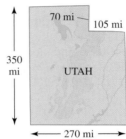

**61.** Find the area of a rectangle that measures 2 *feet* by 8 *inches*. Give the area in square feet and in square inches.

**62.** The Great Pyramid of Khufu at Giza is the largest of the ancient Egyptian pyramids. Its original height was 146.5 meters. The length of each side of its square base was originally 230 meters. Find the volume of the Great Pyramid of Khufu as it was originally built. Round to the nearest whole cubic meter. (*Source:* PBS *NOVA* Online)

**51.** 25

**52.** 49

**53.** 9

**54.** 400

**55.** 5

**56.** 34

**57.** 20

**58.** 37

**59.** 12-in. pizza

**60.** 87,150 sq. mi

**61.** $1\frac{1}{3}$ sq. ft; 192 sq. in.

**62.** 2,583,283 cu. m

**713**

**63.** The second largest pyramid at Giza is Khafre's Pyramid. Its original height was 471 feet. The length of each side of its square base was originally 704 feet. Find the volume of Khafre's Pyramid as it was originally built. (*Source:* PBS *NOVA* Online)

**64.** Menkaure's Pyramid, the smallest of the three great pyramids at Giza, was originally 65.5 meters tall. Each of the sides of its square base was originally 344 meters long. What was the volume of Menkaure's Pyramid as it was originally built? Round to the nearest whole cubic meter. (*Source:* PBS *NOVA* Online)

**65.** Due to factors such as weathering and loss of outer stones, the Great Pyramid of Khufu now stands only 137 meters tall. Its square base is now only 227 meters on a side. Find the current volume of the Great Pyramid of Khufu to the nearest whole cubic meter. How much has its volume decreased since it was built? See Exercise 62 for comparison. (*Source:* PBS *NOVA* Online)

**66.** The centerpiece of the New England Aquarium in Boston is its Giant Ocean Tank. This exhibit is a four-story cylindrical saltwater tank containing sharks, sea turtles, stingrays, and tropical fish. The radius of the tank is 16.3 feet and its height is 32 feet (assuming that a story is 8 feet). What is the volume of the Giant Ocean Tank? Use $\pi \approx 3.14$ and round to the nearest tenth of a cubic foot. (*Source:* New England Aquarium)

**67.** Except for service dogs for guests with disabilities, Walt Disney World does not allow pets in its parks or hotels. However, the resort does make pet-boarding services available to guests. The pet-care kennels at Walt Disney World offer three different sizes of indoor kennels. Of these, the smaller two kennels measure
**a.** 2′1″ × 1′8″ × 1′7″ and
**b.** 1′1″ × 2′ × 8″. What is the volume of each kennel rounded to the nearest cubic foot? Which is larger? (*Source:* Walt Disney World Resort)

**68.** In your own words, explain the difference between the perimeter and the area of a polygon.

**69.** Can you compute the volume of a rectangle? Why or why not?

**Name** _____ **Section** _____ **Date** _____

## INTEGRATED REVIEW—GEOMETRY CONCEPTS

△ **1.** Find the supplement and the complement of a 27° angle.

*Find the measures of angles x, y, and z in each figure.*

△ **2.**

△ **3.** $m \parallel n$

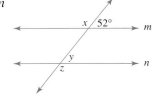

△ **4.** Find the measure of $\angle x$.

*Convert each measurement of length as indicated.*

**5.** 36 in. = _____ ft

**6.** 10,560 ft = _____ mi

**7.** 20 ft = _____ yd

**8.** 6 yd = _____ ft

**9.** 2.1 mi = _____ ft

**10.** 3.2 ft = _____ in.

**11.** 30 m = _____ cm

**12.** 24 mm = _____ cm

**13.** 2000 mm = _____ m

**14.** 18 m = _____ cm

**15.** 7.2 cm = _____ mm

**16.** 600 m = _____ km

17. perimeter = 20 m;
    area = 25 sq. m

18. perimeter = 12 ft;
    area = 6 sq. ft

19. circumference =
    6π cm ≈ 18.84 cm;
    area = 9π sq. cm
    ≈ 28.26 sq. cm

20. perimeter = 32 mi;
    area = 44 sq. mi

21. perimeter = 62 ft;
    area = 238 sq. ft

22. 64 cu. in.

23. 30.6 cu. ft

24. 400 cu. cm

25. $\dfrac{2048}{3}$ π cu. mi ≈
    $2145\dfrac{11}{21}$ cu. mi

*Find the perimeter (or circumference) and area of each figure. For the circle, give an exact circumference and area, then use π ≈ 3.14 to approximate each. Don't forget to attach correct units.*

△ **17.**

Square | 5 m

△ **18.**

4 ft
3 ft
5 ft

△ **19.**

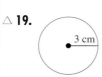
3 cm

△ **20.**

11 mi
Parallelogram | 5 mi
4 mi

△ **21.** The smallest cathedral is in Highlandville, Missouri, and measures 14 feet by 17 feet. Find its perimeter and its area. (*Source: The Guinness Book of Records*, 1998)

*Find the volume of each solid. Don't forget to attach correct units.*

△ **22.** A cube with edges of 4 inches each

△ **23.** A rectangular box 2 feet by 3 feet by 5.1 feet

△ **24.** A pyramid with a square base 10 centimeters on a side and a height of 12 centimeters

△ **25.** A sphere with a diameter of 16 miles; give the exact volume and then use π ≈ $\dfrac{22}{7}$ to approximate.

# 9.5 WEIGHT AND MASS

## A CONVERTING U.S. SYSTEM UNITS OF WEIGHT

Whenever we talk about how heavy an object is, we are concerned with the object's **weight**. We discuss weight when we refer to a 12-ounce box of Rice Krispies, an overweight 19-pound tabby cat, or a barge hauling 24 tons of garbage.

The most common units of weight in the U.S. measurement system are the **ounce**, the **pound**, and the **ton**. The following is a summary of equivalencies between units of weight.

| U.S. UNITS OF WEIGHT | UNIT FRACTIONS |
|---|---|
| 16 ounces (oz) = 1 pound (lb) | $\dfrac{16 \text{ oz}}{1 \text{ lb}} = \dfrac{1 \text{ lb}}{16 \text{ oz}} = 1$ |
| 2000 pounds = 1 ton | $\dfrac{2000 \text{ lb}}{1 \text{ ton}} = \dfrac{1 \text{ ton}}{2000 \text{ lb}} = 1$ |

## TRY THE CONCEPT CHECK IN THE MARGIN.

Unit fractions that equal 1 will be used to convert between units of weight in the U.S. system. When converting using unit fractions, recall that the numerator of a unit fraction should contain units we are converting to and the denominator should contain original units.

To convert 40 ounces to pounds, multiply by $\dfrac{1 \text{ lb}}{16 \text{ oz}}$.    ← Units converting to
            ← Original units

$$40 \text{ oz} = \frac{40 \, \cancel{oz}}{1} \cdot \overbrace{\frac{1 \text{ lb}}{16 \, \cancel{oz}}}^{\text{Unit fraction}}$$

$$= \frac{40 \text{ lb}}{16} \quad \text{Multiply.}$$

$$= \frac{5}{2} \text{ lb} \quad \text{or} \quad 2\frac{1}{2} \text{ lb, as a mixed number}$$

**Example 1**    Convert 9000 pounds to tons.

*Solution:*    We multiply 9000 lb by the unit fraction

$$\frac{1 \text{ ton}}{2000 \text{ lb}} \quad \begin{array}{l} \leftarrow \text{ Units converting to} \\ \leftarrow \text{ Original units} \end{array}$$

**TEACHING TIP**    Classroom Activity
Have students begin this lesson by working in groups to make a list or poster of items having the following estimated weights: 1 ounce, 10 ounces, 1 pound, 10 pounds, 100 pounds, 1000 pounds, 1 ton, and 10 tons.

## ✓ CONCEPT CHECK

If you were describing the weight of the engine car of a train, which type of unit would you use: ounce, pound, or ton? Why?

## Practice Problem 1

Convert 4500 pounds to tons.

**Answers**

**1.** $2\dfrac{1}{4}$ tons

✓ Concept Check: Ton.

$$9000 \text{ lb} = \frac{9000 \text{ lb}}{1} \cdot \frac{1 \text{ ton}}{2000 \text{ lb}} = \frac{9000 \text{ tons}}{2000} = \frac{9}{2} \text{ tons or } 4\frac{1}{2} \text{ tons}$$

**Practice Problem 2**

Convert 56 ounces to pounds.

**Example 2**   Convert 3 pounds to ounces.

*Solution:*   We multiply by the unit fraction $\frac{16 \text{ oz}}{1 \text{ lb}}$ to convert from pounds to ounces.

$$3 \text{ lb} = \frac{3 \text{ lb}}{1} \cdot \frac{16 \text{ oz}}{1 \text{ lb}} = 3 \cdot 16 \text{ oz} = 48 \text{ oz}$$

As with length, it is sometimes useful to simplify a measurement of weight by writing it in terms of mixed units.

33 ounces = _____ lb _____ oz

Because 16 oz = 1 lb, divide 16 into 33 to see how many pounds are in 33 ounces. The quotient is the number of pounds and the remainder is the number of ounces. To see why we divide 16 into 33, notice that

$$33 \text{ oz} = 33 \text{ oz} \cdot \frac{1 \text{ lb}}{16 \text{ oz}} = \frac{33 \text{ lb}}{16}$$

$$\begin{array}{r} 2 \text{ lb } 1 \text{ oz} \\ 16\overline{)33} \\ -32 \\ \hline 1 \end{array}$$

Thus 33 ounces is the same as 2 lb 1 oz.

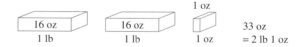

## B   OPERATIONS ON U.S. SYSTEM UNITS OF WEIGHT

Performing arithmetic operations on units of weight works the same way as performing arithmetic operations on units of length.

**Practice Problem 3**

Subtract 5 tons 1200 lb from 8 tons 100 lb.

**Example 3**   Subtract 3 tons 1350 lb from 8 tons 1000 lb.

*Solution:*   To subtract, we line up similar units.

$$\begin{array}{r} 8 \text{ tons } 1000 \text{ lb} \\ -3 \text{ tons } 1350 \text{ lb} \\ \hline \end{array}$$

Since we cannot subtract 1350 lb from 1000 lb, we borrow 1 ton from the 8 tons. To do so, we write 1 ton as 2000 lb and combine it with the 1000 lb.

**Answers**

**2.** $3\frac{1}{2}$ lb   **3.** 2 tons 900 lb

7 tons + (1 ton) 2000 lb

8̸ tons 1000 lb        becomes        7 tons 3000 lb
− 3 tons 1350 lb                              −3 tons 1350 lb
                                                      4 tons 1650 lb

To check, see that the sum of 4 tons 1650 lb and 3 tons 1350 lb is 8 tons 1000 lb.

**Example 4**    Multiply 5 lb 9 oz by 6.

*Solution:*    We multiply 5 lb by 6 and 9 oz by 6.

        5 lb   9 oz
×          6
30 lb 54 oz

To write 54 oz as mixed units, we divide by 16 (1 lb = 16 oz):

        3 lb 6 oz
16)‾5‾4‾
   −48
      6

Thus,

30 lb 54 oz = 30 lb + 3 lb 6 oz = 33 lb 6 oz

**Practice Problem 4**

Multiply 4 lb 11 oz by 8.

**Example 5**    Divide 9 lb 6 oz by 2.

*Solution:*    We divide each of the units by 2.

        4 lb      11 oz
2)‾9‾ ‾l‾b‾    6 oz
 −8
   1 lb = 16 oz
          22 oz    Divide 2 into 22 oz to get 11 oz.

To check, multiply 4 pounds 11 ounces by 2. The result is 9 pounds 6 ounces.

**Practice Problem 5**

Divide 5 lb 8 oz by 4.

**Example 6**    **Finding the Weight of a Child**

Bryan Gimbel weighed 8 lb 8 oz at birth. By the time he was 1 year old, he had gained 11 lb 14 oz. Find his weight at age 1 year.

*Solution:*
birth weight → 8 lb 8 oz
+ weight gained → +11 lb 14 oz
total weight → 19 lb 22 oz

Since 22 oz equals 1 lb 6 oz,

19 lb 22 oz = 19 lb + 1 lb 6 oz
            = 20 lb 6 oz

Bryan weighed 20 lb 6 oz on his first birthday.

**Practice Problem 6**

A 5-lb 14-oz batch of cookies is packed in a 6-oz container before it is mailed. Find the total weight.

**Answers**

**4.** 37 lb 8 oz  **5.** 1 lb 6 oz  **6.** 6 lb 4 oz

## C CONVERTING METRIC SYSTEM UNITS OF MASS

In scientific and technical areas, a careful distinction is made between **weight** and **mass**. **Weight** is really a measure of the pull of gravity. The farther from Earth an object gets, the less it weighs. However, **mass** is a measure of the amount of substance in the object and does not change. Astronauts orbiting Earth weigh much less than they weigh on Earth, but they have the same mass in orbit as they do on Earth. On Earth's surface, weight and mass are the same, so either term may be used.

The basic unit of mass in the metric system is the **gram**. It is defined as the mass of water contained in a cube 1 centimeter (cm) on each side.

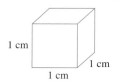

The following examples may help you get a feeling for metric masses.

A tablet contains 200 milligrams of ibuprofen.

A large paper clip weighs approximately 1 gram.

A box of crackers weighs 453 grams.

A kilogram is slightly over 2 pounds. An adult woman may weigh 60 kilograms.

The prefixes for units of mass in the metric system are the same as for units of length, as shown in the following table.

| Prefix | Meaning | Metric Unit of Mass |
|---|---|---|
| kilo | 1000 | 1 kilogram (kg) = 1000 grams (g) |
| hecto | 100 | 1 hectogram (hg) = 100 g |
| deka | 10 | 1 dekagram (dag) = 10g |
| | | **1 gram (g) = 1 g** |
| deci | 1/10 | 1 decigram (dg) = 1/10 g or 0.1 g |
| centi | 1/100 | 1 centigram (cg) = 1/100 g or 0.01 g |
| milli | 1/1000 | 1 milligram (mg) = 1/1000 g or 0.001 g |

TRY THE CONCEPT CHECK IN THE MARGIN.

✓ CONCEPT CHECK

True or false? A decigram is larger than a dekagram. Explain.

The **milligram**, the **gram**, and the **kilogram** are the three most commonly used units of mass in the metric system.

As with lengths, all units of mass are powers of 10 of the gram, so converting from one unit of mass to another only involves moving the decimal point. To convert from one unit of mass to another in the metric system, list the units of mass in order from largest to smallest.

Let's convert 4300 milligrams to grams. To convert from milligrams to grams, we move along the chart 3 units to the left.

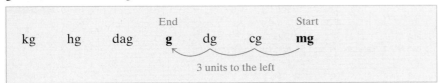

This means that we move the decimal point 3 places to the left to convert from milligrams to grams.

4300 mg = 4.3 g

The same conversion can be done with unit fractions.

$$4300 \text{ mg} = \frac{4300 \text{ mg}}{1} \cdot \frac{0.001 \text{ g}}{1 \text{ mg}}$$

$$= 4300 \cdot 0.001 \text{ g}$$

$$= 4.3 \text{ g} \quad \text{To multiply by 0.001, move the decimal point 3 places to the left.}$$

To see that this is reasonable, study the diagram.

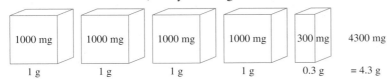

Thus 4300 mg = 4.3 g.

## Example 7

Convert 3.2 kg to grams.

*Solution:*  First we will convert by using a unit fraction.

$$3.2 \text{ kg} = 3.2 \text{ kg} \cdot \overbrace{\frac{1000 \text{ g}}{1 \text{ kg}}}^{\text{Unit fraction}} = 3200 \text{ g}$$

Now let's list the units of mass in a chart and move from kilograms to grams.

Start ⟶ kg  hg  dag  End g  dh  cg  mg

3 units to the right

3.200 kg = 3200. g

3 places to the right

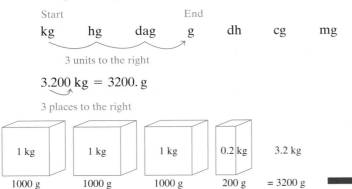

Practice Problem 7

Convert 3.41 g to milligrams.

Answers

**7.** 3410 mg

✓ Concept Check: False.

## Practice Problem 8

Convert 56.2 cg to grams.

### ✓ CONCEPT CHECK

What is wrong with the following conversion? Convert 422.225 g to kilograms.

$$422.225 \text{ g} = 422{,}225 \text{ kg}$$

## Practice Problem 9

Subtract 3.1 dg from 2.5 g.

## Practice Problem 10

Multiply 12.6 kg by 4.

## Practice Problem 11

Twenty-four bags of cement weigh a total of 550 kg. Find the average weight of 1 bag, rounded to the nearest kilogram.

### Answers

**8.** 0.562 g    **9.** 2.19 g or 21.9 dg    **10.** 50.4 kg
**11.** 23 kg
✓ **Concept Check:** Move decimal point 3 places to the left, not right.

---

**Example 8**    Convert 2.35 cg to grams.

*Solution:*    We list the units of mass in a chart and move from centigrams to grams.

|     |     |     | End |     | Start |     |
| --- | --- | --- | --- | --- | --- | --- |
| kg  | hg  | dag | g   | dg  | cg  | mg  |

2 units to the left

$$02.35\text{cg} = 0.0235 \text{ g}$$

2 places to the left

**TRY THE CONCEPT CHECK IN THE MARGIN.**

### D   OPERATIONS ON METRIC SYSTEM UNITS OF MASS

Arithmetic operations can be performed with metric units of mass just as we performed operations with metric units of length. We convert each number to the same unit of mass and add, subtract, multiply, or divide as with decimals.

**Example 9**    Subtract 5.4 dg from 1.6 g.

*Solution:*    We convert both numbers to decigrams or to grams before subtracting.

$$5.4 \text{ dg} = 0.54 \text{ g} \quad \text{or} \quad 1.6 \text{ g} = 16 \text{ dg}$$

$$\begin{array}{r} 1.60 \text{ g} \\ -0.54 \text{ g} \\ \hline 1.06 \text{ g} \end{array} \qquad \begin{array}{r} 16.0 \text{ dg} \\ -\ 5.4 \text{ dg} \\ \hline 10.6 \text{ dg} \end{array}$$

The difference is 1.06 g or 10.6 dg.

**Example 10**    Multiply 15.4 kg by 5.

*Solution:*    We multiply the two numbers together.

$$\begin{array}{r} 15.4 \text{ kg} \\ \times \quad 5 \\ \hline 77.0 \text{ kg} \end{array}$$

The result is 77.0 kg.

**Example 11**    **Calculating Allowable Weight in an Elevator**

An elevator has a weight limit of 1400 kg. A sign posted in the elevator indicates that the maximum capacity of the elevator is 17 persons. What is the average allowable weight for each passenger, rounded to the nearest kilogram?

*Solution:*    To solve, notice that the total weight of 1400 kilograms ÷ 17 = average weight.

$$
\begin{array}{r}
82.3 \text{ kg} \approx 82 \text{ kg} \\
17\overline{)1400.0 \text{ kg}} \\
-136 \phantom{00000} \\
\hline
40 \phantom{0000} \\
-34 \phantom{000} \\
\hline
60 \phantom{00} \\
-51 \phantom{0} \\
\hline
9 \phantom{}
\end{array}
$$

Each passenger can weigh an average of 82 kg. (Recall that a kilogram is slightly over 2 pounds, so 82 kilograms is over 164 pounds. For a better approximation, see the Calculator Explorations box.)

---

## CALCULATOR EXPLORATIONS
### METRIC TO U.S. SYSTEM CONVERSIONS IN WEIGHT

To convert between the two systems of measurement in weight, the following **approximations** can be used.

grams × 0.035 ≈ ounces          ounces × 28.35 ≈ grams

kilograms × 2.20 ≈ pounds        pounds × 0.454 ≈ kilograms

grams × 0.0022 ≈ pounds          pounds × 454 ≈ grams

**Example**    A bulldog weighs 35 pounds. How many kilograms is this?

*Solution:*    From the above approximations,

pounds × 0.454 ≈ kilograms

↓

35 × 0.454 ≈ kilograms

To multiply on your calculator, press the keys

| 35 | × | 0.454 | = | (or | ENTER | ).

The display will read | 15.89 |. Thus 35 pounds ≈ 15.89 kilograms.

*Convert as indicated.*

**1.** 15 oz to grams.    ≈425.25 g

**2.** 11.2 g to ounces.    ≈0.392 oz

**3.** 7 kg to pounds.    ≈15.4 lb

**4.** 23 lb to kilograms.    ≈10.442 kg

**5.** A piece of candy weighs 5 oz grams. How many ounces is this?    ≈0.175 oz

**6.** If a person weighs 82 kilograms, how many pounds is this?    ≈180.4 lb

TEACHING TIP

Ask students to use the approximations to determine which unit is heavier.
gram or ounce
kilogram or pound
gram or pound

# Focus On History

## THE DEVELOPMENT OF UNITS OF MEASURE

The earliest units of measure were based on the human body. The ancient Egyptians, Babylonians, Hebrews, and Mesopotamians used a unit of length called the cubit, which represents the distance between a human elbow and fingertips. For instance, in the book of Genesis in the Bible, Noah's ark is described as having a length of "three hundred cubits, its width fifty cubits, and its height thirty cubits." Other commonly used measures found in documents of these ancient cultures include the digit, hand, span, and foot. These measures are shown at the right.

Several thousand years later, the English system of measurement also consisted of a mixed bag of body- and nature-related units of measure. A rod was the combined length of the left feet of 16 men. An inch was the distance spanned by three grains of barley. A foot was the length of the foot of the king currently in power. Around 1100 A.D., King Henry I of England decreed that a yard was the distance between the king's nose and the thumb of his outstretched arm.

Although the English system became somewhat standardized by the 13th century, the problem with it and the ancient systems of measurement was that the relationship between the various units was not necessarily easy to remember. This made it difficult to convert from one type of unit to another.

The metric system grew out of the French Revolution during the 1790s. As a reaction against the lack of consistency and utility of existing systems of measurement, the French Academy of Sciences set out to develop an easy-to-use, internationally standardized system of measurements. Originally, the basic unit of the metric system, the meter, was to be one ten-millionth of the distance between the North Pole and the Equator on a line through Paris. However, it was soon discovered that this distance was nearly impossible to measure accurately. The meter has been redefined several times since the end of the French Revolution. Most recently, the meter was redefined in 1983 as the distance that light travels in a vacuum during $\dfrac{1}{299,792,458}$ of a second.

Although the definition of the meter has evolved over time, the original idea of developing an easy-to-use system of measurements has endured. The metric system uses a standard set of prefixes for all basic unit types (length, mass, capacity, etc.) and all conversions within a unit type are based on powers of 10.

## CRITICAL THINKING

Develop your own unit of measure for **(a)** length and **(b)** area. Explain how you chose your units and for what types of relative sizes of measurements they would be most useful. Discuss the advantages and disadvantages of your units of measurement. Then use your units to measure the width of your classroom desk and the area of the front cover of this workbook.

digit

hand

span

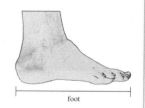

foot

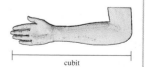

cubit

Name _____  Section _____ Date _____

## MENTAL MATH

*Convert.*

**1.** 16 oz to pounds

**2.** 32 oz to pounds

**3.** 1 ton to pounds

**4.** 2 tons to pounds

**5.** 1 lb to ounces

**6.** 3 lb to ounces

**7.** 2000 lb to tons

**8.** 4000 lb to tons

*Determine whether the measurement in each statement is reasonable.*

**9.** The doctor prescribed a pill containing 2 kg of medication.

**10.** A full-grown cat weighs approximately 15 g.

**11.** A bag of flour weighs 4.5 kg.

**12.** A staple weighs 15 mg.

**13.** A professor weighs less than 150 g.

**14.** A car weighs 2000 mg.

## EXERCISE SET 9.5

**A** *Convert as indicated. See Examples 1 and 2.*

**1.** 2 lb to ounces

**2.** 4 lb to ounces

**3.** 5 tons to pounds

**4.** 3 tons to pounds

**5.** 12,000 lb to tons

**6.** 32,000 lb to tons

**7.** 60 oz to pounds

**8.** 90 oz to pounds

MENTAL MATH ANSWERS

**1.** 1 lb

**2.** 2 lb

**3.** 2000 lb

**4.** 4000 lb

**5.** 16 oz

**6.** 48 oz

**7.** 1 ton

**8.** 2 tons

**9.** no

**10.** no

**11.** yes

**12.** yes

**13.** no

**14.** no

ANSWERS

**1.** 32 oz

**2.** 64 oz

**3.** 10,000 lb

**4.** 6000 lb

**5.** 6 tons

**6.** 16 tons

**7.** $3\frac{3}{4}$ lb

**8.** $5\frac{5}{8}$ lb

**9.** $1\dfrac{3}{4}$ tons _____

**10.** $4\dfrac{1}{2}$ tons _____

**11.** 260 oz _____

**12.** 232 oz _____

**13.** 9800 lb _____

**14.** 16,600 lb _____

**15.** 76 oz _____

**16.** 146 oz _____

**17.** 1.5 tons _____

**18.** 3.2 lb _____

**19.** 53 lb 10 oz _____

**20.** 17 lb 2 oz _____

**21.** 9 tons 390 lb _____

**22.** 4 tons 55 lb _____

**23.** 3 tons 175 lb _____

**24.** 3 tons 590 lb _____

**25.** 8 lb 11 oz _____

**26.** 18 lb 12 oz _____

**27.** 31 lb 2 oz _____

**28.** 11 lb 9 oz _____

**29.** 1 ton 700 lb _____

**30.** 1 ton 600 lb _____

**9.** 3500 lb to tons

**10.** 9000 lb to tons

**11.** 16.25 lb to ounces

**12.** 14.5 lb to ounces

**13.** 4.9 tons to pounds

**14.** 8.3 tons to pounds

**15.** $4\dfrac{3}{4}$ lb to ounces

**16.** $9\dfrac{1}{8}$ lb to ounces

**17.** 2950 lb to the nearest tenth of a ton

**18.** 51 oz to the nearest tenth of a pound

**B** *Perform each indicated operation. See Examples 3 through 5.*

**19.** 34 lb 12 oz + 18 lb 14 oz

**20.** 6 lb 10 oz + 10 lb 8 oz

**21.** 6 tons 1540 lb + 2 tons 850 lb

**22.** 2 tons 1575 lb + 1 ton 480 lb

**23.** 5 tons 1050 lb − 2 tons 875 lb

**24.** 4 tons 850 lb − 1 ton 260 lb

**25.** 12 lb 4 oz − 3 lb 9 oz

**26.** 45 lb 6 oz − 26 lb 10 oz

**27.** 5 lb 3 oz × 6

**28.** 2 lb 5 oz × 5

**29.** 6 tons 1500 lb ÷ 5

**30.** 5 tons 400 lb ÷ 4

**Name** _____

*Solve. Remember to insert units when writing your answers. See Example 6.*

**31.** Doris Johnson has two open containers of Uncle Ben's rice. If she combines 1 lb 10 oz from one container with 3 lb 14 oz from the other container, how much total rice does she have?

**32.** Dru Mizel maintains the records of the amount of coal delivered to his department in the steel mill. In January, 3 tons 1500 lb were delivered. In February, 2 tons 1200 lb were delivered. Find the total amount delivered in these two months.

**33.** Clay Cromer was amazed when he grew a 28 lb 10 oz zucchini in his garden, but later he learned that the heaviest zucchini ever grown weighed 64 lb 8 oz. It was grown in Llanharry, Wales by B. Lavery in 1990. How far below the record weight was Clay's squash? (*Source: The Guinness Book of Records,* 2000)

**34.** The heaviest baby born in good health weighed an incredible 22 lb 8 oz. He was born in Italy in September, 1955. How much heavier is this than a 7 lb 12 oz baby? (*Source: The Guinness Book of Records,* 1996)

**35.** The Shop 'n Bag supermarket chain ships hamburger meat by placing 10 packages of hamburger in a box, with each package weighing 3 lb 4 oz. How much will 4 boxes of hamburger weigh?

**36.** The Quaker Company ships its 1-lb 2-oz boxes of oatmeal in cartons containing 12 boxes of oatmeal. How much will 3 such cartons weigh?

**37.** A carton of Del Monte pineapple weighs 55 lb 4 oz, but 2 lb 8 oz of this weight is due to packaging. Subtract the weight of the packaging to find the actual weight of the pineapple in 4 cartons.

**38.** The Hormel Corporation ships cartons of canned ham weighing 43 lb 2 oz each. Of this weight, 3 lb 4 oz is due to packaging. Find the actual weight of the ham found in 3 cartons.

**39.** One bag of Pepperidge Farm Bordeaux cookies weighs $6\frac{3}{4}$ ounces. How many pounds will a dozen bags weigh?

**40.** One can of Payless red beets weighs $8\frac{1}{2}$ ounces. How much will 8 cans weigh?

**31.** 5 lb 8 oz
**32.** 6 tons 700 lb
**33.** 35 lb 14 oz
**34.** 14 lb 12 oz
**35.** 130 lb
**36.** 40 lb 8 oz
**37.** 211 lb
**38.** 119 lb 10 oz
**39.** 5 lb 1 oz
**40.** 4 lb 4 oz

**Name** _____

**C** *Convert as indicated. See Examples 7 and 8.*

**41.** 500 g to kilograms

**42.** 650 g to kilograms

**43.** 4 g to milligrams

**44.** 9 g to milligrams

**45.** 25 kg to grams

**46.** 18 kg to grams

**47.** 48 mg to grams

**48.** 112 mg to grams

**49.** 6.3 g to kilograms

**50.** 4.9 g to kilograms

**51.** 15.14 g to milligrams

**52.** 16.23 g to milligrams

**53.** 4.01 kg to grams

**54.** 3.16 kg to grams

**D** *Perform each indicated operation. See Examples 9 and 10.*

**55.** 3.8 mg + 9.7 mg

**56.** 41.6 g + 9.8 g

**57.** 205 mg + 5.61 g

**58.** 2.1 g + 153 mg

**59.** 9 g − 7150 mg

**60.** 4 kg − 2410 g

**61.** 1.61 kg − 250 g

**62.** 6.13 g − 418 mg

**63.** 5.2 kg × 2.6

**64.** 4.8 kg × 9.3

**65.** 17 kg ÷ 8

**66.** 8.25 g ÷ 6

*Solve. Remember to insert units when writing your answers. See Example 11.*

**67.** A can of 7-Up weighs 336 grams. Find the weight in kilograms of 24 cans.

**68.** Guy Green normally weighs 73 kg, but he lost 2800 grams after being sick with the flu. Find Guy's new weight.

**69.** Sudafed is a decongestant that comes in two strengths. The regular strength contains 60 mg of medication. The extra strength contains 0.09 g of medication. How much extra medication is in the extra-strength tablet?

**70.** A small can of Planters sunflower nuts weighs 177 grams. If each can contains 6 servings, find the weight of 1 serving.

**71.** Tim Caucutt's doctor recommends that Tim limit his daily intake of sodium to 0.6 gram. A one-ounce serving of Cheerios with $\frac{1}{2}$ cup of fortified skim milk contains 350 mg of sodium. How much more sodium can Tim have after he eats a bowl of Cheerios for breakfast, assuming he intends to follow the doctor's orders?

**72.** A large bottle of Hire's root beer weighs 1900 g. If a carton contains 6 large bottles of root beer, find the weight in kilograms of 5 cartons.

**73.** Three milligrams of preservatives are added to a 0.5-kg box of dried fruit. How many milligrams of preservatives are in 3 cartons of dried fruit if each carton contains 16 boxes?

**74.** One box of Swiss Miss Cocoa Mix weighs 0.385 kg, but 39 grams of this weight is the packaging. Find the actual weight of the cocoa in 8 boxes.

**75.** A carton of 12 boxes of Quaker Oats oatmeal weighs 6.432 kg. Each box includes 26 grams of packaging material. What is the actual weight of the oatmeal in the carton?

**76.** The supermarket prepares hamburger in 875-gram market packages. When Leo Gonzalas gets home, he divides the package in half before refrigerating the meat. How much will each package weigh?

**77.** A package of Trailway's Gorp, a high-energy hiking trail mix, contains 0.3 kg of nuts, 0.15 kg of chocolate bits, and 400 g of raisins. Find the total weight of the package.

**78.** The manufacturer of a pain reliever wants to reduce the caffeine content of its aspirin by $\frac{1}{4}$. Currently, each regular tablet contains 32 mg of caffeine. How much caffeine should be removed from each tablet?

**79.** A regular-size bag of Lay's potato chips weighs 198 grams. Find the weight of a dozen bags, rounded to the nearest hundredth of a kilogram.

**80.** Clarissa Patterson's cat weighs a hefty 9 kg. The vet has recommended that the cat lose 1500 grams. How much should the cat weigh?

**69.** 30 mg

**70.** 29.5 g

**71.** 250 mg

**72.** 57 kg

**73.** 144 mg

**74.** 2.768 kg

**75.** 6.12 kg

**76.** 437.5 g

**77.** 850 g or 0.85 kg

**78.** 8 mg

**79.** 2.38 kg

**80.** 7.5 kg

**81.** 0.25

**82.** 0.05

**83.** 0.16

**84.** 0.6

**85.** 0.875

**86.** 0.1875

**87.** answers may vary

**88.** answers may vary

**Name** _____

## REVIEW AND PREVIEW

*Write each fraction as a decimal See Section 5.6.*

**81.** $\dfrac{1}{4}$

**82.** $\dfrac{1}{20}$

**83.** $\dfrac{4}{25}$

**84.** $\dfrac{3}{5}$

**85.** $\dfrac{7}{8}$

**86.** $\dfrac{3}{16}$

## COMBINING CONCEPTS

**87.** Why is the decimal point moved to the right when grams are converted to milligrams?

**88.** To change 8 pounds to ounces, multiply by 16. Why is this the correct procedure?

# 9.6 Capacity

## A Converting U.S. System Units of Capacity

Units of **capacity** are generally used to measure liquids. The number of gallons of gasoline needed to fill a gas tank in a car, the number of cups of water needed in a bread recipe, and the number of quarts of milk sold each day at a supermarket are all examples of using units of capacity. The following summary shows equivalencies between units of capacity.

---

### U.S. Units of Capacity

$$8 \text{ fluid ounces (fl oz)} = 1 \text{ cup (c)}$$
$$2 \text{ cups} = 1 \text{ pint (pt)}$$
$$2 \text{ pints} = 1 \text{ quart (qt)}$$
$$4 \text{ quarts} = 1 \text{ gallon (gal)}$$

---

Just as with units of length and weight, we can form unit fractions to convert between different units of capacity. For instance,

$$\frac{2 \text{ c}}{1 \text{ pt}} = \frac{1 \text{ pt}}{2 \text{ c}} = 1 \quad \text{and} \quad \frac{2 \text{ pt}}{1 \text{ qt}} = \frac{1 \text{ qt}}{2 \text{ pt}} = 1$$

**Example 1**    Convert 9 quarts to gallons.

*Solution:*    We multiply by the unit fraction $\dfrac{1 \text{ gal}}{4 \text{ qt}}$.

$$9 \text{ qt} = \frac{9 \cancel{\text{ qt}}}{1} \cdot \frac{1 \text{ gal}}{4 \cancel{\text{ qt}}}$$

$$= \frac{9 \text{ gal}}{4}$$

$$= 2\frac{1}{4} \text{ gal}$$

Thus, 9 quarts is the same as $2\dfrac{1}{4}$ gallons, as shown in the diagram.

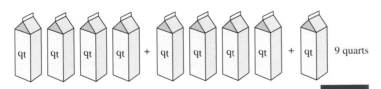

**Example 2**    Convert 14 cups to quarts.

*Solution:*    Our equivalency table contains no direct conversion from cups to quarts. However, from this table we know that

$$1 \text{ qt} = 2 \text{ pt} = 4 \text{ c}$$

so 1 qt = 4 c. Now we have the unit fraction $\dfrac{1 \text{ qt}}{4 \text{ c}}$. Thus,

---

Objectives

**A** Define U.S. units of capacity and convert from one unit to another.

**B** Perform arithmetic operations on U.S. units of capacity.

**C** Define metric units of capacity and convert from one unit to another.

**D** Perform arithmetic operations on metric units of capacity.

Study Guide    SSM    CD-ROM    Video 9.6

**TEACHING TIP**

Point out to students that ounces which are used to measure weight are not the same as fluid ounces which are used to measure capacity.

**Practice Problem 1**

Convert 43 pints to quarts.

**TEACHING TIP**

Ask your students to suggest a method for remembering that there are
    8 fluid ounces in 1 cup
    2 cups in a pint
    2 pints in a quart
    4 quarts in a gallon.

**Practice Problem 2**

Convert 26 quarts to cups.

**Answers**

**1.** $21\dfrac{1}{2}$ qt    **2.** 104 c

$$14\ c = \frac{14\ \cancel{c}}{1} \cdot \frac{1\ qt}{4\ \cancel{c}} = \frac{7}{2}\ qt \quad or \quad 3\frac{1}{2}\ qt$$

14 cups

1 qt     1 qt     1 qt     $\frac{1}{2}$ qt   $= 3\frac{1}{2}$ qt

**TRY THE CONCEPT CHECK IN THE MARGIN.**

### B   OPERATIONS ON U.S. SYSTEM UNITS OF CAPACITY

As is true of units of length and weight, units of capacity can be added, subtracted, multiplied, and divided.

**Example 3**   Subtract 3 qt from 4 gal 2 qt.

*Solution:*   To subtract, we line up similar units.

    4 gal 2 qt
−       3 qt

We cannot subtract 3 qt from 2 qt. We need to borrow 1 gallon from the 4 gallons, convert it to 4 quarts, and then combine it with the 2 quarts.

Borrow 1 gal = 4 qt

3 gal + (1 gal) 4 qt

   $\cancel{4}$ gal 2 qt     or     3 gal 6 qt
−       3 qt           −       3 qt
                              3 gal 3 qt

To check, see that the sum of 3 gal 3 qt and 3 qt is 4 gal 2 qt.

**Example 4**   Multiply 3 qt 1 pt by 3.

*Solution:*   We multiply each of the units of capacity by 3.

    3 qt 1 pt
×         3
    9 qt 3 pt

Since 3 pints is the same as 1 quart and 1 pint, we have

9 qt 3 pt = 9 qt + 1 qt 1 pt = 10 qt 1 pt

The 10 quarts can be changed to gallons by dividing by 4, since there are 4 quarts in a gallon. To see why we divide, notice that

$$10\ qt = \frac{10\ \cancel{qt}}{1} \cdot \frac{1\ gal}{4\ \cancel{qt}} = \frac{10}{4}\ gal$$

$$\begin{array}{r} 2\ gal\ 2\ qt \\ 4\overline{)10\ qts} \\ \underline{-8} \\ 2 \end{array}$$

Then the product is 10 qt 1 pt or 2 gal 2 qt 1 pt.

---

### ✓ CONCEPT CHECK

If 85 quarts are converted to fluid ounces, will the equivalent number of fluid ounces be less than or greater than 85?

### Practice Problem 3

Subtract 2 qt from 1 gal 1 qt.

### TEACHING TIP

After going over Example 3, have students repeat the problem after writing the units as gallons or quarts. Then have them compare their answer to make sure they are equivalent.

### Practice Problem 4

Multiply 2 gal 3 qt by 2.

**Answers**

**3.** 3 qt   **4.** 5 gal 2 qt
✓ **Concept Check:** Greater than 85.

## Example 5

Divide 3 gal 2 qt by 2.

*Solution:*   We divide each unit of capacity by 2.

$$
\begin{array}{r}
1\text{ gal} \quad 3\text{ qt} \\
\overline{2)3\text{ gal} \quad 2\text{ qt}} \\
\underline{-2} \\
1\text{ gal} = 4\text{ qt} \\
6\text{ qt} \qquad 6\text{ qt} \div 2 = 3\text{ qt}
\end{array}
$$

**Practice Problem 5**

Divide 6 gal 1 qt by 2.

## Example 6   Finding the Amount of Water in an Aquarium

An aquarium contains 6 gal 3 qt of water. If 2 gal 2 qt of water is added, what is the total amount of water in the aquarium?

*Solution:*

$$
\begin{array}{rl}
\text{beginning water} \rightarrow & 6\text{ gal } 3\text{ qt} \\
+ \quad \text{water added} \rightarrow & +2\text{ gal } 2\text{ qt} \\
\hline
\text{total water} \rightarrow & 8\text{ gal } 5\text{ qt}
\end{array}
$$

Since 5 qt = 1 gal 1 qt, we have

$$
= \overbrace{8\text{ gal}}^{8\text{ gal}} + \overbrace{1\text{ gal } 1\text{ qt}}^{5\text{ qt}}
$$

$$
= 9\text{ gal } 1\text{ qt}
$$

The total amount of water is 9 gallons 1 quart.

**Practice Problem 6**

A large oil drum contains 15 gal 3 qt of oil. How much will be in the drum if an additional 4 gal 3 qt of oil is poured into it?

## C   CONVERTING METRIC SYSTEM UNITS OF CAPACITY

Thus far, we know that the basic unit of length in the metric system is the meter and the basic unit of mass in the metric system is the gram. What is the basic unit of capacity? The **liter** is the basic unit of capacity in the metric system. By definition, a **liter** is the capacity or volume of a cube measuring 10 centimeters on each side.

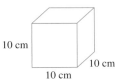

10 cm   10 cm   10 cm

The following examples may help you get a feeling for metric capacities.

One liter of liquid is slightly more than one quart.

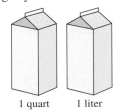

1 quart   1 liter

Many soft drinks are packaged in 2-liter bottles.

COLA

The metric system was designed to be a consistent system. Once again, the prefixes for metric units of capacity are the same as for metric units of length and mass, as summarized in the following table.

| Prefix | Meaning | Metric Unit of Capacity |
|--------|---------|-------------------------|
| kilo | 1000 | 1 kiloliter (kl) = 1000 liters (L) |
| hecto | 100 | 1 hectoliter (hl) = 100 L |
| deka | 10 | 1 dekaliter (dal) = 10 L |
| | | **1 liter (L) = 1 L** |
| deci | 1/10 | 1 deciliter (dl) = 1/10 L or 0.1 L |
| centi | 1/100 | 1 centiliter (cl) = 1/100 L or 0.01 L |
| milli | 1/1000 | 1 milliliter (ml) = 1/1000 L or 0.001 L |

The **milliliter** and the **liter** are the two most commonly used metric units of capacity.

Converting from one unit of capacity to another involves multiplying by powers of 10 or moving the decimal point to the left or to the right. Listing units of capacity in order from largest to smallest helps to keep track of how many places to move the decimal point when converting.

Let's convert 2.6 liters to milliliters. To convert from liters to milliliters, we move along the chart 3 units to the right.

Start           End

kl     hl     dal     **L**     dl     cl     **ml**

3 units to the right

This means that we move the decimal point 3 places to the right to convert from liters to milliliters.

$$2.600 \text{ L} = 2600. \text{ ml}$$

This same conversion can be done with unit fractions.

$$2.6 \text{ L} = \frac{2.6 \text{ L}}{1} \cdot \frac{1000 \text{ ml}}{1 \text{ L}}$$

$$= 2.6 \cdot 1000 \text{ ml} \qquad \text{To multiply by 1000, move the decimal point}$$

$$= 2600 \text{ ml} \qquad \qquad \text{3 places to the right.}$$

To visualize the result, study the diagram below.

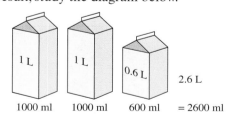

1 L      1 L      0.6 L      2.6 L

1000 ml    1000 ml    600 ml    = 2600 ml

Thus 2.6 L = 2600 ml.

## Example 7
Convert 3210 ml to liters.

*Solution:*    Let's use the unit fraction method first.

Unit fraction

$$3210 \text{ ml} = 3210 \text{ ml} \cdot \frac{1 \text{ L}}{1000 \text{ ml}} = 3.21 \text{ L}$$

**TEACHING TIP**

Before beginning conversions, have students fill in the blanks with "more" or "fewer":
There are _____ liters in a liquid volume than milliliters.
There are _____ milliliters in a liquid volume than kiloliters.
There are _____ kiloliters in a liquid volume than liters.

**Practice Problem 7**

Convert 2100 ml to liters.

**Answer**
**7.** 2.1 L

Now let's list the unit measures in a chart and move from milliliters to liters.

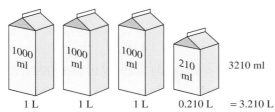

                                    End                    Start
kl        hl        dal        L        dl        cl        ml
                                   ⌣_____⌣_____⌣
                                   3 units to the left

3210 ml = 3.210 L
      ↖
3 places to the left

3210 ml    = 3.210 L

1 L        1 L        1 L        0.210 L    = 3.210 L

**Example 8**    Convert 0.185 dl to milliliters.

*Solution:*    We list the unit measures in a chart and move from deciliters, to milliliters.

                                    Start                    End
kl        hl        dal        L        dl        cl        ml
                                        ⌣_____⌣_____⌣
                                        2 units to the right

0.185 dl = 18.5 ml
         ↘↗
2 places to the right

## D OPERATIONS ON METRIC SYSTEM UNITS OF CAPACITY

As was true for length and weight, arithmetic operations involving metric units of capacity can also be performed. Make sure the metric units of capacity are the same before adding, subtracting, multiplying, or dividing.

**Example 9**    Add 2400 ml to 8.9 L.

*Solution:*    We must convert both to liters or both to milliliters before adding the capacities together.

2400 ml = 2.4 L ⌐    or    8.9 L = 8900 ml ⌐

             2.4 L ←                    2400 ml
            +8.9 L                    + 8900 ml ←
            ─────                     ─────────
            11.3 L                     11,300 ml

The total is 11.3 L or 11,300 ml. They both represent the same capacity.

TRY THE CONCEPT CHECK IN THE MARGIN.

---

**Practice Problem 8**

Convert 2.13 dal to liters.

**Practice Problem 9**

Add 1250 ml to 2.9 L.

## ✓ CONCEPT CHECK

How could you estimate the following operation? Subtract 950 ml from 7.5 L.

**Answers**

**8.** 21.3 L    **9.** 4150 ml or 4.15 L

✓ **Concept Check:** 950 ml = 0.95 L; round 0.95 to 1; 7.5 − 1 = 6.5 L.

**Practice Problem 10**

Divide 146.9 L by 13.

**Practice Problem 11**

If 28.6 L of water can be pumped every minute, how much water can be pumped in 8.5 minutes?

**Example 10**   Divide 18.08 ml by 16.

*Solution:*

$$
\begin{array}{r}
1.13 \text{ ml} \\
16\overline{)18.08 \text{ ml}} \\
\underline{-16} \\
2\,0 \\
\underline{-1\,6} \\
48 \\
\underline{-48} \\
0
\end{array}
$$

The solution is 1.13 ml.

▬▬▬

**Example 11**   Finding the Amount of Medication a Person Has Received

A patient hooked up to an IV unit in the hospital is to receive 12.5 ml of medication every hour. How much medication does the patient receive in 3.5 hours?

*Solution:*   We multiply 12.5 ml by 3.5.

$$
\begin{array}{rr}
\text{medication per hour} \rightarrow & 12.5 \text{ ml} \\
\times \qquad\qquad \text{hours} \rightarrow & \times\ \ 3.5 \\
\hline
\text{total medication} & 625 \\
& 3750 \\
\hline
& 43.75 \text{ ml}
\end{array}
$$

The patient receives 43.75 ml of medication.

▬▬▬

## CALCULATOR EXPLORATIONS
### METRIC TO U.S. SYSTEM CONVERSIONS IN CAPACITY

To convert between the two systems of measurement in capacity, the following *approximations* can be used.

| | |
|---|---|
| liters × 1.06 ≈ quarts | quarts × 0.946 ≈ liters |
| liters × 0.264 ≈ gallons | gallons × 3.785 ≈ liters |

**Example**   How many quarts are there in a 2-liter bottle of cola?

*Solution:*   From the above approximations,

$$\text{liters} \times 1.06 \approx \text{quarts}$$
$$\downarrow$$
$$2 \times 1.06 \approx \text{quarts}$$

To multiply on your calculator, press the keys

[ 2 ] [ × ] [ 1.06 ] [ = ] (or [ ENTER ]).

The display should read [ 2.12 ]. 2 liters ≈ 2.12 quarts

▬▬▬

*Convert as indicated.*

1. 5 qt to liters   ≈4.73 L       2. 26 gal to liters   ≈98.41 L
3. 17.5 L to gallons   ≈4.62 gal  4. 7.8 L to quarts   ≈8.268 qt
5. A 1-gallon container holds how many liters?   ≈3.785 L
6. How many quarts are contained in a 3-liter bottle of cola?
   ≈3.18 qt

**Name** _____ **Section** _____ **Date** _____

## MENTAL MATH

*Convert as indicated.*

**1.** 2 c to pints

**2.** 4 c to pints

**3.** 4 qt to gallons

**4.** 8 qt to gallons

**5.** 2 pt to quarts

**6.** 6 pt to quarts

**7.** 8 fl oz to cups

**8.** 24 fl oz to cups

**9.** 1 pt to cups

**10.** 3 pt to cups

**11.** 1 gal to quarts

**12.** 2 gal to quarts

*Determine whether the measurement in each statement is reasonable.*

**13.** Clair Verdin took a dose of 2 L of cough medicine to cure her cough.

**14.** John Shermak drank 250 ml of milk for lunch.

**15.** Jeannie Ballord likes to relax in a tub filled with 3000 ml of hot water.

**16.** Sarah Fisher pumped 20 L of gasoline into her car yesterday.

## EXERCISE SET 9.6

**A** *Convert each measurement as indicated. See Examples 1 and 2.*

**1.** 32 fl oz to cups

**2.** 16 qt to gallons

**3.** 8 qt to pints

**4.** 9 pt to quarts

**5.** 10 qt to gallons

**6.** 15 c to pints

**7.** 80 fl oz to pints

**8.** 18 pt to gallons

**9.** 2 qt to cups

**10.** 3 pt to fluid ounces

**11.** 120 fl oz to quarts

**12.** 20 c to gallons

**13.** 6 gal to fluid ounces

**14.** 5 qt to cups

**15.** $4\frac{1}{2}$ pt to cups

**16.** $6\frac{1}{2}$ gal to quarts

**17.** $2\frac{3}{4}$ gal to pints

**18.** $3\frac{1}{4}$ qt to cups

MENTAL MATH ANSWERS

**1.** 1 pt
**2.** 2 pt
**3.** 1 gal
**4.** 2 gal
**5.** 1 qt
**6.** 3 qt
**7.** 1 c
**8.** 3 c
**9.** 2 c
**10.** 6 c
**11.** 4 qt
**12.** 8 qt
**13.** no
**14.** yes
**15.** no
**16.** yes

ANSWERS

**1.** 4 c
**2.** 4 gal
**3.** 16 pt
**4.** $4\frac{1}{2}$ qt
**5.** $2\frac{1}{2}$ gal
**6.** $7\frac{1}{2}$ pt
**7.** 5 pt
**8.** $2\frac{1}{4}$ gal
**9.** 8 c
**10.** 48 fl oz
**11.** $3\frac{3}{4}$ qt
**12.** $1\frac{1}{4}$ gal
**13.** 768 fl oz
**14.** 20 c
**15.** 9 c
**16.** 26 qt
**17.** 22 pt
**18.** 13 c

**19.** 10 gal 1 qt

**20.** 11 gal 2 qt

**21.** 4 c 4 fl oz

**22.** 5 c 1 fl oz

**23.** 1 gal 1 qt

**24.** 1 c

**25.** 2 gal 3 qt 1 pt

**26.** 2 qt 1 pt 1 c 4 fl oz

**27.** 2 qt 1 c

**28.** 3 qt

**29.** 17 gal

**30.** 12 gal 1 qt

**31.** 4 gal 3 qt

**32.** 2 gal 2 qt 3 fl oz

**33.** 48 fl oz

**34.** $2\frac{5}{8}$ c

**35.** 2 qt

**36.** 10 fl oz

**37.** yes

**38.** 4 fl oz

**Name** _____

**B** *Perform each indicated operation. See Examples 3 through 5.*

**19.** 4 gal 3 qt + 5 gal 2 qt    **20.** 2 gal 3 qt + 8 gal 3 qt    **21.** 1 c 5 fl oz + 2 c 7 fl oz

**22.** 2 c 3 fl oz + 2 c 6 fl oz    **23.** 3 gal − 1 gal 3 qt    **24.** 2 pt − 1 pt 1 c

**25.** 3 gal 1 qt − 1 qt 1 pt    **26.** 3 qt 1 c − 1 c 4 fl oz    **27.** 1 pt 1 c × 3

**28.** 1 qt 1 pt × 2    **29.** 8 gal 2 qt × 2    **30.** 6 gal 1 pt × 2

**31.** 9 gal 2 qt ÷ 2    **32.** 5 gal 6 fl oz ÷ 2

*Solve. Remember to insert units when writing your answers. See Example 6.*

**33.** A can of Hawaiian punch holds $1\frac{1}{2}$ quarts of liquid. How many fluid ounces is this?

**34.** Weight Watchers double fudge bars contain 21 fluid ounces of ice cream. How many cups of ice cream is this?

**35.** Many diet experts advise individuals to drink 64 ounces of water each day. How many quarts of water is this?

**36.** A recipe for walnut fudge cake calls for $1\frac{1}{4}$ cups of water. How many fluid ounces is this?

**37.** Can 5 pt 1 c of fruit punch and 2 pt 1 c of ginger ale be poured into a 1-gal container without it overflowing?

**38.** Three cups of prepared jello are poured into 6 dessert dishes. How many fluid ounces of jello are in each dish?

**39.** How much punch has been prepared if 1 qt 1 pt of Ocean Spray Cranapple drink is mixed with 1 pt 1 c of ginger ale?

**40.** Henning's Supermarket sells home-made soup in 1-qt-1-pt containers. How much soup is contained in 3 such containers?

**41.** A case of Pepsi Cola holds 24 cans, each of which contains 12 fluid ounces of Pepsi Cola. How many *quarts* are there in a case of Pepsi?

**42.** Manuela's Service Station has a drum that holds 40 gallons of oil. If 6 gallons and 3 quarts have been used, how much oil remains?

**C** *Convert as indicated. See Examples 7 and 8.*

**43.** 5 L to milliliters

**44.** 8 L to milliliters

**45.** 4500 ml to liters

**46.** 3100 ml to liters

**47.** 410 L to kiloliters

**48.** 250 L to kiloliters

**49.** 64 ml to liters

**50.** 39 ml to liters

**51.** 0.16 kl to liters

**52.** 0.48 kl to liters

**53.** 3.6 L to milliliters

**54.** 1.9 L to milliliters

**55.** 0.16 L to kiloliters

**56.** 0.127 L to kiloliters

**D** *Perform each indicated operation. See Examples 9 and 10.*

**57.** 2.9 L + 19.6 L

**58.** 18.5 L + 4.6 L

**59.** 2700 ml + 1.8 L

**60.** 4.6 L + 1600 ml

**61.** 8.6 L − 190 ml

**62.** 4.8 L − 283 ml

**63.** 11,400 ml − 0.8 L

**64.** 6850 ml − 0.3 L

**65.** 480 ml × 8

**66.** 290 ml × 6

**67.** 81.2 L ÷ 0.5

**68.** 5.4 L ÷ 3.6

**39.** 2 qt 1 c

**40.** 1 gal 1 pt

**41.** 9 qt

**42.** 33 gal 1 qt

**43.** 5000 ml

**44.** 8000 ml

**45.** 4.5 L

**46.** 3.1 L

**47.** 0.41 kl

**48.** 0.25 kl

**49.** 0.064 L

**50.** 0.039 L

**51.** 160 L

**52.** 480 L

**53.** 3600 ml

**54.** 1900 ml

**55.** 0.00016 kl

**56.** 0.000127 kl

**57.** 22.5 L

**58.** 23.1 L

**59.** 4.5 L or 4500 ml

**60.** 6.2 L or 6200 ml

**61.** 8410 ml or 8.41 L

**62.** 4.517 L or 4517 ml

**63.** 10,600 ml or 10.6 L

**64.** 6.55 L or 6550 ml

**65.** 3840 ml

**66.** 1740 ml

**67.** 162.4 L

**68.** 1.5 L

*Solve. Remember to insert units when writing your answers. See Example 11.*

**69.** Mike Schaferkotter drank 410 ml of Mountain Dew from a 2-liter bottle. How much Mountain Dew remains in the bottle?

**70.** The Werners' Volvo has a 54.5-L gas tank. Only 3.8 liters of gasoline still remain in the tank. How much is needed to fill it?

**71.** Margie Phitts added 354 ml of Prestone dry gas to the 18.6 L of gasoline in her car's tank. Find the total amount of gasoline in the tank.

**72.** Chris Peckaitis wishes to share a 2-L bottle of Coca Cola equally with 7 of her friends. How much will each person get?

**73.** Stanley Fisher paid $14.00 to fill his car with 44.3 liters of gasoline. Find the price per liter of gasoline to the nearest tenth of a cent.

**74.** A student carelessly misread the scale on a cylinder in the chemistry lab and added 40 cl of water to a mixture instead of 40 ml. Find the excess amount of water.

**75.** A large bottle of Ocean Spray Cranicot drink contains 1.42 L of beverage. The smaller bottle contains only 946 ml. How much more is in the larger bottle?

**76.** In a lab experiment, Melissa Martin added 400 ml of salt water to 1.65 L of water. Later 320 ml of the solution was drained off. How much of the solution still remains?

### REVIEW AND PREVIEW

*Write each decimal as a fraction. See Section 5.6.*

**77.** 0.7

**78.** 0.9

**79.** 0.03

**80.** 0.007

**81.** 0.006

**82.** 0.08

### COMBINING CONCEPTS

**83.** Explain how to borrow to subtract 1 gal 2 qt from 3 gal 1 qt.

# 9.7  TEMPERATURE

When Gabriel Fahrenheit and Anders Celsius independently established units for temperature scales, each based his unit on the heat of water the moment it boils compared to the moment it freezes. One degree Celsius is 1/100 of the difference in heat. One degree Fahrenheit is 1/180 of the difference in heat. Celsius arbitrarily labeled the temperature at the freezing point at 0°C, making the boiling point 100°C; Fahrenheit labeled the freezing point 32°F, making the boiling point 212°F. Water boils at 212°F or 100°C.

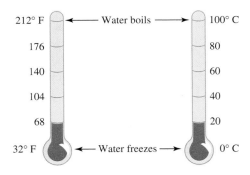

By comparing the two scales in the figure, we see that a 20°C day is as warm as a 68°F day. Similarly, a sweltering 104°F day in the Mojave desert corresponds to a 40°C day.

**TRY THE CONCEPT CHECK IN THE MARGIN.**

## A  CONVERTING DEGREES CELSIUS TO DEGREES FAHRENHEIT

To convert from Celsius temperatures to Fahrenheit temperatures, see the box below. In this box, we use the symbol F to represent degrees Fahrenheit and the symbol C to represent degrees Celsius.

---

**CONVERTING CELSIUS TO FAHRENHEIT**

$$F = \frac{9}{5}C + 32 \qquad \text{or} \qquad F = 1.8\,C + 32$$

(To convert to Fahrenheit temperature, multiply the Celsius temperature by $\frac{9}{5}$ or 1.8, and then add 32.)

---

**Objectives**

**A** Convert temperatures from degrees Celsius to degrees Fahrenheit.

**B** Convert temperatures from degrees Fahrenheit to degrees Celsius.

Study     SSM    CD-ROM    Video
Guide                        9.7

**TEACHING TIP**

Ask students to check the Celsius to Fahrenheit formula by using the known values of F = 32 when C = 0 and F = 212 when C = 100.

### ✓ CONCEPT CHECK

Which of the following statements is correct? Explain.

a. 6°C is below the freezing point of water.

b. 6°F is below the freezing point of water.

**Answer**

✓ Concept Check: **b.** is correct.

## Practice Problem 1

Convert 50°C to degrees Fahrenheit.

## Practice Problem 2

Convert 18°C to degrees Fahrenheit.

## Example 1

Convert 15°C to degrees Fahrenheit.

*Solution:*

$$F = \frac{9}{5}C + 32$$

$$= \frac{9}{5} \cdot 15 + 32 \qquad \text{Replace C with 15.}$$

$$= 27 + 32 \qquad \text{Multiply.}$$

$$= 59 \qquad \text{Add.}$$

Thus 15°C is equivalent to 59°F.

## Example 2

Convert 29°C to degrees Fahrenheit.

*Solution:*

$$F = 1.8\,C + 32$$

$$= 1.8 \cdot 29 + 32 \qquad \text{Replace C with 29.}$$

$$= 52.2 + 32 \qquad \text{Multiply 1.8 by 29.}$$

$$= 84.2 \qquad \text{Add.}$$

Therefore, 29°C is the same as 84.2°F.

### B CONVERTING DEGREES FAHRENHEIT TO DEGREES CELSIUS

To convert from Fahrenheit temperatures to Celsius temperatures, see the box below. The symbol C represents degrees Celsius and F represents degrees Fahrenheit.

CONVERTING FAHRENHEIT TO CELSIUS

$$C = \frac{5}{9}(F - 32)$$

(To convert to Celsius temperature, subtract 32 from the Fahrenheit temperature, and then multiply by $\frac{5}{9}$.)

**Example 3**    Convert 59°F to degrees Celsius.

*Solution:*    We evaluate the formula $C = \frac{5}{9}(F - 32)$ when F is 59.

$$C = \frac{5}{9}(F - 32)$$

$$= \frac{5}{9} \cdot (59 - 32) \quad \text{Replace F with 59.}$$

$$= \frac{5}{9} \cdot (27) \quad \text{Subtract in parentheses.}$$

$$= 15 \quad \text{Multiply.}$$

Therefore, 59°F is the same temperature as 15°C. ▬

**Example 4**    Convert 114°F to degrees Celsius. If necessary, round to the nearest tenth of a degree.

*Solution:*    $$C = \frac{5}{9}(F - 32)$$

$$= \frac{5}{9} \cdot (114 - 32) \quad \text{Replace F with 114.}$$

$$= \frac{5}{9} \cdot (82) \quad \text{Subtract in parentheses.}$$

$$\approx 45.6 \quad \text{Multiply.}$$

Therefore, 114°F is approximately 45.6°C. ▬

**Practice Problem 3**

Convert 86°F to degrees Celsius.

**Practice Problem 4**

Convert 113°F to degrees Celsius. If necessary, round to the nearest tenth of a degree.

Answers

**3.** 30°C   **4.** 45°C

**Practice Problem 5**

During a bout with the flu, Ronald Albert's temperature reaches 102.8°F. What is his temperature measured in degrees Celsius? Round to the nearest tenth of a degree.

✓ **CONCEPT CHECK**

Which of the following statements is correct? Explain.

a. 150°F is cooler than the boiling point of water.

b. 150°C is cooler than the boiling point of water.

**Example 5**    The normal body temperature is 98.6°F. What is this temperature in degrees Celsius?

*Solution:*    We evaluate the formula $C = \dfrac{5}{9}(F - 32)$ when F is 98.6.

$$C = \frac{5}{9}(F - 32)$$

$$= \frac{5}{9} \cdot (98.6 - 32) \quad \text{Replace F with 98.6.}$$

$$= \frac{5}{9} \cdot (66.6) \quad \text{Subtract in parentheses.}$$

$$= 37 \quad \text{Multiply.}$$

Therefore, normal body temperature is 37°C.    ■

**TRY THE CONCEPT CHECK IN THE MARGIN.**

Answers

**5.** 39.3°C

✓ Concept Check: **a.** is correct.

## MENTAL MATH

*Determine whether the measurement in each statement is reasonable.*

**1.** A 72°F room feels comfortable.

**2.** Water heated to 110°F will boil.

**3.** Josiah Jones has a fever if a thermometer reads his temperature as 40°F.

**4.** An air temperature of 20°F on a Vermont ski slope can be expected in the winter.

**5.** When the temperature is 30°C outside an overcoat is needed.

**6.** An air-conditioned room at 60°C feels quite chilly.

**7.** Barbara Smith has a fever if a thermometer records her temperature at 40°C.

**8.** Water cooled to 32°C will freeze.

# EXERCISE SET 9.7

**A** **B** *Convert as indicated. When necessary, round to the nearest tenth of a degree. See Examples 1 through 5.*

**1.** 41°F to degrees Celsius

**2.** 68°F to degrees Celsius

**3.** 104°F to degrees Celsius

**4.** 86°F to degrees Celsius

**5.** 60°C to degrees Fahrenheit

**6.** 80°C to degrees Fahrenheit

**7.** 115°C to degrees Fahrenheit

**8.** 35°C to degrees Fahrenheit

**9.** 62°F to degrees Celsius

**10.** 182°F to degrees Celsius

**11.** 142.1°F to degrees Celsius

**12.** 43.4°F to degrees Celsius

**Name** _____

**13.** 92°C to degrees Fahrenheit

**14.** 75°C to degrees Fahrenheit

**15.** 16.3°C to degrees Fahrenheit

**16.** 48.6°C to degrees Fahrenheit

**17.** The highest temperature ever record-ed in Death Valley was 134°F. Convert this measurement to degrees Celsius. (*Source:* National Climatic Data Center)

**18.** The hottest day in Pennsylvania reached 111°F. Convert this measure-ment to degrees Celsius. (*Source:* National Climatic Data Center)

**19.** A weather forecaster in Caracas pre-dicts a high temperature of 27°C. Find this measurement in degrees Fahren-heit.

**20.** While driving to work, Alan Olda no-tices a temperature of 18°C flash on the local bank's temperature display. Find the corresponding temperature in degrees Fahrenheit.

**21.** At Mack Trucks' headquarters the room temperature is to be set at 70°F, but the thermostat is calibrated in de-grees Celsius. Find the temperature to be set.

**22.** The computer room at Merck, Sharp, and Dohm is normally cooled to 66°F. Find the corresponding temperature in degrees Celsius.

**23.** Najib Tan is running a fever of 100.2°F. Find his temperature as it would be shown on a Celsius thermometer.

**24.** William Saylor generally has a subnor-mal temperature of 98.2°F. Find this temperature on a Celsius thermometer.

**25.** In a European cookbook, the recipe requires the ingredients for caramels to be heated to 118°C, but the cook only has access to a Fahrenheit ther-mometer. Find the temperature in de-grees Fahrenheit that should be used to make the caramels.

**26.** The ingredients for divinity should be heated to 127°C, but the candy ther-mometer that Myung Kim has is cali-brated to degrees Fahrenheit. Find how hot he should heat the ingredients.

Name _____

**27.** 260°C

**27.** Mark Tabbey's recipe for Yorkshire pudding calls for a very hot, 500°F oven. Find the temperature setting he should use with an oven having Celsius controls.

**28.** The temperature of Earth's core is estimated to be 4000°C. Find the corresponding temperature in degrees Fahrenheit.

**28.** 7232°F

**29.** 462.2°C

**29.** The surface temperature of Venus can reach 864°F. Find this temperature in degrees Celsius.

**30.** 12 in.

## REVIEW AND PREVIEW

*Find the perimeter of each figure. See Section 1.2.*

△ **30.**

3 in.

3 in. | Square

△ **31.**

25 m

6 m | Rectangle

**31.** 62 m

**32.** 12 cm

△ **32.**

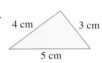

4 cm   3 cm

5 cm

△ **33.**

3 ft   3 ft

3 ft   3 ft

3 ft

**33.** 15 ft

△ **34.**

2 ft 8 in.

1 ft 6 in. | Rectangle

△ **35.**

2.6 m

2.6 m | Square

**34.** 8 ft 4 in.

**35.** 10.4 m

## COMBINING CONCEPTS

**36.** On May 27, 1994, in the Tokamak Fusion Test Reactor at Princeton University, the highest temperature produced in a laboratory was achieved. This temperature was 950,000,000°F. Convert this temperature to degrees Celsius. (*Source: Guiness Book of Records*, 2000)

**37.** The hottest-burning substance known is carbon subnitride. Its flame at one atmosphere pressure reaches 9010°F. Convert this temperature to degrees Celsius. (*Source: Guiness Book of Records*, 2000)

**36.** 527,777,760°C

**37.** 4988°C

**38.** A comfortable room temperature is measured in degrees Celsius and degrees Fahrenheit. Which degree number will be greater? Explain why?

Fahrenheit;
**38.** answers may vary

**39. a.** 33.3°C

**b.** 14°F

**40. a.** 17.4 knots

**b.** 40.3 mph

## Internet Excursions

Go to http://www.prenhall.com/martin-gay

This World Wide Web address will provide you with access to National Weather Service weather calculator tools, or a related site. One of the available calculators converts a temperature reading in either degrees Fahrenheit or degrees Celsius to several other temperature measurement units. Another of the calculators will convert a wind speed in knots, miles per hour, or meters per second to several other wind speed units.

**39.** Select the Temperature Unit Conversion option. Use the converter to convert **(a)** 92°F to degrees Celsius, and **(b)** −10°C to degrees Fahrenheit. Use the formulas from this section to verify the results.

**40.** Select the Wind Speed Unit Conversion option. Use the converter to convert **(a)** a wind speed of 20 miles per hour to knots, and **(b)** a wind speed of 35 knots to miles per hour.

# CHAPTER 9 ACTIVITY
## MAP READING

*This activity may be completed by working in groups or individually.*

According to the American Automobile Association, more and more people are taking driving vacations. To make the most of travel by car, reading road maps effectively is a plus. Use the given map to investigate the driving route you would take from Santa Rosa, New Mexico, to San Antonio, New Mexico.

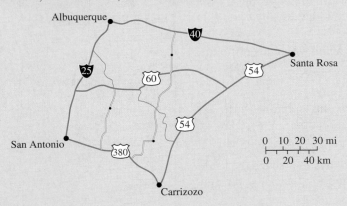

1. Use string to outline the route via Interstate 40 and Interstate 25 from Santa Rosa to San Antonio. Use the given mileage key with the string to estimate the distance between Santa Rosa and San Antonio in miles. Convert this distance to kilometers.

2. Using the method described in Question 1, how many miles is it from Santa Rosa to San Antonio via U.S. 54 and U.S. 380? Convert this distance to kilometers.

3. Assume that the speed limit on Interstates 40 and 25 is 65 miles per hour. How long would the trip take if you took this route?

4. At what average speed would you have to travel on the U.S. routes to make the trip from Santa Rosa to San Antonio in the same amount of time that it would take on the interstate routes? Do you think this speed is reasonable on this route? Explain your reasoning.

5. Discuss in general the factors that might affect your decision among different routes.

6. Explain which route you would choose in this case and why.

# CHAPTER 9 HIGHLIGHTS

| DEFINITIONS AND CONCEPTS | EXAMPLES |
|---|---|

### SECTION 9.1   LINES AND ANGLES

A **line** is a set of points extending indefinitely in two directions. A line has no width or height, but it does have length. We can name a line by any two of its points.

A **line segment** is a piece of a line with two endpoints.

Line $AB$ or $\overleftrightarrow{AB}$

Line segment $AB$ or $\overline{AB}$

A **ray** is a part of a line with one endpoint. A ray extends indefinitely in one direction.

Ray $AB$ or $\overrightarrow{AB}$

An **angle** is made up of two rays that share the same endpoint. The common endpoint is called the **vertex**.

Vertex

An angle that measures 180° is called a **straight angle**.

180°

$\angle RST$ is a straight angle.

An angle that measures 90° is called a **right angle**. The symbol ⌐ is used to denote a right angle.

$\angle ABC$ is a right angle.

An angle whose measure is between 0° and 90° is called an **acute angle**.

Acute angles

An angle whose measure is between 90° and 180° is called an **obtuse angle.**

Obtuse angles

Two angles that have a sum of 90° are called **complementary angles**. We say that each angle is the **complement** of the other.

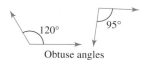

Complementary angles
60° + 30° = 90°

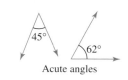

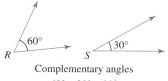

*(continued)*

## SECTION 9.1    LINES AND ANGLES (CONTINUED)

Two angles that have a sum of $180°$ are called **supplementary** angles. We say that each angle is the **supplement** of the other.

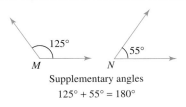

Supplementary angles
$125° + 55° = 180°$

When two lines intersect, four angles are formed. Two of these angles that are opposite each other are called vertical angles. **Vertical angles** have the same measure.

Two of these angles that share a common side are called **adjacent angles**. Adjacent angles formed in intersecting lines are supplementary.

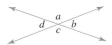

Vertical angles:
$\angle a$ and $\angle c$
$\angle d$ and $\angle b$
Adjacent angles:
$\angle a$ and $\angle b$
$\angle b$ and $\angle c$
$\angle c$ and $\angle d$
$\angle d$ and $\angle a$

A line that intersects two or more lines at different points is called a **transversal**. Line *l* is a transversal that intersects lines *m* and *n*. The eight angles formed have special names. Some of these names are:

Corresponding Angles: $\angle a$ and $\angle e$, $\angle c$ and $\angle g$, $\angle b$ and $\angle f$, $\angle d$ and $\angle h$
Alternate Interior Angles: $\angle c$ and $\angle f$, $\angle d$ and $\angle e$

### PARALLEL LINES CUT BY A TRANSVERSAL

If two parallel lines are cut by a transversal, then the measures of **corresponding angles are equal** and the measures of **alternate interior angles are equal.**

The **sum of the measures** of the angles of a triangle is $180°$.

Find the measure of $\angle x$.

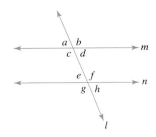

The measure of $\angle x = 180° - 85° - 45° = 50°$

A **right triangle** is a triangle with a right angle. The side opposite the right angle is called the **hypotenuse** and the other two sides are called **legs**.

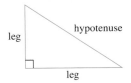

---

## SECTION 9.2　LINEAR MEASUREMENT

To convert from one unit of length to another, unit **fractions** may be used.
The unit fraction should be in the form

$$\frac{\text{units converting to}}{\text{original units}}$$

LENGTH: U.S. SYSTEM OF MEASUREMENT

12 inches (in.) = 1 foot (ft)
3 feet = 1 yard (yd)
5280 feet = 1 mile (mi)

The basic unit of length in the metric system is the **meter**. A meter is slightly longer than a yard.

LENGTH: METRIC SYSTEM OF MEASUREMENT

$$\frac{12 \text{ inches}}{1 \text{ foot}}, \quad \frac{1 \text{ foot}}{12 \text{ inches}}, \quad \frac{3 \text{ feet}}{1 \text{ yard}}$$

Convert 6 feet to inches.

$$6 \text{ feet} = \frac{6 \text{ feet}}{1} \cdot \frac{12 \text{ inches}}{1 \text{ foot}} \quad \begin{matrix} \leftarrow \text{units converting to} \\ \leftarrow \text{original units} \end{matrix}$$
$$= 6 \cdot 12 \text{ inches}$$
$$= 72 \text{ inches}$$

Convert 3650 centimeters to meters.

$$3650 \text{ cm} = \frac{3650 \text{ cm}}{1} \cdot \frac{0.01 \text{ m}}{1 \text{ cm}} = 36.5 \text{ m}$$

or

| | | | End | | Start | |
| km | hm | dam | m | dm | cm | mm |

2 units to the left

36.50 cm = 36.5 m

2 places to the left

| Prefix | Meaning | Metric Unit of Length |
|--------|---------|----------------------|
| kilo | 1000 | 1 **kilo**meter (km) = 1000 meters (m) |
| hecto | 100 | 1 **hecto**meter (hm) = 100 m |
| deka | 10 | 1 **deka**meter (dam) = 10 m |
| | | **1 meter (m) = 1 m** |
| deci | 1/10 | 1 **deci**meter (dm) = 1/10 m or 0.1 m |
| centi | 1/100 | 1 **centi**meter (cm) = 1/100 m or 0.01 m |
| milli | 1/1000 | 1 **milli**meter (mm) = 1/1000 m or 0.001 m |

---

## SECTION 9.3　PERIMETER

PERIMETER FORMULAS

**Rectangle:**　$P = 2l + 2w$

**Square:**　$P = 4s$

**Triangle:**　$P = a + b + c$

**Circumference of a Circle:**

$$C = 2\pi r \quad \text{or} \quad C = \pi d$$

where $\pi \approx 3.14$　or　$\pi \approx \dfrac{22}{7}$

Find the perimeter of the rectangle.

28 m
15 m

$$P = 2l + 2w$$
$$= 2 \cdot 28 \text{ meters} + 2 \cdot 15 \text{ meters}$$
$$= 56 \text{ meters} + 30 \text{ meters}$$
$$= 86 \text{ meters}$$

The perimeter is 86 meters.

## SECTION 9.4    AREA AND VOLUME

AREA FORMULAS

**Rectangle:**   $A = lw$

**Square:**   $A = s^2$

**Triangle:**   $A = \dfrac{1}{2}bh$

**Parallelogram:**   $A = bh$

**Trapezoid:**   $A = \dfrac{1}{2}(b + B)h$

**Circle:**   $A = \pi r^2$

VOLUME FORMULAS

**Rectangular Solid:**   $V = lwh$

**Cube:**   $V = s^3$

**Sphere:**   $V = \dfrac{4}{3}\pi r^3$

**Right Circular Cylinder:**   $V = \pi r^2 h$

**Cone:**   $V = \dfrac{1}{3}\pi r^2 h$

**Square-Based Pyramid:**   $V = \dfrac{1}{3}s^2 h$

Find the area of the square.

8 cm

$$A = s^2$$
$$= (8 \text{ centimeters})^2$$
$$= 64 \text{ square centimeters}$$

The area of the square is 64 square centimeters.

Find the volume of the sphere. Use $\dfrac{22}{7}$ for $\pi$.

4 in.

$$V = \frac{4}{3}\pi r^3$$
$$\approx \frac{4}{3} \cdot \frac{22}{7} \cdot (4 \text{ inches})^3$$
$$= \frac{4 \cdot 22 \cdot 64}{3 \cdot 7} \text{ cubic inches}$$
$$= \frac{5632}{21} \quad \text{or} \quad 268\frac{4}{21} \text{ cubic inches}$$

## SECTION 9.5    WEIGHT AND MASS

**Weight** is really a measure of the pull of gravity.
**Mass** is a measure of the amount of substance in the object and does not change.

WEIGHT: U.S. SYSTEM OF MEASUREMENT

   16 ounces (oz) = 1 pound (lb)
   2000 pounds = 1 ton

A **gram** is the basic unit of mass in the metric system. It is the mass of water contained in a cube 1 centimeter on each side. A large paper clip weighs about 1 gram.

Convert 5 pounds to ounces.

$$5 \text{ lb} = \frac{5 \text{ lb}}{1} \cdot \frac{16 \text{ oz}}{1 \text{ lb}} = 80 \text{ oz}$$

Convert 260 grams to kilograms.

$$260 \text{ g} = \frac{260 \text{ g}}{1} \cdot \frac{1 \text{ kg}}{1000 \text{ g}} = 0.26 \text{ kg}$$

End                                      Start

kg    hg    dag    g    dg    cg    mg

3 units to the left

260 g = 0.260 kg

3 places to the left                          *(continued)*

---

## SECTION 9.5   WEIGHT AND MASS (CONTINUED)

MASS: METRIC SYSTEM OF MEASUREMENT

| Prefix | Meaning | Metric Unit of Mass |
|--------|---------|---------------------|
| kilo | 1000 | 1 kilogram (kg) = 1000 grams (g) |
| hecto | 100 | 1 hectogram (hg) = 100 g |
| deka | 10 | 1 dekagram (dag) = 10g |
| | | **1 gram (g) = 1 g** |
| deci | 1/10 | 1 decigram (dg) = 1/10 g or 0.1 g |
| centi | 1/100 | 1 centigram (cg) = 1/100 g or 0.01 g |
| milli | 1/1000 | 1 milligram (mg) = 1/1000 g or 0.001 g |

---

## SECTION 9.6   CAPACITY

CAPACITY: U.S. SYSTEM OF MEASUREMENT

$$8 \text{ fluid ounces (fl oz)} = 1 \text{ cup (c)}$$
$$2 \text{ cups} = 1 \text{ pint (pt)}$$
$$2 \text{ pints} = 1 \text{ quart (qt)}$$
$$4 \text{ quarts} = 1 \text{ gallon (gal)}$$

The **liter** is the basic unit of capacity in the metric system. It is the capacity or volume of a cube measuring 10 centimeters on each side. A liter of liquid is slightly more than 1 quart.

Convert 5 pints to gallons.

$$1 \text{ gal} = 4 \text{ qt} = 8 \text{ pt}$$

$$5 \text{ pt} = \frac{5 \text{ pt}}{1} \cdot \frac{1 \text{ gal}}{8 \text{ pt}} = \frac{5}{8} \text{ gal}$$

Convert 1.5 liters to milliliters.

$$1.5 \text{ L} = \frac{1.5 \text{ L}}{1} \cdot \frac{1000 \text{ ml}}{1 \text{ L}} = 1500 \text{ ml}$$

or

Start                                End

kl    hl    dal    L    dl    cl    ml

3 units to the right

1.500 L = 1500 ml

3 places to the right

CAPACITY: METRIC SYSTEM OF MEASUREMENT

| Prefix | Meaning | Metric Unit of Capacity |
|--------|---------|-------------------------|
| kilo | 1000 | 1 kiloliter (kl) = 1000 liters (L) |
| hecto | 100 | 1 hectoliter (hl) = 100 L |
| deka | 10 | 1 dekaliter (dal) = 10 L |
| | | **1 liter (L) = 1 L** |
| deci | 1/10 | 1 deciliter (dl) = 1/10 L or 0.1 L |
| centi | 1/100 | 1 centiliter (cl) = 1/100 L or 0.01 L |
| milli | 1/1000 | 1 milliliter (ml) = 1/1000 L or 0.001 L |

---

## SECTION 9.7   TEMPERATURE

CELSIUS TO FAHRENHEIT

$$F = \frac{9}{5}C + 32 \quad \text{or} \quad F = 1.8C + 32$$

FAHRENHEIT TO CELSIUS

$$C = \frac{5}{9}(F - 32)$$

Convert 35°C to degrees Fahrenheit.

$$F = \frac{9}{5} \cdot 35 + 32 = 63 + 32 = 95$$

$$35°C = 95°F$$

Convert 50°F to degrees Celsius.

$$C = \frac{5}{9} \cdot (50 - 32) = \frac{5}{9} \cdot 18 = 10$$

$$50°F = 10°C$$

# CHAPTER 9 REVIEW

△ **(9.1)** *Classify each angle as acute, right, obtuse, or straight.*

**1.**

*A*

right

**2.**

*B*

straight

**3.**

*C*

acute

**4.**

*D*

obtuse

**5.** Find the complement of a 25° angle.    65°

**6.** Find the supplement of a 105° angle.    75°

**7.** Find the supplement of a 72° angle.    108°

**8.** Find the complement of a 1° angle.    89°

*Find the measure of x in each figure.*

**9.**     58°

32°

*x*

**10.**     98°

*x*    82°

**11.** 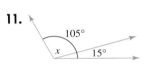    90°

105°

*x*    15°

**12.**    25°

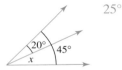

20°
45°
*x*

**13.** Identify the pairs of supplementary angles.
133° and 47°

133°
47°    47°
133°

**14.** Identify the pairs of complementary angles.
43° and 47°; 58° and 32°

58° 32°
47°    43°

*Find the measures of angles x, y, and z in each figure.*

**15.**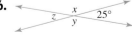

*y*
*x*    80°
*z*

∠*x* = 80°; ∠*y* = 100°; ∠*z* = 100°

**16.**

*z*    *x*    25°
*y*

∠*x* = 155°; ∠*y* = 155°; ∠*z* = 25°

**Name** _____

**17.** *m‖n*

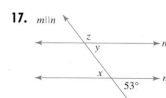

$\angle x = 53°; \angle y = 53°; \angle z = 127°$

**18.** *m‖n*

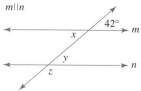

$\angle x = 42°; \angle y = 42°; \angle z = 138°$

*Find the measure of ∠x in each figure.*

**19.**

103°

**20.**

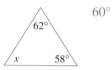

60°

**21.**

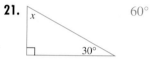

60°

**22.**

65°

**(9.2)** *Convert.*

**23.** 108 in. to feet    9 ft

**24.** 72 ft to yards    24 yd

**25.** 2.5 mi to feet    13,200 ft

**26.** 6.25 ft to inches    75 in.

**27.** 52 ft = _____ yd _____ft    17 yd 1 ft

**28.** 46 in. = _____ ft _____in.    3 ft 10 in.

**29.** 42 m to centimeters    4200 cm

**30.** 82 cm to millimeters    820 mm

**31.** 12.18 mm to meters    0.01218 m

**32.** 2.31 m to kilometers    0.00231 km

**756**

*Perform each indicated operation.*

**33.** 4 yd 2 ft + 16 yd 2 ft   21 yd 1 ft

**34.** 12 ft 1 in. − 4 ft 8 in.   7 ft 5 in.

**35.** 8 ft 3 in. × 5   41 ft 3 in.

**36.** 7 ft 4 in. ÷ 2   3 ft 8 in.

**37.** 8 cm + 15 mm   9.5 cm or 95 mm

**38.** 4 m + 126 cm   5.26 m or 526 cm

**39.** 9.3 km − 183 m   9117 m or 9.117 km

**40.** 4100 mm − 3 m   1.1 m or 1100 mm

*Solve.*

**41.** A bolt of cloth contains 333 yd 1 ft of cotton ticking. Find the amount of material that remains after 163 yd 2 ft is removed from the bolt.   169 yd 2 ft

**42.** The local ambulance corps plans to award 20 framed certificates of valor to some of its outstanding members. If each frame requires 6 ft 4 in. of framing material, how much material is needed for all the frames?   126 ft 8 in.

**43.** The trip from Philadelphia to Washington, D.C., is 217 km one way. Four friends agree to share the driving equally. How far must each drive on this round-trip vacation?   108.5 km

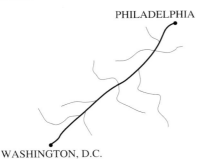

PHILADELPHIA

WASHINGTON, D.C.

△ **44.** The college has ordered that NO SMOKING signs be placed above the doorway to each classroom. Each sign is 0.8 m long and 30 cm wide. Find the area of each sign.   (*Hint:* Recall that the area of a rectangle = width · length.)   0.24 sq. m

NO SMOKING

30 cm

0.8 m

△ **(9.3)** *Find the perimeter of each figure.*

**45.**   88 m

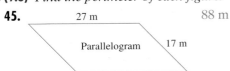

27 m

Parallelogram   17 m

**46.**   30 cm

11 cm

7 cm

12 cm

**47.** 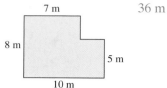 36 m

**48.** 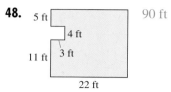 90 ft

*Solve.*

**49.** Find the perimeter of a rectangular sign that measures 6 feet by 10 feet.   32 ft

**50.** Find the perimeter of a town square that measures 110 feet on a side.   440 ft

*Find the circumference of each circle. Use π ≈ 3.14.*

**51.**  5.338 in.

**52.** 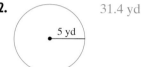 31.4 yd

△ **(9.4)** *Find the area of each figure. For the circles, find the exact area and then use π ≈ 3.14 to approximate the area.*

**53.**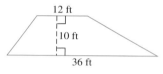

240 sq. ft

**54.**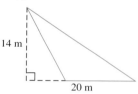

140 sq. m

**55.**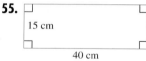

600 sq. cm

**56.**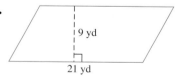

189 sq. yd

**57.**

49 π sq. ft ≈ 153.86 sq. ft

**58.**

4 π sq. in. ≈ 12.56 sq. in.

**59.**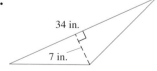

119 sq. in.

**60.**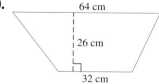

1248 sq. cm

**61.**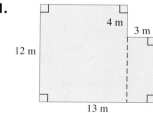

144 sq. m

**62.** The amount of sealer necessary to seal a driveway depends on the area. Find the area of a rectangular driveway 36 feet by 12 feet.   432 sq. ft

**63.** Find how much carpet is needed to cover the floor of the room shown.   130 sq. ft

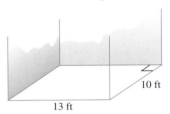

10 ft

13 ft

*Find the volume of each solid. For Exercises 66 and 67, use* $\pi \approx \frac{22}{7}$.

**64.**   $15\frac{5}{8}$ cu. in.

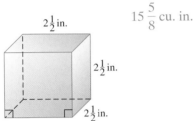

$2\frac{1}{2}$ in.

$2\frac{1}{2}$ in.

$2\frac{1}{2}$ in.

**65.**   84 cu. ft

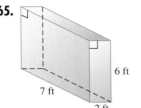

6 ft

7 ft

2 ft

**66.**   $62,857\frac{1}{7}$ cu. cm

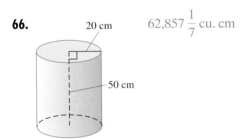

20 cm

50 cm

**67.**   $346\frac{1}{2}$ cu. in.

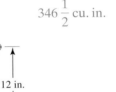

$5\frac{1}{4}$ in.

12 in.

**68.** Find the volume of a pyramid with a square base 2 feet on a side and a height of 2 feet.   $2\frac{2}{3}$ cu. ft

**69.** Approximate the volume of a tin can 8 inches high and 3.5 inches in radius. Use 3.14 for $\pi$.
307.72 cu. in.

**70.** A chest has 3 drawers. If each drawer has inside measurements of $2\frac{1}{2}$ feet by $1\frac{1}{2}$ feet by $\frac{2}{3}$ feet, find the total volume of the 3 drawers.   $7\frac{1}{2}$ cu. ft

**71.** A cylindrical canister for a shop vacuum is 2 feet tall and 1 foot in *diameter*. Find its exact volume.
$0.5\pi$ cu. ft

**72.** Find the volume of air in a rectangular room 15 feet by 12 feet with a 7-foot ceiling.   1260 cu. ft

**73.** A mover has two boxes left for packing. Both are cubical, one 3 feet on a side and the other 1.2 feet on a side. Find their combined volume.   28.728 cu. ft

**(9.5)** *Convert.*

**74.** 66 oz to pounds    4.125 lb

**75.** 2.3 tons to pounds    4600 lb

**76.** 52 oz = _____ lb_____ oz    3 lb 4 oz

**77.** 8200 lb = _____ tons _____ lb    4 tons 200 lb

**78.** 1400 mg to grams    1.4 g

**79.** 40 kg to grams    40,000g

**80.** 2.1 hg to dekagrams    21 dag

**81.** 0.03 mg to decigrams    0.0003 dg

*Perform each indicated operation.*

**82.** 6 lb 5 oz − 2 lb 12 oz    3 lb 9 oz

**83.** 5 tons 1600 lb + 4 tons 1200 lb    10 tons 800 lb

**84.** 6 tons 2250 lb ÷ 3    2 tons 750 lb

**85.** 8 lb 6 oz × 4    33 lb 8 oz

**86.** 1300 mg + 3.6 g    4.9 g or 4900 mg

**87.** 4.8 kg + 4200 g    9 kg or 9000 g

**88.** 9.3 g − 1200 mg    8.1 g or 8100 mg

**89.** 6.3 kg × 8    50.4 kg

*Solve the following.*

**90.** Donshay Berry ordered 1 lb 12 oz of soft-center candies and 2 lb 8 oz of chewy-center candies for his party. Find the total weight of the candy ordered.
4 lb 4 oz

**91.** Four local townships jointly purchase 38 tons 300 lb of cinders to spread on their roads during an ice storm. Determine the weight of the cinders each township receives if they share the purchase equally.
9 tons 1075 lb

**92.** Linda Holden ordered 8.3 kg of whole wheat flour from the health store, but she received 450 g less. How much flour did she actually receive?    7.85 kg

**93.** Eight friends spent a weekend in the Poconos tapping maple trees and preparing 9.3 kg of maple syrup. Find the weight each friend receives if they share the syrup equally.    1.1625 kg

**(9.6)** *Convert.*

**94.** 16 pt to quarts   8 qt

**95.** 40 fl oz to cups   5 c

**96.** 6.75 gal to quarts   27 qt

**97.** 8.5 pt to cups   17 c

**98.** 9 pt = _____ qt _____ pt   4 qt 1 pt

**99.** 15 qt = _____ gal _____ qt   3 gal 3 qt

**100.** 3.8 L to milliliters   3800 ml

**101.** 4.2 ml to deciliters   0.042 dl

**102.** 14 hl to kiloliters   1.4 kl

**103.** 30.6 L to centiliters   3060 cl

*Perform each indicated operation.*

**104.** 1 qt 1 pt + 3 qt 1 pt   1 gal 1 qt

**105.** 3 gal 2 qt 1 pt × 2   7 gal 1 qt

**106.** 0.946 L − 210 ml   736 ml or 0.736 L

**107.** 6.1 L + 9400 ml   15.5 L or 15,500 ml

*Solve.*

**108.** Carlos Perez prepares 4 gal 2 qt of iced tea for a block party. During the first 30 minutes of the party, 1 gal 3 qt of the tea is consumed. How much iced tea remains?   2 gal 3 qt

**109.** A recipe for soup stock calls for 1 c 4 fl oz of beef broth. How much should be used if the recipe is cut in half?   6 fl oz

**110.** Each bottle of Kiwi liquid shoe polish holds 85 ml of the polish. Find the number of liters of shoe polish contained in 8 boxes if each box contains 16 bottles.   10.88 L

**111.** Ivan Miller wants to pour three separate containers of saline solution into a single vat with a capacity of 10 liters. Will 6 liters of solution in the first container combined with 1300 milliliters in the second container and 2.6 liters in the third container fit into the larger vat?   yes, 9.9 L

**(9.7)** *Convert. Round to the nearest tenth of a degree, if necessary.*

**112.** 245°C to degrees Fahrenheit    473°F

**113.** 160°C to degrees Fahrenheit    320°F

**114.** 42°C to degrees Fahrenheit    107.6°F

**115.** 86°C to degrees Fahrenheit    186.8°F

**116.** 93.2°F to degrees Celsius    34°C

**117.** 51.8°F to degrees Celsius    11°C

**118.** 41.3°F to degrees Celsius    5.2°C

**119.** 80°F to degrees Celsius    26.7°C

*Solve. Round to the nearest tenth of a degree, if necessary.*

**120.** A sharp dip in the jet stream caused the temperature in New Orleans to drop to 35°F. Find the corresponding temperature in degrees Celsius.    1.7°C

**121.** The recipe for meat loaf calls for a 165°C oven. Find the setting used if the oven has a Fahrenheit thermometer.    329°F

**ANSWERS**

# CHAPTER 9 TEST

 **1.** Find the complement of a 78° angle.

**2.** Find the supplement of a 124° angle.

**1.** 12° _____

**2.** 56° _____

**3.** Find the measure of ∠x.

**3.** 50° _____

*Find the measure of ∠x, ∠y, and ∠z in each figure.*

**4.**

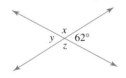

**5.** m‖n

**4.** ∠x = 118°
∠y = 62°; ∠z = 118° _____

**5.** ∠x = 73°; ∠y = 73°;
∠z = 73° _____

**6.** Find the measure of ∠x.

**6.** 26° _____

*Find the perimeter (or circumference) and area of each figure. For the circle, give the exact value and then use π ≈ 3.14 for an approximation.*

**7.**

9 in.

**8.**

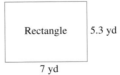

Rectangle   5.3 yd

7 yd

**7.** circumference =
18π in. ≈ 56.52 in.;
area =
81π sq. in. ≈
254.34 sq. in. _____

**8.** perimeter = 24.6 yd;
area = 37.1 sq. yd _____

**9.**

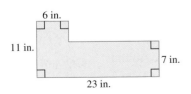

6 in.

11 in.

7 in.

23 in.

**9.** perimeter = 68 in.;
area = 185 sq. in. _____

**Name** _____

*Find the volume of each solid. For the cylinder, use $\pi \approx \frac{22}{7}$.*

△ **10.**

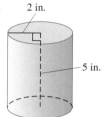
2 in.

5 in.

△ **11.**

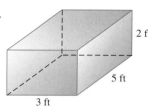
2 ft

5 ft

3 ft

*Solve.*

△ **12.** Find the perimeter of a square frame with a side length of 4 inches.

4 in.

△ **13.** How much soil is needed to fill a rectangular hole 3 feet by 3 feet by 2 feet?

△ **14.** Find how much baseboard is needed to go around a rectangular room that measures 18 feet by 13 feet.

△ **15.** Vivian Thomas is going to put insecticide on her lawn to control grubworms. The lawn is a rectangle measuring 123.8 feet by 80 feet. The amount of insecticide required is 0.02 ounces per square foot. Find how much insecticide Vivian needs to purchase.

*Convert.*

**16.** 280 in. to feet and inches

**17.** $2\frac{1}{2}$ gal to quarts

**18.** 30 oz to pounds

**19.** 2.8 tons to pounds

**20.** 38 pt to gallons

**21.** 40 mg to grams

**22.** 2.4 kg to grams

**23.** 3.6 cm to millimeters

**24.** 4.3 dg to grams

**25.** 0.83 L to milliliters

*Perform each indicated operation.*

**26.** 3 qt 1 pt + 2 qt 1 pt

**27.** 8 lb 6 oz − 4 lb 9 oz

**28.** 2 ft 9 in. × 3

**29.** 5 gal 2 qt ÷ 2

**30.** 8 cm − 14 mm

**31.** 1.8 km + 456 m

*Convert. Round to the nearest tenth of a degree, if necessary.*

**32.** 84°F to degrees Celsius

**33.** 12.6°C to degrees Fahrenheit

**34.** The sugar maples in front of Bette MacMillan's house are 8.4 meters tall. Because they interfere with the phone lines, the telephone company plans to remove the top third of the trees. How tall will the maples be after they are cut back?

**35.** A total of 15 gal 1 qt of oil has been removed from a 20-gallon drum. How much oil still remains in the container?

**36.** The doctors are quite concerned about Lucia Gillespie, who is running a 41°C fever. Find Lucia's temperature in degrees Fahrenheit.

**37.** Gordan Cooper, the engineer in charge of the bridge construction, said that the span of the bridge would be 88 m. But the actual construction required it to be 340 cm longer. Find the span of the bridge.

**20.** $4\frac{3}{4}$ gal

**21.** 0.04 g

**22.** 2400 g

**23.** 36 mm

**24.** 0.43 g

**25.** 830 ml

**26.** 1 gal 2 qt

**26.** 3 lb 13 oz

**28.** 8 ft 3 in.

**29.** 2 gal 3 qt

**30.** 66 mm or 6.6 cm

**31.** 2.256 km or 2256 m

**32.** 28.9°C

**33.** 54.7°F

**34.** 5.6 m

**35.** 4 gal 3 qt

**36.** 105.8°F

**37.** 91.4 m

**38.** $16 \text{ ft } 6 \text{ in. or } 16\frac{1}{2} \text{ ft}$

**38.** If 2 ft 9 in. of material is used to manufacture one scarf, how much material is needed for 6 scarves?

**39.** In December 1998, Nkem Chukwu gave birth to the world's first surviving octuplets in Houston, Texas. The largest octuplet weighed nearly 29 ounces at birth. Convert this birth weight to pounds and ounces. (*Source:* Texas Children's Hospital Houston, Texas)

**39.** 1 lb 13 oz

**40.** 0.320 kg

**40.** The smallest of the world's first surviving octuplets weighed 320 grams at birth. Convert this birth weight to kilograms. (*Source:* Texas Children's Hospital, Houston, Texas)

**41.** The largest ice cream sundae ever made in the U.S. was assembled in Anaheim, California, in 1985. This giant sundae used 4667 gallons of ice cream. How many pints of ice cream were used?

**41.** 37,336 pt

**Name** _____ **Section** _____ **Date** _____

## CUMULATIVE REVIEW

**1.** Solve: $3a - 6 = a + 4$

**2.** Evaluate:

   **a.** $\left(\dfrac{2}{5}\right)^4$    **b.** $\left(-\dfrac{1}{4}\right)^2$

**3.** Add: $1\dfrac{4}{5} + 4 + 2\dfrac{1}{2}$

**4.** Simplify by combining like terms:

   $11.1x - 6.3 + 8.9x - 4.6$

**5.** Simplify: $\dfrac{0.7 + 1.84}{0.4}$

**6.** Insert $<, >,$ or $=$ to form a true

   statement.

   $0.\overline{7}$    $\dfrac{7}{9}$

**7.** Solve for $y$: $0.5y + 2.3 = 1.65$

△ **8.** An inner city park is in the shape of a square that measures 300 feet on a side. Find the length of the diagonal of the park rounded to the nearest whole foot.

△ **9.** Given the rectangle shown:

   7 feet

   5 feet

   **a.** Find the ratio of its width to its length.
   **b.** Find the ratio of its length to its perimeter.

**10.** Write the rate as a fraction in simplest form: 10 nails every 6 feet.

**11.** Solve for $x$: $\dfrac{1.6}{1.1} = \dfrac{x}{0.3}$

   Round the solution to the nearest hundredth.

**12.** The standard dose of an antibiotic is 4 cc (cubic centimeters) for every 25 pounds (lb) of body weight. At this rate, find the standard dose for a 140-lb woman.

**ANSWERS**

**1.** 5 (Sec. 3.4, Ex. 3)

**2. a.** $\dfrac{16}{625}$

   **b.** $\dfrac{1}{16}$ (Sec. 4.3, Ex. 9)

**3.** $8\dfrac{3}{10}$ (Sec. 4.8, Ex. 13)

**4.** $20x - 10.9$ (Sec. 5.2, Ex. 11)

**5.** 6.35 (Sec. 5.5, Ex. 7)

**6.** $=$ (Sec. 5.6, Ex. 5)

**7.** $-1.3$ (Sec. 5.7, Ex. 4)

**8.** 424 feet (Sec. 5.8, Ex. 9)

**9. a.** $\dfrac{5}{7}$

   **b.** $\dfrac{7}{24}$ (Sec. 6.1, Ex. 7)

**10.** $\dfrac{5 \text{ nails}}{3 \text{ feet}}$ (Sec. 6.2, Ex. 1)

**11.** 0.44 (Sec. 6.3, Ex. 10)

**12.** 22.4 cc (Sec. 6.4, Ex. 2)

**Name** _____

**13.** In a survey of 100 people, 17 people drive blue cars. What percent of people drive blue cars?

**14.** 13 is $6\frac{1}{2}$ % of what number?

**15.** What number is 30% of 9?

**16.** The number of applications for a mathematics scholarship at Yale increased from 34 to 45 in one year. What is the percent increase? Round to the nearest whole percent.

**17.** Find the sales tax and the total price on a purchase of an $85.50 trench coat in a city where the sales tax rate is 7.5%.

**18.** Find the ordered pair corresponding to each point plotted on the rectangular coordinate system.

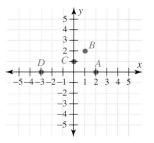

**19.** Graph $y = 4$

**20.** Find the median of the list of numbers: 24, 54, 56, 57, 60, 71, 98

**21.** Find the probability of choosing a red marble from a box containing 1 red, 1 yellow, and 2 blue marbles.

△ **22.** Find the complement of a 48° angle.

**23.** Convert 8 feet to inches.

**24.** Subtract 3 tons 1350 lb from 8 tons 1000 lb.

**25.** Convert 59°F to degrees Celsius.

# Polynomials

Recall that an exponent is a shorthand way of representing repeated multiplication. In this chapter, we learn more about exponents and a special type of expression containing exponents, called a *polynomial*. Studying polynomials is a major part of algebra. Polynomials are also useful for modeling many real-world situations. This chapter serves as an introduction to polynomials and some operations that can be performed on them.

The Royal Gorge Bridge is the world's highest suspension bridge. Located on U.S. Highway 50 near Cañon City, Colorado, the bridge spans the Arkansas River, which lies 1053 feet below the bridge. The Royal Gorge Bridge was built in 1929 at a cost of $350,000. Today it would cost over $10,000,000 to replace the bridge. It is 1260 feet long, 18 feet wide, and can support more than 2,000,000 pounds. In Exercises 43 and 44 on page 777 we will see how a polynomial can be used to describe the height of an object above the Arkansas River after being dropped from the Royal Gorge Bridge.

Name _____ Section _____ Date _____

# CHAPTER 10 PRETEST

*Perform each indicated operation.*

**1.** $(7y^2 - 15y + 9) + (-8y + 17)$

**2.** $(9b - 5) - (-8b + 7)$

**3.** Subtract $(-3z^4 + 6z^2 + 2z)$ from $(2z^4 - z^3 + 5z)$.

**4.** Find the value of the polynomial $6t^3 - 5t + 18$ when $t = -2$.

*Multiply.*

**5.** $9m^6 \cdot 4m^{12}$

**6.** $4x \cdot 5x \cdot 6x$

**7.** $(t^{18})^3$

**8.** $(3n^2)^4$

**9.** $(3a^2bc^3)^5 \cdot (8ab^4c)^2$

**10.** $2d(9d^4 - 5d^2 + 11)$

**11.** $(x + 6)(x + 3)$

**12.** $(y - 4)(2y + 5)$

**13.** $(3x - 2)^2$

**14.** $(n + 10)(n - 10)$

**15.** $(4a + 1)(2a^2 - a + 7)$

*Find the greatest common factor of each list of terms.*

**16.** 18 and 45

**17.** $y^9, y^3, y^8$

**18.** $6m^5, 14m, 18m^4$

*Factor*

**19.** $10y^2 + 6y - 14$

**20.** $8n^6 - 12n^5 + 20n^3$

# 10.1    ADDING AND SUBTRACTING POLYNOMIALS

Before we add and subtract polynomials, let's first review some definitions presented in Section 3.1. Recall that the *addends* of an algebraic expression are the *terms* of the expression.

**Expression**

$3x + 5$
↑    ↑
└────┴────2 terms

$7y^2 + (-6y) + 4$
↑        ↑        ↑
└────────┴────────┴────3 terms

Also, recall that *like terms* can be added or subtracted by using the distributive property. For example,

$7x + 3x = (7 + 3)x = 10x$

## A    ADDING POLYNOMIALS

Some terms are also called **monomials**. A monomial is a term that contains only whole-number exponents and no variable in the denominator.

| **Monomials** | **Not Monomials** | |
|---|---|---|
| $3x^2$ | $\dfrac{2}{y}$ | Variable in denominator |
| $-\dfrac{1}{2}a^2bc^3$ | $-2x^{-5}$ | Not a whole number exponent |
| $7$ | | |

A sum or difference of monomials is called a **polynomial**.

> **POLYNOMIAL**
>
> A **polynomial** is a monomial or a sum or difference of monomials.

**Examples of Polynomials**

$5x^3 - 6x^2 + 2x + 10, \quad -1.2y^3 + 0.7y, \quad z, \quad \dfrac{1}{3}r - \dfrac{1}{2}, \quad 0$

To add polynomials, we use the commutative and associative properties to rearrange and group like terms. Then we combine like terms.

> **ADDING POLYNOMIALS**
>
> To add polynomials, combine like terms.

**Example 1**    Add: $(3x - 1) + (-6x + 2)$

*Solution:*    $(3x - 1) + (-6x + 2) = (3x - 6x) + (-1 + 2)$    Group like terms.

$= (-3x) + (1)$    Combine like terms.

$= -3x + 1$    ▬▬

**Example 2**    Add: $(9y^2 - 6y) + (7y^2 + 10y + 2)$

*Solution:*

$(9y^2 - 6y) + (7y^2 + 10y + 2) = 9y^2 + 7y^2 - 6y + 10y + 2$    Group like terms.

$= 16y^2 + 4y + 2$    ▬▬

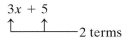
**Practice Problem 1**

Add: $(2y + 7) + (9y - 14)$

**Practice Problem 2**

Add: $(5x^2 + 4x - 3) + (x^2 - 6x)$

**Answers**

**1.** $11y - 7$    **2.** $6x^2 - 2x - 3$

Find the sum of $(7z^2 - 4.2z + 11)$ and $(-9z^2 - 1.9z + 4)$.

**Example 3**     Find the sum of $(-y^2 + 2y + 1.7)$ and $(12y^2 - 6y - 3.6)$.

*Solution:*     Recall that "sum" means addition.

$$(-y^2 + 2y + 1.7) + (12y^2 - 6y - 3.6)$$

$$= \underbrace{-y^2 + 12y^2} + \underbrace{2y - 6y} + \underbrace{1.7 - 3.6} \quad \text{Group like terms.}$$

$$= 11y^2 - 4y - 1.9 \qquad\qquad\qquad \text{Combine like terms.}$$

Polynomials can also be added vertically. To do so, line up like terms underneath one another. Let's vertically add the polynomials in Example 3.

Add the polynomials in Practice Problem 3 vertically.

**Example 4**     Find the sum of $(-y^2 + 2y + 1.7)$ and $(12y^2 - 6y - 3.6)$. Use a vertical format.

*Solution:*     Line up like terms underneath one another.

$$\begin{array}{r} -y^2 + 2y + 1.7 \\ +12y^2 - 6y - 3.6 \\ \hline 11y^2 - 4y - 1.9 \end{array}$$

Notice that we are finding the same sum in Example 3 as in Example 4. Of course, the results are the same.

## B  SUBTRACTING POLYNOMIALS

To subtract one polynomial from another, recall how we subtract numbers. To subtract a number, we add its opposite: $a - b = a + (-b)$.
For example,

$$7 - 10 = 7 + (-10)$$
$$= -3$$

To subtract a polynomial, we also add its opposite. Just as the opposite of 3 is $-3$, the opposite of $(2x^2 - 5x + 1)$ is $-(2x^2 - 5x + 1)$. Let's practice simplifying the opposite of a polynomial.

Simplify: $-(7y^2 + 4y - 6)$

**Example 5**     Simplify: $-(2x^2 - 5x + 1)$

*Solution:*     Rewrite $-(2x^2 - 5x + 1)$ as $-1(2x^2 - 5x + 1)$ and use the distributive property.

$$-(2x^2 - 5x + 1) = -1(2x^2 - 5x + 1)$$
$$= -1(2x^2) + (-1)(-5x) + (-1)(1)$$
$$= -2x^2 + 5x - 1$$

Notice the result of Example 5.

$$-(2x^2 - 5x + 1) = -2x^2 + 5x - 1$$

This means that the opposite of a polynomial can be found by changing the signs of the terms of the polynomial. This leads to the following.

TEACHING TIP

Before Example 7, ask students to translate the following: "Subtract 6 from 10." "Subtract $a$ from $b$."

---

**SUBTRACTING POLYNOMIALS**

To subtract polynomials, change the signs of the terms of the polynomial being subtracted, then add.

**Example 6**   Subtract: $(5a + 7) - (2a - 10)$

*Solution:*

$(5a + 7) - (2a - 10) = (5a + 7) + (-2a + 10)$   Add the opposite of $2a - 10$.
$= 5a - 2a + 7 + 10$   Group like terms.
$= 3a + 17$

**Example 7**   Subtract $(-6z^2 - 2z + 13)$ from $(4z^2 - 20z)$.

*Solution:*   Be careful when arranging the polynomials in this example.

$(4z^2 - 20z) - (-6z^2 - 2z + 13) = (4z^2 - 20z) + (6z^2 + 2z - 13)$   Group like terms.
$= 4z^2 + 6z^2 - 20z + 2z - 13$
$= 10z^2 - 18z - 13$

**TRY THE CONCEPT CHECK IN THE MARGIN.**

Just as with adding polynomials, we can subtract polynomials using a vertical format. Let's subtract the polynomials in Example 7 using a vertical format.

**Example 8**   Subtract $(-6z^2 - 2z + 13)$ from $(4z^2 - 20z)$. Use a vertical format.

*Solution:*   Line up like terms underneath one another.

$$4z^2 - 20z \qquad\qquad 4z^2 - 20z$$
$$\underline{-(-6z^2 - \phantom{0}2z + 13)} \qquad \underline{6z^2 + \phantom{0}2z - 13}$$
$$10z^2 - 18z - 13$$

**C   EVALUATING POLYNOMIALS**

Polynomials have different values depending on replacement values for the variables.

**Example 9**   Find the value of the polynomial $3t^3 - 2t + 5$ when $t = 1$.

*Solution:*   Replace $t$ with 1 and simplify.

$3t^3 - 2t + 5 = 3(1)^3 - 2(1) + 5$   Let $t = 1$.
$= 3(1) - 2 + 5$   $(1)^3 = 1$.
$= 3 - 2 + 5$
$= 6$

The value of $3t^3 - 2t + 5$ when $t = 1$ is 6.

Many real-world applications are modeled by polynomials.

**Practice Problem 6**

Subtract: $(3b - 2) - (7b + 23)$

**Practice Problem 7**

Subtract $(3x^2 - 12x)$ from $(-4x^2 + 20x + 17)$.

**✓ CONCEPT CHECK**

Find and explain the error in the following subtraction.
$(3x^2 + 4) - (x^2 - 3x)$
$= (3x^2 + 4) + (-x^2 - 3x)$
$= 3x^2 - x^2 - 3x + 4$
$= 2x^2 - 3x + 4$

**Practice Problem 8**

Subtract $(3x^2 - 12x)$ from $(-4x^2 + 20x + 17)$. Use a vertical format.

**Practice Problem 9**

Find the value of the polynomial $2y^3 + y^2 - 6$ when $y = 3$.

**Answers**

**6.** $-4b - 25$   **7.** and   **8.** $-7x^2 + 32x + 17$
**9.** 57
**✓ Concept Check:** $(3x^2 + 4) - (x^2 - 3x)$
$= (3x^2 + 4) + (-x^2 + 3x)$
$= 3x^2 - x^2 + 3x + 4$
$= 2x^2 + 3x + 4$

**Practice Problem 10**

An object is dropped from the top of a 530-foot cliff. Its height in feet at time $t$ seconds is given by the polynomial $-16t^2 + 530$. Find the height of the object when $t = 1$ second and when $t = 4$ seconds.

**HELPFUL HINT**

Don't forget to insert units, if appropriate.

**Example 10**    Finding the Height of an Object

An object is dropped from the top of an 800-foot-tall building. Its height at time $t$ seconds is given by the polynomial $-16t^2 + 800$. Find the height of the object when $t = 1$ second and when $t = 3$ seconds.

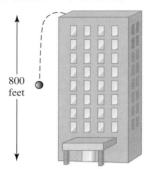

*Solution:* To find each height, we evaluate the polynomial when $t = 1$ and when $t = 3$.

$$-16t^2 + 800 = -16(1)^2 + 800$$
$$= -16 + 800$$
$$= 784$$

The height of the object at 1 second is 784 feet.

$$-16t^2 + 800 = -16(3)^2 + 800$$
$$= -16(9) + 800$$
$$= -144 + 800$$
$$= 656$$

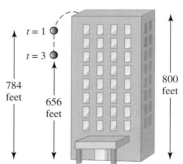

The height of the object at 3 seconds is 656 feet. ▬▬

**Answer**

**10.** 514 feet; 274 feet

## EXERCISE SET 10.1

**A** *Add the polynomials. See Examples 1 through 4.*

**1.** $(2x + 3) + (-7x - 27)$

**2.** $(9y - 16) + (-43y + 16)$

**3.** $(-4z^2 - 6z + 1) + (-5z^2 + 4z + 5)$

**4.** $(17a^2 - 6a + 3) + (16a^2 - 6a - 10)$

**5.** $(12y - 20) + (9y^2 + 13y - 20)$

**6.** $(5x^2 - 6) + (-3x^2 + 17x - 2)$

**7.** $(4.3a^4 + 5) + (-8.6a^4 - 2a^2 + 4)$

**8.** $(-12.7z^3 - 14z) + (-8.9z^3 + 12z + 2)$

**B** *Subtract the polynomials. See Examples 5 through 8.*

**9.** $(5a - 6) - (a + 2)$

**10.** $(12b + 7) - (-b - 5)$

**11.** $(3x^2 - 2x + 1) - (5x^2 - 6x)$

**12.** $(-9z^2 + 6z + 2) - (3z^2 + 1)$

**13.** $(10y^2 - 7) - (20y^3 - 2y^2 - 3)$

**14.** $(11x^3 + 15x - 9) - (-x^3 + 10x^2 - 9)$

**15.** Subtract $(3x - 4)$ from $(2x + 12)$.

**16.** Subtract $(6a + 1)$ from $(-7a + 7)$.

**17.** Subtract $(5y^2 + 4y - 6)$ from $(13y^2 - 6y - 14)$.

**18.** Subtract $(16x^2 - x + 1)$ from $(12x^2 - 3x - 12)$.

**ANSWERS**

**1.** $-5x - 24$

**2.** $-34y$

**3.** $-9z^2 - 2z + 6$

**4.** $33a^2 - 12a - 7$

**5.** $9y^2 + 25y - 40$

**6.** $2x^2 + 17x - 8$

**7.** $-4.3a^4 - 2a^2 + 9$

**8.** $-21.6z^3 - 2z + 2$

**9.** $4a - 8$

**10.** $13b + 12$

**11.** $-2x^2 + 4x + 1$

**12.** $-12z^2 + 6z + 1$

**13.** $-20y^3 + 12y^2 - 4$

**14.** $12x^3 - 10x^2 + 15x$

**15.** $-x + 16$

**16.** $-13a + 6$

**17.** $8y^2 - 10y - 8$

**18.** $-4x^2 - 2x - 13$

**19.** $4x^2 + x - 16$

**20.** $12a^2 - 9a - 9$

**21.** $-15y + 3.6$

**22.** $-1.1x + 97$

**23.** $b^3 - b^2 + 7b - 1$

**24.** $-2z^3 + 8z^2 - 20z$

**25.** $\dfrac{9}{7}$

**26.** $-13y^2 + y$

**27.** 1

**28.** $-17$

**29.** $-5$

**30.** $-72$

**31.** $-8$

**32.** 7

**33.** 20

**34.** $-31$

**35.** 25

**36.** 125

**37.** 50

**38.** 85

Name _____

**A** **B** *Perform each indicated operation.*

**19.** $(9x^2 - 6) + (-5x^2 + x - 10)$      **20.** $(12a^2 - 4a - 4) + (-5a - 5)$

**21.** $(21y - 4.6) - (36y - 8.2)$      **22.** $(8.6x + 4) - (9.7x - 93)$

**23.** $(b^3 - 2b^2 + 10b + 11) + (b^2 - 3b - 12)$

**24.** $(-2z^3 + 5z^2 - 13z + 6) + (3z^2 - 7z - 6)$

**25.** Subtract $\left(3z - \dfrac{3}{7}\right)$ from $\left(3z + \dfrac{6}{7}\right)$

**26.** Subtract $\left(8y^2 - \dfrac{7}{10}\,y\right)$ from $\left(-5y^2 + \dfrac{3}{10}\,y\right)$

**C** *Find the value of each polynomial when $x = 2$. See Examples 9 and 10.*

**27.** $-3x + 7$      **28.** $-5x - 7$

**29.** $x^2 - 6x + 3$      **30.** $5x^2 + 4x - 100$

**31.** $\dfrac{3x^2}{2} - 14$      **32.** $\dfrac{7x^3}{14} - x + 5$

*Find the value of each polynomial when $x = 5$. See Examples 9 and 10.*

**33.** $2x + 10$      **34.** $-5x - 6$      **35.** $x^2$

**36.** $x^3$      **37.** $2x^2 + 4x - 20$      **38.** $4x^2 - 5x + 10$

**Name** _____

*Solve. See Example 10.*

The distance in feet traveled by a free-falling object in $t$ seconds is given by the polynomial

$$16t^2$$

Use this polynomial for Exercises 39 and 40.

**39.** Find the distance traveled by an object that falls for 6 seconds.

**40.** It takes 8 seconds for a hard hat to fall from the top of a building. How high is the building?

Office Supplies, Inc. manufactures office products. They determine that the total cost for manufacturing $x$ file cabinets is given by the polynomial

$$3000 + 20x$$

Use this polynomial for Exercises 41 and 42.

**41.** Find the total cost to manufacture 10 file cabinets.

**42.** Find the total cost to manufacture 100 file cabinets.

An object is dropped from the deck of the Royal Gorge Bridge, which stretches across Royal Gorge at a height of 1053 feet above the Arkansas River. The height of the object above the river after $t$ seconds is given by the polynomial

$$1053 - 16t^2$$

Use this polynomial for Exercises 43 and 44. (*Source:* Royal Gorge Bridge Co.)

**43.** How far above the river is an object that has been falling for 3 seconds?

**44.** How far above the river is an object that has been falling for 8 seconds?

## REVIEW AND PREVIEW

*Evaluate. See Sections 1.7 and 2.4.*

**45.** $3^4$      **46.** $(-2)^5$      **47.** $(-5)^2$      **48.** $4^3$

*Write using exponential notation. See Section 1.8.*

**49.** $x \cdot x \cdot x$      **50.** $y \cdot y \cdot y \cdot y \cdot y$      **51.** $2 \cdot 2 \cdot a \cdot a \cdot a \cdot a$      **52.** $5 \cdot 5 \cdot 5 \cdot b \cdot b$

**39.** 576 ft

**40.** 1024 ft

**41.** $3200

**42.** $5000

**43.** 909 ft

**44.** 29 ft

**45.** 81

**46.** −32

**47.** 25

**48.** 64

**49.** $x^3$

**50.** $y^5$

**51.** $2^2 a^4$

**52.** $5^3 b^2$

**53.** $(8x + 2)$ in.

**Name** _____

**54.** $(6x^2 + 6x - 15)$ m

*Find the perimeter of each figure.*

△ **53.**

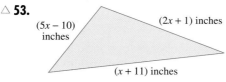

$(5x - 10)$ inches   $(2x + 1)$ inches

$(x + 11)$ inches

△ **54.**

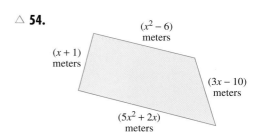

$(x^2 - 6)$ meters

$(x + 1)$ meters

$(3x - 10)$ meters

$(5x^2 + 2x)$ meters

**55.** $(4x - 15)$ units

*Given the lengths in the figure below, we find the unknown length by subtracting. Use the information to find the unknown lengths in Exercises 55 and 56.*

**56.** $(-7x + 4)$ units

←————— 8 units —————→

←— 3 units —→←——— ? ———→
              $(8 - 3)$ units

**55.** ←——— $(7x - 10)$ units ———→

←— $(3x + 5)$ —→←— ? —→
      units        units

**56.** ←——— $(x^2 - 7x + 6)$ units ———→

←— $(x^2 + 2)$ —→←— ? —→
       units          units

**57.** 20;   6;   2

*Fill in the blanks.*

**57.** $\left(3x^2 + \underline{\hspace{1cm}} x - \underline{\hspace{1cm}}\right) + \left(\underline{\hspace{1cm}} x^2 - 6x + 2\right) = 5x^2 + 14x - 4$

**58.** 1;   2;   10

**58.** $\left(\underline{\hspace{1cm}} y^2 + 4y - 3\right) + \left(8y^2 - \underline{\hspace{1cm}} y + \underline{\hspace{1cm}}\right) = 9y^2 + 2y + 7$

**59.** 7.2752

▥ **59.** Find the value of $7a^4 - 6a^2 + 2a - 1$ when $a = 1.2$.

▥ **60.** Find the value of $3b^3 + 4b^2 - 100$ when $b = -2.5$.

**60.** $-121.875$

✎ **61.** For Exercises 43 and 44, the polynomial $1053 - 16t^2$ was used to give the height of an object above the river after $t$ seconds. Find the height when $t = 8$ seconds and $t = 9$ seconds. Explain what happened and why.

29 ft; $-243$ ft;
**61.** answers may vary

**778**

# 10.2 MULTIPLICATION PROPERTIES OF EXPONENTS

## A USING THE PRODUCT RULE

Recall from Section 1.8 that an exponent has the same meaning whether the base is a number or a variable. For example,

$$5^3 = \underbrace{5 \cdot 5 \cdot 5}_{3 \text{ factors of } 5} \quad \text{and} \quad x^3 = \underbrace{x \cdot x \cdot x}_{3 \text{ factors of } x}$$

We can use this definition of an exponent to discover properties that will help us to simplify products and powers of exponential expressions.

For example, let's use the definition of an exponent to find the product of $x^3$ and $x^4$.

$$x^3 \cdot x^4 = (x \cdot x \cdot x)(x \cdot x \cdot x \cdot x)$$
$$= \underbrace{x \cdot x \cdot x \cdot x \cdot x \cdot x \cdot x}_{7 \text{ factors of } x}$$
$$= x^7$$

Notice that the result is the same if we add the exponents.

$$x^3 \cdot x^4 = x^{3+4} = x^7$$

This suggests the following product property for exponents.

> ### PRODUCT PROPERTY FOR EXPONENTS
>
> If $m$ and $n$ are positive integers and $a$ is a real number, then
>
> $$a^m \cdot a^n = a^{m+n}$$

In other words, to multiply two exponential expressions with the same base, keep the base and add the exponents.

### Example 1
Multiply: $y^7 \cdot y^2$

*Solution:* 
$$y^7 \cdot y^2 = y^{7+2} \quad \text{Use the product property for exponents.}$$
$$= y^9 \quad \text{Simplify.}$$

### Example 2
Multiply: $3x^5 \cdot 6x^3$

*Solution:* 
$$3x^5 \cdot 6x^3 = (3 \cdot 6)(x^5 \cdot x^3) \quad \begin{array}{l}\text{Apply the commutative and}\\\text{associative properties.}\end{array}$$
$$= 18x^{5+3} \quad \text{Use the product property for exponents.}$$
$$= 18x^8 \quad \text{Simplify.}$$

### Example 3
Multiply: $(-2a^4b^{10})(9a^5b^3)$

*Solution:* Use properties of multiplication to group numbers and like variables together.

$$(-2a^4b^{10})(9a^5b^3) = (-2 \cdot 9)(a^4 \cdot a^5)(b^{10} \cdot b^3)$$
$$= -18a^{4+5}b^{10+3}$$
$$= -18a^9b^{13}$$

## Objectives

A Use the product rule for exponents.
B Use the power rule for exponents.
C Use the power of a product rule for exponents.

Study Guide     SSM     CD-ROM     Video 10.2

**TEACHING TIP**

Before giving students the product property, try a few more examples such as:
a) $x^2 \cdot x^3$
b) $x \cdot x^5$
c) $x^6 \cdot x^4$
Then see if they can write a shortcut procedure for simplifying these expressions.

**Practice Problem 1**

Multiply: $z^4 \cdot z^8$

**Practice Problem 2**

Multiply: $7y^5 \cdot 3y^9$

**Practice Problem 3**

Multiply: $(-7r^6s^2)(-3r^2s^5)$

**Answers**

**1.** $z^{12}$   **2.** $21y^{14}$   **3.** $21r^8s^7$

**Practice Problem 4**

Multiply: $9y^4 \cdot 3y^2 \cdot y$. (Recall that $y = y^1$.)

> **HELPFUL HINT**
>
> Don't forget that if an exponent is not written, it is assumed to be 1.

**TEACHING TIP**

Before giving students the power property, try a few more examples such as:
 a) $(x^3)^2$
 b) $(x^4)^3$
 c) $(y^2)^5$
Then see if they can write a shortcut procedure for simplifying these expressions.

**Example 4**    Multiply: $2x^3 \cdot 3x \cdot 5x^6$

*Solution:*    First notice the factor $3x$. Since there is one factor of $x$ in $3x$, it can also be written as $3x^1$.

$$2x^3 \cdot 3x^1 \cdot 5x^6 = (2 \cdot 3 \cdot 5)(x^3 \cdot x^1 \cdot x^6)$$
$$= 30x^{10}$$

---

> **Helpful Hint**
>
> These examples will remind you of the difference between adding and multiplying terms.
>
> **Addition**
>
> $$5x^3 + 3x^3 = (5 + 3)x^3 = 8x^3$$
> $$7x + 4x^2 = 7x + 4x^2$$
>
> **Multiplication**
>
> $$(5x^3)(3x^3) = 5 \cdot 3 \cdot x^3 \cdot x^3 = 15x^{3+3} = 15x^6$$
> $$(7x)(4x^2) = 7 \cdot 4 \cdot x \cdot x^2 = 28x^{1+2} = 28x^3$$

### B  USING THE POWER RULE

Next suppose that we want to simplify an exponential expression raised to a power. To see how we simplify $(x^2)^3$, we again use the definition of an exponent.

$$(x^2)^3 = \underbrace{(x^2) \cdot (x^2) \cdot (x^2)}_{3 \text{ factors of } x^2}$$   Apply the definition of an exponent.

$$= x^{2+2+2}$$   Use the product property for exponents.

$$= x^6$$   Simplify.

Notice the result is exactly the same if we multiply the exponents.

$$(x^2)^3 = x^{2 \cdot 3} = x^6$$

This suggests the following power property for exponents.

---

**POWER PROPERTY FOR EXPONENTS**

If $m$ and $n$ are positive integers and $a$ is a real number, then

$$(a^m)^n = a^{m \cdot n}$$

---

In other words, to raise a power to a power, keep the base and multiply the exponents.

**Answer**

**4.** $27y^7$

## Helpful Hint

Take a moment to make sure that you understand when to apply the product rule and when to apply the power rule.

| Product Rule → Add Exponents | Power Rule → Multiply Exponents |
|---|---|
| $x^5 \cdot x^7 = x^{5+7} = x^{12}$ | $\left(x^5\right)^7 = x^{5 \cdot 7} = x^{35}$ |
| $y^6 \cdot y^2 = y^{6+2} = y^8$ | $\left(y^6\right)^2 = y^{6 \cdot 2} = y^{12}$ |

TEACHING TIP

Make sure that students read the Helpful Hint in the text. Students often forget when to add exponents and when to multiply them.

**Example 5**    Simplify: $\left(y^8\right)^2$

*Solution:*    $\left(y^8\right)^2 = y^{8 \cdot 2}$    Use the power property.

$\qquad = y^{16}$

**Practice Problem 5**

Simplify: $\left(z^3\right)^{10}$

**Example 6**    Simplify: $\left(a^3\right)^4 \cdot \left(a^2\right)^9$

*Solution:*    $\left(a^3\right)^4 \cdot \left(a^2\right)^9 = a^{12} \cdot a^{18}$    Use the power property.

$\qquad = a^{12+18}$    Use the product property.

$\qquad = a^{30}$    Simplify.

**Practice Problem 6**

Simplify: $\left(z^4\right)^5 \cdot \left(z^3\right)^7$

### C USING THE POWER OF A PRODUCT RULE

Next, let's simplify the power of a product.

$(xy)^3 = xy \cdot xy \cdot xy$    Apply the definition of an exponent.

$\qquad = (x \cdot x \cdot x)(y \cdot y \cdot y)$    Group like bases.

$\qquad = x^3 y^3$    Simplify.

Notice that the power of a product can be written as the product of powers. This leads to the following power of a product property.

**POWER OF A PRODUCT PROPERTY FOR EXPONENTS**

If $n$ is a positive integer and $a$ and $b$ are real numbers, then

$(ab)^n = a^n b^n$

In other words, to raise a product to a power, raise each factor to the power.

**Example 7**    Simplify: $(5t)^3$

*Solution:*    $(5t)^3 = 5^3 t^3$    Apply the power of a product property.

$\qquad = 125t^3$    Write $5^3$ as 125.

### ✓ CONCEPT CHECK

Which property is needed to simplify $\left(x^6\right)^3$? Explain.

a. Product Property for Exponents
b. Power Property for Exponents
c. Power of a Product Property for Exponents

**Practice Problem 7**

Simplify: $(3b)^4$

**Answers**

**5.** $z^{30}$   **6.** $z^{41}$   **7.** $81b^4$
✓ Concept Check: b

**Practice Problem 8**

Simplify: $\left(4x^2y^6\right)^3$

**Practice Problem 9**

Simplify: $\left(2x^2y^4\right)^4 \cdot \left(3x^6y^9\right)^2$

**Example 8**     Simplify: $\left(2a^5b^3\right)^3$

*Solution:*     $\left(2a^5b^3\right)^3 = 2^3\left(a^5\right)^3 \cdot \left(b^3\right)^3$     Apply the power of a product property.

$\qquad\qquad\qquad = 8a^{15}b^9$     Apply the power property.     ▬▬▬

**Example 9**     Simplify: $\left(3y^4z^2\right)^4 \cdot \left(2y^3z^5\right)^5$

*Solution:*

$\left(3y^4z^2\right)^4 \cdot \left(2y^3z^5\right)^5 = 3^4\left(y^4\right)^4\left(z^2\right)^4 \cdot 2^5\left(y^3\right)^5\left(z^5\right)^5$     Apply the power of a product property.

$\qquad\qquad\qquad = 81y^{16}z^8 \cdot 32y^{15}z^{25}$     Apply the power property.

$\qquad\qquad\qquad = (81 \cdot 32)\left(y^{16} \cdot y^{15}\right)\left(z^8 \cdot z^{25}\right)$     Group like bases.

$\qquad\qquad\qquad = 2592y^{31}z^{33}$     Apply the product property.     ▬▬▬

**Answers**

**8.** $64x^6y^{18}$     **9.** $144x^{20}y^{34}$

## EXERCISE SET 10.2

**A** *Multiply. See Examples 1 through 4.*

**1.** $x^5 \cdot x^9$

**2.** $y^4 \cdot y^7$

**3.** $a^6 \cdot a$

**4.** $b \cdot b^8$

**5.** $3z^3 \cdot 5z^2$

**6.** $8r^2 \cdot 2r^{15}$

**7.** $-4x \cdot 10x$

**8.** $-9y \cdot 3y$

**9.** $(-5x^2y^3)(-5x^4y)$

**10.** $(-2xy^4)(-6x^3y^7)$

**11.** $(7ab)(4a^4b^5)$

**12.** $(3a^3b^6)(12a^2b^9)$

**13.** $2x \cdot 3x \cdot 5x$

**14.** $4y \cdot 3y \cdot 5y$

**15.** $a \cdot 4a^{11} \cdot 3a^5$

**16.** $b \cdot 7b^{10} \cdot 5b^8$

**B** **C** *Simplify. See Examples 5 through 9.*

**17.** $(x^5)^3$

**18.** $(y^4)^7$

**19.** $(z^2)^{10}$

**20.** $(a^6)^9$

**21.** $(b^7)^6 \cdot (b^2)^{10}$

**22.** $(x^2)^9 \cdot (x^5)^3$

**23.** $(3a)^4$

**24.** $(2y)^5$

**25.** $(a^{11}b^8)^3$

**26.** $(x^7y^4)^8$

ANSWERS

1. $x^{14}$

2. $y^{11}$

3. $a^7$

4. $b^9$

5. $15z^5$

6. $16r^{17}$

7. $-40x^2$

8. $-27y^2$

9. $25x^6y^4$

10. $12x^4y^{11}$

11. $28a^5b^6$

12. $36a^5b^{15}$

13. $30x^3$

14. $60y^3$

15. $12a^{17}$

16. $35b^{19}$

17. $x^{15}$

18. $y^{28}$

19. $z^{20}$

20. $a^{54}$

21. $b^{62}$

22. $x^{33}$

23. $81a^4$

24. $32y^5$

25. $a^{33}b^{24}$

26. $x^{56}y^{32}$

**Name** _____

**27.** $\left(11x^3y^6\right)^2$  　　　　　　**28.** $\left(9a^4b^3\right)^2$

**29.** $(-3y)\left(2y^7\right)^3$  　　　　　**30.** $(-2x)\left(5x^2\right)^4$

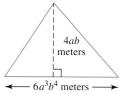 **31.** $(4xy)^3\left(2x^3y^5\right)^2$  　　　　**32.** $(2xy)^4\left(3x^4y^3\right)^3$

## REVIEW AND PREVIEW

*Multiply. See Section 3.1.*

**33.** $7(x - 3)$  　　　**34.** $4(y + 2)$  　　　**35.** $-2(3a + 2b)$

**36.** $-3(8r + 3s)$  　　　**37.** $9(x + 2y - 3)$  　　　**38.** $5(a + 7b - 3)$

## COMBINING CONCEPTS

*Find the area of each figure.*

△ **39.**

square  　$4x^6$ inches

△ **40.**

$9y^2$ centimeters

rectangle  　$9y$ centimeters

△ **41.**

$4ab$ meters

$6a^3b^4$ meters

△ **42.**

$30y^{12}$ feet  　parallelogram

$50y^{15}$ feet

(*Hint:* Area = base · height)

*Multiply and simplify.*

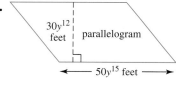 **43.** $\left(14a^7b^6\right)^3\left(9a^6b^3\right)^4$  　　　　**44.** $\left(5x^{14}y^6\right)^7\left(3x^{20}y^{19}\right)^5$

**45.** $\left(8.1x^{10}\right)^5$  　　　　**46.** $\left(4.6a^{14}\right)^4$

**47.** In your own words, explain why $x^2 \cdot x^3 = x^5$ and $\left(x^2\right)^3 = x^6$.

**784**

## INTEGRATED REVIEW—OPERATIONS ON POLYNOMIALS

*Add or subtract the polynomials as indicated.*

**1.** $(7x + 1) + (-3x - 2)$

**2.** $(14y - 6) + (19y - 2)$

**3.** $(7x + 1) - (-3x - 2)$

**4.** $(14y - 6) - (19y - 2)$

**5.** $(a^3 + 1) + (2a^3 + 5a - 9)$

**6.** $(1.2y^2 - 3.6y) + (0.6y^2 + 1.2y - 5.6)$

**7.** $(3.5x^2 - 0.5x) - (5.3x^2 - 2.9x + 1.7)$

**8.** $(2a^3 - 6a^2 + 11) - (6a^3 + 6a^2 + 11)$

**9.** Subtract $(2x - 6)$ from $(8x + 1)$.

**10.** Subtract $(3x^2 - x + 2)$ from $(5x^2 + 2x - 10)$.

*Find the value of each polynomial when $x = -3$.*

**11.** $2x - 7$

**12.** $x^2 + 5x + 2$

*Multiply.*

**13.** $x^7 \cdot x^{11}$

**14.** $x^6 \cdot x^6$

**15.** $y^3 \cdot y$

**1.** $4x - 1$

**2.** $33y - 8$

**3.** $10x + 3$

**4.** $-5y - 4$

**5.** $3a^3 + 5a - 8$

**6.** $1.8y^2 - 2.4y - 5.6$

**7.** $-1.8x^2 + 2.4x - 1.7$

**8.** $-4a^3 - 12a^2$

**9.** $6x + 7$

**10.** $2x^2 + 3x - 12$

**11.** $-13$

**12.** $-4$

**13.** $x^{18}$

**14.** $x^{12}$

**15.** $y^4$

**16.** $a^{11}$

**17.** $x^{77}$

**18.** $x^{36}$

**19.** $16x^4$

**20.** $27y^3$

**21.** $-12x^2y^7$

**22.** $12a^3b^4$

**23.** $x^{36}y^{20}$

**24.** $a^{20}b^{24}$

**25.** $300x^4y^3$

**26.** $128y^6z^7$

**16.** $a \cdot a^{10}$

**17.** $\left(x^7\right)^{11}$

**18.** $\left(x^6\right)^6$

**19.** $(2x)^4$

**20.** $(3y)^3$

**21.** $\left(-6xy^2\right)\left(2xy^5\right)$

**22.** $\left(-4a^2b^3\right)(-3ab)$

**23.** $\left(x^9y^5\right)^4$

**24.** $\left(a^{10}b^{12}\right)^2$

**25.** $\left(10x^2y\right)^2 \cdot (3y)$

**26.** $\left(8y^3z\right)^2 \cdot \left(2z^5\right)$

# 10.3 MULTIPLYING POLYNOMIALS

## A MULTIPLYING A MONOMIAL AND A POLYNOMIAL

Recall that a polynomial that consists of one term is called a **monomial**. For example, $5x$ is a monomial. To multiply a monomial and any polynomial, we use the distributive property $a(b + c) = a \cdot b + a \cdot c$ and properties of exponents.

**Example 1**   Multiply: $5x(3x^2 + 2)$

*Solution:*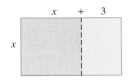

$$5x(3x^2 + 2) = 5x \cdot 3x^2 + 5x \cdot 2 \quad \text{Apply the distributive property.}$$
$$= 15x^3 + 10x$$

**Example 2**   Multiply: $2z(4z^2 + 6z - 9)$

*Solution:*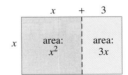

$$2z(4z^2 + 6z - 9) = 2z \cdot 4z^2 + 2z \cdot 6z + 2z(-9)$$
$$= 8z^3 + 12z^2 - 18z$$

To visualize multiplication by a monomial, let's look at two ways we can represent the area of the same rectangle.

The width of the rectangle is $x$ and its length is $x + 3$. One way to calculate the area of the rectangle is

Another way to calculate the area of the rectangle is to find the sum of the areas of the smaller figures.

$$\text{area} = \text{width} \cdot \text{length}$$
$$= x(x + 3)$$

$$\text{area} = x^2 + 3x$$

Since the areas must be equal, we have that

$$x(x + 3) = x^2 + 3x \qquad \text{As expected by the distributive property}$$

## B MULTIPLYING BINOMIALS

A polynomial that consists of exactly two terms is called a **binomial**. To multiply two binomials, we use a version of the distributive property:

$$(b + c)a = b \cdot a + c \cdot a$$

**Example 3**   Multiply: $(x + 2)(x + 3)$

*Solution:*

$$(x + 2)(x + 3) = x(x + 3) + 2(x + 3) \qquad \text{Apply the distributive property.}$$
$$= x \cdot x + x \cdot 3 + 2 \cdot x + 2 \cdot 3 \qquad \text{Apply the distributive property.}$$
$$= x^2 + 3x + 2x + 6 \qquad \text{Multiply.}$$
$$= x^2 + 5x + 6 \qquad \text{Combine like terms.}$$

---

**Objectives**

**A** Multiply a monomial and any polynomial.
**B** Multiply two binomials.
**C** Squaring a binomial.
**D** Multiply any two polynomials.

Study Guide   SSM   CD-ROM   Video 10.3

**Practice Problem 1**

Multiply: $3y(7y^2 + 5)$

**Practice Problem 2**

Multiply: $5r(8r^2 - r + 11)$

TEACHING TIP

You may want to show students how to multiply polynomials using a grid. For example,

|       | $x$    | $+$ $3$ |
|-------|--------|---------|
| $x$   | $x^2$  | $3x$    |
| $+$ $2$ | $2x$ | $6$     |

**Practice Problem 3**

Multiply: $(b + 7)(b + 5)$

**Answers**

**1.** $21y^3 + 15y$   **2.** $40r^3 - 5r^2 + 55r$
**3.** $b^2 + 12b + 35$

## Practice Problem 4

Multiply: $(5x - 1)(5x + 4)$

**TEACHING TIP**

Squaring a binomial incorrectly is a *common* mistake. Make sure that you spend adequate time with students on this concept.

## Practice Problem 5

Multiply: $(6y - 1)^2$

## Practice Problem 6

Multiply: $(2x + 5)(x^2 + 4x - 1)$

**TEACHING TIP**

Before Example 7, you may want to review vertical multiplication of whole numbers. For example,

$$\begin{array}{r} 362 \\ \times\ 47 \end{array}$$

**Answers**

**4.** $25x^2 + 15x - 4$  **5.** $36y^2 - 12y + 1$
**6.** $2x^3 + 13x^2 + 18x - 5$

**Example 4**     Multiply: $(4y + 9)(3y - 2)$

*Solution:*

$$\begin{aligned}
(4y + 9)(3y - 2) &= 4y(3y - 2) + 9(3y - 2) && \text{Apply the distributive property.} \\
&= 4y \cdot 3y + 4y(-2) + 9 \cdot 3y + 9(-2) && \text{Apply the distributive property.} \\
&= 12y^2 - 8y + 27y - 18 && \text{Multiply.} \\
&= 12y^2 + 19y - 18 && \text{Combine like terms.}
\end{aligned}$$

## C  SQUARING A BINOMIAL

Raising a binomial to the power of 2 is also called squaring a binomial. To square a binomial, we use the definition of an exponent, and then multiply.

**Example 5**     Multiply: $(2x + 1)^2$

*Solution:*

$$\begin{aligned}
(2x + 1)^2 &= (2x + 1)(2x + 1) && \text{Apply the definition of an exponent.} \\
&= 2x(2x + 1) + 1(2x + 1) && \text{Apply the distributive property.} \\
&= 2x \cdot 2x + 2x \cdot 1 + 1 \cdot 2x + 1 \cdot 1 && \text{Apply the distributive property.} \\
&= 4x^2 + 2x + 2x + 1 && \text{Multiply.} \\
&= 4x^2 + 4x + 1 && \text{Combine like terms.}
\end{aligned}$$

## D  MULTIPLYING POLYNOMIALS

A polynomial that consists of exactly three terms is called a **trinomial**. Next, we multiply a binomial by a trinomial.

**Example 6**     Multiply: $(3a + 2)(a^2 - 6a + 3)$

*Solution:*     Use the distributive property to multiply $3a$ by the trinomial $(a^2 - 6a + 3)$ and then 2 by the trinomial.

$$\begin{aligned}
(3a + 2)(a^2 - 6a + 3) &= 3a(a^2 - 6a + 3) + 2(a^2 - 6a + 3) && \text{Apply the distributive property.} \\
&= 3a \cdot a^2 + 3a(-6a) + 3a \cdot 3 + && \text{Apply the distributive property.} \\
&\quad\ 2 \cdot a^2 + 2(-6a) + 2 \cdot 3 \\
&= 3a^3 - 18a^2 + 9a + 2a^2 - 12a + 6 && \text{Multiply.} \\
&= 3a^3 - 16a^2 - 3a + 6 && \text{Combine like terms.}
\end{aligned}$$

In general, we have the following.

> **TO MULTIPLY TWO POLYNOMIALS**
>
> Multiply each term of the first polynomial by each term of the second polynomial, and then combine like terms.

A convenient method of multiplying polynomials is to use a vertical format similar to multiplying real numbers.

**TRY THE CONCEPT CHECK IN THE MARGIN.**

**Example 7**  Find the product of $(a^2 - 6a + 3)$ and $(3a + 2)$ vertically.

*Solution:*

$$
\begin{array}{r}
a^2 - 6a + 3 \\
\times \qquad 3a + 2 \\
\hline
2a^2 - 12a + 6 \\
3a^3 - 18a^2 + 9a \\
\hline
3a^3 - 16a^2 - 3a + 6
\end{array}
$$

Multiply $a^2 - 6a + 3$ by 2.

Multiply $a^2 - 6a + 3$ by $3a$. Line up like terms.

Combine like terms

Notice that this example is the same as Example 6 and of course the products are the same. ▬▬▬

---

# Focus On History

### EXPONENTIAL NOTATION

The French mathematician and philosopher René Descartes (1596–1650) is generally credited with devising the system of exponents that we use in math today. His book *La Géométrie* was the first to show successive powers of an unknown quantity $x$ as $x$, $xx$, $x^3$, $x^4$, $x^5$, and so on. No one knows why Descartes preferred to write $xx$ instead of $x^2$. However, the use of $xx$ for the square of the quantity $x$ continued to be popular. Those who used the notation defended it by saying that $xx$ takes up no more space when written than $x^2$ does.

Before Descartes popularized the use of exponents to indicate powers, other less convenient methods were used. Some mathematicians preferred to write out the Latin words *quadratus* and *cubus* whenever they wanted to indicate that a quantity was to be raised to the second power or the third power. Other mathematicians used the abbreviations of *quadratus* and *cubus*, $Q$ and $C$, to indicate second and third powers of a quantity.

## Exercise Set 10.3

**A** *Multiply. See Examples 1 and 2.*

**1.** $3x(9x^2 - 3)$

**2.** $4y(10y^3 + 2y)$

**3.** $-5a(4a^2 - 6a + 1)$

**4.** $-2b(3b^2 - 2b + 5)$

**5.** $7x^2(6x^2 - 5x + 7)$

**6.** $6z^2(-3z^2 - z + 4)$

**B** **C** *Multiply. See Examples 3 through 5.*

**7.** $(x + 3)(x + 10)$

**8.** $(y + 5)(y + 9)$

**9.** $(2x - 6)(x + 4)$

**10.** $(7z + 1)(z - 6)$

**11.** $(6a + 4)^2$

**12.** $(8b - 3)^2$

**D** *Multiply. See Examples 6 and 7.*

**13.** $(a + 6)(a^2 - 6a + 3)$

**14.** $(y + 4)(y^2 + 8y - 2)$

**15.** $(4x - 5)(2x^2 + 3x - 10)$

**16.** $(9z - 2)(2z^2 + z + 1)$

**17.** $(x^3 + 2x + x^2)(3x + 1 + x^2)$

**18.** $(y^2 - 2y + 5)(y^3 + 2 + y)$

**19.** $-30r^2 + 20r$

**20.** $20x^3 + 25x$

**21.** $-6y^3 - 2y^4 + 12y^2$

**22.** $12z^7 - 6z^4 + 3z^6$

**23.** $x^2 + 14x + 24$

**24.** $y^2 - 49$

**25.** $4a^2 - 9$

**26.** $18s^2 - 3s - 1$

**27.** $x^2 + 10x + 25$

**28.** $x^2 + 6x + 9$

**29.** $b^2 + \dfrac{7}{5}b + \dfrac{12}{25}$

**30.** $a^2 - \dfrac{2}{5}a - \dfrac{21}{100}$

**31.** $6x^3 + 25x^2 + 10x + 1$

**32.** $9y^3 + 26y^2 - 48y + 5$

**33.** $49x^2 + 70x + 25$

**34.** $25x^2 + 90x + 81$

**35.** $4x^2 - 4x + 1$

**36.** $16a^2 - 24a + 9$

**792**

---

**Name** _____

**A** **B** **C** **D** *Multiply.*

**19.** $10r(-3r + 2)$

**20.** $5x(4x^2 + 5)$

**21.** $-2y^2(3y + y^2 - 6)$

**22.** $3z^3(4z^4 - 2z + z^3)$

**23.** $(x + 2)(x + 12)$

**24.** $(y + 7)(y - 7)$

**25.** $(2a + 3)(2a - 3)$

**26.** $(6s + 1)(3s - 1)$

**27.** $(x + 5)^2$

**28.** $(x + 3)^2$

**29.** $\left(b + \dfrac{3}{5}\right)\left(b + \dfrac{4}{5}\right)$

**30.** $\left(a - \dfrac{7}{10}\right)\left(a + \dfrac{3}{10}\right)$

**31.** $(6x + 1)(x^2 + 4x + 1)$

**32.** $(9y - 1)(y^2 + 3y - 5)$

**33.** $(7x + 5)^2$

**34.** $(5x + 9)^2$

**35.** $(2x - 1)^2$

**36.** $(4a - 3)^2$

**Name** _____

**37.** $(2x^2 - 3)(4x^3 + 2x - 3)$

**38.** $(3y^2 + 2)(5y^2 - y + 2)$

**39.** $(x^3 + x^2 + x)(x^2 + x + 1)$

**40.** $(a^4 + a^2 + 1)(a^4 + a^2 - 1)$

**41.** $(2z^2 - z + 1)(5z^2 + z - 2)$

**42.** $(2b^2 - 4b + 3)(b^2 - b + 2)$

## REVIEW AND PREVIEW

*Write each number as a product of prime numbers. See Section 4.2.*

**43.** 50

**44.** 48

**45.** 72

**46.** 36

**47.** 200

**48.** 300

## COMBINING CONCEPTS

*Find the area of each figure.*

 **49.**

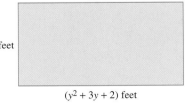

$(y - 6)$ feet

$(y^2 + 3y + 2)$ feet

△ **50.**

$(2x + 11)$ centimeters

---

**37.** $8x^5 - 8x^3 - 6x^2 - 6x + 9$

**38.** $15y^4 - 3y^3 + 16y^2 - 2y + 4$

**39.** $x^5 + 2x^4 + 3x^3 + 2x^2 + x$

**40.** $a^8 + 2a^6 + a^4 - 1$

**41.** $10z^4 - 3z^3 + 3z - 2$

**42.** $2b^4 - 6b^3 + 11b^2 - 11b + 6$

**43.** $2 \cdot 5^2$

**44.** $2^4 \cdot 3$

**45.** $2^3 \cdot 3^2$

**46.** $2^2 \cdot 3^2$

**47.** $2^3 \cdot 5^2$

**48.** $2^2 \cdot 3 \cdot 5^2$

**49.** $(y^3 - 3y^2 - 16y - 12)$ sq. ft

**50.** $(4x^2 + 44x + 121)$ sq. cm

**793**

*Find the area of the shaded figure. To do so, subtract the area of the smaller square from the area of the larger geometric figure.*

△ **51.**

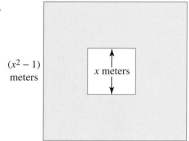

$(x^2 - 1)$
meters

$x$ meters

△ **52.**

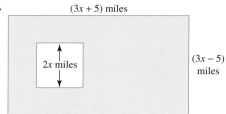

$(3x + 5)$ miles

$2x$ miles

$(3x - 5)$
miles

**53.** Suppose that a classmate asked you why $(2x + 1)^2$ is not $4x^2 + 1$. Write down your response to this classmate.

# 10.4 INTRODUCTION TO FACTORING POLYNOMIALS

Recall that when an integer is written as the product of two or more integers, each of these integers is called a *factor* of the product. This is true of polynomials also. When a polynomial is written as the product of two or more other polynomials, each of these polynomials is called a factor of the product.

$$-2 \cdot 4 = -8 \qquad x^3 \cdot x^7 = x^{10} \qquad 5(x + 2) = 5x + 10$$

factor  factor  product  factor  factor  product  factor  factor  product

The process of writing a polynomial as a product is called **factoring**. Notice that factoring is the reverse process of multiplying.

$$\overbrace{5x + 10}^{\text{factoring}} = \underbrace{5(x + 2)}_{\text{multiplying}}$$

## A FINDING THE GCF OF A LIST OF INTEGERS

Before we factor polynomials, let's practice finding the greatest common factor of a list of integers. The **greatest common factor (GCF)** of a list of integers is the largest integer that is a factor of all the integers in the list. For example,

the GCF of 30 and 18 is 6

because 6 is the largest integer that is a factor of both 30 and 18.
If the GCF cannot be found by inspection, the following steps can be used.

---

**TO FIND THE GCF OF A LIST OF INTEGERS**

**Step 1.** Write each number as a product of prime numbers.
**Step 2.** Identify the common prime factors.
**Step 3.** The product of all common prime factors found in *Step 2* is the greatest common factor. If there are no common prime factors, the greatest common factor is 1.

---

**Objectives**

**A** Find the greatest common factor of a list of integers.

**B** Find the greatest common factor of a list of terms.

**C** Factor the greatest common factor from the terms of a polynomial.

Study   SSM   CD-ROM   Video
Guide                         10.5

Which of the following is the prime factorization of 36?

a. $4 \cdot 9$

b. $2 \cdot 2 \cdot 3 \cdot 3$

c. $6 \cdot 6$

**Practice Problem 1**

Find the GCF of 42 and 28.

**TRY THE CONCEPT CHECK IN THE MARGIN.**

Recall from Section 4.2 that a prime number is a whole number other than 1, whose only factors are 1 and itself.

**Example 1**     Find the GCF of 12 and 20.

*Solution:*     **Step 1.** Write each number as a product of primes.

$$12 = 2 \cdot 2 \cdot 3$$
$$20 = 2 \cdot 2 \cdot 5$$

**Step 2.** $12 = \boxed{2} \cdot \boxed{2} \cdot 3$
$20 = \boxed{2} \cdot \boxed{2} \cdot 5$

$$\downarrow \quad \downarrow$$

$2 \cdot 2$     Identify the common factors.

**Step 3.** The GCF is $2 \cdot 2 = 4$

### B   FINDING THE GCF OF A LIST OF TERMS

How do we find the GCF of a list of variables raised to powers? For example, what is the GCF of $y^3$, $y^5$, and $y^{10}$? Notice that each variable term contains a factor of $y^3$ and no higher power of $y$ is a factor of each term.

$$y^3 = y^3$$
$$y^5 = y^3 \cdot y^2 \quad \text{Recall the product rule for exponents.}$$
$$y^{10} = y^3 \cdot y^7$$

The GCF of $y^3$, $y^5$, and $y^{10}$ is $y^3$. From this example, we can see that **the GCF of a list of variables raised to powers is the variable raised to the smallest exponent in the list.**

**Practice Problem 2**

Find the GCF of $z^7$, $z^8$, and $z$.

**Example 2**     Find the GCF of $x^{11}$, $x^4$, and $x^6$.

*Solution:*     The GCF is $x^4$ since 4 is the smallest exponent to which $x$ is raised.

In general, **the GCF of a list of terms is the product of all common factors.**

**Answers**

**1.** 14   **2.** $z$

✓ Concept Check:  b

**Example 3**    Find the GCF of $4x^3$, $12x$, and $10x^5$.

Solution:    The GCF of 4, 12, and 10 is 2.
The GCF of $x^3$, $x^1$, and $x^5$ is $x^1$.
Thus, the GCF of $4x^3$, $12x$, and $10x^5$ is $2x^1$ or $2x$. ■

## C  Factoring Out the GCF

Next, we practice factoring a polynomial by factoring the GCF from its terms. To do so, we write each term of the polynomial as a product of the GCF and another factor, and then apply the distributive property.

**Example 4**    Factor: $7x^3 + 14x^2$

Solution:    The GCF of $7x^3$ and $14x^2$ is $7x^2$.

$$7x^3 + 14x^2 = 7x^2 \cdot x^1 + 7x^2 \cdot 2$$

$$= 7x^2(x + 2) \qquad \text{Apply the distributive property.}$$

Notice in Example 4 that we factored $7x^3 + 14x^2$ by writing it as the product $7x^2(x + 2)$. Also notice that to check factoring, we multiply

$$7x^2(x + 2) = 7x^2 \cdot x + 7x^2 \cdot 2$$

$$= 7x^3 + 14x^2$$

which is the original binomial.

**Example 5**    Factor: $6x^2 - 24x + 6$

Solution:    The GCF of the terms is 6.

$$6x^2 - 24x + 6 = 6 \cdot x^2 - 6 \cdot 4x + 6 \cdot 1$$

$$= 6(x^2 - 4x + 1)$$

**HELPFUL HINT**
Don't forget to include the term 1.

---

**Practice Problem 3**

Find the GCF of $6a^4$, $3a^5$, and $15a^2$.

**Practice Problem 4**

Factor: $10y^7 + 5y^9$

TEACHING TIP

Remind students that a polynomial and its factored form will always give the same value for a particular value of the variable(s).

**Practice Problem 5**

Factor: $4z^2 - 12z + 2$

Answers

**3.** $3a^2$  **4.** $5y^7(2 + y^2)$  **5.** $2(2z^2 - 6z + 1)$

## Practice Problem 6

Factor: $-3y^2 - 9y + 15x^2$

## ✓ CONCEPT CHECK

Check both factorizations given in Example 6.

**Example 6**    Factor: $-2a + 20b - 4b^2$

*Solution:*
$$-2a + 20b - 4b^2 = 2 \cdot -a + 2 \cdot 10b - 2 \cdot 2b^2$$
$$= 2(-a + 10b - 2b^2)$$

When the coefficient of the first term is a negative number, we often factor out a negative common factor.
$$-2a + 20b - 4b^2 = (-2) \cdot a + (-2)(-10b) + (-2)(2b^2)$$
$$= -2(a - 10b + 2b^2)$$

Both $2(-a + 10b - 2b^2)$ and $-2(a - 10b + 2b^2)$ are factorizations of $-2a + 20b - 4b^2$. ▬▬

## TRY THE CONCEPT CHECK IN THE MARGIN.

**Answers**

**6.** $-3(y^2 + 3y - 5x^2)$

✓ Concept Check:  Answers may vary.

**Name** _____ **Section** _____ **Date** _____

## Exercise Set 10.4

**A** *Find the greatest common factor of each list of numbers. See Example 1.*

**1.** 48 and 15      **2.** 36 and 20      📼 **3.** 60 and 72

**4.** 96 and 45      **5.** 12, 20, and 36      **6.** 18, 24, and 60

**7.** 8, 32, and 100      **8.** 30, 50, and 200

**B** *Find the greatest common factor of each list of terms. See Examples 2 and 3.*

📼 **9.** $y^7, y^2, y^{10}$      **10.** $x^3, x, x^5$      **11.** $a^5, a^5, a^5$

**12.** $b^6, b^6, b^4$      📼 **13.** $x^3y^2, xy^2, x^4y^2$      **14.** $a^5b^3, a^5b^2, a^5b$

**15.** $3x^4, 5x^7, 10x$      **16.** $9z^6, z^5, 2z^3$      **17.** $2z^3, 14z^5, 18z^3$

**18.** $6y^7, 9y^6, 15y^5$

**C** *Factor. Check by multiplying. See Examples 4 through 6.*

📼 **19.** $3y^2 + 18y$      **20.** $2x^2 + 18x$      **21.** $10a^6 - 5a^8$

**22.** $21y^5 + y^{10}$      **23.** $4x^3 + 12x^2 + 20x$      **24.** $9b^3 - 54b^2 + 9b$

**25.** $z^7 - 6z^5$      **26.** $y^{10} + 4y^5$      📼 **27.** $-35 + 14y - 7y^2$

**28.** $-20x + 4x^2 - 2$      **29.** $12a^5 - 36a^6$      **30.** $25z^3 - 20z^2$

**31.** 36

**32.** 119.25

**33.** $\dfrac{4}{5}$

**34.** $\dfrac{13}{20}$

**35.** 37.5%

**36.** 75%

**37.** $x^2 + 2x$

**38.** $3x^2(5x + 2)$ or $15x^3 + 6x^2$

**39.** answers may vary

**40.** answers may vary

**Name** _____

## REVIEW AND PREVIEW

*Solve. See Sections 7.1 and 7.2.*

**31.** Find 30% of 120.

**32.** Find 45% of 265.

**33.** Write 80% as a fraction.

**34.** Write 65% as a fraction.

**35.** Write $\dfrac{3}{8}$ as a percent.

**36.** Write $\dfrac{3}{4}$ as a percent.

 ## COMBINING CONCEPTS

△ **37.** The area of the larger rectangle below is $x(x + 2)$. Find another expression for the area by writing the sum of the areas of the smaller rectangles.

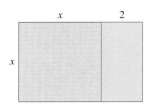

△ **38.** Write an expression for the area of the largest rectangle in two different ways.

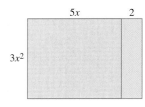

↘ **39.** In your own words, define the greatest common factor of a list of numbers.

↘ **40.** Suppose that a classmate asks you why $4x^2 + 6x + 2$ does not factor as $2(2x^2 + 3x)$. Write down your response to this classmate.

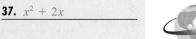 **Internet Excursions**

Go to http://www.prenhall.com/martin-gay

This World Wide Web address will provide you with access to a Webmath calculator for Factoring Polynomials by Finding the Greatest Common Factor (GCF), or a related site. The instructions for using this calculator are included on the Web page. Use this calculator to check your work for the following exercises.

**41.** Exercise 24

**42.** Exercise 26

**43.** Exercise 28

**44.** Exercise 30

**Name** _____ **Section** _____ **Date** _____

# CHAPTER 10 ACTIVITY
## BUSINESS ANALYSIS

*This activity may be completed by working in groups or individually.*

Suppose you own a business that manufactures baskets. You need to decide how many baskets to make. The more baskets you make, the lower the price you will have to charge to sell them all. Naturally, each basket you make costs you money because you must buy the materials to make each basket. The following table summarizes some factors you must consider in deciding how many baskets to make, along with algebraic representations of those factors.

| | Description | | Algebraic Expression |
|---|---|---|---|
| Number of baskets | Unknown | | $x$ |
| Total manufacturing expenses | This is the total amount that it will cost to manufacture all the baskets. It will cost $100 to buy special equipment to manufacture the baskets in addition to basket materials costing $0.50 per basket | | $100 + 0.50x$ |
| Price charged per basket | For each additional basket produced, the price that must be charged per basket decreases from $40 by an additional $0.05 | | $40 - 0.05x$ |

**1.** *Revenue* is the amount of money collected from selling the baskets. Revenue can be found by multiplying the price charged per basket by the number of baskets sold. Use the algebraic expressions given in the table above to find a polynomial that represents the revenue from sales of baskets. Then write this polynomial in the Polynomial column next to "Revenue" in the table to the right.

**2.** *Profit* is the amount of money you make from selling the baskets after deducting the expenses for making the baskets. Profit can be found by subtracting total manufacturing expenses from revenue. Find a polynomial that represents the profit from the sales of baskets. Then write this polynomial in the Polynomial column next to "Profit" in the table to the right.

**3.** Complete the following table by evaluating each polynomial for each of the numbers of baskets given in the table.

| | Polynomial | Number of Baskets, $x$ | | | | |
|---|---|---|---|---|---|---|
| | | 200 | 300 | 400 | 500 | 600 |
| Revenue | | | | | | |
| Total manufacturing expenses | $100 + 0.50x$ | | | | | |
| Profit | | | | | | |

**4.** Study the table. Which number of baskets will give you the largest profit from making and selling baskets?

# CHAPTER 10 HIGHLIGHTS

| DEFINITIONS AND CONCEPTS | EXAMPLES |
|---|---|
| **SECTION 10.1   ADDING AND SUBTRACTING POLYNOMIALS** | |

| | |
|---|---|
| A **polynomial** is a monomial or a sum or difference of monomials. | Polynomials $$5x^2 - 6x + 2, \quad -\frac{9}{10}y, \quad 7$$ |
| **To add polynomials**, combine like terms. | Add: $(7z^2 - 6z + 2) + (5z^2 - 4z + 5)$ <br> $(7z^2 - 6z + 2) + (5z^2 - 4z + 5)$ <br> $= \underbrace{7z^2 + 5z^2} - \underbrace{6z - 4z} + \underbrace{2 + 5}$   Group like terms. <br> $= 12z^2 - 10z + 7$   Combine like terms. |
| **To subtract polynomials,** change the signs of the terms being subtracted, then add. | Subtract: $(20x - 6) - (30x - 6)$ <br> $(20x - 6) - (30x - 6)$ <br> $= (20x - 6) + (-30x + 6)$ <br> $= \underbrace{20x - 30x} - 6 + 6$   Group like terms. <br> $= -10x$   Combine like terms. |

| **SECTION 10.2   MULTIPLICATION PROPERTIES OF EXPONENTS** | |
|---|---|

| | |
|---|---|
| **Product property for exponents** <br>    $a^m \cdot a^n = a^{m+n}$ | $x^3 \cdot x^{11} = x^{3+11} = x^{14}$ |
| **Power property for exponents** <br>    $(a^m)^n = a^{m \cdot n}$ | $(y^5)^3 = y^{5 \cdot 3} = y^{15}$ |
| **Power of a product property for exponents** <br>    $(ab)^n = a^n b^n$ | $(2z^5)^4 = 2^4(z^5)^4 = 16z^{20}$ |

| **SECTION 10.3   MULTIPLYING POLYNOMIALS** | |
|---|---|

| | |
|---|---|
| A **monomial** is a polynomial with one term. <br> A **binomial** is a polynomial with two terms. <br> A **trinomial** is a polynomial with three terms. | Monomial: $-2x^2y^3$ <br> Binomial: $5x - y$ <br> Trinomial: $7z^3 + 0.5z + 1$ |
| **To multiply two polynomials**, multiply each term of the first polynomial by each term of the second polynomial, and then combine like terms. | $(x + 2)(x^2 + 5x - 1)$ <br> $= x(x^2 + 5x - 1) + 2(x^2 + 5x - 1)$ <br> $= x \cdot x^2 + x \cdot 5x + x(-1) + 2 \cdot x^2 + 2 \cdot 5x + 2(-1)$ <br> $= x^3 + 5x^2 - x + 2x^2 + 10x - 2$ <br> $= x^3 + 7x^2 + 9x - 2$ |

## Section 10.4    Introduction to Factoring Polynomials

To Find the Greatest Common Factor of a List of Integers

**Step 1.** Write each number as a product of prime numbers.

**Step 2.** Identify the common prime factors.

**Step 3.** The product of all common prime factors found in *Step 2* is the greatest common factor. If there are no common prime factors, the greatest common factor is 1.

The **GCF of a list of variables** raised to powers is the variable raised to the smallest exponent in the list.

The **GCF of a list of terms** is the product of all common factors.

**To factor the GCF from the terms of a polynomial**, write each term as a product of the GCF and another factor, then apply the distributive property.

Find the GCF of 18 and 30.

$$18 = \boxed{2} \cdot \boxed{3} \cdot 3$$
$$30 = \boxed{2} \cdot \boxed{3} \cdot 5$$
$$\downarrow \quad \downarrow$$

The GCF is $2 \cdot 3$ or 6.

The GCF of $x^6$, $x^8$, and $x^3$ is $x^3$.

Find the GCF of $6y^3$, $12y$, and $4y^7$.

The GCF of 6, 12, and 4 is 2.
The GCF of $y^3$, $y$, and $y^7$ is $y$.

The GCF of $6y^3$, $12y$, and $4y^7$ is $2y$.

Factor $4y^6 + 6y^5$.
The GCF of $4y^6$ and $6y^5$ is $2y^5$.

$$4y^6 + 6y^5 = 2y^5 \cdot 2y + 2y^5 \cdot 3$$
$$= 2y^5(2y + 3)$$

# CHAPTER 10 REVIEW

**(10.1)** *Perform each indicated operation.*

**1.** $(2b + 7) + (8b - 10)$  $10b - 3$

**2.** $(7s - 6) + (14s - 9)$  $21s - 15$

**3.** $(3x + 0.2) - (4x - 2.6)$  $-x + 2.8$

**4.** $(10y - 6) - (11y + 6)$  $-y - 12$

**5.** $(4z^2 + 6z - 1) + (5z - 5)$  $4z^2 + 11z - 6$

**6.** $(17a^3 + 11a^2 + a) + (14a^2 - a)$  $17a^3 + 25a^2$

**7.** $\left(9y^2 - y + \dfrac{1}{2}\right) - \left(20y^2 - \dfrac{1}{4}\right)$  $-11y^2 - y + \dfrac{3}{4}$

**8.** Subtract $(x - 2)$ from $(x^2 - 6x + 1)$.

$x^2 - 7x + 3$

*Find the value of each polynomial when x = 3.*

**9.** $5x^2$  $45$

**10.** $2 - 7x$  $-19$

△ **11.** Find the perimeter of the given rectangle.  $(26x + 28)\,\text{ft}$

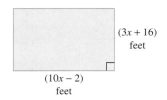

(3x + 16)
feet

(10x − 2)
feet

**Name** _____

**(10.2)** *Multiply and simplify.*

**12.** $x^{10} \cdot x^{14}$   $x^{24}$

**13.** $y \cdot y^6$   $y^7$

**14.** $4z^2 \cdot 6z^5$   $24z^7$

**15.** $(-3x^2y)(5xy^4)$   $-15x^3y^5$

**16.** $(a^5)^7$   $a^{35}$

**17.** $(x^2)^4 \cdot (x^{10})^2$   $x^{28}$

**18.** $(9b)^2$   $81b^2$

**19.** $(a^4b^2c)^5$   $a^{20}b^{10}c^5$

**20.** $(7x)(2x^5)^3$   $56x^{16}$

**21.** $(3x^6y^5)^3(2x^6y^5)^2$   $108x^{30}y^{25}$

△ **22.** Find the area of the square.   $81a^{14}$ sq. mi

$(9a^7)$ miles

**(10.3)** *Multiply.*

**23.** $2a(5a^2 - 6)$   $10a^3 - 12a$

**24.** $-3y^2(y^2 - 2y + 1)$   $-3y^4 + 6y^3 - 3y^2$

**25.** $(x + 2)(x + 6)$   $x^2 + 8x + 12$

**26.** $(3x - 1)(5x - 9)$   $15x^2 - 32x + 9$

**27.** $(y - 5)^2$   $y^2 - 10y + 25$

**28.** $(7a + 1)^2$   $49a^2 + 14a + 1$

**29.** $(x + 1)(x^2 - 2x + 3)$   $x^3 - x^2 + x + 3$

**30.** $(4y^2 - 3)(2y^2 + y + 1)$   $8y^4 + 4y^3 - 2y^2 - 3y - 3$

**31.** $(3z^2 + 2z + 1)(z^2 + z + 1)$   $3z^4 + 5z^3 + 6z^2 + 3z + 1$   △ **32.** Find the area of the given rectangle.

$(a^3 + 5a^2 - 5a + 6)$ sq. cm

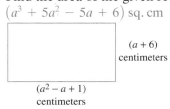

$(a + 6)$
centimeters

$(a^2 - a + 1)$
centimeters

**(10.4)** *Find the greatest common factor (GCF) of each list.*

**33.** 20 and 35   5

**34.** 12 and 32   4

**35.** 24, 30, and 60   6

**36.** 10, 20, and 25   5

**37.** $x^3, x^2, x^{10}$   $x^2$

**38.** $y^{10}, y^7, y^7$   $y^7$

**39.** $xy^2, xy, x^3y^3$   $xy$

**40.** $a^5b^4, a^6b^3, a^7 b^2$   $a^5b^2$

**41.** $5a^3, 10a, 20a^4$   $5a$

**42.** $12y^2z, 20y^2z, 24y^5z$   $4y^2z$

**Name** _____

*Factor.*

**43.** $2x^2 + 12x$   $2x(x + 6)$

**44.** $6a^2 - 12a$   $6a(a - 2)$

**45.** $6y^4 - y^6$   $y^4(6 - y^2)$

**46.** $7x^2 - 14x + 7$   $7(x^2 - 2x + 1)$

**47.** $5a^7 - a^4 + a^3$   $a^3(5a^4 - a + 1)$

**48.** $10y^6 - 10y$   $10y(y^5 - 1)$

Name _____ Section _____ Date _____

# CHAPTER 10 TEST

*Add or subtract as indicated.*

**1.** $(11x - 3) + (4x - 1)$    **2.** $(11x - 3) - (4x - 1)$

**2.** $7x - 2$

**3.** $3.4y^2 + 2y - 3$

**3.** $(1.3y^2 + 5y) + (2.1y^2 - 3y - 3)$    **4.** Subtract $(8a^2 + a)$ from $(6a^2 + 2a + 1)$.

**4.** $-2a^2 + a + 1$

**5.** $17$

**5.** Find the value of $x^2 - 6x + 1$ when $x = 8$.

**6.** $y^{14}$

**7.** $y^{33}$

*Multiply and simplify.*

**6.** $y^3 \cdot y^{11}$    **7.** $(y^3)^{11}$

**8.** $16x^8$

**9.** $-12a^{10}$

**8.** $(2x^2)^4$    **9.** $(6a^3)(-2a^7)$

**10.** $p^{54}$

**10.** $(p^6)^7(p^2)^6$    **11.** $(3a^4b)^2(2ba^4)^3$

**11.** $72a^{20}b^5$

**12.** $10x^3 + 6.5x$

**12.** $5x(2x^2 + 1.3)$    **13.** $-2y(y^3 + 6y^2 - 4)$

**13.** $-2y^4 - 12y^3 + 8y$

**14.** $x^2 - x - 6$

**14.** $(x - 3)(x + 2)$    **15.** $(5x + 2)^2$

**15.** $25x^2 + 20x + 4$

**16.** $a^3 + 8$

perimeter: $(14x - 4)$
in.; area:
$(5x^2 + 33x - 14)$
**17.** sq. in.

**18.** 15

**19.** $3y^3$

**20.** $3y(y - 5)$

**21.** $2a(5a + 6)$

**22.** $6(x^2 - 2x - 5)$

**23.** $x^3(7x^3 - 6x + 1)$

**Name** _____

**16.** $(a + 2)(a^2 - 2a + 4)$

△ **17.** Find the area and the perimeter of the parallelogram. (*Hint:* $A = b \cdot h$).

$(x + 7)$ in.  $2x$ in.

$(5x - 2)$ in.

*Find the greatest common factor of each list.*

**18.** 45 and 60

**19.** $6y^3, 9y^5, 18y^4$

*Factor.*

**20.** $3y^2 - 15y$

**21.** $10a^2 + 12a$

**22.** $6x^2 - 12x - 30$

**23.** $7x^6 - 6x^4 + 1$

**Name** _____ **Section** _____ **Date** _____

# CUMULATIVE REVIEW

△ **1.** The state of Colorado is in the shape of a rectangle whose length is 380 miles and whose width is 280 miles Find its area.

**2.** Add: $1 + (-10) + (-8) + 9$

*Subtract.*

**3.** $8 - 15$

**4.** $-4 - (-5)$

**5.** Solve: $7x = 6x + 4$

**6.** Write $\dfrac{8}{3x}$ as an equivalent fraction whose denominator is $12x$.

**7.** Subtract: $12 - 8\dfrac{3}{7}$

**8.** Round 736.2359 to the nearest tenth.

**9.** Add: $23.85 + 1.604$

**10.** Is 9 a solution of the equation $3.7y = 3.33$?

*Divide*

**11.** $\dfrac{786}{10,000}$

**12.** $\dfrac{0.12}{10}$

**13.** Evaluate $-2x + 5$ when $x = 3.8$

**14.** Write $\dfrac{22}{7}$ as a decimal. Round to the nearest hundreth.

**15.** Find: $\sqrt{\dfrac{1}{36}}$

**16.** Mel Rose is a 6-foot-tall park ranger who needs to know the height of a particular tree. He notices that when the shadow of the tree is 69 feet long, his own shadow is 9 feet long. Find the height of the tree.

**1.** 106,400 sq. mi (Sec. 1.5, Ex. 6)

**2.** $-8$ (Sec. 2.2, Ex. 12)

**3.** $-7$ (Sec. 2.3, Ex. 6)

**4.** 1 (Sec. 2.3, Ex. 7)

**5.** 4 (Sec. 3.2, Ex. 5)

**6.** $\dfrac{32}{12x}$ (Sec. 4.1, Ex. 13)

**7.** $3\dfrac{4}{7}$ (Sec. 4.8, Ex. 16)

**8.** 736.2 (Sec. 5.1, Ex. 13)

**9.** 25.454 (Sec. 5.2, Ex. 1)

**10.** no (Sec. 5.3, Ex. 13)

**11.** 0.0786 (Sec. 5.4, Ex. 4)

**12.** 0.012 (Sec. 5.4, Ex. 5)

**13.** $-2.6$ (Sec. 5.5, Ex. 8)

**14.** 3.14 (Sec. 5.6, Ex. 3)

**15.** $\dfrac{1}{6}$ (Sec. 5.8, Ex. 2)

**16.** 46 ft (Sec. 6.5, Ex. 3)

**Name** _____

**17.** Translate to an equation: 1.2 is 30% of what number?

**18.** What percent of 50 is 8?

**19.** Mr. Buccaran, the principal at Slidell High School, counted 31 freshmen absent during a particular day. If this is 4% of the total number of freshmen, how many freshmen are there at Slidell High School?

**20.** Ivan Borski borrowed $2400 at 10% simple interest for 8 months to buy a used Chevy S-10. Find the simple interest he paid.

**21.** Using the circle graph shown, determine the percent of schools with CD-ROM technology that are either junior or senior high schools.

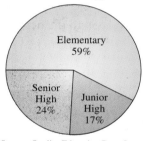

Type of U.S. Public Schools with
CD-ROM Technology in 1999

Elementary 59%
Senior High 24%
Junior High 17%

*Source:* Quality Education Data, Inc.

△ **22.** Find the perimeter of a rectangle with a length of 11 inches and a width of 3 inches.

△ **23.** Find the area of the parallelogram.

1.5 mi
3.4 mi

**24.** Multiply 5 lb 9 oz by 6.

**25.** Convert 3210 ml to liters.

# APPENDIX A

## Addition Table and One Hundred Addition Facts

| + | 0 | 1 | 2 | 3 | 4 | 5 | 6 | 7 | 8 | 9 |
|---|---|---|---|---|---|---|---|---|---|---|
| **0** | 0 | 1 | 2 | 3 | 4 | 5 | 6 | 7 | 8 | 9 |
| **1** | 1 | 2 | 3 | 4 | 5 | 6 | 7 | 8 | 9 | 10 |
| **2** | 2 | 3 | 4 | 5 | 6 | 7 | 8 | 9 | 10 | 11 |
| **3** | 3 | 4 | 5 | 6 | 7 | 8 | 9 | 10 | 11 | 12 |
| **4** | 4 | 5 | 6 | 7 | 8 | 9 | 10 | 11 | 12 | 13 |
| **5** | 5 | 6 | 7 | 8 | 9 | 10 | 11 | 12 | 13 | 14 |
| **6** | 6 | 7 | 8 | 9 | 10 | 11 | 12 | 13 | 14 | 15 |
| **7** | 7 | 8 | 9 | 10 | 11 | 12 | 13 | 14 | 15 | 16 |
| **8** | 8 | 9 | 10 | 11 | 12 | 13 | 14 | 15 | 16 | 17 |
| **9** | 9 | 10 | 11 | 12 | 13 | 14 | 15 | 16 | 17 | 18 |

## Appendix A: One Hundred Addition Facts

*Knowledge of the basic addition facts found above is an important prerequisite for a course in prealgebra. Study the table above and then perform the additions. Check your answers either by comparing them with those found in the back-of-the-book answer section or by using the table. Review any facts that you missed.*

**1.** $\begin{array}{r} 1 \\ +4 \\ \hline 5 \end{array}$
**2.** $\begin{array}{r} 5 \\ +6 \\ \hline 11 \end{array}$
**3.** $\begin{array}{r} 2 \\ +3 \\ \hline 5 \end{array}$
**4.** $\begin{array}{r} 7 \\ +8 \\ \hline 15 \end{array}$
**5.** $\begin{array}{r} 3 \\ +9 \\ \hline 12 \end{array}$
**6.** $\begin{array}{r} 6 \\ +1 \\ \hline 7 \end{array}$
**7.** $\begin{array}{r} 4 \\ +4 \\ \hline 8 \end{array}$
**8.** $\begin{array}{r} 0 \\ +6 \\ \hline 6 \end{array}$
**9.** $\begin{array}{r} 9 \\ +5 \\ \hline 14 \end{array}$
**10.** $\begin{array}{r} 8 \\ +2 \\ \hline 10 \end{array}$

**11.** $\begin{array}{r} 5 \\ +7 \\ \hline 12 \end{array}$
**12.** $\begin{array}{r} 3 \\ +2 \\ \hline 5 \end{array}$
**13.** $\begin{array}{r} 5 \\ +5 \\ \hline 10 \end{array}$
**14.** $\begin{array}{r} 1 \\ +1 \\ \hline 2 \end{array}$
**15.** $\begin{array}{r} 8 \\ +1 \\ \hline 9 \end{array}$
**16.** $\begin{array}{r} 6 \\ +6 \\ \hline 12 \end{array}$
**17.** $\begin{array}{r} 2 \\ +9 \\ \hline 11 \end{array}$
**18.** $\begin{array}{r} 3 \\ +5 \\ \hline 8 \end{array}$
**19.** $\begin{array}{r} 9 \\ +9 \\ \hline 18 \end{array}$
**20.** $\begin{array}{r} 5 \\ +2 \\ \hline 7 \end{array}$

| | | | | | | | | | |
|---|---|---|---|---|---|---|---|---|---|
| **21.** 6 <br> +4 <br> 10 | **22.** 0 <br> +0 <br> 0 | **23.** 1 <br> +9 <br> 10 | **24.** 3 <br> +7 <br> 10 | **25.** 9 <br> +8 <br> 17 | **26.** 0 <br> +8 <br> 8 | **27.** 4 <br> +9 <br> 13 | **28.** 3 <br> +0 <br> 3 | **29.** 7 <br> +5 <br> 12 | **30.** 8 <br> +9 <br> 17 |
| **31.** 9 <br> +7 <br> 16 | **32.** 2 <br> +6 <br> 8 | **33.** 4 <br> +3 <br> 7 | **34.** 8 <br> +5 <br> 13 | **35.** 3 <br> +1 <br> 4 | **36.** 0 <br> +3 <br> 3 | **37.** 7 <br> +1 <br> 8 | **38.** 3 <br> +4 <br> 7 | **39.** 8 <br> +0 <br> 8 | **40.** 6 <br> +3 <br> 9 |
| **41.** 2 <br> +4 <br> 6 | **42.** 0 <br> +9 <br> 9 | **43.** 8 <br> +8 <br> 16 | **44.** 5 <br> +3 <br> 8 | **45.** 3 <br> +6 <br> 9 | **46.** 6 <br> +9 <br> 15 | **47.** 4 <br> +8 <br> 12 | **48.** 0 <br> +1 <br> 1 | **49.** 2 <br> +5 <br> 7 | **50.** 6 <br> +0 <br> 6 |
| **51.** 2 <br> +0 <br> 2 | **52.** 4 <br> +2 <br> 6 | **53.** 8 <br> +3 <br> 11 | **54.** 7 <br> +4 <br> 11 | **55.** 1 <br> +7 <br> 8 | **56.** 4 <br> +6 <br> 10 | **57.** 0 <br> +5 <br> 5 | **58.** 9 <br> +1 <br> 10 | **59.** 8 <br> +6 <br> 14 | **60.** 5 <br> +1 <br> 6 |
| **61.** 6 <br> +7 <br> 13 | **62.** 4 <br> +0 <br> 4 | **63.** 1 <br> +6 <br> 7 | **64.** 4 <br> +5 <br> 9 | **65.** 0 <br> +7 <br> 7 | **66.** 5 <br> +8 <br> 13 | **67.** 7 <br> +6 <br> 13 | **68.** 7 <br> +0 <br> 7 | **69.** 4 <br> +1 <br> 5 | **70.** 5 <br> +4 <br> 9 |
| **71.** 0 <br> +4 <br> 4 | **72.** 1 <br> +2 <br> 3 | **73.** 7 <br> +9 <br> 16 | **74.** 3 <br> +8 <br> 11 | **75.** 7 <br> +7 <br> 14 | **76.** 9 <br> +4 <br> 13 | **77.** 1 <br> +0 <br> 1 | **78.** 4 <br> +7 <br> 11 | **79.** 2 <br> +2 <br> 4 | **80.** 1 <br> +3 <br> 4 |
| **81.** 2 <br> +8 <br> 10 | **82.** 5 <br> +9 <br> 14 | **83.** 6 <br> +2 <br> 8 | **84.** 9 <br> +6 <br> 15 | **85.** 5 <br> +0 <br> 5 | **86.** 8 <br> +7 <br> 15 | **87.** 7 <br> +3 <br> 10 | **88.** 0 <br> +2 <br> 2 | **89.** 9 <br> +2 <br> 11 | **90.** 3 <br> +3 <br> 6 |
| **91.** 9 <br> +3 <br> 12 | **92.** 1 <br> +5 <br> 6 | **93.** 2 <br> +7 <br> 9 | **94.** 6 <br> +5 <br> 11 | **95.** 7 <br> +2 <br> 9 | **96.** 1 <br> +8 <br> 9 | **97.** 6 <br> +8 <br> 14 | **98.** 8 <br> +4 <br> 12 | **99.** 9 <br> +0 <br> 9 | **100.** 2 <br> +1 <br> 3 |

# APPENDIX B

## Multiplication Table and One Hundred Multiplication Facts

| × | 1 | 2 | 3 | 4 | 5 | 6 | 7 | 8 | 9 |
|---|---|---|---|---|---|---|---|---|---|
| 1 | 1 | 2 | 3 | 4 | 5 | 6 | 7 | 8 | 9 |
| 2 | 2 | 4 | 6 | 8 | 10 | 12 | 14 | 16 | 18 |
| 3 | 3 | 6 | 9 | 12 | 15 | 18 | 21 | 24 | 27 |
| 4 | 4 | 8 | 12 | 16 | 20 | 24 | 28 | 32 | 36 |
| 5 | 5 | 10 | 15 | 20 | 25 | 30 | 35 | 40 | 45 |
| 6 | 6 | 12 | 18 | 24 | 30 | 36 | 42 | 48 | 54 |
| 7 | 7 | 14 | 21 | 28 | 35 | 42 | 49 | 56 | 63 |
| 8 | 8 | 16 | 24 | 32 | 40 | 48 | 56 | 64 | 72 |
| 9 | 9 | 18 | 27 | 36 | 45 | 54 | 63 | 72 | 81 |

### Appendix B: One Hundred Multiplication Facts

*Knowledge of the basic multiplication facts found above is an important prerequisite for a course in prealgebra. Study the table above and then perform the multiplications. Check your answers either by comparing them with those found in the back-of-the-book answer section or by using the table. Review any facts that you missed.*

**1.** $\begin{array}{r} 1 \\ \times 1 \\ \hline 1 \end{array}$ **2.** $\begin{array}{r} 5 \\ \times 7 \\ \hline 35 \end{array}$ **3.** $\begin{array}{r} 7 \\ \times 8 \\ \hline 56 \end{array}$ **4.** $\begin{array}{r} 3 \\ \times 3 \\ \hline 9 \end{array}$ **5.** $\begin{array}{r} 8 \\ \times 4 \\ \hline 32 \end{array}$ **6.** $\begin{array}{r} 9 \\ \times 5 \\ \hline 45 \end{array}$ **7.** $\begin{array}{r} 4 \\ \times 7 \\ \hline 28 \end{array}$ **8.** $\begin{array}{r} 7 \\ \times 1 \\ \hline 7 \end{array}$ **9.** $\begin{array}{r} 2 \\ \times 2 \\ \hline 4 \end{array}$ **10.** $\begin{array}{r} 0 \\ \times 5 \\ \hline 0 \end{array}$

**11.** $\begin{array}{r} 9 \\ \times 7 \\ \hline 63 \end{array}$ **12.** $\begin{array}{r} 8 \\ \times 8 \\ \hline 64 \end{array}$ **13.** $\begin{array}{r} 3 \\ \times 2 \\ \hline 6 \end{array}$ **14.** $\begin{array}{r} 6 \\ \times 0 \\ \hline 0 \end{array}$ **15.** $\begin{array}{r} 5 \\ \times 6 \\ \hline 30 \end{array}$ **16.** $\begin{array}{r} 2 \\ \times 5 \\ \hline 10 \end{array}$ **17.** $\begin{array}{r} 4 \\ \times 6 \\ \hline 24 \end{array}$ **18.** $\begin{array}{r} 0 \\ \times 7 \\ \hline 0 \end{array}$ **19.** $\begin{array}{r} 6 \\ \times 3 \\ \hline 18 \end{array}$ **20.** $\begin{array}{r} 8 \\ \times 9 \\ \hline 72 \end{array}$

21.  5     22.  7     23.  4     24.  1     25.  9     26.  3     27.  8     28.  2     29.  6     30.  5
   × 8        × 2        × 8        × 2        × 6        × 1        × 7        × 8        × 9        × 5
   ‾‾‾‾       ‾‾‾‾       ‾‾‾‾       ‾‾‾‾       ‾‾‾‾       ‾‾‾‾       ‾‾‾‾       ‾‾‾‾       ‾‾‾‾       ‾‾‾‾
    40         14         32          2         54          3         56         16         54         25

31.  2     32.  8     33.  4     34.  8     35.  6     36.  4     37.  9     38.  2     39.  3     40.  1
   × 1        × 0        × 9        × 3        × 2        × 5        × 4        × 9        × 4        × 6
   ‾‾‾‾       ‾‾‾‾       ‾‾‾‾       ‾‾‾‾       ‾‾‾‾       ‾‾‾‾       ‾‾‾‾       ‾‾‾‾       ‾‾‾‾       ‾‾‾‾
     2          0         36         24         12         20         36         18         12          6

41.  8     42.  9     43.  1     44.  5     45.  9     46.  7     47.  9     48.  0     49.  3     50.  6
   × 6        × 8        × 8        × 1        × 0        × 4        × 3        × 3        × 5        × 8
   ‾‾‾‾       ‾‾‾‾       ‾‾‾‾       ‾‾‾‾       ‾‾‾‾       ‾‾‾‾       ‾‾‾‾       ‾‾‾‾       ‾‾‾‾       ‾‾‾‾
    48         72          8          5          0         28         27          0         15         48

51.  5     52.  2     53.  1     54.  3     55.  9     56.  5     57.  0     58.  1     59.  5     60.  6
   × 9        × 6        × 0        × 9        × 9        × 4        × 6        × 9        × 0        × 1
   ‾‾‾‾       ‾‾‾‾       ‾‾‾‾       ‾‾‾‾       ‾‾‾‾       ‾‾‾‾       ‾‾‾‾       ‾‾‾‾       ‾‾‾‾       ‾‾‾‾
    45         12          0         27         81         20          0          9          0          6

61.  9     62.  1     63.  1     64.  7     65.  6     66.  4     67.  7     68.  4     69.  7     70.  2
   × 2        × 7        × 3        × 3        × 6        × 0        × 9        × 3        × 5        × 0
   ‾‾‾‾       ‾‾‾‾       ‾‾‾‾       ‾‾‾‾       ‾‾‾‾       ‾‾‾‾       ‾‾‾‾       ‾‾‾‾       ‾‾‾‾       ‾‾‾‾
    18          7          3         21         36          0         63         12         35          0

71.  6     72.  0     73.  8     74.  2     75.  0     76.  3     77.  9     78.  7     79.  5     80.  4
   × 7        × 8        × 5        × 4        × 1        × 8        × 1        × 0        × 3        × 4
   ‾‾‾‾       ‾‾‾‾       ‾‾‾‾       ‾‾‾‾       ‾‾‾‾       ‾‾‾‾       ‾‾‾‾       ‾‾‾‾       ‾‾‾‾       ‾‾‾‾
    42          0         40          8          0         24          9          0         15         16

81.  1     82.  6     83.  3     84.  1     85.  3     86.  4     87.  0     88.  7     89.  8     90.  6
   × 5        × 5        × 0        × 4        × 7        × 2        × 2        × 7        × 2        × 4
   ‾‾‾‾       ‾‾‾‾       ‾‾‾‾       ‾‾‾‾       ‾‾‾‾       ‾‾‾‾       ‾‾‾‾       ‾‾‾‾       ‾‾‾‾       ‾‾‾‾
     5         30          0          4         21          8          0         49         16         24

91.  0     92.  2     93.  4     94.  0     95.  2     96.  8     97.  3     98.  5     99.  0    100.  7
   × 0        × 7        × 1        × 4        × 3        × 1        × 6        × 2        × 9        × 6
   ‾‾‾‾       ‾‾‾‾       ‾‾‾‾       ‾‾‾‾       ‾‾‾‾       ‾‾‾‾       ‾‾‾‾       ‾‾‾‾       ‾‾‾‾       ‾‾‾‾
     0         14          4          0          6          8         18         10          0         42

# Appendix C

## Review of Geometric Figures

| PLANE FIGURES HAVE LENGTH AND WIDTH BUT NO THICKNESS OR DEPTH. | | |
|---|---|---|
| Name | Description | Figure |
| POLYGON | Union of three or more coplanar line segments that intersect with each other only at each endpoint, with each endpoint shared by two segments | |
| TRIANGLE | Polygon with three sides (sum of measures of three angles is 180°) | |
| SCALENE TRIANGLE | Triangle with no sides of equal length | |
| ISOSCELES TRIANGLE | Triangle with two sides of equal length | |
| EQUILATERAL TRIANGLE | Triangle with all sides of equal length | |
| RIGHT TRIANGLE | Triangle that contains a right angle | leg, hypotenuse, leg |
| QUADRILATERAL | Polygon with four sides (sum of measures of four angles is 360°) | |
| TRAPEZOID | Quadrilateral with exactly one pair of opposite sides parallel | base, leg, parallel sides, leg, base |
| ISOSCELES TRAPEZOID | Trapezoid with legs of equal length | |
| PARALLELOGRAM | Quadrilateral with both pairs of opposite sides parallel | |
| RHOMBUS | Parallelogram with all sides of equal length | |

*(continued)*

817

| Name | Description | Figure |
|------|-------------|--------|
| RECTANGLE | Parallelogram with four right angles | |
| SQUARE | Rectangle with all sides of equal length | |
| CIRCLE | All points in a plane the same distance from a fixed point called the **center** | |

| SOLID FIGURES HAVE LENGTH, WIDTH, AND HEIGHT OR DEPTH. | | |
|------|-------------|--------|
| Name | Description | Figure |
| RECTANGULAR SOLID | A solid with six sides, all of which are rectangles | |
| CUBE | A rectangular solid whose six sides are squares | |
| SPHERE | All points the same distance from a fixed point, called the **center** | |
| RIGHT CIRCULAR CYLINDER | A cylinder having two circular bases that are perpendicular to its altitude | |
| RIGHT CIRCULAR CONE | A cone with a circular base that is perpendicular to its altitude | |

# APPENDIX D

## Percents, Decimals, and Fractions

| PERCENT, DECIMAL, AND FRACTION EQUIVALENTS | | |
|---|---|---|
| Percent | Decimal | Fraction |
| 1% | 0.01 | $\frac{1}{100}$ |
| 5% | 0.05 | $\frac{1}{20}$ |
| 10% | 0.1 | $\frac{1}{10}$ |
| 12.5% or $12\frac{1}{2}$% | 0.125 | $\frac{1}{8}$ |
| $16.\overline{6}$% or $16\frac{2}{3}$% | $0.1\overline{6}$ | $\frac{1}{6}$ |
| 20% | 0.2 | $\frac{1}{5}$ |
| 25% | 0.25 | $\frac{1}{4}$ |
| 30% | 0.3 | $\frac{3}{10}$ |
| $33.\overline{3}$% or $33\frac{1}{3}$% | $0.\overline{3}$ | $\frac{1}{3}$ |
| 37.5% or $37\frac{1}{2}$% | 0.375 | $\frac{3}{8}$ |
| 40% | 0.4 | $\frac{2}{5}$ |
| 50% | 0.5 | $\frac{1}{2}$ |
| 60% | 0.6 | $\frac{3}{5}$ |
| 62.5% or $62\frac{1}{2}$% | 0.625 | $\frac{5}{8}$ |
| $66.\overline{6}$% or $66\frac{2}{3}$% | $0.\overline{6}$ | $\frac{2}{3}$ |
| 70% | 0.7 | $\frac{7}{10}$ |
| 75% | 0.75 | $\frac{3}{4}$ |
| 80% | 0.8 | $\frac{4}{5}$ |
| $83.\overline{3}$% or $83\frac{1}{3}$% | $0.8\overline{3}$ | $\frac{5}{6}$ |
| 87.5% or $87\frac{1}{2}$% | 0.875 | $\frac{7}{8}$ |
| 90% | 0.9 | $\frac{9}{10}$ |
| 100% | 1.0 | 1 |
| 110% | 1.1 | $1\frac{1}{10}$ |
| 125% | 1.25 | $1\frac{1}{4}$ |
| $133.\overline{3}$% or $133\frac{1}{3}$% | $1.\overline{3}$ | $1\frac{1}{3}$ |
| 150% | 1.5 | $1\frac{1}{2}$ |
| $166.\overline{6}$% or $166\frac{2}{3}$% | $1.\overline{6}$ | $1\frac{2}{3}$ |
| 175% | 1.75 | $1\frac{3}{4}$ |
| 200% | 2.0 | 2 |

# APPENDIX E

## Table of Squares and Square Roots

| SQUARES AND SQUARE ROOTS | | | | | |
|---|---|---|---|---|---|
| $n$ | $n^2$ | $\sqrt{n}$ | $n$ | $n^2$ | $\sqrt{n}$ |
| 1 | 1 | 1.000 | 51 | 2601 | 7.141 |
| 2 | 4 | 1.414 | 52 | 2704 | 7.211 |
| 3 | 9 | 1.732 | 53 | 2809 | 7.280 |
| 4 | 16 | 2.000 | 54 | 2916 | 7.348 |
| 5 | 25 | 2.236 | 55 | 3025 | 7.416 |
| 6 | 36 | 2.449 | 56 | 3136 | 7.483 |
| 7 | 49 | 2.646 | 57 | 3249 | 7.550 |
| 8 | 64 | 2.828 | 58 | 3364 | 7.616 |
| 9 | 81 | 3.000 | 59 | 3481 | 7.681 |
| 10 | 100 | 3.162 | 60 | 3600 | 7.746 |
| 11 | 121 | 3.317 | 61 | 3721 | 7.810 |
| 12 | 144 | 3.464 | 62 | 3844 | 7.874 |
| 13 | 169 | 3.606 | 63 | 3969 | 7.937 |
| 14 | 196 | 3.742 | 64 | 4096 | 8.000 |
| 15 | 225 | 3.873 | 65 | 4225 | 8.062 |
| 16 | 256 | 4.000 | 66 | 4356 | 8.124 |
| 17 | 289 | 4.123 | 67 | 4489 | 8.185 |
| 18 | 324 | 4.243 | 68 | 4624 | 8.246 |
| 19 | 361 | 4.359 | 69 | 4761 | 8.307 |
| 20 | 400 | 4.472 | 70 | 4900 | 8.367 |
| 21 | 441 | 4.583 | 71 | 5041 | 8.426 |
| 22 | 484 | 4.690 | 72 | 5184 | 8.485 |
| 23 | 529 | 4.796 | 73 | 5329 | 8.544 |
| 24 | 576 | 4.899 | 74 | 5476 | 8.602 |
| 25 | 625 | 5.000 | 75 | 5625 | 8.660 |
| 26 | 676 | 5.099 | 76 | 5776 | 8.718 |
| 27 | 729 | 5.196 | 77 | 5929 | 8.775 |
| 28 | 784 | 5.292 | 78 | 6084 | 8.832 |
| 29 | 841 | 5.385 | 79 | 6241 | 8.888 |
| 30 | 900 | 5.477 | 80 | 6400 | 8.944 |
| 31 | 961 | 5.568 | 81 | 6561 | 9.000 |
| 32 | 1024 | 5.657 | 82 | 6724 | 9.055 |
| 33 | 1089 | 5.745 | 83 | 6889 | 9.110 |
| 34 | 1156 | 5.831 | 84 | 7056 | 9.165 |
| 35 | 1225 | 5.916 | 85 | 7225 | 9.220 |
| 36 | 1296 | 6.000 | 86 | 7396 | 9.274 |
| 37 | 1369 | 6.083 | 87 | 7569 | 9.327 |
| 38 | 1444 | 6.164 | 88 | 7744 | 9.381 |
| 39 | 1521 | 6.245 | 89 | 7921 | 9.434 |
| 40 | 1600 | 6.325 | 90 | 8100 | 9.487 |
| 41 | 1681 | 6.403 | 91 | 8281 | 9.539 |
| 42 | 1764 | 6.481 | 92 | 8464 | 9.592 |
| 43 | 1849 | 6.557 | 93 | 8649 | 9.644 |
| 44 | 1936 | 6.633 | 94 | 8836 | 9.695 |
| 45 | 2025 | 6.708 | 95 | 9025 | 9.747 |
| 46 | 2116 | 6.782 | 96 | 9216 | 9.798 |
| 47 | 2209 | 6.856 | 97 | 9409 | 9.849 |
| 48 | 2304 | 6.928 | 98 | 9604 | 9.899 |
| 49 | 2401 | 7.000 | 99 | 9801 | 9.950 |
| 50 | 2500 | 7.071 | 100 | 10,000 | 10.000 |

# APPENDIX F

## Compound Interest Table

### Compounded Annually

| | 5% | 6% | 7% | 8% | 9% | 10% | 11% | 12% | 13% | 14% | 15% | 16% | 17% | 18% |
|---|---|---|---|---|---|---|---|---|---|---|---|---|---|---|
| 1 year | 1.05000 | 1.06000 | 1.07000 | 1.08000 | 1.09000 | 1.10000 | 1.11000 | 1.12000 | 1.13000 | 1.14000 | 1.15000 | 1.16000 | 1.17000 | 1.18000 |
| 5 years | 1.27628 | 1.33823 | 1.40255 | 1.46933 | 1.53862 | 1.61051 | 1.68506 | 1.76234 | 1.84244 | 1.92541 | 2.01136 | 2.10034 | 2.19245 | 2.28776 |
| 10 years | 1.62889 | 1.79085 | 1.96715 | 2.15892 | 2.36736 | 2.59374 | 2.83942 | 3.10585 | 3.39457 | 3.70722 | 4.04556 | 4.41144 | 4.80683 | 5.23384 |
| 15 years | 2.07893 | 2.39656 | 2.75903 | 3.17217 | 3.64248 | 4.17725 | 4.78459 | 5.47357 | 6.25427 | 7.13794 | 8.13706 | 9.26552 | 10.53872 | 11.97375 |
| 20 years | 2.65330 | 3.20714 | 3.86968 | 4.66096 | 5.60441 | 6.72750 | 8.06231 | 9.64629 | 11.52309 | 13.74349 | 16.36654 | 19.46076 | 23.10560 | 27.39303 |

### Compounded Semiannually

| | 5% | 6% | 7% | 8% | 9% | 10% | 11% | 12% | 13% | 14% | 15% | 16% | 17% | 18% |
|---|---|---|---|---|---|---|---|---|---|---|---|---|---|---|
| 1 year | 1.05063 | 1.06090 | 1.07123 | 1.08160 | 1.09203 | 1.10250 | 1.11303 | 1.12360 | 1.13423 | 1.14490 | 1.15563 | 1.16640 | 1.17723 | 1.18810 |
| 5 years | 1.28008 | 1.34392 | 1.41060 | 1.48024 | 1.55297 | 1.62889 | 1.70814 | 1.79085 | 1.87714 | 1.96715 | 2.06103 | 2.15892 | 2.26098 | 2.36736 |
| 10 years | 1.63862 | 1.80611 | 1.98979 | 2.19112 | 2.41171 | 2.65330 | 2.91776 | 3.20714 | 3.52365 | 3.86968 | 4.24785 | 4.66096 | 5.11205 | 5.60441 |
| 15 years | 2.09757 | 2.42726 | 2.80679 | 3.24340 | 3.74532 | 4.32194 | 4.98395 | 5.74349 | 6.61437 | 7.61226 | 8.75496 | 10.06266 | 11.55825 | 13.26768 |
| 20 years | 2.68506 | 3.26204 | 3.95926 | 4.80102 | 5.81636 | 7.03999 | 8.51331 | 10.28572 | 12.41607 | 14.97446 | 18.04424 | 21.72452 | 26.13302 | 31.40942 |

### Compounded Quarterly

| | 5% | 6% | 7% | 8% | 9% | 10% | 11% | 12% | 13% | 14% | 15% | 16% | 17% | 18% |
|---|---|---|---|---|---|---|---|---|---|---|---|---|---|---|
| 1 year | 1.05095 | 1.06136 | 1.07186 | 1.08243 | 1.09308 | 1.10381 | 1.11462 | 1.12551 | 1.13648 | 1.14752 | 1.15865 | 1.16986 | 1.18115 | 1.19252 |
| 5 years | 1.28204 | 1.34686 | 1.41478 | 1.48595 | 1.56051 | 1.63862 | 1.72043 | 1.80611 | 1.89584 | 1.98979 | 2.08815 | 2.19112 | 2.29891 | 2.41171 |
| 10 years | 1.64362 | 1.81402 | 2.00160 | 2.20804 | 2.43519 | 2.68506 | 2.95987 | 3.26204 | 3.59420 | 3.95926 | 4.36038 | 4.80102 | 5.28497 | 5.81636 |
| 15 years | 2.10718 | 2.44322 | 2.83182 | 3.28103 | 3.80013 | 4.39979 | 5.09225 | 5.89160 | 6.81402 | 7.87809 | 9.10513 | 10.51963 | 12.14965 | 14.02741 |
| 20 years | 2.70148 | 3.29066 | 4.00639 | 4.87544 | 5.93015 | 7.20957 | 8.76085 | 10.64089 | 12.91828 | 15.67574 | 19.01290 | 23.04980 | 27.93091 | 33.83010 |

### Compounded Daily

| | 5% | 6% | 7% | 8% | 9% | 10% | 11% | 12% | 13% | 14% | 15% | 16% | 17% | 18% |
|---|---|---|---|---|---|---|---|---|---|---|---|---|---|---|
| 1 year | 1.05127 | 1.06183 | 1.07250 | 1.08328 | 1.09416 | 1.10516 | 1.11626 | 1.12747 | 1.13880 | 1.15024 | 1.16180 | 1.17347 | 1.18526 | 1.19716 |
| 5 years | 1.28400 | 1.34983 | 1.41902 | 1.49176 | 1.56823 | 1.64861 | 1.73311 | 1.82194 | 1.91532 | 2.01348 | 2.11667 | 2.22515 | 2.33918 | 2.45906 |
| 10 years | 1.64866 | 1.82203 | 2.01362 | 2.22535 | 2.45933 | 2.71791 | 3.00367 | 3.31946 | 3.66845 | 4.05411 | 4.48031 | 4.95130 | 5.47178 | 6.04696 |
| 15 years | 2.11689 | 2.45942 | 2.85736 | 3.31968 | 3.85678 | 4.48077 | 5.20569 | 6.04786 | 7.02625 | 8.16288 | 9.48335 | 11.01738 | 12.79950 | 14.86983 |
| 20 years | 2.71810 | 3.31979 | 4.05466 | 4.95216 | 6.04831 | 7.38703 | 9.02202 | 11.01883 | 13.45751 | 16.43582 | 20.07316 | 24.51533 | 29.94039 | 36.56577 |

# Answers to Selected Exercises

## Chapter 1 Whole Numbers and Introduction to Algebra

### Chapter 1 Pretest

**1.** hundreds; 1.1A **2.** twenty-three thousand, four hundred ninety; 1.1B **3.** 87; 1.2A **4.** 3717; 1.5B **5.** 626; 1.3A **6.** 32; 1.3B
**7.** 136 pages; 1.3C **8.** 9050; 1.4A **9.** 3100; 1.4A **10.** $9 \cdot 3 + 9 \cdot 11$; 1.5A **11.** 25 in.; 1.2C **12.** 184 sq. yd; 1.5C
**13.** 576 seats; 1.5D **14.** 243; 1.6B **15.** 446 R 9; 1.6B **16.** 39; 1.6D **17.** $9^7$; 1.7A **18.** 2401; 1.7B **19.** 39; 1.7C **20.** 6; 1.8A
**21.** $10 - x$; 1.8B

### Exercise Set 1.1

**1.** tens **3.** thousands **5.** hundred-thousands **7.** millions **9.** five thousand, four hundred twenty **11.** twenty-six thousand,
nine hundred ninety **13.** one million, six hundred twenty thousand **15.** fifty-three million, five hundred twenty thousand,
one hundred seventy **17.** four million, nine hundred ninety-two thousand, eight hundred thirty-eight **19.** six hundred twenty thousand
**21.** three thousand, eight hundred ninety-three **23.** 6508 **25.** 29,900 **27.** 6,504,019 **29.** 3,000,014 **31.** 1821
**33.** 63,100,000 **35.** 100,000,000 **37.** $400 + 6$ **39.** $5000 + 200 + 90$ **41.** $60,000 + 2000 + 400 + 7$ **43.** $30,000 + 600 + 80$
**45.** $30,000,000 + 9,000,000 + 600,000 + 80,000$ **47.** $1000 + 6$ **49.** < **51.** > **53.** > **55.** > **57.** four thousand
**59.** $4000 + 100 + 40 + 5$ **61.** Nile **63.** Pomeranian; thirty-eight thousand, five hundred forty **65.** Rottweiler **67.** 55,543
**69.** Answers may vary.

### Calculator Explorations

**1.** 134 **3.** 340 **5.** 2834

### Mental Math

**1.** 12 **3.** 9000 **5.** 1620

### Exercise Set 1.2

**1.** 36 **3.** 92 **5.** 49 **7.** 5399 **9.** 117 **11.** 71 **13.** 117 **15.** 25 **17.** 62 **19.** 212 **21.** 94 **23.** 910 **25.** 8273
**27.** 11,926 **29.** 1884 **31.** 16,717 **33.** 1110 **35.** 8999 **37.** 35,901 **39.** 612,389 **41.** 29 in. **43.** 25 ft **45.** 24 in.
**47.** 8 yd **49.** 383 mi **51.** 340 ft **53.** 123,100 people **55.** 6949 performances **57.** 13,139 **59.** 2425 ft **61.** 13,255 mi
**63.** Texas **65.** 589 **67.** 1663 **69.** Answers may vary. **71.** 1,044,473,765

### Calculator Explorations

**1.** 770 **3.** 109 **5.** 8978

### Mental Math

**1.** 7 **3.** 5 **5.** 0 **7.** 400 **9.** 500

### Exercise Set 1.3

**1.** 44 **3.** 60 **5.** 265 **7.** 254 **9.** 545 **11.** 600 **13.** 25 **15.** 45 **17.** 146 **19.** 288 **21.** 168 **23.** 6 **25.** 447
**27.** 5723 **29.** 504 **31.** 89 **33.** 79 **35.** 39,914 **37.** 32,711 **39.** 5041 **41.** 31,213 **43.** 4 **45.** 20 **47.** 7
**49.** 56 crawfish **51.** 264 pages **53.** 6065 ft **55.** $175 **57.** 358 mi **59.** $389 **61.** $448 **63.** 173 men **65.** Jo; 107 votes
**67.** 5920 sq. ft **69.** 497,000 soldiers **71.** 6679 **73.** Atlanta, Hartsfield International **75.** 733 thousand or 733,000
**77.** General Motors, Procter & Gamble **79.** $454,474,500 **81.** $7,667,991,700 **83.** 9, 3, 8 **85.** Answers may vary.

### Exercise Set 1.4

**1.** 630 **3.** 640 **5.** 790 **7.** 400 **9.** 1100 **11.** 43,000 **13.** 248,700 **15.** 36,000 **17.** 100,000 **19.** 60,000,000
**21.** 5280; 5300; 5000 **23.** 9440; 9400; 9000 **25.** 14,800; 14,900; 15,000 **27.** 11,000 **29.** 73,000 **31.** 170,000,000
**33.** $372,000,000 **35.** 130 **37.** 380 **39.** 5500 **41.** 300 **43.** 8500 **45.** correct **47.** incorrect **49.** correct
**51.** correct **53.** $3100 **55.** 80 mi **57.** 25,000 ft **59.** 5,600,000 **61.** 14,000,000 votes **63.** 253,000 children
**65.** $1,300,000,000 **67.** $1,411,000,000 **69.** 8550 **71.** 1,549,999

### Calculator Explorations

**1.** 3456 **3.** 15,322 **5.** 272,291

### Mental Math

**1.** 24 **3.** 0 **5.** 0 **7.** 87

## Exercise Set 1.5

**1.** $4 \cdot 3 + 4 \cdot 9$ **3.** $2 \cdot 4 + 2 \cdot 6$ **5.** $10 \cdot 11 + 10 \cdot 7$ **7.** 252 **9.** 1872 **11.** 1662 **13.** 5310 **15.** 4172 **17.** 10,857
**19.** 11,326 **21.** 24,800 **23.** 0 **25.** 5900 **27.** 59,232 **29.** 142,506 **31.** 1,821,204 **33.** 456,135 **35.** 64,790
**37.** 240,000 **39.** 300,000 **41.** 63 sq. m **43.** 390 sq. ft **45.** 375 calories **47.** $1295 **49.** 192 cans **51.** 9900 sq. ft
**53.** 495,864 sq. m **55.** 5828 pixels **57.** 1500 characters **59.** 1280 calories **61.** 71,343 mi **63.** 22,464 acres
**65.** 21,700,000 qt **67.** $3,760,000,000 **69.** 50 students **71.** apple and orange **73.** 2, 9 **75.** Answers may vary.
**77.** 686 points **79.** Answers may vary.

## Calculator Explorations

**1.** 53 **3.** 62 **5.** 261 **7.** 0

## Mental Math

**1.** 5 **3.** 9 **5.** 0 **7.** 9 **9.** 1 **11.** 5 **13.** undefined **15.** 7 **17.** 0 **19.** 8

## Exercise Set 1.6

**1.** 12 **3.** 37 **5.** 338 **7.** 16 R 2 **9.** 563 R 1 **11.** 37 R 1 **13.** 265 R 1 **15.** 49 **17.** 13 **19.** 97 R 40 **21.** 206
**23.** 506 **25.** 202 R 7 **27.** 45 **29.** 98 R 100 **31.** 202 R 15 **33.** 202 **35.** 58 students **37.** $252,000
**39.** 415 bushels **41.** 88 bridges **43.** Yes, she needs 176 ft; she has 9 ft left over. **45.** 17 touchdowns **47.** 1760 yd
**49.** 6000 kw-hr **51.** 26 **53.** 498 **55.** 79 **57.** 16° **59.** $1,630,100,625 **61.** increase **63.** no

## Integrated Review

**1.** 148 **2.** 6555 **3.** 1620 **4.** 562 **5.** 79 **6.** undefined **7.** 9 **8.** 1 **9.** 0 **10.** 0 **11.** 0 **12.** 3 **13.** 2433
**14.** 9826 **15.** 213 R 3 **16.** 79,317 **17.** 27 **18.** 9 **19.** 138 **20.** 276 **21.** 663 R6 **22.** 1076 R 60 **23.** tens
**24.** thousands **25.** ones **26.** ten-thousands **27.** eight hundred fifty **28.** eight hundred five **29.** twenty-one thousand, sixty
**30.** four thousand, forty-four **31.** 5612 **32.** 73,001 **33.** 100,306 **34.** 4,000,020 **35.** 3270; 3300 **36.** 46,820; 46,800
**37.** 60,510; 60,500 **38.** 670; 700

## Calculator Explorations

**1.** 729 **3.** 1024 **5.** 2048 **7.** 2526 **9.** 4295 **11.** 8 **13.** 80 **15.** 27

## Exercise Set 1.7

**1.** $3^4$ **3.** $7^8$ **5.** $12^3$ **7.** $6^2 \cdot 5^3$ **9.** $9^3 \cdot 8$ **11.** $3 \cdot 2^5$ **13.** $3 \cdot 2^2 \cdot 5^3$ **15.** 25 **17.** 125 **19.** 64 **21.** 1024 **23.** 7
**25.** 243 **27.** 256 **29.** 64 **31.** 81 **33.** 729 **35.** 100 **37.** 10,000 **39.** 10 **41.** 1920 **43.** 729 **45.** 21 **47.** 8
**49.** 29 **51.** 4 **53.** 17 **55.** 46 **57.** 28 **59.** 10 **61.** 7 **63.** 4 **65.** 14 **67.** 72 **69.** 2 **71.** 35 **73.** 4
**75.** undefined **77.** 52 **79.** 44 **81.** 12 **83.** 13 **85.** 400 sq. mi **87.** 64 sq. cm **89.** 10,000 sq. m **91.** $(2 + 3) \cdot 6 - 2$
**93.** $24 \div (3 \cdot 2) + 2 \cdot 5$ **95.** 1260 ft **97.** 6,384,814

## Exercise Set 1.8

**1.** 9 **3.** 26 **5.** 6 **7.** 3 **9.** 117 **11.** 94 **13.** 5 **15.** 626 **17.** 20 **19.** 4 **21.** 4 **23.** 0 **25.** 33 **27.** 121
**29.** 121 **31.** 100 **33.** 60 **35.** 4 **37.** 16, 64, 144, 256 **39.** $x + 5$ **41.** $x + 8$ **43.** $20 - x$ **45.** $512x$ **47.** $\frac{x}{2}$
**49.** $5x + (17 + x)$ **51.** $5x$ **53.** $11 - x$ **55.** $50 - 8x$ **57.** 274,657 **59.** 777 **61.** $5x$ **63.** As $t$ gets larger, $16t^2$ gets larger.

## Chapter 1 Review

**1.** hundreds **2.** ten-millions **3.** five thousand, four hundred eighty **4.** forty-six million, two hundred thousand, one hundred twenty
**5.** $6000 + 200 + 70 + 9$ **6.** $400,000,000 + 3,000,000 + 200,000 + 20,000 + 5000$ **7.** 59,800 **8.** 6,304,000,000 **9.** 1,595,138
**10.** 2,811,801 **11.** 193,699 **12.** 250,925 **13.** 13 **14.** 17 **15.** 3 **16.** 10 **17.** 38 **18.** 68 **19.** 56 **20.** 40
**21.** 110 **22.** 120 **23.** 950 **24.** 1250 **25.** 1711 **26.** 9867 **27.** $197,699 **28.** 8032 mi **29.** 276 ft **30.** 66 km
**31.** 33 **32.** 43 **33.** 14 **34.** 33 **35.** 362 **36.** 65 **37.** 304 **38.** 476 **39.** 2114 **40.** 321 **41.** $15,626
**42.** 397 pages **43.** May **44.** August **45.** July, August, September **46.** April, May **47.** 90 **48.** 50 **49.** 470 **50.** 500
**51.** 4800 **52.** 58,000 **53.** 50,000,000 **54.** 800,000 **55.** 7400 **56.** 4100 **57.** 11,000 **58.** 1,750,000 **59.** 42
**60.** 24 **61.** 0 **62.** 0 **63.** 1410 **64.** 2898 **65.** 800 **66.** 900 **67.** 3696 **68.** 1694 **69.** 0 **70.** 0 **71.** 16,994
**72.** 8954 **73.** 13,634 **74.** 44,763 **75.** 411,426 **76.** 636,314 **77.** 1500 **78.** 240,000 **79.** 1,040,000 **80.** 7,020,000
**81.** $4,897,341 **82.** 3920 calories **83.** 60 sq. mi **84.** 500 sq. cm **85.** 3 **86.** 4 **87.** 6 **88.** 5 **89.** 5 R 2 **90.** 4 R 2
**91.** undefined **92.** 0 **93.** 1 **94.** 10 **95.** undefined **96.** 0 **97.** 15 **98.** 19 R 7 **99.** 24 R 2 **100.** 56 **101.** 1 R 17
**102.** 35 R 15 **103.** 500 **104.** 21 R 6 **105.** 506 **106.** 16 **107.** 199 R 8 **108.** 200 **109.** 458 ft **110.** 51 **111.** $7^4$
**112.** $3^3$ **113.** $4 \cdot 2^3$ **114.** $5^2 \cdot 7^3$ **115.** 49 **116.** 64 **117.** 1125 **118.** 19,600 **119.** 13 **120.** 10 **121.** 3 **122.** 1
**123.** 32 **124.** 33 **125.** 49 sq. m **126.** 9 sq. in. **127.** 5 **128.** 17 **129.** undefined **130.** 0 **131.** 121 **132.** 2
**133.** 4 **134.** 20 **135.** $x - 5$ **136.** $x + 7$ **137.** $10 \div (x + 1)$ **138.** $5(x + 3)$ **139.** 0, 8, 32, 72

## Chapter 1 Test

**1.** 141 **2.** 113 **3.** 14,880 **4.** 766 R 42 **5.** 200 **6.** 48 **7.** 98 **8.** 0 **9.** undefined **10.** 33 **11.** 21 **12.** 36
**13.** 52,000 **14.** 13,700 **15.** 1600 **16.** $17 **17.** $119 **18.** $126 **19.** 7 billion **20.** 170,000 cards **21.** 30 **22.** 1
**23. a.** $17x$ **b.** $20 - 2x$ **24.** 20 cm, 25 sq. cm **25.** 60 yd, 200 sq. yd **26.** 511 **27.** 1

# Chapter 2 INTEGERS

## CHAPTER 2 PRETEST

**1.** −22; 2.1A    **2.** >; 2.1 C    **3.** 8; 2.1D    **4.** 12; 2.1E    **5.** −11; 2.2A    **6.** −17; 2.2B    **7.** 3°F; 2.2C    **8.** 5; 2.3A    **9.** 8; 2.3B
**10.** −15; 2.3C    **11.** −4; 2.3D    **12.** −104; 2.4A    **13.** 9; 2.4B    **14.** 48; 2.4C    **15.** −20; 2.4C    **16.** −60; 2.4D    **17.** −25; 2.5A
**18.** 2; 2.5A    **19.** −8; 2.5A    **20.** −34; 2.5B

## EXERCISE SET 2.1

**1.** −1445    **3.** +14,494    **5.** −15    **7.** −317    **9.** −5049 thousand    **11.** −10; cooler    **13.** −16

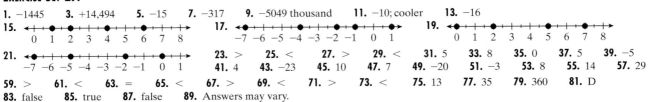

**15.** (number line 0 to 8)    **17.** (number line −7 to 1)    **19.** (number line 0 to 8)
**21.** (number line −7 to 1)    **23.** >    **25.** <    **27.** >    **29.** <    **31.** 5    **33.** 8    **35.** 0    **37.** 5    **39.** −5
**41.** 4    **43.** −23    **45.** 10    **47.** 7    **49.** −20    **51.** −3    **53.** 8    **55.** 14    **57.** 29
**59.** >    **61.** <    **63.** =    **65.** <    **67.** >    **69.** <    **71.** >    **73.** <    **75.** 13    **77.** 35    **79.** 360    **81.** D
**83.** false    **85.** true    **87.** false    **89.** Answers may vary.

## CALCULATOR EXPLORATIONS

**1.** −159    **3.** 44    **5.** −894,855

## MENTAL MATH

**1.** 5    **3.** −35

## EXERCISE SET 2.2

**1.**    **3.**    **5.**

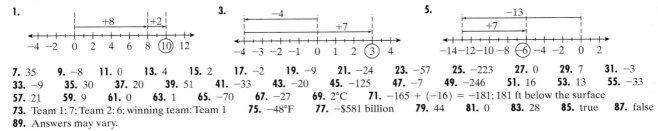

**7.** 35    **9.** −8    **11.** 0    **13.** 4    **15.** 2    **17.** −2    **19.** −9    **21.** −24    **23.** −57    **25.** −223    **27.** 0    **29.** 7    **31.** −3
**33.** −9    **35.** 30    **37.** 20    **39.** 51    **41.** −33    **43.** −20    **45.** −125    **47.** −7    **49.** −246    **51.** 16    **53.** 13    **55.** −33
**57.** 21    **59.** 9    **61.** 0    **63.** 1    **65.** −70    **67.** −27    **69.** 2°C    **71.** −165 + (−16) = −181; 181 ft below the surface
**73.** Team 1: 7; Team 2: 6; winning team: Team 1    **75.** −48°F    **77.** −$581 billion    **79.** 44    **81.** 0    **83.** 28    **85.** true    **87.** false
**89.** Answers may vary.

## EXERCISE SET 2.3

**1.** 0    **3.** 5    **5.** −5    **7.** 14    **9.** 3    **11.** −18    **13.** −14    **15.** 0    **17.** 0    **19.** −4    **21.** −15    **23.** −14    **25.** −1100
**27.** 10    **29.** −4    **31.** 7    **33.** −38    **35.** −17    **37.** 13    **39.** 2    **41.** 0    **43.** −1    **45.** −27    **47.** 40    **49.** −22
**51.** −8    **53.** 14    **55.** −8    **57.** 36    **59.** 0    **61.** 19    **63.** 265°F    **65.** −$16    **67.** −12°C    **69.** 652 ft    **71.** 144 ft
**73.** 382°C    **75.** Saturn; 39°C    **77.** −$249 billion    **79.** 0    **81.** 8    **83.** 1058    **85.** −22    **87.** 20    **89.** −4
**91.** 0    **93.** false    **95.** Answers may vary.    **97.** Answers may vary.

## INTEGRATED REVIEW

**1.** +29,028    **2.** −35,840    **3.** (number line −4 to 4)    **4.** >    **5.** <    **6.** <    **7.** >    **8.** 1    **9.** −4    **10.** 8
**11.** 5    **12.** −6    **13.** 3    **14.** −89    **15.** 0    **16.** 5
**17.** −20    **18.** −10    **19.** −2    **20.** 52    **21.** −3    **22.** 84    **23.** 6    **24.** 1    **25.** −19    **26.** 12    **27.** −4    **28.** D
**29.** A, B, C

## EXERCISE SET 2.4

**1.** 6    **3.** −36    **5.** −64    **7.** 0    **9.** −48    **11.** −8    **13.** 80    **15.** 0    **17.** −15    **19.** 4    **21.** −27    **23.** 25    **25.** −8
**27.** −4    **29.** −5    **31.** 8    **33.** 0    **35.** undefined    **37.** −13    **39.** 0    **41.** −12    **43.** −54    **45.** 42    **47.** −24    **49.** 16
**51.** −2    **53.** −7    **55.** −4    **57.** 48    **59.** −1080    **61.** 0    **63.** −5    **65.** −6    **67.** 3    **69.** 1    **71.** −243    **73.** 180
**75.** 1    **77.** −20    **79.** −966    **81.** −2050    **83.** −28    **85.** −6    **87.** 25    **89.** −1    **91.** undefined    **93.** 6    **95.** 8; 2
**97.** 0; 0    **99.** (−4)(3) = −12; a loss of 12 yd    **101.** (−20)(5) = −100; a depth of 100 ft    **103.** −2    **105.** −$186 million
**107. a.** −8 condors    **b.** −1 condor per yr    **109.** 225    **111.** 109    **113.** 8    **115.** false    **117.** true
**119. a.** −19,024 Saturns    **b.** −38,048 Saturns    **c.** 193,738 Saturns    **121.** false    **123.** false    **125.** Answers may vary.

## CALCULATOR EXPLORATIONS

**1.** 48    **3.** −258

## MENTAL MATH

**1.** base: 3; exponent: 2    **3.** base: 2; exponent: 3    **5.** base: −7; exponent: 5    **7.** base: 5; exponent: 7

## EXERCISE SET 2.5

**1.** 3    **3.** −1    **5.** −7    **7.** −14    **9.** −43    **11.** −8    **13.** 7    **15.** −1    **17.** 4    **19.** −3    **21.** −55    **23.** 24    **25.** 16
**27.** 15    **29.** −65    **31.** 64    **33.** 452    **35.** 129    **37.** −3    **39.** −3    **41.** 4    **43.** 16    **45.** −27    **47.** 34    **49.** 65

**51.** −59  **53.** −7  **55.** −61  **57.** −11  **59.** 36  **61.** −117  **63.** 30  **65.** −3  **67.** 1  **69.** −20  **71.** 0  **73.** −20
**75.** 9  **77.** −16  **79.** 1  **81.** −54  **83.** 4050  **85.** 45  **87.** 32 in.  **89.** 30 ft  **91.** $2 \cdot (7 - 5) \cdot 3$  **93.** $-6 \cdot (10 - 4)$
**95.** 20,736  **97.** 8900  **99.** 9  **101.** Answers may vary.

## Chapter 2 Review

**1.** −1435  **2.** +7562  **3.**  **4.**  **5.** >  **6.** <

**7.** >  **8.** 12  **9.** 0  **10.** −6  **11.** 12  **12.** −3  **13.** false  **14.** true  **15.** true  **16.** true  **17.** 2  **18.** 14  **19.** 4
**20.** 17  **21.** −23  **22.** −22  **23.** −21  **24.** −70  **25.** 0  **26.** 0  **27.** −151  **28.** −606  **29.** −7  **30.** −9
**31.** −20°C  **32.** 150 ft below the surface  **33.** 8  **34.** −16  **35.** −11  **36.** −27  **37.** 20  **38.** 8  **39.** 0  **40.** −32
**41.** 0  **42.** −7  **43.** −10  **44.** −9  **45.** −25  **46.** 692 ft  **47.** true  **48.** false  **49.** true  **50.** true  **51.** 21
**52.** −18  **53.** −64  **54.** 60  **55.** −5  **56.** 3  **57.** 0  **58.** undefined  **59.** 38  **60.** −5  **61.** $(-5)(2) = -10$
**62.** $(-50)(4) = -200$  **63.** 49  **64.** −49  **65.** −32  **66.** −32  **67.** 0  **68.** −8  **69.** −16  **70.** 35  **71.** −28  **72.** −44
**73.** 3  **74.** −1  **75.** 7  **76.** −17  **77.** 39  **78.** −26  **79.** 7  **80.** −80  **81.** −2  **82.** −12  **83.** −5  **84.** 5
**85.** −1  **86.** −7  **87.** 4  **88.** −4  **89.** 3  **90.** 108

## Chapter 2 Test

**1.** 3  **2.** −6  **3.** −100  **4.** 4  **5.** −30  **6.** 12  **7.** 65  **8.** 5  **9.** 12  **10.** −6  **11.** 50  **12.** −2  **13.** −11
**14.** −46  **15.** −117  **16.** 3456  **17.** 28  **18.** −213  **19.** −1  **20.** −2  **21.** 2  **22.** −5  **23.** −3  **24.** 5  **25.** −1
**26.** 8  **27.** 14,893 + (−147) = 14,746; 14,746 ft  **28.** $41  **29.** 31,642 ft  **30.** 3820 ft below sea level

## Cumulative Review

**1.** ten-thousands; Sec. 1.1, Ex. 1  **2.** tens; Sec. 1.1, Ex. 2  **3.** millions; Sec. 1.1, Ex. 3  **4. a.** <  **b.** >  **c.** >; Sec. 1.1, Ex. 12
**5.** 39; Sec. 1.2, Ex. 3  **6.** 7321; Sec. 1.3, Ex. 2  **7.** 2440 km; Sec. 1.3, Ex. 5  **8.** 570; Sec. 1.4, Ex. 1  **9.** 1800; Sec. 1.4, Ex. 5
**10. a.** $3 \cdot 4 + 3 \cdot 5$  **b.** $10 \cdot 6 + 10 \cdot 8$  **c.** $2 \cdot 7 + 2 \cdot 8$; Sec. 1.5, Ex. 2  **11.** 78,875; Sec. 1.5, Ex. 5
**12. a.** 6  **b.** 9  **c.** 6; Sec. 1.6, Ex. 1  **13.** 741; Sec. 1.6, Ex. 4  **14.** 7 boxes; Sec. 1.6, Ex. 11  **15.** 64; Sec. 1.7, Ex. 5
**16.** 7; Sec. 1.7, Ex. 6  **17.** 180; Sec. 1.7, Ex. 8  **18.** 2; Sec. 1.7, Ex. 12  **19.** 15; Sec. 1.8, Ex. 1  **20. a.** 2  **b.** 5  **c.** 0; Sec. 2.1, Ex. 4
**21.** 3; Sec. 2.2, Ex. 6  **22.** 14; Sec. 2.3, Ex. 12  **23.** −21; Sec. 2.4, Ex. 1  **24.** 0; Sec. 2.4, Ex. 3  **25.** −16; Sec. 2.5, Ex. 10

# Chapter 3 SOLVING EQUATIONS AND PROBLEM SOLVING

## Chapter 3 Pretest

**1.** $15x + 4$; 3.1A  **2.** $5x + 5$; 3.1C  **3.** $56b$; 3.1B  **4.** $-10y + 35$; 3.1B  **5.** $(13a + 17)$ in.; 3.1D  **6.** $(12x - 3)$ sq. ft; 3.1D
**7.** not a solution; 3.2A  **8.** −12; 3.2B  **9.** −9; 3.2B  **10.** −7; 3.3A  **11.** 11; 3.3A  **12.** $2(x - 12)$; 3.3B  **13.** $3x$; 3.3B

**14.** 9; 3.4A  **15.** 2; 3.4A  **16.** 1; 3.4B  **17.** −3; 3.4B  **18.** $\dfrac{54}{-6} = -9$; 3.4C  **19.** $x - 5 = 12$; 3.5A  **20.** 7; 3.5B

## Mental Math

**1.** 5  **3.** 1  **5.** 11

## Exercise Set 3.1

**1.** $8x$  **3.** $-4n$  **5.** $-2c$  **7.** $-4x$  **9.** $13a - 8$  **11.** $30x$  **13.** $-22y$  **15.** $72a$  **17.** $2y + 4$  **19.** $5a - 40$
**21.** $-12x - 28$  **23.** $2x + 15$  **25.** $-21n + 20$  **27.** $15c + 3$  **29.** $7w + 15$  **31.** $11x - 8$  **33.** $-5x - 9$  **35.** $-2y$
**37.** $-7z$  **39.** $8d - 3c$  **41.** $6y - 14$  **43.** $-q$  **45.** $2x + 22$  **47.** $-3x - 35$  **49.** $-3z - 15$  **51.** $-6x + 6$
**53.** $3x - 30$  **55.** $-r + 8$  **57.** $-7n + 3$  **59.** $9z - 14$  **61.** −6  **63.** $-4x + 10$  **65.** $2xy + 20$  **67.** $7a + 12$
**69.** $3y + 5$  **71.** $(-25x + 55)$ in.  **73.** $(14y + 22)$ m  **75.** $(11a + 12)$ ft  **77.** $16z^2$ sq. cm  **79.** −3  **81.** 8  **83.** 0
**85.** $4824q + 12,274$  **87.** Answers may vary.  **89.** $(20x + 16)$ sq. mi  **91.** Answers may vary.

## Exercise Set 3.2

**1.** yes  **3.** yes  **5.** yes  **7.** yes  **9.** yes  **11.** yes  **13.** 18  **15.** 26  **17.** 9  **19.** −16  **21.** 11  **23.** −4  **25.** 6
**27.** 0  **29.** 8  **31.** −1  **33.** −6  **35.** 24  **37.** 2  **39.** 0  **41.** −28  **43.** 73  **45.** 0  **47.** −4  **49.** −65  **51.** 13
**53.** −3  **55.** −22  **57.** −16  **59.** 3  **61.** 28  **63.** 1900  **65.** about 700  **67.** 1  **69.** 1  **71.** Answers may vary.
**73.** 162,964  **75.** 3900 yd  **77.** $106,146 million

## Exercise Set 3.3

**1.** 4  **3.** −4  **5.** 0  **7.** −17  **9.** 5  **11.** −4  **13.** 8  **15.** −1  **17.** 2  **19.** −8  **21.** −1  **23.** −7  **25.** 0  **27.** 6
**29.** −9  **31.** 9  **33.** 4  **35.** −6  **37.** 1  **39.** −1  **41.** −25  **43.** −2  **45.** −2  **47.** 0  **49.** 4  **51.** 7  **53.** 3

**55.** 36  **57.** 1  **59.** $4x + 7$  **61.** $2x - 17$  **63.** $-6(x + 15)$  **65.** $\dfrac{45}{-5x}$  **67.** −5  **69.** −5  **71.** 11  **73.** Answers may vary.

**75.** −36  **77.** 12 hr  **79.** 58 mph

## Integrated Review

**1.** 4   **2.** −3   **3.** 1   **4.** −4   **5.** 8x   **6.** −4y   **7.** −2a − 2   **8.** −8x − 14   **9.** 4y + 14   **10.** (12x − 6) sq. m   **11.** 13
**12.** −9   **13.** 5   **14.** 0   **15.** −4   **16.** 5   **17.** −10   **18.** −3   **19.** 6   **20.** 8   **21.** −1   **22.** −24

## Calculator Explorations

**1.** yes   **3.** no   **5.** yes

## Exercise Set 3.4

**1.** 3   **3.** −2   **5.** −4   **7.** −3   **9.** −12   **11.** 5   **13.** −5   **15.** 8   **17.** 5   **19.** 1   **21.** −4   **23.** 2   **25.** 5   **27.** −3
**29.** 2   **31.** −2   **33.** 3   **35.** 6   **37.** −10   **39.** −4   **41.** −1   **43.** 4   **45.** −4   **47.** 3   **49.** −1   **51.** 0   **53.** −22

**55.** 4   **57.** 1   **59.** −30   **61.** −42 + 16 = −26   **63.** −5(−29) = 145   **65.** 3(−14 − 2) = −48   **67.** $\dfrac{100}{2(50)} = 1$   **69.** 33

**71.** −37   **73.** 49   **75.** −6   **77.** 0   **79.** −4   **81.** Answers may vary.

## Exercise Set 3.5

**1.** −5 + x = −7   **3.** 3x = 27   **5.** −20 − x = 104   **7.** $2\big(x + (-1)\big) = 50$   **9.** 8   **11.** 6   **13.** 9   **15.** 6   **17.** 24
**19.** 8   **21.** 5   **23.** 5   **25.** 12   **27.** 5   **29.** Dole: 159 votes; Clinton: 379 votes   **31.** 80 coins   **33.** truck, 35 mph; car, 70 mph
**35.** $225   **37.** Michigan Stadium, 107,501; Neyland Stadium, 102,854   **39.** California, 309 thousand; Washington, 103 thousand
**41.** 71 points   **43.** 590   **45.** 1000   **47.** 3000   **49.** Answers may vary.   **51.** $8250   **53.** $17   **55.** Answers may vary.

## Chapter 3 Review

**1.** 10y − 15   **2.** −6y − 10   **3.** −6a − 7   **4.** −8y + 2   **5.** 2x + 10   **6.** −3y − 24   **7.** 11x − 12   **8.** −4m − 12
**9.** −5a + 4   **10.** 12y − 9   **11.** (4x + 6) yd   **12.** 20y m   **13.** (6x − 3) sq. yd   **14.** (45x + 8) sq. cm   **15.** 600,822x − 9180
**16.** 3292y − 3840   **17.** yes   **18.** no   **19.** −2   **20.** 10   **21.** −6   **22.** −1   **23.** −25   **24.** −8   **25.** 7   **26.** −9   **27.** 1

**28.** −5   **29.** 0   **30.** −2   **31.** 0   **32.** −128   **33.** 2x + 11   **34.** −5x − 50   **35.** $\dfrac{70}{x + 6}$   **36.** 2(x − 13)   **37.** 5

**38.** 12   **39.** 17   **40.** 7   **41.** −2   **42.** 8   **43.** 11   **44.** −5   **45.** 2   **46.** 2   **47.** 20 − (−8) = 28

**48.** 5(2 + (−6)) = −20   **49.** $\dfrac{-75}{5 + 20} = -3$   **50.** −2 − 19 = −21   **51.** 2x − 8 = 40   **52.** $\dfrac{x}{2} - 12 = 10$   **53.** x − 3 = x ÷ 4

**54.** 6x = x + 2   **55.** 5   **56.** −16   **57.** 2386 votes   **58.** 84 CDs

## Chapter 3 Test

**1.** −5x + 5   **2.** −6y − 14   **3.** −8z − 20   **4.** (15x + 15) in.   **5.** (9x − 3) sq. m   **6.** −6   **7.** −6   **8.** 10   **9.** 8
**10.** 24   **11.** 1   **12.** −2   **13.** −2   **14.** 6   **15.** 0   **16.** −2   **17.** 7   **18.** 4   **19.** 0   **20.** 6000 sq. ft   **21. a.** 17x
**b.** 20 − 2x   **22.** −2   **23.** 8 free throws   **24.** 244 women

## Cumulative Review

**1.** one hundred six million, fifty-two thousand, four hundred forty-seven; Sec. 1.1, Ex. 6   **2.** 13 in.; Sec. 1.2, Ex. 5   **3.** 726; Sec. 1.3, Ex. 4
**4.** 249,000; Sec. 1.4, Ex. 3   **5.** 200; Sec. 1.5, Ex. 3   **6.** 208; Sec. 1.6, Ex. 5   **7.** 7; Sec. 1.7, Ex. 9   **8.** 26; Sec. 1.8, Ex. 4
**9. a.** <   **b.** >   **c.** >; Sec. 2.1, Ex. 3   **10.** −7; Sec. 2.2, Ex. 1   **11.** −6; Sec. 2.2, Ex. 4   **12.** 8; Sec. 2.2, Ex. 5   **13.** −14; Sec. 2.3, Ex. 2
**14.** 11; Sec. 2.3, Ex. 3   **15.** −4; Sec. 2.3, Ex. 4   **16.** −2; Sec. 2.4, Ex. 7   **17.** 5; Sec. 2.4, Ex. 8   **18.** −16; Sec. 2.4, Ex. 9
**19.** 9; Sec. 2.5, Ex. 1   **20.** −9; Sec. 2.5, Ex. 2   **21.** 6y + 2; Sec. 3.1, Ex. 2   **22.** not a solution; Sec. 3.2, Ex. 2   **23.** 3; Sec. 3.3, Ex. 3
**24.** 12; Sec. 3.4, Ex. 1   **25.** software, $420; computer system, $1680; Sec. 3.5, Ex. 4

# Chapter 4 Fractions

## Chapter 4 Pretest

**1.** $\dfrac{3}{8}$; 4.1B   **2.** $\dfrac{15}{18}$; 4.1D   **3.** 0; 4.1E   **4.** 2 · 2 · 5 · 7 or $2^2 \cdot 5 \cdot 7$; 4.2A   **5.** $\dfrac{5}{9}$; 4.2B   **6.** $\dfrac{9}{20}$ of a gram; 4.2C   **7.** $\dfrac{6}{5}$; 4.3A

**8.** $\dfrac{1}{14x}$; 4.3C   **9.** $\dfrac{5}{11}$; 4.4A   **10.** $-\dfrac{1}{10}$; 4.4A   **11.** $-\dfrac{8}{27}$; 4.3B   **12.** $-\dfrac{1}{3}$; 4.3D   **13.** $\dfrac{23}{36}$; 4.5A   **14.** $\dfrac{1}{14}$; 4.5B   **15.** $\dfrac{3x}{7}$; 4.6A

**16.** $-\dfrac{46}{25}$; 4.6B   **17.** $-\dfrac{23}{2}$; 4.7A   **18.** $\dfrac{13}{5}$; 4.8B   **19.** $\dfrac{44}{5}$ or $8\dfrac{4}{5}$; 4.8D   **20.** $9\dfrac{5}{6}$; 4.8E

## Mental Math

**1.** numerator = 1; denominator = 2   **3.** numerator = 10; denominator = 3   **5.** numerator = 3z; denominator = 7

## Exercise Set 4.1

**1.** $\dfrac{1}{3}$   **3.** $\dfrac{4}{7}$   **5.** $\dfrac{7}{12}$   **7.** $\dfrac{3}{7}$   **9.** $\dfrac{4}{9}$   **11.** $\dfrac{5}{8}$   **13.** $\dfrac{11}{4}$   **15.** $\dfrac{11}{3}$   **17.** $\dfrac{3}{2}$   **19.** $\dfrac{4}{3}$   **21.** $\dfrac{17}{6}$   **23.** $\dfrac{14}{9}$   **25.** $\dfrac{42}{131}$ of the students

**27.** 89; $\dfrac{89}{131}$ of the students   **29.** $\dfrac{4}{10}$ of the visits   **31.** $\dfrac{8}{42}$ of the presidents   **33.** $\dfrac{5}{12}$ of a foot   **35.** $\dfrac{11}{24}$ of a day   **37.** $\dfrac{7}{11}$ of the children

**39. a.** $\frac{40}{50}$ of the states  **b.** 10 states  **c.** $\frac{10}{50}$ of the states  **41.**

(number line: point at $\frac{1}{4}$ between 0 and 1)

**43.**
(number line: point at $\frac{4}{7}$ between 0 and 1)

**45.**
(number line: point at $\frac{8}{5}$ between 1 and 2)

**47.**
(number line: point at $\frac{7}{3}$ between 2 and 3)

**49.**
(number line: point at $\frac{3}{8}$ between 0 and 1)

**51.** $\frac{20}{35}$  **53.** $\frac{14}{21}$  **55.** $\frac{10y}{25}$  **57.** $\frac{15}{30}$  **59.** $\frac{30}{21x}$  **61.** $\frac{10}{5}$  **63.** $\frac{9}{12}$  **65.** $\frac{8y}{12}$

**67.** $\frac{6}{12}$  **69.** $\frac{48x}{36x}$  **71.** $\frac{20x}{36x}$  **73.** $\frac{36x}{36x}$  **75.** $\frac{35}{100}, \frac{57}{100}, \frac{43}{100}, \frac{37}{100}, \frac{36}{100}, \frac{31}{100}, \frac{48}{100}, \frac{32}{100}, \frac{30}{100}, \frac{46}{100}, \frac{26}{100}$  **77.** United States  **79.** 1

**81.** $-5$  **83.** 0  **85.** 1  **87.** undefined  **89.** 3  **91.** 9  **93.** 125  **95.** 49  **97.** 24  **99.** $\frac{464}{2088}$  **101.** Answers may vary.

**103.** $\frac{1300}{1585}$ of the affiliates  **105.** $\frac{56}{172}$ of the licensees

## Calculator Explorations

**1.** $\frac{4}{7}$  **3.** $\frac{20}{27}$  **5.** $\frac{15}{8}$  **7.** $\frac{9}{2}$

## Mental Math

**1.** yes, yes, yes  **3.** $2 \cdot 5$  **5.** $3 \cdot 7$  **7.** $3^2$

## Exercise Set 4.2

**1.** $2^2 \cdot 5$  **3.** $2^4 \cdot 3$  **5.** $2^6$  **7.** $2^4 \cdot 3 \cdot 5$  **9.** $\frac{1}{4}$  **11.** $\frac{x}{5}$  **13.** $\frac{7}{8}$  **15.** $\frac{4}{5}$  **17.** $\frac{5}{6}$  **19.** $\frac{5x}{6}$  **21.** $\frac{2}{3}$  **23.** $\frac{3x}{4}$  **25.** $\frac{3b}{2a}$

**27.** $\frac{7}{8}$  **29.** $\frac{3}{7}$  **31.** $\frac{3y}{5}$  **33.** $\frac{4}{7}$  **35.** $\frac{4x^2y}{5}$  **37.** $\frac{4}{5}$  **39.** $\frac{5x}{8}$  **41.** $\frac{3x}{10}$  **43.** $\frac{3a^2}{2b^3}$  **45.** $\frac{5}{8z}$  **47.** $\frac{3}{4}$ of a shift  **49.** $\frac{1}{2}$ mi

**51.** $\frac{257}{408}$ of individuals  **53.** $\frac{9}{20}$ of the students  **55. a.** $\frac{3}{10}$ of the states  **b.** 35 states  **c.** $\frac{7}{10}$ of the states  **57.** $\frac{5}{12}$ of the width

**59.** $\frac{1}{10}$  **61.** Answers may vary.  **63.** $-3$  **65.** $-14$  **67.** 0  **69.** 29  **71.** d  **73.** false  **75.** true  **77.** $\frac{3}{5}$

**79.** $\frac{9}{25}$ of the donors  **81.** $\frac{1}{25}$ of the donors  **83.** $\frac{3}{20}$ of the donors

## Mental Math

**1.** $\frac{2}{15}$  **3.** $\frac{6}{35}$  **5.** $\frac{9}{8}$

## Exercise Set 4.3

**1.** $\frac{7}{12}$  **3.** $-\frac{5}{28}$  **5.** $\frac{1}{15}$  **7.** $\frac{18x}{55}$  **9.** $\frac{3a^2}{4}$  **11.** $\frac{x^2}{y}$  **13.** $\frac{1}{125}$  **15.** $\frac{4}{9}$  **17.** $-\frac{4}{27}$  **19.** $\frac{4}{5}$  **21.** $-\frac{1}{6}$  **23.** $\frac{16}{9x}$

**25.** $\frac{121y}{60}$  **27.** $-\frac{1}{6}$  **29.** $\frac{x}{25}$  **31.** $\frac{10}{27}$  **33.** $\frac{18x^2}{35}$  **35.** $-\frac{1}{4}$  **37.** $\frac{3}{4}$  **39.** $\frac{9}{16}$  **41.** $xy^2$  **43.** $\frac{77}{2}$  **45.** $-\frac{36}{x}$  **47.** $\frac{3}{49}$

**49.** $-\frac{19y}{7}$  **51.** $\frac{4}{11}$  **53.** $\frac{8}{3}$  **55.** $\frac{15x}{4}$  **57.** $\frac{8}{9}$  **59.** $\frac{1}{60}$  **61.** $b$  **63.** $\frac{2}{5}$  **65.** $-\frac{5}{3}$  **67. a.** $\frac{1}{3}$  **b.** $\frac{12}{25}$

**69. a.** $-\frac{36}{55}$  **b.** $-\frac{44}{45}$  **71.** yes  **73.** no  **75.** 30 gal  **77.** $\frac{3}{2}$ in.  **79.** 21,735 tornados  **81.** $\frac{17}{2}$ in.  **83.** 3 ft  **85.** $\frac{1}{14}$ sq. ft

**87.** $2 \cdot 3 \cdot 3 \cdot 5$ or $2 \cdot 3^2 \cdot 5$  **89.** $5 \cdot 13$  **91.** $2 \cdot 3 \cdot 3 \cdot 7$ or $2 \cdot 3^2 \cdot 7$  **93.** 7,210,000 households  **95.** Answers may vary.  **97.** 15,432 species

## Mental Math

**1.** unlike  **3.** like  **5.** like  **7.** unlike  **9.** $\frac{5}{7}$  **11.** $\frac{6}{11}$  **13.** $\frac{7}{11}$  **15.** $\frac{8}{15}$

## Exercise Set 4.4

**1.** 0  **3.** $\frac{2}{3x}$  **5.** $-\frac{1}{13}$  **7.** $\frac{2}{3}$  **9.** $-\frac{3}{y}$  **11.** $\frac{7a-3}{4}$  **13.** $-\frac{3}{4}$  **15.** $-\frac{1}{3}$  **17.** $\frac{2x}{3}$  **19.** $-\frac{x}{2}$  **21.** $\frac{3}{4z}$  **23.** $-\frac{3}{10}$

**25.** 2  **27.** $-\frac{2}{3}$  **29.** $\frac{3x}{4}$  **31.** $\frac{5}{4}$  **33.** $-\frac{4}{5}$  **35.** $-\frac{3}{4}$  **37.** $-\frac{2}{3}$  **39.** $\frac{1}{13}$  **41.** $\frac{4}{5}$  **43.** 1 in.  **45.** 2 m  **47.** $\frac{3}{2}$ hr

**49.** Traditional fee-for-service, Point of Service, Health Maintenance Organizations, Preferred Provider Organizations

**51.** $\frac{13}{20}$ of the employees  **53.** $\frac{23}{50}$ of the states  **55.** 12  **57.** 45  **59.** 36  **61.** $24x$  **63.** 150  **65.** 126  **67.** 75  **69.** 24

**71.** 50  **73.** $12a$  **75.** 168  **77.** 363  **79.** $\frac{12}{35}$  **81.** 8  **83.** $\frac{4}{5}$  **85.** $-28$  **87.** $\frac{8}{11}$  **89.** $\frac{1}{10}$ of men  **91.** Answers may vary.

## CALCULATOR EXPLORATIONS

**1.** $\dfrac{37}{80}$  **3.** $\dfrac{95}{72}$  **5.** $\dfrac{394}{323}$

## MENTAL MATH

**1.** 6  **3.** 12  **5.** 56  **7.** 12

## EXERCISE SET 4.5

**1.** $\dfrac{5}{6}$  **3.** $\dfrac{1}{6}$  **5.** $-\dfrac{4}{33}$  **7.** $\dfrac{3x-6}{14}$  **9.** $\dfrac{3}{5}$  **11.** $\dfrac{24y-5}{12}$  **13.** $\dfrac{11}{36}$  **15.** $\dfrac{12}{7}$  **17.** $\dfrac{89a}{99}$  **19.** $\dfrac{4y-1}{6}$  **21.** $\dfrac{x+6}{2x}$

**23.** $-\dfrac{8}{33}$  **25.** $\dfrac{3}{14}$  **27.** $\dfrac{11y-10}{35}$  **29.** $-\dfrac{11}{36}$  **31.** $\dfrac{1}{20}$  **33.** $\dfrac{33}{56}$  **35.** $\dfrac{17}{16}$  **37.** $\dfrac{8}{9}$  **39.** $\dfrac{15+11y}{33}$  **41.** $-\dfrac{53}{42}$  **43.** $\dfrac{11}{18}$

**45.** $\dfrac{44a}{39}$  **47.** $-\dfrac{11}{60}$  **49.** $\dfrac{5y+9}{9y}$  **51.** $\dfrac{56}{45}$  **53.** $\dfrac{40+9x}{72x}$  **55.** 1  **57.** $-\dfrac{11}{30}$  **59.** $\dfrac{7x}{8}$  **61.** $-\dfrac{13}{16}$  **63.** $\dfrac{19}{20}$  **65.** $-\dfrac{5}{24}$

**67.** $\dfrac{57+56x}{144}$  **69.** $\dfrac{37x-20}{56}$  **71.** $\dfrac{13}{12}$  **73.** $\dfrac{1}{4}$  **75.** $\dfrac{11}{6}$  **77.** $\dfrac{11}{12}$  **79.** $-\dfrac{7}{10}$  **81.** $\dfrac{11}{16}$  **83.** $\dfrac{11}{18}$  **85.** $\dfrac{34}{15}$ cm  **87.** $\dfrac{17}{10}$ m

**89.** $\dfrac{61}{264}$ mi  **91.** $\dfrac{49}{100}$ of students  **93.** $\dfrac{77}{100}$ of Americans  **95.** $\dfrac{29}{100}$ of adults  **97.** $\dfrac{25}{36}$  **99.** $\dfrac{25}{36}$  **101.** 57,200  **103.** 330

**105.** $\dfrac{49}{44}$  **107.** Answers may vary.  **109.** $\dfrac{212}{579}$ of the land area  **111.** 21,200,000 sq. mi

## INTEGRATED REVIEW

**1.** $\dfrac{1}{4}$  **2.** $\dfrac{3}{6}$ or $\dfrac{1}{2}$  **3.** $\dfrac{7}{4}$  **4.** $\dfrac{73}{85}$  **5.** $2\cdot 3$  **6.** $2\cdot 5\cdot 7$  **7.** $2^2\cdot 3^2\cdot 7$  **8.** $\dfrac{1}{7}$  **9.** $\dfrac{5}{6}$  **10.** $\dfrac{9}{19}$  **11.** $\dfrac{21}{55}$  **12.** $\dfrac{1}{2}$  **13.** $\dfrac{9}{10}$

**14.** $\dfrac{1}{25}$  **15.** $\dfrac{4}{5}$  **16.** $-\dfrac{2}{5}$  **17.** $\dfrac{3}{25}$  **18.** $\dfrac{1}{3}$  **19.** $\dfrac{4}{5}$  **20.** $\dfrac{5}{9}$  **21.** $-\dfrac{1}{6}$  **22.** $\dfrac{3}{2}$  **23.** $\dfrac{1}{18}$  **24.** $\dfrac{4}{21}$  **25.** $-\dfrac{7}{48}$  **26.** $-\dfrac{9}{50}$

## EXERCISE SET 4.6

**1.** $\dfrac{1}{6}$  **3.** $\dfrac{3}{7}$  **5.** $\dfrac{x}{6}$  **7.** $\dfrac{23}{22}$  **9.** $\dfrac{2x}{13}$  **11.** $\dfrac{35}{9}$  **13.** $-\dfrac{17}{45}$  **15.** $\dfrac{11}{8}$  **17.** $\dfrac{7}{6}$  **19.** $-\dfrac{2}{5}$  **21.** $-\dfrac{2}{9}$  **23.** $\dfrac{5}{2}$  **25.** $\dfrac{7}{2}$  **27.** $\dfrac{4}{9}$

**29.** $-\dfrac{13}{2}$  **31.** $\dfrac{9}{25}$  **33.** $-\dfrac{5}{32}$  **35.** 1  **37.** $\dfrac{1}{10}$  **39.** $-\dfrac{11}{40}$  **41.** $\dfrac{x+6}{16}$  **43.** 8  **45.** 25  **47.** $1x$ or $x$  **49.** $1a$ or $a$

**51.** $-\dfrac{77}{16}$  **53.** $-\dfrac{55}{16}$  **55.** $\dfrac{5}{8}$  **57.** $\dfrac{11}{56}$  **59.** Halfway between $a$ and $b$  **61.** false  **63.** false  **65.** true

## EXERCISE SET 4.7

**1.** $\dfrac{2}{7}$  **3.** 12  **5.** $-27$  **7.** $\dfrac{27}{8}$  **9.** $\dfrac{1}{21}$  **11.** $\dfrac{2}{11}$  **13.** 1  **15.** $\dfrac{15}{2}$  **17.** $-1$  **19.** $-15$  **21.** $\dfrac{3x-28}{21}$  **23.** $\dfrac{y+10}{2}$

**25.** $\dfrac{7x}{15}$  **27.** $\dfrac{4}{3}$  **29.** 2  **31.** $\dfrac{21}{10}$  **33.** $-\dfrac{1}{14}$  **35.** $-3$  **37.** 50  **39.** $-\dfrac{1}{9}$  **41.** $-6$  **43.** 4  **45.** $\dfrac{3}{5}$  **47.** $-\dfrac{1}{24}$

**49.** $-\dfrac{5}{14}$  **51.** 4  **53.** $-36$  **55.** $\dfrac{7}{2}$  **57.** $\dfrac{59}{10}$  **59.** $\dfrac{49}{6}$  **61.** Answers may vary.  **63.** $-2$

## CALCULATOR EXPLORATIONS

**1.** $\dfrac{280}{11}$  **3.** $\dfrac{3776}{35}$  **5.** $26\dfrac{1}{14}$  **7.** $92\dfrac{3}{10}$

## EXERCISE SET 4.8

**1. a.** $\dfrac{11}{4}$  **b.** $2\dfrac{3}{4}$  **3. a.** $\dfrac{11}{3}$  **b.** $3\dfrac{2}{3}$  **5. a.** $\dfrac{3}{2}$  **b.** $1\dfrac{1}{2}$  **7. a.** $\dfrac{4}{3}$  **b.** $1\dfrac{1}{3}$  **9.** $\dfrac{7}{3}$  **11.** $\dfrac{27}{8}$  **13.** $\dfrac{83}{7}$  **15.** $1\dfrac{6}{7}$  **17.** $3\dfrac{2}{15}$

**19.** $4\dfrac{5}{8}$  **21.** $\dfrac{8}{21}$  **23.** $4\dfrac{2}{3}$  **25.** $5\dfrac{1}{2}$  **27.** $18\dfrac{2}{3}$  **29.** $6\dfrac{4}{5}$  **31.** $25\dfrac{5}{14}$  **33.** $13\dfrac{13}{24}$  **35.** $2\dfrac{3}{5}$  **37.** $7\dfrac{5}{14}$  **39.** $\dfrac{24}{25}$  **41.** 4

**43.** $5\dfrac{11}{14}$  **45.** $6\dfrac{2}{9}$  **47.** $\dfrac{25}{33}$  **49.** $35\dfrac{13}{18}$  **51.** $2\dfrac{1}{2}$  **53.** $72\dfrac{19}{30}$  **55.** $\dfrac{11}{14}$  **57.** $5\dfrac{4}{7}$  **59.** $13\dfrac{16}{33}$  **61.** $10\dfrac{43}{60}$ min  **63.** $\dfrac{49}{60}$ min

**65.** $\dfrac{39}{2}$ or $19\dfrac{1}{2}$ in.  **67.** 12 shares  **69.** $\dfrac{1}{16}$ in.  **71.** no, she will be $\dfrac{1}{12}$ of a foot short  **73.** $7\dfrac{5}{6}$ gal  **75.** $7\dfrac{13}{20}$ in.  **77.** $\dfrac{7}{2}$ or $3\dfrac{1}{2}$ sq. yd

**79.** $\dfrac{121}{4}$ or $30\dfrac{1}{4}$ sq. in.  **81.** 7 mi  **83.** $21\dfrac{5}{24}$ m  **85.** $1\dfrac{5}{12}$ hr  **87.** $306\dfrac{2}{3}$ ft  **89.** $\dfrac{19}{30}$ in.  **91.** $9x-13$  **93.** $-3y-6$  **95.** 17

**97.** 1  **99.** Supreme is heavier by $\dfrac{1}{8}$ lb  **101.** Answers may vary.  **103.** Answers may vary.

## Chapter 4 Review

1. $\dfrac{3}{4}$  2. $\dfrac{2}{5}$  3. (number line, point at $\dfrac{7}{9}$)  4. (number line, point at $\dfrac{4}{7}$)  5. (number line, point at $1\,\dfrac{5}{4}$)

6. (number line, point at $\dfrac{7}{5}$)  7. $\dfrac{35}{242}$ of job specialties  8. $\dfrac{17}{20}$ of soldiers  9. 20  10. 35  11. $49a$  12. $45b$

13. 40  14. 10  15. $\dfrac{3}{7}$  16. $\dfrac{5}{9}$  17. $\dfrac{1}{3x}$  18. $\dfrac{y^2}{2}$  19. $\dfrac{29}{32c}$  20. $\dfrac{18z}{23}$  21. $\dfrac{5x}{3y^2}$  22. $\dfrac{7b}{5c^2}$  23. $\dfrac{2}{3}$ of a foot

24. $\dfrac{3}{5}$ of the cars  25. $\dfrac{3}{10}$  26. $-\dfrac{5}{14}$  27. $-\dfrac{7}{12x}$  28. $\dfrac{y}{4}$  29. $\dfrac{9}{x^2}$  30. $\dfrac{y}{2}$  31. $-\dfrac{1}{27}$  32. $\dfrac{25}{144}$  33. $x^2y^2$  34. $\dfrac{b}{a^2}$

35. $\dfrac{2}{15}$  36. $-\dfrac{63}{10}$  37. $-2$  38. $\dfrac{15}{4}$  39. $9x^2$  40. $\dfrac{27}{2}$  41. $-\dfrac{5}{6y}$  42. $\dfrac{y^2}{2x}$  43. $\dfrac{12}{7}$  44. $-\dfrac{15}{2}$  45. $\dfrac{77}{48}$ sq. ft

46. $\dfrac{4}{9}$ sq. m  47. $\dfrac{10}{11}$  48. $\dfrac{2}{3}$  49. $-\dfrac{1}{3}$  50. $\dfrac{2}{y}$  51. $\dfrac{4x}{5}$  52. $\dfrac{4y-3}{21}$  53. $\dfrac{1}{3}$  54. $-\dfrac{1}{15}$  55. $3x$  56. 24  57. yes

58. yes  59. $\dfrac{1}{2}$ hr  60. $\dfrac{3}{8}$ gal  61. $\dfrac{3}{4}$ of his homework  62. $\dfrac{3}{2}$ mi  63. $\dfrac{11}{18}$  64. $\dfrac{7}{26}$  65. $-\dfrac{1}{12}$  66. $-\dfrac{5}{12}$  67. $\dfrac{25x+2}{55}$

68. $\dfrac{4+3b}{15}$  69. $\dfrac{7y}{36}$  70. $\dfrac{11x}{18}$  71. $\dfrac{4y+45}{9y}$  72. $-\dfrac{15}{14}$  73. $\dfrac{91}{150}$  74. $\dfrac{5}{18}$  75. $\dfrac{5}{6}$  76. $-\dfrac{9}{10}$  77. $\dfrac{19}{9}$ m  78. $\dfrac{3}{2}$ ft

79. no  80. yes  81. $\dfrac{21}{50}$  82. $\dfrac{1}{4}$ yd  83. $\dfrac{4x}{7}$  84. $\dfrac{3y}{11}$  85. $\dfrac{22}{7}$  86. $\dfrac{17}{16}$  87. $-2$  88. $-7y$  89. $\dfrac{1}{3}$  90. $\dfrac{15}{4}$

91. $-\dfrac{5}{24}$  92. $\dfrac{19}{30}$  93. $\dfrac{4}{9}$  94. $-\dfrac{3}{10}$  95. $-10$  96. $-6$  97. 15  98. 1  99. 5  100. $-4$  101. $3\dfrac{3}{4}$  102. 3

103. 1  104. $31\dfrac{1}{4}$  105. $\dfrac{11}{5}$  106. $\dfrac{5}{1}$  107. $\dfrac{35}{9}$  108. $\dfrac{3}{1}$  109. $45\dfrac{16}{21}$  110. 60  111. $32\dfrac{13}{22}$  112. $3\dfrac{19}{60}$  113. $111\dfrac{5}{18}$

114. $20\dfrac{7}{24}$  115. $1\dfrac{1}{12}$  116. $1\dfrac{4}{11}$  117. 10  118. $12\dfrac{3}{4}$  119. $5\dfrac{1}{4}$  120. $2\dfrac{29}{46}$  121. $2\dfrac{1}{3}$  122. $6\dfrac{2}{5}$  123. $6\dfrac{7}{20}$ lb

124. $44\dfrac{1}{2}$ yd  125. $7\dfrac{4}{5}$ in.  126. $7\dfrac{1}{2}$ sq. ft  127. 5 ft  128. $\dfrac{119}{80}$ or $1\dfrac{39}{80}$ sq. in.  129. $5\dfrac{1}{12}$ m  130. 203 calories  131. $13\dfrac{1}{3}$ g

132. $\dfrac{21}{20}$ or $1\dfrac{1}{20}$ mi

## Chapter 4 Test

1. $\dfrac{23}{3}$  2. $\dfrac{39}{11}$  3. $4\dfrac{3}{5}$  4. $18\dfrac{3}{4}$  5. $\dfrac{9}{35}$  6. $\dfrac{3}{5}$  7. $\dfrac{4}{3}$  8. $-\dfrac{4}{3}$  9. $\dfrac{8x}{9}$  10. $\dfrac{x-21}{7x}$  11. $y^2$  12. $\dfrac{16}{45}$  13. $\dfrac{9a+4}{10}$

14. $-\dfrac{2}{3y}$  15. $24y^2$  16. 9  17. $14\dfrac{1}{40}$  18. $\dfrac{1}{a^2}$  19. $\dfrac{64}{3}$  20. $22\dfrac{1}{2}$  21. $3\dfrac{3}{5}$  22. $\dfrac{1}{3}$  23. $\dfrac{3}{4}$  24. $\dfrac{3}{4x}$  25. $\dfrac{76}{21}$

26. $-2$  27. $-4$  28. 1  29. $\dfrac{5}{2}$  30. $\dfrac{4}{31}$  31. $\dfrac{1}{2}$ of the calories  32. $3\dfrac{3}{4}$ ft  33. $\dfrac{5}{16}$ of the sales  34. 125,000

35. 8800 sq. yd  36. $\dfrac{34}{27}$ or $1\dfrac{7}{27}$ sq. mi  37. 24 mi  38. \$90 per share  39. \$200

## Cumulative Review

1. eighty-five; Sec. 1.1, Ex. 4    2. one hundred twenty-six; Sec. 1.1, Ex. 5    3. 159; Sec. 1.2, Ex. 1    4. 14; Sec. 1.3, Ex. 3
5. 278,000; Sec. 1.4, Ex. 2    6. 15,540 thousand bytes; Sec. 1.5, Ex. 7    7. 7089 R 5; Sec. 1.6, Ex. 7    8. $4^3$; Sec. 1.7, Ex. 1
9. $6^3 \cdot 8^5$; Sec. 1.7, Ex. 4    10. 8; Sec. 1.8, Ex. 2    11. $-150$; Sec. 2.1, Ex. 1    12. $-4$; Sec. 2.2, Ex. 2    13. 3; Sec. 2.3, Ex. 9
14. 25; Sec. 2.4, Ex. 6    15. $-2$; Sec. 2.5, Ex. 7    16. $15y$; Sec. 3.1, Ex. 6    17. $-8x$; Sec. 3.1, Ex. 7    18. $-9$; Sec. 3.2, Ex. 4

19. $-1$; Sec. 3.3, Ex. 6    20. a. $7 \cdot 6 = 42$    b. $2(3 + 5) = 16$    c. $\dfrac{-45}{5} = -9$; Sec. 3.4, Ex. 6    21. $-9$; Sec. 3.5, Ex. 2

22. $\dfrac{36x}{44}$; Sec. 4.1, Ex. 11    23. $3 \cdot 3 \cdot 5$ or $3^2 \cdot 5$; Sec. 4.2, Ex. 1    24. $\dfrac{10}{33}$; Sec. 4.3, Ex. 1    25. $\dfrac{1}{8}$; Sec. 4.3, Ex. 2

# Chapter 5 Decimals

## Chapter 5 Pretest

1. $\dfrac{27}{100}$; 5.1C    2. $<$; 5.1D    3. 54.7; 5.1E    4. 60.042; 5.2A    5. 7.14; 5.3A    6. 201.6; 5.3B    7. 40.6; 5.4A    8. 0.0891; 5.4B

9. 11.69; 5.2B    10. $-17.7 - 4.8x$; 5.2C    11. 0.126; 5.3C    12. $12\pi$ in. $\approx 37.68$ in.; 5.3D    13. \$28.80; 5.4D    14. 0.778; 5.5B

15. 0.375; 5.6A    16. $<$; 5.6B    17. $-2.14$; 5.7A    18. $\dfrac{6}{7}$; 5.8A    19. 6.78; 5.8B    20. 1.5 in.; 5.8C

## Mental Math

1. tens    3. tenths

## EXERCISE SET 5.1

**1.** six and fifty-two hundredths　　**3.** sixteen and twenty-three hundredths　　**5.** three and two hundred five thousandths
**7.** one hundred sixty-seven and nine thousandths　　**9.** 6.5　　**11.** 9.08　　**13.** 5.625　　**15.** 0.0064　　**17.** 20.33　　**19.** 5.9
**21.** $\frac{3}{10}$　　**23.** $\frac{27}{100}$　　**25.** $5\frac{47}{100}$　　**27.** $\frac{6}{125}$　　**29.** $7\frac{7}{100}$　　**31.** $15\frac{401}{500}$　　**33.** $\frac{601}{2000}$　　**35.** $487\frac{8}{25}$　　**37.** <　　**39.** >　　**41.** <
**43.** =　　**45.** <　　**47.** >　　**49.** 0.6　　**51.** 0.23　　**53.** 0.594　　**55.** 98,210　　**57.** 12.3　　**59.** 17.67　　**61.** 0.5　　**63.** $0.07
**65.** $27　　**67.** $0.20　　**69.** 0.26499; 0.25786　　**71.** 40,000 people　　**73.** 2.39 hr　　**75.** 24.623 hr　　**77.** 136 mph　　**79.** 27 points
**81.** category 3　　**83.** 5766　　**85.** 71　　**87.** 243　　**89.** 226.130; Walter Ray Williams, Jr.
**91.** 226.130; 225.490; 225.370; 222.980; 222.830; 222.008; 219.702; 218.158; 215.432　　**93.** Answers may vary.

## CALCULATOR EXPLORATIONS

**1.** 328.742　　**3.** 5.2414　　**5.** 865.392

## MENTAL MATH

**1.** 0.5　　**3.** 1.26　　**5.** 8.9　　**7.** 0.6

## EXERCISE SET 5.2

**1.** 3.5　　**3.** 6.83　　**5.** 27.0578　　**7.** 56.432　　**9.** −8.57　　**11.** 11.16　　**13.** 6.5　　**15.** 15.3　　**17.** 598.23　　**19.** 16.3　　**21.** −6.32
**23.** −6.4　　**25.** 3.1　　**27.** 450.738　　**29.** −9.9　　**31.** 25.67　　**33.** −5.62　　**35.** 776.89　　**37.** 11,983.32　　**39.** 465.56　　**41.** −549.8
**43.** 861.6　　**45.** 115.123　　**47.** 876.6　　**49.** 0.088　　**51.** 465.902　　**53.** 3.81　　**55.** 3.39　　**57.** 1.61　　**59.** no　　**61.** yes　　**63.** no
**65.** $6.9x + 6.9$　　**67.** $-10.97 + 3.47y$　　**69.** $454.71　　**71.** $0.06　　**73.** $7.52　　**75.** 197.8 lb　　**77.** 15.81 in.　　**79.** 763.035 mph
**81.** $322.7 million　　**83.** 240.8 in.　　**85.** 67.44 ft　　**87.** 29.614 mph　　**89.** 715.05 hr　　**91.** Switzerland　　**93.** 8.1 lb

**95.**

| Country | Pounds of Chocolate Per Person |
| --- | --- |
| Switzerland | 22.0 |
| Norway | 16.0 |
| Germany | 15.8 |
| United Kingdom | 14.5 |
| Belgium | 13.9 |

**97.** $\frac{1}{125}$　　**99.** $\frac{5}{12}$　　**101.** Answers may vary.　　**103.** $-109.544x + 15.604$

## EXERCISE SET 5.3

**1.** 0.12　　**3.** 0.6　　**5.** −17.595　　**7.** 39.273　　**9.** 65　　**11.** 0.65　　**13.** −7093　　**15.** 0.0983　　**17.** 43.274　　**19.** 8.23854
**21.** 14,790　　**23.** −9.3762　　**25.** 1.12746　　**27.** −0.14444　　**29.** 5,500,000,000 chocolate bars　　**31.** 36,400,000 households
**33.** 1,600,000 hr　　**35.** −0.6　　**37.** 17.1　　**39.** 114　　**41.** no　　**43.** no　　**45.** yes　　**47.** $8\pi$ m ≈ 24.12 m　　**49.** $10\pi$ cm ≈ 31.4 cm
**51.** $18.2\pi$ yd ≈ 57.148 yd　　**53.** $250\pi$ ft ≈ 785 ft　　**55. a.** 62.8 m; 125.6 m　　**b.** yes　　**57.** 24.8 g　　**59.** $2700　　**61.** 64.9605 in.
**63.** $555.20　　**65.** 72,900 yen　　**67.** 518 Canadian dollars　　**69.** 26　　**71.** 36　　**73.** 8　　**75.** 9　　**77.** 3,831,600 mi
**79.** Answers may vary.　　**81.** Answers may vary.

## EXERCISE SET 5.4

**1.** 0.094　　**3.** −300　　**5.** 5.8　　**7.** −6.6　　**9.** 200　　**11.** 23.87　　**13.** 110　　**15.** 0.54982　　**17.** −0.0129　　**19.** 8.7　　**21.** 0.413
**23.** −7　　**25.** 4.8　　**27.** 2100　　**29.** 30　　**31.** 7000　　**33.** −0.69　　**35.** 0.024　　**37.** 65　　**39.** 0.0003　　**41.** 120,000　　**43.** 5.65
**45.** 7.0625　　**47.** 0.4　　**49.** yes　　**51.** no　　**53.** yes　　**55.** 24 mo　　**57.** $1245.69　　**59.** 202.1 lb　　**61.** 5.1 m　　**63.** $0.596/lb
**65.** 128.6 mph　　**67.** 22.1 points/game　　**69.** 65.2–82.6 knots　　**71.** 345.22　　**73.** 1001.0　　**75.** 20　　**77.** 18　　**79.** 85　　**81.** 8.6 ft
**83.** Answers may vary.

## INTEGRATED REVIEW

**1.** 2.57　　**2.** 4.05　　**3.** 8.9　　**4.** 3.5　　**5.** 0.16　　**6.** 0.24　　**7.** 0.27　　**8.** 0.52　　**9.** −4.8　　**10.** 6.09　　**11.** 75.56　　**12.** 289.12
**13.** −24.974　　**14.** −43.875　　**15.** −8.6　　**16.** 5.4　　**17.** −280　　**18.** 1600　　**19.** 224.938　　**20.** 145.079　　**21.** 0.56　　**22.** −0.63
**23.** 27.6092　　**24.** 145.6312　　**25.** 5.4　　**26.** −17.74　　**27.** −414.44　　**28.** −1295.03　　**29.** −34　　**30.** −28　　**31.** 116.81
**32.** 18.79　　**33.** yes　　**34.** 7640.25　　**35.** 26.3

## EXERCISE SET 5.5

**1.** 2.8　　**3.** 2898.66　　**5.** 4.2　　**7.** 149.87　　**9.** 22.89　　**11.** 36 in.　　**13.** 39 ft　　**15.** 43.96 m　　**17.** 52 gal　　**19.** $12,000
**21.** 53 mi　　**23.** $2600 million　　**25.** 132,300 people　　**27.** 0.16　　**29.** −3　　**31.** 0.28　　**33.** 2.3　　**35.** 5.29　　**37.** 7.6　　**39.** 0.2025
**41.** −1.29　　**43.** 5.76　　**45.** 5.7　　**47.** 3.6　　**49.** yes　　**51.** no　　**53.** $\frac{5}{16}$　　**55.** $\frac{3}{4}$　　**57.** $\frac{1}{12}$　　**59.** 43.388569　　**61.** Answers may vary.

## EXERCISE SET 5.6

**1.** 0.2　　**3.** 0.5　　**5.** 0.75　　**7.** 0.08　　**9.** 0.375　　**11.** $0.91\overline{6}$　　**13.** 0.425　　**15.** 0.45　　**17.** $0.\overline{3}$　　**19.** 0.4375　　**21.** $0.\overline{2}$　　**23.** $1.\overline{6}$
**25.** 0.33　　**27.** 0.44　　**29.** 0.2　　**31.** 1.7　　**33.** 0.194　　**35.** 0.44　　**37.** 0.8　　**39.** <　　**41.** >　　**43.** <　　**45.** <　　**47.** <
**49.** >　　**51.** <　　**53.** <　　**55.** 0.32, 0.34, 0.35　　**57.** 0.49, 0.491, 0.498　　**59.** 0.73, $\frac{3}{4}$, 0.78　　**61.** 0.412, 0.453, $\frac{4}{7}$　　**63.** 5.23, $\frac{42}{8}$, 5.34

**65.** $\dfrac{17}{8}, 2.37, \dfrac{12}{5}$   **67.** 25.65 sq. in.   **69.** 9.36 sq. cm   **71.** 0.248 sq. yd   **73.** 8   **75.** 72   **77.** $\dfrac{1}{81}$   **79.** $\dfrac{9}{25}$   **81.** $\dfrac{5}{2}$
**83.** 0.221   **85.** 8300   **87.** 0.625   **89.** Answers may vary.   **91.** $-47.25$   **93.** 3.37   **95.** $-0.45$

## EXERCISE SET 5.7

**1.** 5.9   **3.** 0.43   **5.** 0.45   **7.** 4.2   **9.** $-4$   **11.** 1.8   **13.** 10   **15.** 7.6   **17.** 60   **19.** $-0.07$   **21.** 0.0148   **23.** $-8.13$
**25.** $-1$   **27.** 7   **29.** 7   **31.** $6x - 16$   **33.** $3x - 5$   **35.** $-2y + 6.8$   **37.** Answers may vary.   **39.** 7.683   **41.** 4.683

## CALCULATOR EXPLORATIONS

**1.** 32   **3.** 3.873   **5.** 9.849

## EXERCISE SET 5.8

**1.** 2   **3.** 25   **5.** $\dfrac{1}{9}$   **7.** $\dfrac{12}{8} = \dfrac{3}{2}$   **9.** 16   **11.** $\dfrac{3}{2}$   **13.** 1.732   **15.** 3.873   **17.** 3.742   **19.** 6.856   **21.** 2.828   **23.** 5.099
**25.** 8.426   **27.** 2.646   **29.** 13 in.   **31.** 6.633 cm   **33.** 5   **35.** 8   **37.** 17.205   **39.** 16.125   **41.** 12   **43.** 44.822
**45.** 42.426   **47.** 1.732   **49.** 141.42 yd   **51.** 25.0 ft   **53.** 340 ft   **55.** $\dfrac{5}{6}$   **57.** $\dfrac{2}{5}$   **59.** $\dfrac{5}{12}$   **61.** 6, 7   **63.** 10, 11
**65.** Answers may vary.

## CHAPTER 5 REVIEW

**1.** tenths   **2.** hundred-thousandths   **3.** twenty-three and forty-five hundredths   **4.** three hundred forty-five hundred-thousandths
**5.** one hundred nine and twenty-three hundredths   **6.** two hundred and thirty-two millionths   **7.** 2.15   **8.** 503.102   **9.** 16,025.0014
**10.** $\dfrac{4}{25}$   **11.** $12\dfrac{23}{1000}$   **12.** $1\dfrac{9}{2000}$   **13.** $\dfrac{231}{100,000}$   **14.** $25\dfrac{1}{4}$   **15.** $>$   **16.** $=$   **17.** $>$   **18.** $<$   **19.** 0.6   **20.** 0.94
**21.** 42.90   **22.** 16.349   **23.** 13,500 people   **24.** $10\dfrac{3}{4}$ tsp   **25.** 9.5   **26.** 5.1   **27.** $-7.28$   **28.** $-12.04$   **29.** 320.312
**30.** 148.74236   **31.** 1.7   **32.** 2.49   **33.** $-1324.5$   **34.** $-10.136$   **35.** 65.02   **36.** 199.99802   **37.** 10,923.55 points   **38.** $-5.7$
**39.** $4.2x + 12.8$   **40.** $9.89y - 4.17$   **41.** 72   **42.** 9345   **43.** $-78.246$   **44.** 73,246.446   **45.** $14\pi$ m $\approx 43.96$ m   **46.** 63.8 mi
**47.** 70   **48.** $-0.21$   **49.** $-4900$   **50.** 23.904   **51.** 8.059   **52.** 158.25   **53.** 7.3 m   **54.** 45 mo   **55.** 18.2   **56.** 50
**57.** 99.05   **58.** 54.1   **59.** 35.5782   **60.** 0.3526   **61.** 32.7   **62.** 30.4   **63.** $1070 million   **64.** 8932 sq. ft
**65.** Yes, $5 is enough.   **66.** $-11.94$   **67.** 3.89   **68.** 0.1024   **69.** 3.6   **70.** 0.8   **71.** 0.923   **72.** 0.429   **73.** 0.217
**74.** 0.113   **75.** 51.057   **76.** $=$   **77.** $<$   **78.** $<$   **79.** $<$   **80.** 0.832, 0.837, 0.839   **81.** $0.42, \dfrac{3}{7}, 0.43$   **82.** $\dfrac{19}{12}, 1.63, \dfrac{18}{11}$
**83.** $\dfrac{3}{4}, \dfrac{6}{7}, \dfrac{8}{9}$   **84.** 6.9 sq. ft   **85.** 5.46 sq. in.   **86.** 0.3   **87.** 47.19   **88.** 8.6   **89.** 80   **90.** 0.48   **91.** 34   **92.** 0.42
**93.** 0.28   **94.** 20   **95.** 1   **96.** 8   **97.** 12   **98.** 6   **99.** 1   **100.** $\dfrac{2}{5}$   **101.** $\dfrac{1}{10}$   **102.** 13   **103.** 29   **104.** 10.7
**105.** 93   **106.** 86.6   **107.** 28.28 cm   **108.** 88.2 ft

## CHAPTER 5 TEST

**1.** forty-five and ninety-two thousandths   **2.** 3000.059   **3.** 17.595   **4.** $-51.20$   **5.** $-20.42$   **6.** 40.902   **7.** 0.037   **8.** 34.9
**9.** 0.862   **10.** $<$   **11.** $<$   **12.** $\dfrac{69}{200}$   **13.** $24\dfrac{73}{100}$   **14.** 0.5   **15.** 0.941   **16.** 1.93   **17.** $-6.2$   **18.** $0.5x - 13.4$   **19.** 7
**20.** 12.530   **21.** $\dfrac{8}{10} = \dfrac{4}{5}$   **22.** $-3$   **23.** 3.7   **24.** 5.66 cm   **25.** 2.31 sq. mi   **26.** 198.08 oz   **27.** $18\pi$ mi $\approx 56.52$ mi
**28.** 86 high-density disks   **29.** 54 mi   **30.** decrease; 1.9 pounds per person

## CUMULATIVE REVIEW

**1.** 184,046; Sec. 1.2, Ex. 2   **2.** 2300; Sec. 1.4, Ex. 4   **3.** 401 R 2; Sec. 1.6, Ex. 8   **4.** 2; Sec. 1.8, Ex. 3
**5. a.** $-11$   **b.** 2   **c.** 0; Sec. 2.1, Ex. 5   **6.** $-23$; Sec. 2.2, Ex. 3   **7.** 50; Sec. 2.5, Ex. 3   **8.** $-64$; Sec. 2.5, Ex. 4   **9.** $-16$; Sec. 2.5, Ex. 5
**10. a.** $5x$   **b.** $-6y$   **c.** $8x^2 - 2$; Sec. 3.1, Ex. 1   **11.** $-4$; Sec. 3.2, Ex. 7   **12.** $-3$; Sec. 3.3, Ex. 1   **13.** $-4$; Sec. 3.4, Ex. 4
**14.** 4066 votes; Sec. 3.5, Ex. 3   **15.** numerator: 3; denominator: 7; Sec. 4.1, Ex. 1   **16.** numerator: $13x$; denominator: 5; Sec. 4.1, Ex. 2
**17.** $\dfrac{3}{5}$; Sec. 4.2, Ex. 4   **18.** $-\dfrac{1}{8}$; Sec. 4.3, Ex. 5   **19.** $\dfrac{5}{7}$; Sec. 4.4, Ex. 1   **20.** 2; Sec. 4.4, Ex. 3   **21.** $\dfrac{2}{3}$; Sec. 4.5, Ex. 1
**22.** $\dfrac{x}{6}$; Sec. 4.6, Ex. 1   **23.** 15; Sec. 4.7, Ex. 2   **24.** 829.6561; Sec. 5.2, Ex. 2   **25.** 18.408; Sec. 5.3, Ex. 1

# Chapter 6 Ratio and Proportion

## Chapter 6 Pretest

1. $\frac{15}{19}$; 6.1A   2. $\frac{3\frac{4}{5}}{9\frac{1}{8}}$; 6.1A   3. $\frac{51}{79}$; 6.1B   4. $\frac{3}{2}$; 6.1B   5. $\frac{\$2}{1\,\text{lb}}$; 6.2A   6. 19.5 mi/gal; 6.2B   7. \$1.09 per box; 6.2C

8. $\frac{18}{6} = \frac{15}{5}$; 6.3A   9. $\frac{49}{7} = \frac{35}{5}$; 6.3A   10. no; 6.3B   11. yes; 6.3B   12. 4; 6.3C   13. 2; 6.3C   14. 48; 6.3C   15. 10; 6.3C

16. $107\frac{1}{2}$ miles; 6.4A   17. $\frac{1}{2}$; 6.5A   18. 6; 6.5B   19. 5.5; 6.5B   20. 19.8 ft; 6.5C

## Exercise Set 6.1

1. $\frac{11}{14}$   3. $\frac{23}{10}$   5. $\frac{151}{201}$   7. $\frac{2.8}{7.6}$   9. $\frac{5}{7\frac{1}{2}}$   11. $\frac{3\frac{3}{4}}{1\frac{1}{3}}$   13. $\frac{2}{3}$   15. $\frac{77}{100}$   17. $\frac{463}{821}$   19. $\frac{3}{4}$   21. $\frac{5}{12}$   23. $\frac{8}{25}$   25. $\frac{12}{7}$

27. $\frac{16}{23}$   29. $\frac{3}{50}$   31. $\frac{5}{3}$   33. $\frac{17}{40}$   35. $\frac{4}{9}$   37. $\frac{5}{4}$   39. $\frac{3}{1}$   41. $\frac{15}{1}$   43. $\frac{103}{5}$   45. $\frac{5}{41}$   47. 2.3   49. 0.15   51. $\frac{3}{7}$

53. Answers may vary.   55. No, the shipment should not be refused.   57. Answers may vary.

## Exercise Set 6.2

1. $\frac{1\,\text{shrub}}{13\,\text{ft}}$   3. $\frac{3\,\text{returns}}{20\,\text{sales}}$   5. $\frac{2\,\text{phone lines}}{9\,\text{employees}}$   7. $\frac{9\,\text{gal}}{2\,\text{acres}}$   9. $\frac{3\,\text{flight attendants}}{100\,\text{passengers}}$   11. $\frac{71\,\text{cal}}{2\,\text{fl. oz}}$

13. 75 riders/car   15. 110 cal/oz   17. 6 diapers/baby   19. \$50,000/yr   21. $\approx$6.67 km/min   23. 1,225,000 voters/senator

25. 300 good/defective   27. 0.48 tons/acre   29. \$50,000/species   31. \$5,134,125/library

33. **a.** 31.25 computer boards/hr   **b.** $\approx$33.3 computer boards/hr   **c.** Lamont   35. \$11.50/compact disk   37. \$0.17/banana

39. 8 oz: \$0.149/oz; 12 oz: \$0.133/oz; 12-oz size   41. 16 oz: \$0.106/oz; 6 oz: \$0.115/oz; 16-oz size   43. 12 oz: \$0.191/oz; 8 oz: \$0.186/oz; 8-oz size

45. 100: \$0.006/napkin; 180: \$0.005/napkin; 180 napkins   47. 10.2   49. 4.44   51. 1.9

53. miles driven: 257, 352, 347; miles per gallon: 19.2, 22.3, 21.6   55. 1.5 steps/ft   57. 24 students/teacher   59. Answers may vary.

## Mental Math

1. true   3. false   5. true

## Exercise Set 6.3

1. $\frac{10\,\text{diamonds}}{6\,\text{opals}} = \frac{5\,\text{diamonds}}{3\,\text{opals}}$   3. $\frac{3\,\text{printers}}{12\,\text{computers}} = \frac{1\,\text{printer}}{4\,\text{computers}}$   5. $\frac{6\,\text{eagles}}{58\,\text{sparrows}} = \frac{3\,\text{eagles}}{29\,\text{sparrows}}$   7. $\frac{2\frac{1}{4}\,\text{c flour}}{24\,\text{cookies}} = \frac{6\frac{3}{4}\,\text{c flour}}{72\,\text{cookies}}$

9. $\frac{22\,\text{vanilla wafers}}{1\,\text{c cookie crumbs}} = \frac{55\,\text{vanilla wafers}}{2.5\,\text{c cookie crumbs}}$   11. true   13. false   15. true   17. true   19. false   21. true   23. true

25. false   27. 3   29. 5   31. 4   33. 3.2   35. 19.2   37. $\frac{3}{4}$   39. 3   41. 12   43. 0.0025   45. 25   47. 14.9

49. 0.07   51. 3.3   53. 3.163   55. <   57. >   59. <   61. >   63. 0   65. 1400   67. 252.5

69. Answers may vary.

## Integrated Review

1. $\frac{9}{10}$   2. $\frac{9}{25}$   3. $\frac{43}{50}$   4. $\frac{8}{23}$   5. $\frac{173}{139}$   6. $\frac{6}{7}$   7. $\frac{7}{26}$   8. $\frac{20}{33}$   9. $\frac{2}{3}$   10. $\frac{1}{8}$   11. $\frac{60}{47}$   12. $\frac{779}{118}$

13. $\frac{1\,\text{office}}{4\,\text{graduate assistants}}$   14. $\frac{2\,\text{lights}}{5\,\text{ft}}$   15. $\frac{2\,\text{Senators}}{1\,\text{state}}$   16. $\frac{1\,\text{teacher}}{28\,\text{students}}$   17. 55 mph   18. 140 ft/sec

19. 21 employees/fax line   20. 17 phone calls/teenager   21. 8 lb: \$0.27/lb; 18 lb: \$0.28/lb; 8-lb size

22. 100: \$0.020/plate; 500: \$0.018/plate; 500 paper plates   23. 3 packs: \$0.80/pack; 8 packs: \$0.75/pack; 8 packs

24. 4: \$0.92/battery; 10: \$0.99/battery; 4 batteries

## Exercise Set 6.4

1. 12 passes   3. 165 min   5. 180 students   7. 23 ft   9. 270 sq. ft   11. 56 mi   13. 450 km   15. 24 oz   17. 16 bags

19. \$162,000   21. 15 hits   23. 27 people   25. 86 weeks   27. 6 people   29. 112 ft; 11-in. difference

31. $2\frac{2}{3}$ lb   33. 434 emergency room visits   35. 427.5 cal   37. 2.4 c   39. $3 \cdot 5$   41. $2^2 \cdot 5$   43. $2^3 \cdot 5^2$   45. $2^5$   47. $4\frac{2}{3}$ ft

49. Answers may vary.

## Exercise Set 6.5

1. $\frac{2}{1}$   3. $\frac{3}{2}$   5. 4.5   7. 6   9. 5   11. 13.5   13. 17.5   15. 8   17. 21.25   19. 50 ft   21. $x = 4.4$ ft; $y = 5.6$ ft

23. 14.4 ft   25. 0.25   27. 0.65   29. 0.9   31. 8.4   33. Answers may vary.

## CHAPTER 6 REVIEW

**1.** $\dfrac{5}{4}$ **2.** $\dfrac{11}{13}$ **3.** $\dfrac{9}{40}$ **4.** $\dfrac{14}{5}$ **5.** $\dfrac{4}{15}$ **6.** $\dfrac{1}{2}$ **7.** $\dfrac{1}{2}$ **8.** $\dfrac{7}{150}$ **9.** $\dfrac{1 \text{ stillborn birth}}{125 \text{ live births}}$ **10.** $\dfrac{3 \text{ professors}}{10 \text{ assistants}}$ **11.** $\dfrac{5 \text{ pages}}{2 \text{ min}}$

**12.** $\dfrac{4 \text{ computers}}{3 \text{ hr}}$ **13.** 52 mph **14.** 15 ft/sec **15.** $0.31/pear **16.** $1.74/diskette **17.** 65 km/hr **18.** $1\dfrac{1}{3}$ gal/acre

**19.** $36.80/course **20.** 13 bushels/tree **21.** 8 oz: $0.124 per oz; 12 oz: $0.141 per oz; 8-oz size

**22.** 18 oz: $0.828 per oz; 28 oz: $0.854 per oz; 18-oz size **23.** 16 oz: $0.037 per oz; $\dfrac{1}{2}$ gal: $0.026 per oz; 1 gal: $0.018 per oz; 1-gal size

**24.** 12 oz: $0.0492 per oz; 16 oz: $0.0494 per oz; 32 oz: $0.0372 per oz; 32-oz size **25.** $\dfrac{20 \text{ men}}{14 \text{ women}} = \dfrac{10 \text{ men}}{7 \text{ women}}$ **26.** $\dfrac{50 \text{ tries}}{4 \text{ successes}} = \dfrac{25 \text{ tries}}{2 \text{ successes}}$

**27.** $\dfrac{16 \text{ sandwiches}}{8 \text{ players}} = \dfrac{2 \text{ sandwiches}}{1 \text{ player}}$ **28.** $\dfrac{12 \text{ tires}}{3 \text{ cars}} = \dfrac{4 \text{ tires}}{1 \text{ car}}$ **29.** false **30.** true **31.** false **32.** true **33.** 5 **34.** 15

**35.** 32.5 **36.** 5.625 **37.** 32 **38.** 13.5 **39.** 60 **40.** $7\dfrac{1}{5}$ **41.** 0.63 **42.** 30.9 **43.** 14 **44.** 35 **45.** 8 bags

**46.** 16 bags **47.** no **48.** 79 gal **49.** $54,600 **50.** $1023.50 **51.** $40\dfrac{1}{2}$ ft **52.** $8\dfrac{1}{4}$ in. **53.** 37.5 **54.** $13\dfrac{1}{3}$ **55.** 17.4

**56.** $6\dfrac{1}{2}$ **57.** 33 ft **58.** $x = \dfrac{5}{6}$ in., $y = 2\dfrac{1}{6}$ in.

## CHAPTER 6 TEST

**1.** $\dfrac{9}{13}$ **2.** $\dfrac{15}{2}$ **3.** $\dfrac{7 \text{ men}}{1 \text{ woman}}$ **4.** $\dfrac{3 \text{ in.}}{10 \text{ days}}$ **5.** 81.25 km/hr **6.** $\dfrac{2}{3}$ in./hr **7.** 28 students/teacher

**8.** 8 oz: $0.149 per oz; 12 oz: $0.158 per oz; 8-oz size **9.** 16 oz: $0.093 per oz; 24 oz: $0.100 per oz; 16-oz size **10.** true **11.** false **12.** 5

**13.** $4\dfrac{4}{11}$ **14.** $\dfrac{7}{3}$ **15.** 8 **16.** $49\dfrac{1}{2}$ ft **17.** $3\dfrac{3}{4}$ hr **18.** $53\dfrac{1}{3}$ g **19.** $144\dfrac{2}{3}$ cartons **20.** 14,880 adults **21.** 7.5 **22.** 69 ft

## CUMULATIVE REVIEW

**1.** 128 R 6; Sec. 1.6, Ex. 9 **2. a.** $7 + x$ **b.** $15 - x$ **c.** $2x$ **d.** $x \div 5$ or $\dfrac{x}{5}$ **e.** $x - 2$; Sec. 1.8, Ex. 6

**3. a.** 4 **b.** $-5$ **c.** $-6$; Sec. 2.1, Ex. 6 **4.** $-15$; Sec. 2.3, Ex. 5 **5.** 0; Sec. 2.3, Ex. 10 **6.** $8x - 5$; Sec. 3.1, Ex. 3

**7.** $6x - 7y + 6$; Sec. 3.1, Ex. 5 **8.** yes; Sec. 3.2, Ex. 1 **9.** $-4$; Sec. 3.3, Ex. 2 **10.** $-1$; Sec. 3.4, Ex. 2 **11.** $\dfrac{6}{15}$; Sec. 4.1, Ex. 10

**12.** $2^4 \cdot 5$; Sec. 4.2, Ex. 2 **13.** $\dfrac{6}{5}$; Sec. 4.3, Ex. 7 **14.** $\dfrac{7}{9}$; Sec. 4.4, Ex. 4 **15.** $\dfrac{1}{4}$; Sec. 4.4, Ex. 5 **16.** $\dfrac{6 - x}{3}$; Sec. 4.5, Ex. 4

**17.** $\dfrac{2y - 20}{3}$; Sec. 4.6, Ex. 4 **18.** 21; Sec. 4.7, Ex. 1 **19.** $\dfrac{35}{12}$ or $2\dfrac{11}{12}$; Sec. 4.8, Ex. 5 **20.** $\dfrac{43}{100}$; Sec. 5.1, Ex. 6 **21.** 4.68; Sec. 5.4, Ex. 1

**22.** $-3.7$; Sec. 5.5, Ex. 5 **23.** $\dfrac{4}{9}, \dfrac{9}{20}, 0.456$; Sec. 5.6, Ex. 6 **24.** 9.5; Sec. 5.7, Ex. 1 **25.** $\dfrac{7}{6}$; Sec. 6.3, Ex. 7

# Chapter 7 PERCENT

## CHAPTER 7 PRETEST

**1.** 12%; 7.1A **2.** 0.57; 7.1B **3.** 275%; 7.1C **4.** $\dfrac{3}{40}$; 7.1D **5.** 15%; 7.1E **6.** $18\% \cdot 50 = x$; 7.2A **7.** $40\% \cdot x = 89$; 7.2A

**8.** $\dfrac{82}{b} = \dfrac{90}{100}$; 7.3A **9.** $\dfrac{48}{112} = \dfrac{p}{100}$; 7.3A **10.** 20%; 7.2B, 7.3B **11.** 3.3; 7.2B, 7.3B **12.** 200; 7.2B, 7.3B **13.** 3 light bulbs; 7.4A
**14.** decrease of 84 students; current enrollment: 4116 students; 7.4B **15.** 7%; 7.5A **16.** $22,400; 7.5B
**17.** discount: $78, sales price: $572; 7.5C **18.** $144; 7.6A **19.** $7147.50; 7.6B **20.** $37.33; 7.6C

## MENTAL MATH

**1.** 13% **3.** 87% **5.** 1%

## EXERCISE SET 7.1

**1.** 81% **3.** 9% **5.** chocolate chip: 52% **7.** 12% **9.** 0.48 **11.** 0.06 **13.** 1 **15.** 0.613 **17.** 0.028 **19.** 0.006
**21.** 3 **23.** 0.3258 **25.** 0.737 **27.** 0.25 **29.** 0.802 **31.** 0.462 **33.** 310% **35.** 2900% **37.** 0.3% **39.** 22%

**41.** 5.6% **43.** 33.28% **45.** 300% **47.** 70% **49.** 10% **51.** 75.3% **53.** 38% **55.** $\dfrac{1}{25}$ **57.** $\dfrac{9}{200}$ **59.** $1\dfrac{3}{4}$ **61.** $\dfrac{73}{100}$

**63.** $\dfrac{1}{8}$ **65.** $\dfrac{1}{16}$ **67.** $\dfrac{31}{300}$ **69.** $\dfrac{179}{800}$ **71.** 75% **73.** 70% **75.** 40% **77.** 59% **79.** 34% **81.** $37\dfrac{1}{2}\%$ **83.** $31\dfrac{1}{4}\%$

**85.** $66\dfrac{2}{3}\%$ **87.** 250% **89.** 190% **91.** 63.64% **93.** 26.67% **95.** 14.29% **97.** 91.67% **99.** $0.35, \dfrac{7}{20}$; 20%, 0.2; 50%, $\dfrac{1}{2}$; 0.7, $\dfrac{7}{10}$;

37.5%, 0.375 **101.** 0.4, $\dfrac{2}{5}$; $23\dfrac{1}{2}\%, \dfrac{47}{200}$; 80%, 0.8; $0.333\overline{3}, \dfrac{1}{3}$; 87.5%, 0.875; 0.075, $\dfrac{3}{40}$ **103.** $\dfrac{47}{250}$ **105.** 8.5% **107.** 0.674 **109.** 30%

**111.** 15    **113.** −10    **115.** −12    **117.** 1.155; 115.5%    **119.** greater    **121.** Answers may vary.
**123.** Database administrators, computer support specialists, and other computer scientists    **125.** 0.85    **127.** Answers may vary.

## Mental Math

**1.** percent: 42; base: 50; amount: 21    **3.** percent: 125; base: 86; amount: 107.5

## Exercise Set 7.2

**1.** $15\% \cdot 72 = x$    **3.** $30\% \cdot x = 80$    **5.** $x \cdot 90 = 20$    **7.** $1.9 = 40\% \cdot x$    **9.** $x = 9\% \cdot 43$    **11.** 3.5    **13.** 7.28    **15.** 600
**17.** 10    **19.** 110%    **21.** 32%    **23.** 1    **25.** 45    **27.** 500    **29.** 5.16%    **31.** 25.2    **33.** 45%    **35.** 35    **37.** $n = 30$
**39.** $n = 3\frac{7}{11}$    **41.** $\frac{17}{12} = \frac{n}{20}$    **43.** $\frac{8}{9} = \frac{14}{n}$    **45.** 686.625    **47.** 12,285

## Mental Math

**1.** amount: 12.6; base: 42; percent 30    **3.** amount: 102; base: 510; percent: 20

## Exercise Set 7.3

**1.** $\frac{a}{65} = \frac{32}{100}$    **3.** $\frac{75}{b} = \frac{40}{100}$    **5.** $\frac{70}{200} = \frac{p}{100}$    **7.** $\frac{2.3}{b} = \frac{58}{100}$    **9.** $\frac{a}{130} = \frac{19}{100}$    **11.** 5.5    **13.** 18.9    **15.** 400    **17.** 10
**19.** 125%    **21.** 28%    **23.** 29    **25.** 1.92    **27.** 1000    **29.** 210%    **31.** 55.18    **33.** 45%    **35.** 85    **37.** $\frac{7}{8}$    **39.** $3\frac{2}{15}$
**41.** 0.7    **43.** 2.19    **45.** 12,011.2    **47.** 7270.6

## Integrated Review

**1.** 12%    **2.** 68%    **3.** 25%    **4.** 50%    **5.** 520%    **6.** 780%    **7.** 6%    **8.** 44%    **9.** 250%    **10.** 325%    **11.** 3%    **12.** 5%
**13.** 0.65    **14.** 0.31    **15.** 0.08    **16.** 0.07    **17.** 1.42    **18.** 5.38    **19.** 0.029    **20.** 0.066    **21.** $\frac{3}{100}$    **22.** $\frac{2}{25}$    **23.** $\frac{21}{400}$
**24.** $\frac{51}{400}$    **25.** $\frac{19}{50}$    **26.** $\frac{9}{20}$    **27.** $\frac{37}{300}$    **28.** $\frac{1}{6}$    **29.** 8.4    **30.** 100    **31.** 250    **32.** 120%    **33.** 28%    **34.** 76    **35.** 11
**36.** 130%    **37.** 86%    **38.** 37.8    **39.** 150    **40.** 62

## Exercise Set 7.4

**1.** 1600 bolts    **3.** 30 hr    **5.** 15%    **7.** 295 components    **9.** 15.3%    **11.** 100,102    **13.** 29.2%    **15.** 3,974,600; 4,729,774
**17.** $136    **19.** $867.87; $20,153.87    **21.** 8.844 million; 35.644 million    **23.** 10; 25%    **25.** 102; 120%    **27.** 2; 25%    **29.** 120; 75%
**31.** 44%    **33.** 21.5%    **35.** 1.3%    **37.** 86.2%    **39.** 300%    **41.** 81.3%    **43.** 4.56    **45.** 11.18    **47.** 58.54    **49.** 8.2%
**51.** The increased number is double the original number.    **53.** Answers may vary.

## Exercise Set 7.5

**1.** $7.50    **3.** $858.93    **5.** 9%    **7.** $238.40    **9.** $1917    **11.** $11,500    **13.** $112.35    **15.** 6%    **17.** $49,474.24    **19.** 14%
**21.** $1888.50    **23.** 3%    **25.** $6.80; $61.20    **27.** $48.25; $48.25    **29.** $75.25; $139.75    **31.** $3255.00; $18,445.00    **33.** $45; $255
**35.** 1200    **37.** 132    **39.** 16    **41.** $26,838.45    **43.** A discount of 60% is better.

## Calculator Explorations

**1.** 1.56051    **3.** 8.06231    **5.** $634.50

## Exercise Set 7.6

**1.** $32    **3.** $73.60    **5.** $750    **7.** $33.75    **9.** $700    **11.** $78,125    **13.** $5562.50    **15.** $12,580    **17.** $46,815.40
**19.** $2327.15    **21.** $58,163.60    **23.** $240.75    **25.** $938.66    **27.** $971.90    **29.** $260.31    **31.** $637.26    **33.** −29    **35.** 50
**37.** −1    **39.** Answers may vary.    **41.** Answers may vary.

## Chapter 7 Review

**1.** 37%    **2.** 77%    **3.** 0.83    **4.** 0.75    **5.** 0.735    **6.** 0.015    **7.** 1.25    **8.** 1.45    **9.** 0.005    **10.** 0.007    **11.** 2.00 or 2
**12.** 4.00 or 4    **13.** 0.2625    **14.** 0.8534    **15.** 260%    **16.** 5.5%    **17.** 35%    **18.** 102%    **19.** 72.5%    **20.** 25%    **21.** 7.6%
**22.** 8.5%    **23.** 75%    **24.** 65%    **25.** 400%    **26.** 900%    **27.** $\frac{1}{100}$    **28.** $\frac{1}{10}$    **29.** $\frac{1}{4}$    **30.** $\frac{17}{200}$    **31.** $\frac{51}{500}$    **32.** $\frac{1}{6}$    **33.** $\frac{1}{3}$
**34.** $1\frac{1}{10}$    **35.** 20%    **36.** 70%    **37.** $83\frac{1}{3}\%$    **38.** 62.5%    **39.** $166\frac{2}{3}\%$    **40.** 125%    **41.** 60%    **42.** 6.25%    **43.** $\frac{9}{10}, \frac{1}{10}$
**44.** $\frac{24}{25}$    **45.** 84%    **46.** 1.5    **47.** 100,000    **48.** 8000    **49.** 23%    **50.** 114.5    **51.** 3000    **52.** 150%    **53.** 418    **54.** 300
**55.** 64.8    **56.** 180%    **57.** 110%    **58.** 165    **59.** 66%    **60.** 16%    **61.** 106.25%    **62.** 20.9%    **63.** $206,400    **64.** $13.23
**65.** $263.75    **66.** $1.15    **67.** $5000    **68.** $300.38    **69.** discount: $900; sale price: $2100    **70.** discount: $9; sale price: $81
**71.** $120    **72.** $1320    **73.** $30,104.64    **74.** $17,506.56    **75.** $80.61    **76.** $32,830.10

## CHAPTER 7 TEST

**1.** 0.85 **2.** 5 **3.** 0.006 **4.** 5.6% **5.** 610% **6.** 35% **7.** $1\frac{1}{5}$ **8.** $\frac{77}{200}$ **9.** $\frac{1}{500}$ **10.** 55% **11.** 37.5% **12.** 175%

**13.** 20% **14.** $\frac{16}{25}$ **15.** 33.6 **16.** 1250 **17.** 75% **18.** 38.4 lb **19.** $56,750 **20.** $358.43 **21.** 5%

**22.** discount: $18; sale price: $102 **23.** $395 **24.** 1% **25.** $647.50 **26.** $2005.64 **27.** $427 **28.** 16.9%

## CUMULATIVE REVIEW

**1.** 20,296; Sec. 1.5, Ex. 4 **2.** −10; Sec. 2.3, Ex. 8 **3.** 3; Sec. 3.2, Ex. 3 **4.** 2; Sec. 3.4, Ex. 5 **5.** $\frac{21}{7}$; Sec. 4.1, Ex. 12

**6.** $\frac{14x}{15}$; Sec. 4.2, Ex. 6 **7.** $\frac{5}{12}$; Sec. 4.3, Ex. 12 **8.** $-\frac{1}{2}$; Sec. 4.4, Ex. 9 **9.** $\frac{1}{28}$; Sec. 4.5, Ex. 5 **10.** $\frac{3}{2}$; Sec. 4.6, Ex. 2

**11.** 3; Sec. 4.7, Ex. 7 **12. a.** $\frac{38}{9}$ **b.** $\frac{19}{11}$; Sec. 4.8, Ex. 3 **13.** $\frac{1}{8}$; Sec. 5.1, Ex. 8 **14.** $105\frac{83}{1000}$; Sec. 5.1, Ex. 10

**15.** 67.69; Sec. 5.2, Ex. 6 **16.** 76.8; Sec. 5.3, Ex. 5 **17.** −76,300; Sec. 5.3, Ex. 7 **18.** yes; Sec. 5.4, Ex. 7 **19.** $\frac{2}{5}$; Sec. 5.8, Ex. 3

**20.** $\frac{26}{31}$; Sec. 6.1, Ex. 5 **21.** $0.21/oz; Sec. 6.2, Ex. 6 **22.** no; Sec. 6.3, Ex. 3 **23.** 17.5 miles; Sec. 6.4, Ex. 1 **24.** $\frac{19}{1000}$; Sec. 7.1, Ex. 12

**25.** $\frac{1}{3}$; Sec. 7.1, Ex. 14

# Chapter 8 GRAPHING AND INTRODUCTION TO STATISTICS

## CHAPTER 8 PRETEST

**1.** ; 8.1B **2.** April; 8.2D **3.** July; 8.2D **4.** 700; 8.2D **5.** no; 8.3B
**6. a.** (0, −6) **b.** (−3, −12) **c.** (3, 0); 8.3C
**7.** ; 8.4A & 8.4ABC **8.** mean: 53.5; median: 64; mode: 74;

**9.** 2.65; 8.5A

**10.** $\frac{1}{6}$; 8.6B

**11.** $\frac{1}{2}$; 8.6B

**12.** $\frac{1}{3}$; 8.6B

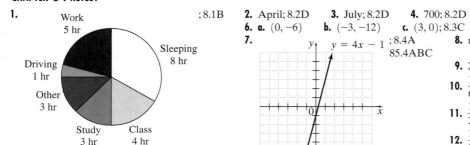

## EXERCISE SET 8.1

**1.** Parent or guardian's home **3.** $\frac{9}{35}$ **5.** $\frac{9}{16}$ **7.** less than $20 **9.** 53% **11.** 1410 teenagers **13.** 1316 teenagers **15.** 55%

**17.** nonfiction **19.** 31,400 books **21.** 27,632 books **23.** 25,120 books
**25.**
**27.** $2^2 \cdot 5$
**29.** $2^3 \cdot 5$
**31.** $5 \cdot 17$
**33.** 29,600,002 sq. km
**35.** 55,542,858 sq. km
**37.** Answers may vary.

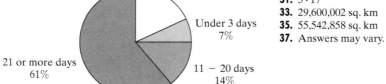

## EXERCISE SET 8.2

**1.** 1998 **3.** 4000 cars **5.** 1994, 1995, 1999 **7.** 1993, 1996 **9.** 21 oz **11.** 1994, 1996, 1998 **13.** 3 oz/week
**15.** consumption of chicken is increasing **17.** April **19.** 14 **21.** February, March, April, May, June
**23.** Tokyo, 26.5 million or 26,500,000 **25.** New York City, 16.2 million or 16,200,000 **27.** 12 million or 12,000,000

**29.**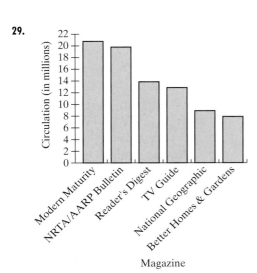

Magazine

**31.**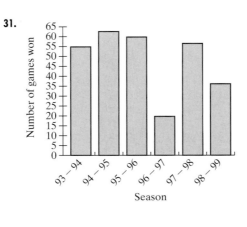

Season

**33.** 54.5 field goals
**35.** 1975
**37.** increase
**39.** $101 million
**41.** 1st week
**43.** $94 million
**45.** During week 8, *The Phantom Menace* grossed $12 million at the box office.
**47.** 3.6
**49.** 6.2
**51.** 25%
**53.** 34%
**55.** 83°F
**57.** Sunday, 68°F
**59.** Tuesday, 13°F
**61.** North America
**63.** North America
**65.** Answers may vary.

## Exercise Set 8.3

**1.**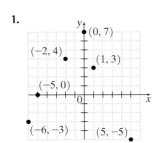

**3.**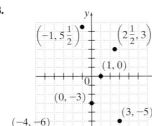

**5.** $A(0,0); B(3\frac{1}{2},0); C(3,2); D(-1,3); E(-2,-2); F(0,-1); G(2,-1)$
**7.** yes   **9.** no   **11.** yes   **13.** yes   **15.** yes   **17.** no
**19.** 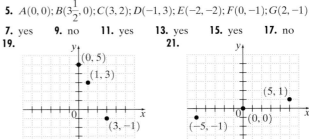   **21.**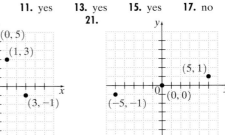

**23.**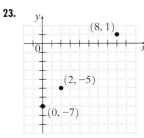

**25.** $(1, 8), (0, 0), (-2, -16)$   **27.** $(2, 12), (22, -8), (0, 14)$   **29.** $(1, 6), (2, 7), (3, 8)$
**31.** $(1, -2), (2, 1), (3, 4)$   **33.** $(0, 0), (2, -2), (-2, 2)$   **35.** $(-2, 0), (1, -3), (-7, 5)$
**37.** 1.7   **39.** 21.84   **41.** −23.6   **43.** true   **45.** true   **47.** false   **49.** false
**51.** Answers may vary.

## Integrated Review

**1.** 50 lb   **2.** 52.5 lb   **3.** 1999   **4.** 1997   **5.** Oroville Dam, 755 ft   **6.** New Bullards Bar Dam, 635 ft   **7.** 15 ft
**8.** 4 dams   **9.** Thursday and Saturday, 100°F   **10.** Monday, 82°F   **11.** Sunday, Monday, and Tuesday
**12.** Wednesday, Thursday, Friday, and Saturday   **13.** 70 qt containers   **14.** 52 qt containers   **15.** 2 qt containers   **16.** 6 qt containers
**17.** 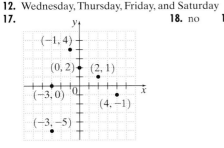   **18.** no   **19.** yes   **20.** $(0, -6), (6, 0), (2, -4)$

**1.**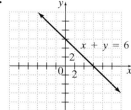

**3.**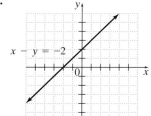

**5.**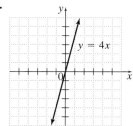

**7.**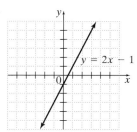

**9.**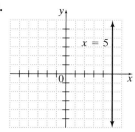

**11.**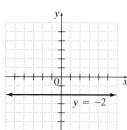

**13.**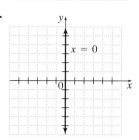

**15.**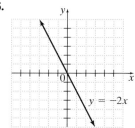

**17.**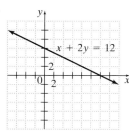

**19.**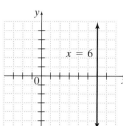

**21.**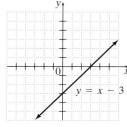

**23.**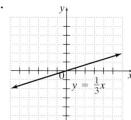

**25.**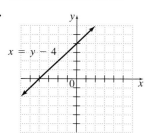

**27.**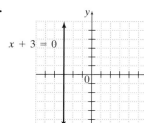

**29.**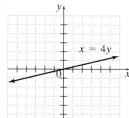

**31.**

**33.**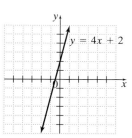

**35.**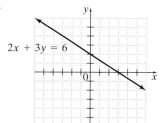

**37.**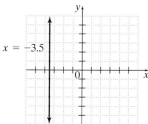

**39.**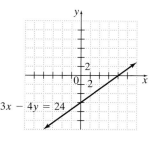

**41.** 90   **43.** 20   **45.** 27   **47.** 3, 2, 1, 0, 1, 2, 3

**49.** Answers may vary.

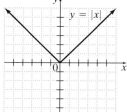

**MENTAL MATH**

**1.** 4   **3.** 3

## EXERCISE SET 8.5

**1.** mean: 29; median: 28; no mode　　**3.** mean: 8.1; median: 8.2; mode: 8.2　　**5.** mean: 0.6; median: 0.6; mode: 0.2 and 0.6
**7.** mean: 370.9; median: 313.5; no mode　　**9.** 1313.2 ft　　**11.** 1131.5 ft　　**13.** 2.79　　**15.** 3.46　　**17.** 6.8　　**19.** 6.9　　**21.** 85.5
**23.** 73　　**25.** 70 and 71　　**27.** 9　　**29.** $\frac{1}{3}$　　**31.** $\frac{3}{5}$　　**33.** $\frac{11}{15}$　　**35.** 35, 35, 37, 43

## MENTAL MATH

**1.** $\frac{1}{2}$　　**3.** $\frac{1}{2}$

## EXERCISE SET 8.6

**1.**

**3.**

**5.**
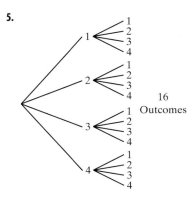

**7.**
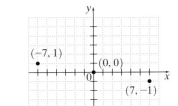

**9.**
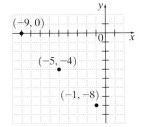

**11.** $\frac{1}{6}$　　**13.** $\frac{1}{3}$　　**15.** $\frac{1}{2}$　　**17.** $\frac{1}{3}$
**19.** $\frac{2}{3}$　　**21.** $\frac{1}{7}$　　**23.** $\frac{2}{7}$　　**25.** $\frac{5}{6}$
**27.** $\frac{1}{6}$　　**29.** $\frac{20}{3}$ or $6\frac{2}{3}$　　**31.** $\frac{1}{52}$
**33.** $\frac{1}{13}$　　**35.** $\frac{1}{4}$　　**37.** $\frac{1}{12}$　　**39.** 0
**41.** No; no; answers may vary.

## CHAPTER 8 REVIEW

**1.** Mortgage　　**2.** Utilities　　**3.** $1225　　**4.** $700　　**5.** $\frac{39}{160}$　　**6.** $\frac{7}{40}$　　**7.** 20 states　　**8.** 10 states　　**9.** 26 states　　**10.** 23 states
**11.** 11,500 homes　　**12.** 11,500 homes　　**13.** Alabama　　**14.** Delaware　　**15.** Utah and Alabama　　**16.** Maine and Delaware　　**17.** 8%
**18.** 1998　　**19.** 1980, 1990, 1998　　**20.** Answers may vary.　　**21.** 10 billion pounds　　**22.** 39.5 billion pounds　　**23.** 1980–1990　　**24.** 1940–1950
**25.** $(0, 0), (7, -1), (-7, 1)$　　**26.** $(0, -2), (1, 1), (-2, -8)$　　**27.** $(-1, -8), (-9, 0), (-5, -4)$　　**28.** $(4, 1), (0, -3), (6, 3)$

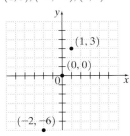
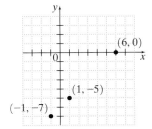
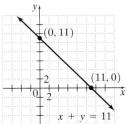
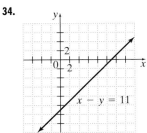

**29.** $(1, 3), (-2, -6), (0, 0)$　　**30.** $(1, -5), (6, 0), (-1, -7)$　　**33.**　　**34.**

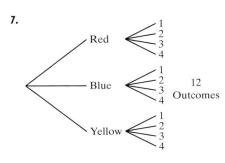
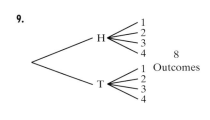

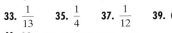
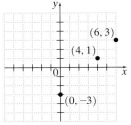
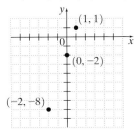

**35.**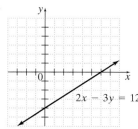

**36.**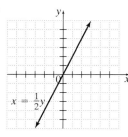

**37.**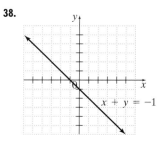

**38.**

**39.**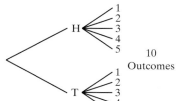

**40.**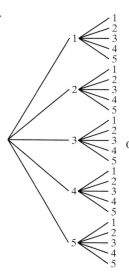

**41.** mean: 17.8; median: 14; no mode
**42.** mean: 55.2; median: 60; no mode
**43.** mean: $24,500; median: $20,000; mode: $20,000
**44.** mean: 447.3; median: 420; mode: 400
**45.** 3.25
**46.** 2.57

**47.**

H — 1, 2, 3, 4, 5
T — 1, 2, 3, 4, 5
10 Outcomes

**48.**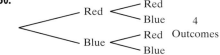

Red — H, T
Blue — H, T
4 Outcomes

**49.**

1 — 1, 2, 3, 4, 5
2 — 1, 2, 3, 4, 5
3 — 1, 2, 3, 4, 5
4 — 1, 2, 3, 4, 5
5 — 1, 2, 3, 4, 5
25 Outcomes

**50.**

Red — Red, Blue
Blue — Red, Blue
4 Outcomes

**51.**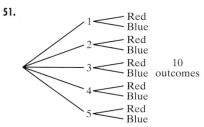

1 — Red, Blue
2 — Red, Blue
3 — Red, Blue
4 — Red, Blue
5 — Red, Blue
10 outcomes

**52.** $\frac{1}{6}$   **53.** $\frac{1}{6}$   **54.** $\frac{1}{5}$   **55.** $\frac{1}{5}$   **56.** $\frac{3}{5}$   **57.** $\frac{2}{5}$

## Chapter 8 Test

**1.** $225   **2.** 3rd week, $350   **3.** $1100   **4.** June, August, September   **5.** February; 3 cm   **6.** March and November

**7.** $110 million   **8.** 1999   **9.** 1996 and 1997   **10.** $\frac{17}{40}$   **11.** $\frac{31}{22}$   **12.** 2,034,750 small cars

**13.** 1,302,240 luxury cars   **14.** $(4, 0)$   **15.** $(0, -3)$   **16.** $(-3, 4)$   **17.** $(-2, -1)$
**18.** $(0, 0), (-6, 1), (12, -2)$   **19.** $(2, 10), (-1, -11), (0, -4)$   **20.**   **21.**

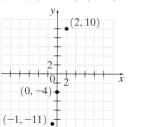

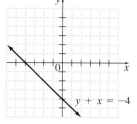

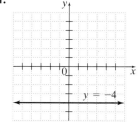

**22.**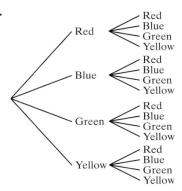
$y = 3x - 5$

**23.**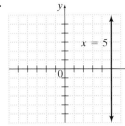
$x = 5$

**24.**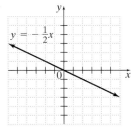
$y = -\frac{1}{2}x$

**25.**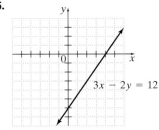
$3x - 2y = 12$

**26.** mean: 38.4; median: 42; no mode    **27.** mean: 12.625; median: 12.5; mode: 16 and 12    **28.** 3.07

**29.**
```
              Red
Red     <     Blue
              Green
              Yellow
              Red
Blue    <     Blue
              Green
              Yellow
              Red
Green   <     Blue
              Green
              Yellow
              Red
Yellow  <     Blue
              Green
              Yellow
```

**30.**
```
        H
H   <   
        T
        H
T   <   
        T
```

**31.** $\frac{1}{10}$    **32.** $\frac{1}{5}$

## CUMULATIVE REVIEW

**1.** 47; Sec. 1.7, Ex. 11    **2.** −12; Sec. 2.3, Ex. 11    **3.** −3; Sec. 3.3, Ex. 4    **4.** 2; Sec. 4.7, Ex. 5    **5.** $7\frac{17}{24}$; Sec. 4.8, Ex. 11

**6.** $5\frac{3}{5}$; Sec. 5.1, Ex. 7    **7.** 3.432; Sec. 5.2, Ex. 5    **8.** 0.0849; Sec. 5.3, Ex. 2    **9.** −0.052; Sec. 5.4, Ex. 2    **10.** 1.69; Sec. 5.5, Ex. 6

**11.** 0.25; Sec. 5.6, Ex. 1    **12.** 0.7; Sec. 5.7, Ex. 3    **13.** 5.657; Sec. 5.8, Ex. 5    **14.** $\frac{12}{17}$; Sec. 6.1, Ex. 1    **15.** 24.5 mi/gal; Sec. 6.2, Ex. 5

**16.** 5; Sec. 6.3, Ex. 6    **17.** $\frac{12}{17}$; Sec. 6.5, Ex. 1    **18.** 0.046; Sec. 7.1, Ex. 4    **19.** 0.0074; Sec. 7.1, Ex. 6    **20.** 21; Sec. 7.2, Ex. 7

**21.** 52; Sec. 7.3, Ex. 9    **22.** 7.5%; Sec. 7.5, Ex. 2    **23.** $101.50; Sec. 7.6, Ex. 6    **24.** 80.5; Sec. 8.5, Ex. 4    **25.** $\frac{1}{3}$; Sec. 8.6, Ex. 4

# Chapter 9 GEOMETRY AND MEASUREMENT

## CHAPTER 9 PRETEST

**1.** acute; 9.1B    **2.** straight; 9.1B    **3.** 88°; 9.1C    **4.** $x = 55°$, $y = 125°$, $z = 125°$; 9.1D    **5.** 110°; 9.1D    **6.** $3\frac{2}{3}$ yd; 9.2A

**7.** 62.5 km; 9.2D    **8.** 60 in.; 9.3A    **9.** 56.52 yd; 9.3B    **10.** 6 sq. cm; 9.4A    **11.** 7 lb 12 oz; 9.5B    **12.** 8 gal 3 qt; 9.6B
**13.** 25,000 ml; 9.6C    **14.** 47.8°C; 9.7B

## EXERCISE SET 9.1

**1.** line; line $yz$ or $\overleftrightarrow{yz}$    **3.** line segment; line segment $LM$ or $\overline{LM}$    **5.** line segment; line segment $PQ$ or $\overline{PQ}$    **7.** ray; ray $UW$ or $\overrightarrow{UW}$
**9.** 15°    **11.** 50°    **13.** 65°    **15.** 95°    **17.** 90°    **19.** 0°; 90°    **21.** straight    **23.** right    **25.** obtuse    **27.** right    **29.** 73°
**31.** 163°    **33.** 45°    **35.** 55°    **37.** $\angle MNP$ and $\angle RNO$; $\angle PNQ$ and $\angle QNR$    **39.** $\angle SPT$ and $\angle TPQ$; $\angle SPR$ and $\angle RPQ$; $\angle SPT$ and
$\angle SPR$; $\angle TPQ$ and $\angle QPR$    **41.** 32°    **43.** 75°    **45.** $\angle x = 35°$; $\angle y = 145°$; $\angle z = 145°$    **47.** $\angle x = 77°$; $\angle y = 103°$; $\angle z = 77°$
**49.** $\angle x = 100°$; $\angle y = 80°$; $\angle z = 100°$    **51.** $\angle x = 134°$; $\angle y = 46°$; $\angle z = 134°$    **53.** 25°    **55.** 13°    **57.** 40°
**59.** $\frac{9}{8}$ or $1\frac{1}{8}$    **61.** $\frac{7}{32}$    **63.** $\frac{5}{6}$    **65.** $\frac{4}{3}$ or $1\frac{1}{3}$    **67.** 54.8°    **69.** $\angle a = 60°$; $\angle b = 50°$; $\angle c = 110°$; $\angle d = 70°$; $\angle e = 120°$
**71.** The sum of the measures of the angles of a triangle is 180°.

## CALCULATOR EXPLORATIONS

**1.** ≈ 22.96 ft    **3.** ≈ 21.59 cm    **5.** ≈ 3.1 mi

## MENTAL MATH

**1.** 1 ft    **3.** 2 ft    **5.** 1 yd    **7.** no    **9.** yes    **11.** no

## Exercise Set 9.2

**1.** 5 ft   **3.** 36 ft   **5.** 8 mi   **7.** $8\frac{1}{2}$ ft   **9.** $3\frac{1}{3}$ yd   **11.** 33,792 ft   **13.** 13 yd 1 ft   **15.** 3 ft 5 in.   **17.** 1 mi 4720 ft   **19.** 62 in.
**21.** 17 ft   **23.** 84 in.   **25.** 12 ft 3 in.   **27.** 22 yd 1 ft   **29.** 8 ft 5 in.   **31.** 5 ft 6 in.   **33.** 3 ft 4 in.   **35.** 50 yd 2 ft   **37.** 10 ft 6 in.
**39.** 18 ft 5 in.   **41.** 15 ft 9 in.   **43.** 3 ft 1 in.   **45.** 86 ft 6 in.   **47.** $105\frac{1}{3}$ yd   **49.** 4000 cm   **51.** 4 cm   **53.** 0.3 km   **55.** 1.4 m
**57.** 15 m   **59.** 83 mm   **61.** 0.201 dm   **63.** 40 mm   **65.** 8.94 m   **67.** 2.94 m or 2940 mm   **69.** 1.29 cm or 12.9 mm
**71.** 12.640 km or 12,640 m   **73.** 54.9 m   **75.** 1.55 km   **77.** 9.12 m   **79.** 26.7 mm   **81.** 41.25 m or 4125 cm   **83.** 3.35 m
**85.** 6.009 km or 6009 m   **87.** 15 tiles   **89.** 32 yd   **91.** 35 m   **93.** Answers may vary.   **95.** 6.575 m

## Exercise Set 9.3

**1.** 64 ft   **3.** 36 cm   **5.** 21 in.   **7.** 120 cm   **9.** 48 ft   **11.** 66 in.   **13.** 21 ft   **15.** 60 ft   **17.** 346 yd   **19.** 22 ft   **21.** $66
**23.** 36 in.   **25.** 28 in.   **27.** $24.08   **29.** 96 m   **31.** 66 ft   **33.** 128 mi   **35.** $17\pi$ cm; 53.38 cm   **37.** $16\pi$ mi; 50.24 mi
**39.** $26\pi$ m; 81.64 m   **41.** $31\frac{3}{7}$ ft   **43.** 12,560 ft   **45.** 23   **47.** 1   **49.** 10   **51.** 216   **53.** perimeter   **55.** area   **57.** area
**59.** perimeter   **61. a.** 62.8 m; 125.6 m   **b.** yes   **63.** $44 + 10\pi \approx 75.4$ m

## Exercise Set 9.4

**1.** 7 sq. m   **3.** $9\frac{3}{4}$ sq. yd   **5.** 15 sq. yd   **7.** $2.25\pi$ sq. in. $\approx 7.065$ sq. in.   **9.** 36.75 sq. ft   **11.** 28 sq. m   **13.** 22 sq. yd
**15.** $36\frac{3}{4}$ sq. ft   **17.** $22\frac{1}{2}$ sq. in.   **19.** 25 sq. cm   **21.** 86 sq. mi   **23.** 24 sq. cm   **25.** $36\pi$ sq. in. $\approx 113\frac{1}{7}$ sq. in.   **27.** 72 cu. in.
**29.** 512 cu. cm   **31.** $12\frac{4}{7}$ cu. yd   **33.** $523\frac{17}{21}$ cu. in.   **35.** 75 cu. cm   **37.** $2\frac{10}{27}$ cu. in.   **39.** 8.4 cu. ft   **41.** 5000 sq. ft
**43.** 168 sq. ft   **45.** 960 cu. cm   **47.** 9200 sq. ft   **49.** 381 sq. ft   **51.** 25   **53.** 9   **55.** 5   **57.** 20   **59.** 12-in. pizza
**61.** $1\frac{1}{3}$ sq. ft; 192 sq. in.   **63.** 77,811,712 cu. ft   **65.** 2,353,158 cu. m; 230,125 cu. m   **67. a.** 5 cu. ft   **b.** 1 cu. ft; (a) is larger
**69.** Answers may vary.

## Integrated Review

**1.** 153°; 63°   **2.** $\angle x = 75°$; $\angle y = 105°$; $\angle z = 75°$   **3.** $\angle x = 128°$; $\angle y = 52°$; $\angle z = 128°$   **4.** $\angle x = 52°$   **5.** 3 ft   **6.** 2 mi
**7.** $6\frac{2}{3}$ yd   **8.** 18 ft   **9.** 11,088 ft   **10.** 38.4 in.   **11.** 3000 cm   **12.** 2.4 cm   **13.** 2 m   **14.** 1800 cm   **15.** 72 mm
**16.** 0.6 km   **17.** perimeter = 20 m; area = 25 sq. m   **18.** perimeter = 12 ft; area = 6 sq. ft
**19.** circumference = $6\pi$ cm $\approx 18.84$ cm; area = $9\pi$ sq. cm $\approx 28.26$ sq. cm   **20.** perimeter = 32 mi; area = 44 sq. mi
**21.** perimeter = 62 ft; area = 238 sq. ft   **22.** 64 cu. in.   **23.** 30.6 cu. ft   **24.** 400 cu. cm   **25.** $\frac{2048}{3}\pi$ cu. mi $\approx 2145\frac{11}{21}$ cu. mi

## Calculator Explorations

**1.** $\approx 425.25\ g$   **3.** $\approx 15.4$ lb   **5.** $\approx 0.175$ oz

## Mental Math

**1.** 1 lb   **3.** 2000 lb   **5.** 16 oz   **7.** 1 ton   **9.** no   **11.** yes   **13.** no

## Exercise Set 9.5

**1.** 32 oz   **3.** 10,000 lb   **5.** 6 tons   **7.** $3\frac{3}{4}$ lb   **9.** $1\frac{3}{4}$ tons   **11.** 260 oz   **13.** 9800 lb   **15.** 76 oz   **17.** 1.5 tons
**19.** 53 lb 10 oz   **21.** 9 tons 390 lb   **23.** 3 tons 175 lb   **25.** 8 lb 11 oz   **27.** 31 lb 2 oz   **29.** 1 ton 700 lb   **31.** 5 lb 8 oz
**33.** 35 lb 14 oz   **35.** 130 lb   **37.** 211 lb   **39.** 5 lb 1 oz   **41.** 0.5 kg   **43.** 4000 mg   **45.** 25,000 g   **47.** 0.048 g
**49.** 0.0063 kg   **51.** 15,140 mg   **53.** 4010 g   **55.** 13.5 mg   **57.** 5.815 g or 5815 mg   **59.** 1850 mg or 1.85 g
**61.** 1360 g or 1.36 kg   **63.** 13.52 kg   **65.** 2.125 kg   **67.** 8.064 kg   **69.** 30 mg   **71.** 250 mg   **73.** 144 mg   **75.** 6.12 kg
**77.** 850 g or 0.85 kg   **79.** 2.38 kg   **81.** 0.25   **83.** 0.16   **85.** 0.875   **87.** Answers may vary.

## Calculator Explorations

**1.** $\approx 4.73$ L   **3.** $\approx 4.62$ gal   **5.** $\approx 3.785$ L

## Mental Math

**1.** 1 pt   **3.** 1 gal   **5.** 1 qt   **7.** 1 c   **9.** 2 c   **11.** 4 qt   **13.** no   **15.** no

## Exercise Set 9.6

**1.** 4 c   **3.** 16 pt   **5.** $2\frac{1}{2}$ gal   **7.** 5 pt   **9.** 8 c   **11.** $3\frac{3}{4}$ qt   **13.** 768 fl oz   **15.** 9 c   **17.** 22 pt   **19.** 10 gal 1 qt
**21.** 4 c 4 fl oz   **23.** 1 gal 1 qt   **25.** 2 gal 3 qt 1 pt   **27.** 2 qt 1 c   **29.** 17 gal   **31.** 4 gal 3 qt   **33.** 48 fl oz   **35.** 2 qt   **37.** yes
**39.** 2 qt 1 c   **41.** 9 qt   **43.** 5000 ml   **45.** 4.5 L   **47.** 0.41 kl   **49.** 0.064 L   **51.** 160 L   **53.** 3600 ml   **55.** 0.00016 kl
**57.** 22.5 L   **59.** 4.5 L or 4500 ml   **61.** 8410 ml or 8.41 L   **63.** 10,600 ml or 10.6 L   **65.** 3840 ml   **67.** 162.4 L   **69.** 1.59 L

**71.** 18.954 L   **73.** $0.316   **75.** 474 ml   **77.** $\frac{7}{10}$   **79.** $\frac{3}{100}$   **81.** $\frac{3}{500}$   **83.** Answers may vary.

## MENTAL MATH

**1.** yes   **3.** no   **5.** no   **7.** yes

## EXERCISE SET 9.7

**1.** 5°C   **3.** 40°C   **5.** 140°F   **7.** 239°F   **9.** 16.7°C   **11.** 61.2°C   **13.** 197.6°F   **15.** 61.3°F   **17.** 56.7°C   **19.** 80.6°F
**21.** 21.1°C   **23.** 37.9°C   **25.** 244.4°F   **27.** 260°C   **29.** 462.2°C   **31.** 62 m   **33.** 15 ft   **35.** 10.4 m   **37.** 4988°C
**39. a.** 33.3°C   **b.** 14°F

## CHAPTER 9 REVIEW

**1.** right   **2.** straight   **3.** acute   **4.** obtuse   **5.** 65°   **6.** 75°   **7.** 108°   **8.** 89°   **9.** 58°   **10.** 98°   **11.** 90°   **12.** 25°
**13.** 133° and 47°   **14.** 43° and 47°; 58° and 32°   **15.** $\angle x = 80°$; $\angle y = 100°$; $\angle z = 100°$   **16.** $\angle x = 155°$; $\angle y = 155°$; $\angle z = 25°$
**17.** $\angle x = 53°$; $\angle y = 53°$; $\angle z = 127°$   **18.** $\angle x = 42°$; $\angle y = 42°$; $\angle z = 138°$   **19.** 103°   **20.** 60°   **21.** 60°   **22.** 65°   **23.** 9 ft
**24.** 24 yd   **25.** 13,200 ft   **26.** 75 in.   **27.** 17 yd 1 ft   **28.** 3 ft 10 in.   **29.** 4200 cm   **30.** 820 mm   **31.** 0.01218 m
**32.** 0.00231 km   **33.** 21 yd 1 ft   **34.** 7 ft 5 in.   **35.** 41 ft 3 in.   **36.** 3 ft 8 in.   **37.** 9.5 cm or 95 mm   **38.** 5.26 m or 526 cm
**39.** 9117 m or 9.117 km   **40.** 1.1 m or 1100 mm   **41.** 169 yd 2 ft   **42.** 126 ft 8 in.   **43.** 108.5 km   **44.** 0.24 sq. m   **45.** 88 m
**46.** 30 cm   **47.** 36 m   **48.** 90 ft   **49.** 32 ft   **50.** 440 ft   **51.** 5.338 in.   **52.** 31.4 yd   **53.** 240 sq. ft   **54.** 140 sq. m
**55.** 600 sq. cm   **56.** 189 sq. yd   **57.** $49\pi$ sq. ft ≈ 153.86 sq. ft   **58.** $4\pi$ sq. in. ≈ 12.56 sq. in.   **59.** 119 sq. in.   **60.** 1248 sq. cm
**61.** 144 sq. m   **62.** 432 sq. ft   **63.** 130 sq. ft   **64.** $15\frac{5}{8}$ cu. in.   **65.** 84 cu. ft   **66.** $62{,}857\frac{1}{7}$ cu. cm   **67.** $346\frac{1}{2}$ cu. in.
**68.** $2\frac{2}{3}$ cu. ft   **69.** 307.72 cu. in.   **70.** $7\frac{1}{2}$ cu. ft   **71.** $0.5\pi$ cu. ft   **72.** 1260 cu. ft   **73.** 28.728 cu. ft   **74.** 4.125 lb   **75.** 4600 lb
**76.** 3 lb 4 oz   **77.** 4 tons 200 lb   **78.** 1.4 g   **79.** 40,000 g   **80.** 21 dag   **81.** 0.0003 dg   **82.** 3 lb 9 oz   **83.** 10 tons 800 lb
**84.** 2 tons 750 lb   **85.** 33 lb 8 oz   **86.** 4.9 g or 4900 mg   **87.** 9 kg or 9000 g   **88.** 8.1 g or 8100 mg   **89.** 50.4 kg   **90.** 4 lb 4 oz
**91.** 9 tons 1075 lb   **92.** 7.85 kg   **93.** 1.1625 kg   **94.** 8 qt   **95.** 5 c   **96.** 27 qt   **97.** 17 c   **98.** 4 qt 1 pt   **99.** 3 gal 3 qt
**100.** 3800 ml   **101.** 0.042 dl   **102.** 1.4 kl   **103.** 3060 cl   **104.** 1 gal 1 qt   **105.** 7 gal 1 qt   **106.** 736 ml or 0.736 L
**107.** 15.5 L or 15,500 ml   **108.** 2 gal 3 qt   **109.** 6 fl oz   **110.** 10.88 L   **111.** yes   **112.** 473°F   **113.** 320°F   **114.** 107.6°F
**115.** 186.8°F   **116.** 34°C   **117.** 11°C   **118.** 5.2°C   **119.** 26.7°C   **120.** 1.7°C   **121.** 329°F

## CHAPTER 9 TEST

**1.** 12°   **2.** 56°   **3.** 50°   **4.** $\angle x = 118°$; $\angle y = 62°$; $\angle z = 118°$   **5.** $\angle x = 73°$; $\angle y = 73°$; $\angle z = 73°$   **6.** 26°
**7.** circumference $= 18\pi$ in. ≈ 56.52 in.; area $= 81\pi$ sq. in. ≈ 254.34 sq. in.   **8.** perimeter $= 24.6$ yd; area $= 37.1$ sq. yd
**9.** perimeter $= 68$ in.; area $= 185$ sq. in.   **10.** $62\frac{6}{7}$ cu. in.   **11.** 30 cu. ft   **12.** 16 in.   **13.** 18 cu. ft   **14.** 62 ft   **15.** 198.08 oz
**16.** 23 ft 4 in.   **17.** 10 qt   **18.** 1.875 lb   **19.** 5600 lb   **20.** $4\frac{3}{4}$ gal   **21.** 0.04 g   **22.** 2400 g   **23.** 36 mm   **24.** 0.43 g
**25.** 830 ml   **26.** 1 gal 2 qt   **27.** 3 lb 13 oz   **28.** 8 ft 3 in.   **29.** 2 gal 3 qt   **30.** 66 mm or 6.6 cm   **31.** 2.256 km or 2256 m
**32.** 28.9°C   **33.** 54.7°F   **34.** 5.6 m   **35.** 4 gal 3 qt   **36.** 105.8°F   **37.** 91.4 m   **38.** 16 ft 6 in. or $16\frac{1}{2}$ ft   **39.** 1 lb 13 oz
**40.** 0.320 kg   **41.** 37,336 pt

## CUMULATIVE REVIEW

**1.** 5; Sec. 3.4, Ex. 3   **2. a.** $\frac{16}{625}$   **b.** $\frac{1}{16}$; Sec. 4.3, Ex. 9   **3.** $8\frac{3}{10}$; Sec. 4.8, Ex. 13   **4.** $20x - 10.9$; Sec. 5.2, Ex. 11   **5.** 6.35; Sec. 5.5, Ex. 7
**6.** =; Sec. 5.6, Ex. 5   **7.** −1.3; Sec. 5.7, Ex. 4   **8.** 424 ft; Sec. 5.8, Ex. 9   **9. a.** $\frac{5}{7}$   **b.** $\frac{7}{24}$; Sec. 6.1, Ex. 7   **10.** $\frac{5 \text{ nails}}{3 \text{ feet}}$; Sec. 6.2, Ex. 1
**11.** 0.44; Sec. 6.3, Ex. 10   **12.** 22.4 cc; Sec. 6.4, Ex. 2   **13.** 17%; Sec. 7.1, Ex. 1   **14.** 200; Sec. 7.2, Ex. 10   **15.** 2.7; Sec. 7.3, Ex. 7
**16.** 32%; Sec. 7.4, Ex. 4   **17.** Sales tax: $6.41, total price: $91.91; Sec. 7.5; Ex. 1   **18.** $A(2, 0)$, $B(1, 2)$, $C(0, 1)$, $D(-3, 0)$; Sec. 8.3, Ex. 2
**19.** ; Sec. 8.4, Ex. 4

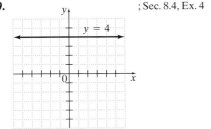

**20.** 57; Sec. 8.5, Ex. 3
**21.** $\frac{1}{4}$; Sec. 8.6, Ex. 5
**22.** 42°; Sec. 9.1, Ex. 4
**23.** 96 in.; Sec. 9.2, Ex. 1
**24.** 4 tons 1650 lb; Sec. 9.5, Ex. 3
**25.** 15°C; Sec. 9.7, Ex. 3

# Chapter 10 POLYNOMIALS

## CHAPTER 10 PRETEST

**1.** $7y^2 - 23y + 26$; 10.1A　**2.** $17b - 12$; 10.1B　**3.** $5z^4 - z^3 - 6z^2 + 3z$; 10.1B　**4.** $-20$; 10.1C　**5.** $36n^{18}$; 10.2A
**6.** $120x^3$; 10.2A　**7.** $t^{54}$; 10.2B　**8.** $81n^8$; 10.2C　**9.** $15{,}552a^{12}b^{13}c^{17}$; 10.2C　**10.** $18d^5 - 10d^3 + 22d$; 10.3A　**11.** $x^2 + 9x + 18$; 10.3B
**12.** $2y^2 - 3y - 20$; 10.3B　**13.** $9x^2 - 12x + 4$; 10.3C　**14.** $n^2 - 100$; 10.3B　**15.** $8a^3 - 2a^2 + 27a + 7$; 10.3D　**16.** $9$; 10.4A
**17.** $y^3$; 10.4B　**18.** $2m$; 10.4B　**19.** $2(5y^2 + 3y - 7)$; 10.4C　**20.** $4n^3(2n^3 - 3n^2 + 5)$; 10.4C

## EXERCISE SET 10.1

**1.** $-5x - 24$　**3.** $-9z^2 - 2z + 6$　**5.** $9y^2 + 25y - 40$　**7.** $-4.3a^4 - 2a^2 + 9$　**9.** $4a - 8$　**11.** $-2x^2 + 4x + 1$
**13.** $-20y^3 + 12y^2 - 4$　**15.** $-x + 16$　**17.** $8y^2 - 10y - 8$　**19.** $4x^2 + x - 16$　**21.** $-15y + 3.6$　**23.** $b^3 - b^2 + 7b - 1$
**25.** $\dfrac{9}{7}$　**27.** $1$　**29.** $-5$　**31.** $-8$　**33.** $20$　**35.** $25$　**37.** $50$　**39.** $576$ ft　**41.** \$3200　**43.** $909$ ft　**45.** $81$　**47.** $25$
**49.** $x^3$　**51.** $2^2a^4$　**53.** $(8x + 2)$ in.　**55.** $(4x - 15)$ units　**57.** $20, 6, 2$　**59.** $7.2752$　**61.** $29$ ft; $-243$ ft; Answers may vary.

## EXERCISE SET 10.2

**1.** $x^{14}$　**3.** $a^7$　**5.** $15z^5$　**7.** $-40x^2$　**9.** $25x^6y^4$　**11.** $28a^5b^6$　**13.** $30x^3$　**15.** $12a^{17}$　**17.** $x^{15}$　**19.** $z^{20}$　**21.** $b^{62}$
**23.** $81a^4$　**25.** $a^{33}b^{24}$　**27.** $121x^6y^{12}$　**29.** $-24y^{22}$　**31.** $256x^9y^{13}$　**33.** $7x - 21$　**35.** $-6a - 4b$　**37.** $9x + 18y - 27$
**39.** $16x^{12}$ sq. in.　**41.** $12a^4b^5$ sq. m　**43.** $18{,}003{,}384a^{45}b^{30}$　**45.** $34{,}867.84401x^{50}$　**47.** Answers may vary.

## INTEGRATED REVIEW

**1.** $4x - 1$　**2.** $33y - 8$　**3.** $10x + 3$　**4.** $-5y - 4$　**5.** $3a^3 + 5a - 8$　**6.** $1.8y^2 - 2.4y - 5.6$　**7.** $-1.8x^2 + 2.4x - 1.7$
**8.** $-4a^3 - 12a^2$　**9.** $6x + 7$　**10.** $2x^2 + 3x - 12$　**11.** $-13$　**12.** $-4$　**13.** $x^{18}$　**14.** $x^{12}$　**15.** $y^4$　**16.** $a^{11}$　**17.** $x^{77}$
**18.** $x^{36}$　**19.** $16x^4$　**20.** $27y^3$　**21.** $-12x^2y^7$　**22.** $12a^3b^4$　**23.** $x^{36}y^{20}$　**24.** $a^{20}b^{24}$　**25.** $300x^4y^3$　**26.** $128y^6z^7$

## EXERCISE SET 10.3

**1.** $27x^3 - 9x$　**3.** $-20a^3 + 30a^2 - 5a$　**5.** $42x^4 - 35x^3 + 49x^2$　**7.** $x^2 + 13x + 30$　**9.** $2x^2 + 2x - 24$　**11.** $36a^2 + 48a + 16$
**13.** $a^3 - 33a + 18$　**15.** $8x^3 + 2x^2 - 55x + 50$　**17.** $x^5 + 4x^4 + 6x^3 + 7x^2 + 2x$　**19.** $-30r^2 + 20r$　**21.** $-6y^3 - 2y^4 + 12y^2$
**23.** $x^2 + 14x + 24$　**25.** $4a^2 - 9$　**27.** $x^2 + 10x + 25$　**29.** $b^2 + \dfrac{7}{5}b + \dfrac{12}{25}$　**31.** $6x^3 + 25x^2 + 10x + 1$　**33.** $49x^2 + 70x + 25$
**35.** $4x^2 - 4x + 1$　**37.** $8x^5 - 8x^3 - 6x^2 - 6x + 9$　**39.** $x^5 + 2x^4 + 3x^3 + 2x^2 + x$　**41.** $10z^4 - 3z^3 + 3z - 2$　**43.** $2 \cdot 5^2$
**45.** $2^3 \cdot 3^2$　**47.** $2^3 \cdot 5^2$　**49.** $(y^3 - 3y^2 - 16y - 12)$ sq. ft　**51.** $(x^4 - 3x^2 + 1)$ sq. m　**53.** Answers may vary.

## EXERCISE SET 10.4

**1.** $3$　**3.** $12$　**5.** $4$　**7.** $4$　**9.** $y^2$　**11.** $a^5$　**13.** $xy^2$　**15.** $x$　**17.** $2z^3$　**19.** $3y(y + 6)$　**21.** $5a^6(2 - a^2)$
**23.** $4x(x^2 + 3x + 5)$　**25.** $z^5(z^2 - 6)$　**27.** $-7(5 - 2y + y^2)$ or $7(-5 + 2y - y^2)$　**29.** $12a^5(1 - 3a)$　**31.** $36$　**33.** $\dfrac{4}{5}$
**35.** $37.5\%$　**37.** $x^2 + 2x$　**39.** Answers may vary.

## CHAPTER 10 REVIEW

**1.** $10b - 3$　**2.** $21s - 15$　**3.** $-x + 2.8$　**4.** $-y - 12$　**5.** $4z^2 + 11z - 6$　**6.** $17a^3 + 25a^2$　**7.** $-11y^2 - y + \dfrac{3}{4}$
**8.** $x^2 - 7x + 3$　**9.** $45$　**10.** $-19$　**11.** $(26x + 28)$ ft　**12.** $x^{24}$　**13.** $y^7$　**14.** $24z^7$　**15.** $-15x^3y^5$　**16.** $a^{35}$　**17.** $x^{28}$
**18.** $81b^2$　**19.** $a^{20}b^{10}c^5$　**20.** $56x^{16}$　**21.** $108x^{30}y^{25}$　**22.** $81a^{14}$ sq. mi　**23.** $10a^3 - 12a$　**24.** $-3y^4 + 6y^3 - 3y^2$
**25.** $x^2 + 8x + 12$　**26.** $15x^2 - 32x + 9$　**27.** $y^2 - 10y + 25$　**28.** $49a^2 + 14a + 1$　**29.** $x^3 - x^2 + x + 3$
**30.** $8y^4 + 4y^3 - 2y^2 - 3y - 3$　**31.** $3z^4 + 5z^3 + 6z^2 + 3z + 1$　**32.** $(a^3 + 5a^2 - 5a + 6)$ sq. cm　**33.** $5$　**34.** $4$　**35.** $6$
**36.** $5$　**37.** $x^2$　**38.** $y^7$　**39.** $xy$　**40.** $a^5b^2$　**41.** $5a$　**42.** $4y^2z$　**43.** $2x(x + 6)$　**44.** $6a(a - 2)$　**45.** $y^4(6 - y^2)$
**46.** $7(x^2 - 2x + 1)$　**47.** $a^3(5a^4 - a + 1)$　**48.** $10y(y^5 - 1)$

## CHAPTER 10 TEST

**1.** $15x - 4$　**2.** $7x - 2$　**3.** $3.4y^2 + 2y - 3$　**4.** $-2a^2 + a + 1$　**5.** $17$　**6.** $y^{14}$　**7.** $y^{33}$　**8.** $16x^8$　**9.** $-12a^{10}$　**10.** $p^{54}$
**11.** $72a^{20}b^5$　**12.** $10x^3 + 6.5x$　**13.** $-2y^4 - 12y^3 + 8y$　**14.** $x^2 - x - 6$　**15.** $25x^2 + 20x + 4$　**16.** $a^3 + 8$
**17.** perimeter: $(14x - 4)$ in.; area: $(5x^2 + 33x - 14)$ sq. in.　**18.** $15$　**19.** $3y^3$　**20.** $3y(y - 5)$　**21.** $2a(5a + 6)$
**22.** $6(x^2 - 2x - 5)$　**23.** $x^3(7x^3 - 6x + 1)$

## CUMULATIVE REVIEW

**1.** $106{,}400$ sq. mi; Sec. 1.5, Ex. 6　**2.** $-8$; Sec. 2.2, Ex. 12　**3.** $-7$; Sec. 2.3, Ex. 6　**4.** $1$; Sec. 2.3, Ex. 7　**5.** $4$; Sec. 3.2, Ex. 5
**6.** $\dfrac{32}{12x}$; Sec. 4.1, Ex. 13　**7.** $3\dfrac{4}{7}$; Sec. 4.8, Ex. 16　**8.** $736.2$; Sec. 5.1, Ex. 13　**9.** $25.454$; Sec. 5.2, Ex. 1　**10.** no; Sec. 5.3, Ex. 13
**11.** $0.0786$; Sec. 5.4, Ex. 4　**12.** $0.012$; Sec. 5.4, Ex. 5　**13.** $-2.6$; Sec. 5.5, Ex. 8　**14.** $3.14$; Sec. 5.6, Ex. 3　**15.** $\dfrac{1}{6}$; Sec. 5.8, Ex. 2
**16.** $46$ ft; Sec. 6.5, Ex. 3　**17.** $1.2 = 30\% \cdot x$; Sec. 7.2, Ex. 2　**18.** $16\%$; Sec. 7.3, Ex. 10　**19.** $775$ freshmen; Sec. 7.4, Ex. 1
**20.** \$160; Sec. 7.6, Ex. 2　**21.** $41\%$; Sec. 8.1, Ex. 2　**22.** $28$ in.; Sec. 9.3, Ex. 2　**23.** $5.1$ sq. mi; Sec. 9.4, Ex. 2
**24.** $33$ lb $6$ oz; Sec. 9.5, Ex. 4　**25.** $3.21$ L; Sec. 9.6, Ex. 7

# Appendix A

**1.** 5 **2.** 11 **3.** 5 **4.** 15 **5.** 12 **6.** 7 **7.** 8 **8.** 6 **9.** 14 **10.** 10 **11.** 12 **12.** 5 **13.** 10 **14.** 2 **15.** 9
**16.** 12 **17.** 11 **18.** 8 **19.** 18 **20.** 7 **21.** 10 **22.** 0 **23.** 10 **24.** 10 **25.** 17 **26.** 8 **27.** 13 **28.** 3
**29.** 12 **30.** 17 **31.** 16 **32.** 8 **33.** 7 **34.** 13 **35.** 4 **36.** 3 **37.** 8 **38.** 7 **39.** 8 **40.** 9 **41.** 6 **42.** 9
**43.** 16 **44.** 8 **45.** 9 **46.** 15 **47.** 12 **48.** 1 **49.** 7 **50.** 6 **51.** 2 **52.** 6 **53.** 11 **54.** 11 **55.** 8 **56.** 10
**57.** 5 **58.** 10 **59.** 14 **60.** 6 **61.** 13 **62.** 4 **63.** 7 **64.** 9 **65.** 7 **66.** 13 **67.** 13 **68.** 7 **69.** 5 **70.** 9
**71.** 4 **72.** 3 **73.** 16 **74.** 11 **75.** 14 **76.** 13 **77.** 1 **78.** 11 **79.** 4 **80.** 4 **81.** 10 **82.** 14 **83.** 8
**84.** 15 **85.** 5 **86.** 15 **87.** 10 **88.** 2 **89.** 11 **90.** 6 **91.** 12 **92.** 6 **93.** 9 **94.** 11 **95.** 9 **96.** 9 **97.** 14
**98.** 12 **99.** 9 **100.** 3

# Appendix B

**1.** 1 **2.** 35 **3.** 56 **4.** 9 **5.** 32 **6.** 45 **7.** 28 **8.** 7 **9.** 4 **10.** 0 **11.** 63 **12.** 64 **13.** 6 **14.** 0 **15.** 30
**16.** 10 **17.** 24 **18.** 0 **19.** 18 **20.** 72 **21.** 40 **22.** 14 **23.** 32 **24.** 2 **25.** 54 **26.** 3 **27.** 56 **28.** 16
**29.** 54 **30.** 25 **31.** 2 **32.** 0 **33.** 36 **34.** 24 **35.** 12 **36.** 20 **37.** 36 **38.** 18 **39.** 12 **40.** 6 **41.** 48
**42.** 72 **43.** 8 **44.** 5 **45.** 0 **46.** 28 **47.** 27 **48.** 0 **49.** 15 **50.** 48 **51.** 45 **52.** 12 **53.** 0 **54.** 27
**55.** 81 **56.** 20 **57.** 0 **58.** 9 **59.** 0 **60.** 6 **61.** 18 **62.** 7 **63.** 3 **64.** 21 **65.** 36 **66.** 0 **67.** 63 **68.** 12
**69.** 35 **70.** 0 **71.** 42 **72.** 0 **73.** 40 **74.** 8 **75.** 0 **76.** 24 **77.** 9 **78.** 0 **79.** 15 **80.** 16 **81.** 5 **82.** 30
**83.** 0 **84.** 4 **85.** 21 **86.** 8 **87.** 0 **88.** 49 **89.** 16 **90.** 24 **91.** 0 **92.** 14 **93.** 4 **94.** 0 **95.** 6 **96.** 8
**97.** 18 **98.** 10 **99.** 0 **100.** 42

# SOLUTIONS TO SELECTED EXERCISES

## Chapter 1

### EXERCISE SET 1.1

**1.** The place value of the 5 in 352 is tens.

**5.** The place value of the 5 in 62,500,000 is hundred-thousands.

**9.** 5420 is written as five thousand, four hundred twenty.

**13.** 1,620,000 is written as one million, six hundred twenty thousand.

**17.** 4,992,838 is written as four million, nine hundred ninety-two thousand, eight hundred thirty-eight.

**21.** 3893 is written as three thousand, eight hundred ninety-three.

**25.** Twenty-nine thousand, nine hundred in standard form is 29,900.

**29.** Three million, fourteen in standard form is 3,000,014.

**33.** Sixty-three million, one hundred thousand in standard form is 63,100,000.

**37.** $406 = 400 + 6$

**41.** $62,407 = 60,000 + 2000 + 400 + 7$

**45.** $39,680,000 = 30,000,000 + 9,000,000$
$+ 600,000 + 80,000$

**49.** $3 < 8$

**53.** $6 > 2$

**57.** 4000 is written as four thousand.

**61.** The Nile is the longest river in the world.

**65.** There are fewer Rottweilers than German Shepherds registered.

**69.** Answers may vary.

### EXERCISE SET 1.2

**1.**
$$\begin{array}{r} 14 \\ + 22 \\ \hline 36 \end{array}$$

**5.**
$$\begin{array}{r} 12 \\ 13 \\ + 24 \\ \hline 49 \end{array}$$

**9.**
$$\begin{array}{r} \overset{1}{5}3 \\ + 64 \\ \hline 117 \end{array}$$

**13.**
$$\begin{array}{r} \overset{1\,1}{3}8 \\ + 79 \\ \hline 117 \end{array}$$

**17.**
$$\begin{array}{r} \overset{2}{\phantom{0}}6 \\ 21 \\ 14 \\ 9 \\ + 12 \\ \hline 62 \end{array}$$

**21.**
$$\begin{array}{r} \overset{1}{6}2 \\ 18 \\ + 14 \\ \hline 94 \end{array}$$

**25.**
$$\begin{array}{r} \overset{1\,1\,1}{7}542 \\ 49 \\ + 682 \\ \hline 8273 \end{array}$$

**29.**
$$\begin{array}{r} \overset{1\ \ 2}{6}27 \\ 628 \\ + 629 \\ \hline 1884 \end{array}$$

**33.**
$$\begin{array}{r} \overset{1\,1\,1}{5}07 \\ 593 \\ + 10 \\ \hline 1110 \end{array}$$

**37.**
$$\begin{array}{r} \overset{1\,1\,2\,2}{\phantom{00}}49 \\ 628 \\ 5\ 762 \\ + 29,462 \\ \hline 35,901 \end{array}$$

**41.** $8 + 3 + 5 + 7 + 5 + 1 = 8 + 1 + 3 + 7 + 5 + 5$
$= 9 + 10 + 10$
$= 29$
The perimeter is 29 inches.

**45.** Opposite sides of a rectangle have the same length.
$4 + 8 + 4 + 8 = 12 + 12 = 24$
The perimeter is 24 inches.

**49.**
$$\begin{array}{r} \overset{1\,1}{2}85 \\ + 98 \\ \hline 383 \end{array}$$
It is 383 miles from Kansas City to Colby.

**53.**
$$\begin{array}{r} 105,600 \\ + 17,500 \\ \hline 123,100 \end{array}$$

**57.**
$$\begin{array}{r} \overset{1\ \ 1}{12},166 \\ + 973 \\ \hline 13,139 \end{array}$$
There were 13,139 transplants involving a kidney performed in 1998.

**61.**
$$\begin{array}{r} 1795 \\ +\,11{,}460 \\ \hline 13{,}255 \end{array}$$
The total highway mileage in Alaska is 13,255 miles.

**65.**
$$\begin{array}{r} \overset{1}{1}66 \\ 130 \\ +\,293 \\ \hline 589 \end{array}$$
Texas, Florida, and Illinois have the most Wal-Mart stores, with a total of 589 stores.

**69.** Answers may vary.

### EXERCISE SET 1.3

**1.**
$$\begin{array}{r} 67 \\ -\,23 \\ \hline 44 \end{array}$$
*Check:*
$$\begin{array}{r} 44 \\ +\,23 \\ \hline 67 \end{array}$$

**5.**
$$\begin{array}{r} 389 \\ -\,124 \\ \hline 265 \end{array}$$
*Check:*
$$\begin{array}{r} 265 \\ +\,124 \\ \hline 389 \end{array}$$

**9.**
$$\begin{array}{r} 998 \\ -\,453 \\ \hline 545 \end{array}$$
*Check:*
$$\begin{array}{r} 545 \\ +\,453 \\ \hline 998 \end{array}$$

**13.**
$$\begin{array}{r} 62 \\ -\,37 \\ \hline 25 \end{array}$$
*Check:*
$$\begin{array}{r} \overset{1}{2}5 \\ +\,37 \\ \hline 62 \end{array}$$

**17.**
$$\begin{array}{r} 938 \\ -\,792 \\ \hline 146 \end{array}$$
*Check:*
$$\begin{array}{r} \overset{1}{1}46 \\ +\,792 \\ \hline 938 \end{array}$$

**21.**
$$\begin{array}{r} 600 \\ -\,432 \\ \hline 168 \end{array}$$
*Check:*
$$\begin{array}{r} \overset{1\,1}{1}68 \\ +\,432 \\ \hline 600 \end{array}$$

**25.**
$$\begin{array}{r} 923 \\ -\,476 \\ \hline 447 \end{array}$$
*Check:*
$$\begin{array}{r} \overset{1\,1}{4}47 \\ +\,476 \\ \hline 923 \end{array}$$

**29.**
$$\begin{array}{r} 533 \\ -\,29 \\ \hline 504 \end{array}$$
*Check:*
$$\begin{array}{r} \overset{1}{5}04 \\ +\,29 \\ \hline 533 \end{array}$$

**33.**
$$\begin{array}{r} 1983 \\ -\,1904 \\ \hline 79 \end{array}$$
*Check:*
$$\begin{array}{r} \overset{1}{7}9 \\ +\,1904 \\ \hline 1983 \end{array}$$

**37.**
$$\begin{array}{r} 50{,}000 \\ -\,17{,}289 \\ \hline 32{,}711 \end{array}$$
*Check:*
$$\begin{array}{r} \overset{1\,1\,\,\,1\,1}{32}{,}711 \\ +\,17{,}289 \\ \hline 50{,}000 \end{array}$$

**41.**
$$\begin{array}{r} 51{,}111 \\ -\,19{,}898 \\ \hline 31{,}213 \end{array}$$
*Check:*
$$\begin{array}{r} \overset{1\,1\,\,\,1\,1}{31}{,}213 \\ +\,19{,}898 \\ \hline 51{,}111 \end{array}$$

**45.**
$$\begin{array}{r} 41 \\ -\,21 \\ \hline 20 \end{array}$$
*Check:*
$$\begin{array}{r} 20 \\ +\,21 \\ \hline 41 \end{array}$$

**49.**
$$\begin{array}{r} 63 \\ -\,7 \\ \hline 56 \end{array}$$
Ronnie ate 56 crawfish.

**53.**
$$\begin{array}{r} 20{,}320 \\ -\,14{,}255 \\ \hline 6{,}065 \end{array}$$
Mt. McKinley is 6065 feet higher than Long's Peak.

**57.**
$$\begin{array}{r} 645 \\ -\,287 \\ \hline 358 \end{array}$$
The distance between Hays and Denver is 358 miles.

**61.**
$$\begin{array}{r} 547 \\ -\,99 \\ \hline 448 \end{array}$$
The sale price is $448.

**65.** The total number of votes cast for Jo was:

$$
\begin{array}{r}
^{1\ 2\ 1}\\
276\\
362\\
201\\
+\ 179\\
\hline
1018
\end{array}
$$

The total number of votes cast for Trudy was:

$$
\begin{array}{r}
^{2\ 1}\\
295\\
122\\
312\\
+\ 182\\
\hline
911
\end{array}
$$

Since more votes were cast for Jo than for Trudy, Jo won the election.

$$
\begin{array}{r}
1018\\
-911\\
\hline
107
\end{array}
$$

Jo won by 107 votes.

**69.**
$$
\begin{array}{r}
1,130,000\\
-633,000\\
\hline
497,000
\end{array}
$$

North Korea's fighting force is 497,000 soldiers larger than South Korea's fighting force.

**73.** Atlanta, Hartsfield International is busiest.

**77.** General Motors and Procter & Gamble spent more than $1500 million on ads.

**81.**
$$
\begin{array}{r}
2,121,040,900\\
1,724,259,700\\
1,410,748,700\\
1,264,353,200\\
+\ 1,147,589,200\\
\hline
7,667,991,700
\end{array}
$$

The top five companies spent $7,667,991,700 on ads.

**85.** Answers may vary.

## EXERCISE SET 1.4

**1.** To round 632 to the nearest ten, observe that the digit in the ones place is 2. Since this digit is less than 5, we do not add 1 to the digit in the tens place. The number 632 rounded to the nearest ten is 630.

**5.** To round 792 to the nearest ten, observe that the digit in the ones place is 2. Since this digit is less than 5, we do not add 1 to the digit in the tens place. The number 792 rounded to the nearest ten is 790.

**9.** To round 1096 to the nearest ten, observe that the digit in the ones place is 6. Since this digit is at least 5, we need to add 1 to the digit in the tens place. The number 1096 rounded to the nearest ten is 1100.

**13.** To round 248,695 to the nearest hundred, observe that the digit in the tens place is 9. Since this digit is at least 5, we need to add 1 to the digit in the hundreds place. The number 248,695 rounded to the nearest hundred is 248,700.

**17.** To round 99,995 to the nearest ten, observe that the digit in the ones place is 5. Since this digit is at least 5, we need to add 1 to the digit in the tens place. The number 99,995 rounded to the nearest ten is 100,000.

**21.** Estimate 5281 to a given place value by rounding it to that place value. 5281 estimated to the tens place value is 5280, to the hundreds place value is 5300, and to the thousands place value is 5000.

**25.** Estimate 14,876 to a given place value by rounding it to that place value. 14,876 estimaed to the tens place value is 14,880, to the hundreds place value is 14,900, and to the thousands place value is 15,000.

**29.** To round 72,625 to the nearest thousand, observe that the digit in the hundreds place is 6. Since this digit is as least 5, we need to add 1 to the digit in the thousands place. The number 72,625 rounded to the nearest thousand is 73,000.

**33.** To round $371,971,500 to the nearest hundred-thousand, observe that the digit in the ten-thousands place is 7. Since this digit is at least 5, we need to add 1 to the digit in the hundred-thousands place. The number $371,971,500 rounded to the nearest hundred-thousand is $372,000,000.

**37.**
$$
\begin{array}{llr}
649 & \text{rounds to} & 650\\
-272 & \text{rounds to} & -270\\
\hline
& & 380
\end{array}
$$
The estimated difference is 380.

**41.**
$$
\begin{array}{llr}
1774 & \text{rounds to} & 1800\\
-1492 & \text{rounds to} & -1500\\
\hline
& & 300
\end{array}
$$
The estimated difference is 300.

**45.** $362 + 419$ is approximately $360 + 420 = 780$.
The answer of 781 is correct.

**49.** $7806 + 5150$ is approximately $7800 + 5200 = 13,000$.
The answer of 12,956 is correct.

**53.**
$$
\begin{array}{llr}
799 & \text{rounds to} & 800\\
1299 & \text{rounds to} & 1300\\
+999 & \text{rounds to} & +1000\\
\hline
& & 3100
\end{array}
$$
The total cost is approximately $3100.

**57.**
$$
\begin{array}{llr}
29,028 & \text{rounds to} & 29,000\\
-4039 & \text{rounds to} & -4\ 000\\
\hline
& & 25,000
\end{array}
$$
The difference in elevation is approximately 25,000 feet.

**61.**
$$
\begin{array}{llr}
41,126,233 & \text{rounds to} & 41,000,000\\
-27,174,898 & \text{rounds to} & -27,000,000\\
\hline
& & 14,000,000
\end{array}
$$
Johnson won the election by approximately 14,000,000 votes.

**65.** 1,264,353,200 rounds to 1,300,000,000. Philip Morris spent approximately $1,300,000,000.

**69.** The smallest possible number that rounds to 8600 is 8550.

## EXERCISE SET 1.5

**1.** $4(3 + 9) = 4 \cdot 3 + 4 \cdot 9$

**5.** $10(11 + 7) = 10 \cdot 11 + 10 \cdot 7$

**9.**  $\begin{array}{r} 624 \\ \times\ \ 3 \\ \hline 1872 \end{array}$

**13.**  $\begin{array}{r} 1062 \\ \times\ \ 5 \\ \hline 5310 \end{array}$

**17.**  $\begin{array}{r} 231 \\ \times\ \ 47 \\ \hline 1617 \\ 9240 \\ \hline 10,857 \end{array}$

**21.**  $\begin{array}{r} 620 \\ \times\ \ 40 \\ \hline 0 \\ 24,800 \\ \hline 24,800 \end{array}$

**25.**  $(590)(1)(10) = 5900$

**29.**  $\begin{array}{r} 609 \\ \times\ \ 234 \\ \hline 2\ 436 \\ 18,270 \\ 121,800 \\ \hline 142,506 \end{array}$

**33.**  $\begin{array}{r} 1941 \\ \times\ \ 235 \\ \hline 9\ 705 \\ 58,230 \\ 388,200 \\ \hline 456,135 \end{array}$

**37.**  $\begin{array}{r} 576 \\ \times\ \ 354 \end{array}$ rounds to $\begin{array}{r} 600 \\ \times\ \ 400 \\ \hline 240,000 \end{array}$

$576 \times 354$ is approximately 240,000.

**41.** Area = length · width
= (9 meters)(7 meters)
= 63 square meters
The area is 63 square meters.

**45.**  $\begin{array}{r} 125 \\ \times\ \ 3 \\ \hline 375 \end{array}$

There are 375 calories in 3 tablespoons of olive oil.

**49.**  $\begin{array}{r} 12 \\ \times\ \ 8 \\ \hline 96 \end{array}$

$2 \times 96 = 192$
There are 192 cans in a case.

**53.**  $\begin{array}{r} 776 \\ \times\ \ 639 \\ \hline 6\ 984 \\ 23,280 \\ 465,600 \\ \hline 495,864 \end{array}$

The floor area is 495,864 square meters.

**57.**  $\begin{array}{r} 60 \\ \times\ \ 25 \\ \hline 300 \\ 1200 \\ \hline 1500 \end{array}$

There are 1500 characters in 25 lines.

**61.**  $\begin{array}{r} 7927 \\ \times\ \ 9 \\ \hline 71,343 \end{array}$

Saturn has a diameter of 71,343 miles.

**65.**  $\begin{array}{r} 700,000 \\ \times\ \ 31 \\ \hline 700,000 \\ 21,000,000 \\ \hline 21,700,000 \end{array}$

21,700,000 quarts of milk would be used in March.

**69.**  $5 \times 10 = 50$
50 students chose grapes as their favorite fruit.

**73.** The result of multiplying 3 by the digit in the first blank is a number ending in 6. Only 2 works. The result of multiplying the digit in the second blank by 42 is 378, so the digit in the second blank is 9.

$\begin{array}{r} 42 \\ \times\ \ 93 \\ \hline 126 \\ 3780 \\ \hline 3906 \end{array}$

**77.**  $154 \times 2 = 308$ and $58 \times 3 = 174$.

$\begin{array}{r} 204 \\ 308 \\ +\ 174 \\ \hline 686 \end{array}$

Cynthia Cooper scored 686 points in the 1999 season.

### EXERCISE SET 1.6

**1.**
$$\begin{array}{r} 12 \\ 9\overline{)108} \\ \underline{-9} \\ 18 \\ \underline{-18} \\ 0 \end{array}$$
Check: $12 \cdot 9 = 108$

**5.**
$$\begin{array}{r} 338 \\ 3\overline{)1014} \\ \underline{-9} \\ 11 \\ \underline{-9} \\ 24 \\ \underline{-24} \\ 0 \end{array}$$
Check: $338 \cdot 3 = 1014$

**9.**
$$\begin{array}{r} 563\ R\ 1 \\ 2\overline{)1127} \\ \underline{-10} \\ 12 \\ \underline{-12} \\ 07 \\ \underline{-6} \\ 1 \end{array}$$
Check: $563 \cdot 2 + 1 = 1127$

**13.**

$$
\begin{array}{r}
265 \text{ R } 1 \\
8\overline{)2121} \\
-16 \phantom{00} \\
\hline
52 \phantom{0} \\
-48 \phantom{0} \\
\hline
41 \\
-40 \\
\hline
1
\end{array}
$$

Check: $265 \cdot 8 + 1 = 2121$

**17.**

$$
\begin{array}{r}
13 \\
55\overline{)715} \\
-55 \phantom{0} \\
\hline
165 \\
-165 \\
\hline
0
\end{array}
$$

Check: $13 \cdot 55 = 715$

**21.**

$$
\begin{array}{r}
206 \\
18\overline{)3708} \\
-36 \phantom{00} \\
\hline
10 \phantom{0} \\
-0 \phantom{0} \\
\hline
108 \\
-108 \\
\hline
0
\end{array}
$$

Check: $206 \cdot 18 = 3708$

**25.**

$$
\begin{array}{r}
202 \text{ R } 7 \\
46\overline{)9299} \\
-92 \phantom{00} \\
\hline
09 \phantom{0} \\
-0 \phantom{0} \\
\hline
99 \\
-92 \\
\hline
7
\end{array}
$$

Check: $202 \cdot 46 + 7 = 9299$

**29.**

$$
\begin{array}{r}
98 \text{ R } 100 \\
103\overline{)10194} \\
-927 \phantom{0} \\
\hline
924 \\
-824 \\
\hline
100
\end{array}
$$

Check: $98 \cdot 103 + 100 = 10,194$

**33.**

$$
\begin{array}{r}
202 \\
223\overline{)45046} \\
-446 \phantom{00} \\
\hline
44 \phantom{0} \\
-0 \phantom{0} \\
\hline
446 \\
-446 \\
\hline
0
\end{array}
$$

Check: $202 \cdot 223 = 45,046$

**37.**

$$
\begin{array}{r}
252000 \\
21\overline{)5292000} \\
-42 \phantom{00000} \\
\hline
109 \phantom{0000} \\
-105 \phantom{0000} \\
\hline
42 \phantom{000} \\
-42 \phantom{000} \\
\hline
00 \phantom{00} \\
-0 \phantom{00} \\
\hline
00 \phantom{0} \\
-0 \phantom{0} \\
\hline
00 \\
-0 \\
\hline
0
\end{array}
$$

Each person receives $252,000.

**41.**

$$
\begin{array}{r}
88 \text{ R } 1 \\
3\overline{)265} \\
-24 \phantom{0} \\
\hline
25 \\
-24 \\
\hline
1
\end{array}
$$

There are 88 bridges in 265 miles.

**45.**

$$
\begin{array}{r}
17 \\
6\overline{)102} \\
-6 \phantom{0} \\
\hline
42 \\
-42 \\
\hline
0
\end{array}
$$

He scored 17 touchdowns during 1999.

**49.**

$$
\begin{array}{r}
6,000 \\
3,500,000\overline{)21,000,000,000} \\
-21,000,000 \phantom{000} \\
\hline
0\,0 \phantom{0000} \\
-0 \phantom{0000} \\
\hline
00 \phantom{000} \\
-0 \phantom{000} \\
\hline
00 \phantom{00} \\
-0 \phantom{00} \\
\hline
0
\end{array}
$$

A single home uses 6000 kilowatt-hours of electricity during one year on average.

**53.** There are four numbers.

$$
\begin{array}{r}
204 \\
968 \\
552 \\
+268 \\
\hline
1992
\end{array}
$$

$$
\begin{array}{r}
498 \\
4\overline{)1992} \\
-16 \phantom{00} \\
\hline
39 \phantom{0} \\
-36 \phantom{0} \\
\hline
32 \\
-32 \\
\hline
0
\end{array}
$$

Average $= \dfrac{1992}{4} = 498$

**57.** Add those month's temperatures, then divide by 3, the number of temperatures.

$$\begin{array}{r} 18 \\ 12 \\ +18 \\ \hline 48 \end{array}$$

$$\begin{array}{r} 16 \\ 3{\overline{\smash{\big)}\,48}} \\ \underline{-3\phantom{0}} \\ 18 \\ \underline{-18} \\ 0 \end{array}$$

The average temperature is $\dfrac{48}{3} = 16$ degrees.

**61.** Increases, because 71 lies below the average, which is 83.

## EXERCISE SET 1.7

**1.** $3 \cdot 3 \cdot 3 \cdot 3 = 3^4$

**5.** $12 \cdot 12 \cdot 12 = 12^3$

**9.** $9 \cdot 9 \cdot 9 \cdot 8 = (9 \cdot 9 \cdot 9)8 = 9^3 \cdot 8$

**13.** $3 \cdot 2 \cdot 2 \cdot 5 \cdot 5 \cdot 5 = 3(2 \cdot 2)(5 \cdot 5 \cdot 5) = 3 \cdot 2^2 \cdot 5^3$

**17.** $5^3 = 5 \cdot 5 \cdot 5 = 125$

**21.** $2^{10} = 2 \cdot 2 \cdot 2 \cdot 2 \cdot 2 \cdot 2 \cdot 2 \cdot 2 \cdot 2 \cdot 2 = 1024$

**25.** $3^5 = 3 \cdot 3 \cdot 3 \cdot 3 \cdot 3 = 243$

**29.** $4^3 = 4 \cdot 4 \cdot 4 = 64$

**33.** $9^3 = 9 \cdot 9 \cdot 9 = 729$

**37.** $10^4 = 10 \cdot 10 \cdot 10 \cdot 10 = 10,000$

**41.** $1920^1 = 1920$

**45.** $15 + 3 \cdot 2 = 15 + 6 = 21$

**49.** $5 \cdot 9 - 16 = 45 - 16 = 29$

**53.** $14 + \dfrac{24}{8} = 14 + 3 = 17$

**57.** $0 \div 6 + 4 \cdot 7 = 0 + 4 \cdot 7 = 0 + 28 = 28$

**61.** $(6 + 8) \div 2 = 14 \div 2 = 7$

**65.** $(3 + 5^2) \div 2 = (3 + 25) \div 2 = 28 \div 2 = 14$

**69.** $\dfrac{18 + 6}{2^4 - 4} = \dfrac{24}{16 - 4} = \dfrac{24}{12} = 2$

**73.** $\dfrac{7(9 - 6) + 3}{3^2 - 3} = \dfrac{7 \cdot 3 + 3}{9 - 3} = \dfrac{21 + 3}{6} = \dfrac{24}{6} = 4$

**77.** $3^4 - [35 - (12 - 6)] = 3^4 - [35 - 6]$
$$= 81 - 29$$
$$= 52$$

**81.** $8 \cdot [4 + (6 - 1) \cdot 2] - 50 \cdot 2 = 8 \cdot [4 + 5 \cdot 2] - 50 \cdot 2$
$$= 8 \cdot [4 + 10] - 100$$
$$= 8 \cdot 14 - 100$$
$$= 112 - 100$$
$$= 12$$

**85.** Area of a square $=$ (side)$^2$
$$= (20 \text{ miles})^2$$
$$= 400 \text{ square miles}$$

**89.** Area of base $=$ (side)$^2$
$$= (100 \text{ meters})^2$$
$$= 10,000 \text{ square meters}$$

**93.** $24 \div (3 \cdot 2) + 2 \cdot 5 = 24 \div 6 + 2 \cdot 5$
$$= 4 + 10$$
$$= 14$$

**97.** $(7 + 2^4)^5 - (3^5 - 2^4)^2 = (7 + 16)^5 - (3^5 - 2^4)^2$
$$= 23^5 - (243 - 16)^2$$
$$= 6,436,343 - 227^2$$
$$= 6,436,343 - 51,529$$
$$= 6,384,814$$

## EXERCISE SET 1.8

**1.** $3 + 2z = 3 + 2(3)$
$$= 3 + 6$$
$$= 9$$

**5.** $z - x + y = 3 - 2 + 5$
$$= 6$$

**9.** $y^3 - 4x = 5^3 - 4(2)$
$$= 125 - 4(2)$$
$$= 125 - 8$$
$$= 117$$

**13.** $8 - (y - x) = 8 - (5 - 2)$
$$= 8 - 3$$
$$= 5$$

**17.** $\dfrac{6xy}{z} = \dfrac{6 \cdot 2 \cdot 5}{3}$
$$= \dfrac{60}{3}$$
$$= 20$$

**21.** $\dfrac{x + 2y}{z} = \dfrac{2 + 2 \cdot 5}{3}$
$$= \dfrac{2 + 10}{3}$$
$$= \dfrac{12}{3}$$
$$= 4$$

**25.** $2y^2 - 4y + 3 = 2 \cdot 5^2 - 4 \cdot 5 + 3$
$$= 2 \cdot 25 - 4 \cdot 5 + 3$$
$$= 50 - 20 + 3$$
$$= 33$$

**29.** $(xy + 1)^2 = (2 \cdot 5 + 1)^2$
$$= (10 + 1)^2$$
$$= (11)^2$$
$$= 121$$

**33.** $xy(5 + z - x) = 2 \cdot 5(5 + 3 - 2)$
$$= 2 \cdot 5(8 - 2)$$
$$= 2 \cdot 5(6)$$
$$= 10(6)$$
$$= 60$$

**37.**

| $t$ | 1 | 2 | 3 | 4 |
|---|---|---|---|---|
| $16t^2$ | $16(1)^2$ | $16(2)^2$ | $16(3)^2$ | $16(4)^2$ |
| | $16 \cdot 1$ | $16 \cdot 4$ | $16 \cdot 9$ | $16 \cdot 16$ |
| | 16 | 64 | 144 | 256 |

**41.** $x + 8$  **45.** $512x$

**49.** $5x + (17 + x)$  **53.** $11 - x$

**57.** $x^4 - y^2 = (23)^4 - (72)^2$
$$= 279,841 - 5184$$
$$= 274,657$$

**61.** Compare expressions:

$$\frac{x}{3} = \left(\frac{1}{3}\right)x$$

$$\left(\frac{1}{3}\right)x < 2x < 5x$$

$5x$ is the largest.

## CHAPTER 1 TEST

**1.** $59 + 82 = 141$

**5.** $2^3 \cdot 5^2 = 8 \cdot 25 = 200$

**9.** $62 \div 0$ is undefined.

**13.** 52,369 rounded to the nearest thousand is 52,000.

**17.**
$$\begin{array}{r} 17 \\ \times\ 7 \\ \hline 119 \end{array}$$

**21.** $5[(2)^3 - 2] = 5[8 - 2] = 5 \cdot 6 = 30$

**25.** Opposite sides of a rectangle have the same length.
Find the perimeter by adding the lengths of the sides.
$20 + 10 + 20 + 10 = 60$.
The perimeter is 60 yards.
Area $=$ length $\cdot$ width
$\phantom{Area} = (20\text{ yards})(10\text{ yards})$
$\phantom{Area} = 200$ square yards

# Chapter 2

## EXERCISE SET 2.1

**1.** If 0 represents ground level, then 1445 feet underground is $-1445$.

**5.** If 0 represents the line of scrimmage, a loss of 15 yards is $-15$.

**9.** $-5,049$ thousand

**13.** If 0 represents 0%, a loss of 16% is $-16$.

**17.**
```
←•─┼─┼─•─┼─•─•─┼─┼─•→
 -7 -6 -5 -4 -3 -2 -1  0  1
```

**21.**
```
←•─┼─•─•─┼─┼─•─┼─•─┼→
 -7 -6 -5 -4 -3 -2 -1  0  1
```

**25.** $-7 < -5$

**29.** $-26 < 26$

**33.** $|-8| = 8$, because $-8$ is 8 units from 0.

**37.** $|-5| = 5$, because $-5$ is 5 units from 0.

**41.** The opposite of $-4$ is $-(-4) = 4$.

**45.** The opposite of $-10$ is $-(-10) = 10$.

**49.** $-|20| = -20$
The opposite of the absolute value of 20 is the opposite of 20.

**53.** $-(-8) = 8$
The opposite of negative 8 is 8.

**57.** $-(-29) = 29$
The opposite of negative 29 is 29.

**61.** $|-9|\ ?\ |-14|$
$\phantom{xx} 9\ ?\ 14$
$\phantom{xx} 9 < 14$

**65.** $-|-10|\ ?\ -(-10)$
$\phantom{xx} -10\ ?\ 10$
$\phantom{xx} -10 < 10$

**69.** $|0|\ ?\ |-9|$
$\phantom{xx} 0\ ?\ 9$
$\phantom{xx} 0 < 9$

**73.** $-(-12)\ ?\ -(-18)$
$\phantom{xxxx} 12 < 18$

**77.**
$$\begin{array}{r} 15 \\ +20 \\ \hline 35 \end{array}$$

**81.** $-12 < -8$; D

**85.** True; consider the values on a number line.

**89.** Answers will vary.

## EXERCISE SET 2.2

**1.** $8 + 2 = 10$

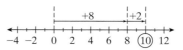

**5.** $-13 + 7 = -6$

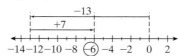

**9.** $|-6| + |-2| = 8$
Common sign is negative, so answer is $-8$.

**13.** $|6| - |-2| = 4$
$6 > 2$ so answer is $+4$.

**17.** $|-5| - |3| = 2$
$5 > 3$ so answer is $-2$.

**21.** $|-12| + |-12| = 24$
Common sign is negative, so answer is $-24$.

**25.** $|-123| + |-100| = 223$
Common sign is negative, so answer is $-223$.

**29.** $|12| - |-5| = 7$
$12 > 5$, so answer is $+7$.

**33.** $|-12| - |3| = 9$
$12 > 3$, so answer is $-9$.

**37.** $|57| - |-37| = 20$
$57 > 37$, so answer is $+20$.

**41.** $|-67| - |34| = 33$
$67 > 34$, so answer is $-33$.

**45.** $|-82| + |-43| = 125$
Common sign is negative, so $-125$.

**49.** $-52 + (-77) + (-117)$ $\quad$ add from left to right
$\phantom{x} = -129 + (-117)$ $\quad$ add from left to right
$\phantom{x} = -246$

**53.** $(-10) + 14 + 25 + (-16)$ $\quad$ add from left to right
$\phantom{x} = 4 + 25 + (-16)$ $\quad\quad$ add from left to right
$\phantom{x} = 29 + (-16)$ $\quad\quad\quad$ add from left to right
$\phantom{x} = 13$

**57.** $5 + (-1) + 17$ $\quad$ add from left to right
$\phantom{x} = 4 + 17$ $\quad\quad$ add from left to right
$\phantom{x} = 21$

**61.** $-3 + (-8) + 12 + (-1)$     add from left to right
    $= -11 + 12 + (-1)$     add from left to right
    $= 1 + (-1)$     add from left to right
    $= 0$

**65.** $x + y$     substitute values
    $= (-20) + (-50)$     add from left to right
    $= -70$

**69.** $-10 + 12 = 2$
The temperature at 11 p.m. was 2°C.

**73.** Team 1 Total $= -2 + (-13) + 20 + 2$
             $= 7$
Team 2 Total $= 5 + 11 + (-7) + (-3)$
             $= 6$
$7 > 6$ so Team 1 is the winning team.

**77.** $-170 + (-181) + (-230)$
    $= -351 + (-230)$
    $= -581$
The total U.S. trade balance was $-\$581$ billion.

**81.** $52 - 52 = 0$

**85.** True; add any two negative numbers on a number line to verify.

**89.** Answers will vary.

### EXERCISE SET 2.3

**1.** $5 - 5 = 5 + (-5) = 0$

**5.** $3 - 8 = 3 + (-8) = -5$

**9.** $-5 - (-8) = -5 + 8 = 3$

**13.** $2 - 16 = 2 + (-16) = -14$

**17.** $-15 - (-15) = -15 + 15 = 0$

**21.** $30 - 45 = 30 + (-45) = -15$

**25.** $-230 - 870 = -230 + (-870) = -1100$

**29.** $-7 - (-3) = -7 + 3 = -4$

**33.** $-20 - 18 = -20 + (-18) = -38$

**37.** $2 - (-11) = 2 + 11 = 13$

**41.** $12 - 5 - 7 = 12 + (-5) + (-7)$
              $= 7 + (-7)$
              $= 0$

**45.** $-10 + (-5) - 12 = -10 + (-5) + (-12)$
                     $= -15 + (-12)$
                     $= -27$

**49.** $-(-6) - 12 + (-16) = 6 + (-12) + (-16)$
                       $= -6 + (-16)$
                       $= -22$

**53.** $-3 + 4 - (-23) - 10 = -3 + 4 + 23 + (-10)$
                        $= 1 + 23 + (-10)$
                        $= 24 + (-10)$
                        $= 14$

**57.** $x - y = 6 - (-30)$     show substitution exactly
          $= 6 + (30)$     change subtraction to addition
          $= 36$

**61.** $x - y = (1) - (-18)$     show substitution exactly
          $= 1 + (18)$     change subtraction to addition
          $= 19$

**65.** Subtract check amounts from the checking account, and add the deposit.
$125 - 117 + 45 - 69$
  $= 125 + (-117) + 45 + (-69)$
  $= 8 + 45 + (-69)$
  $= 53 + (-69)$
  $= -16$
Aaron has overdrawn his checking account by $16.

**69.** $600 - (-52) = 600 + 52 = 652$ feet

**73.** $167 - (-215) = 167 + 215 = 382$
The surface of Mercury is 382°C warmer than the surface of Neptune.

**77.** $663 - 912 = 663 + (-912) = -249$
The U.S. trade balance in 1998 was $-\$249$ billion.

**81.** $1 \cdot 8 = 8$

**85.** $x - y - z = (-4) - (3) - (15)$     show substitution exactly
         $= -4 + (-3) + (-15)$     change subtraction to addition
         $= -7 + (-15)$     add left to right
         $= -22$

**89.** $|-3| - |-7| = 3 - 7 = 3 + (-7) = -4$

**93.** False. $|-8 - 3| = |-8 + (-3)| = |-11| = 11$

**97.** Answers may vary.

### EXERCISE SET 2.4

**1.** $-2(-3) = 6$

**5.** $8(-8) = -64$

**9.** $6(-4)(2) = -24(2) = -48$

**13.** $-4(4)(-5) = -16(-5) = 80$

**17.** $-5(3)(-1)(-1) = -15(-1)(-1)$
                 $= 15(-1)$
                 $= -15$

**21.** $(-3)^3 = (-3)(-3)(-3) = 9(-3) = -27$

**25.** $(-2)^3 = (-2)(-2)(-2) = 4(-2) = -8$

**29.** $\dfrac{-30}{6} = -5$

**33.** $\dfrac{0}{14} = 0$

**37.** $\dfrac{39}{-3} = -13$

**41.** $-4(3) = -12$

**45.** $-7(-6) = 42$

**49.** $(-4)^2 = (-4)(-4) = 16$

**53.** $-\dfrac{56}{8} = -7$

**57.** $4(-4)(-3) = -16(-3) = 48$

**61.** $3 \cdot (-2) \cdot 0 = (-6) \cdot 0 = 0$

**65.** $240 \div (-40) = -6$

**69.** $(-1)^4 = (-1)(-1)(-1)(-1)$
          $= 1 \cdot (-1)(-1)$
          $= (-1) \cdot (-1)$
          $= 1$

**73.** $-2(3)(5)(-6) = (-6)(5)(-6)$
$\qquad\qquad\quad = (-30)(-6)$
$\qquad\qquad\quad = 180$

**77.** $-2(-2)(-5) = 4(-5) = -20$

**81.**
$\qquad 25$
$\underline{\times\ 82}$
$\qquad 50$
$\underline{2000}$
$\ 2050$
$25 \cdot (-82) = -2050$

**85.** $a \cdot b = (3) \cdot (-2) = -6$

**89.** $\dfrac{x}{y} = \dfrac{5}{-5}$ $\qquad$ show substitution exactly
$\qquad = -1$

**93.** $\dfrac{x}{y} = \dfrac{-36}{-6}$ $\qquad$ show substitution exactly
$\qquad = 6$

**97.** $x \cdot y = (0)(-6)$ $\qquad$ show substitution exactly
$\qquad\quad = 0$ $\qquad\qquad$ multiplication property of zero
$\dfrac{x}{y} = \dfrac{0}{-6}$ $\qquad$ show substitution exactly
$\qquad = 0$ $\qquad$ zero division property

**101.** $5(-20) = -100$
He is at a depth of 100 feet.

**105.** $-62 \cdot 3 = -186$
After three years the income would be $-\$186$ million.

**109.** $(3 \cdot 5)^2 = (3 \cdot 5)(3 \cdot 5) = (15)(15) = 225$

**113.** $12 \div\ 4 - 2 + 7 = 3 - 2 + 7 = 8$

**117.** True

**121.** False

**125.** Answers may vary.

## EXERCISE SET 2.5

**1.** $-1(-2) + 1$ $\qquad$ multiply before adding
$\quad = 2 + 1$
$\quad = 3$

**5.** $9 - 12 - 4$ $\qquad$ change to addition
$\quad = 9 + (-12) + (-4)$
$\quad = -3 + (-4)$
$\quad = -7$

**9.** $5(-9) + 2$ $\qquad$ multiply before adding
$\quad = -45 + 2$
$\quad = -43$

**13.** $25 \div\ (-5) + 12$ $\qquad$ divide before adding
$\quad = -5 + 12$
$\quad = 7$

**17.** $\dfrac{24}{10 + (-4)}$ $\qquad$ simplify the bottom of the
$\qquad\qquad\qquad$ fraction bar first
$\quad = \dfrac{24}{6}$
$\quad = 4$

**21.** $(-19) - 12(3)$ $\qquad$ multiply before adding
$\quad = (-19) - 36$ $\qquad$ change to addition
$\quad = (-19) + (-36)$
$\quad = -55$

**25.** $\left(8 + (-4)\right)^2$ $\qquad$ grouping symbols first
$\quad = (4)^2$
$\quad = 16$

**29.** $16 - (-3)^4$ $\qquad$ exponents first
$\quad = 16 - 81$ $\qquad$ change to addition
$\quad = 16 + (-81)$
$\quad = -65$

**33.** $7 \cdot 8^2 + 4$ $\qquad$ exponents first
$\quad = 7 \cdot 64 + 4$ $\qquad$ multiply before adding
$\quad = 448 + 4$
$\quad = 452$

**37.** $(3 - 12) \div\ 3$ $\qquad$ grouping symbols first
$\quad = (-9) \div\ 3$
$\quad = -3$

**41.** $(5 - 9)^2 \div\ (4 - 2)^2$ $\qquad$ grouping symbols first
$\quad = (-4)^2 \div\ (2)^2$ $\qquad\quad$ exponents next
$\quad = 16 \div\ 4$
$\quad = 4$

**45.** $(-12 - 20) \div\ 16 - 25$ $\qquad$ grouping symbols first
$\quad = (-32) \div\ 16 - 25$ $\qquad$ divide before subtracting
$\quad = -2 - 25$
$\quad = -27$

**49.** $(2 - 7) \cdot (6 - 19)$ $\qquad$ grouping symbols first
$\quad = (-5) \cdot (-13)$
$\quad = 65$

**53.** $(-36 \div\ 6) - (4 \div\ 4)$ $\qquad$ divide before subtracting
$\quad = -6 - 1$
$\quad = -7$

**57.** $(-5)^2 - 6^2$ $\qquad$ exponents first (watch the signs)
$\quad = 25 - 36$
$\quad = -11$

**61.** $2(8 - 10)^2 - 5(1 - 6)^2$
$\quad = 2(-2)^2 - 5(-5)^2$
$\quad = 2(-2)(-2) - 5(-5)(-5)$
$\quad = 2(-2)(-2) + (-5)(-5)(-5)$
$\quad = (-4)(-2) + (25)(-5)$
$\quad = 8 + (-125)$
$\quad = -117$

**65.** $\dfrac{(-7)(-3) - (4)(3)}{3\left(7 \div\ (3 - 10)\right)} = \dfrac{(-7)(-3) + (-4)(3)}{3\left(7 \div\ (3 + (-10))\right)}$

$\qquad\qquad\qquad\qquad = \dfrac{21 + (-12)}{3\left(7 \div\ (-7)\right)}$

$\qquad\qquad\qquad\qquad = \dfrac{9}{3(-1)}$

$\qquad\qquad\qquad\qquad = \dfrac{9}{-3}$

$\qquad\qquad\qquad\qquad = -3$

**69.** $2x - y^2$ $\qquad$ show substitution exactly
$\quad = 2(-2) - (4)^2$ $\qquad$ exponents first
$\quad = 2(-2) - 16$ $\qquad$ multiplication before
$\qquad\qquad\qquad\qquad$ subtraction
$\quad = -4 - 16$
$\quad = -20$

**73.** $\dfrac{5y}{z}$      show substitution exactly

$\quad = \dfrac{5 \cdot 4}{-1}$      simplify the top of the fraction bar first

$\quad = \dfrac{20}{-1}$

$\quad = -20$

**77.** $-z^2 = -(-4)^2 = -16$

**81.** $2x^3 = 2(-3)^3 = 2(-27) = -54$

**85.** $90 - 45 = 45$

**89.** $p = 2(9) + 2(6) = 18 + 12 = 30$ feet

**93.** $-6 \cdot (10 - 4) = -6 \cdot 6 = -36$

**97.** $x^3 - y^2 = 21^3 - (-19)^2$      substitute
$\quad = 21 \cdot 21 \cdot 21 - (-19)(-19)$
$\quad = 21 \cdot 21 \cdot 21 + (19)(-19)$
$\quad = 441 \cdot 21 + (-361)$
$\quad = 9261 + (-361)$
$\quad = 8900$

**101.** Answers may vary.

### Chapter 2 test

**1.** $-5 + 8 = 3$

**5.** $(-18) + (-12) = -30$

**9.** $|-25| + (-13) = 25 + (-13) = 12$

**13.** $(-8) + 9 \div (-3) = -8 + (-3) = -11$

**17.** $-(-7)^2 \div 7 \cdot (-4) = -49 \div 7 \cdot (-4)$
$\quad = -7 \cdot (-4)$
$\quad = 28$

**21.** $\dfrac{(-3)(-2) + 12}{-1(-4 - 5)} = \dfrac{6 + 12}{-1(-9)} = \dfrac{18}{9} = 2$

**25.** $\dfrac{3(2)}{2(-3)} = \dfrac{6}{-6} = -1$

**29.** $6288 - (-25{,}354) = 6288 + 25{,}354 = 31{,}642$
The difference in elevation is 31,642 feet.

# Chapter 3

### Exercise Set 3.1

**1.** $3x + 5x = (3 + 5)x = 8x$

**5.** $4c + c - 7c = (4 + 1 - 7)c = -2c$

**9.** $4a + 3a + 6a - 8 = (4 + 3 + 6)a - 8 = 13a - 8$

**13.** $-2(11y) = (-2 \cdot 11)y = -22y$

**17.** $2(y + 2) = 2 \cdot y + 2 \cdot 2 = 2y + 4$

**21.** $-4(3x + 7) = -4 \cdot 3x + (-4) \cdot 7$
$\quad = -12x - 28$

**25.** $-4(6n - 5) + 3n = -4 \cdot 6n + (-4) \cdot (-5) + 3n$
$\quad = -24n + 20 + 3n$
$\quad = -21n + 20$

**29.** $3 + 6(w + 2) + w = 3 + 6 \cdot w + 6 \cdot 2 + w$
$\quad = 3 + 6w + 12 + w$
$\quad = 6w + w + 3 + 12$
$\quad = 7w + 15$

**33.** $-(5x - 1) - 10 = -1(5x - 1) - 10$
$\quad = -1 \cdot 5x + (-1) \cdot (-1) - 10$
$\quad = -5x + 1 - 10$
$\quad = -5x - 9$

**37.** $z - 8z = (1 - 8)z = -7z$

**41.** $2y - 6 + 4y - 8 = 2y + 4y - 6 - 8$
$\quad = 6y - 14$

**45.** $2(x + 1) + 20 = 2 \cdot x + 2 \cdot 1 + 20$
$\quad = 2x + 2 + 20$
$\quad = 2x + 22$

**49.** $-5(z + 3) + 2z = -5 \cdot z + (-5) \cdot 3 + 2z$
$\quad = -5z - 15 + 2z$
$\quad = -5z + 2z - 15$
$\quad = -3z - 15$

**53.** $-7(x + 5) + 5(2x + 1)$
$\quad = -7 \cdot x + (-7) \cdot 5 + 5 \cdot 2x + 5 \cdot 1$
$\quad = -7x - 35 + 10x + 5$
$\quad = -7x + 10x - 35 + 5$
$\quad = 3x - 30$

**57.** $-3(n - 1) - 4n = -3 \cdot n - (-3) \cdot 1 - 4n$
$\quad = -3n + 3 - 4n$
$\quad = -3n - 4n + 3$
$\quad = -7n + 3$

**61.** $6(2x - 1) - 12x = 6 \cdot 2x - 6 \cdot 1 - 12x$
$\quad = 12x - 6 - 12x$
$\quad = 12x - 12x - 6$
$\quad = -6$

**65.** $-(4xy - 10) + 2(3xy + 5)$
$\quad = -1(4xy - 10) + 2 \cdot 3xy + 2 \cdot 5$
$\quad = -1 \cdot 4xy - (-1) \cdot 10 + 6xy + 10$
$\quad = -4xy + 10 + 6xy + 10$
$\quad = -4xy + 6xy + 10 + 10$
$\quad = 2xy + 20$

**69.** $5y - 2(y - 1) + 3 = 5y - 2y - 2(-1) + 3$
$\quad = 3y + 2 + 3$
$\quad = 3y + 5$

**73.** $5y + 16 + 3y + 4y + 2y + 6$
$\quad = 5y + 3y + 4y + 2y + 16 + 6$
$\quad = (14y + 22)$ meters

**77.** $\text{Area} = (\text{side})^2$
$\quad = (4z)^2$
$\quad = (4z)(4z)$
$\quad = (4)(4)(z)(z)$
$\quad = 16z^2$ square centimeters

**81.** $-4 - (-12) = -4 + 12 = 8$

**85.** $9684q - 686 - 4860q + 12{,}960$
$\quad = (9684 - 4860)q + (12{,}960 - 686)$
$\quad = 4824q + 12{,}274$

**89.** Add the areas of the two rectangles.

$\text{Area} = \dfrac{\text{Area of Left}}{\text{Rectangle}} + \dfrac{\text{Area of Right}}{\text{Rectangle}}$

$\quad = 7(2x + 1) + 3(2x + 3)$
$\quad = 7(2x) + 7(1) + 3(2x) + 3(3)$
$\quad = 14x + 7 + 6x + 9$
$\quad = (14 + 6)x + 16$
$\quad = (20x + 16)$

The area is $(20x + 16)$ square miles.

## EXERCISE SET 3.2

**1.** $x - 8 = 2$
$10 - 8 \stackrel{?}{=} 2$
$2 \stackrel{?}{=} 2$ True
Yes, 10 is a solution.

**5.** $x + 12 = 7$
$-5 + 12 \stackrel{?}{=} 7$
$7 \stackrel{?}{=} 7$ True
Yes, $-5$ is a solution.

**9.** $h - 8 = -8$
$0 - 8 \stackrel{?}{=} -8$
$-8 \stackrel{?}{=} -8$ True
Yes, 0 is a solution.

**13.** $a + 5 = 23$
$a + 5 - 5 = 23 - 5$
$a = 18$
*Check:*
$a + 5 = 23$
$18 + 5 \stackrel{?}{=} 23$
$23 \stackrel{?}{=} 23$ True
The solution is 18.

**17.** $7 = y - 2$
$7 + 2 = y - 2 + 2$
$9 = y$
*Check:*
$7 = y - 2$
$7 \stackrel{?}{=} 9 - 2$
$7 \stackrel{?}{=} 7$ True
The solution is 9.

**21.** $3x = 2x + 11$
$3x - 2x = 2x + 11 - 2x$
$x = 11$
*Check:*
$3x = 2x + 11$
$3(11) \stackrel{?}{=} 2(11) + 11$
$33 \stackrel{?}{=} 22 + 11$
$33 \stackrel{?}{=} 33$ True
The solution is 11.

**25.** $x - 3 = -1 + 4$
$x - 3 = 3$
$x - 3 + 3 = 3 + 3$
$x = 6$
*Check:*
$x - 3 = -1 + 4$
$6 - 3 \stackrel{?}{=} -1 + 4$
$3 \stackrel{?}{=} 3$ True
The solution is 6.

**29.** $-7 + 10 = m - 5$
$3 = m - 5$
$3 + 5 = m - 5 + 5$
$8 = m$
*Check:*
$-7 + 10 = m - 5$
$-7 + 10 \stackrel{?}{=} 8 - 5$
$3 \stackrel{?}{=} 3$ True
The solution is 8.

**33.** $2(5x - 3) = 11x$
$2 \cdot 5x - 2 \cdot 3 = 11x$
$10x - 6 = 11x$
$10x - 6 - 10x = 11x - 10x$
$-6 = x$
*Check:*
$2(5x - 3) = 11x$
$2(5 \cdot (-6) - 3) \stackrel{?}{=} 11 \cdot (-6)$
$2(-30 - 3) \stackrel{?}{=} -66$
$2(-33) \stackrel{?}{=} -66$
$-66 \stackrel{?}{=} -66$ True
The solution is $-6$.

**37.** $-8x + 4 + 9x = -1 + 7$
$x + 4 = 6$
$x + 4 - 4 = 6 - 4$
$x = 2$
*Check:*
$-8x + 4 + 9x = -1 + 7$
$-8(2) + 4 + 9(2) \stackrel{?}{=} -1 + 7$
$-16 + 4 + 18 \stackrel{?}{=} 6$
$6 \stackrel{?}{=} 6$ True
The solution is 2.

**41.** $7x + 14 - 6x = -4 - 10$
$x + 14 = -14$
$x + 14 - 14 = -14 - 14$
$x = -28$
*Check:*
$7x + 14 - 6x = -4 - 10$
$7(-28) + 14 - 6(-28) \stackrel{?}{=} -4 - 10$
$-196 + 14 + 168 \stackrel{?}{=} -14$
$-14 \stackrel{?}{=} -14$ True
The solution is $-28$.

**45.** $67 = z + 67$
$67 + (-67) = z + 67 + (-67)$
$0 = z$

**49.** $z - 23 = -88$
$z - 23 + 23 = -88 + 23$
$z = -65$

**53.** $-12 + x = -15$
$-12 + x + 12 = -15 + 12$
$x = -3$

**57.** $8(3x - 2) = 25x$
$8 \cdot 3x - 8 \cdot 2 = 25x$
$24x - 16 = 25x$
$24x - 16 - 24x = 25x - 24x$
$-16 = x$

**61.** $50y = 7(7y + 4)$
$50y = 7 \cdot 7y + 7 \cdot 4$
$50y = 49y + 28$
$50y - 49y = 49y + 28 - 49y$
$y = 28$

**65.** In 1995 there were about 900 trumpeter swans. In 1985 there were about 200 trumpeter swans. Thus, there were about $900 - 200 = 700$ more trumpeter swans in 1995 than in 1985.

**69.** $\dfrac{-3}{-3} = 1$

**73.**
$$x - 76{,}862 = 86{,}102$$
$$x - 76{,}862 + 76{,}862 = 86{,}102 + 76{,}862$$
$$x = 162{,}964$$

**77.**
$$I = R - E$$
$$3056 = R - 103{,}090$$
$$3056 + 103{,}090 = R - 103{,}090 + 103{,}090$$
$$106{,}146 = R$$
The total revenues were $106,146 million.

## EXERCISE SET 3.3

**1.** $5x = 20$
$$\frac{5x}{5} = \frac{20}{5}$$
$$x = 4$$

**5.** $4y = 0$
$$\frac{4y}{4} = \frac{0}{4}$$
$$y = 0$$

**9.** $-3x = -15$
$$\frac{-3x}{-3} = \frac{-15}{-3}$$
$$x = 5$$

**13.** $16 = 10t - 8t$
$$16 = 2t$$
$$\frac{16}{2} = \frac{2t}{2}$$
$$8 = t$$

**17.** $4 - 10 = -3z$
$$-6 = -3z$$
$$\frac{-6}{-3} = \frac{-3z}{-3}$$
$$2 = z$$

**21.** $-10x = 10$
$$\frac{-10x}{-10} = \frac{10}{-10}$$
$$x = -1$$

**25.** $0 = 3x$
$$\frac{0}{3} = \frac{3x}{3}$$
$$0 = x$$

**29.** $10z - 3z = -63$
$$7z = -63$$
$$\frac{7z}{7} = \frac{-63}{7}$$
$$z = -9$$

**33.** $12 = 13y - 10y$
$$12 = 3y$$
$$\frac{12}{3} = \frac{3y}{3}$$
$$4 = y$$

**37.** $18 - 11 = 7x$
$$7 = 7x$$
$$\frac{7}{7} = \frac{7x}{7}$$
$$1 = x$$

**41.** $10p - 11p = 25$
$$-1p = 25$$
$$\frac{-1p}{-1} = \frac{25}{-1}$$
$$p = -25$$

**45.** $10 = 7t - 12t$
$$10 = -5t$$
$$\frac{10}{-5} = \frac{-5t}{-5}$$
$$-2 = t$$

**49.** $4r - 9r = -20$
$$-5r = -20$$
$$\frac{-5r}{-5} = \frac{-20}{-5}$$
$$r = 4$$

**53.** $3w - 12w = -27$
$$-9w = -27$$
$$\frac{-9w}{-9} = \frac{-27}{-9}$$
$$w = 3$$

**57.** $23x - 25x = 7 - 9$
$$-2x = -2$$
$$\frac{-2x}{-2} = \frac{-2}{-2}$$
$$x = 1$$

**61.** Let $x$ represent a number. Twice a number is $2x$. Thus, twice a number, decreased by 17 is $2x - 17$.

**65.** Let $x$ represent a number. The product of a number and $-5$ is $-5x$. Thus, the quotient of 45 and the product of a number and $-5$ is $\frac{45}{-5x}$.

**69.** $\dfrac{x - 5}{2} = \dfrac{-5 - 5}{2} = \dfrac{-10}{2} = -5$

**73.** Answers may vary.

**77.**
$$d = r \cdot t$$
$$780 = 65t$$
$$\frac{780}{65} = \frac{65t}{65}$$
$$12 = t$$
It will take 12 hours.

## EXERCISE SET 3.4

**1.**
$$2x - 6 = 0$$
$$2x - 6 + 6 = 0 + 6$$
$$2x = 6$$
$$\frac{2x}{2} = \frac{6}{2}$$
$$x = 3$$

**5.**
$$6 - n = 10$$
$$6 - n - 6 = 10 - 6$$
$$-n = 4$$
$$\frac{-n}{-1} = \frac{4}{-1}$$
$$n = -4$$

**9.**
$$3x - 7 = 4x + 5$$
$$3x - 7 + 7 = 4x + 5 + 7$$
$$3x = 4x + 12$$
$$3x - 4x = 4x + 12 - 4x$$
$$-x = 12$$
$$\frac{-x}{-1} = \frac{12}{-1}$$
$$x = -12$$

**13.**
$$-2(y + 4) = 2$$
$$-2y - 8 = 2$$
$$-2y - 8 + 8 = 2 + 8$$
$$-2y = 10$$
$$\frac{-2y}{-2} = \frac{10}{-2}$$
$$y = -5$$

**17.**
$$8 - t = 3$$
$$8 - t - 8 = 3 - 8$$
$$-t = -5$$
$$\frac{-t}{-1} = \frac{-5}{-1}$$
$$t = 5$$

**21.**
$$2n + 8 = 0$$
$$2n + 8 - 8 = 0 - 8$$
$$2n = -8$$
$$\frac{2n}{2} = \frac{-8}{2}$$
$$n = -4$$

**25.**
$$3r + 4 = 19$$
$$3r + 4 - 4 = 19 - 4$$
$$3r = 15$$
$$\frac{3r}{3} = \frac{15}{3}$$
$$r = 5$$

**29.**
$$2 = 3z - 4$$
$$2 + 4 = 3z - 4 + 4$$
$$6 = 3z$$
$$\frac{6}{3} = \frac{3z}{3}$$
$$2 = z$$

**33.**
$$-7c + 1 = -20$$
$$-7c + 1 - 1 = -20 - 1$$
$$-7c = -21$$
$$\frac{-7c}{-7} = \frac{-21}{-7}$$
$$c = 3$$

**37.**
$$8m + 79 = -1$$
$$8m + 79 - 79 = -1 - 79$$
$$8m = -80$$
$$\frac{8m}{8} = \frac{-80}{8}$$
$$m = -10$$

**41.**
$$-5 = -13 - 8k$$
$$-5 + 13 = -13 - 8k + 13$$
$$8 = -8k$$
$$\frac{8}{-8} = \frac{-8k}{-8}$$
$$-1 = k$$

**45.**
$$-2y - 10 = 5y + 18$$
$$-2y - 10 + 10 = 5y + 18 + 10$$
$$-2y = 5y + 28$$
$$-2y - 5y = 5y + 28 - 5y$$
$$-7y = 28$$
$$\frac{-7y}{-7} = \frac{28}{-7}$$
$$y = -4$$

**49.**
$$9 - 3x = 14 + 2x$$
$$9 - 3x - 9 = 14 + 2x - 9$$
$$-3x = 5 + 2x$$
$$-3x - 2x = 5 + 2x - 2x$$
$$-5x = 5$$
$$\frac{-5x}{-5} = \frac{5}{-5}$$
$$x = -1$$

**53.**
$$2t - 1 = 3(t + 7)$$
$$2t - 1 = 3t + 21$$
$$2t - 1 + 1 = 3t + 21 + 1$$
$$2t = 3t + 22$$
$$2t - 3t = 3t + 22 - 3t$$
$$-1t = 22$$
$$\frac{-t}{-1} = \frac{22}{-1}$$
$$t = -22$$

**57.**
$$10 + 5(z - 2) = 4z + 1$$
$$10 + 5z - 10 = 4z + 1$$
$$5z = 4z + 1$$
$$5z - 4z = 4z + 1 - 4z$$
$$z = 1$$

**61.** The sum of $-42$ and 16 is $-26$ translates to
$$-42 + 16 = -26.$$

**65.** Three times the difference of $-14$ and 2 amounts to $-48$ translates to $3(-14 - 2) = -48$.

**69.** $x^3 - 2xy = 3^3 - 2(3)(-1)$
$$= 27 - (-6)$$
$$= 27 + 6$$
$$= 33$$

**73.** $(2x - y)^2 = \left(2(3) - (-1)\right)^2$
$$= (6 + 1)^2$$
$$= (7)^2$$
$$= 49$$

**77.**
$$(-8)^2 + 3x = 5x + 4^3$$
$$64 + 3x = 5x + 64$$
$$64 + 3x - 5x = 5x + 64 - 5x$$
$$64 - 2x = 64$$
$$64 - 2x - 64 = 64 - 64$$
$$-2x = 0$$
$$\frac{-2x}{-2} = \frac{0}{-2}$$
$$x = 0$$

**81.** Answers may vary.

### EXERCISE SET 3.5

**1.** A number added to $-5$ is $-7$ translates to
$$-5 + x = -7.$$

**5.** A number subtracted from $-20$ amounts to 104 translates to $-20 - x = 104$.

**9.** $3x + 9 = 33$
$$3x = 24$$
$$x = 8$$

**13.** $3 + 4 + x = 16$
$$7 + x = 16$$
$$x = 9$$

**17.** $x - 3 = 45 - x$
$$2x = 48$$
$$x = 24$$

**21.** $8 - x = \dfrac{15}{5}$
$$8 - x = 3$$
$$-x = -5$$
$$x = 5$$

**25.** $5x - 40 = x + 8$
$$4x = 48$$
$$x = 12$$

**29.** Let $x$ be the number of electoral votes for Bob Dole. Then $x + 220$ is the number of electoral votes for Bill Clinton.
$$x + x + 220 = 538$$
$$2x + 220 = 538$$
$$2x = 318$$
$$x = 159$$
Dole received 159 votes and Clinton received $159 + 220 = 379$ votes.

**33.** Let $x$ be the speed of the Dodge truck. Then $2x$ is the speed of the Toyota Camry.
$$x + 2x = 105$$
$$3x = 105$$
$$x = 35$$
The truck's speed is 35 mph and the car's speed is $2 \cdot 35 = 70$ mph.

**37.** Let $x$ be the capacity of Neyland Stadium. Then $x + 4647$ is the capacity of Michigan Stadium.
$$x + x + 4647 = 210,355$$
$$2x + 4647 = 210,355$$
$$2x = 205,708$$
$$x = 102,854$$
Neyland Stadium has a capacity of 102,854 and Michigan Stadium has a capacity of $102,854 + 4647 = 107,501$.

**41.** Let $x$ be the points scored by the Tennessee Lady Volunteers. Then $x + 19$ are the points scored by the Connecticut Huskies.
$$x + x + 19 = 123$$
$$2x + 19 = 123$$
$$2x = 104$$
$$x = 52$$
The Connecticut Huskies scored $52 + 19 = 71$ points.

**45.** To round 1026 to the nearest hundred, observe that the digit in the tens place is 2. Since this digit is less than 5, we do not add 1 to the digit in the hundreds place. The number 1026 rounded to the nearest hundred is 1000.

**49.** Answers may vary.

**53.** $P = C + M$
$$29 = 12 + M$$
$$29 - 12 = 12 + M - 12$$
$$17 = M$$
The markup on the blouse is $17.

## CHAPTER 3 TEST

**1.** $7x - 5 - 12x + 10 = (7 - 12)x + (-5 + 10)$
$$= -5x + 5$$

**5.** $A = 3(3x - 1) = 9x - 3$
The area is $(9x - 3)$ square meters.

**9.** $5 + 4z = 37$
$$5 + 4z - 5 = 37 - 5$$
$$4z = 32$$
$$\dfrac{4z}{4} = \dfrac{32}{4}$$
$$z = 8$$

**13.** $-4x + 7 = 15$
$$-4x + 7 - 7 = 15 - 7$$
$$-4x = 8$$
$$\dfrac{-4x}{-4} = \dfrac{8}{-4}$$
$$x = -2$$

**17.** $10y - 1 = 7y + 20$
$$10y - 1 + 1 = 7y + 20 + 1$$
$$10y = 7y + 21$$
$$10y - 7y = 7y + 21 - 7y$$
$$3y = 21$$
$$\dfrac{3y}{3} = \dfrac{21}{3}$$
$$y = 7$$

**21. a.** Let $x$ represent "a number." The product of a number and 17 is $17x$.

**b.** Let $x$ represent "a number." Twice a number is $2x$. Twice a number subtracted from 20 is $20 - 2x$.

# Chapter 4

## EXERCISE SET 4.1

**1.** 1 out of 3 equal parts is shaded: $\dfrac{1}{3}$.

**5.** 7 out of 12 equal parts are shaded: $\dfrac{7}{12}$.

**9.** 4 out of 9 equal parts are shaded: $\dfrac{4}{9}$.

**13.** $\dfrac{11}{4}$      **17.** $\dfrac{3}{2}$      **21.** $\dfrac{17}{6}$

**25.** $\dfrac{\text{Freshman} \rightarrow 42}{\text{Students} \rightarrow 131} = \dfrac{42}{131}$

Thus, $\dfrac{42}{131}$ of the students are freshmen.

**29.** $\dfrac{\text{Injury-related visits} \rightarrow 4}{\text{Total visits} \rightarrow 10} = \dfrac{4}{10}$

Thus, $\dfrac{4}{10}$ of the visits are injury-related.

**33.** $\dfrac{\text{Number of inches} \rightarrow 5}{\text{Inches in a foot} \rightarrow 12} = \dfrac{5}{12}$

Thus, 5 inches represents $\dfrac{5}{12}$ of a foot.

**37.** $\dfrac{\text{Number of girls} \rightarrow 7}{\text{Number of children} \rightarrow 11} = \dfrac{7}{11}$

Thus the girls are $\dfrac{7}{11}$ of the children.

**41.** To graph $\dfrac{1}{4}$ on a number line, divide the distance from 0 to 1 into 4 equal parts. Then start at 0 and count over part 1.

**45.** To graph $\dfrac{8}{5}$ on a number line, divide the distance from 0 to 1 into 5 equal parts and divide the distance from 1 to 2 into 5 equal parts. Then start at 0 and count over 8 parts.

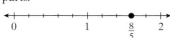

**49.** To graph $\dfrac{3}{8}$ on a number line, divide the distance from 0 to 1 into 8 equal parts. Then start at 0 and count over 3 parts.

**53.** $\dfrac{2}{3} = \dfrac{?}{21}$

$\dfrac{2 \cdot 7}{3 \cdot 7} = \dfrac{14}{21}$

**57.** $\dfrac{1}{2} = \dfrac{?}{30}$

$\dfrac{1 \cdot 15}{2 \cdot 15} = \dfrac{15}{30}$

**61.** $2 = \dfrac{?}{5}$

$\dfrac{2}{1} = \dfrac{?}{5}$

$\dfrac{2 \cdot 5}{1 \cdot 5} = \dfrac{10}{5}$

**65.** $\dfrac{2y}{3} = \dfrac{?}{12}$

$\dfrac{2y \cdot 4}{3 \cdot 4} = \dfrac{8y}{12}$

**69.** $\dfrac{4}{3} = \dfrac{?}{36x}$

$\dfrac{4 \cdot 12x}{3 \cdot 12x} = \dfrac{48x}{36x}$

**73.** $1 = \dfrac{?}{36x}$

$\dfrac{1}{1} = \dfrac{?}{36x}$

$\dfrac{1 \cdot 36x}{1 \cdot 36x} = \dfrac{36x}{36x}$

**77.** The smallest numerator of all the fractions with a denominator of 100 is 26, which is the United States.

**81.** $\dfrac{-5}{1} = -5 \div 1 = -5$

**85.** $\dfrac{-8}{-8} = -8 \div (-8) = 1$

**89.** $\dfrac{3}{1} = 3 \div 1 = 3$

**93.** $5^3 = 5 \cdot 5 \cdot 5 = 125$

**97.** $2^3 \cdot 3 = 2 \cdot 2 \cdot 2 \cdot 3 = 24$

**101.** Answers may vary.

**105.** Total number of licenses $= 87 + 56 + 21 + 8$
$\qquad\qquad\qquad\qquad\qquad = 172$

$\dfrac{56}{172}$ of the licensees are colleges or universities.

### Exercise Set 4.2

**1.** $20 = 2 \cdot 10$

$\qquad 2 \cdot 2 \cdot 5 = 2^2 \cdot 5$

**5.** $64 = 8 \cdot 8$

$\qquad 2 \cdot 4 \cdot \quad 2 \cdot 4$

$\qquad 2 \cdot 2 \cdot 2 \cdot 2 \cdot 2 \cdot 2 = 2^6$

**9.** $\dfrac{3}{12} = \dfrac{3}{3 \cdot 4} = \dfrac{1}{4}$

**13.** $\dfrac{14}{16} = \dfrac{2 \cdot 7}{2 \cdot 8} = \dfrac{7}{8}$

**17.** $\dfrac{35}{42} = \dfrac{7 \cdot 5}{7 \cdot 2 \cdot 3} = \dfrac{5}{6}$

**21.** $\dfrac{16}{24} = \dfrac{2 \cdot 2 \cdot 2 \cdot 2}{2 \cdot 2 \cdot 2 \cdot 3} = \dfrac{2}{3}$

**25.** $\dfrac{39ab}{26a^2} = \dfrac{3 \cdot 13 \cdot a \cdot b}{2 \cdot 13 \cdot a \cdot a} = \dfrac{3b}{2a}$

**29.** $\dfrac{21}{49} = \dfrac{3 \cdot 7}{7 \cdot 7} = \dfrac{3}{7}$

**33.** $\dfrac{36z}{63z} = \dfrac{3 \cdot 3 \cdot 2 \cdot 2 \cdot z}{3 \cdot 3 \cdot 7 \cdot z} = \dfrac{4}{7}$

**37.** $\dfrac{12}{15} = \dfrac{2 \cdot 2 \cdot 3}{3 \cdot 5} = \dfrac{4}{5}$

**41.** $\dfrac{27xy}{90y} = \dfrac{3 \cdot 3 \cdot 3 \cdot x \cdot y}{2 \cdot 3 \cdot 3 \cdot 5 \cdot y} = \dfrac{3x}{10}$

**45.** $\dfrac{40xy}{64xyz} = \dfrac{2 \cdot 2 \cdot 2 \cdot 5 \cdot x \cdot y}{2 \cdot 2 \cdot 2 \cdot 2 \cdot 2 \cdot 2 \cdot x \cdot y \cdot z} = \dfrac{5}{8z}$

**49.** $\dfrac{2640 \text{ feet}}{5280 \text{ feet}} = \dfrac{2\cdot2\cdot2\cdot2\cdot3\cdot5\cdot11}{2\cdot2\cdot2\cdot2\cdot2\cdot3\cdot5\cdot11} = \dfrac{1}{2}$

2640 feet represents $\dfrac{1}{2}$ mile.

**53.** $16{,}000 - 8800 = 7200$

$\dfrac{\text{Number of male students} \to 7200}{\text{Number of female students} \to 16{,}000}$

$\dfrac{7200}{16{,}000} = \dfrac{9\cdot800}{20\cdot800} = \dfrac{9}{20}$

$\dfrac{9}{20}$ of the students are male.

**57.** $\dfrac{10}{24} = \dfrac{2\cdot5}{2\cdot2\cdot2\cdot3} = \dfrac{5}{12}$

$\dfrac{5}{12}$ of the width is concrete.

**61.** Answers will vary.

**65.** $2y = 2(-7) = -14$

**69.** $a^2 + 2b + 3 = 4^2 + 2(5) + 3 = 16 + 10 + 3 = 29$

**73.** False; $\dfrac{14}{42} = \dfrac{2\cdot7}{2\cdot3\cdot7} = \dfrac{1}{3}$

**77.** $\dfrac{372}{620} = \dfrac{2\cdot2\cdot3\cdot31}{2\cdot2\cdot5\cdot31} = \dfrac{3}{5}$

**81.** $3 + 1 = 4$

$\dfrac{4}{100} = \dfrac{2\cdot2}{2\cdot2\cdot5\cdot5} = \dfrac{1}{5\cdot5} = \dfrac{1}{25}$

$\dfrac{1}{25}$ of the donors have AB blood type (either Rh-positive or Rh-negative).

## EXERCISE SET 4.3

**1.** $\dfrac{7}{8}\cdot\dfrac{2}{3} = \dfrac{7\cdot2}{8\cdot3} = \dfrac{7\cdot2}{2\cdot2\cdot2\cdot3} = \dfrac{7}{2\cdot2\cdot3} = \dfrac{7}{12}$

**5.** $-\dfrac{1}{2}\cdot-\dfrac{2}{15} = \dfrac{1\cdot2}{2\cdot15} = \dfrac{1}{15}$

**9.** $3a^2\cdot\dfrac{1}{4} = \dfrac{3\cdot a^2\cdot1}{4} = \dfrac{3a^2}{4}$

**13.** $\left(\dfrac{1}{5}\right)^3 = \dfrac{1}{5}\cdot\dfrac{1}{5}\cdot\dfrac{1}{5} = \dfrac{1\cdot1\cdot1}{5\cdot5\cdot5} = \dfrac{1}{125}$

**17.** $\left(-\dfrac{2}{3}\right)^3\cdot\dfrac{1}{2} = \left(-\dfrac{2}{3}\right)\cdot\left(-\dfrac{2}{3}\right)\cdot\left(-\dfrac{2}{3}\right)\cdot\left(\dfrac{1}{2}\right)$

$\qquad = -\dfrac{2\cdot2\cdot2\cdot1}{3\cdot3\cdot3\cdot2}$

$\qquad = -\dfrac{4}{27}$

**21.** $-\dfrac{6}{15}\div\dfrac{12}{5} = -\dfrac{6}{15}\cdot\dfrac{5}{12}$

$\qquad = -\dfrac{6\cdot5}{3\cdot5\cdot2\cdot6}$

$\qquad = -\dfrac{1}{6}$

**25.** $\dfrac{11y}{20}\div\dfrac{3}{11} = \dfrac{11y}{20}\cdot\dfrac{11}{3} = \dfrac{11\cdot y\cdot11}{2\cdot2\cdot5\cdot3} = \dfrac{121y}{60}$

**29.** $\dfrac{1}{5x}\div\dfrac{5}{x^2} = \dfrac{1}{5x}\cdot\dfrac{x^2}{5} = \dfrac{x\cdot x}{5\cdot x\cdot5} = \dfrac{x}{25}$

**33.** $\dfrac{3x}{7}\div\dfrac{5}{6x} = \dfrac{3x}{7}\cdot\dfrac{6x}{5}$

$\qquad = \dfrac{3\cdot x\cdot2\cdot3\cdot x}{7\cdot5}$

$\qquad = \dfrac{18x^2}{35}$

**37.** $-\dfrac{3}{5}\div-\dfrac{4}{5} = \left(-\dfrac{3}{5}\right)\cdot\left(-\dfrac{5}{4}\right) = \dfrac{3\cdot5}{5\cdot2\cdot2} = \dfrac{3}{4}$

**41.** $\dfrac{x^2}{y}\cdot\dfrac{y^3}{x} = \dfrac{x\cdot x\cdot y\cdot y^2}{y\cdot x} = \dfrac{x\cdot y^2}{1} = xy^2$

**45.** $-3x\div\dfrac{x^2}{12} = -\dfrac{3x}{1}\cdot\dfrac{12}{x^2} = -\dfrac{3\cdot x\cdot12}{1\cdot x\cdot x} = -\dfrac{36}{x}$

**49.** $-\dfrac{19}{63y}\cdot9y^2 = -\dfrac{19}{63y}\cdot\dfrac{9y^2}{1} = -\dfrac{19\cdot9\cdot y\cdot y}{7\cdot9\cdot y\cdot1} = -\dfrac{19y}{7}$

**53.** $\dfrac{4}{8}\div\dfrac{3}{16} = \dfrac{4}{8}\cdot\dfrac{16}{3} = \dfrac{4\cdot8\cdot2}{8\cdot3} = \dfrac{8}{3}$

**57.** $\left(1\div\dfrac{3}{4}\right)\cdot\dfrac{2}{3} = \left(\dfrac{1}{1}\cdot\dfrac{4}{3}\right)\cdot\dfrac{2}{3} = \dfrac{1\cdot2\cdot2\cdot2}{1\cdot3\cdot3} = \dfrac{8}{9}$

**61.** $\dfrac{ab^2}{c}\cdot\dfrac{c}{ab} = \dfrac{a\cdot b\cdot b\cdot c}{c\cdot a\cdot b} = \dfrac{b}{1} = b$

**65.** $-\dfrac{4}{7}\div\left(\dfrac{4}{5}\cdot\dfrac{3}{7}\right) = -\dfrac{4}{7}\div\left(\dfrac{2\cdot2\cdot3}{5\cdot7}\right)$

$\qquad = -\dfrac{4}{7}\cdot\dfrac{5\cdot7}{2\cdot2\cdot3}$

$\qquad = -\dfrac{4\cdot5\cdot7}{7\cdot4\cdot3} = -\dfrac{5}{3}$

$\qquad = -\dfrac{5}{3}$

**69.** $xy = -\dfrac{4}{5}\cdot\dfrac{9}{11} = -\dfrac{2\cdot2\cdot3\cdot3}{5\cdot11} = -\dfrac{36}{55}$

**73.** $-\dfrac{1}{2}z = \dfrac{1}{10}$

$-\dfrac{1}{2}\cdot\dfrac{2}{5} = \dfrac{1}{10}$      replace $z$ with $\dfrac{2}{5}$

$-\dfrac{1\cdot2}{2\cdot5} = \dfrac{1}{10}$

$-\dfrac{1}{5} = \dfrac{1}{10}$ False

No, it is not a solution.

**77.** $8\left(\dfrac{3}{16}\right) = \dfrac{24}{16} = \dfrac{3}{2}$

The screw is $\dfrac{3}{2}$ inches deep.

**81.** Let $x$ be the size of Jorge's wrist.

(wrist size) = (fraction of waist) $\cdot$ (waist size)

$x = \dfrac{1}{4}\cdot\dfrac{34}{1} = \dfrac{1\cdot2\cdot17}{2\cdot2\cdot1} = \dfrac{17}{2}$

Jorge's wrist is about $\dfrac{17}{2}$ inches.

**85.** $\dfrac{5}{14}\cdot\dfrac{1}{5} = \dfrac{5\cdot1}{14\cdot5} = \dfrac{1}{14}$

The area is $\dfrac{1}{14}$ square foot.

**89.** $65 = 5\cdot13$

**93.** $10{,}300{,}000 \cdot \dfrac{7}{10} = 7{,}210{,}000$ households

**97.** $5144 \div \dfrac{1}{3} = 5144 \cdot 3 = 15{,}432$

There are 15,432 flowering plant species that are native to the United States.

## Exercise Set 4.4

**1.** $-\dfrac{1}{2} + \dfrac{1}{2} = \dfrac{-1 + 1}{2} = \dfrac{0}{2} = 0$

**5.** $-\dfrac{4}{13} + \dfrac{2}{13} + \dfrac{1}{13} = \dfrac{-4 + 2 + 1}{13} = -\dfrac{1}{13}$

**9.** $\dfrac{1}{y} - \dfrac{4}{y} = \dfrac{1 - 4}{y} = \dfrac{-3}{y} = -\dfrac{3}{y}$

**13.** $\dfrac{1}{8} - \dfrac{7}{8} = \dfrac{1 - 7}{8} = \dfrac{-6}{8} = -\dfrac{3 \cdot 2}{4 \cdot 2} = -\dfrac{3}{4}$

**17.** $\dfrac{9x}{15} + \dfrac{1x}{15} = \dfrac{9x + 1x}{15} = \dfrac{10x}{15} = \dfrac{5 \cdot 2x}{5 \cdot 3} = \dfrac{2x}{3}$

**21.** $\dfrac{15}{16z} - \dfrac{3}{16z} = \dfrac{15 - 3}{16z} = \dfrac{12}{16z} = \dfrac{4 \cdot 3}{4 \cdot 4 \cdot z} = \dfrac{3}{4z}$

**25.** $\dfrac{15}{17} + \dfrac{5}{17} + \dfrac{14}{17} = \dfrac{15 + 5 + 14}{17}$

$$= \dfrac{34}{17}$$
$$= \dfrac{17 \cdot 2}{17}$$
$$= \dfrac{2}{1}$$
$$= 2$$

**29.** $\dfrac{x}{4} + \dfrac{3x}{4} - \dfrac{2x}{4} + \dfrac{x}{4} = \dfrac{x + 3x - 2x + x}{4} = \dfrac{3x}{4}$

**33.** $x - y = -\dfrac{1}{5} - \dfrac{3}{5} = \dfrac{-1 - 3}{5} = \dfrac{-4}{5} = -\dfrac{4}{5}$

**37.** $x + \dfrac{1}{3} = -\dfrac{1}{3}$

$x + \dfrac{1}{3} - \dfrac{1}{3} = -\dfrac{1}{3} - \dfrac{1}{3}$

$x = \dfrac{-1 - 1}{3} = \dfrac{-2}{3} = -\dfrac{2}{3}$

Check:

$x + \dfrac{1}{3} = -\dfrac{1}{3}$

$-\dfrac{2}{3} + \dfrac{1}{3} \overset{?}{=} -\dfrac{1}{3}$

$-\dfrac{1}{3} \overset{?}{=} -\dfrac{1}{3}$ True

**41.** $3x - \dfrac{1}{5} - 2x = \dfrac{1}{5} + \dfrac{2}{5}$

$(3x - 2x) - \dfrac{1}{5} = \dfrac{1 + 2}{5}$

$x - \dfrac{1}{5} = \dfrac{3}{5}$

$x - \dfrac{1}{5} + \dfrac{1}{5} = \dfrac{3}{5} + \dfrac{1}{5}$

$x = \dfrac{4}{5}$

Check:

$3x - \dfrac{1}{5} - 2x = \dfrac{1}{5} + \dfrac{2}{5}$

$3 \cdot \dfrac{4}{5} - \dfrac{1}{5} - 2 \cdot \dfrac{4}{5} \overset{?}{=} \dfrac{1}{5} + \dfrac{2}{5}$

$\dfrac{3}{1} \cdot \dfrac{4}{5} - \dfrac{1}{5} - \dfrac{2}{1} \cdot \dfrac{4}{5} \overset{?}{=} \dfrac{3}{5}$

$\dfrac{3 \cdot 4}{1 \cdot 5} - \dfrac{1}{5} - \dfrac{2 \cdot 4}{1 \cdot 5} \overset{?}{=} \dfrac{3}{5}$

$\dfrac{12}{5} - \dfrac{1}{5} - \dfrac{8}{5} \overset{?}{=} \dfrac{3}{5}$

$\dfrac{12 - 1 - 8}{5} \overset{?}{=} \dfrac{3}{5}$

$\dfrac{3}{5} \overset{?}{=} \dfrac{3}{5}$ True

**45.** The perimeter is the distance around. A rectangle has 2 sets of equal sides. Add the lengths of the sides.

$\dfrac{5}{12} + \dfrac{7}{12} + \dfrac{5}{12} + \dfrac{7}{12} = \dfrac{5 + 7 + 5 + 7}{12}$

$$= \dfrac{24}{12}$$
$$= \dfrac{2 \cdot 12}{1 \cdot 12}$$
$$= 2$$

The perimeter is 2 meters.

**49.** The fraction of employees enrolled in each plan in order from smallest to largest are $\dfrac{3}{20}, \dfrac{4}{20}, \dfrac{6}{20},$ and $\dfrac{7}{20}$.

Thus, the health plans in order from the smallest fraction of employees to the largest fraction of employees are Traditional fee-for-service, Point-of-Service, Health Maintenance Organization, and Preferred Provider Organization.

**53.** Subtract $\dfrac{16}{50}$ from $\dfrac{39}{50}$.

$\dfrac{39}{50} - \dfrac{16}{50} = \dfrac{39 - 16}{50} = \dfrac{23}{50}$

$\dfrac{23}{50}$ of the states had maximum speed limits that were less than 70 mph.

**57.** $9 = 3 \cdot 3$
$15 = 3 \cdot 5$
$\text{LCD} = 3 \cdot 3 \cdot 5 = 45$

**61.** $24 = 2 \cdot 2 \cdot 2 \cdot 3$
$x = x$
$\text{LCD} = 2 \cdot 2 \cdot 2 \cdot 3 \cdot x = 24x$

**65.** $18 = 2 \cdot 3 \cdot 3$
$21 = 3 \cdot 7$
$\text{LCD} = 2 \cdot 3 \cdot 3 \cdot 7 = 126$

**69.** $8 = 2 \cdot 2 \cdot 2$
$24 = 2 \cdot 2 \cdot 2 \cdot 3$
$\text{LCD} = 2 \cdot 2 \cdot 2 \cdot 3 = 24$

**73.** $a = a$
$12 = 2 \cdot 2 \cdot 3$
$\text{LCD} = 2 \cdot 2 \cdot 3 \cdot a = 12a$

**77.** $11 = 11$
$33 = 3 \cdot 11$
$121 = 11 \cdot 11$
$\text{LCD} = 3 \cdot 11 \cdot 11 = 363$

**81.** $-2 + 10 = 8$

**85.** $-12 - 16 = -12 + (-16) = -28$

**89.** Subtract the sum of $\dfrac{38}{50}$ and $\dfrac{7}{50}$ from the total of $\dfrac{50}{50}$.

$$\frac{50}{50} - \left(\frac{38}{50} + \frac{7}{50}\right) = \frac{50}{50} - \frac{45}{50}$$
$$= \frac{50 - 45}{45}$$
$$= \frac{5}{50}$$
$$= \frac{1 \cdot 5}{10 \cdot 5}$$
$$= \frac{1}{10}$$

$\dfrac{1}{10}$ of American men over age 65 are either single or divorced.

### EXERCISE SET 4.5

**1.** The LCD of 3 and 6 is 6.
$$\frac{2}{3} + \frac{1}{6} = \frac{2 \cdot 2}{3 \cdot 2} + \frac{1}{6} = \frac{4}{6} + \frac{1}{6} = \frac{5}{6}$$

**5.** The LCD of 11 and 33 is 33.
$$-\frac{2}{11} + \frac{2}{33} = -\frac{2 \cdot 3}{11 \cdot 3} + \frac{2}{33} = -\frac{6}{33} + \frac{2}{33} = -\frac{4}{33}$$

**9.** The LCD of 35 and 7 is 35.
$$\frac{11}{35} + \frac{2}{7} = \frac{11}{35} + \frac{2 \cdot 5}{7 \cdot 5} = \frac{11}{35} + \frac{10}{35} = \frac{21}{35} = \frac{3 \cdot 7}{5 \cdot 7} = \frac{3}{5}$$

**13.** The LCD of 12 and 9 is 36.
$$\frac{5}{12} - \frac{1}{9} = \frac{5 \cdot 3}{12 \cdot 3} - \frac{1 \cdot 4}{9 \cdot 4} = \frac{15}{36} - \frac{4}{36} = \frac{11}{36}$$

**17.** The LCD of 11 and 9 is 99.
$$\frac{5a}{11} + \frac{4a}{9} = \frac{5a \cdot 9}{11 \cdot 9} + \frac{4a \cdot 11}{9 \cdot 11}$$
$$= \frac{45a}{99} + \frac{44a}{99}$$
$$= \frac{89a}{99}$$

**21.** The LCD of 2 and $x$ is $2x$.
$$\frac{1}{2} + \frac{3}{x} = \frac{1 \cdot x}{2 \cdot x} + \frac{3 \cdot 2}{x \cdot 2} = \frac{x}{2x} + \frac{6}{2x} = \frac{x + 6}{2x}$$

**25.** The LCD of 14 and 7 is 14.
$$\frac{9}{14} - \frac{3}{7} = \frac{9}{14} \cdot \frac{3 \cdot 2}{7 \cdot 2} = \frac{9}{14} - \frac{6}{14} = \frac{3}{14}$$

**29.** The LCD of 9 and 12 is 36.
$$\frac{1}{9} - \frac{5}{12} = \frac{1 \cdot 4}{9 \cdot 4} - \frac{5 \cdot 3}{12 \cdot 3} = \frac{4}{36} - \frac{15}{36} = -\frac{11}{36}$$

**33.** The LCD is 7 and 8 is 56.
$$\frac{5}{7} - \frac{1}{8} = \frac{5 \cdot 8}{7 \cdot 8} - \frac{1 \cdot 7}{8 \cdot 7} = \frac{40}{56} - \frac{7}{56} = \frac{33}{56}$$

**37.** $\dfrac{5}{9} + \dfrac{3}{9} = \dfrac{8}{9}$

**41.** The LCD of 6 and 7 is 42.
$$-\frac{5}{6} - \frac{3}{7} = -\frac{5 \cdot 7}{6 \cdot 7} - \frac{3 \cdot 6}{7 \cdot 6}$$
$$= -\frac{35}{42} - \frac{18}{42}$$
$$= -\frac{53}{42}$$

**45.** The LCD of 3 and 13 is 39.
$$\frac{2a}{3} + \frac{6a}{13} = \frac{2a \cdot 13}{3 \cdot 13} + \frac{6a \cdot 3}{13 \cdot 3}$$
$$= \frac{26a}{39} + \frac{18a}{39}$$
$$= \frac{44a}{39}$$

**49.** The LCD of 9 and $y$ is $9y$.
$$\frac{5}{9} + \frac{1}{y} = \frac{5 \cdot y}{9 \cdot y} + \frac{1 \cdot 9}{y \cdot 9} = \frac{5y}{9y} + \frac{9}{9y} = \frac{5y + 9}{9y}$$

**53.** The LCD of $9x$ and 8 is $72x$.
$$\frac{5}{9x} + \frac{1}{8} = \frac{5 \cdot 8}{9x \cdot 8} + \frac{1 \cdot 9x}{8 \cdot 9x}$$
$$= \frac{40}{72x} + \frac{9x}{72x}$$
$$= \frac{40 + 9x}{72x}$$

**57.** The LCD of 5, 3, and 10 is 30.
$$-\frac{2}{5} + \frac{1}{3} - \frac{3}{10} = -\frac{2 \cdot 6}{5 \cdot 6} + \frac{1 \cdot 10}{3 \cdot 10} - \frac{3 \cdot 3}{10 \cdot 3}$$
$$= -\frac{12}{30} + \frac{10}{30} - \frac{9}{30}$$
$$= -\frac{11}{30}$$

**61.** The LCD of 2, 4, and 16 is 16.
$$-\frac{1}{2} - \frac{1}{4} - \frac{1}{16} = -\frac{1 \cdot 8}{2 \cdot 8} - \frac{1 \cdot 4}{4 \cdot 4} - \frac{1}{16}$$
$$= -\frac{8}{16} - \frac{4}{16} - \frac{1}{16}$$
$$= -\frac{13}{16}$$

**65.** The LCD of 12, 24, and 6 is 24.
$$-\frac{9}{12} + \frac{17}{24} - \frac{1}{6} = -\frac{9 \cdot 2}{12 \cdot 2} + \frac{17}{24} - \frac{1 \cdot 4}{6 \cdot 4}$$
$$= -\frac{18}{24} + \frac{17}{24} - \frac{4}{24}$$
$$= -\frac{5}{24}$$

**69.** The LCD of 8, 7 and 14 is 56.

$$\frac{3x}{8} + \frac{2x}{7} - \frac{5}{14} = \frac{3x \cdot 7}{8 \cdot 7} + \frac{2x \cdot 8}{7 \cdot 8} - \frac{5 \cdot 4}{14 \cdot 4}$$

$$= \frac{21x}{56} + \frac{16x}{56} - \frac{20}{56}$$

$$= \frac{37x - 20}{56}$$

**73.** $xy = \dfrac{1}{3} \cdot \dfrac{3}{4} = \dfrac{1 \cdot 3}{3 \cdot 4} = \dfrac{1}{4}$

**77.**
$$x - \frac{1}{12} = \frac{5}{6}$$

$$x - \frac{1}{12} + \frac{1}{12} = \frac{5}{6} + \frac{1}{12}$$

$$x = \frac{5 \cdot 2}{6 \cdot 2} + \frac{1}{12}$$

$$x = \frac{10}{12} + \frac{1}{12}$$

$$x = \frac{11}{12}$$

Check:
$$x - \frac{1}{12} = \frac{5}{6}$$

$$\frac{11}{12} - \frac{1}{12} \stackrel{?}{=} \frac{5}{6}$$

$$\frac{10}{12} \stackrel{?}{=} \frac{5}{6}$$

$$\frac{2 \cdot 5}{2 \cdot 6} \stackrel{?}{=} \frac{5}{6}$$

$$\frac{5}{6} \stackrel{?}{=} \frac{5}{6} \text{ True}$$

**81.**
$$7z + \frac{1}{16} - 6z = \frac{3}{4}$$

$$(7z - 6z) + \frac{1}{16} = \frac{3}{4}$$

$$z + \frac{1}{16} = \frac{3}{4}$$

$$z + \frac{1}{16} - \frac{1}{16} = \frac{3}{4} - \frac{1}{16}$$

$$z = \frac{3 \cdot 4}{4 \cdot 4} - \frac{1}{16}$$

$$z = \frac{12}{16} - \frac{1}{16}$$

$$z = \frac{11}{16}$$

Check:
$$7z + \frac{1}{16} - 6z = \frac{3}{4}$$

$$7 \cdot \frac{11}{16} + \frac{1}{16} - 6 \cdot \frac{11}{16} \stackrel{?}{=} \frac{3}{4}$$

$$\frac{7}{1} \cdot \frac{11}{16} + \frac{1}{16} - \frac{6}{1} \cdot \frac{11}{16} \stackrel{?}{=} \frac{3}{4}$$

$$\frac{7 \cdot 11}{1 \cdot 16} + \frac{1}{16} - \frac{6 \cdot 11}{1 \cdot 16} \stackrel{?}{=} \frac{3}{4}$$

$$\frac{77}{16} + \frac{1}{16} - \frac{66}{16} \stackrel{?}{=} \frac{3}{4}$$

$$\frac{77 + 1 - 66}{16} \stackrel{?}{=} \frac{3}{4}$$

$$\frac{12}{16} \stackrel{?}{=} \frac{3}{4}$$

$$\frac{4 \cdot 3}{4 \cdot 4} \stackrel{?}{=} \frac{3}{4}$$

$$\frac{3}{4} \stackrel{?}{=} \frac{3}{4} \text{ True}$$

**85.** Add the lengths of the 4 sides. A parallelogram has 2 sets of equal sides.

$$\frac{1}{3} + \frac{4}{5} + \frac{1}{3} + \frac{4}{5} = \frac{1 \cdot 5}{3 \cdot 5} + \frac{4 \cdot 3}{5 \cdot 3} + \frac{1 \cdot 5}{3 \cdot 5} + \frac{4 \cdot 3}{5 \cdot 3}$$

$$= \frac{5}{15} + \frac{12}{15} + \frac{5}{15} + \frac{12}{15}$$

$$= \frac{34}{15}$$

The perimeter is $\dfrac{34}{15}$ centimeters.

**89.** Subtract $\dfrac{5}{264}$ from $\dfrac{1}{4}$.

$$\frac{1}{4} - \frac{5}{264} = \frac{1 \cdot 66}{4 \cdot 66} - \frac{5}{264} = \frac{66 - 5}{264} = \frac{61}{264}$$

A killer bee will chase a person $\dfrac{61}{264}$ mile farther.

**93.** Add the fractions for 1 or 2 times per week and 3 times per week.

$$\frac{23}{50} + \frac{31}{100} = \frac{23 \cdot 2}{50 \cdot 2} + \frac{31}{100}$$

$$= \frac{46}{100} + \frac{31}{100}$$

$$= \frac{77}{100}$$

$\frac{77}{100}$ of Americans eat pasta 1, 2, or 3 times per week.

**97.** $\left(\dfrac{5}{6}\right)^2 = \dfrac{5}{6} \cdot \dfrac{5}{6} = \dfrac{5 \cdot 5}{6 \cdot 6} = \dfrac{25}{36}$

**101.** 57,236 rounded to the nearest hundred is 57,200.

**105.** $\dfrac{30}{55} + \dfrac{1000}{1760} = \dfrac{30 \cdot 32}{55 \cdot 32} + \dfrac{1000}{1760}$

$$= \frac{960}{1760} + \frac{1000}{1760}$$

$$= \frac{1960}{1760}$$

$$= \frac{49 \cdot 40}{44 \cdot 40}$$

$$= \frac{49}{44}$$

**109.** Find the sum of the fractions for Asia and Europe.

$$\frac{58}{193} + \frac{38}{579} = \frac{58 \cdot 3}{193 \cdot 3} + \frac{38}{579}$$

$$= \frac{174}{579} + \frac{38}{579}$$

$$= \frac{212}{579}$$

$\frac{212}{579}$ of the world's land area is accounted for by Asia and Europe.

### EXERCISE SET 4.6

**1.** $\dfrac{\frac{1}{8}}{\frac{3}{4}} = \dfrac{1}{8} \div \dfrac{3}{4} = \dfrac{1}{8} \cdot \dfrac{4}{3} = \dfrac{1 \cdot 4}{2 \cdot 4 \cdot 3} = \dfrac{1}{6}$

**5.** $\dfrac{\frac{2x}{27}}{\frac{4}{9}} = \dfrac{2x}{27} \div \dfrac{4}{9} = \dfrac{2x}{27} \cdot \dfrac{9}{4} = \dfrac{2 \cdot x \cdot 3 \cdot 3}{3 \cdot 3 \cdot 3 \cdot 2 \cdot 2} = \dfrac{x}{6}$

**9.** $\dfrac{\frac{3x}{4}}{5 - \frac{1}{8}} = \dfrac{8 \cdot \left(\frac{3x}{4}\right)}{8 \cdot \left(\frac{5}{1} - \frac{1}{8}\right)}$

$$= \frac{6x}{8 \cdot \left(\frac{5}{1}\right) - 8\left(\frac{1}{8}\right)}$$

$$= \frac{6x}{40 - 1}$$

$$= \frac{6x}{39}$$

$$= \frac{3 \cdot 2 \cdot x}{3 \cdot 13}$$

$$= \frac{2x}{13}$$

**13.** $\left(\dfrac{2}{9} + \dfrac{4}{9}\right)\left(\dfrac{1}{3} - \dfrac{9}{10}\right) = \left(\dfrac{6}{9}\right)\left(\dfrac{1 \cdot 10}{3 \cdot 10} - \dfrac{9 \cdot 3}{10 \cdot 3}\right)$

$$= \left(\frac{6}{9}\right)\left(\frac{10}{30} - \frac{27}{30}\right)$$

$$= \frac{6}{9} \cdot \left(-\frac{17}{30}\right)$$

$$= -\frac{6 \cdot 17}{9 \cdot 5 \cdot 6}$$

$$= -\frac{17}{45}$$

**17.** $5y - z = 5\left(\dfrac{2}{5}\right) - \left(\dfrac{5}{6}\right)$

$$= 2 - \frac{5}{6}$$

$$= \frac{2 \cdot 6}{6} - \frac{5}{6}$$

$$= \frac{12}{6} - \frac{5}{6}$$

$$= \frac{7}{6}$$

**21.** $x^2 - yz = \left(-\dfrac{1}{3}\right)^2 - \left(\dfrac{2}{5}\right)\left(\dfrac{5}{6}\right)$

$$= \frac{1}{9} - \left(\frac{2}{5}\right)\left(\frac{5}{6}\right)$$

$$= \frac{1}{9} - \frac{2 \cdot 5}{5 \cdot 2 \cdot 3}$$

$$= \frac{1}{9} - \frac{1}{3}$$

$$= \frac{1}{9} - \frac{3}{9}$$

$$= -\frac{2}{9}$$

**25.** $\left(\dfrac{3}{2}\right)^3 + \left(\dfrac{1}{2}\right)^3 = \dfrac{27}{8} + \dfrac{1}{8} = \dfrac{28}{8} = \dfrac{7 \cdot 4}{2 \cdot 4} = \dfrac{7}{2}$

**29.** $\dfrac{2 + \frac{1}{6}}{1 - \frac{4}{3}} = \dfrac{6\left(2 + \frac{1}{6}\right)}{6\left(1 - \frac{4}{3}\right)}$

$$= \frac{6 \cdot 2 + 6 \cdot \frac{1}{6}}{6 \cdot 1 - 6 \cdot \frac{4}{3}}$$

$$= \frac{12 + 1}{6 - 8}$$

$$= \frac{13}{-2}$$

$$= -\frac{13}{2}$$

**33.** $\left(\dfrac{3}{4} - 1\right)\left(\dfrac{1}{8} + \dfrac{1}{2}\right) = \left(\dfrac{3}{4} - \dfrac{4}{4}\right)\left(\dfrac{1}{8} + \dfrac{4}{8}\right)$

$$= \left(-\frac{1}{4}\right)\left(\frac{5}{8}\right)$$

$$= -\frac{1 \cdot 5}{4 \cdot 8}$$

$$= -\frac{5}{32}$$

**37.** $\dfrac{\left(\frac{1}{2} - \frac{3}{8}\right)}{\left(\frac{3}{4} + \frac{1}{2}\right)} = \dfrac{8\left(\frac{1}{2} - \frac{3}{8}\right)}{8\left(\frac{3}{4} + \frac{1}{2}\right)}$

$\qquad = \dfrac{8 \cdot \frac{1}{2} - 8 \cdot \frac{3}{8}}{8 \cdot \frac{3}{4} + 8 \cdot \frac{1}{2}}$

$\qquad = \dfrac{4 - 3}{6 + 4} = \dfrac{1}{10}$

$\qquad = \dfrac{1}{10}$

**41.** $\dfrac{\frac{x}{3} + 2}{5 + \frac{1}{3}} = \dfrac{3\left(\frac{x}{3} + 2\right)}{3\left(5 + \frac{1}{3}\right)}$

$\qquad = \dfrac{3 \cdot \frac{x}{3} + 3 \cdot 2}{3 \cdot 5 + 3 \cdot \frac{1}{3}}$

$\qquad = \dfrac{x + 6}{15 + 1}$

$\qquad = \dfrac{x + 6}{16}$

**45.** $5^2 = 5 \cdot 5 = 25$

**49.** $\dfrac{2}{3}\left(\dfrac{3}{2}a\right) = \dfrac{2}{3} \cdot \dfrac{3}{2} \cdot \dfrac{a}{1} = \dfrac{2 \cdot 3 \cdot a}{3 \cdot 2 \cdot 1} = \dfrac{a}{1} = a$

**53.** $x^2 + 7y = \left(\dfrac{3}{4}\right)^2 + 7\left(-\dfrac{4}{7}\right)$

$\qquad = \dfrac{9}{16} + \dfrac{7}{1} \cdot \left(-\dfrac{4}{7}\right)$

$\qquad = \dfrac{9}{16} + \left(-\dfrac{7 \cdot 4}{1 \cdot 7}\right)$

$\qquad = \dfrac{9}{16} - \dfrac{4}{1}$

$\qquad = \dfrac{9}{16} - \dfrac{4 \cdot 16}{1 \cdot 16}$

$\qquad = \dfrac{9}{16} - \dfrac{64}{16}$

$\qquad = -\dfrac{55}{16}$

**57.** $\dfrac{\frac{1}{4} + \frac{2}{14}}{2} = \dfrac{28\left(\frac{1}{4} + \frac{2}{14}\right)}{28 \cdot 2}$

$\qquad = \dfrac{28 \cdot \frac{1}{4} + 28 \cdot \frac{2}{14}}{56}$

$\qquad = \dfrac{7 + 4}{56}$

$\qquad = \dfrac{11}{56}$

**61.** False, the average of two numbers is between the two numbers.

**65.** True, consider $\dfrac{9}{4} - \dfrac{5}{4} = \dfrac{4}{4} = 1$.

### EXERCISE SET 4.7

**1.** $7x = 2$

$\quad \dfrac{7x}{7} = \dfrac{2}{7}$

$\qquad x = \dfrac{2}{7}$

**5.** $\dfrac{2}{9}y = -6$

$\quad \dfrac{9}{2} \cdot \dfrac{2}{9}y = \dfrac{9}{2} \cdot (-6)$

$\qquad y = -27$

**9.** $7a = \dfrac{1}{3}$

$\quad \dfrac{1}{7} \cdot 7a = \dfrac{1}{7} \cdot \dfrac{1}{3}$

$\qquad a = \dfrac{1}{21}$

**13.** Multiply both sides of the equation by 3.

$\qquad \dfrac{x}{3} + 2 = \dfrac{7}{3}$

$\quad 3\left(\dfrac{x}{3} + 2\right) = 3 \cdot \dfrac{7}{3}$

$\qquad x + 6 = 7$

$\quad x + 6 - 6 = 7 - 6$

$\qquad x = 1$

**17.** Multiply both sides of the equation by the LCD of 2, 5, and 10: 10.

$\qquad \dfrac{1}{2} - \dfrac{3}{5} = \dfrac{x}{10}$

$\quad 10\left(\dfrac{1}{2} - \dfrac{3}{5}\right) = 10\left(\dfrac{x}{10}\right)$

$\qquad 5 - 6 = x$

$\qquad -1 = x$

**21.** $\dfrac{x}{7} - \dfrac{4}{3} = \dfrac{x \cdot 3}{7 \cdot 3} - \dfrac{4 \cdot 7}{3 \cdot 7}$

$\qquad = \dfrac{3x}{21} - \dfrac{28}{21}$

$\qquad = \dfrac{3x - 28}{21}$

**25.** $\dfrac{3x}{10} + \dfrac{x}{6} = \dfrac{3x \cdot 3}{10 \cdot 3} + \dfrac{x \cdot 5}{6 \cdot 5}$

$\qquad = \dfrac{9x}{30} + \dfrac{5x}{30}$

$\qquad = \dfrac{14x}{30}$

$\qquad = \dfrac{2 \cdot 7x}{2 \cdot 15}$

$\qquad = \dfrac{7x}{15}$

**29.** $\qquad \dfrac{2}{3} - \dfrac{x}{5} = \dfrac{4}{15}$

$\quad 15\left(\dfrac{2}{3} - \dfrac{x}{5}\right) = 15\left(\dfrac{4}{15}\right)$

$\qquad 10 - 3x = 4$

$\quad 10 - 10 - 3x = 4 - 10$

$\qquad \dfrac{-3x}{-3} = \dfrac{-6}{-3}$

$\qquad x = 2$

**33.** $-3m - 5m = \dfrac{4}{7}$

$$-8m = \dfrac{4}{7}$$

$$\left(-\dfrac{1}{8}\right)(-8m) = \left(-\dfrac{1}{8}\right)\left(\dfrac{4}{7}\right)$$

$$m = -\dfrac{1 \cdot 4}{2 \cdot 4 \cdot 7}$$

$$m = -\dfrac{1}{14}$$

**37.** $\dfrac{1}{5}y = 10$

$$5\left(\dfrac{1}{5}y\right) = 5 \cdot 10$$

$$y = 50$$

**41.** $-\dfrac{3}{4}x = \dfrac{9}{2}$

$$-\dfrac{4}{3} \cdot \left(-\dfrac{3}{4}\right) = -\dfrac{4}{3} \cdot \left(\dfrac{9}{2}\right)$$

$$x = -\dfrac{2 \cdot 2 \cdot 3 \cdot 3}{3 \cdot 2}$$

$$x = -6$$

**45.** $-\dfrac{5}{8}y = \dfrac{3}{16} - \dfrac{9}{16}$

$$-\dfrac{5}{8}y = -\dfrac{6}{16}$$

$$-\dfrac{5}{8}y = -\dfrac{2 \cdot 3}{2 \cdot 8}$$

$$-\dfrac{5}{8}y = -\dfrac{3}{8}$$

$$\left(-\dfrac{8}{5}\right)\left(-\dfrac{5}{8}y\right) = \left(-\dfrac{8}{5}\right)\left(-\dfrac{3}{8}\right)$$

$$y = \dfrac{8 \cdot 3}{5 \cdot 8}$$

$$y = \dfrac{3}{5}$$

**49.** $\dfrac{7}{6}x = \dfrac{1}{4} - \dfrac{2}{3}$

$$12\left(\dfrac{7}{6}x\right) = 12\left(\dfrac{1}{4} - \dfrac{2}{3}\right)$$

$$14x = 3 - 8$$

$$14x = -5$$

$$\dfrac{14x}{14} = \dfrac{-5}{14}$$

$$x = -\dfrac{5}{14}$$

**53.** $\dfrac{x}{3} + 2 = \dfrac{x}{2} + 8$

$$6\left(\dfrac{x}{3} + 2\right) = 6\left(\dfrac{x}{2} + 8\right)$$

$$2x + 12 = 3x + 48$$

$$2x + 12 - 2x = 3x + 48 - 2x$$

$$12 = x + 48$$

$$12 - 48 = x + 48 - 48$$

$$-36 = x$$

**57.** $5 + \dfrac{9}{10} = \dfrac{5}{1} + \dfrac{9}{10}$

$$= \dfrac{5 \cdot 10}{1 \cdot 10} + \dfrac{9}{10}$$

$$= \dfrac{50}{10} + \dfrac{9}{10}$$

$$= \dfrac{50 + 9}{10}$$

$$= \dfrac{59}{10}$$

**61.** Answers will vary.

### EXERCISE SET 4.8

**1.** Each part is $\dfrac{1}{4}$, and there are 11 parts shaded, or 2 wholes and 3 more parts.

**a.** $\dfrac{11}{4}$  **b.** $2\dfrac{3}{4}$

**5.** Each part is $\dfrac{1}{2}$, and there are 3 parts shaded or 1 whole part and 1 more part.

**a.** $\dfrac{3}{2}$  **b.** $1\dfrac{1}{2}$

**9.** $2\dfrac{1}{3} = \dfrac{2 \cdot 3 + 1}{3} = \dfrac{7}{3}$

**13.** $11\dfrac{6}{7} = \dfrac{11 \cdot 7 + 6}{7} = \dfrac{83}{7}$

**17.**
$$15\overline{)47}$$
$$\underline{-45}$$
$$2$$

$$\dfrac{47}{15} = 3\dfrac{2}{15}$$

**21.** $2\dfrac{2}{3} \cdot \dfrac{1}{7} = \dfrac{8}{3} \cdot \dfrac{1}{7} = \dfrac{8 \cdot 1}{3 \cdot 7} = \dfrac{8}{21}$

**25.** $3\dfrac{2}{3} \cdot 1\dfrac{1}{2} = \dfrac{11}{3} \cdot \dfrac{3}{2} = \dfrac{11 \cdot 3}{3 \cdot 2} = \dfrac{11}{2} = 5\dfrac{1}{2}$

**29.** $4\dfrac{7}{10} + 2\dfrac{1}{10} = 6\dfrac{8}{10} = 6\dfrac{4}{5}$

**33.** $3\dfrac{5}{8} = 3\dfrac{15}{24}$

$$2\dfrac{1}{6} = 2\dfrac{4}{24}$$

$$+ 7\dfrac{3}{4} = 7\dfrac{18}{24}$$

$$12\dfrac{37}{24} = 13\dfrac{13}{24}$$

**37.** $10\dfrac{13}{14} = 10\dfrac{13}{14}$

$$-3\dfrac{4}{7} = -3\dfrac{8}{14}$$

$$7\dfrac{5}{14}$$

**41.**

$$2\frac{3}{4}$$
$$+1\frac{1}{4}$$
$$\overline{\phantom{+1}3\frac{4}{4}} = 3 + 1 = 4$$

**45.** $3\frac{1}{9} \cdot 2 = \frac{28}{9} \cdot \frac{2}{1} = \frac{56}{9} = 6\frac{2}{9}$

**49.** $22\frac{4}{9} + 13\frac{5}{18} = 22\frac{8}{18} + 13\frac{5}{18} = 35\frac{13}{18}$

**53.**

$$15\frac{1}{5} = 15\frac{6}{30}$$
$$20\frac{3}{10} = 20\frac{9}{30}$$
$$+37\frac{2}{15} = 37\frac{4}{30}$$
$$\overline{\phantom{+37}72\frac{19}{30}}$$

**57.** $4\frac{2}{7} \cdot 1\frac{3}{10} = \frac{30}{7} \cdot \frac{13}{10} = \frac{3 \cdot 10 \cdot 13}{7 \cdot 10} = \frac{39}{7} = 5\frac{4}{7}$

**61.** The phrase "total duration" tells us to add. Find the sum of the three durations.

$$4\frac{14}{15} = 4\frac{56}{60}$$
$$4\frac{7}{60} = 4\frac{7}{60}$$
$$+1\frac{2}{3} = 1\frac{40}{60}$$
$$\overline{\phantom{+1}9\frac{103}{60}} = 10\frac{43}{60}$$

**65.** $6 \cdot 3\frac{1}{4} = \frac{6}{1} \cdot \frac{13}{4}$

$$= \frac{2 \cdot 3 \cdot 13}{1 \cdot 2 \cdot 2}$$
$$= \frac{39}{2} = 19\frac{1}{2}$$

The sidewalk is $19\frac{1}{2}$ inches wide.

**69.** Subtract $1\frac{1}{2}$ inches from $1\frac{9}{16}$ inches.

$$1\frac{9}{16} = 1\frac{9}{16}$$
$$-1\frac{1}{2} = -1\frac{8}{16}$$
$$\overline{\phantom{-1}\frac{1}{16}}$$

The entrance holes for Mountain Bluebirds should be $\frac{1}{16}$ inch wider than the entrance for Eastern Bluebirds.

**73.** $58\frac{3}{4} \div 7\frac{1}{2} = \frac{235}{4} \div \frac{15}{2}$

$$= \frac{235}{4} \cdot \frac{2}{15}$$
$$= \frac{5 \cdot 47 \cdot 2}{2 \cdot 2 \cdot 3 \cdot 5}$$
$$= \frac{47}{6} = 7\frac{5}{6}$$

$7\frac{5}{6}$ gallons were used each hour.

**77.** $2 \cdot 1\frac{3}{4} = \frac{2}{1} \cdot \frac{7}{4}$

$$= \frac{2 \cdot 7}{1 \cdot 2 \cdot 2}$$
$$= \frac{7}{2} \text{ or } 3\frac{1}{2}$$

The area is $\frac{7}{2}$ or $3\frac{1}{2}$ square yards.

**81.** Find the distance around. Add the lengths of the three sides.

$$2\frac{1}{3}$$
$$2\frac{1}{3}$$
$$+2\frac{1}{3}$$
$$\overline{\phantom{+2}6\frac{3}{3}} = 7$$

The perimeter is 7 miles.

**85.** Subtract $1\frac{1}{3}$ hours from $2\frac{3}{4}$ hours.

$$2\frac{3}{4} = 2\frac{9}{12}$$
$$-1\frac{1}{3} = -1\frac{4}{12}$$
$$\overline{\phantom{-1}1\frac{5}{12}}$$

He will have to wait $1\frac{5}{12}$ hours.

**89.** $15\frac{1}{5} \div 24 = \frac{76}{5} \div \frac{24}{1}$

$= \frac{76}{5} \cdot \frac{1}{24}$

$= \frac{4 \cdot 19 \cdot 1}{5 \cdot 4 \cdot 6}$

$= \frac{19}{30}$

On average, $\frac{19}{30}$ inch of rain fell each hour.

**93.** $3(y - 2) - 6y = 3 \cdot y - 3 \cdot 2 - 6y$

$= 3y - 6 - 6y$

$= -3y - 6$

**97.** $\frac{7 - 3}{2^2} = \frac{4}{4} = 1$

**101.** Answers will vary.

## CHAPTER 4 TEST

**1.** $7\frac{2}{3} = \frac{7 \cdot 3 + 2}{3} = \frac{23}{3}$

**5.** $\frac{54}{210} = \frac{2 \cdot 3 \cdot 3 \cdot 3}{2 \cdot 3 \cdot 5 \cdot 7} = \frac{3 \cdot 3}{5 \cdot 7} = \frac{9}{35}$

**9.** $\frac{7x}{9} + \frac{x}{9} = \frac{7x + x}{9} = \frac{8x}{9}$

**13.** $\frac{9a}{10} + \frac{2}{5} = \frac{9a}{10} + \frac{2 \cdot 2}{5 \cdot 2} = \frac{9a}{10} + \frac{4}{10} = \frac{9a + 4}{10}$

**17.** $3\frac{7}{8} = 3\frac{35}{40}$

$7\frac{2}{5} = 7\frac{16}{40}$

$+2\frac{3}{4} = 2\frac{30}{40}$

$12\frac{81}{40} = 14\frac{1}{40}$

**21.** $12 \div 3\frac{1}{3} = 12 \div \frac{10}{3}$

$= 12 \cdot \frac{3}{10}$

$= \frac{2 \cdot 6 \cdot 3}{2 \cdot 5}$

$= \frac{18}{5}$

$= 3\frac{3}{5}$

**25.** $\frac{5 + \frac{3}{7}}{2 - \frac{1}{2}} = \frac{14(5 + \frac{3}{7})}{14(2 - \frac{1}{2})}$

$= \frac{14 \cdot 5 + 14 \cdot \frac{3}{7}}{14 \cdot 2 + 14 \cdot \frac{1}{2}}$

$= \frac{70 + 6}{28 - 7}$

$= \frac{76}{21}$

**29.** $-5x = -5\left(-\frac{1}{2}\right) = -\frac{5}{1} \cdot \left(-\frac{1}{2}\right) = \frac{5 \cdot 1}{1 \cdot 2} = \frac{5}{2}$

**33.** Find the sum of the fractions representing Back Woods and Westward.

$\frac{3}{16} + \frac{1}{8} = \frac{3}{16} + \frac{1 \cdot 2}{8 \cdot 2}$

$= \frac{3}{16} + \frac{2}{16}$

$= \frac{5}{16}$

$\frac{5}{16}$ of backpack sales go the Back Woods and Westward combined.

**37.** $258 \div 10\frac{3}{4} = \frac{258}{1} \div \frac{43}{4}$

$= \frac{258}{1} \cdot \frac{4}{43}$

$= \frac{258 \cdot 4}{1 \cdot 43}$

$= \frac{43 \cdot 6 \cdot 4}{1 \cdot 43}$

$= \frac{24}{1}$

$= 24$

The car will travel about 24 miles on one gallon of gas.

# Chapter 5

## EXERCISE SET 5.1

**1.** 6.52 is six and fifty-two hundredths.

**5.** 3.205 is three and two hundred five thousandths.

**9.** Six and five-tenths is 6.5

**13.** Five and six hundred twenty-five thousandths is 5.625.

**17.** Twenty and thirty-three hundredths is 20.33.

**21.** $0.3 = \frac{3}{10}$      **25.** $5.47 = 5\frac{47}{100}$

**29.** $7.07 = 7\frac{7}{100}$

**33.** $0.3005 = \frac{3005}{10,000} = \frac{601}{2000}$

**37.** 0.15  0.16

↑    ↑

5 < 6

so 0.15 < 0.16

**41.** 0.098  0.1

↑    ↑

0 < 1

so 0.098 < 0.1

**45.** 167.908  167.980

↑      ↑

0 < 8

so 167.908 < 169.980

**49.** To round 0.57 to the nearest tenth, observe that the digit in the hundredths place is 7. Since this digit is at least 5, we need to add 1 to the digit in the tenths place. The number 0.57 rounded to the nearest tenth is 0.6.

**53.** To round 0.5942 to the nearest thousandth, observe that the digit in the ten-thousandths place is 2. Since this digit is less than 5, we do not add 1 to the digit in the thousandths place. The number 0.5942 rounded to the nearest thousandth is 0.594.

**57.** To round 12,342 to the nearest tenth, observe that the digit in the hundredths place is 4. Since this digit is less than 5, we do not need to add 1 to the digit in the tenths place. The number 12.342 rounded to the nearest tenth is 12.3.

**61.** To round 0.501 to the nearest tenth, observe that the digit in the hundredths place is 0. Since the digit is less than 5, we do not add 1 to the digit in the tenths place. The number 0.501 rounded to the nearest tenth is 0.5.

**65.** To round 26.95 to the nearest one, observe that the digit in the tenths place is 9. Since this digit is at least 5, we need to add 1 to the digit in the ones place. The number 26.95 rounded to the nearest one is 27. The amount is $27.

**69.** 0.26559 rounds to 0.27
0.26499 rounds to 0.26
0.25786 rounds to 0.26
0.25186 rounds to 0.25
Therefore, 0.26499 and 0.25786 round to 0.26.

**73.** To round 2.39027 to the nearest hundredth, observe that the digit in the thousandths place is 0. Since this digit is less than 5, we do not add 1 to the digit in the hundredths place. The number 2.39027 rounded to the nearest hundredth is 2.39. The time is 2.39 hours.

**77.** To round 135.74 to the nearest one, observe that the digit in the tenths place is 7. Since this digit is at least 5, we need to add 1 to the digit in the ones place. The number 135.74 rounded to the nearest one is 136. The record is 136 mph.

**81.** Since 28.21 > 27.91 and 28.21 < 28.49, the hurricane is a Category 3.

**85.**
$$\begin{array}{r} 94 \\ -23 \\ \hline 71 \end{array}$$

**89.** All the hundreds places are the same. When comparing the tens places, 6 of the values have 2 and 2 > 1. So continue to compare the ones places for these numbers. Notice then that 6 > 5. So the highest average score is 226.130 achieved by Walter Ray Williams Jr. in 1998.

**93.** Answers may vary.

## EXERCISE SET 5.2

**1.**
$$\begin{array}{r} 1.3 \\ +2.2 \\ \hline 3.5 \end{array}$$

**5.**
$$\begin{array}{r} 24.6000 \\ 2.3900 \\ +0.0678 \\ \hline 27.0578 \end{array}$$

**9.** $-2.6 + (-5.97)$
Add the absolute values.
$$\begin{array}{r} 2.60 \\ +5.97 \\ \hline 8.57 \end{array}$$
Attach the common sign.
$-8.57$

**13.** $8.8 - 2.3$
$$\begin{array}{r} 8.8 \\ -2.3 \\ \hline 6.5 \end{array}$$

**17.**
$$\begin{array}{r} 654.90 \\ -56.67 \\ \hline 598.23 \end{array}$$

**21.** $-1.12 - 5.2 = -1.12 + (-5.2)$
Add the absolute values.
$$\begin{array}{r} 1.12 \\ +5.20 \\ \hline 6.32 \end{array}$$
Attach the common sign.
$-6.32$

**25.**
$$\begin{array}{r} 0.9 \\ +2.2 \\ \hline 3.1 \end{array}$$

**29.** $-5.9 - 4 = -5.9 + (-4)$
Add the absolute values.
$$\begin{array}{r} 5.9 \\ +4.0 \\ \hline 9.9 \end{array}$$
Attach the common sign.
$-9.9$

**33.** $-6.06 + 0.44$
Subtract the absolute values.
$$\begin{array}{r} 6.06 \\ -0.44 \\ \hline 5.62 \end{array}$$
Attach the sign of the larger absolute value.
$-5.62$

**37.**
$$\begin{array}{r} 3490.23 \\ +8493.09 \\ \hline 11,983.32 \end{array}$$

**41.** $50.2 - 600 = 50.2 + (-600)$
Subtract the absolute values.
$$\begin{array}{r} 600.0 \\ -50.2 \\ \hline 549.8 \end{array}$$
Attach the sign of the larger absolute value.
$-549.8$

**45.** 
$$
\begin{array}{r}
100.009 \\
6.080 \\
+9.034 \\
\hline
115.123
\end{array}
$$

**49.** $-0.003 + 0.091$

Subtract the absolute values.
$$
\begin{array}{r}
0.091 \\
-0.003 \\
\hline
0.088
\end{array}
$$

Attach the sign of the larger absolute value.

0.088

**53.** $x + z = 3.6 + 0.21 = 3.81$

**57.** $y - x + z = 5 - 3.6 + 0.21$
$$
\begin{aligned}
&= 5.00 - 3.60 + 0.21 \\
&= 1.40 + 0.21 \\
&= 1.61
\end{aligned}
$$

**61.** $27.4 - y = 16$

$27.4 - 11.4 \overset{?}{=} 16$

$16 = 16$ True

Yes, it is a solutuion.

**65.** $30.7x + 17.6 - 23.8x - 10.7$
$$
\begin{aligned}
&= 30.7x + (-23.8x) + 17.6 + (-10.7) \\
&= 6.9x + 6.9
\end{aligned}
$$

**69.** The phrase "total monthly cost" tells us to add. Find the sum of the four expenses.
$$
\begin{array}{r}
\overset{2\ 1\ 1\ \ 1}{275.36} \\
83.00 \\
81.60 \\
+14.75 \\
\hline
454.71
\end{array}
$$

The total monthly cost is $454.71.

**73.** Subtract the cost of the book from what she paid: ($20 + $20 = $40).
$$
\begin{array}{r}
40.00 \\
-32.48 \\
\hline
7.52
\end{array}
$$

Check:
$$
\begin{array}{r}
\overset{1\ 1\ \ 1}{7.52} \\
+32.48 \\
\hline
40.00 \text{ or } 40
\end{array}
$$

Her change was $7.52.

**77.** The phrase "How much more" tells us to subtract. Subtract 46.07 from 61.88.
$$
\begin{array}{r}
61.88 \\
-46.07 \\
\hline
15.81
\end{array}
$$

Check:
$$
\begin{array}{r}
\overset{1}{15.81} \\
+46.07 \\
\hline
61.88
\end{array}
$$

New Orleans receives 15.81 more inches of rain annually than Houston.

**81.** The phrase "total amount" tells us to add. Find the sum of the 3 concert's earnings.
$$
\begin{array}{r}
\overset{1\ 1}{121.2} \\
103.5 \\
+\ 98.0 \\
\hline
322.7
\end{array}
$$

The total amount of money these three concerts have earned is $322.7 million.

**85.** Find the sum of the lengths of the three sides.
$$
\begin{array}{r}
12.40 \\
29.34 \\
+25.70 \\
\hline
67.44
\end{array}
$$

The architect needs 67.44 feet of border material.

**89.** Add the durations of all four missions.
$$
\begin{array}{r}
\overset{2\ 1\ 2\ \ 1\ 1}{330.583} \\
94.567 \\
147.000 \\
+142.900 \\
\hline
715.050 \text{ or } 715.05
\end{array}
$$

James A. Lovell has spent 715.05 hours in spaceflight.

**93.** Subtract 13.9 from 22.
$$
\begin{array}{r}
22.0 \\
-13.9 \\
\hline
8.1
\end{array}
$$

Check:
$$
\begin{array}{r}
\overset{1\ 1}{8.1} \\
+13.9 \\
\hline
22.0 \text{ or } 22
\end{array}
$$

The difference in consumption is 8.1 pounds per person.

**97.** $\left(\dfrac{1}{5}\right)^3 = \dfrac{1}{5} \cdot \dfrac{1}{5} \cdot \dfrac{1}{5} = \dfrac{1}{125}$

**101.** Answers may vary.

### EXERCISE SET 5.3

**1.**
$$
\begin{array}{rl}
0.2 & \text{1 decimal place} \\
\times\ 0.6 & \text{1 decimal place} \\
\hline
0.12 & \text{2 decimal places}
\end{array}
$$

**5.** The product, $(-2.3)(7.65)$, is negative.
$$
\begin{array}{rl}
7.65 & \text{2 decimal places} \\
\times\ \ 2.3 & \text{1 decimal place} \\
\hline
2295 & \\
15300 & \\
\hline
-17.595 & \text{3 decimal places and include negative sign.}
\end{array}
$$

**9.** $6.5 \times 10$

10 has 1 zero, so move decimal point 1 place to the right.

$6.5 \times 10 = 65$

**13.** $(-7.093)(1000)$

1000 has 3 zeros, so move decimal point 3 places to the right.

$(-7.093)(1000) = -7093$

**17.**
$$
\begin{array}{rl}
5.62 & \text{2 decimal places} \\
\times\ 7.7 & \text{1 decimal place} \\
\hline
3934 & \\
39340 & \\
\hline
43.274 & \text{3 decimal places}
\end{array}
$$

**21.** $(147.9)(100)$

100 has 2 zeros, so move decimal point 2 places to the right (a 0 is inserted).

$(147.9)(100) = 14,790$

**25.**
$$\begin{array}{r} 49.02 \\ \times\ 0.023 \\ \hline 14706 \\ 98040 \\ \hline 1.12746 \end{array}$$

2 decimal places
3 decimal places

**29.** 5.5 billion $= 5.5 \times 1,000,000,000$
$$= 5,500,000,000$$

**33.** 1.6 million $= 1.6 \times 1,000,000$
$$= 1,600,000$$

**37.** Recall that $xz$ means $x \cdot z$.

$xz = (3)(5.7) = 17.1$
$$\begin{array}{r} 5.7 \\ \times\ 3 \\ \hline 17.1 \end{array}$$

**41.**
$$\begin{array}{r} 0.6x = 4.92 \\ 0.6(14.2) = 4.92 \\ 8.52 = 4.92\ \ \text{False} \end{array}$$
No, it is not a solution.

**45.**
$$\begin{array}{r} 3.5y = -14 \\ 3.5(-4) = -14 \\ -14.0 = -14 \\ -14 = -14\ \ \text{True} \end{array}$$
Yes, it is a solution.

**49.** Circumference $= \pi \cdot \text{diameter}$

$C = \pi \cdot 10 = 10\pi$

$C \approx 10(3.14) = 31.4$

The circumference is 101 centimeters, which is approximately 31.4 centimeters.

**53.** Circumference $= \pi \cdot \text{diameter}$

$C = \pi \cdot 250 = 250\pi$

$C \approx 250(3.14) = 785$

The circumference of the Ferris wheel is $250\pi$ feet which is approximately 785 feet.

**57.** Multiply the number of ounces by the number of grams of fat in 1 ounce to get the total amount of fat.
$$\begin{array}{r} 6.2 \\ \times\ 4 \\ \hline 24.8 \end{array}$$
There are 24.8 grams of fat in a 4-ounce serving of cream cheese.

**61.** Multiply 39.37 by 1.65.
$$\begin{array}{r} 39.37 \\ \times\ 1.65 \\ \hline 19685 \\ 236220 \\ 393700 \\ \hline 64.9605 \end{array}$$
She is about 64.9605 inches tall.

**65.** $675 \times 108 = 72,900$ yen

**69.**
$$\begin{array}{r} 26 \\ 5)\overline{130} \\ -10 \\ \hline 30 \\ -30 \\ \hline 0 \end{array}$$

**73.**
$$\begin{array}{r} 8 \\ 365)\overline{2920} \\ -2920 \\ \hline 0 \end{array}$$

**77.** $(20.6)(1.86)(100,000) = 3,831,600$ miles

## Exercise Set 5.4

**1.** $0.47 \div 5$
$$\begin{array}{r} 0.094 \\ 5)\overline{0.470} \\ -45 \\ \hline 20 \\ -20 \\ \hline 0 \end{array}$$

**5.** $4.756 \div 0.82$

$0.82)\overline{4.756}$  Move the decimal point 2 places.
$$\begin{array}{r} 5.8 \\ 82.)\overline{475.6} \\ -410 \\ \hline 656 \\ -656 \\ \hline 0 \end{array}$$

**9.** $2.4)\overline{429.34}$  Move the decimal points 1 place.
$$\begin{array}{r} 178.8 \\ 24.)\overline{4293.4} \\ -24 \\ \hline 189 \\ -168 \\ \hline 213 \\ -192 \\ \hline 214 \\ -192 \\ \hline 22 \end{array}$$
178.8 rounds to 200.

**13.** $0.4)\overline{45.23}$  Move the decimal points 1 place.
$$\begin{array}{r} 113.0 \\ 4.)\overline{452.300} \\ -4 \\ \hline 05 \\ -4 \\ \hline 12 \\ -12 \\ \hline 03 \\ -0 \\ \hline 3 \end{array}$$
113.0 rounds to 110.

**17.** $12.9 \div (-1000)$

1000 has 3 zeros so move the decimal point 3 places to the left and attach the negative sign.

$12.9 \div (-1000) = -0.0129$

**21.** $1.239 \div 3$
$$\begin{array}{r} 0.413 \\ 3)\overline{1.239} \\ -12 \\ \hline 03 \\ -3 \\ \hline 09 \\ -9 \\ \hline 0 \end{array}$$

**25.** $1.296 \div \ 0.27$

$0.27)\overline{1.296}$   Move the decimal points 2 places.

$$
\begin{array}{r}
4.8 \\
27.)\overline{129.6} \\
-108 \\
\hline
21\,6 \\
-21\,6 \\
\hline
0
\end{array}
$$

**29.** $-18 \div \ -0.6$

$0.6)\overline{18.0}$   Move the decimal points 1 place.

$$
\begin{array}{r}
30. \\
6.)\overline{180.} \\
-18 \\
\hline
00 \\
-\ 0 \\
\hline
0
\end{array}
$$

**33.** $-1.104 \div \ 1.6$

$$
\begin{array}{r}
0.69 \\
16)\overline{11.04} \\
-9\,6 \\
\hline
1\,44 \\
-1\,44 \\
\hline
0
\end{array}
$$

Thus $-1.104 \div \ 1.6 = -0.69$

**37.** $\dfrac{4.615}{0.071} = \dfrac{4615}{71} = 65$

$$
\begin{array}{r}
65 \\
71)\overline{4615} \\
-426 \\
\hline
355 \\
-355 \\
\hline
0
\end{array}
$$

**41.** $0.0043)\overline{500.}$   Move the decimal point 4 places.

$$
\begin{array}{r}
116279. \\
43)\overline{5000000.} \\
-43 \\
\hline
70 \\
-43 \\
\hline
270 \\
-258 \\
\hline
120 \\
-86 \\
\hline
340 \\
-301 \\
\hline
390 \\
-387 \\
\hline
3
\end{array}
$$

116,279 rounded to nearest ten-thousand is 120,000.

**45.** $x \div \ y$

$5.65 \div \ 0.8$

$0.8)\overline{5.65}$   Move the decimal points 1 place.

$$
\begin{array}{r}
7.0625 \\
8.)\overline{56.5000} \\
-56 \\
\hline
0\,5 \\
-\ 0 \\
\hline
50 \\
-48 \\
\hline
20 \\
-16 \\
\hline
40 \\
-40 \\
\hline
0
\end{array}
$$

**49.** $\dfrac{x}{4} = 3.04$

$\dfrac{12.16}{4} = 3.04$

$3.04 = 3.04$  True

Yes, it is a solution.

**53.** $\dfrac{z}{10} = 0.8$

$\dfrac{8}{10} = 0.8$

$0.8 = 0.8$  True

Yes, it is a solution.

**57.** There are 52 weeks per year and 40 hours per week. Therefore, there are $52 \times 40 = 2080$ hours per year.

$$
\begin{array}{r}
1\,245.687 \\
2080)\overline{2,591,031.000} \\
-2\,080 \\
\hline
511\,0 \\
-416\,0 \\
\hline
95\,03 \\
-83\,20 \\
\hline
11\,831 \\
-10\,400 \\
\hline
1\,431\,0 \\
-1\,248\,0 \\
\hline
183\,00 \\
-166\,40 \\
\hline
16\,600 \\
-14\,560 \\
\hline
2\,040
\end{array}
$$

His hourly wage was \$1245.69.

**61.** $39.37\overline{)200}$  Move the decimal points 2 places.

$$\begin{array}{r} 5.08 \approx 5.1 \\ 3937.\overline{)20{,}000.00} \\ \underline{-19\,685} \\ 315\,0 \\ \underline{-0} \\ 315\,00 \\ \underline{-314\,96} \\ 4 \end{array}$$

There are about 5.1 meters in 200 inches.

**65.**
$$\begin{array}{r} 128.63 \approx 128.6 \\ 24\overline{)3087.12} \\ \underline{-24} \\ 68 \\ \underline{-48} \\ 207 \\ \underline{-192} \\ 15\,1 \\ \underline{-14\,4} \\ 72 \\ \underline{-72} \\ 0 \end{array}$$

Their average speed was about 128.6 mph.

**69.** 75 mph $\div$ 1.15

$1.15\overline{)75}$  Move the decimal points 2 places.

$$\begin{array}{r} 65.21 \approx 65.2 \\ 115.\overline{)7500.00} \\ \underline{-690} \\ 600 \\ \underline{-575} \\ 25\,0 \\ \underline{-23\,0} \\ 2\,00 \\ \underline{-1\,15} \\ 85 \end{array}$$

95 mph $\div$ 1.15

$1.15\overline{)95}$  Move the decimal points 2 places.

$$\begin{array}{r} 82.60 \approx 82.6 \\ 115.\overline{)9500.00} \\ \underline{-920} \\ 300 \\ \underline{-230} \\ 70\,0 \\ \underline{-69\,0} \\ 1\,00 \\ \underline{-\,0} \\ 1\,00 \end{array}$$

The wind speed range for a Category 1 hurricane is from 65.2 knots to 82.6 knots.

**73.** To round 1000.994 to the nearest tenth, observe that the digit in the hundredths place is 9. Since this digit is at least 5, we need to add 1 to the digit in the tenths place. The number 1000.994 rounded to the nearest tenth is 1001.0.

**77.** $20 - 10 \div 5 = 20 - 2 = 18$

**81.** $4.5\overline{)38.7}$  Move the decimal points 1 place.

$$\begin{array}{r} 8.6 \\ 45.\overline{)387.0} \\ \underline{-360} \\ 27\,0 \\ \underline{-27\,0} \\ 0 \end{array}$$

The length is 8.6 feet.

### Exercise Set 5.5

**1.** $4.9 - 2.1 = 2.8$

$5 - 2 = 3$

3 is close to 2.8, so the answer is reasonable.

**5.** $62.16 \div 14.8 = 4.2$

$60 \div 15 = 4$

4 is close to 4.2 so the answer is reasonable.

**9.**
$$\begin{array}{rr} 34.92 & 35 \\ -12.03 & -12 \\ \hline 22.89 & 23 \end{array}$$

23 is close to 22.89 so the answer is reasonable.

**13.** $11.8 + 12.9 + 14.2 \approx 12 + 13 + 14 = 39$ ft.

**17.**
$$\begin{array}{r} 51.6 \\ 30\overline{)1550.0} \\ \underline{-150} \\ 50 \\ \underline{-30} \\ 20\,0 \\ \underline{-18\,0} \\ 2\,0 \end{array}$$

Their car will use about 52 gallons.

**21.** $19.9 + 15.1 + 10.9 + 6.7 \approx 20 + 15 + 11 + 7$
$$= 53 \text{ miles}$$

The distance is about 53 miles.

**25.** 485 rounded to the nearest 10 is 490.

271 rounded to the nearest 10 is 270.

$$\begin{array}{r} 490 \\ \times\;\;270 \\ \hline 0 \\ 34\,300 \\ 98\,000 \\ \hline 132{,}300 \end{array}$$

The population is about 132,300 people.

**29.** $\dfrac{1 + 0.8}{-0.6} = \dfrac{1.8}{-0.6} = \dfrac{18}{-6} = -3$

**33.** $4.83 \div 2.1 = 2.3$

**37.** $(3.1 + 0.7)(2.9 - 0.9) = (3.8)(2.0) = 7.6$

**41.** $\dfrac{7 + 0.74}{-6} = \dfrac{7.74}{-6} = -1.29$

**45.** $x - y$

$6 - (0.3)$

$= 5.7$

**49.** $7x + 2.1 = -7$

$7(-1.3) + 2.1 = -7$

$-9.1 + 2.1 = -7$

$-7 = -7$  True

Yes, it is a solution.

**53.** $\dfrac{3}{4} \cdot \dfrac{5}{12} = \dfrac{3 \cdot 5}{4 \cdot 3 \cdot 4} = \dfrac{5}{16}$

**57.** $\frac{5}{12} - \frac{1}{3} = \frac{5}{12} - \frac{4}{12} = \frac{1}{12}$

**61.** Answers may vary.

## EXERCISE SET 5.6

**1.** $\frac{1}{5} = 0.2$

$$\begin{array}{r} 0.2 \\ 5\overline{)1.0} \\ \underline{-1\,0} \\ 0 \end{array}$$

**5.** $\frac{3}{4} = 0.75$

$$\begin{array}{r} 0.75 \\ 4\overline{)3.00} \\ \underline{-2\,8} \\ 20 \\ \underline{-20} \\ 0 \end{array}$$

**9.** $\frac{3}{8} = 0.375$

$$\begin{array}{r} 0.375 \\ 8\overline{)3.000} \\ \underline{-2\,4} \\ 60 \\ \underline{-56} \\ 40 \\ \underline{-40} \\ 0 \end{array}$$

**13.** $\frac{17}{40} = 0.425$

$$\begin{array}{r} 0.425 \\ 40\overline{)17.000} \\ \underline{-16\,0} \\ 1\,00 \\ \underline{-80} \\ 200 \\ \underline{-200} \\ 0 \end{array}$$

**17.** $\frac{1}{3} = 0.\overline{3}$

$$\begin{array}{r} .33 \\ 3\overline{)1.00} \\ \underline{-9} \\ 10 \\ \underline{-9} \\ 1 \end{array}$$

**21.** $\frac{2}{9} = 0.\overline{2}$

$$\begin{array}{r} 0.22 \\ 9\overline{)2.00} \\ \underline{-1\,8} \\ 20 \\ \underline{-18} \\ 2 \end{array}$$

**25.** $0.\overline{3} = 0.333 \approx 0.33$

**29.** $0.\overline{2} = 0.222 \approx 0.2$

**33.**

$$\begin{array}{r} 0.1941... \\ 376\overline{)73.0000} \\ \underline{-37\,6} \\ 35\,40 \\ \underline{-33\,84} \\ 1\,560 \\ \underline{-1\,504} \\ 560 \\ \underline{-376} \\ 184 \end{array}$$

$\frac{73}{376} \approx 0.194$

**37.**

$$\begin{array}{r} 0.8 \\ 5\overline{)4.0} \\ \underline{-4\,0} \\ 0 \end{array}$$

$\frac{4}{5} = 0.8$

**41.** $2 > 1$, so $0.823 > 0.813$

**45.** $\frac{2}{3} = \frac{4}{6}$ and $\frac{4}{6} < \frac{5}{6}$, so $\frac{2}{3} < \frac{5}{6}$.

**49.** $\frac{4}{7} \approx 0.5714$ and $0.5714 > 0.14$, so $\frac{4}{7} > 0.14$.

**53.** $\frac{456}{64} = 7.125$ and $7.123 < 7.125$, so $7.123 < \frac{456}{64}$.

**57.** $0.49 = 0.490$

$0.49, 0.491, 0.498$

**61.** $\frac{4}{7} \approx 0.571$

$0.412, 0.453, \frac{4}{7}$

**65.** $\frac{12}{5} = 2.4, \frac{17}{8} = 2.125$

$\frac{17}{8}, 2.37, \frac{12}{5}$

**69.** Area $= \frac{1}{2} \times$ base $\times$ height

$= \frac{1}{2} \times 5.2 \times 3.6$

$= 0.5 \times 5.2 \times 3.6$

$= 9.36$

The area is 9.36 square centimeters.

**73.** $2^3 = (2)(2)(2) = 8$

**77.** $\left(\frac{1}{3}\right)^4 = \frac{1}{3} \cdot \frac{1}{3} \cdot \frac{1}{3} \cdot \frac{1}{3} = \frac{1}{81}$

**81.** $\left(\frac{2}{5}\right)\left(\frac{5}{2}\right)^2 = \frac{2}{5} \cdot \frac{5}{2} \cdot \frac{5}{2} = \frac{5}{2}$

**85.**

| | |
|---|---|
| 2321 | Estimate: 2300 |
| 1576 | 1600 |
| 1396 | 1400 |
| 1109 | 1100 |
| 1088 | 1100 |
| +803 | +800 |
| | 8300 |

**89.** Answers may vary.

**93.** $\left(\dfrac{1}{10}\right)^2 + (1.6)(2.1) = \dfrac{1}{100} + (1.6)(2.1)$

$\qquad\qquad\qquad = 0.01 + 3.36$

$\qquad\qquad\qquad = 3.37$

## EXERCISE SET 5.7

**1.** $\qquad x + 1.2 = 7.1$

$\qquad x + 1.2 - 1.2 = 7.1 - 1.2$

$\qquad\qquad\quad x = 5.9$

**5.** $\qquad\qquad 6x + 8.65 = 3x + 10$

$\qquad 6x + 8.65 - 8.65 = 3x + 10 - 8.65$

$\qquad\qquad\qquad 6x = 3x + 1.35$

$\qquad\qquad 6x - 3x = 3x - 3x + 1.35$

$\qquad\qquad\qquad 3x = 1.35$

$\qquad\qquad\qquad \dfrac{3x}{3} = \dfrac{1.35}{3}$

$\qquad\qquad\qquad\quad x = 0.45$

**9.** $\quad 0.4x + 0.7 = -0.9$

$\qquad\quad 4x + 7 = -9$

$\quad 4x + 7 - 7 = -9 - 7$

$\qquad\qquad 4x = -16$

$\qquad\qquad \dfrac{4x}{4} = \dfrac{-16}{4}$

$\qquad\qquad\quad x = -4$

**13.** $2.1x + 5 - 1.6x = 10$

$\quad 21x + 50 - 16x = 100$

$\qquad\quad 5x + 50 = 100$

$\quad 5x + 50 - 50 = 100 - 50$

$\qquad\qquad 5x = 50$

$\qquad\qquad \dfrac{5x}{5} = \dfrac{50}{5}$

$\qquad\qquad\quad x = 10$

**17.** $\quad -0.02x = -1.2$

$\qquad\quad -2x = -120$

$\qquad\quad \dfrac{-2x}{-2} = \dfrac{-120}{-2}$

$\qquad\qquad\quad x = 60$

**21.** $\qquad 200x - 0.67 = 100x + 0.81$

$\quad 200x - 0.67 + 0.67 = 100x + 0.81 + 0.67$

$\qquad\qquad 200x = 100x + 1.48$

$\quad 200x - 100x = 100x - 100x + 1.48$

$\qquad\qquad 100x = 1.48$

$\qquad\qquad\qquad x = 0.0148$

**25.** $\qquad 1.2 + 0.3x = 0.9$

$\qquad\quad 12 + 3x = 9$

$\quad 12 - 12 + 3x = 9 - 12$

$\qquad\qquad 3x = -3$

$\qquad\qquad \dfrac{3x}{3} = \dfrac{-3}{3}$

$\qquad\qquad\quad x = -1$

**29.** $\qquad\qquad 4x + 7.6 = 2(3x - 3.2)$

$\qquad\qquad\quad 4x + 7.6 = 6x - 6.4$

$\quad 4x - 6x + 7.6 - 7.6 = 6x - 6x - 6.4 - 7.6$

$\qquad\qquad\qquad -2x = -14$

$\qquad\qquad\qquad\quad x = 7$

**33.** $3(x - 5) + 10 = 3x - 15 + 10 = 3x - 5$

**37.** Answers may vary.

**41.** $\quad 1.95y + 6.834 = 7.65y - 19.8591$

$\quad 19.8591 + 6.834 = 7.65y - 1.95y$

$\qquad\qquad 26.6931 = 5.7y$

$\qquad\qquad \dfrac{26.6931}{5.7} = \dfrac{5.7y}{5.7}$

$\qquad\qquad 4.683 = y$

## EXERCISE SET 5.8

**1.** $\sqrt{4} = 2$ because $2^2 = 4.$

**5.** $\sqrt{\dfrac{1}{81}} = \dfrac{1}{9}$ because $\dfrac{1}{9} \cdot \dfrac{1}{9} = \dfrac{1}{81}.$

**9.** $\sqrt{256} = 16$ because $16^2 = 256.$

**13.** $\sqrt{3} \approx 1.732$

**17.** $\sqrt{14} \approx 3.742$

**21.** $\sqrt{8} \approx 2.828$

**25.** $\sqrt{71} \approx 8.426$

**29.** $\quad a^2 + b^2 = c^2$

$\quad 5^2 + 12^2 = c^2$

$\quad 25 + 144 = c^2$

$\qquad\quad 169 = c^2$

$\qquad\qquad c = \sqrt{169}$

$\qquad\qquad c = 13$

The length of the hypotenuse is 13 inches.

**33.**

$\quad a^2 + b^2 = c^2$

$\quad 3^2 + 4^2 = c^2$

$\quad 9 + 16 = c^2$

$\qquad\quad 25 = c^2$

$\qquad\qquad c = \sqrt{25}$

$\qquad\qquad c = 5$

**37.**

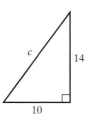

$\quad a^2 + b^2 = c^2$

$\quad 10^2 + 14^2 = c^2$

$\quad 100 + 196 = c^2$

$\qquad\quad 296 = c^2$

$\qquad\qquad c = \sqrt{296}$

$\qquad\qquad c \approx 17.205$

**41.**

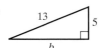

$\quad a^2 + b^2 = c^2$

$\quad 5^2 + b^2 = 13^2$

$\quad 25 + b^2 = 169$

$\qquad\quad b^2 = 144$

$\qquad\quad b = 12$

**45.**

$$a^2 + b^2 = c^2$$
$$30^2 + 30^2 = c^2$$
$$900 + 900 = c^2$$
$$1800 = c^2$$
$$c = \sqrt{1800}$$
$$c \approx 42.426$$

**49.**
$$a^2 + b^2 = c^2$$
$$100^2 + 100^2 = c^2$$
$$10{,}000 + 10{,}000 = c^2$$
$$20{,}000 = c^2$$
$$c = \sqrt{20{,}000}$$
$$c \approx 141.42$$

The length of the diagonal is about 141.42 yards.

**53.**
$$a^2 + b^2 = c^2$$
$$300^2 + 160 = c^2$$
$$90{,}000 + 25{,}600 = c^2$$
$$115{,}600 = c^2$$
$$c = \sqrt{115{,}600}$$
$$c = 340$$

The length of the diagonal is about 340 feet.

**57.** $\dfrac{24}{60} = \dfrac{2 \cdot 2 \cdot 2 \cdot 3}{2 \cdot 3 \cdot 2 \cdot 5} = \dfrac{2}{5}$

**61.** $\sqrt{38}$ is between 6 and 7.
$\sqrt{38} \approx 6.164$

### CHAPTER 5 TEST

**1.** 45.092 is forty-five and ninety-two thousandths.

**5.** $9.83 - 30.25 = 9.83 + (-30.25)$
Subtract the absolute values.
$$\begin{array}{r} 30.25 \\ -9.83 \\ \hline 20.42 \end{array}$$
Attach the sign of the larger absolute value.
$-20.42$

**9.** 0.8623 rounded to the nearest thousandth is 0.862.

**13.** $24.73 = 24\dfrac{73}{100}$

**17.** $\dfrac{0.23 + 1.63}{-0.3} = \dfrac{1.86}{-0.3} = -6.2$

**21.** $\sqrt{\dfrac{64}{100}} = \dfrac{8}{10} = \dfrac{4}{5}$ since $\left(\dfrac{8}{10}\right)^2 = \dfrac{64}{100}$.

**25.** Area $= \dfrac{1}{2} \times$ base $\times$ height
$= 0.5 \times 4.2 \times 1.1$
$= 2.31$
The area is 2.31 square miles.

**29.**
$$\begin{array}{r} 1.42 \\ 16.1 \\ +23.7 \\ \hline \end{array}$$
Estimate:
$$\begin{array}{r} 14 \\ 16 \\ +24 \\ \hline 54 \end{array}$$
The total distance is about 54 miles.

# Chapter 6

### EXERCISE SET 6.1

**1.** The ratio of 11 to 14 is $\dfrac{11}{14}$.

**5.** The ratio of 151 to 201 is $\dfrac{151}{201}$.

**9.** The ratio of 5 to $7\dfrac{1}{2}$ is $\dfrac{5}{7\frac{1}{2}}$.

**13.** $\dfrac{16}{24} = \dfrac{2 \cdot 8}{3 \cdot 8} = \dfrac{2}{3}$

**17.** $\dfrac{4.63}{8.21} = \dfrac{4.63 \times 100}{8.21 \times 100} = \dfrac{463}{821}$

**21.** $\dfrac{10 \text{ hours}}{24 \text{ hours}} = \dfrac{5 \cdot 2}{12 \cdot 2} = \dfrac{5}{12}$

**25.** $\dfrac{24 \text{ days}}{14 \text{ days}} = \dfrac{12 \cdot 2}{7 \cdot 2} = \dfrac{12}{7}$

**29.** $\dfrac{\text{vacation/other miles}}{\text{total miles}} = \dfrac{900 \text{ miles}}{15{,}000 \text{ miles}}$
$= \dfrac{300 \cdot 3}{300 \cdot 50}$
$= \dfrac{3}{50}$

**33.** perimeter $= 8 + 15 + 17 = 40$ feet
$\dfrac{\text{longest side}}{\text{perimeter}} = \dfrac{17 \text{ feet}}{40 \text{ feet}} = \dfrac{17}{40}$

**37.** $\dfrac{\text{women}}{\text{men}} = \dfrac{125}{100} = \dfrac{25 \cdot 5}{25 \cdot 4} = \dfrac{5}{4}$

**41.** $\dfrac{600 \text{ red blood cells}}{40 \text{ platelet cells}} = \dfrac{40 \cdot 15}{40} = \dfrac{15}{1}$

**45.** $\dfrac{25 \text{ medals}}{205 \text{ medals}} = \dfrac{5 \cdot 5}{41 \cdot 5} = \dfrac{5}{41}$

**49.** $3.7\overline{)0.555}$    Move decimal points 1 place.
$$\begin{array}{r} 0.15 \\ 37\overline{)5.55} \\ -3\,7 \\ \hline 1\,85 \\ -1\,85 \\ \hline 0 \end{array}$$

**53.** Answers may vary.

**57.** Answers may vary.

### EXERCISE SET 6.2

**1.** $\dfrac{5 \text{ shrubs}}{15 \text{ feet}} = \dfrac{1 \text{ shrub}}{3 \text{ feet}}$

**5.** $\dfrac{8 \text{ phone lines}}{36 \text{ employees}} = \dfrac{2 \text{ phone lines}}{9 \text{ employees}}$

**9.** $\dfrac{6 \text{ flight attendants}}{200 \text{ passengers}} = \dfrac{3 \text{ flight attendants}}{100 \text{ passengers}}$

**13.** $\dfrac{375 \text{ riders}}{5 \text{ subway cars}} = \dfrac{75 \text{ riders}}{1 \text{ subway car}} = 75$ riders/car

**17.** $\dfrac{144 \text{ diapers}}{24 \text{ babies}} = \dfrac{6 \text{ diapers}}{1 \text{ baby}} = 6$ diapers/baby

**21.** $\dfrac{600 \text{ kilometers}}{90 \text{ minutes}} = \dfrac{20 \text{ kilometers}}{3 \text{ minutes}}$

$\qquad\qquad = 6\dfrac{2}{3} \text{ kilometers per minute}$

$\qquad\qquad \approx 6.67 \text{ kilometers/minute}$

**25.** $\dfrac{12{,}000 \text{ good}}{40 \text{ defective}} = \dfrac{300 \text{ good}}{1 \text{ defective}} = 300 \text{ good/defective}$

**29.** $\dfrac{\$20{,}000{,}000}{400 \text{ species}} = \dfrac{\$50{,}000}{1 \text{ species}}$ or $\$50{,}000$/species

**33. a.** Greer: $\dfrac{250 \text{ boards}}{8 \text{ hours}} = 31.25 \text{ computer boards/hr}$

**b.** Lamont: $\dfrac{400 \text{ boards}}{12 \text{ hours}} \approx 33.3 \text{ boards/hr}$

**c.** Lamont can assemble about 33.3 boards in the time that Greer assembles 31.25 boards. Thus, Lamont assembles them faster.

**37.** $\dfrac{\$1.19}{7 \text{ bananas}} = \dfrac{\$0.17}{1 \text{ banana}}$ or $\$0.17$ per banana

**41.** 16-ounce size: $\dfrac{\$1.69}{16 \text{ ounces}} \approx \$0.106 \text{ per ounce}$

6-ounce size: $\dfrac{\$0.69}{6 \text{ ounces}} = \$0.115 \text{ per ounce}$

The 16-ounce size costs less per ounce thus, it is the better buy.

**45.** 100-pack size: $\dfrac{\$0.59}{100 \text{ napkins}} \approx \$0.006 \text{ per napkin}$

180-pack size: $\dfrac{\$0.93}{180 \text{ napkins}} \approx \$0.005 \text{ per napkin}$

The package of 180 napkins cost less per napkin thus, it is the better buy.

**49.**
$$\begin{array}{r} 3.7 \\ \times\ 1.2 \\ \hline 74 \\ 370 \\ \hline 4.44 \end{array}$$

**53.**

| Miles Driven | Miles Per Gallon (round to the nearest tenth) |
| --- | --- |
| $79{,}543 - 79{,}286 = 257$ | $\dfrac{257}{13.4} = 19.2$ |
| $79{,}895 - 79{,}543 = 352$ | $\dfrac{352}{15.8} = 22.3$ |
| $80{,}242 - 79{,}895 = 347$ | $\dfrac{347}{16.1} = 21.6$ |

**57.** $\dfrac{477{,}121 \text{ students}}{20{,}039 \text{ teachers}} \approx 23.8 \text{ students/teacher}$

$\qquad\qquad\qquad \approx 24 \text{ students/teacher}$

### EXERCISE SET 6.3

**1.** $\dfrac{10 \text{ diamonds}}{6 \text{ opals}} = \dfrac{5 \text{ diamonds}}{3 \text{ opals}}$

**5.** $\dfrac{6 \text{ eagles}}{58 \text{ sparrows}} = \dfrac{3 \text{ eagles}}{29 \text{ sparrows}}$

**9.** $\dfrac{22 \text{ vanilla wafers}}{1 \text{ cup cookie crumbs}} = \dfrac{55 \text{ vanilla wafers}}{2.5 \text{ cups cookie crumbs}}$

**13.** $\dfrac{8}{6} \stackrel{?}{=} \dfrac{9}{7}$

$6 \cdot 9 \stackrel{?}{=} 8 \cdot 7$

$54 \neq 56$

Since the cross products are not equal, the proportion is false.

**17.** $\dfrac{5}{8} \stackrel{?}{=} \dfrac{625}{1000}$

$8 \cdot 625 \stackrel{?}{=} 5 \cdot 1000$

$5000 = 5000$

Since the cross products are equal, the proportion is true.

**21.** $\dfrac{4.2}{8.4} \stackrel{?}{=} \dfrac{5}{10}$

$8.4 \cdot 5 \stackrel{?}{=} 4.2 \cdot 10$

$42 = 42$

Since the cross products are equal, the proportion is true.

**25.** $\dfrac{2\frac{2}{5}}{\frac{2}{3}} \stackrel{?}{=} \dfrac{\frac{10}{9}}{\frac{1}{4}}$

$\dfrac{2}{3} \cdot \dfrac{10}{9} \stackrel{?}{=} 2\dfrac{2}{5} \cdot \dfrac{1}{4}$

$\dfrac{20}{27} \stackrel{?}{=} \dfrac{12}{5} \cdot \dfrac{1}{4}$

$\dfrac{20}{27} \stackrel{?}{=} \dfrac{3}{5}$

$27 \cdot 3 \stackrel{?}{=} 20 \cdot 5$

$81 \neq 100$

Since the cross products are not equal, the proportion is false.

**29.** $\dfrac{30}{10} = \dfrac{15}{y}$

$10 \cdot 15 = 30y$

$150 = 30y$

$\dfrac{150}{30} = \dfrac{30y}{30}$

$5 = y$

**33.** $\dfrac{n}{6} = \dfrac{8}{15}$

$6 \cdot 8 = 15n$

$48 = 15n$

$\dfrac{48}{15} = \dfrac{15n}{15}$

$3.2 = n$

**37.**

$$\frac{n}{\frac{6}{5}} = \frac{4\frac{1}{6}}{6\frac{2}{3}}$$

$$\frac{6}{5} \cdot 4\frac{1}{6} = 6\frac{2}{3} \cdot n$$

$$\frac{6}{5} \cdot \frac{25}{6} = \frac{20}{3}n$$

$$\frac{150}{30} = \frac{20}{3}n$$

$$5 = \frac{20}{3}n$$

$$\frac{3}{20} \cdot \frac{5}{1} = \frac{3}{20} \cdot \frac{20}{3}n$$

$$\frac{15}{20} = n$$

$$\frac{3}{4} = n$$

**41.**

$$\frac{\frac{2}{3}}{\frac{6}{9}} = \frac{12}{z}$$

$$\frac{6}{9} \cdot 12 = \frac{2}{3}z$$

$$\frac{3}{2} \cdot \frac{6}{9} \cdot \frac{12}{1} = \frac{3}{2} \cdot \frac{2}{3}z$$

$$\frac{3 \cdot 2 \cdot 3 \cdot 12}{2 \cdot 3 \cdot 3 \cdot 1} = z$$

$$12 = z$$

**45.**

$$\frac{3.5}{12.5} = \frac{7}{z}$$

$$12.5 \cdot 7 = 3.5 \cdot z$$

$$87.5 = 3.5z$$

$$\frac{87.5}{3.5} = \frac{3.5z}{3.5}$$

$$25 = z$$

**49.**

$$\frac{z}{5.2} = \frac{0.08}{6}$$

$$5.2 \cdot 0.08 = 6z$$

$$0.416 = 6z$$

$$\frac{0.416}{6} = \frac{6z}{6}$$

$$0.07 \approx z$$

**53.**

$$\frac{43}{17} = \frac{8}{z}$$

$$17 \cdot 8 = 43z$$

$$136 = 43z$$

$$\frac{136}{43} = \frac{43z}{43}$$

$$3.163 \approx z$$

**57.** $-2 > -3$

**61.** $-1\frac{1}{2} > -2\frac{1}{2}$

**65.**

$$\frac{n}{1150} = \frac{588}{483}$$

$$1150 \cdot 588 = 483n$$

$$676,200 = 483n$$

$$\frac{676,200}{483} = \frac{483n}{483}$$

$$1400 = n$$

**69.** Answers may vary.

**A58**

---

## Exercise Set 6.4

**1.** Let $x$ be the number of completed passes.

$$\frac{4 \text{ completed}}{9 \text{ attempted}} = \frac{x \text{ completed}}{27 \text{ attempted}}$$

$$9x = 27 \cdot 4$$

$$9x = 108$$

$$\frac{9x}{9} = \frac{108}{9}$$

$$x = 12$$

He completed 12 passes.

**5.** Let $x$ be the number of students accepted.

$$\frac{2 \text{ accepts}}{7 \text{ applicants}} = \frac{x \text{ accepts}}{630 \text{ applicants}}$$

$$7x = 1260$$

$$\frac{7x}{7} = \frac{1260}{7}$$

$$x = 180$$

180 students were accepted.

**9.** Let $x$ be the amount of floor space.

$$\frac{9 \text{ sq. ft}}{1 \text{ student}} = \frac{x \text{ sq. ft}}{30 \text{ students}}$$

$$x = 270$$

270 square feet is required.

**13.** Let $x$ be the distance from Milan to Rome.

$$\frac{1 \text{ cm}}{30 \text{ km}} = \frac{15 \text{ cm}}{x \text{ km}}$$

$$450 = x$$

It is 450 km from Milan to Rome.

**17.** Let $x$ be the number of bags of fertilizer.

$$\text{Area of lawn} = 260 \text{ ft} \cdot 180 \text{ ft}$$

$$= 46,800 \text{ square feet}$$

$$\frac{1 \text{ bag}}{3000 \text{ sq ft}} = \frac{x}{46,800 \text{ sq ft}}$$

$$3000x = 46,800$$

$$\frac{3000x}{3000} = \frac{46,800}{3000}$$

$$x = 15.6$$

Since a whole number of bags must be purchased, 16 bags are needed.

**21.** Let $x$ be the number of hits expected.

$$\frac{3 \text{ hits}}{8 \text{ at bats}} = \frac{x \text{ hits}}{40 \text{ at bats}}$$

$$8x = 120$$

$$\frac{8x}{8} = \frac{120}{8}$$

$$x = 15$$

He would be expected to make 15 hits.

**25.** Let $x$ be the number of weeks it will last.

$$\frac{5 \text{ boxes}}{3 \text{ weeks}} = \frac{144 \text{ boxes}}{x \text{ weeks}}$$

$$432 = 5x$$

$$\frac{432}{5} = \frac{5x}{5}$$

$$86.4 = x$$

The envelopes will last 86 weeks.

**29.** Let $n$ be the estimated height of Statue of Liberty.

$$\frac{\text{statue}}{\text{student}} = \frac{42 \text{ feet}}{2 \text{ feet}} = \frac{n \text{ feet}}{5\frac{1}{3} \text{ feet}}$$

$$2n = 42 \cdot 5\frac{1}{3}$$

$$2n = \frac{42}{1} \cdot \frac{16}{3}$$

$$2n = 224$$

$$\frac{2n}{2} = \frac{224}{2}$$

$$n = 112$$

The estimate is 112 feet. The actual height is 111 feet 1 inch. The difference in the estimated height and the actual height is $112 - 111\frac{1}{12} = \frac{11}{12}$ of a foot or 11 inches.

**33.** Let $n$ be the number of visits including a prescription.

$$\frac{7 \text{ prescriptions}}{10 \text{ visits}} = \frac{n \text{ prescriptions}}{620 \text{ visits}}$$

$$10n = 7 \cdot 620$$

$$10n = 4340$$

$$\frac{10n}{10} = \frac{4340}{10}$$

$$n = 434$$

434 emergency room visits included a prescription.

**37.** Let $n$ be the number of cups of rock salt.

$$\frac{5}{1} = \frac{12}{n}$$

$$1 \cdot 12 = 5n$$

$$12 = 5n$$

$$\frac{12}{5} = \frac{5n}{5}$$

$$2.4 = n$$

Mix 2.4 cups of rock salt with the ice.

**41.** $20 = 4 \cdot 5 = 2 \cdot 2 \cdot 5 = 2^2 \cdot 5$

**45.** $32 = 4 \cdot 8$
$= 2 \cdot 2 \cdot 2 \cdot 4$
$= 2 \cdot 2 \cdot 2 \cdot 2 \cdot 2$
$= 2^5$

**49.** Answers may vary.

### Exercise Set 6.5

**1.** Since the triangles are similar, we can compare any of the corresponding sides to find the ratio.

$$\frac{22}{11} = \frac{2}{1}$$

**5.** $\frac{3}{n} = \frac{6}{9}$

$$6n = 27$$

$$\frac{6n}{6} = \frac{27}{6}$$

$$n = 4.5$$

**9.** $\frac{n}{3.75} = \frac{12}{9}$

$$45 = 9n$$

$$\frac{45}{9} = \frac{9n}{9}$$

$$5 = n$$

**13.** $\frac{n}{3.25} = \frac{17.5}{3.25}$

$$56.875 = 3.25n$$

$$\frac{56.875}{3.25} = \frac{3.25n}{3.25}$$

$$17.5 = n$$

**17.** $\frac{34}{n} = \frac{16}{10}$

$$16n = 340$$

$$\frac{16n}{16} = \frac{340}{16}$$

$$n = 21.25$$

**21.** $\frac{11}{x} = \frac{5}{2}$

$$5x = 22$$

$$\frac{5x}{5} = \frac{22}{5}$$

$$x = 4.4$$

$$\frac{14}{y} = \frac{5}{2}$$

$$5y = 28$$

$$\frac{5y}{5} = \frac{28}{5}$$

$$y = 5.6$$

The lengths are $x = 4.4$ feet and $y = 5.6$ feet.

**25.** $\frac{1}{4} = \frac{1 \cdot 25}{4 \cdot 25} = \frac{25}{100} = 0.25$

**29.** $\frac{9}{10} = 0.9$

**33.** Answers may vary.

### Chapter 6 Test

**1.** $\frac{4500 \text{ trees}}{6500 \text{ trees}} = \frac{9 \cdot 500}{13 \cdot 500} = \frac{9}{13}$

**5.** $\frac{650 \text{ kilometers}}{8 \text{ hours}} = 81.25$ kilometers/hour

$$\begin{array}{r} 81.25 \\ 8\overline{)650.00} \\ -64 \phantom{0000} \\ \hline 10 \phantom{000} \\ -8 \phantom{000} \\ \hline 2\,0 \phantom{00} \\ -1\,6 \phantom{00} \\ \hline 40 \phantom{0} \\ -40 \phantom{0} \\ \hline 0 \phantom{0} \end{array}$$

**9.** 16-ounce size: $\frac{\$1.49}{16 \text{ oz}} \approx \$0.093$ per ounce

24-ounce size: $\frac{\$2.39}{24 \text{ oz}} \approx \$0.100$ per ounce

The 16-ounce size costs less per ounce, thus it is the better buy.

**13.** $\dfrac{8}{x} = \dfrac{11}{6}$

$11 \cdot x = 8 \cdot 6$

$11x = 48$

$\dfrac{11x}{11} = \dfrac{48}{11}$

$x = \dfrac{48}{11} = 4\dfrac{4}{11}$

**17.** Let $x$ be the length of time.

$\dfrac{3}{80} = \dfrac{x}{100}$

$80x = 300$

$\dfrac{80x}{80} = \dfrac{300}{80}$

$x = \dfrac{15}{4} = 3\dfrac{3}{4}$

It will take $3\dfrac{3}{4}$ hours.

**21.** $\dfrac{5}{8} = \dfrac{n}{12}$

$8n = 5 \cdot 12$

$8n = 60$

$\dfrac{8n}{8} = \dfrac{60}{8}$

$n = 7.5$

# Chapter 7

## EXERCISE SET 7.1

**1.** $\dfrac{81}{100} = 81\%$

**5.** The largest section of the circle graph is chocolate chip. Therefore, chocolate chip was the most preferred cookie.

$\dfrac{52}{100} = 52\%$

**9.** $48\% = 48.\% = 0.48$

**13.** $100\% = 100.\% = 1.00 = 1$

**17.** $2.8\% = 0.028$

**21.** $300\% = 300.\% = 3.00 = 3$

**25.** $73.7\% = 0.737$

**29.** $80.2\% = 0.802$

**33.** $3.1 = 3.10 = 310\%$

**37.** $0.003 = 000.3\% = 0.3\%$

**41.** $0.056 = 005.6\% = 5.6\%$

**45.** $3.00 = 300.\% = 300\%$

**49.** $0.10 = 010.\% = 10\%$

**53.** $0.38 = 038.\% = 38\%$

**57.** $4.5\% = \dfrac{4.5}{100} = \dfrac{45}{1000} = \dfrac{5 \cdot 9}{5 \cdot 200} = \dfrac{9}{200}$

**61.** $73\% = \dfrac{73}{100}$

**65.** $6.25\% = \dfrac{6.25}{100} = \dfrac{625}{10,000} = \dfrac{625}{625 \cdot 16} = \dfrac{1}{16}$

**69.** $22\dfrac{3}{8}\% = \dfrac{22\frac{3}{8}}{100}$

$= \dfrac{\frac{179}{8}}{100}$

$= \dfrac{179}{8} \div 100$

$= \dfrac{179}{8} \cdot \dfrac{1}{100}$

$= \dfrac{179}{800}$

**73.** $\dfrac{7}{10} = \dfrac{7}{10} \cdot 100\% = \dfrac{700}{10}\% = 70\%$

**77.** $\dfrac{59}{100} = \dfrac{59}{100} \cdot 100\% = \dfrac{59 \cdot 100}{100}\% = 59\%$

**81.** $\dfrac{3}{8} = \dfrac{3}{8} \cdot 100\% = \dfrac{300}{8}\% = \dfrac{4.75}{4.2}\% = \dfrac{75}{2}\% = 37\dfrac{1}{2}\%$

**85.** $\dfrac{2}{3} = \dfrac{2}{3} \cdot 100\% = \dfrac{200}{3}\% = 66\dfrac{2}{3}\%$

**89.** $1\dfrac{9}{10} = \dfrac{19}{10} \cdot 100\% = \dfrac{1900\%}{10} = 190\%$

**93.** $\dfrac{4}{15} = \dfrac{4}{15} \cdot 100\% = \dfrac{400}{15}\% \approx 26.67\%$

$\begin{array}{r} 26.666 \approx 26.67 \\ 15\overline{)400.000} \\ \underline{-30\phantom{0.000}} \\ 100 \\ \underline{-90} \\ 10\,0 \\ \underline{-9\,0} \\ 1\,00 \\ \underline{-90} \\ 100 \\ \underline{-90} \\ 10 \end{array}$

**97.** $\dfrac{11}{12} = \dfrac{11}{12} \cdot 100\% = \dfrac{1100}{12}\% \approx 91.67\%$

$\begin{array}{r} 91.666 \approx 91.67 \\ 12\overline{)1100.000} \\ \underline{-108\phantom{0.000}} \\ 20 \\ \underline{-12} \\ 8\,0 \\ \underline{-7\,2} \\ 80 \\ \underline{-72} \\ 80 \\ \underline{-72} \\ 8 \end{array}$

**101.**

| Percent | Decimal | Fraction |
|---|---|---|
| 40% | 0.4 | $\dfrac{2}{5}$ |
| 23.5% | 0.235 | $\dfrac{47}{200}$ |
| 80% | 0.8 | $\dfrac{4}{5}$ |
| $33\dfrac{1}{3}\%$ | $0.333\overline{3}$ | $\dfrac{1}{3}$ |
| 87.5% | 0.875 | $\dfrac{7}{8}$ |
| 7.5% | 0.075 | $\dfrac{3}{40}$ |

**105.** $\dfrac{17}{200} = \dfrac{17}{200} \cdot 100\%$

$\qquad = \dfrac{1700}{200}\%$

$\qquad = \dfrac{17}{2}\%$

$\qquad = 0.085$

$\qquad = 8.5\%$

$$\begin{array}{r} 8.5 \\ 2\overline{)17.0} \\ \underline{-16} \\ 1\,0 \\ \underline{-1\,0} \\ 0 \end{array}$$

**109.** $\dfrac{3}{10} = \dfrac{3 \cdot 10}{10 \cdot 10} = \dfrac{30}{100} = 30\%$

**113.** $-8n = 80$

$\qquad \dfrac{-8n}{-8} = \dfrac{80}{-8}$

$\qquad n = -10$

**117.** $\dfrac{850}{736} \approx 1.155 = 115.5\%$

**121.** Answers may vary.

**125.** $85\% = 85.\% = 0.85$

## EXERCISE SET 7.2

**1.** $15\% \cdot 72 = x$

**5.** $x \cdot 90 = 20$

**9.** $x = 9\% \cdot 43$

**13.** $x = 14\% \cdot 52$

$\qquad x = 0.14 \cdot 52$

$\qquad x = 7.28$

**17.** $1.2 = 12\% \cdot x$

$\qquad 1.2 = 0.12x$

$\qquad \dfrac{1.2}{0.12} = \dfrac{0.12x}{0.12}$

$\qquad 10 = x$

**21.** $16 = x \cdot 50$

$\qquad \dfrac{16}{50} = \dfrac{x \cdot 50}{50}$

$\qquad 0.32 = x$

$\qquad 32\% = x$

**25.** $125\% \cdot 36 = x$

$\qquad 1.25 \cdot 36 = x$

$\qquad 45 = x$

**29.** $2.58 = x \cdot 50$

$\qquad \dfrac{2.58}{50} = \dfrac{x \cdot 50}{50}$

$\qquad 0.0516 = x$

$\qquad 5.16\% = x$

**33.** $x \cdot 150 = 67.5$

$\qquad \dfrac{x \cdot 150}{150} = \dfrac{67.5}{150}$

$\qquad x = 0.45$

$\qquad x = 45\%$

**37.** $\dfrac{27}{n} = \dfrac{9}{10}$

$\qquad 9 \cdot n = 27 \cdot 10$

$\qquad \dfrac{9 \cdot n}{9} = \dfrac{270}{9}$

$\qquad n = 30$

**41.** $\dfrac{17}{12} = \dfrac{n}{20}$

**45.** $1.5\% \cdot 45,775 = n$

$\qquad 0.015 \cdot 45,775 = n$

$\qquad 686.625 = n$

## EXERCISE SET 7.3

**1.** $\dfrac{a}{65} = \dfrac{32}{100}$

**5.** $\dfrac{70}{200} = \dfrac{p}{100}$

**9.** $\dfrac{a}{130} = \dfrac{19}{100}$

**13.** $\dfrac{a}{105} = \dfrac{18}{100}$

$\qquad \dfrac{a}{105} = \dfrac{9}{50}$

$\qquad a \cdot 50 = 105 \cdot 9$

$\qquad \dfrac{a \cdot 50}{50} = \dfrac{945}{50}$

$\qquad a = 18.9$

Therefore, 18% of 105 is 18.9.

**17.** $\dfrac{7.8}{b} = \dfrac{78}{100}$

$\qquad \dfrac{7.8}{b} = \dfrac{39}{50}$

$\qquad 7.8 \cdot 50 = b \cdot 39$

$\qquad 390 = b \cdot 39$

$\qquad \dfrac{390}{39} = \dfrac{b \cdot 39}{39}$

$\qquad 10 = b$

Therefore, 78% of 10 is 7.8.

**21.**
$$\frac{14}{50} = \frac{p}{100}$$
$$\frac{7}{25} = \frac{p}{100}$$
$$7 \cdot 100 = 25 \cdot p$$
$$700 = 25 \cdot p$$
$$\frac{700}{25} = \frac{25 \cdot p}{25}$$
$$28 = p$$
Therefore, 28% of 50 is 14.

**25.**
$$\frac{a}{80} = \frac{2.4}{100}$$
$$a \cdot 100 = 80 \cdot 2.4$$
$$a \cdot 100 = 192$$
$$\frac{a \cdot 100}{100} = \frac{192}{100}$$
$$a = 1.92$$
Therefore, 2.4% of 80 is 1.92.

**29.**
$$\frac{348.6}{166} = \frac{p}{100}$$
$$348.6 \cdot 100 = 166 \cdot p$$
$$34,860 = 166 \cdot p$$
$$\frac{34,860}{166} = \frac{166 \cdot p}{166}$$
$$210 = p$$
Therefore, 210% of 166 is 348.6.

**33.**
$$\frac{3.6}{8} = \frac{p}{100}$$
$$3.6 \cdot 100 = 8 \cdot p$$
$$360 = 8 \cdot p$$
$$\frac{360}{8} = \frac{8 \cdot p}{8}$$
$$45 = p$$
Therefore, 45% of 8 is 3.6.

**37.** $\dfrac{11}{16} + \dfrac{3}{16} = \dfrac{11 + 3}{16} = \dfrac{14}{16} = \dfrac{7 \cdot 2}{8 \cdot 2} = \dfrac{7}{8}$

**41.**
$$\overset{1}{0}.41$$
$$+0.29$$
$$\overline{0.70} \text{ or } 0.7$$

**45.**
$$\frac{a}{53,862} = \frac{22.3}{100}$$
$$a \cdot 100 = 53,862 \cdot 22.3$$
$$a \cdot 100 = 1,201,122.6$$
$$\frac{a \cdot 100}{100} = \frac{1,201,122.6}{100}$$
$$a = 12,011.226$$
Therefore, 22.3% of 53,862 rounded to the nearest tenth is 12,011.2.

### EXERCISE SET 7.4

**1.** Let $n$ be the number of bolts inspected.
$$1.5\% \cdot n = 24$$
$$0.015 \cdot n = 24$$
$$\frac{0.015 \cdot n}{0.015} = \frac{24}{0.015}$$
$$n = 1600$$
1600 bolts were inspected.

**5.** Let $x$ be the percent of income spent on food.
$$\$300 = x \cdot \$2000$$
$$\frac{300}{2000} = \frac{x \cdot 2000}{2000}$$
$$0.15 = x$$
$$0.15 \cdot 100\% = x$$
$$15\% = x$$
She spends 15% of her monthly income on food.

**9.** Let $x$ be the percent of members with some direct connection.
$$82 = x \cdot 535$$
$$\frac{82}{535} = \frac{x \cdot 535}{535}$$
$$0.153 = x$$
$$0.153 \cdot 100\% = x$$
$$15.3\% = x$$
15.3% of the members had some direct connection.

**13.** Let $x$ be the percent of calories from fat.
$$35 = x \cdot 120$$
$$\frac{35}{120} = \frac{x \cdot 120}{120}$$
$$0.292 = x$$
$$0.292 \cdot 100\% = x$$
$$29.2\% = x$$
29.2% of the food's total calories is from fat.

**17.** 20% of $170 is what number?
$$20\% \cdot \$170 = n$$
$$0.2 \cdot \$170 = n$$
$$\$34.00 = n$$
new bill $= \$170 - \$34$
$$= \$136$$
Their new bill is $136.

**21.** 33% of 26.8 million is what number?
$$33\% \cdot 26.8 \text{ million} = n$$
$$0.33 \cdot 26,800,00 = n$$
$$8,844,000 = n$$
projected population:
26.8 million + 8.844 million = 35.644 million
The increase is 8.844 million. The projected population is 35.644 million.

**25.**

| Original amount | New amount | Amount of Increase | Percent Increase |
|---|---|---|---|
| 85 | 187 | $187 - 85 = 102$ | $\dfrac{102}{85} = 1.2 = 120\%$ |

**29.**

| Original amount | New amount | Amount of Decrease | Percent Decrease |
|---|---|---|---|
| 160 | 40 | $160 - 40 = 120$ | $\dfrac{120}{160} = 0.75 = 75\%$ |

**33.** percent increase $= \dfrac{23.7 - 19.5}{19.5}$
$$= \frac{4.2}{19.5}$$
$$\approx 0.215$$
$$= 21.5\%$$

**37.** percent decrease $= \dfrac{52.1 - 7.2}{52.1}$

$\qquad\qquad\qquad\quad = \dfrac{44.9}{52.1}$

$\qquad\qquad\qquad\quad \approx 0.862$

$\qquad\qquad\qquad\quad = 86.2\%$

**41.** percent increase $= \dfrac{29{,}000 - 16{,}000}{16{,}000}$

$\qquad\qquad\qquad\quad = \dfrac{13{,}000}{16{,}000}$

$\qquad\qquad\qquad\quad = 0.8125$

$\qquad\qquad\qquad\quad = 81.25\%$

or 81.3% when rounded to the nearest tenth.

**45.** $\begin{array}{r} 9.20 \\ +1.98 \\ \hline 11.18 \end{array}$

**49.** percent increase $= \dfrac{28{,}700 - 26{,}518}{26{,}518}$

$\qquad\qquad\qquad\quad = \dfrac{2182}{26{,}518}$

$\qquad\qquad\qquad\quad \approx 0.082$

$\qquad\qquad\qquad\quad = 8.2\%$

**53.** Answers may vary.

## EXERCISE SET 7.5

**1.** tax $= 5\% \cdot \$150.00$

$\qquad\ = 0.05 \cdot \$150.00$

$\qquad\ = \$7.50$

The sales tax is \$7.50.

**5.** $\$54 = r \cdot \$600$

$\dfrac{\$54}{\$600} = r$

$0.09 = r$

$9\% = r$

The sales tax rate is 9%.

**9.** tax $= 6.5\% \cdot \$1800$

$\qquad\ = 0.065 \cdot \$1800$

$\qquad\ = \$117$

total $= \$1800 + \$117 = \$1917$

The total price is \$1917.

**13.** total purchase $= \$90 + \$15 = \$105$

tax $= 7\% \cdot \$105$

$\qquad = 0.07 \cdot \$105$

$\qquad = \$7.35$

total $= \$105 + \$7.35 = \$112.35$

The total price is \$112.35.

**17.** commission $= 4\% \cdot \$1{,}236{,}856$

$\qquad\qquad\quad = 0.04 \cdot \$1{,}236{,}856$

$\qquad\qquad\quad = \$49{,}474.24$

Her commission was \$49,474.24.

**21.** commission $= 1.5\% \cdot \$125{,}900$

$\qquad\qquad\quad = 0.015 \cdot \$125{,}900$

$\qquad\qquad\quad = \$1888.50$

His commission is \$1888.50.

**25.**

| Original Price | Discount Rate | Amount of Discount | Sale Price |
|---|---|---|---|
| \$68.00 | 10% | $\$68.00 \cdot 10\%$ $= \$68.00 \cdot 0.10$ $= \$6.80$ | $\$68.00 - \$6.80$ $= \$61.20$ |

**29.**

| Original Price | Discount Rate | Amount of Discount | Sale Price |
|---|---|---|---|
| \$215.00 | 35% | $\$215.00 \cdot 35\%$ $= \$215.00 \cdot 0.35$ $= \$75.25$ | $\$215.00 - \$75.25$ $= \$139.75$ |

**33.** discount $= 15\% \cdot \$300$

$\qquad\qquad = 0.15 \cdot \$300$

$\qquad\qquad = \$45$

sale price $= \$300 - \$45$

$\qquad\qquad\ = \$255$

**37.** $400 \cdot 0.03 \cdot 11 = 12 \cdot 11 = 132$

**41.** tax $= 7.5\% \cdot \$24{,}966$

$\qquad\ = 0.075 \cdot \$24{,}966$

$\qquad\ = \$1872.45$

total price $= \$24{,}966 + \$1872.45$

$\qquad\qquad\ = \$26{,}838.45$

The total price is \$26,838.45.

## EXERCISE SET 7.6

**1.** simple interest $=$ principal $\cdot$ rate $\cdot$ time

$\qquad\qquad\qquad\ = \$200 \cdot 8\% \cdot 2$

$\qquad\qquad\qquad\ = \$200 \cdot (0.08) \cdot 2$

$\qquad\qquad\qquad\ = \$32$

**5.** simple interest $=$ principal $\cdot$ rate $\cdot$ time

$\qquad\qquad\qquad\ = \$5000 \cdot 10\% \cdot 1\tfrac{1}{2}$

$\qquad\qquad\qquad\ = \$5000 \cdot (0.10) \cdot 1.5$

$\qquad\qquad\qquad\ = \$750$

**9.** simple interest $=$ principal $\cdot$ rate $\cdot$ time

$\qquad\qquad\qquad\ = \$2500 \cdot 16\% \cdot \dfrac{21}{12}$

$\qquad\qquad\qquad\ = \$2500 \cdot (0.16) \cdot 1.75$

$\qquad\qquad\qquad\ = \$700$

**13.** simple interest $=$ principal $\cdot$ rate $\cdot$ time

$\qquad\qquad\qquad\ = \$5000 \cdot 9\% \cdot \dfrac{15}{12}$

$\qquad\qquad\qquad\ = \$5000 \cdot (0.09) \cdot 1.25$

$\qquad\qquad\qquad\ = \$562.50$

total $= \$5000 + \$562.50 = \$5562.50$

The total amount received is \$5562.50.

**17.** total $=$ principal $\cdot$ compound interest factor

$\qquad\ = \$6150(7.61226)$

$\qquad\ = \$46{,}815.399$

The total amount is \$46,815.40.

**21.** total $=$ principal $\cdot$ compound interest factor

$\qquad\ = \$10{,}000(5.81636)$

$\qquad\ = \$58{,}163.60$

The total amount is \$58,163.60.

**25.** total = principal · compound interest factor
   = $2000(1.46933)
   = $2938.66
   compound interest = total − principal
      = $2938.66 − $2000
      = $938.66

**29.** monthly payment = $\dfrac{\text{principal} + \text{interest}}{\text{number of payments}}$
   = $\dfrac{\$1500 + \$61.88}{6}$
   = $\dfrac{\$1561.88}{6}$
   ≈ $260.31
   The monthly payment is $260.31.

**33.** $-5 + (-24) = -29$

**37.** $\dfrac{7 - 10}{3} = \dfrac{-3}{3} = -1$

**41.** Answers may vary.

## Chapter 7 Test

**1.** $85\% = 85.\% = 0.85$

**5.** $6.1 = 6.10 = 610.\% = 610\%$

**9.** $0.2\% = \dfrac{0.2}{100} = \dfrac{2}{1000} = \dfrac{1}{500}$

**13.** $\dfrac{1}{5} = \dfrac{1}{5} \cdot 100\% = \dfrac{100}{5}\% = 20\%$

**17.** $567 = x \cdot 756$
   $\dfrac{567}{756} = \dfrac{x \cdot 756}{x}$
   $0.75 = x$
   $75\% = x$
   Therefore, 75% of 756 is 567.

**21.** percent increase = $\dfrac{26{,}460 - 25{,}200}{25{,}200}$
   = $\dfrac{1260}{25{,}200}$
   = $0.05$
   = $5\%$
   The population increased 5%.

**25.** simple interest = principal · rate · time
   = $\$2000 \cdot 9.25\% \cdot 3\dfrac{1}{2}$
   = $\$2000 \cdot (0.0925) \cdot 3.5$
   = $\$647.50$

# Chapter 8

## Exercise Set 8.1

**1.** The largest sector, or 320 students, corresponds to where most college students live. Most college students live at parent or guardian's home.

**5.** $\dfrac{\text{Students living in campus housing}}{\text{total students}} = \dfrac{180}{320} = \dfrac{9}{16}$

**9.** The sectors which represent "less than $30" are "less than $20" and "$20–$29."
   percent = 30% + 23% = 53%

**13.** amount = 28% · 4700 = 0.28 · 4700 = 1316
   1316 Central High teenagers spent $30–$49.

**17.** The second-largest sector is 25% which corresponds to "Nonfiction." Nonfiction is the second-largest category of books.

**21.** amount = 22% · 125,600 = 0.22 · 125,600 = 27,632
   This library has 27,632 books in children's fiction.

**25.** 
   Under 3 days       7% · 360° = 25.2°
   3–10 days          18% · 360° = 64.8°
   11–20 days         14% · 360° = 50.4°
   21 or more days    61% · 360° = 219.6°

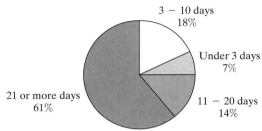

**29.** $40 = 8 \cdot 5 = 2 \cdot 2 \cdot 2 \cdot 5 = 2^3 \cdot 5$

**33.** Pacific Ocean:
   49% · 264,489,800 = 0.49 · 264,489,800 = 129,600,002
   The Pacific Ocean is 129,600,002 square kilometers.

## Exercise Set 8.2

**1.** The year 1998 has the most cars, so the greatest number of automobiles was manufactured in 1998.

**5.** Compare each year with the previous year to see that the production of automobiles decreased from the previous year in 1994, 1995, and 1999.

**9.** The year 1992 has seven chickens and each chicken represents 3 ounces of chicken, so approximately 7 · 3 = 21 ounces of chicken were consumed per person per week in 1992.

**13.** There is one more chicken for 1992 than for 1988, so there was an increase of approximately 1 · 3 = 3 ounces per person per week.

**17.** The tallest bar corresponds to the month of April, so April has the most tornado-related deaths.

**21.** Look for bars that extend above the horizontal line at 5. The months of February, March, April, May and June have over 5 tornado-related deaths.

**25.** The two U.S. cities are New York City and Los Angeles. New York City is the largest, with an estimated population of 16.2 million or 16,200,000 people.

**29.**

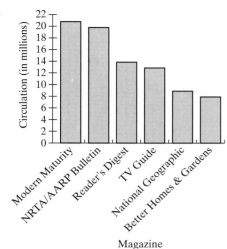
Magazine

**33.** In 1985, there was an average of approximately 54.5 field goals per game.

**37.** The average number of field goals per game increased between 1955 and 1965.

**41.** The receipts were the greatest the first week after release.

**45.** During week 8, *The Phantom Menace* grossed $12 million at the box office.

**49.** $10\% \cdot 62 = 0.1 \cdot 62 = 6.2$

**53.** $\dfrac{17}{50} = 0.34 = 34\%$

**57.** The lowest temperature occurred on Sunday, with 68°F.

**61.** Look at the two bars for each continent. Since the bars representing North America have the largest difference in height, North America had the greatest change in percent of U.S. immigrants.

**65.** Answers may vary.

## EXERCISE SET 8.3

**1.**

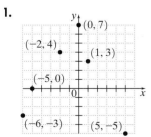

**5.** Point $A$ has coordinates $(0, 0)$.

Point $B$ has coordinates $\left(3\dfrac{1}{2}, 0\right)$.

Point $C$ has coordinates $(3, 2)$.
Point $D$ has coordinates $(-1, 3)$.
Point $E$ has coordinates $(-2, -2)$.
Point $F$ has coordinates $(0, -1)$.
Point $G$ has coordinates $(2, -1)$.

**9.** $x - y = 3$
$1 - 2 = 3$
$\quad -1 = 3$ False
No, $(1, 2)$ is not a solution of $x - y = 3$.

**13.** $y = -4x$
$-8 = -4 \cdot 2$
$-8 = -8$ True
Yes, $(2, -8)$ is a solution of $y = -4x$.

**17.** $x - 5y = -1$
$3 - 5(1) = -1$
$\quad 3 - 5 = -1$
$\quad\quad -2 = -1$ False
No, $(3, 1)$ is not a solution of $x - 5y = -1$.

**21.** $x = 5y$; $(5, 1)$, $(0, 0)$, $(-5, -1)$

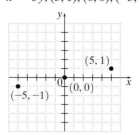

**25.** $y = 8x$
$y = 8(1)$
$y = 8$
The solution is $(1, 8)$.
$y = 8x$
$y = 8(0)$
$y = 0$
The solution is $(0, 0)$, the origin.
$\quad y = 8x$
$-16 = 8x$
$\dfrac{-16}{8} = \dfrac{8x}{8}$
$\quad -2 = x$
The solution is $(-2, -16)$.

**29.** $y = x + 5$
$y = 1 + 5$
$y = 6$
The solution is $(1, 6)$.
$\quad\quad y = x + 5$
$\quad\quad 7 = x + 5$
$7 - 5 = x + 5 - 5$
$\quad\quad 2 = x$
The solution is $(2, 7)$.
$y = x + 5$
$y = 3 + 5$
$y = 8$
The solution is $(3, 8)$.

**33.**
$$y = -x$$
$$0 = -x$$
$$\frac{0}{-1} = \frac{-x}{-1}$$
$$0 = x$$
The solution is $(0, 0)$.
$$y = -x$$
$$y = -(2)$$
$$y = -2$$
The solution is $(2, -2)$.
$$y = -x$$
$$2 = -x$$
$$\frac{2}{-1} = \frac{-x}{-1}$$
$$-2 = x$$
The solution is $(-2, 2)$.

**37.** $5.6 - 3.9 = 1.7$

**41.** $(0.236)(-100) = -23.6$

**45.** To plot the point $(0, b)$, we start at the origin and move 0 units to the right and $b$ units up. Our point will be on the $y$-axis. Thus $(0, b)$ lies on the $y$-axis is a true statement.

**49.** To plot $(-a, b)$, we start at the origin and move $a$ units to the left and $b$ units up. Our point will be in quadrant II. Thus $(-a, b)$ lies in quadrant III is a false statement.

## EXERCISE SET 8.4

**1.** $x + y = 6$
Find any 3 ordered pair solutions. Let $x = 0$.
$$x + y = 6$$
$$0 + y = 6$$
$$y = 6$$
$(0, 6)$
Let $x = 3$.
$$x + y = 6$$
$$3 + y = 6$$
$$3 + y \cdot 3 = 6 - 3$$
$$y = 3$$
$(3, 3)$
Let $x = 6$.
$$x + y = 6$$
$$6 + y = 6$$
$$6 + y - 6 = 6 - 6$$
$$y = 0$$
$(6, 0)$.
Plot $(0, 6)$, $(3, 3)$ and $(6, 0)$. Then draw the line through them.

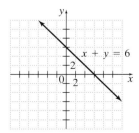

**5.** $y = 4x$
Find any 3 ordered pair solutions. Let $x = 0$.
$$y = 4x$$
$$y = 4(0)$$
$$y = 0$$
$(0, 0)$
Let $x = 1$.
$$y = 4x$$
$$y = 4(1)$$
$$y = 4$$
$(1, 4)$
Let $x = -1$.
$$y = 4x$$
$$y = 4(-1)$$
$$y = -4$$
$(-1, -4)$
Plot $(0, 0)$, $(1, 4)$ and $(-1, -4)$. Then draw the line through them.

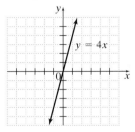

**9.** $x = 5$
No matter what $y$-value we choose, $x$ is always 5.

| $x$ | $y$ |
|---|---|
| 5 | $-4$ |
| 5 | 0 |
| 5 | 4 |

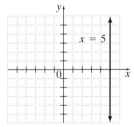

**13.** $x = 0$
No matter what $y$-value we choose, $x$ is always 0.

| $x$ | $y$ |
|---|---|
| 0 | $-3$ |
| 0 | 0 |
| 0 | 3 |

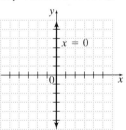

**17.** $y = -2$
No matter what $x$-value we choose, $y$ is always $-2$.

| $x$ | $y$ |
|---|---|
| $-4$ | $-2$ |
| 0 | $-2$ |
| 4 | $-2$ |

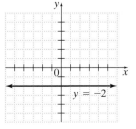

**21.** $x = 6$

No matter what $y$-value we choose, $x$ is always 6.

| $x$ | $y$ |
|---|---|
| 6 | $-5$ |
| 6 | 0 |
| 6 | 5 |

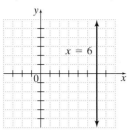

**25.** $x = y - 4$

Find any 3 ordered pair solutions. Let $y = 0$.

$x = y - 4$
$x = 0 - 4$
$x = -4$
$(-4, 0)$
Let $y = 4$.
$x = y - 4$
$x = 4 - 4$
$x = 0$
$(0, 4)$
Let $y = 6$.
$x = y - 4$
$x = 6 - 4$
$x = 2$
$(2, 6)$
Plot $(-4, 0)$, $(0, 4)$, and $(2, 6)$. Then draw the line through them.

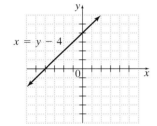

**29.** $x = 4y$

Find any 3 ordered pair solutions. Let $y = -1$.

$x = 4y$
$x = 4(-1)$
$x = -4$
$(-4, -1)$
Let $y = 0$.
$x = 4y$
$x = 4(0)$
$x = 0$
$(0, 0)$
Let $y = 1$.
$x = 4y$
$x = 4(1)$
$x = 4$
$(4, 1)$
Plot $(-4, -1)$, $(0, 0)$ and $(4, 1)$. Then draw the line through them.

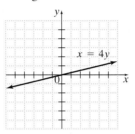

**33.** $y = 4x + 2$

Find any 3 ordered pair solutions. Let $x = -1$.

$y = 4x + 2$
$y = 4(-1) + 2$
$y = -4 + 2$
$y = -2$
$(-1, -2)$
Let $x = 0$.
$y = 4x + 2$
$y = 4(0) + 2$
$y = 0 + 2$
$y = 2$
$(0, 2)$
Let $x = 1$.
$y = 4x + 2$
$y = 4(1) + 2$
$y = 4 + 2$
$y = 6$
$(1, 6)$
Plot $(-1, -2)$, $(0, 2)$ and $(1, 6)$. Then draw the line through them.

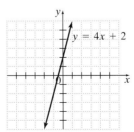

**37.** $x = -3.5$

No matter what $y$-value we choose, $x$ is always $-3.5$.

| $x$ | $y$ |
|------|-----|
| $-3.5$ | $-3$ |
| $-3.5$ | $0$ |
| $-3.5$ | $3$ |

$x = -3.5$

**41.** $\dfrac{86 + 94}{2} = \dfrac{180}{2} = 90$

**45.** $\dfrac{30 + 22 + 23 + 33}{4} = \dfrac{108}{4} = 27$

**49.** Answers may vary.

### Exercise Set 8.5

**1.** Mean: $\dfrac{21 + 28 + 16 + 42 + 38}{5} = \dfrac{145}{5} = 29$

Median: Write the numbers in order.

16, 21, 28, 38, 42

The middle number is 28.

Mode: There is no mode, since there is no number that occurs more often than the others.

**5.** Mean:

$$\dfrac{0.2 + 0.3 + 0.5 + 0.6 + 0.6 + 0.9 + 0.2 + 0.7 + 1.1}{9}$$

$$= \dfrac{5.1}{9} \approx 0.6$$

Median: Write the numbers in order.

0.2, 0.2, 0.3, 0.5, 0.6, 0.6, 0.7, 0.9, 1.1

The middle number is 0.6.

Mode: Since 0.2 and 0.6 occur most often, there are two modes, 0.2 and 0.6.

**9.** Mean: $\dfrac{1450 + 1368 + 1362 + 1250 + 1136}{5} = \dfrac{6566}{5}$

$$= 1313.2$$

The mean height of the five tallest buildings is 1313.2 feet.

**13.**

| Grade | Point Value | Credit Hours | $\left(\begin{array}{c}\text{Point}\\\text{Value}\end{array}\right) \cdot \left(\begin{array}{c}\text{Credit}\\\text{Hours}\end{array}\right)$ |
|-------|-------------|--------------|---------|
| B | 3 | 3 | 9 |
| C | 2 | 3 | 6 |
| A | 4 | 4 | 16 |
| C | 2 | 4 | 8 |
| Totals | | 14 | 39 |

Grade Point Average $= \dfrac{39}{14} \approx 2.79$.

**17.** Mean:

$$\dfrac{7.8 + 6.9 + 7.5 + 4.7 + 6.9 + 7.0}{6} = \dfrac{40.8}{6} = 6.8$$

The mean time is 6.8 seconds.

**21.** Write the numbers in order.

74, 77, 85, 86, 91, 95

Since there is an even number of scores, find the average of the middle two numbers.

median $= \dfrac{85 + 86}{2} = \dfrac{171}{2} = 85.5$

**25.** Since 70 and 71 occur more often than the other numbers, the mode is 70 and 71.

**29.** $\dfrac{6}{18} = \dfrac{2 \cdot 3}{2 \cdot 3 \cdot 3} = \dfrac{1}{3}$

**33.** $\dfrac{55}{75} = \dfrac{5 \cdot 11}{5 \cdot 15} = \dfrac{11}{15}$

### Exercise Set 8.6

**1.**

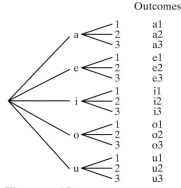

Outcomes

There are 15 outcomes.

**5.**

Outcomes

There are 16 outcomes.

**9.**

Outcomes

There are 8 outcomes.

**13.** possible outcomes: 1, 2, 3, 4, 5, 6

probability of a 1 or a 4 $= \dfrac{2}{6} = \dfrac{1}{3}$

**17.** possible outcomes: 1, 2, 3

probability of a 2 $= \dfrac{1}{3}$

**21.** possible outcomes: red, blue, yellow, yellow, green, green, green

probability of red $= \dfrac{1}{7}$

**25.** $\dfrac{1}{2} + \dfrac{1}{3} = \dfrac{1}{2} \cdot \dfrac{3}{3} + \dfrac{1}{3} \cdot \dfrac{2}{2}$

$\qquad = \dfrac{1 \cdot 3}{2 \cdot 3} + \dfrac{1 \cdot 2}{3 \cdot 2}$

$\qquad = \dfrac{3}{6} + \dfrac{2}{6}$

$\qquad = \dfrac{5}{6}$

**29.** $5 \div \dfrac{3}{4} = \dfrac{5}{1} \cdot \dfrac{4}{3} = \dfrac{5 \cdot 4}{1 \cdot 3} = \dfrac{20}{3}$ or $6\dfrac{2}{3}$

**33.** There are 52 possible outcomes.
There are four kings.

probability of a king $= \dfrac{4}{52} = \dfrac{1}{13}$

**37.**

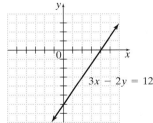

                              Sum

There are 36 possible outcomes. Three outcomes result in a sum of 4.

probability of a sum of 4 $= \dfrac{3}{36} = \dfrac{1}{12}$

**41.** Answers may vary.

## CHAPTER 8 TEST

**1.** The second week has $4\dfrac{1}{2}$ bills and each bill represents $50.

$4.5 \cdot \$50 = \$225$ was collected during the second week.

**5.** The shortest bar is above the month of February. February has the least amount of normal monthly precipitation with 3 centimeters.

**9.** There are two points below the line representing $50 million. 1996 and 1997 had sales less than $50 million.

**13.** amount $= 16\% \cdot 8{,}139{,}000$
$\qquad = 1{,}302{,}240$ luxury cars
In 1998, 1,302,240 luxury cars were sold.

**17.** Point $D$ is 2 units to the left of the origin and 1 unit down. It has coordinates $(-2, -1)$.

**21.** $y = -4$
No matter what $x$-value we choose, $y$ is always $-4$.

| $x$ | $y$ |
|-----|-----|
| $-5$ | $-4$ |
| $0$ | $-4$ |
| $5$ | $-4$ |

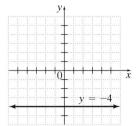

**25.** $3x - 2y = 12$
Find any 3 ordered pair solutions. Let $x = 0$.

$3x - 2y = 12$
$3(0) - 2y = 12$
$0 - 2y = 12$
$-2y = 12$
$\dfrac{-2y}{-2} = \dfrac{12}{-2}$
$y = -6$

$(0, -6)$
Let $y = 0$.

$3x - 2y = 12$
$3x - 2(0) = 12$
$3x - 0 = 12$
$3x = 12$
$\dfrac{3x}{3} = \dfrac{12}{3}$
$x = 4$

$(4, 0)$
Let $x = 2$.

$3x - 2y = 12$
$3(2) - 2y = 12$
$6 - 2y = 12$
$6 - 2y - 6 = 12 - 6$
$-2y = 6$
$\dfrac{-2y}{-2} = \dfrac{6}{-2}$
$y = -3$

$(2, -3)$
Plot $(0, -6)$, $(2, -3)$ and $(4, 0)$.
Then draw the line through them.

**29.**

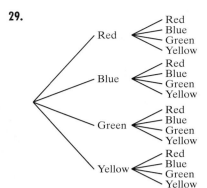

# Chapter 9

## Exercise Set 9.1

**1.** The figure extends indefinitely in two directions. It is line $yz$, or $\overleftrightarrow{yz}$.

**5.** The figure has two endpoints. It is line segment $PQ$, or $\overline{PQ}$.

**9.** $\angle ABC = 15°$

**13.** $\angle DBA = 50° + 15° = 65°$

**17.** 90°

**21.** $\angle S$ is a straight angle.

**25.** Since $\angle Q$ measures between 90° and 180°, it is an obtuse angle.

**29.** The complement of an angle that measures 17° is an angle that measures $90° - 17° = 73°$.

**33.** The complement of a 45° angle is an angle that measures $90° - 45° = 45°$.

**37.** $\angle MNP$ and $\angle RNO$ are complementary angles since $60° + 30° = 90°$. $\angle PNQ$ and $\angle QNR$ are complementary angles since $52° + 38° = 90°$.

**41.** $\angle x = 120° - 88° = 32°$

**45.** Since $\angle x$ and the 35° angle are vertical angles, $\angle x = 35°$. $\angle x$ and $\angle y$ are adjacent angles, so $\angle y = 180° - 35° = 145°$. $\angle y$ and $\angle z$ are vertical angles, so $\angle z = 145°$.

**49.** $\angle x$ and the 80° angle are adjacent angles, so $\angle x = 180° - 80° = 100°$. $\angle y$ and the 80° angle are alternate interior angles, so $\angle y = 80°$. $\angle x$ and $\angle z$ are corresponding angles, so $\angle z = 100°$.

**53.** $\angle x = 180° - 70° - 85° = 25°$

**57.** $\angle x = 180° - 90° - 50° = 40°$

**61.** $\dfrac{7}{8} \cdot \dfrac{1}{4} = \dfrac{7 \cdot 1}{8 \cdot 4} = \dfrac{7}{32}$

**65.** $3\dfrac{1}{3} \div 2\dfrac{1}{2} = \dfrac{10}{3} \div \dfrac{5}{2} = \dfrac{10}{3} \cdot \dfrac{2}{5} = \dfrac{10 \cdot 2}{3 \cdot 5} = \dfrac{20}{15} = \dfrac{4}{3}$ or $1\dfrac{1}{3}$

**69.** $\angle a$ and the 60° angle are alternate interior angles, so $\angle a = 60°$. $\angle a$, $\angle b$ and the 70° angle, form a straight angle, so
$\angle b = 180° - 70° - \angle a$
$= 180° - 70° - 60°$
$= 50°$
$\angle c$ and the angle that is $\angle a$ and $\angle b$ together are alternate interior angles, so
$\angle c = 60° + 50° = 110°$.
$\angle c$ and $\angle d$ are adjacent angles, so
$\angle d = 180° - 110° = 70°$.
$\angle e$ and the 60° angle are adjacent angles, so
$\angle e = 180° - 60° = 120°$.

## Exercise Set 9.2

**1.** 60 in. $= 60$ in. $\cdot \dfrac{1 \text{ ft}}{12 \text{ in.}} = \dfrac{60 \text{ ft}}{12} = 5$ ft
60 inches is 5 feet.

**5.** 42,240 ft $= 42,240$ ft $\cdot \dfrac{1 \text{ mi}}{5280 \text{ ft}} = \dfrac{42,240 \text{ mi}}{5280} = 8$ mi
42,240 feet is 8 miles.

**9.** 10 ft $= 10$ ft $\cdot \dfrac{1 \text{ yd}}{3 \text{ ft}} = \dfrac{10 \text{ yd}}{3} = 3\dfrac{1}{3}$ yd
10 feet is $3\dfrac{1}{3}$ yards.

**13.** 40 ft$\div$ 3 = 13 with remainder 1
40 ft = 13 yd 1 ft

**17.** 10,000 ft$\div$ 5280 = 1 with remainder 4720
10,000 ft = 1 mi 4720 ft

**21.** 5 yd $= 5$ yd $\cdot \dfrac{3 \text{ ft}}{1 \text{ yd}} = 15$ ft
5 yd 2 ft = 15 ft + 2 ft = 17 ft

**25.** $\begin{array}{r} 5 \text{ ft } 8 \text{ in.} \\ + 6 \text{ ft } 7 \text{ in.} \\ \hline 11 \text{ ft } 15 \text{ in.} \end{array}$ = 11 ft + 1 ft 3 in. = 12 ft 3 in.

**29.** $\begin{array}{r} 24 \text{ ft } 8 \text{ in.} \\ -16 \text{ ft } 3 \text{ in.} \\ \hline 8 \text{ ft } 5 \text{ in.} \end{array}$

**33.** $\begin{array}{r} 3 \text{ ft } 4 \text{ in.} \\ 2\overline{)6 \text{ ft } 8 \text{ in.}} \\ \underline{6 \text{ ft}} \\ 0 \quad 8 \text{ in.} \\ \underline{-8 \text{ in.}} \\ 0 \end{array}$

**37.** $\begin{array}{r} 6 \text{ ft } 10 \text{ in.} \\ + 3 \text{ ft } 8 \text{ in.} \\ \hline 9 \text{ ft } 18 \text{ in.} \end{array}$ = 9 ft + 1 ft 6 in. = 10 ft 6 in.
The bamboo is 10 ft 6 in. tall.

**41.** $\begin{array}{r} 1 \text{ ft } 9 \text{ in.} \\ \times \quad\quad 9 \\ \hline 9 \text{ ft } 81 \text{ in.} \end{array}$ = 9 ft + 6 ft 9 in. = 15 ft 9 in.
9 stacks would extend 15 ft 9 in. from the wall.

**45.** perimeter = 24 ft 9 in. + 18 ft. 6 in.
$$+ \ 24 \text{ ft 9 in.} + 18 \text{ ft 6 in.}$$
$$= 84 \text{ ft 30 in.}$$
$$= 84 \text{ ft} + 2 \text{ ft 6 in.}$$
$$= 86 \text{ ft 6 in.}$$
She must purchase 86 ft 6 in. of fencing material.

**49.** To convert meters to centimeters, move the decimal 2 places to the right.
40 m = 4000 cm

**53.** To convert meters to kilometers, move the decimal 3 places to the left.
300 m = 0.300 km = 0.3 km

**57.** To convert centimeters to meters, move the decimal 2 places to the left.
1500 cm = 15 m

**61.** To convert millimeters to decimeters, move the decimal 2 places to the left.
20.1 mm = 0.201 dm

**65.**   8.60 m
+ 0.34 m
‾‾‾‾‾‾‾
  8.94 m

**69.**  24.8 mm =   2.48 cm
−1.19 cm = −1.19 cm
‾‾‾‾‾‾‾‾‾‾‾‾‾‾‾‾‾
               1.29 cm  or 12.9 mm

**73.**  18.3 m
× 3
‾‾‾‾‾
54.9 m

**77.** 3.4 m + 5.8 m − 8 cm = 3.4 m + 5.8 m − 0.08 m
$$= 9.12 \text{ m}$$
The tied ropes are 9.12 m long.

**81.** 25(1 m + 65 cm) = 25(100 cm + 65 cm)
$$= 25(165 \text{ cm})$$
$$= 4125 \text{ cm or } 41.25 \text{ m}$$
4125 cm or 41.25 m of wood must be ordered.

**85.** 5.988 km + 21 m = 5988 m + 21 m
$$= 6009 \text{ m or } 6.009 \text{ km}$$

**89.** perimeter = 10 + 6 + 10 + 6 = 32
The perimeter is 32 yards.

**93.** Answers may vary.

## EXERCISE SET 9.3

**1.** $P = 2l + 2w$
$$= 2 \cdot 17 \text{ feet} + 2 \cdot 15 \text{ feet}$$
$$= 34 \text{ feet} + 30 \text{ feet}$$
$$= 64 \text{ feet}$$

**5.** $P = a + b + c$
$$= 5 \text{ inches} + 7 \text{ inches} + 9 \text{ inches}$$
$$= 21 \text{ inches}$$

**9.** $P = 10 \text{ feet} + 8 \text{ feet} + 8 \text{ feet} + 15 \text{ feet} + 7 \text{ feet}$
$$= 48 \text{ feet}$$

**13.** $P = 5 \text{ feet} + 3 \text{ feet} + 2 \text{ feet} + 7 \text{ feet} + 4 \text{ feet}$
$$= 21 \text{ feet}$$

**17.** $P = 2l + 2w$
$$= 2(120 \text{ yards}) + 2(53 \text{ yards})$$
$$= 240 \text{ yards} + 106 \text{ yards}$$
$$= 346 \text{ yards}$$

**21.** Since 22 feet of stripping is needed, the cost is $3(22) = $66.

**25.** $P = 4s = 4(7 \text{ inches}) = 28 \text{ inches}$

**29.** The missing lengths are:
28 meters − 20 meters = 8 meters
20 meters − 17 meters = 3 meters
$P = 8 \text{ meters} + 3 \text{ meters} + 20 \text{ meters} + 20 \text{ meters}$
$$+ \ 28 \text{ meters} + 17 \text{ meters}$$
$$= 96 \text{ meters}$$

**33.** The missing lengths are:
12 miles + 10 miles = 22 miles
8 miles + 34 miles = 42 miles
$P = 12 \text{ miles} + 34 \text{ miles} + 10 \text{ miles} + 8 \text{ miles}$
$$+ \ 22 \text{ miles} + 42 \text{ miles}$$
$$= 128 \text{ miles}$$

**37.** $C = 2\pi r$
$$= 2\pi \cdot 8 \text{ mi}$$
$$= 16\pi \text{ mi}$$
$$\approx 16 \cdot 3.14 \text{ mi}$$
$$= 50.24 \text{ mi}$$

**41.** $C = 2\pi r$
$$= 2\pi \cdot 5 \text{ feet}$$
$$= 10\pi \text{ feet}$$
$$\approx 10 \cdot \frac{22}{7} \text{ feet}$$
$$= \frac{220}{7} \text{ feet}$$
$$= 31\frac{3}{7} \text{ feet}$$

**45.** $5 + 6 \cdot 3 = 5 + 18 = 23$

**49.** $(18 + 8) - (12 + 4) = 26 - 16 = 10$

**53.** Since a fence goes along the edge of the yard, we are concerned with the perimeter of the yard.

**57.** Since we paint the surface of the wall, we are concerned with the area of the wall.

**61. a.** Circumference of the smaller circle:
$C = 2\pi r$
$$= 2\pi \cdot 10 \text{ meters}$$
$$= 20\pi \text{ meters}$$
$$\approx 20 \cdot 3.14 \text{ meters}$$
$$= 62.8 \text{ meters}$$
Circumference of the larger circle:
$C = 2\pi r$
$$= 2\pi \cdot 20 \text{ meters}$$
$$= 40\pi \text{ meters}$$
$$= 40 \cdot 3.14 \text{ meters}$$
$$= 125.6 \text{ meters}$$

**b.** Yes, the circumference is doubled.

## EXERCISE SET 9.4

**1.** $A = lw$
$$= 3.5 \text{ meters} \cdot 2 \text{ meters}$$
$$= 7 \text{ square meters}$$

**5.** $A = \frac{1}{2}bh$

$= \frac{1}{2} \cdot 6 \text{ yards} \cdot 5 \text{ yards}$

$= \frac{1}{2} \cdot 6 \cdot 5 \text{ square yards}$

$= 15 \text{ square yards}$

**9.** $A = bh$

$= 7 \text{ feet} \cdot 5.25 \text{ feet}$

$= 36.75 \text{ square feet}$

**13.** $A = \frac{1}{2}(b + B)h$

$= \frac{1}{2} \cdot (4 \text{ yards} + 7 \text{ yards}) \cdot 4 \text{ yards}$

$= \frac{1}{2} \cdot 11 \text{ yards} \cdot 4 \text{ yards}$

$= \frac{1}{2} \cdot 11 \cdot 4 \text{ square yards}$

$= 22 \text{ square yards}$

**17.** $A = bh$

$= 5 \text{ inches} \cdot 4\frac{1}{2} \text{ inches}$

$= 5 \cdot \frac{9}{2} \text{ square inches}$

$= \frac{45}{2} \text{ or } 22\frac{1}{2} \text{ square inches}$

**21.**

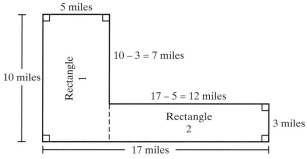

Area of rectangle 1 $= bh$

$= 5 \text{ miles} \cdot 10 \text{ miles}$

$= 50 \text{ square miles}$

Area of rectangle 2 $= bh$

$= 12 \text{ miles} \cdot 3 \text{ miles}$

$= 36 \text{ square miles}$

Total area $= 50 \text{ square miles} + 36 \text{ square miles}$

$= 86 \text{ square miles}$

**25.** $A = \pi r^2$

$= \pi(6 \text{ inches})^2$

$= 36\pi \text{ square inches}$

$= 36 \cdot \frac{22}{7} \text{ square inches}$

$\approx 113.1 \text{ square inches}$

**29.** $V = s^3$

$= (8 \text{ centimeters})^3$

$= 512 \text{ cubic centimeters}$

**33.** $V = \frac{4}{3}\pi r^3$

$= \frac{4}{3} \cdot \pi \left(\frac{10 \text{ inches}}{2}\right)^3$

$= \frac{4}{3} \cdot 125 \cdot \pi \text{ cubic inches}$

$= \frac{500}{3}\pi \text{ cubic inches}$

$\approx \frac{500}{3} \cdot \frac{22}{7} \text{ cubic inches}$

$= \frac{11,000}{21} \text{ cubic inches} = 523\frac{17}{21} \text{ cubic inches}$

**37.** $V = s^3$

$= \left(1\frac{1}{3} \text{ inches}\right)^3$

$= \left(\frac{4}{3}\right)^3 \text{ cubic inches}$

$= \frac{64}{27} \text{ cubic inches} = 2\frac{10}{27} \text{ cubic inches}$

**41.** $A = lw$

$= 100 \text{ feet} \cdot 50 \text{ feet}$

$= 5000 \text{ square feet}$

**45.** $V = \frac{1}{3}s^2h$

$= \frac{1}{3} \cdot (12 \text{ centimeters})^2 \cdot 20 \text{ centimeters}$

$= \frac{2880}{3} \text{ cubic centimeters}$

$= 960 \text{ cubic centimeters}$

**49.** Area of shaded part $= \frac{1}{2}(b + B)h$

$= \frac{1}{2} \cdot (25 \text{ feet} + 36 \text{ feet}) \cdot 12\frac{1}{2} \text{ feet}$

$= \frac{1}{2} \cdot 61 \text{ feet} \cdot \frac{25}{2} \text{ feet}$

$= \frac{1}{2} \cdot \frac{61}{1} \cdot \frac{25}{2} \text{ square feet}$

$= \frac{1525}{4} \text{ square feet} = 381\frac{1}{4} \text{ square feet}$

The area of the shaded part, to the nearest square foot is 381 square feet.

**53.** $3^2 = 9$

**57.** $4^2 + 2^2 = 16 + 4 = 20$

**61.** $8 \text{ inches} = \frac{8}{12} \text{ feet} = \frac{2}{3} \text{ feet}$

$A = lw$

$= 2 \text{ feet} \cdot \frac{2}{3} \text{ feet}$

$= \frac{4}{3} \text{ or } 1\frac{1}{3} \text{ square feet}$

$2 \text{ feet} = 24 \text{ inches}$

$A = lw$

$= (24 \text{ inches}) \cdot (8 \text{ inches})$

$= 192 \text{ square inches}$

**65.** $V = \frac{1}{3}s^2h$

$\quad = \frac{1}{3} \cdot (227 \text{ meters})^2 \cdot (137 \text{ meters})$

$\quad = \frac{7{,}059{,}473}{4} \text{ cubic meters}$

$\quad \approx 2{,}353{,}158 \text{ cubic meters}$

The original volume was approximately 2,583,283 cubic meters, so the volume has decreased

2,583,283 cubic meters − 2,353,158 cubic meters
$\quad = 230{,}125$ cubic meters.

## EXERCISE SET 9.5

**1.** $2 \text{ lb} = 2 \text{ lb} \cdot \dfrac{16 \text{ oz}}{1 \text{ lb}} = 2 \cdot 16 \text{ oz} = 32 \text{ oz}$

**5.** $12{,}000 \text{ lb} = 12{,}000 \text{ lb} \cdot \dfrac{1 \text{ ton}}{2000 \text{ lb}} = \dfrac{12{,}000 \text{ tons}}{2000} = 6 \text{ tons}$

**9.** $3500 \text{ lb} = 3500 \text{ lb} \cdot \dfrac{1 \text{ ton}}{2000 \text{ lb}}$

$\quad = \dfrac{3500 \text{ tons}}{2000}$

$\quad = \dfrac{7}{4} \text{ tons}$

$\quad = 1\dfrac{3}{4} \text{ tons}$

**13.** $4.9 \text{ tons} = 4.9 \text{ tons} \cdot \dfrac{2000 \text{ lb}}{1 \text{ ton}}$

$\quad = 4.9 \cdot 2000 \text{ lb}$

$\quad = 9800 \text{ lb}$

**17.** $2950 \text{ lb} = 2950 \text{ lb} \cdot \dfrac{1 \text{ ton}}{2000 \text{ lb}}$

$\quad = \dfrac{2950 \text{ tons}}{2000}$

$\quad = 1.475 \text{ tons}$

$\quad \approx 1.5 \text{ tons}$

**21.**
$$\begin{array}{r} 6 \text{ tons } 1540 \text{ lb} \\ + \ 2 \text{ tons } \ \ 850 \text{ lb} \\ \hline 8 \text{ tons } 2390 \text{ lb} \end{array} = 8 \text{ tons} + 1 \text{ ton } 390 \text{ lb}$$
$$\qquad\qquad\qquad\quad = 9 \text{ tons } 390 \text{ lb}$$

**25.**
$$\begin{array}{r} 12 \text{ lb } \ \ 4 \text{ oz} = 11 \text{ lb } 20 \text{ oz} \\ -3 \text{ lb } \ \ 9 \text{ oz} = \underline{-3 \text{ lb } \ \ 9 \text{ oz}} \\ 8 \text{ lb } 11 \text{ oz} \end{array}$$

**29.**
$$\begin{array}{r} 1 \text{ ton } \qquad 700 \text{ lb} \\ 5\overline{)6 \text{ tons } \quad 1500 \text{ lb}} \\ \underline{-5 \text{ tons}} \qquad \\ 1 \text{ ton} = 2000 \text{ lb} \\ 3500 \text{ lb} \\ \underline{-3500 \text{ lb}} \\ 0 \end{array}$$

**33.**
$$\begin{array}{r} 64 \text{ lb } \ \ 8 \text{ oz} = \ \ 63 \text{ lb } 24 \text{ oz} \\ -28 \text{ lb } 10 \text{ oz} = \underline{-28 \text{ lb } 10 \text{ oz}} \\ 35 \text{ lb } 14 \text{ oz} \end{array}$$
His zucchini was 35 lb 14 oz below the record.

**37.**
$$\begin{array}{r} 55 \text{ lb } 4 \text{ oz} = \ \ 54 \text{ lb } 20 \text{ oz} \\ -2 \text{ lb } 8 \text{ oz} = \underline{-2 \text{ lb } \ \ 8 \text{ oz}} \\ 52 \text{ lb } 12 \text{ oz} \end{array}$$
$$\begin{array}{r} 52 \text{ lb } 12 \text{ oz} \\ \times \qquad\quad 4 \\ \hline 208 \text{ lb } 48 \text{ oz} \end{array} = 208 \text{ lb} + 3 \text{ lb} = 211 \text{ lb}$$
There are 211 lb of pineapple in 4 boxes.

**41.** To convert grams to kilograms, move the decimal 3 places to the left.
$500 \text{ g} = 0.5 \text{ kg}$

**45.** To convert kilograms to grams, move the decimal 3 places to the right.
$25 \text{ kg} = 25{,}000 \text{ g}$

**49.** To convert grams to kilograms, move the decimal 3 places to the left.
$6.3 \text{ g} = 0.0063 \text{ kg}$

**53.** To convert kilograms to grams, move the decimal 3 places to the right.
$4.01 \text{ kg} = 4010 \text{ g}$

**57.** $205 \text{ mg} = 0.205 \text{ g}$ or $5.61 \text{ g} = 5610 \text{ mg}$
$$\begin{array}{r} 0.205 \text{ g} = \quad 205 \text{ mg} \\ +5.610 \text{ g} = \underline{+ \ 5610 \text{ mg}} \\ 5.815 \text{ g or} \quad 5815 \text{ mg} \end{array}$$

**61.** $1.61 \text{ kg} = 1610 \text{ g}$ or $250 \text{ g} = 0.25 \text{ kg}$
$$\begin{array}{r} 1610 \text{ g} = \quad 1.61 \text{ kg} \\ -250 \text{ g} = \underline{-0.25 \text{ kg}} \\ 1360 \text{ g or} \quad 1.36 \text{ kg} \end{array}$$

**65.**
$$\begin{array}{r} 2.125 \text{ kg} \\ 8\overline{)17.000 \text{ kg}} \\ \underline{-16} \qquad\quad \\ 1\ 0 \qquad\quad \\ \underline{-8} \qquad\quad \\ 20 \qquad \\ \underline{-16} \qquad \\ 40 \quad \\ \underline{-40} \quad \\ 0 \end{array}$$

**69.** $0.09 \text{ g} - 60 \text{ mg} = 90 \text{ mg} - 60 \text{ mg}$
$$\qquad\qquad\qquad\qquad = 30 \text{ mg}$$
The extra-strength tablet contains 30 mg more medication.

**73.** $3 \cdot 16 \cdot 3 \text{ mg} = 144 \text{ mg}$
3 cartons contain 144 mg of preservatives.

**77.** $0.3 \text{ kg} + 0.15 \text{ kg} + 400 \text{ g}$
$\quad = 0.3 \text{ kg} + 0.15 \text{ kg} + 0.4 \text{ kg}$
$\quad = 0.85 \text{ kg or } 850 \text{ g}$
The package weighs 0.85 kg or 850 g.

**81.** $\dfrac{1}{4} = \dfrac{1 \cdot 25}{4 \cdot 25} = \dfrac{25}{100} = 0.25$

**85.** $\dfrac{7}{8} = \dfrac{7 \cdot 125}{8 \cdot 125} = \dfrac{875}{1000} = 0.875$

## EXERCISE SET 9.6

**1.** $32 \text{ fl oz} = 32 \text{ fl oz} \cdot \dfrac{1 \text{ c}}{8 \text{ fl oz}} = \dfrac{32 \text{ c}}{8} = 4 \text{ c}$

**5.** $10 \text{ qt} = 10 \text{ qt} \cdot \dfrac{1 \text{ gal}}{4 \text{ qt}} = \dfrac{10 \text{ gal}}{4} = 2\dfrac{1}{2} \text{ gal}$

**9.** $2 \text{ qt} = 2 \text{ qt} \cdot \dfrac{4 \text{ c}}{1 \text{ qt}} = 2 \cdot 4 \text{ c} = 8 \text{ c}$

**13.** $6 \text{ gal} = 6 \text{ gal} \cdot \dfrac{4 \text{ qt}}{1 \text{ gal}} \cdot \dfrac{2 \text{ pt}}{1 \text{ qt}} \cdot \dfrac{2 \text{ c}}{1 \text{ pt}} \cdot \dfrac{8 \text{ fl oz}}{1 \text{ c}}$

$\qquad = 6 \cdot 4 \cdot 2 \cdot 2 \cdot 8 \text{ fl oz}$

$\qquad = 768 \text{ fl oz}$

**17.** $2\dfrac{3}{4} \text{ gal} = \dfrac{11}{4} \text{ gal} \cdot \dfrac{4 \text{ qt}}{1 \text{ gal}} \cdot \dfrac{2 \text{ pt}}{1 \text{ qt}} = \dfrac{11 \cdot 4 \cdot 2 \text{ pt}}{4} = 22 \text{ pt}$

**21.**
$\quad 1 \text{ c } 5 \text{ fl oz}$
$\underline{+2 \text{ c } 7 \text{ fl oz}}$
$\quad 3 \text{ c } 12 \text{ fl oz} = 3 \text{ c} + 1 \text{ c } 4 \text{ fl oz} = 4 \text{ c } 4 \text{ fl oz}$

**25.**
$3 \text{ gal } 1 \text{ qt} \qquad = 2 \text{ gal } 4 \text{ qt } 2 \text{ pt}$
$\underline{- \qquad 1 \text{ qt } 1 \text{ pt}} = \underline{- \qquad 1 \text{ qt } 1 \text{ pt}}$
$\qquad\qquad\qquad\qquad 2 \text{ gal } \quad 3 \text{ qt } 1 \text{ pt}$

**29.**
$\quad 8 \text{ gal } 2 \text{ qt}$
$\underline{\times \qquad\quad 2}$
$16 \text{ gal } 4 \text{ qt} = 16 \text{ gal} + 1 \text{ gal} = 17 \text{ gal}$

**33.** $1\dfrac{1}{2} \text{ qt} = \dfrac{3}{2} \text{ qt} \cdot \dfrac{2 \text{ pt}}{1 \text{ qt}} \cdot \dfrac{2 \text{ c}}{1 \text{ qt}} \cdot \dfrac{8 \text{ fl oz}}{1 \text{ c}}$

$\qquad = \dfrac{3 \cdot 2 \cdot 2 \cdot 8 \text{ fl oz}}{2}$

$\qquad = 48 \text{ fl oz}$

**37.**
$\quad 5 \text{ pt } 1 \text{ c}$
$\underline{+ 2 \text{ pt } 1 \text{ c}}$
$\quad 7 \text{ pt } 2 \text{ c} = 7 \text{ pt} + 1 \text{ pt} = 8 \text{ pt} = 4 \text{ qt} = 1 \text{ gal}$
Yes, the punch can be poured into the container.

**41.** $12 \text{ fl oz} \cdot 24 = 288 \text{ fl oz}$

$288 \text{ fl oz} = 288 \text{ fl oz} \cdot \dfrac{1 \text{ c}}{8 \text{ fl oz}} \cdot \dfrac{1 \text{ pt}}{2 \text{ c}} \cdot \dfrac{1 \text{ qt}}{2 \text{ pt}}$

$\qquad = \dfrac{288 \text{ qt}}{8 \cdot 2 \cdot 2}$

$\qquad = 9 \text{ qt}$

There are 9 quarts in a case of Pepsi.

**45.** To convert milliliters to liters, move the decimal 3 places to the left.
$4500 \text{ ml} = 4.5 \text{ L}$

**49.** To convert milliliters to liters, move the decimal 3 places to the left.
$64 \text{ ml} = 0.064 \text{ L}$

**53.** To convert liters to milliliters, move the decimal 3 places to the right.
$3.6 \text{ L} = 3600 \text{ ml}$

**57.**
$\quad 2.9 \text{ L}$
$\underline{+ 19.6 \text{ L}}$
$\quad 22.5 \text{ L}$

**61.** $8.6 \text{ L} = 8600 \text{ ml or } 190 \text{ ml} = 0.19 \text{ L}$
$8600 \text{ ml} = \quad 8.60 \text{ L}$
$\underline{-190 \text{ ml} = -0.19 \text{ L}}$
$8410 \text{ ml or} \quad 8.41 \text{ L}$

**65.**
$\quad 480 \text{ ml}$
$\underline{\times \qquad 8}$
$\quad 3840 \text{ ml}$

**69.**
$\quad 2 \text{ L} = \quad 2.00 \text{ L}$
$\underline{-410 \text{ ml} = -0.41 \text{ L}}$
$\qquad\qquad\quad 1.59 \text{ L}$
1.59 liters remain in the bottle.

**73.** $44.3\overline{)\$14.00}$

$\qquad\qquad 0.3160$
$443\overline{)140.0000}$
$\underline{-132\ 9}$
$\qquad 7\ 10$
$\qquad \underline{-4\ 43}$
$\qquad 2\ 670$
$\qquad \underline{-2\ 658}$
$\qquad\qquad 120$

The cost is about \$0.316 per liter.

**77.** $0.7 = \dfrac{7}{10}$

**81.** $0.006 = \dfrac{6}{1000} = \dfrac{3}{500}$

### EXERCISE SET 9.7

**1.** $\quad C = \dfrac{5}{9}(F - 32)$

$\qquad C = \dfrac{5}{9} \cdot (41 - 32)$

$\qquad\quad = \dfrac{5}{9} \cdot (9)$

$\qquad\quad = 5$
$41°F = 5°C$

**5.** $\quad F = \dfrac{9}{5}C + 32$

$\qquad F = \dfrac{9}{5} \cdot (60) + 32$

$\qquad\quad = 108 + 32$

$\qquad\quad = 140$
$60°C = 140°F$

**9.** $\quad C = \dfrac{5}{9}(F - 32)$

$\qquad C = \dfrac{5}{9} \cdot (62 - 32)$

$\qquad\quad = \dfrac{5}{9} \cdot (30)$

$\qquad\quad = \dfrac{150}{9}$

$\qquad\quad \approx 16.7$
$62°F \approx 16.7°C$

**13.** $\quad F = 1.8C + 32$
$\qquad F = 1.8(92) + 32$
$\qquad\quad = 165.6 + 32$
$\qquad\quad = 197.6$
$92°C = 197.6°F$

**17.** $\quad C = \dfrac{5}{9}(F - 32)$

$\qquad C = \dfrac{5}{9} \cdot (134 - 32)$

$\qquad\quad = \dfrac{5}{9} \cdot (102)$

$\qquad\quad = \dfrac{510}{9}$

$\qquad\quad \approx 56.7$
$134°F = 56.7°C$

**21.** $C = \dfrac{5}{9}(F - 32)$

$\quad C = \dfrac{5}{9} \cdot (70 - 32)$

$\quad = \dfrac{5}{9} \cdot (38)$

$\quad = \dfrac{190}{9} \approx 21.1$

$70°F = 21.1°C$

**25.** $\quad F = 1.8C + 32$

$\quad F = 1.8 \cdot 118 + 32$

$\quad = 212.4 + 32$

$\quad = 244.4$

$118°C = 244.4°F$

**29.** $C = \dfrac{5}{9}(F - 32)$

$\quad C = \dfrac{5}{9} \cdot (864 - 32)$

$\quad = \dfrac{5}{9} \cdot (832)$

$\quad = \dfrac{4160}{9}$

$\quad \approx 462.2$

$864°F = 462.2°C$

**33.** $P = 3\text{ ft} + 3\text{ ft} + 3\text{ ft} + 3\text{ ft} + 3\text{ ft} = 15\text{ ft}$

**37.** $C = \dfrac{5}{9}(F - 32)$

$\quad C = \dfrac{5}{9} \cdot (9010 - 32)$

$\quad = \dfrac{5}{9} \cdot (8978)$

$\quad = \dfrac{44,890}{9}$

$\quad \approx 4988$

$9010°F$ is about $4988°C$.

## CHAPTER 9 TEST

**1.** The complement of a 78° angle is $90° - 78° = 12°$.

**5.** $\angle x$ and the 73° angle are vertical angles, so $\angle x = 73°$. $\angle y$ and the 73° angle are corresponding angles, so $\angle y = 73°$. $\angle z$ and $\angle y$ are vertical angles, so $\angle z = 73°$.

**9.**

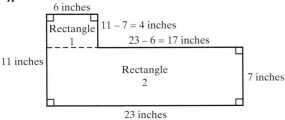

Perimeter
$\quad = 6\text{ inches} + 4\text{ inches} + 17\text{ inches} + 7\text{ inches}$
$\qquad + 23\text{ inches} + 11\text{ inches}$
Area of rectangle 1 $= 6\text{ inches} \cdot 4\text{ inches}$
$\qquad\qquad\qquad\qquad = 24\text{ square inches}$
Area of rectangle 2 $= 7\text{ inches} \cdot 23\text{ inches}$
$\qquad\qquad\qquad\qquad = 161\text{ square inches}$
Total area $= 161\text{ square inches} + 24\text{ square inches}$
$\qquad\qquad = 185\text{ square inches}$

**13.** $V = lwh$

$\quad = 3\text{ feet} \cdot 3\text{ feet} \cdot 2\text{ feet}$

$\quad = 18\text{ cubic feet}$

**17.** $2\dfrac{1}{2}\text{ gal} = \dfrac{5}{2}\text{ gal} \cdot \dfrac{4\text{ qt}}{1\text{ gal}} = \dfrac{20\text{ qt}}{2} = 10\text{ qt}$

**21.** To convert milligrams to grams, move the decimal place 3 places to the left.
$40\text{ mg} = 0.04\text{ g}$

**25.** To convert liters to milliliters, move the decimal place 3 places to the right.
$0.83\text{ L} = 830\text{ ml}$

**29.**
$$\begin{array}{r}
2\text{ gal} \quad 3\text{ qt} \\
\hline
2)\overline{5\text{ gal} \quad 2\text{ qt}} \\
-4\text{ gal} \\
\hline
1\text{ gal} = 4\text{ qt} \\
6\text{ qt} \\
-6\text{ qt} \\
\hline
0
\end{array}$$

**33.** $\quad F = 1.8C + 32$

$\quad F = 1.8(12.6) + 32$

$\quad = 22.68 + 32$

$\quad = 54.68$

$12.6°C \approx 54.7°F$

**37.**
$$\begin{array}{r}
88\text{ m} = 88.0\text{ m} \\
+\ 340\text{ cm} = \underline{3.4\text{ m}} \\
91.4\text{ m}
\end{array}$$
The span is 91.4 m long.

**41.** $4667\text{ gal} = \dfrac{4667}{1}\text{ gal} \cdot \dfrac{4\text{ qt}}{1\text{ gal}} \cdot \dfrac{2\text{ pt}}{1\text{ qt}}$

$\quad = 4667 \cdot 4 \cdot 2\text{ pt}$

$\quad = 37,336\text{ pt}$

Thus, 37,336 pints of ice cream were used.

# Chapter 10

### EXERCISE SET 10.1

**1.** $(2x + 3) + (-7x - 27) = (2x - 7x) + (3 - 27)$
$\qquad\qquad\qquad\qquad\qquad = -5x + (-24)$
$\qquad\qquad\qquad\qquad\qquad = -5x - 24$

**5.** $(12y - 20) + (9y^2 + 13y - 20)$
$\quad = 9y^2 + (12y + 13y) + (-20 - 20)$
$\quad = 9y^2 + 25y - 40$

**9.** $(5a - 6) - (a + 2) = (5a - 6) + (-a - 2)$
$\qquad\qquad\qquad\qquad = (5a - a) + (-6 - 2)$
$\qquad\qquad\qquad\qquad = 4a - 8$

**13.** $(10y^2 - 7) - (20y^3 - 2y^2 - 3)$
$\quad = (10y^2 - 7) + (-20y^3 + 2y^2 + 3)$
$\quad = (-20y^3) + (10y^2 + 2y^2) + (-7 + 3)$
$\quad = -20y^3 + 12y^2 - 4$

**17.**
$$\begin{array}{r}
13y^2 - 6y - 14 \\
-(5y^2 + 4y - 6) \\
\end{array}
\qquad
\begin{array}{r}
13y^2 - 6y - 14 \\
+ -5y^2 - 4y + 6 \\
\hline
8y^2 - 10y - 8
\end{array}$$

**21.** $(21y - 4.6) - (36y - 8.2)$
$\quad = (21y - 4.6) + (-36y + 8.2)$
$\quad = (21y - 36y) + (-4.6 + 8.2)$
$\quad = -15y + 3.6$

**25.** $\left(3z + \dfrac{6}{7}\right) - \left(3z - \dfrac{3}{7}\right)$

$\quad = \left(3z + \dfrac{6}{7}\right) + \left(-3z + \dfrac{3}{7}\right)$

$\quad = (3z - 3z) + \left(\dfrac{6}{7} + \dfrac{3}{7}\right)$

$\quad = \dfrac{9}{7}$

**29.** $x^2 - 6x + 3$

Let $x = 2$.

$(2)^2 - 6(2) + 3 = 4 - 12 + 3 = -5$

**33.** $2x + 10$

Let $x = 5$.

$2(5) + 10 = 10 + 10 = 20$

**37.** $2x^2 + 4x - 20$

Let $x = 5$.

$2(5)^2 + 4(5) - 20 = 2(25) + 20 - 20 = 50$

**41.** $3000 + 20x$

Let $x = 10$.

$3000 + 20(10) = 3000 + 200 = 3200$

It costs $3200 to manufacture 10 file cabinets.

**45.** $3^4 = 3 \cdot 3 \cdot 3 \cdot 3 = 81$

**49.** $x \cdot x \cdot x = x^3$

**53.** $(2x + 1) + (x + 11) + (5x - 10)$

$\quad = (2x + x + 5x) + (1 + 11 - 10)$

$\quad = 8x + 2$

The perimeter is $(8x + 2)$ inches.

**57.**

$$\begin{array}{r} 3x^2 + \phantom{2}x - \phantom{2} \\ + \phantom{3}x^2 - \phantom{2}6x + 2 \\ \hline 5x^2 + 14x - 4 \end{array}$$

$\left[\begin{array}{l} (3 + \_\_)x^2 = 5x \\ (3 + \underline{2})x^2 = 5x^2 \end{array}\right]$

$\left[\begin{array}{l} (\_\_ - 6)x = 14x \\ (\underline{20} - 6)x = 14x \end{array}\right]$

$\left[\begin{array}{l} (-\_\_ + 2) = -4 \\ (\underline{-6} + 2) = -4 \end{array}\right]$

$$\begin{array}{r} 3x^2 + 20x - 6 \\ + \phantom{}2x^2 - \phantom{2}6x + 2 \\ \hline 5x^2 + 14x - 4 \end{array}$$

**61.** When $t = 8$ seconds:

$1053 - 16t^2 = 1053 - 16(8)^2$

$\qquad\qquad = 1053 - 16 \cdot 64$

$\qquad\qquad = 1053 - 1024$

$\qquad\qquad = 29$

The height of the object above the river after 8 seconds is 29 feet.

When $t = 9$ seconds:

$1053 - 16t^2 = 1053 - 16(9)^2$

$\qquad\qquad = 1053 - 16 \cdot 81$

$\qquad\qquad = 1053 - 1296$

$\qquad\qquad = -243$

The height of the object above the river after 9 seconds is $-243$ feet. The object hits the water between 8 and 9 seconds.

## EXERCISE SET 10.2

**1.** $x^5 \cdot x^9 = x^{5+9} = x^{14}$

**5.** $3z^3 \cdot 5z^2 = (3 \cdot 5)(z^3 \cdot z^2) = 15z^5$

**9.** $(-5x^2y^3)(-5x^4y) = (-5)(-5)(x^2 \cdot x^4)(y^3 \cdot y)$

$\qquad\qquad\qquad\qquad = 25x^6y^4$

**13.** $2x \cdot 3x \cdot 5x = (2 \cdot 3 \cdot 5)(x \cdot x \cdot x) = 30x^3$

**17.** $(x^5)^3 = x^{5 \cdot 3} = x^{15}$

**21.** $(b^7)^6 \cdot (b^2)^{10} = b^{7 \cdot 6} \cdot b^{2 \cdot 10}$

$\qquad\qquad\qquad = b^{42} \cdot b^{20}$

$\qquad\qquad\qquad = b^{42 + 20}$

$\qquad\qquad\qquad = b^{62}$

**25.** $(a^{11}b^8)^3 = a^{11 \cdot 3}b^{8 \cdot 3} = a^{33}b^{24}$

**29.** $(-3y)(2y^7)^3 = (-3y) \cdot 2^3(y^7)^3$

$\qquad\qquad\qquad = (-3y) \cdot 8y^{21}$

$\qquad\qquad\qquad = (-3)(8)(y^1 \cdot y^{21})$

$\qquad\qquad\qquad = -24y^{22}$

**33.** $7(x - 3) = 7x - 21$

**37.** $9(x + 2y - 3) = 9x + 18y - 27$

**41.** area $= \dfrac{1}{2}bh$

$\qquad = \dfrac{1}{2} \cdot (6a^3b^4) \cdot (4ab)$

$\qquad = \left(\dfrac{1}{2} \cdot 6 \cdot 4\right)(a^3 \cdot a)(b^4 \cdot b)$

$\qquad = 12a^4b^5$

The area is $12a^4b^5$ square meters.

**45.** $(8.1x^{10})^5 = 8.1^5(x^{10})^5 = 34{,}867.84401x^{50}$

## EXERCISE SET 10.3

**1.** $3x(9x^2 - 3) = 3x \cdot 9x^2 + 3x \cdot (-3)$

$\qquad\qquad\qquad = (3 \cdot 9)(x \cdot x^2) + (3)(-3)(x)$

$\qquad\qquad\qquad = 27x^3 + (-9x)$

$\qquad\qquad\qquad = 27x^3 - 9x$

**5.** $7x^2(6x^2 - 5x + 7)$

$\quad = (7x^2)(6x^2) + (7x^2)(-5x) + (7x^2)(7)$

$\quad = (7 \cdot 6)(x^2 \cdot x^2) + (7)(-5)(x^2 \cdot x) + (7 \cdot 7)x^2$

$\quad = 42x^4 - 35x^3 + 49x^2$

**9.** $(2x - 6)(x + 4) = 2x(x + 4) - 6(x + 4)$

$\qquad\qquad\qquad = 2x \cdot x + 2x \cdot 4 - 6 \cdot x - 6 \cdot 4$

$\qquad\qquad\qquad = 2x^2 + 8x - 6x - 24$

$\qquad\qquad\qquad = 2x^2 + 2x - 24$

**13.** $(a + 6)(a^2 - 6a + 3)$

$\quad = a(a^2 - 6a + 3) + 6(a^2 - 6a + 3)$

$\quad = a \cdot a^2 + a(-6a) + a \cdot 3 + 6 \cdot a^2 + 6(-6a) + 6 \cdot 3$

$\quad = a^3 - 6a^2 + 3a + 6a^2 - 36a + 18$

$\quad = a^3 - 33a + 18$

**17.** $(x^3 + 2x + x^2)(3x + 1 + x^2)$

$\quad = x^3(3x + 1 + x^2) + 2x(3x + 1 + x^2)$

$\qquad + x^2(3x + 1 + x^2)$

$\quad = x^3 \cdot 3x + x^3 \cdot 1 + x^3 \cdot x^2 + 2x \cdot 3x + 2x \cdot 1$

$\qquad + 2x \cdot x^2 + x^2 \cdot 3x + x^2 \cdot 1 + x^2 \cdot x^2$

$\quad = 3x^4 + x^3 + x^5 + 6x^2 + 2x + 2x^3$

$\qquad + 3x^3 + x^2 + x^4$

$\quad = x^5 + 4x^4 + 6x^3 + 7x^2 + 2x$

**21.** $-2y^2(3y + y^2 - 6)$
$= -2y^2 \cdot 3y + (-2y^2) \cdot y^2 + (-2y^2)(-6)$
$= -6y^3 - 2y^4 + 12y^2$

**25.** $(2a + 3)(2a - 3)$
$= 2a(2a - 3) + 3(2a - 3)$
$= 2a \cdot 2a + 2a(-3) + 3 \cdot 2a + 3(-3)$
$= 4a^2 - 6a + 6a - 9$
$= 4a^2 - 9$

**29.** $\left(b + \dfrac{3}{5}\right)\left(b + \dfrac{4}{5}\right) = b\left(b + \dfrac{4}{5}\right) + \dfrac{3}{5}\left(b + \dfrac{4}{5}\right)$
$$= b^2 + \dfrac{4}{5}b + \dfrac{3}{5}b + \dfrac{3}{5} \cdot \dfrac{4}{5}$$
$$= b^2 + \dfrac{7}{5}b + \dfrac{12}{25}$$

**33.** $(7x + 5)^2 = (7x + 5)(7x + 5)$
$= 7x(7x + 5) + 5(7x + 5)$
$= 49x^2 + 35x + 35x + 25$
$= 49x^2 + 70x + 25$

**37.** $(2x^2 - 3)(4x^3 + 2x - 3)$
$= 2x^2(4x^3 + 2x - 3) - 3(4x^3 + 2x - 3)$
$= 8x^5 + 4x^3 - 6x^2 - 12x^3 - 6x + 9$
$= 8x^5 - 8x^3 - 6x^2 - 6x + 9$

**41.**
$$
\begin{array}{r}
2z^2 - z + 1 \\
\times\ 5z^2 + z - 2 \\
\hline
-4z^2 + 2z - 2 \\
2z^3 - z^2 + z \\
10z^4 - 5z^3 + 5z^2 \\
\hline
10z^4 - 3z^3\ \ \ \ \ \ \ + 3z - 2
\end{array}
$$

**45.** $72 = 2 \cdot 2 \cdot 2 \cdot 3 \cdot 3 = 2^3 \cdot 3^2$

**49.** $(y - 6)(y^2 + 3y + 2)$
$= y(y^2 + 3y + 2) - 6(y^2 + 3y + 2)$
$= y^3 + 3y^2 + 2y - 6y^2 - 18y - 12$
$= y^3 - 3y^2 - 16y - 12$
The area is $(y^3 - 3y^2 - 16y - 12)$ square feet.

**53.** Answers may vary.

## EXERCISE SET 10.4

**1.** $48 = 2 \cdot 2 \cdot 2 \cdot 2 \cdot 3$
$15 = 3 \cdot 5$
GCF $= 3$

**5.** $12 = 2 \cdot 2 \cdot 3$
$20 = 2 \cdot 2 \cdot 5$
$36 = 2 \cdot 2 \cdot 3 \cdot 3$
GCF $= 2 \cdot 2 = 4$

**9.** $y^7 = y^2 \cdot y^5$
$y^2 = y^2$
$y^{10} = y^2 \cdot y^8$
GCF $= y^2$

**13.** $x^3y^2 = x \cdot x^2 \cdot y^2$
$xy^2 = x \cdot y^2$
$x^4y^2 = x \cdot x^3y^2$
GCF $= x \cdot y^2 = xy^2$

**17.** $2 = 2$
$14 = 2 \cdot 7$
$18 = 2 \cdot 3 \cdot 3$
GCF $= 2$
$z^3 = z^3$
$z^5 = z^3 \cdot z^2$
$z^3 = z^3$
GCF $= z^3$
GCF $= 2z^3$

**21.** $10a^6 = 5a^6 \cdot 2$
$5a^8 = 5a^6 \cdot a^2$
GCF $= 5a^6$
$10a^6 - 5a^8 = 5a^6 \cdot 2 - 5a^6 \cdot a^2$
$= 5a^6(2 - a^2)$

**25.** $z^7 = z^5 \cdot z^2$
$6z^5 = z^5 \cdot 6$
GCF $= z^5$
$z^7 - 6z^5 = z^5 \cdot z^2 - z^5 \cdot 6 = z^5(z^2 - 6)$

**29.** $12a^5 = 12a^5$
$36a^6 = 12a^5 \cdot 3a$
GCF $= 12a^5$
$12a^5 - 36a^6 = 12a^5 \cdot 1 - 12a^5 \cdot 3a$
$= 12a^5(1 - 3a)$

**33.** $80\% = \dfrac{80}{100} = \dfrac{4}{5}$

**37.** area on the left: $x \cdot x = x^2$
area on the right: $2 \cdot x = 2x$
total area: $x^2 + 2x$
Notice that $x(x + 2) = x^2 + 2x$.

## CHAPTER 10 TEST

**1.** $(11x - 3) + (4x - 1) = (11x + 4x) + (-3 - 1)$
$= 15x + (-4)$
$= 15x - 4$

**5.** $x^2 - 6x + 1 = (8)^2 - 6(8) + 1$
$= 64 - 48 + 1$
$= 17$

**9.** $(6a^3)(-2a^7) = (6)(-2)(a^3 \cdot a^7) = -12a^{10}$

**13.** $-2y(y^3 + 6y^2 - 4)$
$= -2y \cdot y^3 - 2y \cdot 6y^2 - 2y \cdot (-4)$
$= -2y^4 - 12y^3 + 8y$

**17.** area:
$(x + 7)(5x - 2) = x(5x - 2) + 7(5x - 2)$
$= 5x^2 - 2x + 35x - 14$
$= 5x^2 + 33x - 14$
The area is $(5x^2 + 33x - 14)$ square inches.
perimeter:
$2(2x) + 2(5x - 2) = 4x + 10x - 4 = 14x - 4$
The perimeter is $(14x - 4)$ inches.

**21.** $10a^2 = 2a \cdot 5a$
$12a = 2a \cdot 6$
$10a^2 + 12a = 2a \cdot 5a + 2a \cdot 6$
$= 2a(5a + 6)$

# Index

# Photo Credits

Chapter 1 CO Tony Cordoza/Liaison Agency, Inc., (p.32 ) The Schiller Group, Ltd./Dole Plantation, Hawaii, (p.32) Jerry L. Ferrara/Photo Researchers, Inc., (p.55) Hyatt Corporation, (p.82) Jean-Claude LeJeune/Stock Boston

Chapter 2 CO N. Schiller/The Image Works, (p.109) Paul Souders/Allstock/Picture Quest Vienna, (p.116) Voscar—The Maine Photographer, (p.135) Kathy Willens/AP/Wide World Photos, (p.141) © Alon Reininger/Contact Press Images/Picture Quest, (p.156) © AFP/Tom Mihalek/CORBIS

Chapter 3 CO D. Young-Wolff/PhotoEdit, (p.170) The New York Public Library Photographic Services, (p.204) John A. Rizzo/PhotoDisc, Inc., (p.207) Steve Morrell/AP/Wide World Photos, (p.207) F. Hangel/The Image Works, (p.220) Bob Krist/Corbis

Chapter 4 CO Scott Camazine/Photo Researchers, Inc., (p.256) Kevin Horan/Stone, (p.261) P. Lloyd/Weatherstock, (p.285) Richard Hutchings/PhotoEdit, (p.320) Rev. Ronald Royer/Photo Researchers, Inc., (p.324) Robert Harbison, (p.324) Peter Welmann/Animals Animals/Earth Scenes

Chapter 5 CO Tom Salyer/Reuters/Corbis, (p.348) AP/Wide World Photos, (p.357) AFP Photo/Timothy A. Clary/Corbis, (p.358) Tony Gutierrez/AP/Wide World Photos, (p.358) Doug Densinger/Allsport Photography (USA), Inc.

Chapter 6 CO © Bettmann/CORBIS, (p.451) Paul Sakuma/AP/Wide World Photos, (p.456) Grace Davies/Omni-Photo Communications, Inc., (p.456) British Airways, (p.459) Photo-Verlag Gyger/Niesenbahn/Switzerland Tourism, (p.473) Michael Newman/PhotoEdit, (p.475) Stock Boston

Chapter 7 CO M. H. Sharp/Photo Researchers, Inc., (p.505) C. Borland/PhotoDisc, Inc., (p.513) SIU/Photo Researchers, Inc., (p.513) Santokh Kochar/PhotoDisc, Inc., (p.516) Telegraph Colour Library/FPG International LLC, (p.539) Jeff Greenberg/Omni-Photo Communications, Inc., (p.544) John Serafin/Pearson Education Corporate Digital Archive, (p.544) © Richard Hamilton Smith/CORBIS, (p.545) Michael Heron/Pearson Education/PH College

Chapter 8 CO UPI/Corbis, (p.624) Jeff Greenberg/Visuals Unlimited

Chapter 9 CO Paul Conklin, (p.671) Brian K. Diggs/AP/Wide World Photos, (p.671) Will & Deni McIntyre/Photo Researchers, Inc., (p.698) Tom Bean/DRK Photo, (p.713) Will & Deni McIntyre/Photo Researchers, Inc.

Chapter 10 CO © Joseph Sohm; ChromoSoh, Inc./Corbis, (p.790) Archive Photos